Series in *Polymers, Interfaces, and Biomaterials*

Other books in the series:

Alexander Grosberg, Editor, *Theoretical and Mathematical Models in Polymer Research* (1998).

Kaoru Tsujii, *Chemistry and Physics of Surfactants: Principles, Phenomena, and Applications* (1998).

Teruo Okano, Editor, *Biorelated Polymers and Gels: Controlled Release and Applications in Biomedical Engineering* (1998).

Also available:

Alexander Grosberg and Alexei R. Khokhlov, *Giant Molecules: Here, There, and Everywhere* (1997).

Jacob Israelachvili, *Intermolecular and Surface Forces, Second Edition* (1992).

Experimental Methods in Polymer Science

Modern Methods in Polymer Research and Technology

Edited by

Toyoichi Tanaka

DEPARTMENT OF PHYSICS AND CENTER FOR
MATERIALS SCIENCE AND ENGINEERING
MASSACHUSETTS INSTITUTE OF TECHNOLOGY
CAMBRIDGE, MASSACHUSETTS

ACADEMIC PRESS

An Imprint of Elsevier

San Diego San Francisco New York Boston
London Sydney Tokyo

This book is printed on acid-free paper. ♾

ACADEMIC PRESS
An Imprint of Elsevier
525 B Street, Suite 1900, San Diego, CA 92101-4495, USA
http://www.academicpress.com

Academic Press
24-28 Oval Road, London NW1 7DX, UK
http://www.hbuk.co.uk/ap/

Library of Congress Catalog Number: 99-61578
ISBN-13: 978-0-12-683265-5
ISBN-10: 0-12-683265-X

Printed and bound by CPI Group (UK) Ltd, Croydon, CR0 4YY

Transferred to Digital Print 2011

Contents

Contributors

Numbers in parentheses indicate the pages on which the authors' contributions begin.

S. Amiya (261), Analytical Research Center, Kuraray Co., Ltd., Kurashiki, Okayama, Japan

I. Ando (261), Department of Polymer Chemistry, Tokyo Institute of Technology, 2-12-1 Ookayama, Meguro-ku Tokyo, Japan

S. Ando (261), Department of Polymer Chemistry, Tokyo Institute of Technology, 2-12-1 Ookayama, Meguro-ku Tokyo, Japan

B. Chu (1), Department of Chemistry, State University of New York at Stony Brook, Stony Brook, New York 11794

E. R. Edelman (547), Harvard-MIT Division of Health Sciences and Technology, Massachusetts Institute of Technology, Cambridge, Massachusetts 02139

A. English (547), Department of Biomedical Engineering, University of Iowa, Iowa City, Iowa 52242

T. Hashimoto (57), Department of Polymer Science and Engineering, Kyoto Institute of Technology, Matsugasaki, Sakyo-ku, Kyoto 606-8585, Japan

H. Itagaki (155), Department of Chemistry, Faculty of Education, Shizuoka University, 836 Ohya, Shizuoka 422, Japan

H. Jinnai (57), Department of Polymer Science and Engineering, Kyoto Institute of Technology, Matsugasaki, Sakyo-ku, Kyoto 606-8585, Japan

M. Kanekiyo (261), Department of Polymer Chemistry, Tokyo Institute of Technology, 2-12-1 Ookayama, Meguro-ku Tokyo, Japan

M. Kobayashi (261), Department of Polymer Chemistry, Tokyo Institute of Technology, 2-12-1 Ookayama, Meguro-ku Tokyo, Japan

S. Kuroki (261), Department of Polymer Chemistry, Tokyo Institute of Technology, 2-12-1 Ookayama, Meguro-ku Tokyo, Japan

H. Kurosu (261), Department of Textile and Apparel Science, Nara Women's University, Kita-uoya, Nishimachi, Nara, Japan

S. Matsukawa (261), Department of Food Science, Tokyo University of Fisheries, Konan, Minato-ku, Tokyo, Japan

M. Mours (495), BASF AG, D-67056 Ludwigshafen, Germany

M. Shibayama (57), Department of Polymer Science and Engineering, Kyoto Institute of Technology, Matsugasaki, Sakyo-ku, Kyoto 606-8585, Japan

T. Tanaka (547), Department of Physics and Center for Materials Science and Engineering, Massachusetts Institute of Technology, Cambridge, Massachusetts 02139

H. H. Winter (495), Department of Chemical Engineering, University of Massachusetts at Amherst, Amherst, Massachusetts 01003

C. Wu (1), The Chinese University of Hong Kong, Department of Chemistry, Shatin, N.T. Hong Kong

H. Yasunaga (261), Department of Chemistry and Materials Technology, Kyoto Institute of Technology, Goshokaido-machi, Matsugasaki, Kyoto, Japan

Preface by Series Editor

The 20th century will be remembered as the century of science and technology. It began with the advent of relativity and quantum mechanics that revolutionized our concept of time, space, and matter. The atomism of physics that has sought the smallest constituent of matter, elementary particles, for thousands of years is about to end its long journey. Physics now aims at more and more complex systems and has found principles that predict macroscopic behaviors from microscopic interactions. Chemistry determines molecular structures of numerous natural matters and is now able to synthesize practically any molecules. The discovery of DNA structure opened the new world of molecular biology and biotechnology. Electronics and computers converted our everyday life into the modern society of communication and information. All these advances create an impression that everything is known, all major questions have been answered, and that there are no more fundamental challenges left in science and technology.

In reality, quite the opposite is true: the advances of 20th-century science have opened up an unprecedented range of challenges, including the most breathtaking ones such as the origin of universe, the brain function, the origin of life, and the synthesis of life. There are many secrets that nature has not yet disclosed to us. In spite of the dramatic progress in molecular biology during the last several decades, biotechnology still depends entirely on the genetic information provided by nature. The principle by which to design proteins and DNA, the polymers of life, is not in our hand. For this reason, the essence of biotechnology is still in a "black box," and so is the mystery of the origin of life. There are innumerable unsolved questions that remain to be resolved before many major diseases will be conquered. One cannot yet produce artificial organs, muscles, and skin to our satisfaction. Too rapid an advance of modern technologies has brought the earth into an increasingly dangerous and chaotic environment. The 21st century and the third millennium will soon begin with these unsolved problems and issues.

Among the key concepts in science that can crucially address these issues are polymers, interfaces, and biomaterials. They have been a main subject of science for some time, but their true potential value for humans has not been fully explored. Biopolymers are the bearer of life that appeared on this earth and are responsible for all the biological functions of molecular sensing, homeostasis, cleaning, catalysis, and molecular motions. Synthesis of polymers capable of such biological functions is the ultimate challenge to modern science and will lead to the answers to many questions, including the mystery of the origin of life. The medical and technological implications of these accomplishments will be huge and endless. Understanding biomaterials, their interactions, and their interfaces will open the door to development of novel soft and active materials in areas in which hard, and inert materials used to be the key players. They will be particularly useful for the development of artificial organs, tissues, and limbs for both humans and robots. Many environmental problems can be solved by materials capable of molecular sensing, absorption, recovery, as well as responsiveness to stabilized environmental conditions. Here again, polymers, interfaces, and biomaterials will play the key roles.

With all this in mind and in response to the great demand by researchers and students who look at our future from a global perspective, Academic Press has decided to launch a series of books titled *Polymers, Interfaces, and Biomaterials*. The series is aimed at researchers in the field of polymers science, applied chemistry and physics, medicine, pharmacology, agriculture, and other industries. We believe that the series is timely and will help promote the field of modern science and technology for the next century.

Toyoichi Tanaka, Series Editor
Massachusetts Institute of Technology

Preface by Editor

Polymers are considered among the most important materials in science and technology for the 21st century. The uses of polymers in our everyday life are being extended and diversified day by day. The chemical, medical, and agricultural industries as well as many others are heavily dependent on a wide variety of polymers. Moreover, polymers are the materials that nature chose as the vehicle for life that appeared on this earth.

Polymer science may be said to have begun when Staudinger first hypothesized that polymers are giant molecules in which small molecules are connected in the form of a long chain. Numerous experimental works proved that hypothesis, one of which was Carothers' unambiguous proof by indeed creating the first synthetic polymer, nylon. The discovery of synthesis opened the door to the rich industrial fields of polymer technology, completely changing our everyday lives. In parallel with technological developments, the science of polymers has been a great challenge to the human intellect. Works by scientists, including such pioneers as Flory and de Gennes, laid the foundations and established the physical principles of polymer science.

In spite of the dramatic advances in polymer science, biotechnology, and materials science and engineering during the last several decades, our understanding and ability for designing polymers are far from complete. We are far behind nature that so elegantly designed and created biopolymers and utilizes them as the molecular machinery of life. The complete dependence of biotechnology on genetic information makes evident our ignorance of the principles underlying polymer design and our lack of understanding of the mechanisms behind the marvelous biopolymer functions that we see at hand. During the last decade, however, we have begun to unravel the secrets behind biopolymers and have started to see the possibility of synthesizing polymers that can mimic the activities of life. There appears to be an enormous amount of room to improve polymers in the directions of higher performance, higher strength, higher efficiency, and higher functionality.

For this reason, *now* is a crucial moment in science to advance our understanding of polymers in general. The knowledge of the technical methods of characterizing polymer systems is of inevitable importance to modern technologies spanning many different fields. It is essential to master the modern polymer techniques in all areas of experiment, theory, and computation to understand fully the up-to-date research as well as to open the doors to new research areas.

This handbook is designed to fulfill the requirements by many scientists and engineers who wish to be able to carry out experiments on polymers using modern physical techniques to characterize polymer systems. The book is meant to be one volume of a set of handbooks addressing the subjects of theoretical understanding (edited by Alexander Grosberg), characterization (edited by Toyoichi Tanaka), and computations of polymers (edited by Kurtz Kremer). These three volumes will provide a useful guide to the physical characterization and analysis of polymer behavior.

Each chapter in this book describes the principle of a technique along with detailed experimental procedures providing examples of actual applications of the technique. With this book, the authors will be able to start applying the concepts and methods to the systems of their own studies. The subjects covered by this volume are as follows:

- Light scattering (Wu and Chu)
- Neutron scattering (Shibayama *et al.*)
- Fluorescence spectroscopy (Itagaki)
- NMR spectroscopy (Ando *et al.*)
- Mechanical spectroscopy (Mours and Winter)
- Polymer hydrogel phase transitions (English, Edelman, Tanaka)

Although this volume does not span the entire spectrum of the experimental characterizations of polymers, further volumes are under way to cover other techniques. We sincerely hope that this volume will be of help to various researchers and students with ambitions to explore new fields of polymer science and technology with the aim of discovering new concepts and principles and developing novel polymer systems.

Toyoichi Tanaka

Experimental Methods in Polymer Science

Modern Methods in Polymer Research and Technology

Chapter 1 | Light Scattering

C. Wu

Department of Chemistry
The Chinese University of Hong Kong
Shatin, N.T. Hong Kong
The Open Laboratory of Bond-Selective Chemistry
Department of Chemical Physics
University of Science and Technology of China
Hefei, Anhui, China

B. Chu

Department of Chemistry
State University of New York at Stony Brook
Stony Brook, New York
Department of Materials Science and Engineering
State University of New York at Stony Brook
Stony Brook, New York

1.1 Introduction

When a monochromatic, coherent beam of light is incident on a dilute solution of macromolecules or suspension of colloidal particles and the solvent refractive index is different from that of the solute (macromolecules or colloidal particles), the incident light is scattered by each illuminated macromolecule or colloidal particle in all directions. The scattered light waves from different macromolecules or particles mutually interfere, or combine, at a distant, fast detector (e.g., a photomultiplier tube) and produce a net scattered intensity $I(t)$ or photon counts $n(t)$ that are not uniform on the scattering (or detection) plane. If all the macromolecules or particles are stationary, the scattered light intensity at each direction would be a constant (i.e., independent of time). However, in reality, all the scatterers in solution are undergoing constant Brownian motions, and this fact leads to fluctuations of the scattered intensity pattern on the detection plane and the fluctuations in $I(t)$ if the detection area is sufficiently small. The fluctuation rates can be related to different relaxation processes such as translational and rotational diffusions as well as internal motions of the macromolecules. The faster the relaxation process, the faster the intensity fluctuations will be.

EXPERIMENTAL METHODS
IN POLYMER SCIENCE

ISBN 0-12-683265-X/$30.00.

In a broad definition, laser light scattering (LLS) could be classified as inelastic (e.g., Raman, fluorescence, and phosphorescence) and elastic (no absorption) light scattering. However, in polymer and colloid science, light scattering is normally referred to in terms of static (elastic) or dynamic (quasi-elastic) measurements, or both, of the scattered light [1]. Static LLS as a classic and absolute analytical method measures the time-average scattered intensity, and it has been widely used to characterize synthetic and natural macromolecules [2]. On the other hand, dynamic LLS measures the intensity fluctuations instead of the average light intensity (this is where the word *dynamic* comes from), and its essence may be explained as follows: When the incident light is scattered by a moving macromolecule or particle, the detected frequency of the scattered light will be slightly higher or lower than that of the original incident light owing to the Doppler effect, depending on whether the particle moves towards or away from the detector. Thus, the frequency distribution of the scattered light is slightly broader than that of the incident light. This is why dynamic LLS is also called quasi-elastic light scattering (QELS). The frequency broadening ($\approx 10^5$–10^7 Hz) is so small in comparison with the incident light frequency ($\approx 10^{15}$ Hz) that it is very difficult, if not impossible, to detect the broadening directly in the frequency domain. However, it can be effectively recorded in the time domain via a time correlation function. Thus, dynamic light scattering is sometimes known as intensity fluctuation spectroscopy. If we use digital photons to measure the intensity fluctuations, the term photon correlation spectroscopy (PCS) is then used to refer to the technique described here.

In the last two decades, thanks to the advance of stable laser, ultrafast electronics and personal computers, LLS, especially dynamic LLS, has evolved from a very special instrument for physicists and physical chemists to a routine analytical tool in polymer laboratories or even to a daily quality-control device in production lines. Commercially available research-grade LLS instruments are normally capable of making static and dynamic measurements simultaneously for studies of colloidal particles in suspension or macromolecules in solution as well as in gels and viscous media.

1.1.1 ENERGY TRANSFER VERSUS MOMENTUM TRANSFER

Considering the interaction of light (an electromagnetic radiation) with matter, we can describe it in terms of two fundamental quantities: the momentum transfer ($\hbar\mathbf{K}$) and the energy transfer ($\hbar\Delta\omega$) obeying the conservation equations

$$\hbar\mathbf{K} = \hbar(\mathbf{k}_i - \mathbf{k}_s) \tag{1}$$

$$\hbar\Delta\omega = \hbar(\omega_i - \omega_s) \tag{2}$$

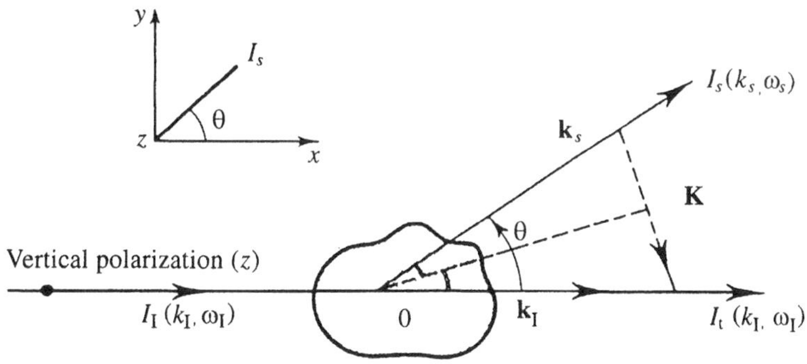

Figure 1.1 Scattering geometry. I_I, I_s, and I_t are, respectively, the incident, the scattered, and the transmitted intensities; θ is the scattering angle; and $K_I = 2\pi/\lambda_I$, $k_s = 2\pi/\lambda_s$ with $\lambda_i \equiv (\lambda_0/n) \cong \lambda_s$, λ_0 being the wavelength *in vacuo*, and n the refractive index of the scattering medium. $\mathbf{K}\,[= (4\pi/\lambda)\sin(\theta/2)]$ is the magnitude of the momentum transfer vector. For visualization of the geometry in Eq. (1), we set the incident beam polarization to be perpendicular to the plane of the paper and the scattering plane defined by I_I and I_s.

where $\hbar = h/2\pi$ with h being Planck's constant; $\mathbf{k}_i$, $\mathbf{k}_s$, and ω_i, ω_s are, respectively, the incident and scattered wave vectors with magnitudes $2\pi/\lambda_i$, $2\pi/\lambda_s$ and angular frequencies $2\pi\nu_i$, $2\pi\nu_s$, as shown in Figure 1.1. For structural and dynamic information, we can use $R \approx \mathbf{K}^{-1}$ as a spatial resolution ruler with which static LLS is able to probe the size of colloidal particles and macromolecules; and $\tau \approx 1/\Delta\nu = 1/(\nu_I - \nu_s)$ as a characteristic time with which dynamic (quasi-elastic) LLS is able to measure the translational or internal motions, or both, of colloidal particles in suspension or macromolecules in solution as well as their cooperative motions in complex fluids. In Table 1.1, typical magnitudes of the momentum and energy transfers in LLS are compared with those of small-angle X-ray scattering (SAXS) [3]. It is worth noting that small-angle neutron scattering (SANS) with neutron wavelength of a few tenths of a nanometer has a ΔK-range as SAXS. Table 1.1 illustrates that LLS is complementary to both SAXS and SANS.

1.1.2 SCOPE OF LASER LIGHT SCATTERING

The magnitude of the scattering vector $\mathbf{K}\,[= 4\pi \sin(\theta/2)/\lambda]$ is a pertinent parameter in all the scattering experiments, not the scattering angle θ or the wavelength λ of the probing radiation in the scattering medium but the ratio of $\sin(\theta/2)/\lambda$. This implies that visible light with wavelengths in the range ≈400–760 nm in vacuum can only have relatively small values of K even at the maximum value for

Table 1.1 Typical magnitudes of momentum and energy transfers in laser light scattering (LLS) and small-angle X-ray scattering (SAXS).

	LLS		***SAXS***	
(λ/nm)	300		0.15	
Scattering angle (θ/rad)	Minimum	Maximum	Minimum	Maximum
	5×10^{-2}	π	5×10^{-4}	2×10^{-1}
Momentum transfer				
($\mathbf{K}$/nm^{-1})	1×10^{-3}	4×10^{-2}	2×10^{-2}	10
($R \approx \mathbf{K}^{-1}$/nm)	1×10^{3}	25	25	10^{-1}
Energy transfer[a]				
$h\Delta\nu(= \hbar\Delta\omega)$/eV	$\approx 10^{-15}$	$\approx 10^{-9}$	No energy transfer	
τ/s	≈ 10	$\approx 10^{-5}$		

[a]Respectively calculated on the basis of the detectable particle size range of 1–1000 nm in dynamic LLS.

$\theta = \pi$, and thus, if using visible light as a probing radiation, we can only measure the size R down to about tens of nanometers. On the other hand, with modern LLS instrumentation, we are able to measure static and dynamic scattering at a scattering angle as small as $\approx 3°$ or $\approx 5 \times 10^{-2}$ radians, $R \approx \mathbf{K}^{-1} \approx 1000$ nm, i.e., small-angle laser light scattering (SALLS) is capable of measuring micron-sized colloidal particles in suspension or macromolecules in solution as well as slow relaxations in gels and viscous media. In comparison with visible light, X-ray and neutron scattering at a very small angle (e.g., 5×10^{-4} radians) can have a value of K as small as 2×10^{-2} nm^{-1}, which is very close to the highest K of visible light. Therefore, it is feasible to overlap the SAXS or SANS pattern experimentally with the one from visible light scattering under favorable conditions. In other words, if making SAXS measurements at very small scattering angles and visible LLS at large scattering angles, we are able to match the SAXS and LLS results experimentally. At higher scattering angles, SAXS or SANS overlaps with its diffraction range and is able to reach atomic dimensions. The visibility of the scattering objects (macromolecules or colloidal particles) in LLS, SAXS, and SANS, respectively, depends on the differences in the refractive index, the electron density, and the scattering cross section between the scattering object and the background.

Strictly speaking, QELS includes Fabry–Perot interferometry and optical mixing spectroscopy; i.e., the frequency broadening due to translational or internal motions of the scattering object can be detected either in the frequency domain or in the time domain. Nowadays, the most commonly used method in QELS is a digital technique of photon correlation spectroscopy in the self-beating mode to measure the intensity fluctuations of the scattered light in the time domain. In dynamic LLS, translational motions of macromolecules or particles within the size range 1–1000 nm can be measured, whereas structures with the same size can be studied experimentally by a combination of static LLS, SAXS, and SANS. The characteristic time of dynamic relaxation in dynamic LLS, which includes translational, rotational, and internal motions, could vary from seconds to tens of nanoseconds [4]. In this chapter, we shall discuss only dynamic LLS—specifically, the self-beating intensity–intensity time correlation spectroscopy—and ignore the interferometry technique, which could be an appropriate method to study the dynamics of complex fluids [5].

Many reviews, books, proceedings, and chapters have been published on the topic. The present chapter can be viewed as a long abstract in that context, discussing only the basic practice and principles of laser light scattering. The interested reader should consult [1] and [2] for details. For those who are interested in a particular application of LLS, Appendix I of [1] could be a good starting point. As a chapter for beginners, readers could also use other books, rather than proceedings or articles, as reference materials. In particular, the first monograph on the theoretical aspects of dynamic laser light scattering by Berne and Pecora [6] is highly recommended because it remains the best source reference in the relation between the basic equations of light scattering and the dynamic physical parameters of macromolecules in solution and colloidal particles in suspension. For the convenience of discussion, hereafter, macromolecules and colloidal particles are referred to as *particles*.

Basic static and dynamic LLS theories are outlined in Sections 1.2 and 1.3, respectively. The emphasis is on the principles of light scattering, not on the theoretical relations between light scattering and the physical parameters of interest via statistical mechanics; namely, we will use much description instead of mathematical equations. In the past, static and dynamic LLS were often used separately, which seriously limited their applications. Section 1.4 specially deals with this problem by using a few examples to show how to combine static and dynamic LLS results to extract more information. Section 1.5 illustrates the practical experimental aspects of light scattering, including the development of laser light sources, the optical and special cell design, sample preparation, and differential refractometry.

1.2 Static Laser Light Scattering

1.2.1 INTENSITY OF SCATTERED LIGHT

For an incident beam I_{INC} having its polarization vertical to a horizontal scattering plane, as shown schematically in Figure 1.2, the scattered intensity of a single

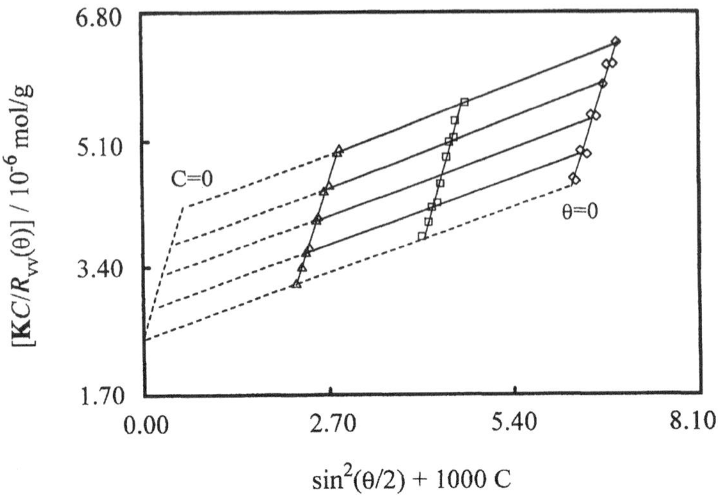

Figure 1.2 Typical Zimm plot for an alternating copolymer of ethylene and tetrafluoroethylene ($M_w = 5.4 \times 10^5$ g/mol, $R_g = 45.4$ nm, and $A_2 = 1.97 \times 10^{-4}$ mol ml/g²) in diisobutyl adipate at 240°C.

small particle i_s is

$$\frac{i_s}{I_{INC}} = \frac{k^4\alpha^2}{d^2} \tag{3}$$

where $k = 2\pi/\lambda_0$, α is the polarizability of the particle, and d is the distance between the particle and the observer. For N identical particles in a solvent of refractive index n_0 and polarizability α_0, the background scattering of the solvent has to be subtracted; namely, α in Eq. (3) has to be replaced by $\alpha_{ex} = \alpha - \alpha_0$ as follows:

$$n^2 - n_0^2 = 4\pi N\alpha_{ex} \tag{4}$$

$$\alpha_{ex} = \frac{n_0}{2\pi}\left(\frac{n - n_0}{C}\right)\frac{M}{N_A} \tag{5}$$

where $N = CN_A/M$ with C being the weight concentration. The excess scattered intensity I_{ex} for a dilute solution with N identical small particles in volume V without both intraparticle interference (i.e., the particles are much smaller than λ) and interparticle interactions (i.e., the particles are sufficiently far apart from each other) is

$$\frac{I_{ex}}{I_{INC}} = \frac{k^4\alpha_{ex}^2 N}{d^2} \tag{6}$$

Correspondingly, the excess Rayleigh ratio $\Delta R(K)$ $(= I_{ex}d^2/I_{INC})$ of the solute particles for the vertically polarized incident and scattering lights has the form

$$\Delta R_{VV}(K) = \frac{k^4 n_0^2}{4\pi^2 N_A}\left(\frac{\partial n}{\partial C}\right)^2_{T,P} MC = HMC \tag{7}$$

where the optical constant $H = 4\pi^2 n_0^2(\partial h/\partial C)^2_{T,P}/(N_A\lambda_0^4)$. In the presence of intraparticle interference (i.e., the particle is not so small that light scattered from two scattering elements within the volume of the same particle has a significant phase difference), the scattered intensity of a single particle i_s of a uniform polarizability α due to a phase shift is

$$\frac{i_s}{I_{INC}} = \frac{k^4\alpha^2}{d^2 V_P}\int_{V_P} p(r)\exp(i\mathbf{K}\cdot\mathbf{r})\,dv = \frac{k^4\alpha^2}{d^2}P(K) \tag{8}$$

or written as

$$\Delta R_{VV}(K) = HMCP(K) \tag{9}$$

where $\exp(i\mathbf{K}\cdot\mathbf{r})$ is the phase factor with $\mathbf{r}\,(= \mathbf{r}_i - \mathbf{r}_j)$ being the vector distance between the two scattering elements inside the particle volume V_P; $p(r)$ is a radial distribution function for the scattering elements inside the particle, which may be defined by the statement that $p(r)\,dv/V_P$ is the probability of finding the ith scattering element within the volume element dv at a distance r from the jth scattering element; and

$$P(K) = \frac{1}{V_P}\int_{V_P} p(r)\exp(i\mathbf{K}\cdot\mathbf{r})\,dv \tag{10}$$

is a normalized intraparticle scattering factor of a single particle of uniform density and finite size. The integration is over all orientations and magnitudes of $\mathbf{r}$ at constant $\mathbf{K}$ and also over all the scattering elements inside the entire particle volume V_P [7].

1.2.2 SCATTERING BY PARTICLES WITH DIFFERENT SHAPES

In general, for a randomly oriented particle with an arbitrary shape, the probability that $\mathbf{K}$ and $\mathbf{r}$ have an angle between α and $\alpha + d\alpha$ is $2\pi \sin\alpha\, d\alpha$ and thus the average of the phase factor,

$$\langle\exp(\mathbf{K}\cdot\mathbf{r})\rangle = \frac{\displaystyle\int_0^{\pi}\cos(Kr\cos\alpha)\sin\alpha\, d\alpha}{\displaystyle\int_0^{\pi}\sin\alpha\, d\alpha} = \frac{\sin(Kr)}{Kr} \tag{11}$$

where $K = |\mathbf{K}|$ and $r = |\mathbf{r}_i - \mathbf{r}_j|$. Therefore, for a single particle with a radial distribution function $p(r)$ and finite size, its normalized intraparticle scattering factor can generally be defined as

$$P(K) = \frac{\int_{V_P} p(r) \frac{\sin(Kr)}{Kr} dv}{\int_{V_P} p(r)\, dv} \tag{12}$$

It is clear that, for a uniform sphere, $p(r) = 1$ and $\int_{V_P} p(r)\, dv = V_P$, and thus

$$P(K) = \frac{1}{V_P} \int_0^{2\pi} d\varphi \int_0^{\pi} \sin\phi\, d\phi \int_0^R r^2 \frac{\sin(Kr)}{Kr} dr = \frac{4\pi}{V_P} \int_0^R r^2 \frac{\sin(Kr)}{Kr} dr \tag{13}$$

where R is the radius of the particle. On the other hand, for a uniform, long, thin, rigid rod (i.e., its diameter is much smaller than not only its length L but also the light wavelength λ) in an equilibrium ensemble, its orientation in all directions is equally probable so that its orientation distribution function is $p(\varphi, \phi) = 1/4\pi$. Therefore, $P(K)$ can be written as

$$P(K) = \frac{1}{4\pi L} \int_0^{2\pi} d\varphi \int_0^{\pi} \sin\phi\, d\phi \int_{-L/2}^{L/2} \exp(i\mathbf{K}\cdot\mathbf{r})\, dr \tag{14}$$

where the integration of r is in one dimension along the length of the rod. If we let $\mathbf{K}$ be along the z axis and use spherical polar coordinates, $\mathbf{K}\cdot\mathbf{r} = Kr\cos\phi$ and P(K) can be expressed in terms of a spherical zero-order Bessel function $j_0(w) = \sin w/w$; i.e.,

$$P(K) = \frac{1}{4\pi} \int_0^{2\pi} d\varphi \int_0^{\pi} \sin\phi\, d\phi \left| j_0\left(\frac{x}{2}\cos\phi\right)\right|^2 \tag{15}$$

Letting $y = \cos\theta$ and integrating over ϕ, we have

$$P(K) = \frac{1}{2} \int_{-1}^{1} \left| j_0\left(\frac{xy}{2}\right)\right|^2 dy \tag{16}$$

where $x = KL$. Equation (16) cannot be solved analytically, but a numerical evaluation can show how $P(K)$ depends on K and more precisely on x. Equation (16) can also be written in a more common form [6]

$$P(K) = \frac{2}{x} \int_0^{\pi} dz \frac{\sin z}{z} - \left[\frac{2}{x}\sin\left(\frac{x}{2}\right)\right]^2 \tag{17}$$

The integral on the left is a tabulated function. Further, for a Gaussian chain with a total number of n statistic segments its mean square end-to-end distance by definition is

$$\langle r^2(n)\rangle = \ell^2 n \tag{18}$$

where ℓ is the length of a statistic segment depending on the nature of the particular chain. The definition is also valid for a portion of the chain with a number of m statistic segments. In the volume element dv, the probability of finding any two segments (i, j, and $m = i - j$) follows a Gaussian distribution [8]

$$w(r)\,dv = \left[\frac{3}{2\pi\,\langle r^2(m)\rangle}\right]^{3/2} \exp\left(-\frac{3r^2}{2\langle r^2(m)\rangle}\right) dv \tag{19}$$

It is worth noting that this probability distribution is derived from the random flight of a particle over a distance r over a large number of steps, implying that the polymer chain is very flexible and that each statistic segment contains a sufficient number of elementary chemical bonds. Integrating $w(r)$ over all possible values of m (i.e., from 0 to n), we have

$$p(r) = \frac{2}{n^2}\int_0^n (n-m)\left[\frac{3}{2\pi\,\langle r^2(m)\rangle}\right]^{3/2} \exp\left(-\frac{3r^2}{2\langle r^2(m)\rangle}\right) dm \tag{20}$$

Placing $p(r)$ in Eq. (10), we have

$$P(K) = \frac{2}{n^2}\int_0^n (n-m)\,dm \int_{V_P}\left[\frac{3}{2\pi\,\langle r^2(m)\rangle}\right]^{3/2} \exp\left(i\mathbf{K}\cdot\mathbf{r} - \frac{3r^2}{2\langle r^2(m)\rangle}\right) dv \tag{21}$$

Integrating r over the particle volume V_P, we finally get

$$P(K) = \frac{2}{n^2}\int_0^n (n-m)\exp\left(-\frac{K\,\langle r^2(m)\rangle}{6}\right) dm \tag{22}$$

or further,

$$P(K) = \frac{2}{\left(K^2 R_g^2\right)^2}\left[K^2 R_g^2 - 1 + \exp\left(-K^2 R_g^2\right)\right] \tag{23}$$

where R_g^2 ($= \ell^2 n/6$) is the square radius of gyration of the Gaussian chain, and its general definition is

$$R_g^2 = \frac{\displaystyle\int_{V_p} \rho(\mathbf{r})|\mathbf{r}-\mathbf{r}_0|^2\,dv}{\displaystyle\int_{V_p} \rho(\mathbf{r})\,dv} \tag{24}$$

where $\rho(\mathbf{r})$ is the mass density at $\mathbf{r}$ with $\mathbf{r}_0$ being the gravity center of the particle and M is the mass of the particle. For a uniform sphere with a radius of R_0,

$$R_g^2 = \frac{\int_0^{R_0} 4\pi r^4 \, dr}{\int_0^{R_0} 4\pi r^2 \, dr} = \frac{3}{5} R_0^2 \tag{25}$$

For a uniform rigid rod with a length of L, the number of mass elements at a distance between x and $x + dx$ is proportional to dx. Therefore,

$$R_g^2 = \frac{\int_{-L/2}^{L/2} x^2 \, dx}{\int_{-L/2}^{L/2} dx} = \frac{L^2}{12} \tag{26}$$

On the basis of Eq. (23), at small values of KR_g, the scattering factor of the polymer chain can be approximated by

$$P(K) \approx 1 + \frac{1}{3} K^2 R_g^2 + \cdots \tag{27}$$

It can be shown that Eq. (23) is not only valid for the Gaussian chain but also for particles or macromolecules of arbitrary shape. This is because Eq. (12) can be approximated as

$$P(K) \approx 1 - \frac{K^2}{6} \frac{\int_{V_P} p(r) r^2 \, dv}{\int_{V_P} p(r) \, dv} + \cdots \tag{28}$$

where $r = |\mathbf{r}_i - \mathbf{r}_j|$. Replacing r with $|(\mathbf{r}_i - \mathbf{r}_0) - (\mathbf{r}_j - \mathbf{r}_0)|$, we can rewrite Eq. (26) as

$$P(K) \approx 1 + \frac{K^2}{6} \left[2 \frac{\int_{V_p} \rho(r)(\mathbf{r} - \mathbf{r}_0)^2 \, dv}{\int_{V_p} \rho(r) \, dv} + 2 \left(\frac{\int_{V_p} \rho(r)(\mathbf{r} - \mathbf{r}_0) \, dv}{\int_{V_p} \rho(r) \, dv} \right)^2 \right] + \cdots \tag{29}$$

where the second term in the bracket is equal to zero because $\int_{V_p} \rho(r)\mathbf{r} \, dv / \int_{V_p} \rho(r) \, dv = \mathbf{r}_0$, and thus we have $P(K) \approx 1 + (1/3)K^2 R_g^2 + \cdots$, as long as $KR_g \ll 1$. A graphic display of $P(K)$ of sphere, thin rod, and Gaussian coil can be found elsewhere [9].

1.2.3 *ZIMM PLOT—CONCENTRATION AND ANGULAR DEPENDENCE*

So far, we have only dealt with monodisperse particles at infinite dilution; namely, we have not considered the interparticle interaction or the interference between the light scattered from different particles. For large particles ($R_0 > \lambda/10$) in a dilute solution, Eq. (9) is simply not a linear function of the particle concentration. Considering the intraparticle and interparticle interference between the scattered light, Debye [10] showed that the concentration dependence can be expanded as a power series in concentration, i.e., the virial expansion

$$\frac{HC}{R_{\mathrm{VV}}(K)} = \frac{1}{MP(K)} + 2A_2C + \cdots \tag{30}$$

where A_2 is the second virial coefficient. Further, by considering that the particles have a size or mass distribution, we can rewrite Eq. (9) as

$$\sum_i [\Delta R_{\mathrm{VV}}(K)]_i = H \sum_i M_i C_i \left(1 - \frac{1}{3}K^2 R_{\mathrm{g},i}^2 + \cdots\right)$$

or

$$\sum_i [\Delta R_{\mathrm{VV}}(K)]_i = HCM_{\mathrm{w}} \left(1 - \frac{1}{3}K^2 \langle R_{\mathrm{g}}^2 \rangle_z + \cdots\right) \tag{31}$$

where $M_{\mathrm{w}}(= \sum M_i C_i / C)$ is the weight average molar mass, and $\langle R_{\mathrm{g}}^2 \rangle_z^{1/2} (= \sum M_i C_i R_{\mathrm{g},i}^2 / \sum M_i C_i)$ (or written as $\langle R_{\mathrm{g}} \rangle$) is the root mean square z-average radius of gyration. Thus, we have the basic equation for a polydisperse sample in dilute solution and measured under the condition $K\langle R_{\mathrm{g}} \rangle < 1$ in static light scattering

$$\frac{HC}{R_{\mathrm{VV}}(K)} \approx \frac{1}{M_{\mathrm{w}}} \left(1 + \frac{1}{3}K^2 \langle R_{\mathrm{g}}^2 \rangle_z\right) + 2A_2C \tag{32}$$

where ΔR_{VV} is now denoted by $\mathrm{R_{VV}}$ because the excess value is obvious. It shows that with $R_{\mathrm{vv}}(K)$ measured over a series of C and K, we are able to determine $\langle R_{\mathrm{g}} \rangle$ from the slope of $[HC/R_{\mathrm{vv}}(K)]_{C\to 0}$ versus K^2; A_2 from the slope of $[HC/R_{\mathrm{vv}}(K)]_{K\to 0}$ versus C; and M_{w} from $[HC/R_{\mathrm{vv}}(K)]_{C\to 0, K\to 0}$. The Zimm plot, i.e., $HC/R_{\mathrm{vv}}(K)$ versus $K^2 + kC$ with k being an adjustable constant, allows both K and C extrapolations to be made on a single grid [11]. Figure 1.3 shows a typical Zimm plot for an alternating copolymer of ethylene and tetrafluoroethylene ($M_{\mathrm{w}} = 5.4 \times 10^5$ g/mol, $\langle R_{\mathrm{g}} \rangle = 45.4$ nm, and $A_2 = 1.97 \times 10^{-4}$ mol ml/g^2) in diisobutyl adipate at 240°C [12]. It should be noted that Eq. (32) is valid under the restriction that the polymer solution exhibits no absorption, no fluorescence, and no depolarized scattering. As for anisotropic rigid and nearly rigid rods that result in

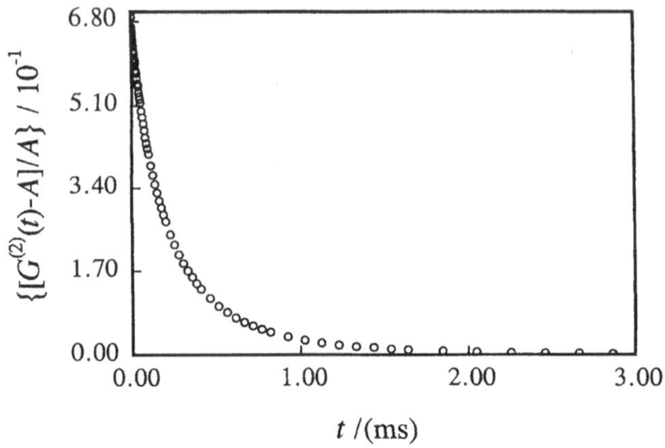

Figure 1.3 Typical normalized intensity–intensity time correlation function for chitosan (M_w = 1.06 × 10^5 g/mol and $\langle\Gamma\rangle$ = 2.19 ms) in 0.2 *M* CH_3COOH/0.1 *M* CH_3COONa aqueous solution at T = 25°C, θ = 45°, and C = 4.96 × 10^{-4} g/mol.

depolarized scattering, readers should refer to the excellent review article of Russo and the references therein [13]. As for the correction of absorption and fluorescence, readers are advised to refer to the characterization of Kevlar (a DuPont trademark) in concentrated sulphuric acid by Chu *et al.* [14, 15] and Ying and Chu [16].

In practice, the Rayleigh ratio is determined by a relative method, namely, by measuring the scattered intensity of a standard, such as benzene or toluene, we can calculate the Rayleigh ratio of a given solution using the expression

$$R_{VV}(K) = R_{VV}^0(K)\frac{\langle I\rangle_{\text{solution}} - \langle I\rangle_{\text{solvent}}}{\langle I^0\rangle}\left(\frac{n}{n^0}\right)^{\gamma} \tag{33}$$

where the superscript 0 denotes the standard and I and n are, respectively, the time-averaged scattered intensity and the refractive index. The term $(n/n^0)^\gamma$ is a refraction correction for the scattering volume and γ is a constant between 1 and 2, depending on the detection geometry of the light-scattering instrument, because we should compare the same scattering volume from the solution and the reference standard. If taking the incident light as the x-direction and the scattered light as the y-direction (i.e., $\theta = 90°$), we only need to have a linear correction of the refraction in the x-direction if a slit is used to determine the scattering volume (i.e., $\gamma = 1$) because we already see all the scattered lights in the z-direction (vertical). On the other hand, if a pinhole with a size much smaller than the diameter of the incident beam at the center of the scattering cell were used, we would have to correct the refraction in both the x- and z-directions (i.e., $\gamma = 2$). However, if the pinhole size is comparable with the beam diameter, $1 < \gamma < 2$. In practice, we should avoid this situation by choosing either a slit or a smaller pinhole.

1.3 Dynamic Light Scattering

1.3.1 SPECTRUM OF SCATTERED LIGHT

The phase integral in Eq. (10) accounts only for the intraparticle interference effect. However, in optical mixing spectroscopy (dynamic light scattering), we now consider the interference of the light scattered by different volume elements within a scattering volume V with local dielectric constant fluctuations. Therefore, the phase integral in this case has the same form as Eq. (10) but is over a scattering volume V (no longer the particle volume V_{p}), and $\mathbf{r}$ is the position vector in V. For a rectangular parallelepiped of dimensions L_x, L_y, and L_z the normalized phase integral as a function of $\mathbf{K}'$ around $\mathbf{K}' = 0$ results in

$$\int_V \exp(i\mathbf{K}\cdot\mathbf{r})\,dv = \begin{cases} V & \text{when } 2\pi/|\mathbf{K}| \text{ is within } V \\ 0 & \text{otherwise} \end{cases} \tag{34}$$

where $\mathbf{K}$ $(= \mathbf{r}_i - \mathbf{r}_s + \mathbf{K}')$ and the variation range of $\mathbf{K}'$ is the uncertainty of the momentum transfer and is related to the finite size (and shape) of V. It is important to keep the angular aperture of the detector *small* in the design of an optical mixing instrument because the farfield observation of the radiation field from the plane wave components of local dielectric constant fluctuations must satisfy the conditions of Eq. (34). A practical question of how small the angular aperture is has been discussed by Berne and Pecora [6]. It has been estimated that the coherence area A_{coh} for a typical optical mixing experiment is

$$A_{\mathrm{coh}} \approx \frac{\lambda^2}{\Omega} \tag{35}$$

where Ω is the solid angle subtended by the scattering volume at the detector. In an optical mixing experiment, the important quantity is the signal per coherence area. By reducing the scattering volume, we can have a smaller Ω and a larger A_{coh}. However, the larger the scattering volume, the stronger the scattered intensity and the smaller the statistical noise. Therefore, there is a trade-off and balance in choosing a proper scattering volume. We will come back to this point later in Section 1.5.

1.3.2 SIEGERT RELATION

Without a local oscillator (i.e., a constant fraction of the incident light reaching the detector from various intentional or unintentional sources, such as surface scratching or reflection), the self-beating of the scattered electric field leads to the intensity–intensity time correlation function, $G^{(2)}(K,t)$ based in essence on the Siegert relation:

$$G^{(2)}(K,t) = \langle I(K,0)I(K,t)\rangle = A\left(1+\beta\left|g^{(1)}(K,t)\right|^2\right) \tag{36}$$

where A ($\equiv \langle I(K,0)I(K,0)\rangle$) is the baseline, t is the delay time, β is a parameter depending on the coherence of the detection optics, $|g^{(1)}(K,t)|$ ($= \langle \mathbf{E}(K,0)\mathbf{E}^*(K,t)\rangle / \langle \mathbf{E}(K,0)\mathbf{E}^*(K,0)\rangle$) is the normalized electric field–field time correlation function, and $I(K,t)$ is the detected scattered intensity or photon counts at time t, including contributions from the solvent and the solute. Therefore, $G^{(2)}(K,t) = \langle [I_{\text{solvent}}(K,0) + I_{\text{solute}}(K,0)][I_{\text{solvent}}(K,t) + I_{\text{solute}}(K,t)]\rangle$ and Eq. (36) becomc

$$G^{(2)}(K,t) = A\left\{1+\beta\left[\frac{I_{\text{solvent}}}{I_{\text{solution}}}\left|g^{(1)}_{\text{solvent}}(K,t)\right| + \frac{I_{\text{solute}}}{I_{\text{solution}}}\left|g^{(1)}_{\text{solute}}(K,t)\right|\right]^2\right\} \tag{37}$$

where all the cross terms have been dropped by assuming that the light scattered by solvent molecules and particles is not correlated. It should be noted that $|g^{(1)}_{\text{solvent}}(K,t)|$ decays much faster than $|g^{(1)}_{\text{solute}}(K,t)|$ because small solvent molecules diffuse much faster than larger particles. Thus, after a very short delay time, Eq. (37) becomes

$$G^{(2)}(K,t) \cong A\left[1+\beta\left(\frac{I_{\text{solute}}}{I_{\text{solution}}}\right)^2\left|g^{(1)}_{\text{solute}}(K,t)\right|^2\right] = A\left[1+\beta_{\text{app}}\left|g^{(1)}_{\text{solute}}(K,t)\right|^2\right] \tag{38}$$

where $\beta_{\text{app}} = \beta(I_{\text{solute}}/I_{\text{solution}})^2$. For a dilute solution, the scattering from solvent molecules could become appreciable (i.e., $I_{\text{solute}} \leq I_{\text{solution}}$) and thus the apparent coherence would be lower; i.e., $G^{(2)}(K,0)$ appears to have a lower value than expected. The reader should be aware of this fact, especially for weakly scattered, dilute low-molar-mass polymer solution. For example, if $I_{\text{solute}} = I_{\text{solvent}}$, $\beta_{\text{app}} = \beta/4$. It should be noted that β is a constant for each particular optical geometry of the scattering instrument. In fact, I_{solute} can be estimated from β_{app} if the values of β at different scattering angles have been precalibrated with a narrowly distributed latex standard whose scattering intensity is much stronger than water (solvent), as was first demonstrated by Sun *et al.* [17]. The beginner in LLS should be aware that such a measurement is not a routine method and is reserved only for some particular experiments in which a direct and accurate measurement of $I_{\text{solution}} - I_{\text{solvent}}$ is difficult.

1.3.3 DIFFUSIONS AND INTERNAL MOTIONS

Generally, the relaxation of $|g^{(1)}(K, t)|$ includes diffusion (translation and rotation) and internal motions. Let us first consider the translational diffusion relaxation of the particles. For a monodisperse sample, $|g^{(1)}(K, t)|$ is theoretically represented by

$$\left|g^{(1)}(K, t)\right| = Ge^{-\Gamma t} \tag{39}$$

where G and Γ are the proportionality factor and the line width, respectively. For a dilute solution, Γ measured at a finite scattering angle is related to C and K by [18]

$$\Gamma = K^2 D(1 + k_d C)\left(1 + f\langle R_g^2\rangle_z K^2\right) \tag{40}$$

Here D is the translational diffusion coefficient of the solute molecule at $C \to 0$, k_d is the diffusion second virial coefficient, and f is a dimensionless parameter depending on polymer chain structure and solvent. Hence, for small C and $\mathbf{K}$, $D \approx \Gamma/K^2$. For a polydisperse polymer sample with a continuous distribution of molar mass M, Eq. (39) may be generalized as

$$\left|g^{(1)}(K, t)\right| = \int_0^\infty G(D)e^{-K^2 Dt}\, dD \tag{41}$$

where $G(D)$ is called the translational diffusion coefficient distribution. It should be noted that by the definition of $|g^{(1)}(K, t)|$, $G(D)$ is an intensity distribution of D. This equation is the basis of some discussion in this chapter. Note that because $|g^{(1)}(K, t)|$ approaches unity as $t \to 0$, we have

$$\left|g^{(1)}(K, t \to 0)\right| = \frac{\langle \mathbf{E}(K, 0)\mathbf{E}^*(K, t \to 0)\rangle}{\langle \mathbf{E}(K, 0)\mathbf{E}^*(K, 0)\rangle} = \int_0^\infty G(D)\, dD = 1 \tag{42}$$

Figure 1.3 illustrates the $g^{(1)}(K, t)$ data for chitosan ($M_w = 1.06 \times 10^5$ g/mol and $\langle D\rangle = 5.92 \times 10^{-8}$ cm^2/s) in 0.2 M CH_3COOH/0.1 M CH_3COONa aqueous solution at 25°C, $\theta = 45°$, and $C = 4.96 \times 10^{-4}$ g/mL [19], where $\langle D\rangle$ is the average diffusion coefficient defined as

$$\langle D\rangle = \int_0^\infty G(D) D\, dD \tag{43}$$

In the consideration of the contribution of the rotational diffusion to the relaxation, the simplest case would be a monodisperse rigid thin rod. For an incident light vertically polarized to the scattering plane, the time correlation function of vertically polarized scattered light in the self-beating mode leads to

$$\left|g^{(1)}(t, K)\right|_{\mathrm{VV}} \propto e^{-K^2 Dt}(1 + \gamma e^{-6D_r t}) \tag{44}$$

and that of horizontally polarized scattered light in the self-beating mode has the form

$$\left|g^{(1)}(t, K)\right|_{\mathrm{VH}} \propto e^{-k^2 Dt} e^{-6D_r t} \tag{45}$$

where D_r is the rotational diffusion coefficient and γ is a constant that may be related to molecular anisotropy or configuration. For details, the reader should refer to Berne and Pecora's book [6] and the chapter written by Russo [13]. Equation (44) shows that if the scattering angle is very small (i.e., $K \ll 1/L$), $|g^{(1)}(t, K)|_{\mathrm{VV}}$ is approximately proportional to $e^{-K^2 Dt}$ because the relaxation of the rotation term is relatively fast in this case. Therefore, to observe the rotational diffusion, one has to measure the time correlation function at a relatively high scattering angle.

As for a long, flexible polymer chain, we have to consider the relaxation associated with the internal motions, which are also known as the normal modes or "breathing modes." For simplicity, we leave the polydispersity out of the following discussion. As was shown by Berne and Pecora, when an infinitely dilute polymer solution is illuminated by a coherent and monochromatic laser light beam, the spectral distribution of the light scattered from a flexible polymer chain can be written as

$$S(\mathbf{K}, \omega) = \frac{1}{2\pi} \int e^{-i\omega t} e^{-q^2 D|t|} J(\mathbf{K}, t)\, dt \tag{46}$$

where ω is the difference between the angular frequency of the scattered light and that of the incident light, $\mathbf{K}$ is the scattering vector as previously defined, and the function

$$J(\mathbf{K}, t) = \left\langle \frac{1}{N^2} \sum_{i=0}^{N} \sum_{m=0}^{N} e^{-i\mathbf{K}\cdot[\mathbf{r}_\ell(0) - \mathbf{r}_m(t)]} \right\rangle \tag{47}$$

is the spatial Fourier transform of the segment–segment time correlation function. It arises from the interference of the scattered light from different segments in a polymer chain with N such segments. It contains all the temporal and spatial information on the intramolecular (or internal) motions of a polymer chain. Here $\mathbf{r}_\ell(0)$ is the position of the lth segment at time 0 and $\mathbf{r}_m(t)$, that of the mth segment at time t, both are referred to the center of mass of the polymer chain.

To perform the ensemble average in $J(K, t)$, an explicit model for the internal motions of a polymer chain is needed. By incorporating the Oseen–Kirkwood–

Riseman hydrodynamic interaction into the bead-and-spring model, Perico *et al.* [20] have shown

$$\begin{aligned} S(K,\omega) &= P_0(x)L(\omega, DK^2) + \sum_{\alpha=1}^{N} P_1(x,\alpha)L(\omega, DK^2+\Gamma_\alpha) \\ &\quad + \sum_{\beta=1}^{N}\sum_{\alpha=1}^{N} P_2(x,\alpha,\beta)L(\omega, DK^2+\Gamma_\alpha+\Gamma_\beta) + \cdots \end{aligned} \tag{48}$$

where $x = (R_g K)^2$ and the functions

$$L(\omega, \Lambda) = \frac{2\Lambda}{2\pi(\omega^2 + \Lambda^2)} \tag{49}$$

represent the ω-normalized Lorentzian distribution with Λ being the half-width at half-height, i.e., the line width, and the $P_n s$ $(n = 0, 1, 2, \ldots)$ determine the contributions of the different Lorentzians to the spectrum of the scattered light. The zeroth-order $P_0(x)$ represents the contribution of the translational diffusion, $P_1(x, \alpha)$ the first-order contribution of the αth internal mode, $P_2(x, \alpha, \beta)$ the second-order contribution of the αth and βth internal modes, and so on.

When $x < 1$, the spectral distribution is measured in the long-wavelength regime, and hence $P_0(x)$ is dominant in $S(K, \omega)$. As x increases, the contributions from $P_1(x, \alpha)$, $P_2(x, \alpha, \beta)$, and other higher-order terms become more and more important. Perico *et al.* [20] have numerically shown that $P_0(x, 1, 1)$ is the largest contribution to $S(K, \omega)$ among all the Lorentzian terms associated with the internal modes. In a modern dynamic laser light-scattering experiment, the intensity–intensity time correlation function of the scattered light is usually measured, from which $S(K, t)$, the Fourier transform of $S(K, \omega)$, is determined.

Figure 1.4 shows typical plots of $G(\Gamma/K^2)$ versus Γ/K^2 for a narrowly distributed high-molar-mass polystyrene standard ($M_w = 1.02 \times 10^7$ g/mol and $M_w/M_n = 1.17$) in toluene at $T = 20°\text{C}$ and at different x values; the insert shows a $\approx$10-times enlargement of the second (smaller) peak of the distribution in the range 10^{-7}–10^{-6} cm^2/s [21]. It clearly shows the following:

1. At $x < 1$, as expected and discussed previously, there exists only one single and narrow peak.
2. At $x \approx 1$, the second peak with a higher Γ appears in $G(\Gamma/K^2)$, whereas the first peak is basically unchanged in position and shape. This second peak is related to the internal motions.
3. At higher x, the first peak is getting broader and shifting to higher Γ. This is because at higher values of x, the measurement scale K^{-1} is

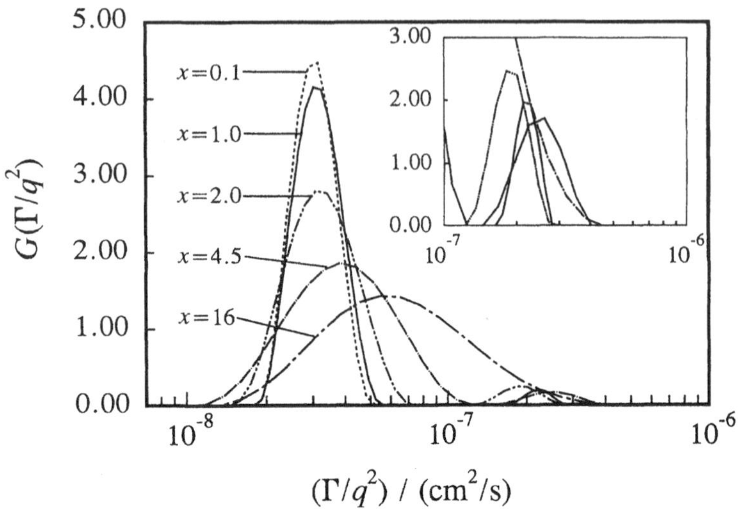

Figure 1.4 The x-dependence of $G(\Gamma/K^2)$ for a high-molar-mass polystyrene standard ($M_w = 1.02 \times 10^7$ g/mol and $M_w/M_n = 1.17$) in toluene at $T = 20°C$, where $x = (R_g K)^2$ and $G(\Gamma/K^2)$ was calculated by using the CONTIN Laplace inversion program.

smaller than the chain dimension $\approx R_g$, and thus the contributions from the translational and internal motions are mixed in the measured spectrum that caused the broadening and shifting of the first peak.

4. At $x > 15$, the first and second peaks in $G(K/\Gamma^2)$ merge into a single and broader distribution because the line width ($K^2 D$) associated with the pure translational diffusion increases with x, but the line widths related to the internal motions are independent of the scattering angle.

1.3.4 METHODS OF CORRELATION FUNCTION PROFILE ANALYSIS

Equation (39) indicates that once $|g^{(1)}(K, t)|$ is determined from $G^{(2)}(K, t)$ through Eq. (36), $G(\Gamma)$ or $G(D)$ can be computed from the Laplace inversion of $|g^{(1)}(K, t)|$ [22–26]. In the last three decades, many computation programs were developed. In the early stage, the computation speed was a very important factor in the program development. This constraint has gradually been removed because the personal computer has become faster and faster in the last 10 years. Among the many programs, the CONTIN program developed by Provencher [27] is still one of the most widely used and accepted for this computation. However, it should be noted that

Eq. (41) is one of the first kind of Fredholm integral equations. Its inversion is an ill-conditioned problem because of the bandwidth limitation of photon correlation instruments, some unavoidable noises in the measured time correlation function, and a limited number of data points. In other words, the data for $g^{(1)}(K, t)$ does not always provide information necessary and sufficient to determine $G(\Gamma)$ uniquely. Thus, in practice, reducing the noises in the measured intensity time correlation function becomes more important than choosing a program for data analysis. It is crucial that the solution be cleaned (i.e., "dust-freed") very thoroughly before it is subjected to laser light-scattering measurements. A common guideline is to keep the relative difference between the measured and the calculated baselines less than 0.1%. The error analysis related to the preceding problem can be found elsewhere [28, 29].

It is worth noting that there is a temptation among the users of dynamic LLS to extract too much information from the measured intensity–intensity time correlation function, actually from experimental noises. In the literature, three or four peaks in $G(D)$ were often reported. It is important to note that even a bimodal distribution of $G(D)$ has to be well justified by other physical evidence or pre-experimental knowledge. This does not mean that many of the Laplace inversion programs developed in the past are useless. On the contrary, they have been quite successful in retrieving the desired information, especially in terms of the average line width $\langle\Gamma\rangle$ ($\equiv \int_0^\infty \Gamma G(\Gamma)\, d\Gamma$) and the relative width of the line-width distribution $\mu_2/\langle\Gamma\rangle^2$ with $\mu_2 = \int_0^\infty (\Gamma - \langle\Gamma\rangle)^2 G(\Gamma)\, d\Gamma$. Therefore, the Laplace inversion is a very helpful method in the analysis of the line-width distribution, but it should be used with a clear understanding of its ill-conditioned nature and its limitations.

The uninitiated reader may wish to consult chapters from two books on polydispersity analysis: *Photon Correlation Spectroscopy of Brownian Motion: Polydispersity Analysis and Studies of Particle Dynamics* edited by Schulz-DuBois [30] and *Essentials of Size Distribution Measurements* edited by Dahneke [31]. In practice, if one is only interested in the determination of $\langle\Gamma\rangle$ and $\mu_2/\langle\Gamma\rangle^2$, a fast but more limited cumulants analysis adopted by Koppel [22] can be used, wherein $[G^{(2)}(K, t) - A]/A$ is expanded as

$$\ln\left[\frac{G^{(2)}(K, t) - A}{A}\right] = \ln\beta - \langle\Gamma\rangle t + \frac{\mu_2 t^2}{2!} - \frac{\mu_3 t^3}{3!} + \cdots \tag{50}$$

where

$$\mu_m = \int_0^\infty (\Gamma - \langle\Gamma\rangle)^m G(\Gamma)\, d\Gamma \tag{51}$$

is the mth moment of the line-width distribution $G(\Gamma)$. An mth-order cumulants fit means that all the terms higher than t^m in Eq. (50) are terminated in the data analysis. It is worth noting that, in practice, the cumulants fit can be used for a relatively narrow characteristic line-width distribution. For $\mu_2/\langle\Gamma\rangle^2 < \approx 0.2$, the second-order cumulants fit is normally sufficient, whereas with $\mu_2/\langle\Gamma\rangle^2$ in the range $\approx$0.2–0.3, the third-order cumulants fit is required. For even higher values of $\mu_2/\langle\Gamma\rangle^2$, higher-order expansions should be used. However, it is often difficult to know how many terms are sufficient to obtain a meaningful result because using too many terms in the cumulants fit might lead to an overfitting of experimental noises. Therefore, for broadly distributed samples, the use of a cumulants fit is very tedious. On the other hand, the use of CONTIN or MEM (maximum entropy method) can yield reliable $\langle\Gamma\rangle$ and $\mu_2/\langle\Gamma\rangle^2$ values under all conditions as long as the measured time correlation function is obtained within a proper bandwidth range and the photon counts have sufficient statistics, e.g., the baseline (A) has a relatively large number of total counts (i.e., over 10^6). However, the reader has to be warned that the line-width distribution obtained from the Laplace inversion is only an estimate, and one should therefore be aware of the limitations. Yet, it should be noted that the Laplace inversion method can provide useful information and distinguish between unimodal and multimodal line-width distributions—especially when the peak positions are separated by a factor of more than 2.

1.4 Methods of Combining Static and Dynamic LLS

Dynamic LLS (DLS) plays a very important role in particle sizing. In a dilute dispersion, if the colloidal particles have a spherical shape, or in other words, for colloidal particles in the absence of rotational and internal motions, $G(\Gamma)$ can be converted to the hydrodynamic size distribution $f(R_h)$ by means of $D = \Gamma/K^2$ and the Stokes–Einstein relation $D = k_B T/6\pi\eta R_h$ with k_B, T, and η being the Boltzmann constant, the absolute temperature, and solvent viscosity, respectively. All the parameters in the conversion are either well-known constants or precisely measurable by other methods. Therefore, particle sizing on the basis of DLS can be considered as an absolute method without calibration. This is why DLS is so successful in particle sizing and many commercial instruments have been developed by using the DLS principle. For details, the reader should refer to the book edited by Gouesbet and Grehan [32].

1.4.1 *MOLAR MASS DISTRIBUTION DETERMINATION*

The determination of molar mass and molar mass distribution is a very important part of polymer characterization. Many experimental methods, including LLS, have been developed and routinely used for this purpose. The absolute methods include end-group chemical analysis, vapor pressure osmometry, membrane osmometry, sedimentation equilibrium, static (classic) LLS, and the very recently developed matrix-assisted time-of-flight mass spectroscopy because they do not require a calibration with a set of narrowly distributed polymer samples with known molar masses. The relative methods include viscometry, gel permeation chromatography (GPC), field flow fractionation (FFF), and DLS.

The average molar mass $\langle M \rangle$ of a polydisperse polymer sample is generally defined by

$$\langle M \rangle = \int_0^\infty f_n(M) M^{\psi} dM \bigg/ \int_0^\infty f_n(M) M^{\psi^{-1}} dM \tag{52}$$

if ψ is an integer, where $f_n(M)$ is the number distribution of molecular weight M. Thus, $\psi = 1$ for the number-average molar mass (M_n), $\psi = 2$ for the weight-average molar mass (M_w), and $\psi = 3$ for z-average molar mass (M_z). For example, M_n is measured by end-group analysis and osmometry; M_w by GPC, sedimentation equilibrium, and static LLS; and M_z, by sedimentation equilibrium and DLS. In practice, the ratio of $M_\mathrm{w}/M_\mathrm{n}$ is called the polydispersity index and is conveniently used to characterize the distribution width of a given polymer sample.

Among these methods, using DLS to characterize polymer molar mass distribution has yet to become popular because it requires a more sophisticated instrumentation, a better understanding of LLS, and, more important, a calibration between D and M. It is worth noting that in the characterization of polymer molar mass distribution, LLS as a nonintrusive and nondestructive method has certain advantages because it can use a strong corrosive solvent, such as concentrated sulphuric acid, and it can be operated at high temperatures (e.g., at 340°C). Though not involving fractionation as in GPC, a combination of static and DLS results allows estimation of the molar mass distribution of a polymer. The principle is as follows: For a polydisperse polymer consisting of n homologous species it is known on the basis of Eq. (41) that $G(D_i)$ for species i at vanishingly small C and K is proportional to its scattered intensity I_i, which is further proportional to $M_i w_i$, and thus

$$G(D_i) = M_i w_i \bigg/ \sum_{j=1}^{n} M_j w_j \tag{53}$$

where w_i denotes the weight fraction of species i with molar mass M_i. For a continuous distribution of molar mass, this gives

$$G(D) dD = M f_w(M)\, dM / M_\mathrm{w} \tag{54}$$

where $f_w(M)$ denotes the differential weight distribution of molar mass M. Thus, we get

$$f_w(M) = \frac{M_w G(D)}{M} \frac{dD}{dM} \tag{55}$$

Empirically we have for a series of homologous polymers [33]

$$D = k_D M^{-\alpha_D} \tag{56}$$

where k_D and α_D are two scaling constants. Experimentally, it has been confirmed that for a flexible polymer, $0.5 < \alpha_D < 0.6$ in a good solvent and $\alpha_D = 0.5$ in a Flory Θ-solvent; for a rigid rodlike chain, $\alpha_D = 1$; and for a semirigid, wormlike chain, $0.6 < \alpha_D < 1$. Therefore, Eq. (5) can be written as

$$f_w(M) \propto D^{1+\frac{2}{\alpha_D}} G(D) \tag{57}$$

because M_w, k_D, and α_D are constants for a given polymer solution. With Eqs. (56) and (57), we are able to transfer D to M and $f_w(M)$ to $G(D)$ if the values of k_D and α_D are available from separate sources. This is the basic idea of the method that allows information about $f_w(M)$ to be derived by dynamic LLS, which is very similar to the particle sizing approach.

The most straightforward method for calibrating the relation between D and M is to measure both D and M for a set of monodisperse samples with different molar masses. In reality, the monodisperse samples have to be replaced by narrowly distributed standards made available either by relevant living polymerization or by fractionation of a broadly distributed sample. However, only very few kinds of polymers such as polystyrene and poly(methyl methacrylate) can actually be prepared so as to have a sufficiently narrow molar mass distribution ($M_w/M_n \approx 1.1$), and the fractionation is very time consuming. Thus, the straightforward calibration of the D versus M relation is not always practical.

Figure 1.5 shows the plot of log(D) versus log(M) for a set of narrowly distributed polystyrene standards in toluene at 20°C [34]. The line represents a least-square fitting to the data points, giving D (cm/s) $= 3.64 \times 10^{-4} M^{-0.577}$. Using this relation, we were able to estimate the molar mass distribution of polystyrene by using only DLS [35]. However, as noted earlier, in reality only a few kinds of polymers can be prepared to have narrow enough distributions of molar mass. Hence, we often have to satisfy ourselves with more broadly distributed samples having different average molar masses. This means that special analytical methods have to be developed to calibrate or scale the translational diffusion coefficient D and molar mass M from broadly distributed samples. An approach to this problem is described as follows. As can be easily shown, it follows from Eqs. (52) and (56)

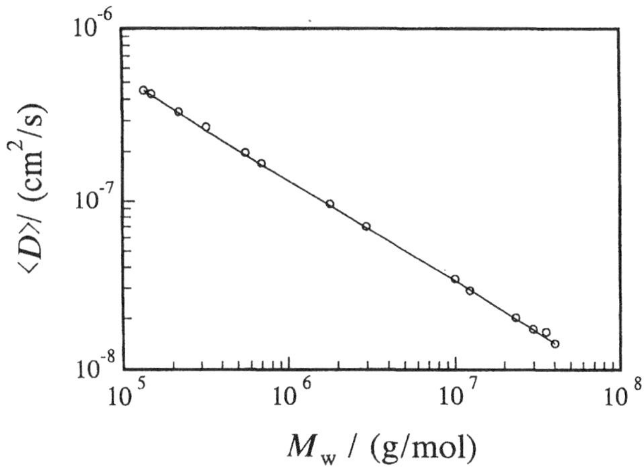

Figure 1.5 Double logarithmic plot D versus M for polystyrene in toluene at 20°C. The line represents a least-square fit of D (cm/s) $= 3.64 \times 10^{-4} M^{-0.577}$. See [42].

that

$$M_{\mathrm{w}} = (k_D)^{1/\alpha_D} \Big/ \int_0^\infty G(D) D^{1/\alpha_D} dD \tag{58}$$

The quantity on the right-hand side is denoted by $M_{\mathrm{w,calc}}^{\mathrm{DLS}}$. For any given set of k_D and α_D, $M_{\mathrm{w,calc}}^{\mathrm{DLS}}$ can be calculated when $G(D)$ is determined from DLS measurements. For two polydisperse samples i and j, one can obtain

$$\frac{\left(M_{\mathrm{w,calc}}^{\mathrm{DLS}}\right)_i}{\left(M_{\mathrm{w,calc}}^{\mathrm{DLS}}\right)_j} = \frac{\int_0^\infty G_j(D) D^{1/\alpha_D} dD}{\int_0^\infty G_i(D) D^{1/\alpha_D} dD} \tag{59}$$

where $G_i(D)$ and $G_j(D)$ are for the samples i and j, respectively. The right-hand side can be calculated from measured $G(D)_i$ and $G(D)_j$ for any α_D, and if the chosen value of α_D happens to be equal to that for the system under study, the resulting value of $(M_{\mathrm{w,calc}}^{\mathrm{DLS}})_i/(M_{\mathrm{w,calc}}^{\mathrm{DLS}})_j$ should agree with the value of $(M_{\mathrm{w}})_i/(M_{\mathrm{w}})_j$ from static LLS. In reality, the desired α_D will be reached by using a computer program that seeks a minimum of ERROR(α_D) defined by

$$\mathrm{ERROR}(\alpha_D) = \sum_i^N \sum_j^N \left[\frac{(M_{\mathrm{w}})_i}{(M_{\mathrm{w}})_j} - \frac{\left(M_{\mathrm{w,calc}}^{\mathrm{DLS}}\right)_i}{\left(M_{\mathrm{w,calc}}^{\mathrm{DLS}}\right)_j} \right]^2 \tag{60}$$

where N is the total number of the polydisperse samples examined. Next, with the α_D value so determined, we can compute $(M_{\mathrm{w,calc}}^{\mathrm{DLS}})_i$ for each of the N samples from Eq. (58) by varying k_D and seeking a k_D value that minimizes ERROR(k_D)

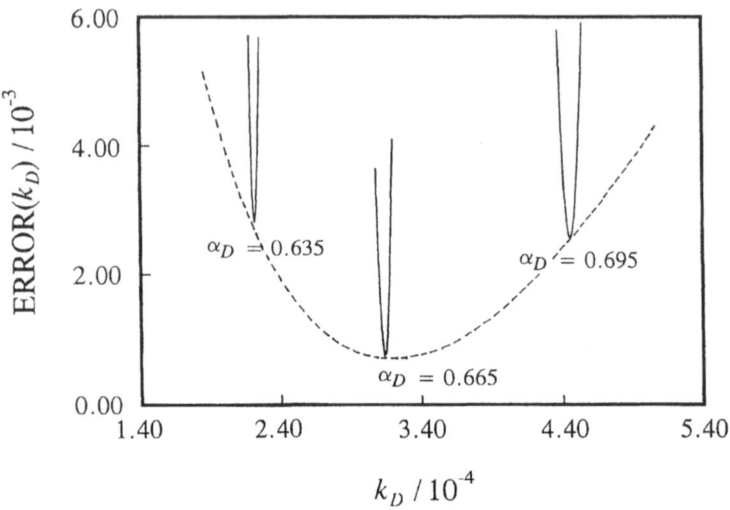

Figure 1.6 Plot of ERROR(k_D) calculated with data for five chitosan samples with different weight average molecular weights. Here the minimum of the dashed curve corresponds to $\alpha_D = 0.665 \pm 0.015$ and $k_D = (3.14 \pm 0.20) \times 10^{-4}$.

defined by

$$\mathrm{ERROR}(k_D) = \sum_i^N \left[\frac{(M_\mathrm{w})_i}{\left(M_{\mathrm{w,calc}}^{\mathrm{DLS}}\right)_i} - 1 \right]^2 \tag{61}$$

Figure 1.6 depicts a plot of ERROR(k_D) at three values of α_D calculated from a combination of static and dynamic LLS results for five chitosan samples with different values of M_w. The figure illustrates that for each chosen α_D, ERROR(k_D) shows a sharp minimum, but the location and height of the minimum vary significantly with α_D, and the minimum becomes the smallest at $\alpha_D = 0.665$ and $k_D = 3.14 \times 10^{-4}$. With these values, the molar mass distributions of chitosan samples have been successfully characterized [19]. It can be seen that the method discussed in the preceding involves only a light-scattering instrument, and it is a self-calibrated method. It is worth noting that unlike GPC, the light-scattering calibration is independent of the LLS instrument used.

If only one broadly distributed sample is available, we have to resort to another method to determine the relation between D and M. One possibility is to estimate α_D from the Mark–Houwink equation for intrinsic viscosity. It is known that the intrinsic viscosity $[\eta]$ can be scaled with M by the Mark–Houwink equation, i.e., $[\eta] = k_\eta M^{\alpha_\eta}$, and, according to Flory and also to de Gennes [33, 36], α_D may be equal to $(\alpha_\eta + 1)/3$. With α_D estimated from α_η by this relation, M_w from static LLS and $G(D)$ from dynamic LLS can be used to determine k_D as described in the preceding. Chu *et al.* and Pope and Chu [37, 38] successfully applied this method to linear polyethylene in 1,2,4-trichlorobenzene at 135°C for which α_D was estimated from the value of $\alpha_\eta = 0.72$ reported by Cervenka [39]. Figure 1.7 shows the resulting cumulative weight distribution $F_{w,cum}(M)\,[=\int_M^\infty f_w(M)\,dM]$ and compares it with the result obtained by high-temperature GPC. The agreement is reasonably good, but it should be noted that the weight distribution from LLS is usually narrower and more skewed toward the high molar mass than that from GPC because the scattered light intensity is proportional to the square of molar mass, and thus higher-molar-mass species weigh more in LLS.

For only one given broadly distributed sample, one can also combine LLS with other methods, such as GPC. The static LLS apparatus used as an on-line GPC detector has been popular for a while (e.g., the multiangle LLS detector from Wyatt Technology Co.). Here, we illustrate another less-known method of combining the results from GPC and off-line dynamics LLS. The basic principle is as follows: There is a similarity between these two tools in that the translational diffusion coefficient D obtained by dynamic LLS and the elution volume V in GPC are related to the hydrodynamic size of a given macromolecule. In a first

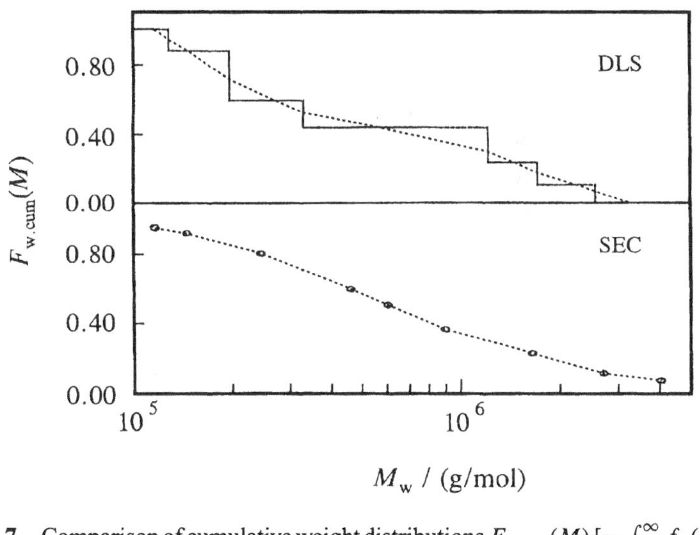

Figure 1.7 Comparison of cumulative weight distributions $F_{w,cum}(M)\,[=\int_M^\infty f_w(M)\,dM]$ obtained by DLS and SEC (size exclusion chromatography) for a linear polyethylene in 1,2,4-trichlorobenzene at 135°C.

approximation, if the hydrodynamic size is proportional to the molar mass, we have

$$V = A + B\log(M) \tag{62}$$

where A and B are constants similar to k_D and α_D. It should be noted that this approximation simplifies but does not affect the following discussion. A combination of Eqs. (56) and (62) leads to

$$V = \boldsymbol{A} + \boldsymbol{B}\log(D) \tag{63}$$

where $\boldsymbol{A} = A + B\log(k_D)/\alpha_D$ and $\boldsymbol{B} = -B/\alpha_D$. Furthermore, we get from Eq. (63)

$$V^2 = \boldsymbol{A}^2 + 2\boldsymbol{AB}\log(D) + \boldsymbol{B}^2\log^2(D) \tag{64}$$

Averaging both sides of Eqs. (63) and (64) over the concentration profile $C(V)$ in GPC, we obtain

$$\langle V\rangle = \boldsymbol{A} + \boldsymbol{B}\langle\log(D)\rangle \tag{65}$$

and

$$\langle V^2\rangle = \boldsymbol{A}^2 + 2\boldsymbol{AB}\langle\log(D)\rangle + \boldsymbol{B}^2\langle\log^2(D)\rangle \tag{66}$$

where

$$\langle V\rangle = \int_0^\infty VC(V)\,dV \tag{67}$$

and

$$\langle V^2\rangle = \int_0^\infty V^2C(V)\,dV \tag{68}$$

whereas

$$\langle\log(D)\rangle = \frac{\int_0^\infty \log(D)C(V)\,dV}{\int_0^\infty C(V)\,dV} \tag{69}$$

and

$$\langle\log^2(D)\rangle = \frac{\int_0^\infty \log^2(D)C(V)\,dV}{\int_0^\infty C(V)\,dV} \tag{70}$$

Though not yet theoretically proved, it is usually assumed that the differential area $C(V)\,dV$ under a GPC curve is proportional to the differential mass of the polymers

dW contained in the differential elution volume dV. Because $dW \propto f_w(M)\,dM$, we have

$$C(V)\,dV \propto f_w(M)\,dM \tag{71}$$

Combining Eqs. (57), (69), and (71), we have

$$\langle \log(D) \rangle = \frac{\int_0^\infty \log(D)G(D)D^{1/a_D}dV}{\int_0^\infty G(D)D^{1/a_D}\,dV} \tag{72}$$

and, correspondingly

$$\langle \log^2(D) \rangle = \frac{\int_0^\infty \log^2(D)G(D)D^{1/a_D}dV}{\int_0^\infty G(D)D^{1/a_D}\,dV} \tag{73}$$

Multiplying both sides of Eq. (71) by M and integrating over the entire range of M, we have

$$M_w = \int_0^\infty MC(V)\,dV \tag{74}$$

Elimination of M from the right-hand side using the relation $D = k_D M^{-\alpha_D}$ and Eq. (63) transforms Eq. (74) to

$$M_w = k_d^{1/a_D} \int_0^\infty 10^{(\mathbf{A}-V)/(a_D\mathbf{B})} C(V)\,dV \tag{75}$$

which is combined with Eq. (58) for M_w to give

$$\frac{\int_0^\infty 10^{(\mathbf{A}-V)/(a_D\mathbf{B})} C(V)\,dV}{\int_0^\infty G(D)D^{1/\alpha_D}dD} = 1 \tag{76}$$

This equation contains only one unknown parameter α_D. For a chosen α_D, one first calculates $\langle \log(D) \rangle$ and $\langle \log^2(D) \rangle$ using Eqs. (72 and 73, respectively), then calculates $\boldsymbol{A}$ and $\boldsymbol{B}$ by solving Eq. (66) with $\langle V \rangle$ and $\langle V^2 \rangle$ computed from the GPC curve, and finally calculates the left side of Eq. (76). By iteration, one can find a value of α_D to minimize the difference between the left and right sides of Eq. (76). For the α_D so obtained, one can calculate k_D from either Eq. (58) or (75) by using M_w determined directly from static LLS and $C(V)$ from GPC or $G(D)$ from dynamic LLS. With $\boldsymbol{A}$, $\boldsymbol{B}$, k_D, and α_D, one is ready to calculate A and B. In this way, one can calibrate not only the M versus V relation but also the M versus

D in a single process by using only one broadly distributed sample. This method has been tested and applied in the characterization of gelatin in water [40, 41].

When the relation between D and M is established, one can easily convert $G(D)$ obtained by dynamic LLS into a corresponding molar mass distribution. The applications of this method in the characterization of special polymers has been reviewed by Wu [42].

1.4.2 CHARACTERIZATION OF BLOCK COPOLYMERS

A combination of static and dynamic LLS can also be used to estimate the composition distribution of a block copolymer. Let us consider a copolymer sample consisting of monomers A and B. The sample is generally polydisperse in both molar mass and chain composition. We suppose that the copolymer species i is characterized by the molar mass M_i and the composition $w_{\mathrm{A}i}$, which is the weight fraction of A in the chain of that species. It is assumed that no compositional heterogeneity exists in the chains of the same length. For homopolymers, the refractive index increment (at infinite dilution) does not depend, in a good approximation, on the molar mass of the chain but is equal to that of the entire sample. For copolymers, this is not the case, and, according to the theory of light scattering, Eq. (53) for G_i may by replaced by

$$G_i = (\nu_i/\nu)^2 M_i w_i \Big/ \sum_j M_j w_j \tag{77}$$

where ν_i is the refractive index increment due to the copolymer species i; ν, that of the entire sample; and w_i, the weight fraction of the copolymer species i. With a continuous distribution of molar mass, Eq. (77) is generalized to

$$G(D)\,dD = (\nu(M)/\nu)^2 M f_w(M) M dM \tag{78}$$

where $\nu(M)$ is the refractive index increment due to the chains of molar mass M. Note that $\nu(M)$ depends on $w_{\mathrm{A}}(M)$, which is the continuous version of $w_{\mathrm{A}i}$. We assume that this dependence is represented by

$$\nu(M) = \nu_{\mathrm{A}} w_{\mathrm{A}}(M) + \nu_{\mathrm{B}}[1 - w_{\mathrm{A}}(M)] \tag{79}$$

where ν_{A} and ν_{B} are the refractive index increments of the homopolymers consisting, respectively, of A and B. Recalling Eq. (55) for homopolymers, we can introduce $f_{\mathrm{w,\,app}}(M)$, an apparent weight distribution function of M, by

$$f_{\mathrm{w,app}}(M) = (M_{\mathrm{w}}/M)G(D)(dD/dM) \tag{80}$$

With Eq. (56) (assumed to also hold for copolymers), this gives

$$f_{\mathrm{w,app}}(M)/M_{\mathrm{w}} = (1/k_D)^{2/\alpha_D} G(D) D^{1+(2/\alpha_D)} \tag{81}$$

Therefore, $f_{\mathrm{w,app}}(M)/M_{\mathrm{w}}$ can be calculated from $G(D)$ measured in dynamic light scattering when k_D and α_D are known separately. On the other hand, substituting Eq. (78) together with Eq. (79) into Eq. (80), we get

$$f_{\mathrm{w,app}}(M)/M_{\mathrm{w}} = \nu^{-2}[\nu_{\mathrm{A}} w_{\mathrm{A}}(M) + \nu_{\mathrm{B}}(1 - w_{\mathrm{A}}(M))]^2 f_w(M) \tag{82}$$

where $f_{\mathrm{w,app}}(M)$, ν, ν_{A}, and ν_{B} vary with solvent. On the basis of Eq. (82), and by choosing two solvents 1 and 2 for a given copolymer, we have

$$\frac{f^{(1)}_{\mathrm{w,app}}(M)}{f^{(2)}_{\mathrm{w,app}}(M)} = \left\{ \frac{n^{(2)}}{n^{(1)}} \frac{w_{\mathrm{A}}(M) n^{(1)}_{\mathrm{A}} + [1 - w_{\mathrm{A}}(M)] n^{(1)}_{\mathrm{B}}}{w_{\mathrm{A}}(M) n^{(2)}_{\mathrm{A}} + [1 - w_{\mathrm{A}}(M)] n^{(2)}_{\mathrm{B}}} \right\}^2 \tag{83}$$

The ratio on the left-hand side can be obtained as a function of M because $f_{\mathrm{w,app}}(M)^{(i)}/M_{\mathrm{w}}$ can be determined as described previously from dynamic LLS. Thus, Eq. (83) allows a determination of $w_{\mathrm{A}}(M)$, the chain composition distribution when all other parameters on its right-hand side are measured by differential refractometry. Once $w_{\mathrm{A}}(M)$ is known, we are ready to compute $v(M)$ from Eq. (79), $f_w(M)$ from Eq. (82), and finally M_{w}.

Figure 1.8 shows two typical apparent molar mass distributions determined from $G(D)$s by Eq. (81) for the low-mass (○) and high-mass (□) segmented copolymers

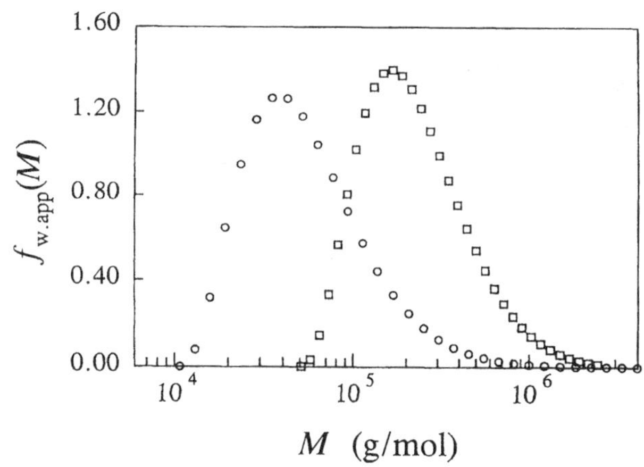

Figure 1.8 Apparent weight distributions calculated from the translational diffusion coefficient distributions corresponding to low-mass (○) and high-mass (□) copolymer segmented poly(ethylene terephthalate-*co*-caprolactone) (PET–PCL) containing 13% PET in tetrahydrofuran (THF) at 25°C.

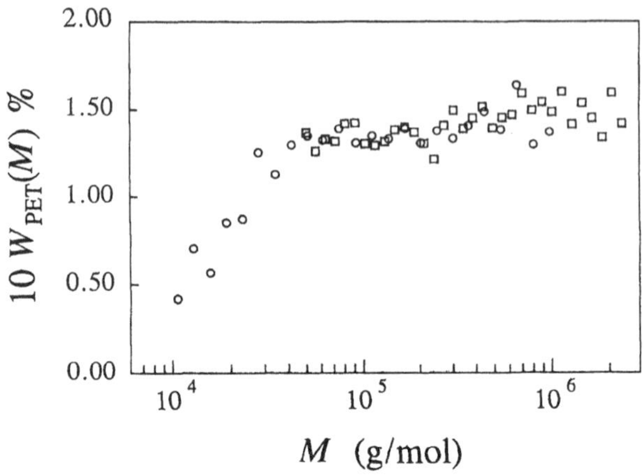

Figure 1.9 Estimate of the chain composition distributions for low mass (○) and high mass (□) 13% PET–PCL samples.

made of ethylene terephthalate and caprolactone (PET–PCL) in tetrahydrofuran (THF) at 25°C. The overall PET contents of the two copolymers are the same: 13%. Repeating the measurements on these samples in another solvent chloroform leads to a new set of $G(D)$s, which allows Eq. (73) to be used to calculate $w_{PET}(M)$.

Figure 1.9 shows the results for the low-mass (○) and high-mass (□)13% PET–PCL samples. We see that the PET content increases as M for $M < \approx 4 \times 10^4$ and approaches a constant value of ≈14% in the high-molar-mass range. For the 58% PET–PCL sample, the composition distribution is nearly constant. The composition of the high-molar-mass 13% PET–PCL sample overlaps with that of the low-mass 13% PET–PCL sample in the same molar mass range. This indirectly indicates that the estimation of the composition distribution is reasonable. The lower content of PET in the low-molar-mass range was attributed to the two-step synthesis [43, 44].

1.4.3 *INVESTIGATION OF AGGREGATION*

A combination of static and dynamic LLS is also useful in the study of the aggregation process in solution. If a mixture is made of individual polymer chains and polymer clusters (or aggregates), the measurement of static LLS will lead to an apparent weight-average molar mass $M_{w,app}$, which is expressed by

$$M_{w,app} = M_{w,L}x_L + M_{w,H}x_H \tag{84}$$

where the subscripts L and H denote low-molar-mass linear polymer chains and high-molar-mass polymer clusters, respectively, and x_L and x_H are their weight fractions with $x_L + x_H = 1$. If the linear chains and clusters are significantly

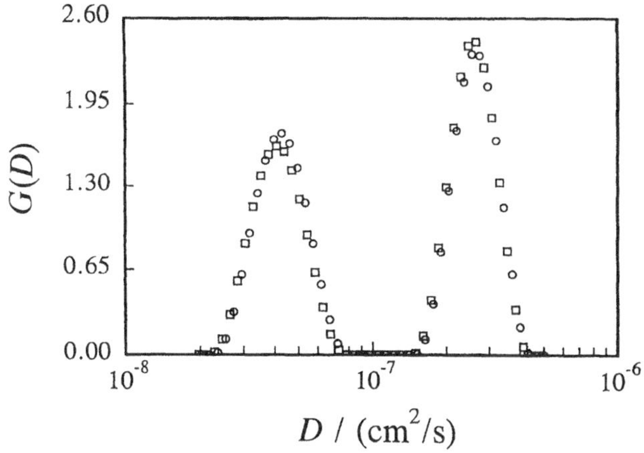

Figure 1.10 Translational diffusion coefficient distributions $G(D)$ of a simulated polymer mixture at two scattering angles: 14° (○) and 17° (□). The mixture contains two polystyrene standards of distinctly different weight average molar masses (3.0×10^5 and 5.9×10^6 g/mol) and a high-mass polystyrene.

different in the hydrodynamic size, dynamic LLS will detect two distinct peaks in the measured $G(D)$, with one peak corresponding to the linear chains and the other to clusters.

Figure 1.10 shows $G(D)$ of a simulated polymer mixture at two scattering angles: 14° (○) and 17° (□). The mixture consists of two polystyrene standards having distinctly different weight-average molar masses (3.0×10^5 and 5.9×10^6 g/mol), and the high-mass polystyrene is used to simulate the polymer cluster [45]. The area ratio A_r of the two peaks is expressed by

$$A_r = \frac{A_L}{A_H} = \frac{\int_0^{D_M} G_L(D)\,dD}{\int_{D_M}^{\infty} G_H(D)\,dD} = \frac{M_{w,L}x_L}{M_{w,H}x_H} \qquad (85)$$

with D_M being the cutoff translational diffusion coefficient between $G_L(D)$ and $G_H(D)$. In practice, the values of A_r at finite scattering angles must be extrapolated to $K \to 0$. On the basis of Eqs. (84) and (85), one can use $M_{w,app}$ and A_r, respectively, from static and dynamic LLS, to calculate $M_{w,L}x_L$ and $M_{w,H}x_H$. In principle, by knowing any one of the four parameters ($M_{w,L}$, $M_{w,H}$, x_L, and x_H), one should be able to determine the remaining three parameters. This method has been thoroughly tested with mixtures of polystyrene standards by Wu *et al.* [45]. As for a particular polymer mixture, we should find a way allowing determination of one of the four parameters. For example, in the study of polymer association, we can determine the $M_{w,L}$ of starting individual polymer chains, and in the study of the gelation process, we can use a filtration method to remove large microgels, and thus the weight fractions of x_L and x_H can subsequently be determined. This method has been used to characterize novel thermoplastic polymers with phenolphthalein in their backbone chains by Wu *et al.* [46].

1.4.4 STUDY OF POLYMER ADSORPTION ON COLLOIDS

Besides the particle sizing, a combination of static and dynamic LLS can provide much more information on the study of colloids, such as particle structures, the thickness of the adsorbed surfactant or polymer layer, and the formation of nanoparticles, as demonstrated by Li *et al.*, Wu, and Wu *et al.* [47–49]. Recently, Gao *et al.*, Wu and Chan, and Wu *et al.* [50–52] used a combination of static and dynamic LLS results to determine the density of polymer particles. The following outline of the method illustrates how a combination of static and dynamic LLS as a powerful analytical tool can be used to investigate more sophisticated colloidal problems.

For a colloidal particle of uniform density, its molar mass M is proportional to the cubic of its radius R, i.e., $M = (4/3)\pi R^3 \rho N_A$, where ρ is the particle density and N_A the Avogadro constant. D is related to R_h through the Stokes–Einstein equation. In general, $R_h \geq R$, and thus one can write down $R_h = R + b$ with b being the thickness of the solvated layer. Therefore,

$$D = F\left(\frac{k_B T}{6\pi\eta}\right)(4\pi\rho N_A)^{1/3} M^{1/3} \tag{86}$$

with

$$F = \frac{1}{1 + b\left(\dfrac{4\pi\rho N_A}{M}\right)^{1/3}} \tag{87}$$

Comparing Eq. (86) with Eq. (56) and considering $b \ll R$, one finds approximately $\alpha = 1/3$ and $k_D = F(k_B T/6\pi\eta)(4\pi\rho N_A)^{1/3}$. Thus, with M in Eq. (87) replaced by M_w, it follows from Eq. (58) that

$$M_w = \frac{1}{[1 + b(4\pi\rho N_A/M_w)^{1/3}]^3}\left(\frac{4\pi\rho N_A}{3}\right)\left(\frac{K_B T}{6\pi\eta}\right)^3 \Big/ \int_0^\infty G(D) D^3 dD \tag{88}$$

This equation contains two unknown parameters (b and ρ), and if we know one of them, the other can be calculated from M_w and $G(D)$. With this idea, Wu *et al.* [53] found that the average density of the polystyrene microspheres made of a few uncross-linked chains is slightly lower than that of bulk polystyrene or conventional polystyrene latex.

1.4.5 STRUCTURES AND DYNAMICS OF POLYMER GELS

There are two different ways of using LLS to study polymer gels. One is the study of strongly scattering probes (usually uniform latex spheres) in a weakly scattering gel network [54], and the other is to study the polymer gel network itself directly [55]. In static LLS, Debye and Bueche [56] showed that on the basis of Eq. (12), the angular dependence of the average scattered intensity can lead to the mean-square average local dielectric constant fluctuations ($\langle \varsigma^2 \rangle$) and the correlation length (ξ) through

$$R_{\mathrm{vv}}(K) = \frac{8\pi^3 \langle \varsigma^2 \rangle \xi^3}{\lambda_0^4 (1 + K^2 \xi^2)^2} \tag{89}$$

where $p(r)$ in Eq. (12) has been defined by

$$p(r) = \frac{\langle \varsigma_1 \varsigma_2 \rangle}{\langle \varsigma^2 \rangle} = \exp\left(-\frac{r}{\xi}\right) \tag{90}$$

with ς_1 and ς_2 being the local dielectric constant fluctuations in volume elements 1 and 2, respectively. Equation (89) shows that $\langle \varsigma^2 \rangle$ and ξ can be calculated from the Rayleigh ratio by plotting $(R_{\mathrm{vv}}(K))^{-1/2}$ versus K^2. Figure 1.11 shows how ξ and $\langle \varsigma^2 \rangle$ change in the gelation process of the epoxy polymerization of 1,4-butanediol diglycidyl ether with *cis*-cyclohexane-dicarboxylic anhydride with different catalyst concentrations [57]. The Figure reveals that a lower catalyst concentration results in a smaller value of $\langle \varsigma^2 \rangle$ and a larger extension of local optical inhomogeneties, suggesting that at lower catalyst concentrations, each polymer molecule has more chance to grow bigger before it reacts with other polymer molecules. It also shows that two different rates are associated with the characteristic length increases with time. By combining this observation with the results from dynamic LLS, Wu *et al.* [58] proposed a four-step reaction process that led to a better understanding of the structure formation in the epoxy polymerization.

However, if the scattering elements are unable to diffuse throughout the medium during dynamic LLS measurements, the time-averaged intensity–intensity correlation function will not be equal to the ensemble-averaged intensity–intensity correlation function but will correspond to a subensemble-averaged intensity–intensity correlation function. Pusey and van Megen [55] have made a fundamental study on this topic and provided methods to study gels or glasses using a time-averaged intensity–intensity correlation function from a single dynamic LLS measurement. Shibayama [59] has summarized how to use dynamic LLS to study spatial

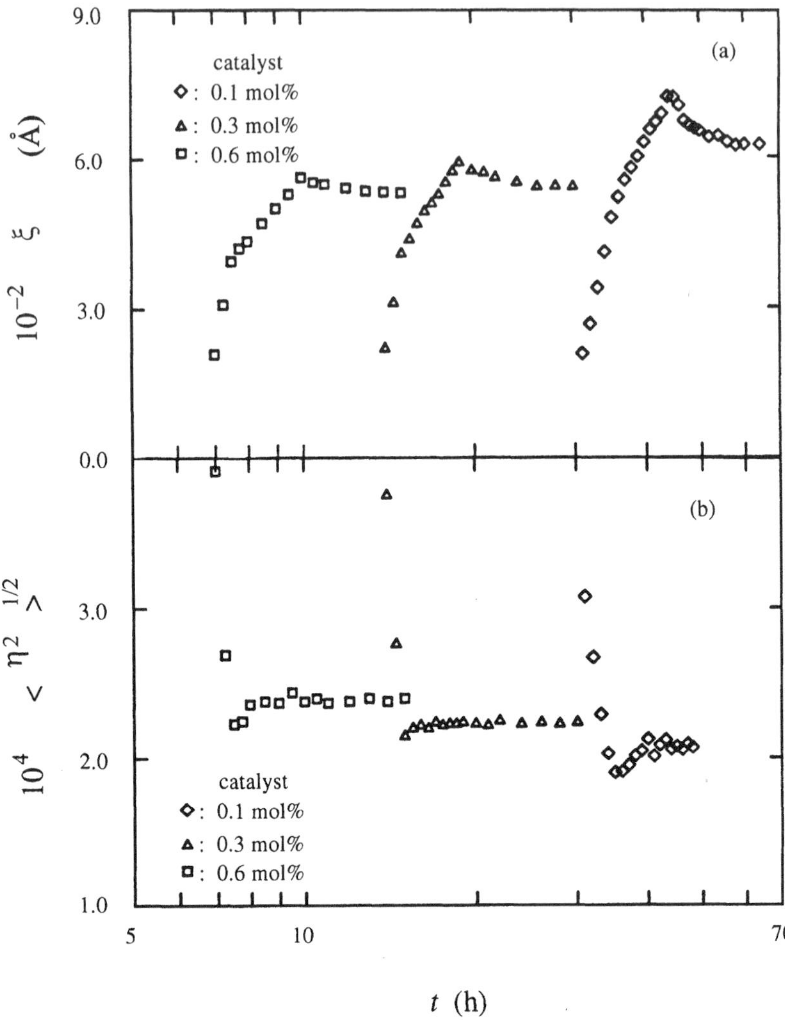

Figure 1.11 (a) Correlation length (ξ) of local optical inhomogeneties versus reaction time t and (b) mean-square average local dielectric constant fluctuation $\langle \eta^2 \rangle$ versus reaction time t at three different catalyst concentrations (0.1, 0.3 and 0.6%) but the same molar ratio of DGEB:CH = 1:2 at 80°C.

inhomogeneity and dynamic fluctuations of polymer gels in his feature article. The basic principle is outlined as follows.

In the gelation process, the inhomogeneity is gradually frozen inside the gel, resulting in a drastic increase and a sample position-dependence of the scattered intensity. By changing the gel position (e.g., rotating the gel), we can observe a speckle pattern, and these speckles can be used to detect the gelation point [60, 61]. Owing to the nonuniformity of the intensity of the scattered light over the sample position, the gel was referred to as a "nonergodic medium." It has been argued that, strictly speaking, the gel should be called a "restricted ergodic medium" because each point in a gel has its own free-energy minimum [59]. The scattered electric field $\mathbf{E}(k, t)$ from a gel consists of a fluctuating component $\mathbf{E}_{\mathrm{F}}(k, t)$ and a time-independent background $\mathbf{E}_{\mathrm{C}}(k, t)$, which is a constant for a given sample position. By relating the field–field time correlation function $|g_{\mathrm{NE}}^{(1)}(K, t)|$ with the amplitude δ of the Brownian excursion along the scattering vector $\mathbf{K}$, one has

$$|g_{\mathrm{NE}}^{(1)}(K, t)| = \exp\left\{-K^2\langle\delta^2\rangle\left[1 - \exp\left(-\frac{D_{\mathrm{NE}}K^2 t}{K^2\langle\delta^2\rangle}\right)\right]\right\} \tag{91}$$

where the subscript NE is used to address the nonergodic medium. In general, the diffusion coefficient D and the density–density correlation function of the gel network $g(r)$ are related by

$$D = \int \frac{k_{\mathrm{B}}T}{6\pi\eta r} g(r)\, dr \Big/ \!\! \int g(r)\, dr \tag{92}$$

It has been shown by de Gennes [33] that a gel in its swelling equilibrium and a semidilute solution have the same spatial correlation function, i.e.,

$$g(r) \propto \frac{\xi}{r} \exp\left(-\frac{r}{\xi}\right) \tag{93}$$

By substituting Eq. (91) into Eq. (90), one has $D = k_{\mathrm{B}}T/6\pi\eta\xi$. In Eq. (91), if $K^2\langle\delta^2\rangle \to \infty$, $|g_{\mathrm{NE}}^{(1)}(K, t)| \to \exp(-D_{\mathrm{NE}}K^2 t)$, as we discussed before for Brownian particles free in solution. Pusey and van Megen [55] showed that for a gel, the intensity–intensity time correlation function can be written as

$$\begin{aligned} G^{(2)}(K, t)/A &= 1 + Y^2\left\{\left[g_{\mathrm{NE}}^{(1)}(K, t)\right]^2 - \left[g_{\mathrm{NE}}^{(1)}(K, \infty)\right]^2\right\} \\ &\quad + 2Y(1 - Y)\left\{\left[g_{\mathrm{NE}}^{(1)}(K, t)\right] - \left[g_{\mathrm{NE}}^{(1)}(K, \infty)\right]\right\} \end{aligned} \tag{94}$$

where Y is the ratio of the assembly-average-scattered intensity $\langle I\rangle_{\mathrm{E}}$ over the time-average-scattered intensity $\langle I\rangle_{\mathrm{T}}$. From the measured $G^{(2)}(K, t)/A$, we can obtain the initial decay rate Γ [$\equiv D_{\mathrm{A}}K^2$] with D_{A} being the apparent diffusion coefficient and the initial amplitude σ_I^2 by

$$D_{\mathrm{A}} = -\frac{1}{2k^2}\lim \frac{\partial}{\partial t}\ln\left\{\left[G^{(2)}(K, t)/A\right] - 1\right\} \tag{95}$$

$$\sigma_I^2 \equiv \left[G^{(2)}(K, 0)/A\right] - 1 \tag{96}$$

where D_A is related to D_{NE} by

$$D_{NE} = \frac{\sigma_I^2}{Y} D_{A\Sigma} \tag{97}$$

It is worth noting that the estimate of D_{NE} is a function of K because both σ_I^2 and Y are dependent on K. For a highly inhomogeneous system (i.e., very small in $\langle\delta^2\rangle$, one has to use a larger K (i.e., $K^2 \geq 2/\langle\delta^2\rangle$), to ensure a reasonable value of of D_{NE} [55]. On the other hand, if using the partial heterodyne approach proposed by Joosten *et al.* [62], we have

$$G^{(2)}(K,t)/A = 1 + X^2\left[g_{HT}^{(1)}(K,t)\right]^2 + 2X(1-X)g_{HT}^{(1)}(K,t) \tag{98}$$

where X is the ratio of the time-average-scattered intensity of the fluctuating part $\langle I_F(K)\rangle_T$ over the time-average total scattered intensity $\langle I(K)\rangle_T$, and $g_{HT}^{(1)}(K,t)$, associated with the relevant dynamic component of the density fluctuations of the gel network, is defined as

$$g_{HT}^{(1)} = \frac{\left\langle \mathbf{E}_F(K,0)\mathbf{E}_F^*(K,t)\right\rangle_E}{\left\langle \mathbf{E}_F(K,0)\mathbf{E}_F^*(K,0)\right\rangle_E} = \frac{\left\langle \mathbf{E}_F(K,0)\mathbf{E}_F^*(K,t)\right\rangle_T}{\left\langle \mathbf{E}_F(K,0)\mathbf{E}_F^*(K,t)\right\rangle_T} = \exp(-D_{HT}K^2t) \tag{99}$$

Note that for the fluctuating part of the scattered light intensity, the assembly and time averages are equal. The relationship between D_{HT}, D_A, and D_{NE} is given by

$$D_{HT} = (2-X)D_A = \frac{\sigma_I^2}{X}D_A = \frac{Y}{X}D_{NE} \tag{100}$$

The equation shows that $\langle I(K)\rangle_T/D_A = (2/D_{HT})\langle I(K)\rangle_T - \langle I_F(K)\rangle_T/D_{HT}$.

Figure 1.12 shows a typical $\langle I\rangle_T$-dependence of D_A for a poly(N-isopropylacrylamide-*co*-acrylic acid) gel at 20°C. The solid line shows a fitting of Eq. (100). The insert shows a plot on the basis of $\langle I\rangle_T/D_A = (2/D)\langle I\rangle_T - \langle I\rangle_F\rangle_T/D$ from which D and $\langle I_F\rangle_T$ were estimated. Fang and Brown [63] modified the partial heterodyne method by introducing a correction of the instrumental coherence factor β, and thus Eq. (98) can be rewritten as

$$G^{(2)}(K,t)/A = 1 + \beta R^2\left[g_{HT}^{(1)}(K,t)\right]^2 + 2\beta R(1-R)g_{HT}^{(1)}(K,t) \tag{101}$$

where $R = 1-(1-\sigma_I^2/\beta)^{1/2}$. This method leads to the diffusion coefficient $D_{HT\beta}$. Another approach is the homodyne method, very similar to Eq. (36), in which the time-independent background intensity B is simply taken as a coherent constant

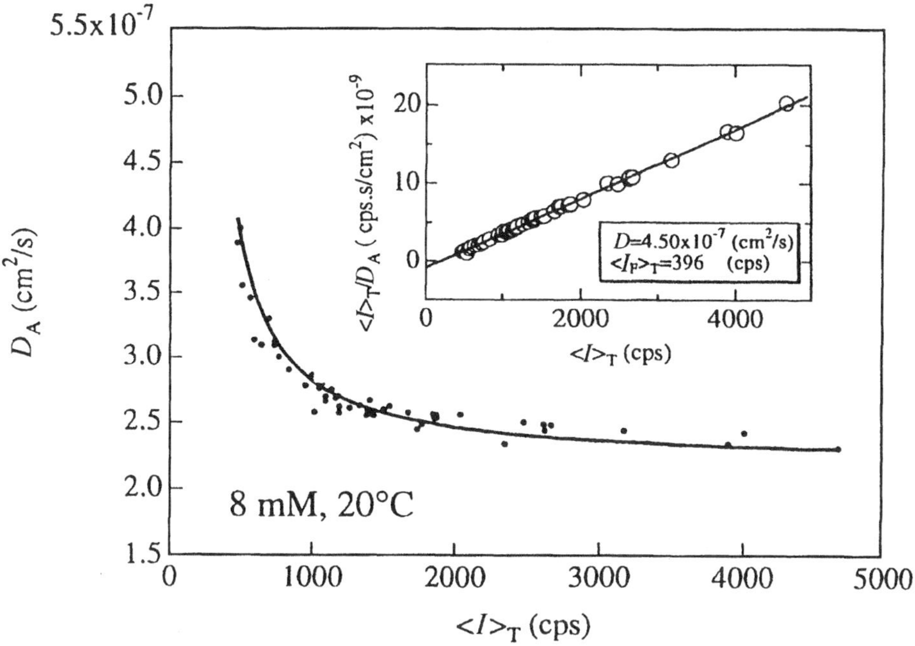

Figure 1.12 $\langle I\rangle_T$-dependence of D_A for a poly(N-isopropylacrylamide-*co*-acrylic acid) gel at 20°C. The solid line shows a fitting of Eq. (90). The insert shows a plot on the basis of $\langle I\rangle_T/D_A = (2/D)\langle I\rangle_T - \langle I_F\rangle_T/D$ from which D and $\langle I_F\rangle_T$ were estimated.

intensity, i.e.,

$$G^{(2)}(K,t)/A = 1 + \beta\left[g_{HT}^{(1)}(K,t) + B\right]^2 \tag{102}$$

which results in another diffusion coefficient D_{HM}. The four types of diffusion coefficients are related to each other by

$$D_{NE} : D_{HT} : D_{HT\beta} : D_{HM} = (2 - Y) : (2 - X) : (2 - R) : (1 + B) \tag{103}$$

Shibayama [59] studied swelling and shrinking kinetics of chemically cross-linked poly(vinyl alcohol) gels and found that the collective diffusion coefficient D_C of a shrinking gel is close to D_{NE}, whereas D_C of a swelling gel is similar to D_{HT}, or in other words D_{HT} and D_{NE} are, respectively, more suitable parameters to describe the swollen and shrunken gels. It should be noted that a full understanding of the physical meaning of these diffusion coefficients is still missing. The reader should refer to the recent review article by Shibayama [59].

1.5 Practice of Laser Light Scattering

An LLS spectrometer contains a limited number of components; namely, the light source, the optics, the cell holder, and the detector. Nowadays, an LLS instrument should have a digital output (single photon counting) from a fast photomultiplier; i.e., the output current pulse has already been treated by a preamplifier–amplifier–discriminator before it is connected to a time correlator, which is often a single plug-in board to a PC computer.

1.5.1 LIGHT SOURCE

Traditionally, the light source is a helium–neon (He–Ne) laser with a wavelength of 632.8 nm and an output power of 5–50 mW or an argon-ion (Ar^+) laser with a wavelength of 488 or 514.5 nm and an output power of 50–400 mW. Krypton lasers have also been used because their wavelength can be longer than the 632.8 nm of the He–Ne laser. The additional cost and somewhat short plasma tube life are drawbacks. Recently, there has been a tendency to replace the gas lasers with solid-state continuous wave (CW) lasers.

Specifically, we like to state that the frequency-doubled Nd–YAG laser (532 nm) is a much better choice nowadays if one does not require two wavelengths in experiments. In comparison with the gas lasers, (1) its beam diameter is $\approx$5 times smaller, making small scattering angles more easily accessible; (2) it is $\approx$1000 times more coherent; (3) it is $\approx$10 times more stable; (4) it has a much smaller overall size; (5) it is air-cooled and uses plug-in electric power; and (6) most important, its running cost is lower by a factor of about 5. It is expected that the Nd–YAG laser will replace most of the Ar-ion lasers in LLS spectrometers within the next few years. The manufacturers are starting to provide a new kind of solid-state CW diode lasers in the visible ($\approx$670 nm) and near visible (780–830 nm) range that are particularly useful in the study of conjugated polymer systems to avoid the light adsorption.

The laser used in dynamic light scattering should have a TEM_{00} mode with a Gaussian intensity profile so that it can be focused to produce a higher power density for the incident beam, which leads to a smaller scattering volume and a higher coherence factor in the optical mixing experiment. Many commercial laser companies can produce stable laser light sources suitable for LLS experiments. The reader should choose a laser with beam amplitude root-mean-square noises less than 0.5%, so that the noise level of the intensity–intensity time correlation function in dynamic LLS will not be affected, and a long-term amplitude stability of better than $\pm 1\%$ for the convenience of time-averaged scattered light intensity measurements will be obtained. It should be noted that in dynamic LLS measurements, long-term stability is usually not very important because the maximum delay time is normally no more than a few minutes (typically less than 1 s).

1.5.2 OPTICS AND CELL DESIGN

It is well known that the laser light follows the Gaussian optics and that its beam in the TEM_{00} mode has a Gaussian profile

$$i = i_0 \exp(-2r^2/r_0^2) \tag{104}$$

where r is the radial distance from the laser cavity axis and i and i_0 are the beam intensities at r and $r = 0$, respectively. The beam intensity decays to $1/e^2$ of its maximum value at $r = r_0$. The total intensity $I(r)$ on the cross-section πr^2 is

$$I(r) = \int_0^r 2\pi r i(r)\, dr = \frac{\pi i r_0^2 i_0}{2}\left[1 - \exp\left(-\frac{2r^2}{r_0^2}\right)\right] \tag{105}$$

The equation shows that $\approx$87% of the light intensity falls in an aperture of πr_0^2. If a laser beam is focused through an aperture by a lens, the diameter d_s of the focused spot will be $\approx 1.22\lambda f/r_0$ with λ and f being the wavelength of the laser light and the focal length of the lens. If $r_0 \approx$ 0.8 mm, $\lambda \approx 532$ nm, and $f \approx 300$ mm, $d_s \approx 0.25$ mm and the incident beam divergence (d_s/f) is less than 1 mrad, which is sufficiently small and typical in a laser light-scattering spectrometer. A polarizer may be placed in the light path before the incident beam strikes the sample cell to define the polarization (normally vertical) of the incident beam. Nowadays, the polarization ratio of lasers is usually better than 100:1, and the polarization of the laser is well aligned. Therefore, the polarizer is really not necessary.

The design of the conventional sample cell holder in LLS has become common knowledge. The cell holder normally consists of a hollow cylindrical brass block with an outside diameter of 50–80 mm and an inside diameter of 10–20 mm, which matches the outside diameter of the scattering cell. The brass block is normally placed inside a cylindrical optical glass cup filled with a refractive index matching fluid (e.g., xylene, toluene, and silicone oil) whose refractive index is very close to that of optical glass ($\approx$1.5) to reduce surface scattering and the curvature of the scattering cell. A proper alignment of the optical path is normally judged by the constant scattered intensity for benzene or toluene after scattering volume correction by $\sin\theta$ to within 1% (if the scattering volume is chosen by a slit) or 2% (if a small pinhole is used) over an angular range $\approx$15–150°. In principle, the scattering cell with an optical quality should be used. However, we found in practice that a selected normal cylindrical sample vial can also be used as the scattering cell, which greatly reduces the experimental cost and makes the scattering cell disposable.

On the other hand, if one is mainly interested in dynamic LLS experiments, a rectangular cell can also be used or may even be preferred. It should also be noted that the scattering volume ($\approx 200 \times 300 \times 300\ \mu m^2$ or $0.02\ \mu L$) is quite small, and thus the scattering cell could have a very small physical dimension. Foord *et al.* [64] have used a melting-point capillary with a microliter-solution volume for time correlation function measurements even though it is difficult to use it for measurements of angular dependence of time-averaged scattered intensity. The scattering cell can also be a flow type (e.g., a flow prism cell has been developed by Chu *et al.* [65]), which could offer not only small solution volume but also *in situ* small angle light scattering measurements as a detector in combination with other analytical separation techniques, such as GPC and electrophoresis. Another challenge in polymer analysis is to characterize polymers soluble only at high temperatures. A very important advancement is the design of a novel light-scattering cell holder by Chu *et al.*, Wu *et al.*, and Wu [66–71]. The cell holder is capable of operating at temperatures as high as 340°C and was first developed at the State University of New York at Stony Brook. Later, this novel technique was transferred to Du Pont and BASF.

Figure 1.13 is a schematic diagram of the high-temperature LLS spectrometer. In it, a thermally controllable plate (13) used as a heat sink isolates the rotary table (12) from the outer thermostat (3) by means of two sets of stainless steel standoffs (14). The outer brass thermostat (3) is isolated from the room with 0.5-in.-thick porous silicone rubber. This arrangement creates an oven that allows the temperature to be kept easily in the 320–340°C range with temperature fluctuations

---

Figure 1.13 Schematic top and side views of the high-temperature thermostat and detection system of the light scattering spectrometer: (1) silicon rubber insulation; (2) heating wires for the brass thermostats; (3) outer brass thermostat with fluid circulation facilities; (4) vacuum glass jacket for thermal isolation made of precision-polished glass of 2.25-in. outer diameter with Kovar seals at both ends of a stainless steel temperature shield with precision-polished glass of 2.25-in. outer diameter; (5) inner brass thermostat, which has a separate temperature controller and a thermometer and can accommodate a light-scattering cell up to 27-mm outer diameter; (6) Glan–Thompson polarizers; (7) fluid circulation paths; (8) lens; (9) field aperture; (10) optical-fiber bundle; (11) rotating plate for multiple detectors; (12) RT-200 Klinger rotary table with 0.01° step size; (13) cooling plate to isolate the outer thermostat from the rotary table; and (14) stainless steel standoffs for thermal isolation.

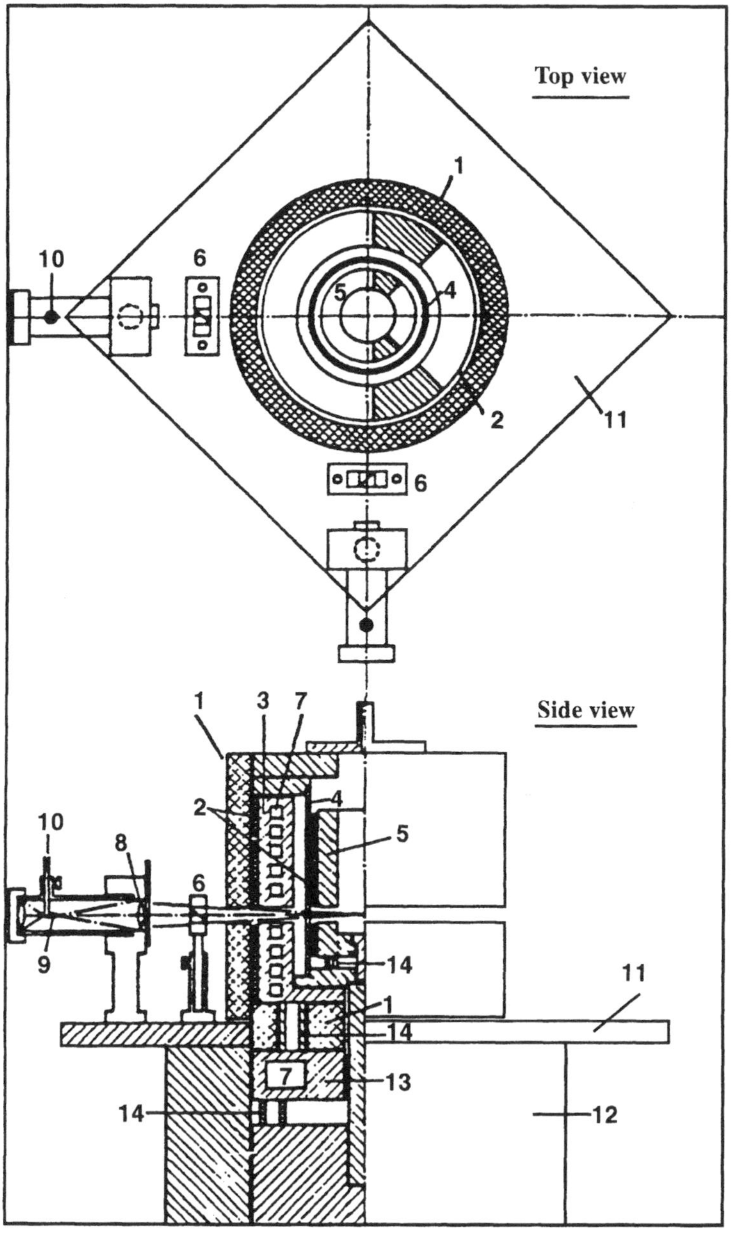
Top view
1
10
6
5
4
2
11
6
Side view
1
3
7
2
4
10
8
5
6
9
14
11
1
14
7
13
14
12

of less than 0.2°C. A glass (vacuum) jacket of 2.25-in outer diameter isolates the inner thermostat from the oven. The vacuum jacket reduces the temperature gradient in the light-scattering cell. The inner thermostat has a separate temperature controller and a miniature platinum resistance thermometer that can be connected to a digital voltmeter through the vacuum jacket by means of ceramic feed-through. With this design, short-term (20 min) control of ±0.05°C, intermediate-term (60 min) control of ±0.1°C, and long-term control of ±0.5°C can be achieved at 340°C even in the absence of a vacuum. Long-term temperature stability depends partially on room temperature fluctuations even in the presence of the outer thermostat and the isolation between the two thermostats. Figure 1.14 shows typical temperature fluctuations of the inner thermostat at 340°C. In addition, a high-temperature LLS detector coupled with GPC has recently been developed, and the determination of the molar mass distribution of poly(phenylene sulphide) in 1-chloro-naphthalene at 220°C has been made possible [36]. The advantage of on-line coupling LLS with GPC is obvious, for GPC is a fractionation method, and LLS allows an absolute molar mass measurement and hence makes the calibration of GPC columns.

It has to be stated that the optics together with the cell design in LLS are undergoing a drastic change because of the development of optical-fiber technology [72–75]. Figure 1.15 shows that a fiber-optic detector probe comprising an optical-fiber (single-mode) and a graded index microlens can form an integral part of the scattering cell. In this cell-detector probe design, the probes can eliminate the need for a goniometer, which is often one of the bulkier components of the spectrometer.

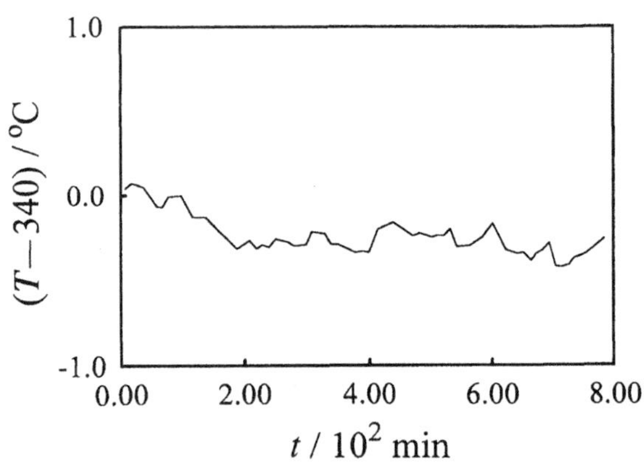

Figure 1.14 Typical temperature fluctuations in the inner thermostat at 340°C. Intermediate-term (1 h) temperature fluctuations were ±0.1°C. Long-term (10 h) temperature fluctuations were ±0.5°C.

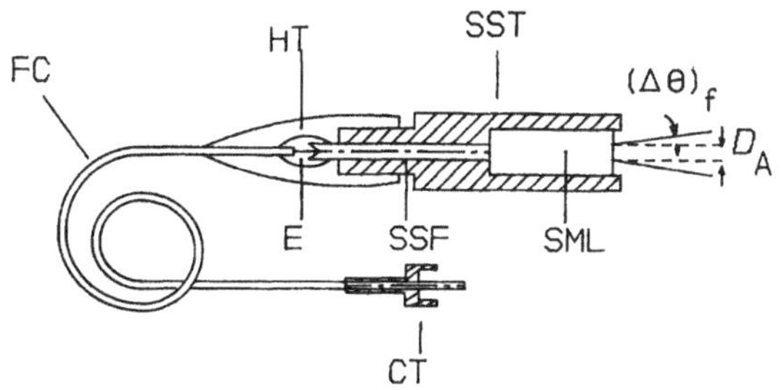

Figure 1.15 Schematic of a typical fiber-optic probe (SST) = matching piece of cylindrical stainless steel, SML = SELFOC microlens, SSF = stainless steel or ceramic ferrule used for mounting the bare optical fiber, E = epoxy used for holding fiber in ferrule, HT = heat shrink tubing, FC = fiber cable, CT = SMA type-II male connector). D_A and $\Delta\theta$ are the effective detector aperture and divergence angle, respectively.

Moreover, the probe can be in contact with the solution or dispersion so that the requirement of a transparent window in the sample chamber can be relaxed.

1.5.3 DETECTORS AND DETECTION

Commercially available standard photomultiplier tubes (PMTs) with low dark count (<30 Hz) and short after-pulsing are normally used to count the photons. When an He–Ne laser light source is used, an S-20 photocathode is preferable because of its higher sensitivity in the red light. PMTs, such as EMI 9863 and the new Hamamatsu miniature PMTs, are more suitable. For photon correlation in dynamic LLS, a low dark count is required, and thus it is advisable to purchase the PMT with a relatively smaller photocathode (typically, 2.54 mm in diameter). If the laser light is in the blue and green range, PMTs with a bialkali photocathode are better because they have a lower dark count at room temperatures. The EMI 9893 PMT with selected low dark count and short after-pulsing has been popular for this purpose. The RCA C31034 PMT, which has a broad spectral response (300–800 nm), is good for the entire visible range, but it is more delicate and expensive.

The silicon avalanche photodiode is another new development and offers promise in making a miniature light-scattering apparatus. With a broad spectral response, the silicon avalanche photodiodes are matched to the diode lasers to simplify the LLS instrumentation—especially if it is combined with the fiber-optic probe.

Figure 1.16 shows two commonly used configurations of the detection optics between the scattering cell and the detector. It should be noted that in static LLS

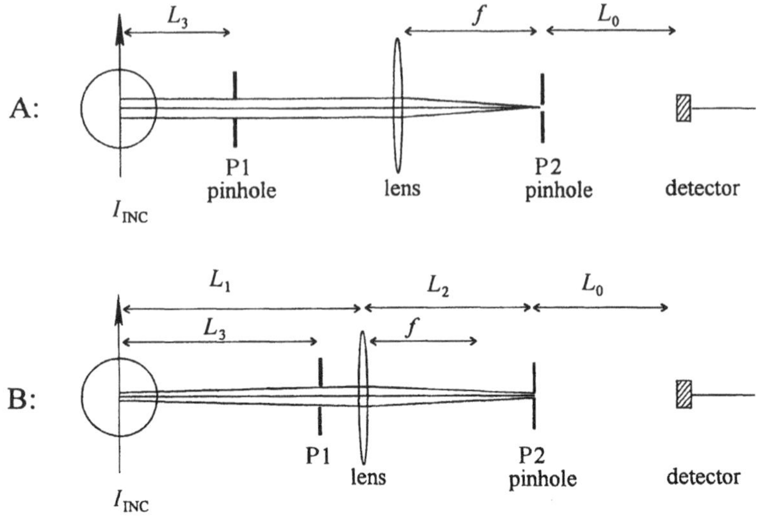

Figure 1.16 Two commonly used configurations of the detection optics for static and dynamic laser light scattering, where I_{INC} is the incident light and f is the focal length.

measurements, the first pinhole (P1) can be replaced by a slit so that the alignment will be easier, but it has to be switched back to pinhole for dynamic LLS measurements, which makes simultaneous static and dynamic LLS measurements impossible. In the **A**-configuration, the scattering volume is mainly determined by the diameter d_{P1} of P1 (or the slit width). The first pinhole should be as close as possible to the cell so that the scattering volume will be well defined. However, owing to the existence of the cell holder and index-matching cuvette, the first pinhole is normally 10–15 cm away from the scattering center. The second pinhole is located exactly on the focal plane of the lens, and the opening angle of d_{P2}/f determines the uncertainty of the scattering angle $\Delta\theta$. The coherent factor β in Eq. (36) is mainly determined by the opening angle of d_{P2}/L_0 but is also influenced by d_{P1}. In this design, the alignment will be easier, and the distance between the cell and the detector can be very small. In practice, $f \approx 10$ cm, $L_0 \approx 10$ cm, $d_{P1} \approx 200$–$400\ \mu$m, and $d_{P2} \approx 100$–$200\ \mu$m.

In the **B**-configuration, L_1, L_2, and f have to be exactly related by $1/L_1 + 1/L_2 = 1/f$; i.e., the scattering center and the second pinhole are located exactly on the imaging planes of the lens. The scattering volume is precisely determined by d_{P2}, and $\Delta\theta$ is determined by the opening angle of d_{P1}/L_3. Therefore, the first pinhole should be placed as close as possible to the lens, which is not difficult in practice. If $L_1 = L_2 = 2f$, one will have the so-called $2f$-$2f$ configuration; i.e., one optically moves the second pinhole to the scattering center. The coherent factor β is still determined by the opening angle of d_{P2}/L_0. In this configuration, $f \approx 5$–10 cm, which is smaller than the previous one. For a given d_{P2}, the scattering volume can be enlarged by a factor of L_1/L_2 (normally, $L_1/L_2 \leq 3$), so that one can simultaneously have a higher scattered intensity, a larger β, and a smaller $\Delta\theta$ for a given distance between the scattering center and the detector. However, the alignment of the lens is more difficult but manageable. Therefore, this design is preferred.

1.5.4 SAMPLE PREPARATION

If a macromolecule can be dissolved in more than one solvent, the choice of the solvent for laser light-scattering measurement should be made generally according to the following three criteria: (1) it should be colorless so that the absorption correction can be avoided, (2) it should have a higher contrast, i.e., a higher value of the specific refractive index increment dn/dC, so that the signal-to-noise ratio is increased, and (3) it should be less polar and less viscous so that the solution may be clarified more easily.

Sometimes in practice, we may have no choice of solvent for a given polymer. For example, poly(1,4-phenyleneterephthalamide) (PPTA or Kevlar) is only soluble in very strong acids that are viscous. In such a case, ultracentrifugation instead of filtration has to be used to remove dust particles from the solution [16, 76]. As for copolymers, the selection of proper solvents is even more difficult because at least two solvents that satisfy the preceding three criteria are needed. For this reason, reported characterization of copolymers is very limited [77–80].

As for polymers that are only soluble at high temperatures, the preparation of polymer solutions for the LLS experiments is challenging. Finding a solvent with a high boiling point is often not easy, but the dissolution of the polymers and the clarification of the solution at high temperatures are even more difficult. Chu and Wu and Chu *et al.* [66, 69] have developed two different apparatuses that can be used to prepare and clarify a polymer solution at temperatures higher than 200°C. Importantly, the two apparatuses are able (1) to dissolve a polymer under an inert atmosphere without losing solvent or building up an inner pressure owing to solvent evaporation, and (2) to transfer the solution into a filtration device by remote control.

Figure 1.17 shows a specially designed dissolution and filtration apparatus that can be placed inside a small oven. Known weights of a polymer sample and a filtered solvent, as well as a small glass-enclosed magnetic stirrer, are placed in A

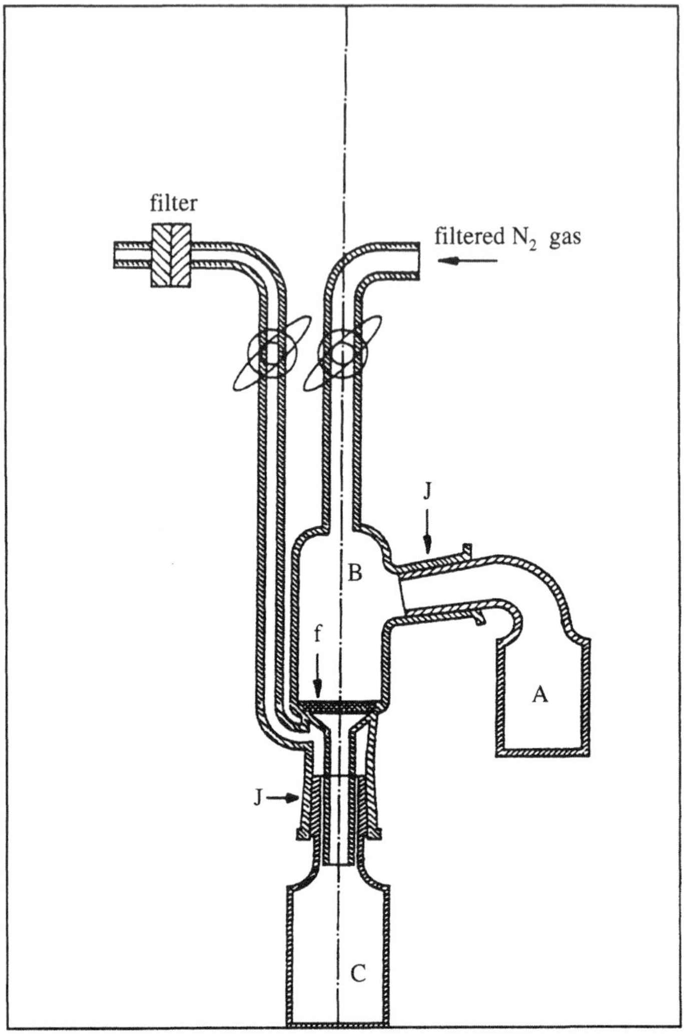

Figure 1.17 High-temperature dissolution and filtration apparatus. The entire apparatus is placed in a high-temperature oven controlled at 250 ± 2°C. (A) Solution vessel, where known weights of polymer and solvent as well as a small glass-enclosed magnetic stirrer are introduced. (B) A fine-grade sintered glass filter, connected to A and C by means of clean seal glass joints. (J) (14/20, Wheaton Scientific). (f) Fine-grade sintered glass filter. (C) Cylindrical light scattering cell of 27-mm outer diameter with a clean seal glass joint.

at room temperature. The solution vessel (A) is then connected to the precleaned filter (B). After degassing followed by introduction of nitrogen, both stopcocks are closed, and the oven is heated to a desired temperature to dissolve the polymer while the solution is stirred. When the polymer is considered to have been completely dissolved, the solution vessel is turned 180° by means of the seal glass joint (J). This allows the polymer solution to be transferred from A into B without exposure to air. A gentle nitrogen pressure is applied to force the polymer solution to pass the fine-grade sintered glass filter (f) and to move directly into the precleaned dust-free cylindrical light scattering cell (C). In this way, dust-free polymer solutions can be successfully prepared, keeping the temperature high.

Figure 1.18 shows another type of dissolution and filtration apparatus. Sleeve A (with no bottom) is joined to the shaded stopper, which is connected to a reflux condenser by means of a greaseless glass joint. Cup (B), with a magnetic stirrer (E) sitting on top of the fine-grade sintered glass filter (F_2), allows solution clarification. First, a dust-free solvent and a polymer are placed in B of the argon-filled apparatus, which has the preattached, dust-free light-scattering cell (D). The reflux condenser flushed with argon is then inserted with care because the apparatus with the polymer and solvent is under an inert atmosphere at room temperature. The entire apparatus is set in a small oven, and the temperature is raised to a desired value. When the polymer is completely dissolved, an argon pressure is applied through the sintered glass filter (F_1) to let the polymer solution move from B to C. The additional pressure difference plus gravity will filter the polymer solution directly into the dust-free light-scattering cell (D). After the filtration process is completed, the additional argon pressure is released. The stopcock above the upper filter is closed during the light-scattering experiment. Needless to say, extreme care must be taken when the test solution is prepared and subjected to light-scattering measurements at temperatures near the boiling point of the solvent because the pressure buildup could cause an explosion. This condition points to the absolute need for the pressure-releasing mechanism.

1.5.5 DIFFERENTIAL REFRACTOMETER

One of the most important parameters in static LLS is the specific refractive index increment dn/dC, which is defined as $\lim_{C\to 0}(\partial n/\partial C)_{\mathrm{T,P},\lambda}$. Because this parameter is not an intrinsic property of the polymer, the conditions of fixing temperature T, pressure P, and wavelength of light in vacuum λ are needed in its definition. Note that, according to Eqs. (7) and (32), an error of E% in dn/dC will lead to an error of 2 E% in the derived M_w.

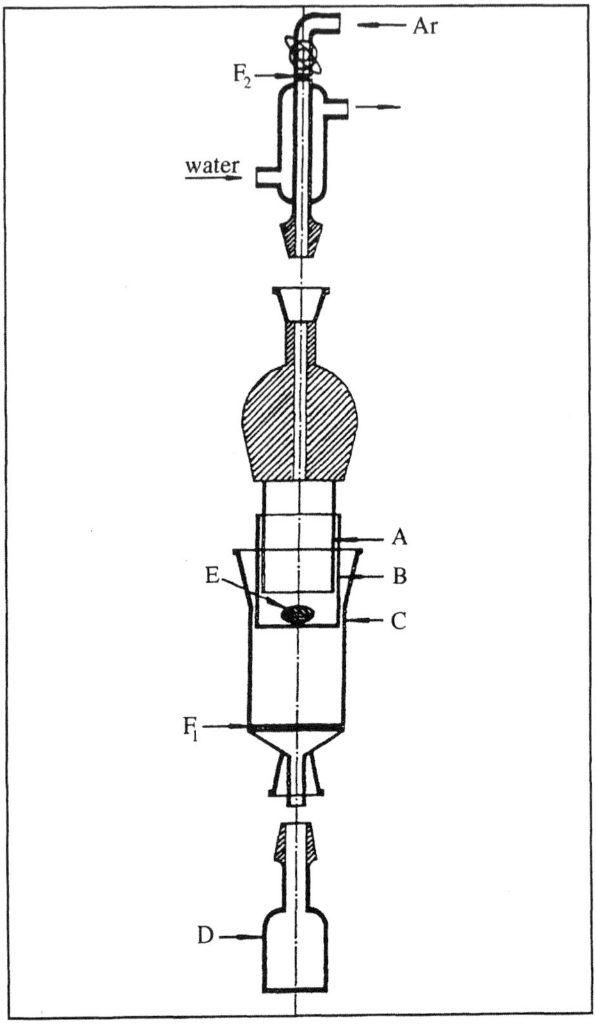
Ar
F_2
water
A
E
B
C
F_1
D

The refractive index increment Δn of a polymer solution is usually measured by using either a differential refractometer or an interferometer. In the former, the light beam is refracted at the boundary between the sample and a reference liquid. Commonly, the beam displacement is directly measured and then converted to Δn after multiplying a calibrated constant, which is normally obtained by using a solution with an accurately known refractive index difference Δn [81, 82]. This method is not an absolute one because the constant has to be calibrated at the same conditions as in the light-scattering measurements. In these measurements, two light beams with identical geometrical paths traverse two different optical paths. One passes through the sample and the other through the reference liquid. This method relies on the interference of the two beams. Its details can be found elsewhere [83, 84]. In a high-temperature LLS measurement, the conventional divided differential refractometer cuvette had to be replaced by a deformed cylindrical light-scattering cell in which the exit laser beam is refracted by the solution–air interface [66].

Figure 1.19a schematically shows a novel differential refractometer that was first designed by Wu *et al.* [85] and later commercialized by ALV GmbH, Langen, Germany. A small pinhole (P) with a diameter of 400 μm is illuminated with a laser light. The illuminated pinhole is imaged to a position-sensitive detector

---

Figure 1.18 Separate components of a high-temperature dissolution and filtration apparatus. The assembled apparatus is placed in a high-temperature oven. (A) Cylindrical insert (without bottom) with a diameter ≈2 mm less than that of the solution vessel (B). (C) Filtration section with a fine-grade sintered glass filter (F_1) and ground glass joints to a light scattering cell (D) and a ground glass joint adapter for the water-cooled condenser, which is located outside the temperature-controlled oven. (E) Magnetic stirring bar. Shaded area denotes volume reduction so that the volume accessible by vapor phase is no more than a few times the fluid phase. The miniature water-cooled condenser has a coarse-grade sintered glass filter (F_2) so that the entire system is always isolated from external dust. The greaseless stopcock above F_2 is for operating the apparatus as a closed system; for introducing low vacuum to degas the solvent before dissolution; for filling the apparatus with inert gas, such as argon, to alleviate chemical decomposition; and for releasing a possible pressure buildup at high temperature if chemical decomposition takes place. The entire apparatus is portable and can be inserted into the high-temperature light-scattering spectrometer with the light-scattering cell (D) and part of the filtration component (C) controlled at a given high temperature.

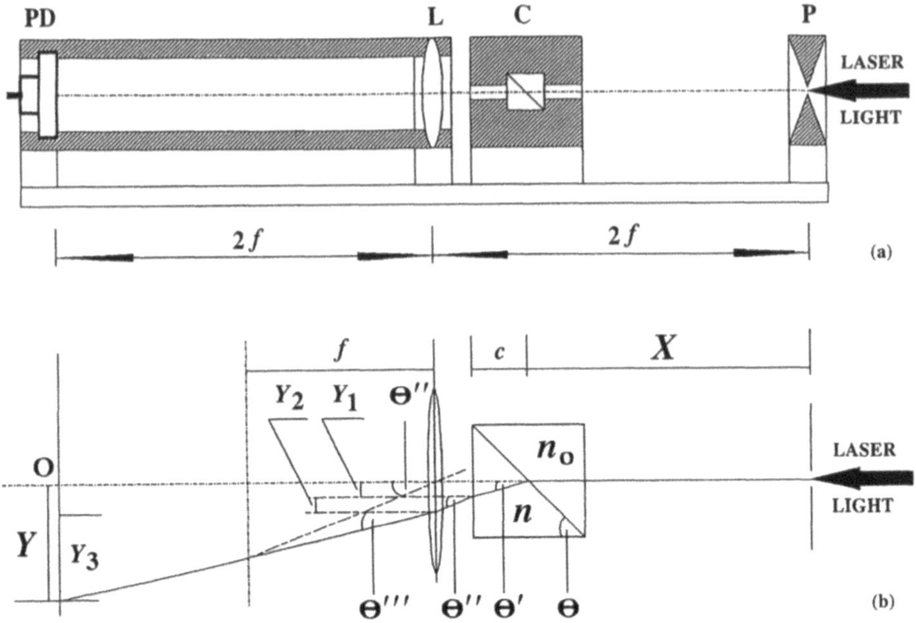

Figure 1.19 (a) Schematic view of a novel differential refractometer (commercialized by ALV GmbH, Langen, Germany), which consists of a pinhole (P), a differential refractometer cuvette (C), a lens (L, $f = 10$ cm), and a position-sensitive detector (PD). All components are rigidly mounted on a 40-cm-long optical rail. (b) Light path in which one compartment of the cuvette contains a solvent with refractive index n and the other contains a solution with slightly different refractive index $n = n_0 + \Delta n$. The cuvette and angles θ', θ'', and θ''' (actually very small, ≈0.01 radian) are enlarged to make the light path distinct.

(PD) (Hamamatsu S 3932) by a lens (L) located at an equal distance from the pinhole and the detector, where the distance is four times the focal length ($f = 100$ mm) of the lens. Thus, this novel refractometer uses a ($2f$-$2f$) design instead of the conventional ($1f$) design, which uses parallel incident light beams and makes the distance between the detector and the lens equal to only one focal length. A temperature-controlled refractometer cuvette (C) (Hellma 590.049-QS) is placed just in front of the lens. It is a flow cell and has a volume of ≈20 mL, which is divided by a glass plate at ≈45° into two compartments. The pinhole, the cuvette, the lens, and the detector are rigidly mounted on a small optical rail. The refractometer has dimensions of only 40 cm in length, 15 cm in width, and 10 cm in height, and the length can easily be reduced to 20 cm with another lens if necessary. The output voltage (-10 to 10 volts) from the position-sensitive detector is proportional to the displacement of the light spot from the center of the detector and can be measured by a digital voltmeter or an analog-to-digital data acquisition system and a personal computer.

Figure 1.19b shows the basic principle and the light path of the refractometer, where θ', θ'', θ''', and the cuvette are drawn enlarged to make the details clearer. If both compartments are filled with a solvent (i.e., $n = n_0$), the illuminated pinhole will be imaged at point O. However, if the solvent in one of the compartments is replaced by a dilute polymer solution with a slightly different refractive index (i.e., $n = n_0 + \Delta n$), the light will be bent first by the glass plate, then by the cuvette wall, and finally by the lens. The image is shifted away from the point O by a distance of Y. Figure 1.19b shows that

$$Y = Y_1 + Y_2 + Y_3 = c\tan(\theta) + (2f - X - c)\tan(\theta') + 2f\tan(\theta'') \quad (106)$$

and

$$f\tan(\theta') = f\tan(\theta'') + c\tan(\theta) + (2f - X - c)\tan(\theta') \quad (107)$$

where c, X, and θ are constants. Snell's law gives

$$n_0 \sin(90 - \theta) = (n_0 + \Delta n)\sin(90^\circ - \theta - \theta) \quad (108)$$

and

$$(n_0 + \Delta n)\sin(\theta) = \sin(\theta') \quad (109)$$

where θ, θ', and θ'' are actually so small because Δn is on the order of 10^{-4} refractive index units that we may set $\sin(\theta) = \theta$, $\sin(\theta') = \theta'$, $\tan(\theta) = \theta$, $\tan(\theta') = \theta'$, and $\tan(\theta'') = \theta''$. Combining Eqs. (106–109) leads to

$$Y = K\Delta n \quad (110)$$

where $K = [X + c''(1 - 1/n_0)]\tan(90^\circ - \theta)$. For a given optical setup and solvent, X, c, θ, n_0, and hence K are constants. Equation (110) shows that the signal is proportional to Δn, and the larger the value of X, the higher the sensitivity ($Y/\Delta n$) is. This means that the cuvette should be placed as close as possible to the lens in the experimental setup.

In the $2f$-$2f$ design, the detector and the pinhole (acting as a light source) are placed at the exact imaging positions along the optical axis of the lens. This configuration is optically equivalent to placing the detector directly behind the pinhole so that the laser beam drift is eliminated. In comparison with the conventional differential refractometer, this novel design has made the measurement of Δn much easier and provides reliable and accurate values for dn/dC because it is stable and

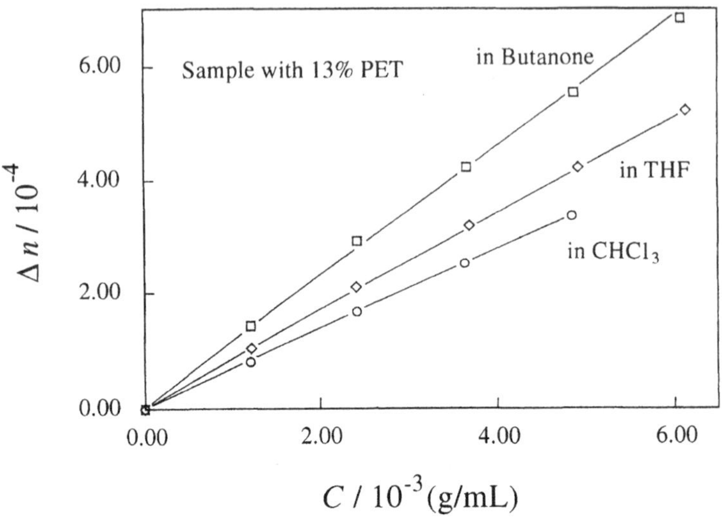

Figure 1.20 Concentration dependence of the refractive index difference (Δn) between the polymer solution and solvent for a 13% PET–PCL copolymer. The lines represent the least-square fits to the measured data.

the results can be recorded and averaged on a computer. Figure 1.20 shows the concentration dependence of Δn for a 13% PET–PCL copolymer in three different solvents. The lines represent the least-square fits to the data points.

The refractometer with its present dimensions can easily be installed into any existing laser light-scattering spectrometer together with the laser source, the thermostat, and the computer, as exemplified in Figure 1.21. The optical glass

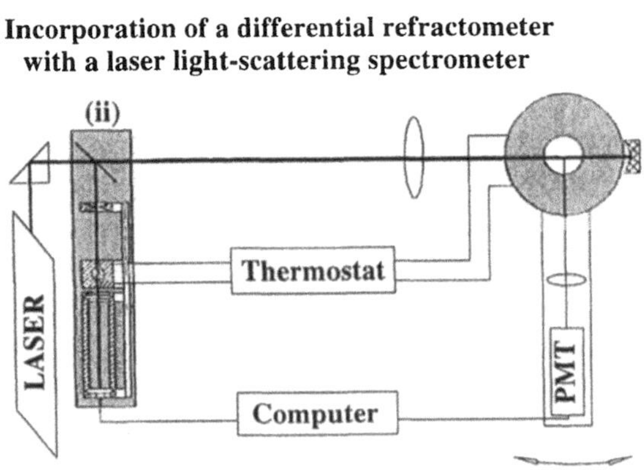

Figure 1.21 Possible arrangement of the novel differential refractometer with an existing laser light-scattering spectrometer.

plate placed in the laser light path at 45° reflects laser light by about 4%, and the reflected light is used as the light source. With this design, the light scattering and the refractive index increment can simultaneously be measured under identical experimental conditions of wavelength and temperature. The details of this novel spectrometer have been described elsewhere [85].

In summary, this chapter has shown that static and dynamic laser light scattering (LLS) combined provide a very powerful method for polymer characterization. LLS has advantages over other polymer characterization methods, which include ultracentrifugation and chromatography, in such features as speed, nonperturbation, and extreme dissolution conditions (high temperature or strong acid). The most important factor is that the calibration is independent of a particular LLS instrument used. However, the LLS method for the determination of mass distributions described in this chapter has its limitations in that its resolution is not as high as the fractionation methods, especially for samples whose mass distributions have closely packed peaks. The LLS method should play a useful role when polymers intractable by conventional characterization methods have to be treated. Finally, dynamic LLS can be used with other characterization methods that take advantage of the dependence of the hydrodynamic volume on molar mass.

Acknowledgments

BC gratefully acknowledges support of this work by the U.S. Army Research Office (DAAG559710022), the National Science Foundation (DMR 9612386), and the U.S. Department of Energy (DEFG0286ER45237.015). The support of the National Distinguished Young Investigator Fund (1996) from the National Science Foundation of China and the Earmarked Grant (1993–1998) from the Research Grants Council of Hong Kong Special Administrative Region are gratefully acknowledged.

References

1. B. Chu, *Laser Light Scattering*, 2nd ed., Academic Press, New York, 1991.
2. M. B. Huglin, Ed., *Light Scattering from Polymer Solution*, Academic Press, New York, 1972. Also, *Topics in Current Chemistry* **73:** 141 (1977).
3. O. Glatter and O. Kratky, *Small-Angle X-ray Scattering*, Academic Press, New York (1982).
4. S. B. Dubin, N. A. Clark, and G. B. Benedek, *J. Chem. Phys.* **54:** 5158 (1971).
5. G. D. Patterson, *Methods of Experimental Physics*, Vol. 16A, 170–204, Academic Press, New York (1980).

6. B. J. Berne and R. Pecora, *Dynamic Light Scattering*, Wiley/Interscience, New York (1976).
7. B. H. Zimm, *J. Chem. Phys.* **16:** 1093 (1948).
8. S. Chandrasekkar, *Rev. Mod. Phys.* **15:** 1 (1943).
9. P. Munk, *Introduction to Macromolecular Science*, 255. John Wiley & Sons, New York (1989).
10. P. Debye, *J. Phys. Coll. Chem.* **51:** 18 (1947).
11. B. H. Zimm, *J. Chem. Phys.* **16:** 1100 (1948).
12. B. Chu and C. Wu, *Macromolecules* **20:** 2642 (1987).
13. P. S. Russo, In *Dynamic Light Scattering*, W. Brown, Ed., Chapter 12, Oxford, New York (1993).
14. B. Chu, Q. Ying, C. Wu, and J. Ford, *Polymer Communications* **25:** 211 (1984).
15. B. Chu, Q. Ying, and C. Wu, *Polymer* **26:** 1408 (1985).
16. Q. Ying and B. Chu, *Die Makromol. Chem., Rapid Communication* **5:** 785 (1984).
17. S. T. Sun, I. Nishio, G. Swislow, and T. Tanaka, *J. Chem. Phys.* **73:** 5971 (1980).
18. W. H. Stockmayer and M. Schmidt, *Macromolecules* **17:** 509 (1984).
19. C. Wu, S. Q. Zhou, and W. Wang, *Biopolymer* **35:** 385 (1995).
20. A. Perico, P. Piaggio, and C. Cuniberti, *J. Chem. Phys.* **62(7):** 2690 (1975).
21. C. Wu, K. K. Chan, and K. Q. Xia, *Macromolecules* **28:** 1032 (1995).
22. D. E. Koppel, *J. Chem. Phys.* **57:** 4814 (1972).
23. A. K. Livesey, P. Licinio, and M. Delaye, *J. Chem. Phys.* **84:** 5102 (1986).
24. J. G. McWhirter and E. R. Pike, *J. Phys. A* **11:** 1729 (1978).
25. B. Chu, J. R. Ford, and H. S. Dhadwal, *Methods Enzymol.* **117:** 256 (1983).
26. B. Chu, C. Wu, and J. R. Ford, *J. Colloid Interface Sci.* **105:** 473 (1985).
27. S. W. Provencher, *J. Chem. Phys.* **64(7):** 2772 (1976).
28. J. Raczek, *Eur. Polym. J.* **19:** 607 (1983).
29. E. Nordmeier and M. D. Lechner, *Polym. J.* **21:** 623 (1989).
30. E. Q. Schulz-DuBois, Ed., *Proceedings of the 5th International Conference on Photon Correlation Techniques in Fluid Mechanics*, Springer-Verlag, New York (1983).
31. B. E. Dahneke, Ed., *Measurements of Suspended Particles by Quasi-Elastic Light Scattering*, Wiley, New York (1983).
32. G. Gouesbet and G. Grehan, *Optical Particle Sizing*, Plenum Press, New York (1988).
33. P. G. de Gennes, *Scaling Concepts in Polymer Physics*, Cornell University Press, Ithaca (1979).
34. B. Appelt and G. Meyerhoff, *Macromolecules* **13:** 657 (1980).
35. C. Wu, Y. B. Zhang, X. H. Yan, and R. S. Cheng, *Acta Polymerica Sinica* **3:** 349 (1995).
36. P. J. Flory, *Principles of Polymer Chemistry*, Cornell University Press, Ithaca (1953).
37. B. Chu, M. Onclin, and J. R. Ford, *J. Phys. Chem.* **88:** 6566 (1984).
38. J. Pope and B. Chu, *Macromolecules* **17:** 2633 (1984).
39. A. Cervenka, *Makromol. Chem.* **170:** 239 (1973).
40. C. Wu, *Macromolecules* **26:** 5423 (1993).

41. C. Wu, *J. Polym. Sci. Polym. Phys. Ed.* **32:** 803 (1994).
42. C. Wu, *Adv. Polym. Sci.* **137:** (1998).
43. C. Wu, D. Ma, X. Luo, K. K. Chan, and K. F. Woo, *J. Appl. Polym. Sci.* **53:** 1323 (1994).
44. C. Wu, K. F. Woo, X. Luo, and D. Ma, *Macromolecules* **27:** 6055 (1994).
45. C. Wu, M. Siddiq, and K. F. Woo, *Macromolecules* **28:** 4914 (1995).
46. C. Wu, M. Siddiq, and S. Q. Bo, *Macromolecules* **29:** 2989, 3157 (1996).
47. M. Li, M. Jiang, and C. Wu, *Macromolecules* **30:** 2201 (1997).
48. C. Wu, *Macromolecules* **27:** 7099 (1994).
49. C. Wu, M. Chen, and M. Akashi, *Macromolecules* **30:** 2187 (1997).
50. J. Gao, S. Q. Zhou, and C. Wu, *Polym. Engineering Sci.* **36:** 2968 (1996).
51. C. Wu and K. K. Chan, *J. Polym. Sci., Polym. Phy., Ed.*, **33:** 919 (1994).
52. C. Wu, R. Y. Qian, and D. H. Napper, *Macromolecules* **28:** 1592 (1994).
53. C. Wu, R. Y. Qian, and D. H. Napper, *Macromolecules* **28:** 1592 (1995).
54. I. Nishio, J. C. Reina, and R. Bansil, *Phys. Rev. Letts.* **59:** 684 (1987).
55. P. N. Pusey and W. van Megen, *Physica A* **157:** 705 (1989).
56. P. Debye and A. M. Bueche, *J. Appl. Phys.* **20:** 518 (1949).
57. C. Wu, J. Zuo, and B. Chu, *Macromolecules* **22:** 838 (1989).
58. C. Wu, J. Zuo, and B. Chu, *Macromolecules* **22:** 633 (1989).
59. M. Shibayama, *Macromol. Chem. Phys.* **199:** 1 (1998).
60. S. Kobayasi, *Rev. Sci. Instrum.* **56:** 160 (1985).
61. C. Allain, M. Drifford, and B. Gauthier-Manuel, *Polymer* **27:** 177 (1986).
62. J. G. H. Joosten, J. L. McCarthy, and P. N. Pusey, *Macromolecules* **24:** 6690 (1991).
63. L. Fang and W. Brown, *Macromolecules* **25:** 6897 (1992).
64. R. Foord, E. Jakeman, C. J. Oliver, E. R. Pike, R. J. Blagrove, F. Wood, and A. R. Peacocke, *Nature* **277:** 242 (1970).
65. B. Chu, R. Xu, T. Maeda, and H. S. Dhadwal, *Rev. Sci. Instrum.* **59:** 716 (1988).
66. B. Chu and C. Wu, *Macromolecules* **20:** 93 (1987).
67. C. Wu, W. Buck, and B. Chu, *Macromolecules* **20:** 98 (1987).
68. B. Chu, C. Wu, and J. Zuo, *Macromolecules* **20:** 700 (1987).
69. B. Chu, C. Wu, and W. Buck, *Macromolecules* **21:** 397 (1988).
70. B. Chu, C. Wu, and W. Buck, *Macromolecules* **22:** 831 (1989).
71. C. Wu, *Makromol. Chem., Makromol. Symp.* **61:** 377 (1992).
72. R. G. Brown, *Appl. Opt.* **26:** 4846 (1987).
73. R. G. Brown and A. P. Jackson, *J. Phys. E* **20:** 1503.
74. H. S. Dhadwal and B. Chu, *Rev. Sci. Instrum.* **60:** 845 (1989).
75. H. S. Dhadwal, C. Wu, and B. Chu, *Applied Optics* **28:** 4199 (1989).
76. B. Chu, C. Wu, and J. Ford, *J. Colloid & Interface Sci.* **105:** 473 (1985).
77. W. Bushuk and H. Benoit, *Can. J. Chem.* **36:** 1616 (1958).
78. B. Chu, Q. Ying, D. C. Lee, and D. Q. Wu, *Macromolecules* **18:**1962 (1985).
79. W. H. Stockmayer, L. D. Moore, M. Fixman, and B. N. Epstein, *J. Polym. Sci.* **16:** 517 (1955).

80. C. Wu, K. F. Woo, X. L. Luo, and D. Z. Ma, *Macromolecules* **27:** 6055 (1994).
81. M. Obasa, H. Nakamura, M. Takasaka, T. Keto, and M. Nagasawa, *Polym. J.* **25:** 301 (1993).
82. Application Notes LS7. Chromatix, Mountain View, California, U.S.A.
83. I. Baltog, C. Ghita, and L. Ghita, *Eur. Polym. J.* **6:** 1299 (1970).
84. Application Notes Optilab 903, Wyatt Technology, Santa Barbara, California, U.S.A.
85. C. Wu and K.Q. Xia, *Review of Scientific Instruments* **65:** 587 (1994).

Chapter 2 | Neutron Scattering

M. Shibayama*, H. Jinnai*, and T. Hashimoto**

**Department of Polymer Science and Engineering*
Kyoto Institute of Technology
Matsugasaki, Sakyo-ku
Kyoto, Japan

***Department of Polymer Chemistry*
Graduate School of Engineering
Kyoto University
Kyoto, Japan

2.1 Introduction

Studies on polymer conformation, morphology, and miscibility are of particular importance as a basic science application of polymers. Because the size of polymer chains is in the range of a few to hundreds of nanometers, scattering techniques such as light scattering [1–3], small-angle X-ray scattering [4], and small-angle neutron scattering [5] have been widely used to characterize the polymer chains in dilute solutions or in bulk. Though a direct observation of individual molecules can be achieved by transmission electron microscopy [6] and scanning probe microscopy [7], these methods are rather limited for this application because of the difficulty of sample preparation. Neutron scattering is particularly useful for investigating the shape and size of these macromolecules because, unlike X-ray scattering, deuterium substitution leads to a remarkable change in the contrast between the labeled and nonlabeled species. Substitution of a hydrogen (^{1}H) to a deuterium (D or ^{2}H) changes the scattering length from -3.739×10^{-13} to $+6.671 \times 10^{-13}$ cm without significantly changing molecular interactions and structures of the systems. Note that the negative value of the scattering length of hydrogen is of particular importance. With the aid of chemistry, we can substitute hydrogen atoms of a target part of a polymer chain by deuterium atoms and "color" (or label) the part. This method is called *molecular labeling*. This colored part becomes exclusively visible with the "eye" of a neutron. Another useful method is *contrast matching*. Because most elements except for ^{1}H (and a few others, such as ^{7}Li, ^{48}Ti, and ^{55}Mn) have positive scattering lengths, polymeric materials consisting mostly of hydrogen (^{1}H) and carbon atoms have an intermediate

ISBN 0-12-683265-X/$30.00.

scattering length between those of the two extremes, namely, fully deuterated and fully nondeuterated ones. Therefore, by substituting some fraction of hydrogen atoms of the matrix component by deuterium, we can tune the scattering length of the system to the value we desire (contrast matching). For example, if one wants to observe the entire shape of a zebra clearly, it is better to tune the background to be gray. But, if the black stripes are focused, the background should be tuned to be white.

Because of the aforementioned advantages, small-angle neutron scattering (SANS) has become one of the most powerful techniques for polymer characterization since its first application to the investigation of the molecular size of molten polystyrene [5]. The random-coil nature of polymer chains in the molten state was proved in 1973 by SANS about a quarter of a century after the prediction by Flory (1949) [8]. This labeling is unique because deuterium substitution does not change the physical properties of the polymers. Although isotope effects became clear in the 1970s (difference in crystallization temperature) [9], the deuterium labeling technique is widely used for studies in polymer science as well as in biophysics. The labeling technique has also been applied to the studies of conformation changes of polymer chains with extension or swelling, stress relaxation in polymers, polymer–polymer miscibility, and so on. These studies were reviewed by Sperling [10], Boué [11], and Sadler [12]. Concurrently, owing to the theoretical development of the structure factor for mixtures of protonated and deuterated polymers, the restriction of dilution for labeled chains has been removed [13], which now makes it possible to study the interaction parameter for polymer blends in a given concentration.

For heterogeneous structures, such as lamellar structures in crystalline polymers and microdomain structures in block copolymers, many studies have been carried out to elucidate chain-folding mechanism, crystallization, chain conformation in a confined space, miscibility, and so forth. Polymer networks and gels are also classified as heterogeneous polymeric systems in which cross-linking generally produces frozen spatial heterogeneities. These heterogeneities have important effects on the physical properties of polymer networks of such as deformation and swelling behavior.

This chapter presents comprehensive guidelines for a SANS experiment, including the background of the theory, the design of an experiment, and data analysis for beginners unfamiliar not only with SANS but with polymer science. Then several examples of the experiments are reviewed to provide suggestions and hints for readers planning a new SANS experiment. For further reading, the book by Higgins and Benoit is highly recommended [14].

2.2 Neutron and Neutron Scattering

2.2.1 *NEUTRON AS AN ELEMENTARY PARTICLE [15–17]*

The neutron is an elementary particle discovered by Chadwick in 1932. The half-life of a neutron is about 10.3 min. The charge of a neutron is less than 10^{-18} e, where e is the charge of an electron; e = 1.6021×10^{-19} C. Therefore, the neutron is substantially a noncharged element. The mass of neutron is $m_n = 1.6749 \times 10^{-27}$ kg (=1.009 u; u is the atomic unit), which is about 2000 times heavier than an electron. The radius of a neutron is about 1.5 fermi (1 fermi = 10^{-13} cm) and is much smaller than the radius of the electron cloud of an atom ($\approx 10^{-8}$ cm), as schematically shown in Figure 2.1 for the case of $^{12}_{6}C$. A neutron has a quantum spin number of 1/2 and hence belongs to the Fermi–Dirac statistics. Like the other particle beams and electromagnetic waves, a neutron has the features of a particle and a wave, i.e., wave–particle duality. The mean wavelength of neutron, λ_{mean}, can be given by combining the Einstein and de Broglie relations,

$$E = \frac{1}{2}mv^2 \text{ (Einstein)}, \qquad E = h\nu; \quad p = mv = \frac{h}{\lambda} \text{ (de Broglie)} \tag{1}$$

where E, m, v, ν, and λ are the energy, mass, velocity, frequency, and wavelength of a neutron, respectively; p is the momentum transfer and h is Planck's constant.

However, because of the characteristic nature of neutrons (i.e., the mass, electroneutrality, and spin), the flux of neutrons (i.e., a neutron beam) exhibits

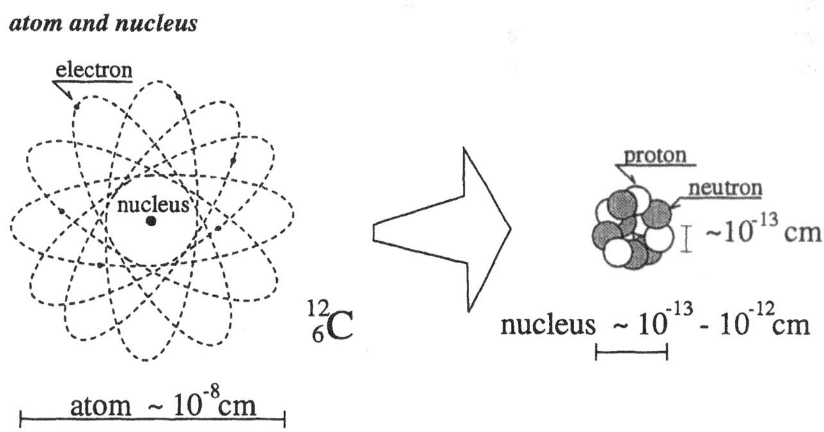

Figure 2.1 Schematic representation of atom and nucleus.

characteristic properties different from other particle beams, such as α and β rays and electron beams, and from electromagnetic waves, such as γ rays, including X rays, which are listed in the following:

1. Maxwell distribution

   The flux of neutrons can be generated in an atomic reactor or pulse source. In an atomic reactor, there exist several kinds of neutrons having different energy levels. However, only thermal neutrons or cold neutrons having lower energies than others are useful for neutron scattering. These neutrons are at thermal equilibrium. Therefore, the population of the neutrons obeys the Maxwell distribution. At the thermal equilibrium, the energy of a neutron is given by

$$E = kT \tag{2}$$

where k is the Boltzmann constant and T is the absolute temperature. By substituting Eq. (2) into (1), the wavelength of the thermal neutrons is obtained as follows:

$$\lambda_{\text{mean}} = \frac{h}{\sqrt{3m_n kT}} \approx T^{-1/2} \tag{3}$$

Note that λ_{mean} is proportional to $T^{-1/2}$. Because the maximum of the population for a Maxwell distribution is given by $\lambda_{\text{max}} = (3/2)^{1/2}\lambda_{\text{mean}}$,

$$\lambda_{\text{max}} = \sqrt{\frac{3}{2}}\lambda_{\text{mean}} = \frac{30.81}{\sqrt{T}}[\text{Å}] \tag{4}$$

For example, when $T = 300$ K, λ_{mean}, is obtained to be 1.78 Å, which is on the same order of the wavelength of the X ray used for diffraction studies (e.g., 1.54 Å for a CuK_α line). It is more convenient to use the electron volt as a unit of energy. In this case, T is given by

$$T[K] = 11.605 \times 10^3 E[\text{eV}] \tag{5}$$

and the wavelength and the velocity of neutrons are given as follows:

$$\lambda[\text{Å}] = \frac{0.2860}{\sqrt{E[\text{eV}]}} \tag{6}$$

$$v[\text{m/s}] = 1.383 \times 10^4 \sqrt{E[\text{eV}]} \tag{7}$$

These relations are shown in Figure 2.2. For example, 1Å-wavelength neutrons have an energy of 0.0818 eV and a velocity of 3.98 km/s.

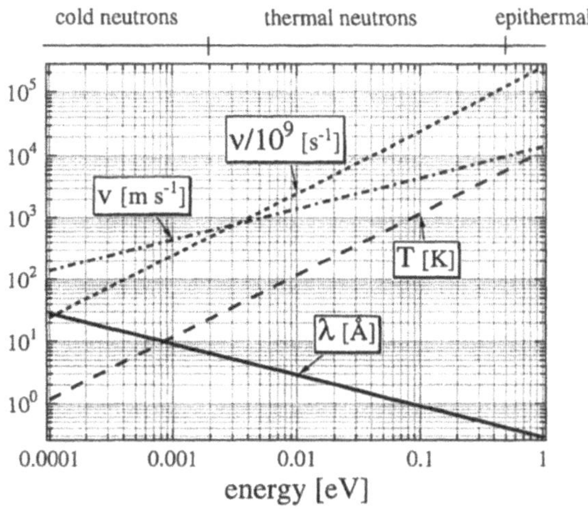

Figure 2.2 Velocity, wave number, temperature for the thermal equilibrium, and wavelengths of neutrons as a function of their energy.

2. Electroneutrality and spin

   Because of electroneutrality, interactions of neutrons with electrons can be ignored. Neutrons interact with nuclei in an atom, with the magnetic moment of an atom, or both. Interaction with nuclei gives rise to *nuclear scattering*, and interaction with the magnetic moment results in *magnetic scattering*. In most cases, however, the nuclear scattering is particularly useful for polymer science.

2.2.2 NEUTRON SCATTERING AND CROSS SECTION

Neutrons are scattered by nuclei because of nuclear force. Because the range of the nuclear force is very short (10^{-13}–10^{-12} cm) and the size of nuclei is much smaller than that of an atom ($\approx 10^{-8}$ cm), most of materials are "very dilute" for the "eye" of neutrons, as schematically illustrated in Figure 2.3. Let us consider a case in which a flux of neutron Φ_0 is irradiated by an object. According to Lambert's law, the transmitted flux Φ is obtained by

$$\Phi = \Phi_0 \exp(-N\sigma_T x/V) \tag{8}$$

where σ_T is the number rate of neutrons that collide with the atoms, N is the number of atoms in the sample volume V, and x is the path length of the object. σ_T is called the total cross section and consists of the absorption σ_a and the scattering

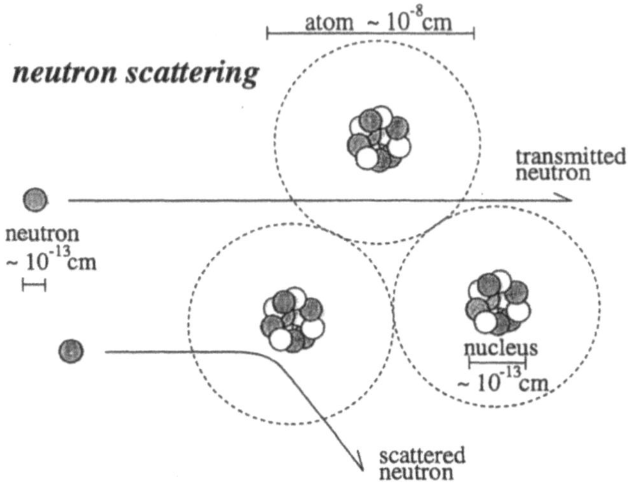

Figure 2.3 Comparison of the sizes of neutrons and atoms for neutron scattering.

cross sections σ, respectively;

$$\sigma_T = \sigma_a + \sigma \tag{9}$$

If the scattering is isotropic, σ is given by

$$\sigma = 4\pi b^2 \tag{10}$$

where b is the scattering length. The scattered intensity for an assembly of atoms of number N can be calculated by taking account of the phase of the scattered waves in the context of the Born approximation [18]

$$\left(\frac{d\sigma}{d\Omega}\right) = \frac{1}{V}\left\langle \sum_{i,j}^{N} b_i b_j \exp[-i\mathbf{q}\cdot(\mathbf{r}_i - \mathbf{r}_j)] \right\rangle \tag{11}$$

where Ω is the solid angle, and b_i and $\mathbf{r}_i$ are respectively the scattering length and the position vector of the atom i; $\langle \ldots \rangle$ denotes the average; and $(d\sigma/d\Omega)$ is called the differential cross section and $\mathbf{q}$ is the momentum transfer given by

$$q = |\mathbf{q}| = \frac{4\pi}{\lambda} \sin\left(\frac{\theta}{2}\right) \tag{12}$$

where q is the magnitude of the momentum transfer (or scattering vector) and θ is the scattering angle. The q is the most important variable describing the structure of matter, of which physical meaning will be discussed later. The differential cross section is divided into the following two terms:

$$\left(\frac{d\sigma}{d\Omega}\right) = \frac{1}{V}\sum_{i}^{N}\left\langle b_i^2\right\rangle + \frac{1}{V}\sum_{i\neq j}^{N}\langle b_i b_j \exp[-i\mathbf{q}\cdot(\mathbf{r}_i - \mathbf{r}_j)]\rangle$$

$$= \frac{N}{V}\langle b^2\rangle + \frac{\langle b\rangle^2}{V}\sum_{i\neq j}^{N}\langle \exp[-i\mathbf{q}\cdot(\mathbf{r}_i - \mathbf{r}_j)]\rangle \qquad (13)$$

where the relation $\langle b_i b_j\rangle = \langle b_i\rangle\langle b_j\rangle \equiv \langle b\rangle^2$ was assumed for $j \neq i$. This means that there is no correlation between b_i and b_j. By defining the dispersion of the scattering length as

$$\langle(\Delta b)^2\rangle \equiv \langle(b - \langle b\rangle)^2\rangle = \langle b^2\rangle - \langle b\rangle^2 \qquad (14)$$

one can rewrite Eq. (13) as

$$\left(\frac{d\sigma}{d\Omega}\right) = \frac{N}{V}\langle b\rangle^2 + \frac{\langle b\rangle^2}{V}\sum_{i\neq j}^{N}\langle \exp[-i\mathbf{q}\cdot(\mathbf{r}_i - \mathbf{r}_j)]\rangle + \frac{N}{V}\langle b^2\rangle - \frac{N}{V}\langle b\rangle^2$$

$$= \frac{\langle b\rangle^2}{V}\sum_{i,j}^{N}\langle \exp[-i\mathbf{q}\cdot(\mathbf{r}_i - \mathbf{r}_j)]\rangle + \frac{N}{V}\langle(\Delta b)^2\rangle \qquad (15)$$

The first term of Eq. (15) is called coherent scattering, and the second term is incoherent scattering. Thus,

$$\left(\frac{d\sigma}{d\Omega}\right) = \left(\frac{d\sigma}{d\Omega}\right)_{\text{coh}} + \left(\frac{d\sigma}{d\Omega}\right)_{\text{inc}} \qquad (16)$$

and

$$\left(\frac{d\sigma}{d\Omega}\right)_{\text{coh}} = \frac{\langle b\rangle^2}{V}\sum_{i,j}^{N}\langle \exp[-i\mathbf{q}\cdot(\mathbf{r}_i - \mathbf{r}_j)]\rangle \qquad (17)$$

$$\left(\frac{d\sigma}{d\Omega}\right)_{\text{inc}} = \frac{N}{V}\langle\Delta b\rangle^2 \qquad (18)$$

Thus, the coherent and incoherent scattering cross sections σ_{coh} and σ_{inc} for an atom are respectively given by

$$\sigma_{\text{coh}} = \int\left(\frac{d\sigma}{d\Omega}\right)_{\text{coh}} d\Omega = 4\pi\langle b\rangle^2 \qquad (19)$$

$$\sigma_{\text{inc}} = \int\left(\frac{d\sigma}{d\Omega}\right)_{\text{inc}} d\Omega = 4\pi\langle\Delta b\rangle^2 \qquad (20)$$

The values of σ_{coh} and σ_{inc} for important elements in polymer science are listed in Table 2.1. The scattering length for molecules (or segments) is calculated by summing the scattering lengths of the component atoms, i.e.,

$$b \equiv b_{\text{molecule}} = \sum_i r_i b_{\text{atom},i}$$

where r_i is the number of atoms of kind i in the molecule (or in the segment). For example, the scattering length of benzene molecule, b_{benzene}, is given by

$$\begin{aligned} b_{\text{benzene}} &= 6b_{\text{H}} + 6b_{\text{C}} = 6 \times (-3.739 \times 10^{-13}) + 6 \times (6.646 \times 10^{-13}) \\ &= 17.442 \times 10^{-13} [\text{cm}] \end{aligned} \tag{21}$$

where b_{H} and b_{C} are the scattering lengths of hydrogen and carbon atoms, respectively.

Equations (15–20) indicate that there are two types of neutron scatterings, i.e., coherent scattering and incoherent scattering. The incoherent scattering is ascribed to the presence of different isotopes for a given element. Because this reduces the coherence of the scattered neutron beam, it is very important to design a SANS experiment to reduce the contribution of the incoherent scattering. The origin of incoherent scattering is discussed in Section 2.2.3. There is another type of classification. Like the Compton (inelastic) and Thomson (elastic) scatterings in X-ray scattering and Stokes (and anti-Stokes; inelastic) and Rayleigh (elastic) scatterings in light scattering [19], neutron scattering is also classified as to elastic and inelastic scattering. In the case of the former, the wavelength of the scattered neutrons is different from the incident neutrons, and in the latter the wavelength is the same because of "elastic scattering." Therefore, there are four different scatterings in neutron scattering, i.e., *elastic coherent scattering*, *elastic incoherent scattering*, *inelastic coherent scattering*, and *inelastic incoherent scattering*. However, we focus on the elastic coherent scattering, which is most useful to the structural investigation of polymeric systems.

2.2.3 *ORIGIN OF INCOHERENT SCATTERING*

The incoherent scattering is, in origin, ascribed to the different isotopes in the nuclei. Therefore, no incoherent scattering is expected for an isotropically pure sample. However, as shown in Table 2.1, hydrogen (^{1}H) has an extraordinarily large incoherent scattering cross section. This is due to the nuclear spin incoherence. A neutron has two quantum spin states, $+1/2$ and $-1/2$, each of which has the scattering length of b^+ and b^-. The neutron interacts with the nuclear spin of the

Table 2.1 Scattering lengths and cross sections of important elements in polymer science. Reprinted from V. E. Sears *et al.* (1992) with permission from the author and the publisher, GORDON AND BREACH PUBLISHERS.

Isotope[a]	I^b	*Concentration*[c] *(%)*	b^d *(fm)*	$b_{inc}{}^e$ *(fm)*	$\sigma_{coh}{}^f$ *(barn)*	$\sigma_{inc}{}^g$ *(barn)*	$\sigma_a{}^h$ *(barn)*
H			−3.7390		1.7568	80.26	0.3326
^{1}H	1/2	99.985	−3.7406	25.274	1.7583	80.27	0.3326
^{2}H	1	0.015	6.671	4.04	5.592	2.05	0.000519
C			6.6460		5.551	0.001	0.0035
^{12}C	0	98.9	6.6511	0	5.559	0	0.00353
^{13}C		1.1	6.19	−0.52	4.81	0.034	0.00137
N			9.36		11.01	0.5	1.9
^{14}N	1	99.63	9.37	2.0	11.03	0.5	1.91
^{15}N		0.37	6.44	−0.02	5.21	0.00005	0.000024
O			5.803		4.232	0.0008	0.00019
^{16}O	0	99.762	5.803	0	4.232	0	0.0001
^{17}O		0.038	5.78	0.18	4.2	0.004	0.236
^{18}O		0.2	5.84	0	4.29	0	0.00016
^{19}F	1/2	100	5.654	−0.082	4.017	0.0008	0.0096
^{23}Na	3/2	100	3.63	3.59	1.66	1.62	0.53
Si			4.1491		2.163	0.004	0.171
P		100	5.13	0.2	3.307	0.005	0.172
S			2.847		1.0186	0.007	0.53
Cl			9.5770		11.5257	5.3	33.5
K			3.67		1.69	0.27	2.1
Ca			4.70		2.78	0.05	0.43

[a] Isotope.
[b] Spin number.
[c] Natural abundance (for radioisotopes the half-life is given instead).
[d] Bound coherent scattering length (fm = 1×10^{-13} cm = 1×10^{-15} m).
[e] Bound incoherent scattering length (fm).
[f] Bound coherent scattering cross section (barn = 1×10^{-24} cm^2).
[g] Bound incoherent scattering cross section (barn).
[h] Absorption cross section for 2200 m/s neutrons (barn).
References:
Acta Cryst. A **28:** 357 (1972); *Neutron Diffraction Newsletter*, March (1976)
Handbook of Chemistry and Physics, CRC (1996).
J. S. Higgins and H. C. Benoit, *Polymers and Neutron Scattering*, Oxford (1994), p. 52.
http://ne43.ne.uiuc.edu/n-scatter/n-lengths/list.html

atom. If the nucleus has the spin number I, the following relations are given:

$$\langle b \rangle = \left(\frac{I+1}{2I+1}\right)b^{+} + \left(\frac{I}{2I+1}\right)b^{-} \tag{22}$$

and

$$\langle b^2 \rangle = \left(\frac{I+1}{2I+1}\right)(b^{+})^2 + \left(\frac{I}{2I+1}\right)(b^{-})^2 \tag{23}$$

Hydrogen is a typical element having a large incoherent cross section. Substituting $b^{+} = 1 \times 10^{-12}$ cm, $b^{-} = -4.7 \times 10^{-12}$ cm, and $I = 1/2$ into Eqs. (22) and (23), we get $\sigma_{\text{inc}} = 4\pi \langle (\Delta b)^2 \rangle = 78 \times 10^{-24}$ cm^2 for hydrogen, which is much larger than $\sigma_{\text{coh}} = 2 \times 10^{-24}$ cm^2. More accurate values of coherent and incoherent cross sections are given in Table 2.1 [20]. These values are also now available via the World Wide Web (http://ne43.ne.uiuc.edu/n-scatter/n-lengths/list.html).

2.3 Experiments

2.3.1 INTRODUCTION

Neutron scattering has been used to explore the static and dynamic properties of materials via the interaction between the neutron and the nucleus. If a neutron particle is scattered by a nucleus with an exchange of energy between the incident neutron and the molecules that are vibrating, rotating, or translating, the neutron is said to be "inelastically" scattered. On the other hand, if no exchange of energy upon scattering occurs, it is called "elastic scattering," i.e., the wavelengths of the incoming and the scattered neutrons are identical. In between these two extreme cases, there is "quasi-elastic scattering": If a scattering element in a macromolecule changes its position in space due to diffusion, the neutron is scattered neither completely elastically nor inelastically.

What kind of information can one get from the neutron scattering? Clearly, inelastic scattering contains information on some kinds of molecular motions. For example, the *backscattering spectrometers* and the *spin-echo spectrometers* are used to study the slow motion of the polymer backbone or of large side groups that can be detected by measuring small changes in energy transfer. However, much more attention has been focused to examine the molecular shapes or morphology of the macromolecules, which can be done by studying the elastic coherent scattering through the static structure factor $S(q)$. This is not only because the q-range (or spatial scale) accessible by the neutron scattering is convenient to investigate the shape and morphology of macromolecules but also results from the excellent feature of deuterium labeling. Thus, as mentioned in the introduction of this chapter, we restrict our attention to SANS. Even so, we believe that application to some classical diffractometers can be understood as an extension of the principles learned in SANS. In the following, we will describe some brief overviews of the SANS instrument and comment about the set up of the instrument and data reductions.

2.3.2 SANS SPECTROMETER

2.3.2.1 Overview

There are more than 10 SANS facilities around the world. The spectrometers mainly used in the polymer society are listed in Table 2.2. They are designed to suit a particular neutron source and scientific demand. To list detailed features of all those spectrometers would not be meaningful. Detailed description of each instrument is always provided by the institute of the neutron source, most of which are open to the public. We listed Internet URLs for those institutes in Table 2.2 for

Table 2.2 Small-angle neutron scattering (SANS) instruments around the world.

Facility	*Location*	*Neutron Source*	*Name of SANS*
BNL[a]	Brookhaven, USA	30-mW reactor	H9B
ILL[b]	Grenoble, France	57-mW reactor	D11, D17
ISIS[c]	Chilton, UK	160-mW accelerator	LOQ
JAERI[d]	Tokai, Japan	20-mW reactor	SANS-J, SANS-U
KEK[e]	Tsukuba, Japan	3.5-kW accelerator	SWAN
KFA[f]	Julich, Germany	23-mW reactor	KWSI, KWSII
NIST[g]	Gaithersburg, USA	20-mW reactor	Two 30-m SANS[j], One 8-m SANS
ORNL[h]	Oak Ridge, USA	85-mW reactor	SANS
RISO[i]	Roskilde, Denmark	10-mW reactor	SANS

[a]http://neutron.chm.bnl.gov/HFBI
[b]http://www.ill.fr/
[c]http://www.isis.rl.ac.uk/largescale/LOQ/loq.htm
[d]http://www.issp.u-tokyo.ac.jp/labs/neutron/inst/sans-u/ and http://rrsys.tokai.jaeri.go.jp:1050/english/cont.html
[e]http://neutron-www.kek.jp/
[f]http://www.kfa-juelich.de/iff/iff_ns1.e.html
[g]http://rrdjazz.nist.gov/sans.html
[h]http://neutrons.ornl.gov/
[i]http://www.risoe.dk/
[j]NG3SANS and NG7SANS

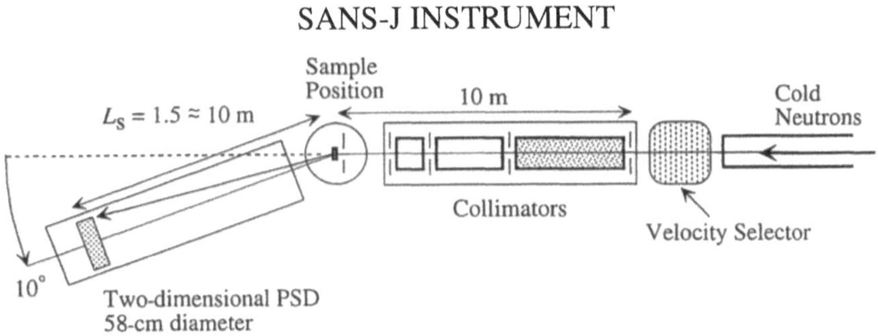

Figure 2.4 Schematic illustration of a small-angle neutron scattering instrument, SANS-J, at JRR3M reactor of Japan Atomic Energy Research Institute (JAERI). Courtesy of Drs. J. Suzuki and S. Koizumi.

the reader's reference. In this chapter, we will describe the SANS spectrometer at the reactor sources. Although some SANS instruments are installed at the pulse source, the principles described here are quite general and can be applied to those instruments at the pulse source.

Figure 2.4 schematically shows a typical SANS spectrometer for a reactor source (SANS-J, Japan Atomic Research Institute, Japan, JAERI). Let us first give an overview of the SANS instrument and then describe some important points of the SANS spectrometer in detail. The slow neutrons from the cold source travel along neutron-guide without losing energy to a region with more space and low background radiation, which is often called a "guide hall." The neutrons guided to the SANS spectrometer have a wavelength distribution. Hence, it is necessary to select a particular wavelength from the "white" beam produced by the reactor source. A velocity selector is usually used for this purpose (see Section 2.3.2.3). After the monochromation, there is a "presample flight path," normally called a collimator, which consists of some diaphragms (in the case of SANS-J in Figure 2.4, four diaphragms). Often, the diameters of each diaphragm are independently controlled from a computer terminal. In some cases, the position of the diaphragms can be changed. Combination of the diaphragm diameters of the collimator should make divergence of the neutron beam on the detector small enough to attain required resolution. The collimated neutron beam goes into the sample, which may be heated, cooled, pressured, etc. After the sample, the scattered neutrons pass through the evacuated tube ("postsample flight path") and arrive at the two-dimensional (2-D) area detector. The evacuation of pre- and postsample flight paths is necessary to prevent the scattering and absorption of neutrons by the air, which would remove most neutrons before they arrived at the detector. The distance between the sample and the detector (sample-to-detector distance L_S) can be arbitrarily controlled, e.g., from 1.5 to 10 m for SANS-J. A beam stop is placed in front of the detector to protect the detector from the full intensity of the unscattered transmitted beam. The beam stop may be mobile via the computer terminal for fine adjustment of the transmitted beam. Several beam stops with different dimensions may be provided so that one can choose the most appropriate size, depending on the collimation of the beam. For a relatively wide-angle measurement, to obtain a large q-range, the postsample evacuated path itself can be "swung" up several degrees (up to 10° for SANS-J). Most of the parameters such as L_S, the beam stop positions and their dimensions, sample environmental parameters, and the dimensions of the collimators diaphragms can normally be controlled from the computer terminal. In the following sections, we describe important parts of the SANS spectrometer as well as the neutron sources in some detail.

2.3.2.2 Neutron Source and Neutron Transportation

Fast neutrons are produced in a fission chain reaction, often from enriched uranium in the reactor core, which is surrounded by a moderator. The moderator is composed of materials rich in light nuclei; for example, D_2O, H_2O, and the fast neutrons suffer many collisions, losing their energies until they reach an average thermal energy. The peak wavelength from a water-moderated reactor at a temperature of 50 °C is about 1.4 Å, which is not ideal for structural and dynamical studies of large molecules. This problem can be solved if a local "cold part" is placed in the moderator. This is called a "cold source." For example, liquid D_2 at 25 K as a cold source provides a Maxwellian distribution with a peak wavelength around 6 Å, and a considerable enhancement in intensity of long-wavelength neutrons can be obtained compared with the moderators without the cold source.

The neutrons thus produced in the reactor then have to be guided to spectrometers. The beam tubes are inserted through the shielding into just outside the core of the reactor, where the maximum flux can be obtained. Normally, many spectrometers are built around the source as close as possible to the outside of the shielding because the neutron flux decreases as the inverse of the square of the distance from the source. This area becomes crowded and, hence, the number of spectrometers that can be built around a given source will be limited. The neutron facility solves this crowding problem by using neutron guides [21], which are "pipes" whose internal surfaces are coated with a very flat reflecting material analogous to the optical fiber for light. The neutrons will be totally reflected just as light when it radiates at an angle less than the critical angle on a medium with higher refractive index. Typical neutron refractive indices are very close to, and usually less than, unity. The critical angle θ_c is very small, e.g., $\theta_c \cong 0.5°$ for 10-Å neutrons if nickel-coated glass is used. Note the critical angle is also wavelength dependent. The guides have to be very well aligned. The primary function of the guides is, of course, to transport the neutrons to accommodate more spectrometers. Moreover, the neutron guides are useful in allowing the spectrometers to be built far from the reactor in region(s) where the background radiation is low.

Another advantage is that the curved neutron guides can be used as a sort of "wavelength selector." The angle of incidence changes if the neutron guide is curved, depending upon the curvature of the guide. In other words, the neutrons reflected at an angle $\theta_1 (\theta_1 < \theta_c)$ would have an angle θ_2 at the following reflection that could be larger than the critical angle and hence could not be reflected. Because the critical angle for short-wavelength neutrons is smaller, it is more difficult for them to satisfy the condition for the total reflection; hence, they are more easily cut off.

The neutrons thus guided from the cold source to the instrument are filtered to remove core gamma rays and fast neutrons from the beam [22, 23]. The filters consist of high-purity bismuth (Bi). Large, randomly oriented single-crystal grains of Bi are cast in polycrystalline beryllium and cooled in a liquid nitrogen. The filtered neutron beams have negligible gamma ray contamination, and the measured transmission is typically 75–80%.

2.3.2.3 Mechanical Velocity Selection

Because the neutron is an elementary particle having a finite velocity (i.e., material wave), monochromation of the neutron beam can be accomplished by selecting neutrons having a given velocity. For example, if one chooses neutrons only with the velocity of 600 ms^{-1}, neutrons with a wavelength λ of 6 Å result. A mechanical selection device called a "velocity selector" is used for this purpose. Figure 2.5(a) shows the principle of mechanical velocity selection for a case in which two rotors

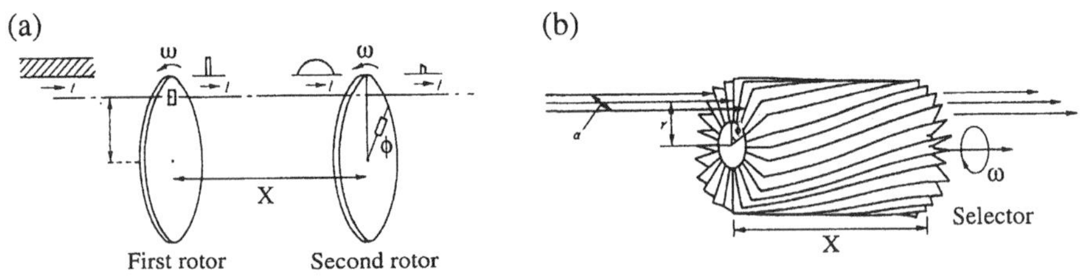

Figure 2.5 Schematic representation of mechanical velocity selection. (a) Principle and (b) helical–slit-type velocity selector.

are turning with an angular velocity ω. The first rotor has an open window through which the neutron can pass. As the rotor turns, the window acts like a gate that opens periodically. The first rotor allows a sharp polychromatic pulse of neutrons to pass. Because the neutrons having different wavelengths have different velocities, only pulsed neutrons that have the right velocity arrive at the second gate. If the distance between the first and the second rotor is X, and the phase difference of the gate of the second rotor with respect to that of the first rotor is ϕ, the selected velocity v of the neutrons is given by

$$v = X\omega/\phi \tag{24}$$

Thus, the velocity (and hence the wavelength) of the neutrons can be selected by changing X and ω. Though this is a basic principle of mechanical velocity selection, not all the *real* velocity selectors actually consist of two rotators; the obtained monochromated neutron beam would be pulsed (i.e., would not have a continuous beam) and would have relatively weak intensity.

To make a continuous monochromated beam with bigger intensity, a single helical-slit (or a multidisk velocity selector that has a set of slotted disks spaced along an axle to form helical pathways) velocity selector has been developed [24–27]. The schematic representation of such a velocity selector is shown in Figure 2.5(b). The velocity of the neutrons monochromated by the helical-slit velocity selector can also be calculated by Eq. (24). We note that the rotation rate of such a helical-slit velocity selector can be large, which is favorable for selecting fast neutrons (i.e., those of short wavelength). To avoid neutrons reflecting from the wall of the selector, the wall is coated by B_4C. Typically, the transmission of the helical-slit velocity selector is more than 80%. A continuous neutron beam in the range of 5 to 10 Å can be selected with a wavelength resolution $\Delta\lambda/\lambda$ of about 10%.

2.3.2.4 Collimator

After the velocity selector one finds the evacuated presample flight path used to collimate the neutron beam [22]. This inner surface of the flight tube is also coated by B_4C to prevent any reflection of the neutron beam from the surface. Figure 2.6 schematically shows a simple collimation geometry that determines the instrument resolution (and hence the minimum q, $q_{\min}$) as well as the intensity on the sample through the diaphragms and distance. At a distance L from the center of the detector, the scattering angle θ is defined by $\tan\theta = L/L_s$. In SANS, because θ is very small, $\tan\theta \to \theta$, and $q = (4\pi/\lambda)\sin(\theta/2) \to (2\pi/\lambda)\theta$. The nominal value of the scattering angle $\theta_{\min}$ at which the measured incident beam intensity

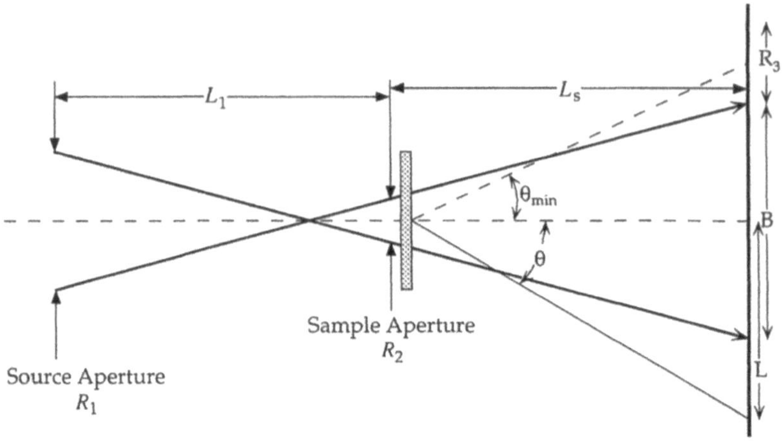

Figure 2.6 Simple collimation geometry. Reprinted from "NIST, Center for Neutron Research NG3 & NG7 SANS Instruments Data Acquisition Manual" by C. J. Glinka *et al.* (1996) with permission from the author.

would go to zero can be expressed as

$$\theta_{\min} = (B/2 + R_3/2)/L_s. \tag{25}$$

Here R_3 and B are, respectively, the FWHM (full width at half maximum) of the detector resolution function and a penumbra beam size at the detector. B depends upon the collimation system of the presample flight path, which consists of two apertures, i.e., the source aperture and the sample aperture, given by

$$B = (L_s/L_1)(R_1/R_2) + R_2. \tag{26}$$

Here R_1 and R_2 are, respectively, the diameters of the source and the sample apertures and L_1 is the source-to-sample distance; thus, $q_{\min}$ for a given configuration can be computed from

$$q_{\min} = (4\pi/\lambda)\sin(\theta_{\min}/2) \approx (2\pi/\lambda)(B/2 + R_3/2)/L_s \tag{27}$$

Therefore, combining Eqs. (26) and (27) gives

$$q_{\min} = (\pi/\lambda)\{(R_1 + R_2)/L_1 + (R_2 + R_3)/L_s\}. \tag{28}$$

There are two ways to reduce $q_{\min}$: (1) using long wavelength and (2) using long sample-to-detector distance. Reduction in the intensity on the sample I_s by using longer wavelength can be estimated knowing that $I_s \approx (1/\lambda^4)$ [23, 28]. On the other hand, changing the sample-to-detector distance by a factor of x results in a change in the solid angle subtended by a factor of x^2. Thus, for a given collimation, doubling the sample-to-detector distance and doubling the wavelength, both making $q_{\min}$ about half, give 1/4 and 1/16 of the initial flux, respectively. Therefore, in general, long flight paths rather than long wavelength should be used first to reduce $q_{\min}$. More detailed and accurate descriptions about how one can optimize the neutron flux at the sample position are given in the literature [23, 29, 30]. Choosing the instrumental configuration will be discussed again later in Section 2.3.5.1.

2.3.2.5 Detector

Unlike X rays, the neutron itself does not generate ions when it enters molecules. Hence, a nuclear reaction between a neutron and a suitable nucleus is often used for neutron detection: The secondary ionization that generates the charged particles is used for the actual detection process. The nucleus used for this purpose is required to have high absorption cross section, and the secondary ionization from the events of reactions should be easily detectable. Light nuclei such as ^{3}He, ^{6}Li, and ^{10}B are often used.

Gas detectors use ^{10}B-enriched BF_3 or ^{3}He-enriched helium gas. They react with neutrons in the following way:

$$^{10}\mathrm{B} + \mathrm{neutron} \rightarrow {}^{7}\mathrm{Li} + {}^{4}\mathrm{He} + 2.78\,\mathrm{MeV}$$
$$^{3}\mathrm{He} + \mathrm{neutron} \rightarrow {}^{3}\mathrm{H} + {}^{1}\mathrm{H} + 0.76\,\mathrm{MeV}$$

The ionized atoms are produced with considerable energy in exothermic reactions. The fast ions thus generated are actually used for the detection process. The ionized atoms produce a trail of secondary ionization that transmits to an electrode and thus is detected as an electric pulse. High pressure can be used for ^{3}He at moderate voltages, and its absorption cross section is higher than ^{10}B, which gives the ^{3}He-enriched detector higher counting efficient.

Scintillation detectors use a solid, and therefore much denser, neutron absorber like ^{6}Li in a glass containing a phosphor such as ZnS(Ag). The ^{4}He is produced according to the following nuclear reaction:

$$^{6}\mathrm{Li} + \mathrm{neutron} \rightarrow {}^{3}\mathrm{He} + {}^{4}\mathrm{He} + 4.77\,\mathrm{MeV}$$

The phosphor emits a flash of light in response to the ^{4}He, which is detected by photomultipliers. High counting efficiency can be achieved because the light flash is very intense. The same principle is used to make a photograph. It is often handy to have a "neutron camera" (especially during the sample stage alignment, which will be described in Section 2.3.5.2). The same principle as the scintillation detector is used. A screen containing a mixture of ^{6}Li and ZnS(Ag) powders is placed in front of a photo plate. The light flash emitted upon incidence of the neutron is recorded on the plate to make photographs.

In these days, state-of-the-art SANS spectrometers are furnished with position-sensitive detectors (PSD) [31, 22, 23]. The same principle as the gas detectors is used for neutron detection in those area detectors. Their area is up to a square meter and has a typical resolution from several millimeters to 1 cm either in one or two dimensions. A similar type of position-sensitive proportional counter (PSPC) is widely used for small-angle X-ray scattering (SAXS) instruments [32]. The grid could be Cartesian (x and y) or radial (circles and spokes). The PSD has three planes of parallel wires: one anode wire plane at the center and two cathode wire planes (for x and y directions) on each side of the anode wire. Each wire is folded with a 5-mm to 1-cm spacing between adjacent wires. The electric pulse induced on the cathode by neutron incidence is transmitted through the wire line to both ends of each wire. According to the position of the incident neutron, the two pulses reached to both ends have a certain time lag, which is used to determine the position of the detection.

As a new two-dimensional area detector, the imaging plate (IP) so far used for X-ray scattering [33], may be used in the future. The IP is made of a thin layer of the photostimulable phosphor crystals such as BaFbr:Eu^{2+} on polyester substrate film [34]. When the IP is exposed to the charged particles produced by the preceding nuclear reaction (^{6}Li), they excite color centers in the photostimulatable materials. The number of excited color centers is proportional to the number of incoming particles (writing process). The IP is scanned by an He–Ne laser beam, which after the exposure to the neutrons, emits a luminescent light that is detected by a photomultiplier (reading process). One of the superior features of the IP is its resolution (ca. 0.1 mm), which is better than the resolution of PSD. However, the IP has drawbacks: it is rather difficult to differentiate whether the observed intensity is due to neutrons or gamma rays; the counting efficiency is rather low compared with the PSD.

2.3.3 PREPARATION FOR THE NEUTRON SCATTERING EXPERIMENT

The reactor starts up, runs for a period of weeks, and then closes down for fuel replacement and maintenance. Normally, the experimental time at an institute is limited because the user's demand for the duration of a SANS experiment is usually more than the total operating time of the reactor. Users who want to perform neutron experiments first have to submit their experimental plans to the neutron institute. If the proposals are approved by the committee of the institute, a certain period of time (often called "beam time"), depending upon the importance and feasibility of the experiment, is assigned for the users to perform their experiments. Therefore, one should keep in mind that beam time is precious and should make the best use of it.

Needless to say, before actually performing the neutron experiments, each experiment has to be well designed so that one can make maximum use of the given beam time. Some items that should be checked in advance are the followings:

1. Does the neutron institute offer basic equipment commonly found in a chemical laboratory (mainly for sample preparation on site)?
2. Does the neutron institute have some kind of a sample chamber for common use? If so, what are the dimensions of the sample container for it?
3. What kind of devices are available for SANS experiments?

Occasionally, users may want to prepare their specimens on site owing to the nature of the scientific problem or the nature of the materials. Or users may want to put some of their samples into the cell just to see the scattering as preliminary data for the preparation of the next experiment. Or the number of the sample cells may not enough for the total number of the samples the user wants to measure. In such cases the users, have to take out samples from the cell and put (or sometimes even make) the new sample into the cell. It is very convenient if there is some basic equipment such as a balance, a vacuum oven, etc., at the institute.

Various types of devices may be offered at the institute. The most common device used in the field of polymer science is probably a heating block, which is sometimes mounted on a translation stage and may have multiple sample positions and be temperature controlled. The temperature and sample positions ("sample changer") may be controlled via the computer terminal so that users do not have to stay in the guide hall all the time during beam time. The heating block is often designed for common use; for example, sample cells may be designed to hold not only solid materials but also liquidlike samples or solutions. Hence, it is convenient if those specifications of the heating block meet the requirements of the user. However, if the specification does not meet the user's need, the user may have to design equipment for his or her own use. It is typical that the sample stage of the SANS instrument has multiple thread holes so that the users can bolt on their own equipment.

Sometimes, as with the sample changer devices mentioned in the preceding paragraph, other apparatuses can be used in collaboration with the scientists at the institute. In these days, shear, pressure, and magnetic devices may be found relatively easily. The devices may differ from institute to institute, and thus users may want to contact the reactor scientists for detailed specifications.

We note here that some neutron institutes may require users to take some classes for safety training beforehand. Therefore, those who are planning to perform SANS experiments may also want to check with the reactor scientists for such requirements.

2.3.4 ANCILLARY MEASUREMENTS REQUIRED FOR A SANS EXPERIMENT

A complete SANS experiment requires not only the scattering measurements of the sample but measurements of some ancillary scattering data. These include empty cell scattering ("empty run") and the blocked beam scattering ("background run" or "blocked beam run") measurements. These should be repeated whenever the instrument configuration is changed because they depend on the instrument configuration. The empty cell scattering is the scattering from the sample container, which sometimes consists of two quartz (or sapphire) plates ("windows") used to sandwich the sample. Quartz container is commercially available, too. It is necessary to subtract the contribution of the windows or the container from the sample scattering (Section 2.3.6.1). In the subtraction, transmissions of the sample with the windows (or the container) with respect to the empty cell are required (see Section 2.3.5.4).

The blocked beam measurement corresponds to the background (electric) noise of the 2-D area detector. Throughout this chapter, the background means either electronic noise of the detector or stray neutrons arriving from surrounding spectrometers and the neutron source itself. Strictly speaking, the background run should be repeated even if users change sample-to-detector distance. Any scattering from parts of the sample that is not of interest can also be referred as the background, which may include the scattering from solvent or the incoherent scattering (Section 2.3.6.1).

Sensitivity of the pixels of the 2-D detector for a given scattering intensity is not necessarily uniform and thus should be corrected in some way. One way to correct this is to use the angularly independent incoherent scattering. Scattering from water is often used for this purpose. We note, however, that the scattering from water does not give complete angularly independent scattering because of the inelastic scattering effect. This effect should be taken into account by considering a factor G

described later in Section 2.3.6.2. In any case, as one of the ancillary measurements, the incoherent scattering from the water sample ("water run") should be measured and is used to correct for detector nonuniformity. The water run for this purpose requires good statistics and is thus often (not always) performed periodically by the reactor scientists and made available to users.

How many neutron counts does each measurement need? According to counting statistics, the noise level of an experiment decreases with the increase of the square root of the number of neutrons N_n or measurement time. Various factors such as the number of samples, counting statistics, and available beam time should be considered to make a sensible choice. Total detector counts necessary for different experiments vary widely depending upon how strongly samples scatter. Empirically speaking, in case the intensity is concentrated in the center of the detector, and one allows several percent error in data analysis, a total of 2×10^4 counts on the detector would be acceptable, whereas 1×10^7 neutrons would be ideal. Radially averaged data consist roughly of 100 data points. Hence, if the intensity is evenly distributed radially, a total of 1×10^6 detector counts is needed to achieve a 1% noise level, i.e., $\sqrt{N_n}/N_n \cong 0.01$.

How is the length of a measurement controlled? One can specify the length of a measurement either by setting a certain time or by setting a certain number of monitor counts. A monitor is normally installed right after the velocity selector. The total neutron intensity of the incident neutron beam is continuously monitored simultaneously with the scattering and recorded. Because the neutron sources may not provide constant neutron flux over a long period of time, time is not always a sensible control parameter. Therefore, it is preferable to use the monitor counts rather than time to control the length of the measurements. Background correction measurements such as the empty run, the blocked beam run, scattering from solvent, etc., should normally have the same length of monitor counts as the sample runs. Only in the case when the sample's weakest scattering (often at the largest q of interest) is much larger than that of the strongest background run, the background runs can be made shorter in length than the sample runs.

In addition, if one needs to put the data on an absolute cross section scale (units of cm^{-1} for example), an absolute standard sample must be measured under the same instrumental configuration as for the sample. The incoherent scattering from water can also be used for the absolute intensity calibration [35] (see Section 2.3.6.2 for details). Another method used to normalize the data is to use the intensity at $q = 0$ from a sample with known scattering intensities. In such a case, water scattering is only used to correct for the detector efficiency but not to normalize the scattering data. Various materials can be used as such secondary standards;

for example, a silica gel standard used at the research reactor at National Institute of Standards and Technology (NIST) that was calibrated against the isotropic incoherent scattering from high purity (99.95%) single crystals of vanadium [35, 36]. Such secondary standards should be stable at least over a period of years and should have very strong scattering intensity in the q-range of the spectrometer. The silica gel, for example, easily gives a million counts within 10 to 20 min at $\lambda \approx 9$ Å on the 8-m SANS instrument ($L_S = 3.5$ m: fixed) and at the intermediate detector distance on the 30-m SANS instrument in which the q-range from about 0.015 to 0.035 Å^{-1} is covered. As secondary standards other than the silica gel, polymer blends, irradiated single-crystal aluminum, and "Lupolene" (polyethylene film) are used, depending upon the q-range of the experiments.

2.3.5 *OPERATING PROCEDURES*

When users arrive at the institute, the first thing they may want to do is to make contact with reactor scientists for getting up-to-date information on the SANS instrument. Normally, all the data acquisition and data reduction will be done in a user account of the computer that may also control the SANS spectrometer. Therefore, unless the users perform a SANS experiment just once, and if, in particular, they regularly do experiments at the institute, it would be a good idea to get their own user accounts assigned. In these days, it is common that the users can log-in to their user accounts over the Internet using TELNET command. In the following sections, some suggestions about actual operations and methods for the SANS experiments will be provided.

At the beginning of the beam time, a couple of hours have to be spent for the initial instrument setup, which includes (1) setting up instrument configuration such as sample-to-detector distance, wavelength of the incident neutron, and collimation; (2) aligning the sample changer; and (3) aligning the beam stop and locating the beam center. After this initial instrument setup, we recommend that the users take scattering from selected samples for short monitor counts to see how "strong" they scatter neutrons, and to derive a sense of the scattering "power" of the samples.

2.3.5.1 Setting Up the Spectrometer

An important factor in setting up the instrument is the choice of the collimation (i.e., the dimension of the diaphragms) and the beam stop dimensions. The main purpose of setting up the spectrometer is to choose an appropriate q-range for a sample with a reasonable flux at the sample position for each measurement. The SANS instruments, as briefly mentioned in Section 2.3.2, with their variable wavelength and its resolution, adjustable lengths of pre- and postsample flight paths, and adjustable diameters and positions of diaphragms in the presample path tubes for collimating the neutron beam offer a wide range of possible configurations. The proper configuration for a given sample depends not only upon those instrumental parameters but also upon the size of the scattering elements to be examined (which determines the q-range required). Users should also consider the available beam time, the number of samples to be measured, etc., as previously mentioned.

Among many instrumental parameters described in the previous paragraph, the main two parameters users have to be particularly concerned about are (1) the wavelength (and sometimes $\Delta\lambda/\lambda$) and (2) the sample-to-detector distance L_s. State-of-the-art SANS spectrometers are furnished with the 2-D area detector that sits on a motor-driven track to permit the users to change the L_s. The sample-to-detector distance can be changed continuously as the detector moves inside the evacuated path. The area detector may be able to move transversely to the beam direction to extend the q-range. In some instruments, however, the distance can be changed only by adding or removing sections of the evacuated tube, which offers only fixed choice of the sample-to-detector distance.

It is best to use the smallest beam stop in order to detect scattered intensities as close to the pixels near the center of the detector as possible. However, in reality, there is always a spillover of neutrons around the beam stop. The spillover of the neutron contaminates the pixels around the beam stop, which makes the quantitative evaluation of intensities in those pixels quite difficult. Hence, this causes $q_{\min}$ to be larger. On the other hand, tighter collimation, which involves smaller diaphragms in collimation for achieving small $q_{\min}$, considerably reduces the neutron flux at the sample position and thus weakens the scattering intensity. Specifically, there is always a compromise between the q-range and the scattered intensity, which is, however, a strong function of instrument design. As mentioned in Section 3.2.4, longer L_s is the primary choice to make $q_{\min}$ smaller rather than longer λ. A computer program to simulate $q_{\min}$ and the neutron flux at the sample position at a given set of the instrumental parameters may be offered. Users may be able to use such a program to find the best instrument configuration.

Thus, it is not an easy job for each polymer scientist to find out the best configuration for his or her experiments. Moreover, the difficult part of setting up instrument depends much on experience. Fortunately, with help from experienced neutron reactor scientists based at the neutron institute (who normally takes care of the SANS spectrometer), most polymer scientists will happily be able to configure the instrument. In collaboration with the neutron reactor scientists, it is the polymer scientists' responsibility to make clear what kind of information is sought and, for that purpose, what q-range should be examined in the experiments. It is not the reactor scientists but the polymer scientists who know the nature of the sample best!

2.3.5.2 Aligning the Sample Stage

A sample translation stage is often offered at the reactor on which a sample holder with multiple sample positions (sample changer) is installed to eliminate the need for users "babysit" the instrument (especially during the night). In the present section, we suppose users use such a sample changer.

Occasionally, a laser is installed for rough centering of the sample translation stage. The light path of the laser should already be aligned to the neutron path. Hence, one can rely on the laser so that it is centered at one of the sample holder positions. A practical method of fine alignment is to take two consecutive exposures of a Polaroid photograph: They are taken with and without a sample holder aperture and by making sure that the camera is not moved. Note that, in this case, each sample holder has its own diaphragm, i.e., the sample holder aperture just in front of the sample and the aperture must be smaller than the last aperture of the collimation system (the sample aperture in Figure 2.6). The photograph shows the outline of both apertures, which should be concentric if the neutron beam is in the middle of the hole of the sample holder position. One should align the sample changer before aligning the beam stop.

2.3.5.3 Beam Stop Alignment and Beam Center Location

After the sample changer is aligned, the beam stop should be placed in the right position on the 2-D area detector to protect it from the radiation damage with the direct strong (unscattered) neutron beam. We note that the beam stop should be roughly positioned to the direct neutron beam before one conducts this fine tuning of the beam stop position. The rough position of the beam stop under similar instrumental setup can be normally found in the log book in which records of the instrumental parameters used by the previous users are kept. By the proper adjustment of the beam stop, one can also minimize the loss of data close to the beam stop, which extends the q-range toward a smaller value. A practical way of aligning the beam stop is to use a sample having very strong isotropic scattering such as SiO_2 gel or teflon. The scattering from such strong scatterer gives scattering around the beam stop, which should be symmetric around it if the beam stop position is appropriate. If not, move the beam stop and repeat the procedure until symmetrical scattering around the beam stop results.

A precise method of determining the beam center involves direct measurement of the neutron beam. *Special care* must be taken to attenuate the neutron beam before moving the beam stop out. This is very important to avoid damage to the detector. There may be a couple of methods to estimate the beam center. One way is to use software that is capable of measuring the intensity contours for finding the center position of the selected region surrounding the direct beam. The other method involves folding of the incident beam profile with respect to either row(s) or column(s) at a certain pixel to find out the center pixel around which the beam profile on both side gets symmetric. The coordinate of the beam center is an important parameter in a radial, angular sector or rectangular average of 2-D data into 1-D data (Section 2.3.6).

2.3.5.4 Transmission Measurement

As already mentioned earlier in Section 2.3.4, to obtain the scattering intrinsic to the sample, transmissions of sample(s) and those of ancillary measurements should be measured. The sample transmissions must be measured for each instrument configuration used (e.g., collimation, wavelength, sample-to-detector distance, etc.).

One of the most common ways to measure the transmissions is as follows: (1) attenuate the incident neutron beam, (2) take the beam stop out from the position of the direct beam, and (3) measure the direct beam for a short time (e.g., typically a couple of minutes) with and without the sample. Attenuators with different thicknesses may be offered to reduce the neutron counts on the detector for the transmission measurements. Intensity of the transmitted direct neutron beam can be estimated by summing the neutron counts of the pixels close to the beam center. The users should take a look at the 2-D image of the detector and decide the area for this sum. In this method, the attenuator(s) *must* be inserted in the optics *before* removing the beam stop from the beam center so as not to accidentally damage the detector. The transmission is then calculated as follows:

$$T = I_s/I_0 \tag{29}$$

where I_s and I_0 represent, respectively, summed intensity with and without the sample.

Besides the method described above, there is another way to measure transmission. Some SANS instruments have a separate small detector (small-diameter ^{3}He proportional counter) specifically installed only for the transmission measurements. The small detector is designed to snap to the position of the beam center so that it counts only transmitted neutrons. However, the diameter of the small detector must be carefully designed so that its area covers only the direct beam and does not count scattered neutrons. This could be a problem with today's state-of-the-art spectrometers that have the capability of choosing different sample-to-detector distances, because, strictly speaking, the diameter of the ^{3}He detector should be scaled by the sample-to-detector distance to keep the solid angle fixed.

2.3.6 DATA REDUCTION

2.3.6.1 Basic Scheme of the Data Reduction

The term *data reduction* refers to a procedure to get a set of data from which all instrumental artifacts should be removed. In contrast, the term *data analysis* includes the process of interpreting the data. Although, sometimes, these two terms seem to be used in the same sense, in this section we show a sequence of data reduction by taking scattering from a phase-separated block copolymer as an example.

Before the data reduction, data such as 2-D scattering from a sample, empty run, blocked beam run, and water run and the corresponding transmissions should already have been taken and calculated. It is often convenient to make list of parameters before the data reduction. It may also be important to measure the ancillary data in the early part of the beam time so that one can start reducing data as soon as a part of the scattering experiment is finished. All of these precautions prevent the user from making mistakes in the data reduction because most users tend to run out of time and become impetuous at the end of their beam time. They have to catch a flight to go home! Of course, users can later connect remotely to the reactor computer via the Internet. Nevertheless, unless they perform data reduction on-site, it is also likely that subtle points of the experiments may be lost and, especially, the data reduction could be put off.

Two-dimensional data should be stored in the reactor computer, which may be the computer that controls the SANS instrument or sometimes a different one, depending on the institute. An operating system (UNIX, Macintosh®, Windows®, MS DOS®, etc.) and program(s) for the data reduction also depend on the institute, and thus one may want to ask reactor scientists what they can offer. Note that a data reduction program may or may not be offered from the institute. The following scheme of data reduction is based on the program offered at NIST. The "SANS Instruments and Data Acquisition Manual" written by the scientists at the NIST center for neutron research would be worth reading. The manual, both in BinHex and PDF formats, can be obtained from http://rrdjazz.nist.gov/sans.html.

Figure 2.7 schematically shows the commonly used sequences of data reduction in which data files are schematically shown as a folder. In Figure 2.7, there are three kinds of data files, namely, WORK.SAM, WORK.EMP, and WORK.BGD, which are, respectively, the work files for sample, empty cell, and blocked beam. As a first step, the user loads the corresponding 2-D data into the data analysis program. In Figure 2.7, a scattering from a microphase-separated block copolymer is taken as an example. In this sequence, the user loads the 2-D scattering data of a microphase-separated block copolymer into WORK.SAM and the 2-D data of two quartz windows into WORK.EMP. Upon loading data, the total counts of those raw data are normalized to a fixed number of monitor counts to facilitate comparisons between different data sets obtained with a different length of the measurement regardless of variation in the reactor power during the measurements. The empty cell and the background are corrected (pixel by pixel) according to the following equation:

$$\mathrm{COR} = (\mathrm{SAM} - \mathrm{BGD}) - [T_{\mathrm{SAM}}/T_{\mathrm{EMP}}](\mathrm{EMP} - \mathrm{BGD}) \tag{30}$$

where T_{EMP} and T_{SAM} correspond, respectively, to the transmissions of the sample and the empty cell taken with respect to the direct beam. Procedure(s) to measure the transmissions have been described earlier (Section 2.3.5.4). The terms showing in three capital letters in Eq. (30) (e.g., SAM) denote the corresponding 2-D data files. The result of the correction shown in Eq. (30) is stored in a data file named WORK.COR.

Now the user can average the 2-D data (WORK.COR) to reduce it to $I(q)$ versus q. However, at this point, we recommend that users look into the corrected 2-D data file for defective pixels, e.g., someone else might have "burned" a part of the detector. Those defective pixels, if any, should be ignored in the subsequent averaging process. It would also be advisable to ignore the center of the detector, i.e., the beam stop where the direct beam falls on the detector and the few rows and columns at the edge of the detector where the detector sensitivity gets highly nonlinear.

In the averaging process, users can usually choose different types of averaging: a radial, angular sector, or rectangular averaging. The radial averaging of the

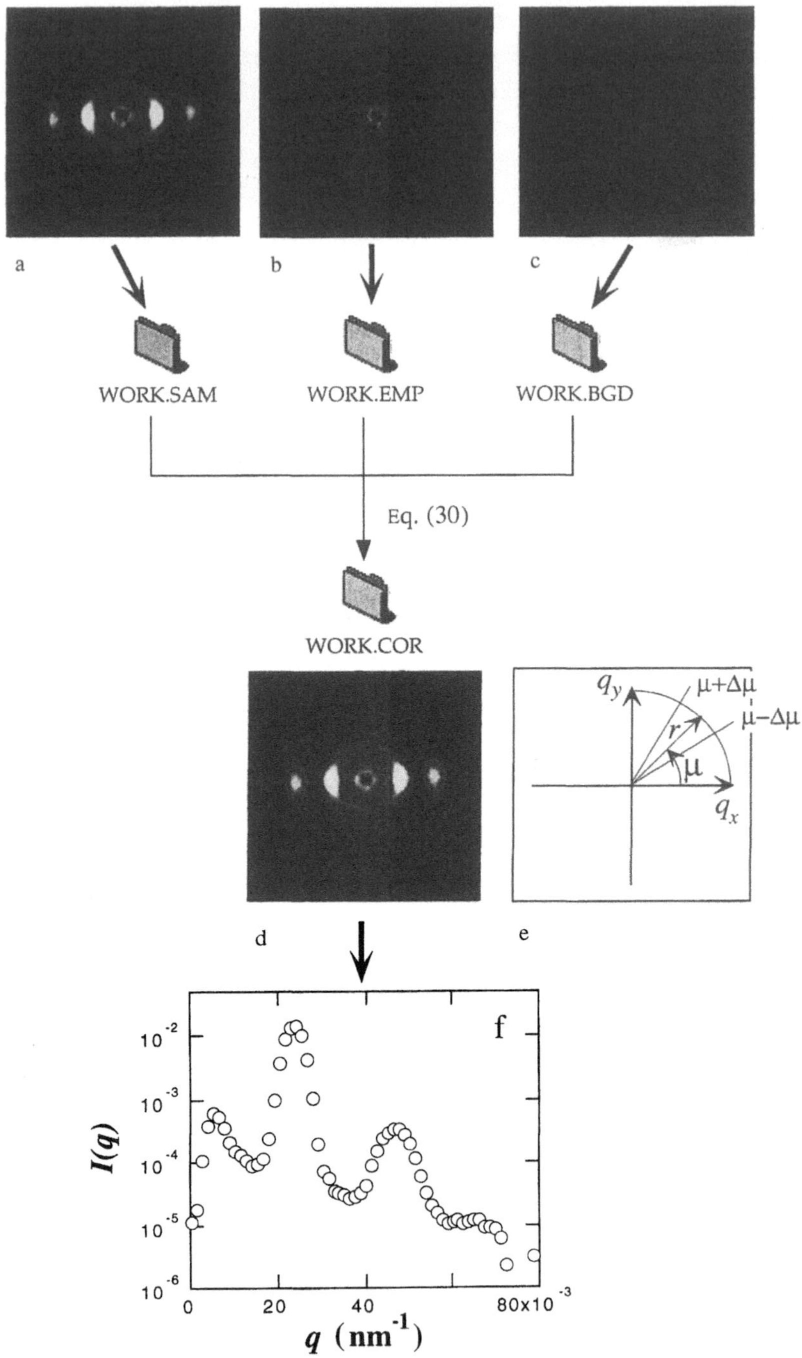

Figure 2.7 A commonly used data reduction sequence.

2-D data set is beneficial in terms of the counting statistics if the scattering from the sample is isotropic. On the other hand, the angular sector averaging is useful if the sample shows anisotropic scattering so that the user can look at the scattering in different area of the 2-D data, i.e., in different azimuthal angles, μ. In Figure 2.7(d) the scattering ($-30^\circ < \mu < 30^\circ$ and $150^\circ < \mu < 210^\circ$) was averaged to obtain $I(q, \mu)$ shown in Figure 7(f). The definition of μ is shown in Figure 2.7(e). The wavenumber of the pixel whose distance is r from the center of the direct beam can be converted by $q = (4\pi/\lambda)\sin(\theta/2)$ and $\theta = \arctan(r/L_s)$. The expression $I(q, \mu)$ is the one-dimensional scattering intensity that depends on the azimuthal angle and is obtained by $I(q, \mu) = (1/2q\Delta\mu)\int_{\mu-\Delta\mu}^{\mu+\Delta\mu} I(q, \mu)q\,d\mu$. We note the one-dimensional data $I(q)$ thus obtained have not yet been converted to an absolute scale. The conversion of the arbitrary intensity scale to the absolute intensity is discussed in Section 2.3.6.2.

A correction related to the detector response may have to be considered: The response of some detector becomes nonlinear towards the edge. A smooth analytic function is used to correct the response [13]. These correction factors are effective for most parts of the detector except for a few rows and columns at the edges where the response is too nonlinear to be corrected. One must neglect the area in the averaging. A data set may also have to be corrected for nonuniformity in the detector sensitivity by dividing, pixel-by-pixel, the scattering data from the sample by a measured scattering from water, for example. The isotropic pattern of water that is counted long enough to give good statistical precision over the entire detector should be used to make a file ("water calibration file") and scaled to an average of 1 count/pixel. Data for all samples should be corrected for the background, empty cell, etc., and then they are divided, pixel-by-pixel, by the water calibration file. After the correction, the 2-D images are subjected to averaging to give 1-D scattering data $I(q)$ or $I(q, \mu)$.

As mentioned in Section 2.2, we are particularly concerned with the elastic coherent scattering, for it is often necessary to subtract the contribution of elastic incoherent scattering from the total scattering intensity. For a mixture of deuterated chains and hydrogenous chains, for example, the incoherent scattering can be estimated from the scattering of each (pure) constituent polymer: The total contribution of the incoherent scattering in the system can be written by weighted sum of the incoherent scattering of each constituent. Some other methods, experimentally or theoretically, are summarized in Higgin's book [14].

In some cases, the 2-D data from samples, ancillary measurements, or both, are averaged to give 1-D scattering data before the background correction. Subsequently, the background subtractions or correction of detector inhomogeniety is done. Basically, this sequence should give the same 1-D scattering profile as the data reduction in 2-D. However, in our opinion, the resulting data using this procedure could be affected if there were damaged parts (hence less sensitive parts for neutrons) in the 2-D detector.

2.3.6.2 Absolute Scattering Correction [35]

Data are placed on the absolute intensity scale by measuring an absolute intensity standard under the same spectrometer configurations as those used to measure the sample. As the absolute intensity standard it is convenient to use the scattering intensity at $q = 0$ from a sample of known scattering intensities or known molecular weight. The scattered intensity $I(q)$ is related to the absolute scattering cross section $d\sigma(q)/d\Omega$ by the following expression:

$$I(q) = \Phi S d_{\text{sam}} T_{\text{sam}} (d\sigma(q)/d\Omega) \Delta\Omega \varepsilon t, \tag{31}$$

where Φ is the flux at the sample and S, d_{sam}, T_{sam} are sample-related quantities; they are, respectively, the sample area, sample thickness, and sample transmission. The expressions $\Delta\Omega$ and ε are, respectively, the solid angle subtended by one pixel of the detector and the detector sensitivity, and t is the effective counting time. Note that, if one follows the data reduction sequences described in the previous section, t has already been normalized to some monitor counts. Equation (31) is also valid for the standard at $q = 0$, i.e.,

$$I_{\text{std}}(q = 0) = \Phi S d_{\text{std}} T_{\text{std}} (d\sigma_{\text{std}}(q = 0)/d\Omega) \Delta\Omega \varepsilon t_{\text{std}}, \tag{32}$$

The subscript std in Eq. (32) means the corresponding quantities for the standard. The reason to consider the scattering intensity at $q = 0$ is that most of the secondary standard is calibrated against the primary standard such as vanadium at $q = 0$. From Eqs. (31) and (32) we now obtain

$$\frac{d\sigma(q)}{d\Omega} = \frac{I(q)}{I_{\text{std}}(q = 0)} \frac{t_{\text{std}}}{t} \frac{d_{\text{std}}}{d_{\text{sam}}} \frac{T_{\text{std}}}{T_{\text{sam}}} \frac{d\sigma_{\text{std}}(q = 0)}{d\Omega} \tag{33}$$

In Eq. (33), $t_{\text{std}} = t$. Measured quantities in Eq. (33) are $I(q)$, d_{sam}, and T_{sam}. The expressions d_{std}, T_{std}, and $d\sigma_{\text{std}}(q = 0)/d\Omega$ are quantities related to the standard, and thus they are known. One has to measure the zero-wavevector scattering intensity from the standard under the same spectrometer configuration. The method to determine this quantity may depend on a standard due to the nature of it, which should be provided from the reactor scientists. Thus, one can calculate a rescale factor ReF by which the measured intensity $I(q)$ should be multiplied as follows:

$$\frac{d\sigma(q)}{d\Omega} = \text{ReF}\, I(q). \tag{33a}$$

The factor ReF is now given by

$$\text{ReF} = \frac{d_{\text{std}}}{d_{\text{sam}}} \frac{T_{\text{std}}}{T_{\text{sam}}} \frac{d\sigma_{\text{std}}(q=0)/d\Omega}{I_{\text{std}}(q=0)}. \tag{33b}$$

Another method for normalizing the data is to use the scattering from a standard incoherent scatterer [35]. Thin (≈1-mm) water is most commonly used as such a standard. The scattering of water should be measured under the condition in which the scattering intensity of the provided standard is too strong to be measured or the condition in which the q-range of the standard does not match the spectrometer configuration used. In this case, water is used not only to correct for detector sensitivity but also to normalize the data. However, we note that water can effectively be used only if the sample-to-detector distance is short, the incident wavelength of the neutron is short, or both, so that the total scattering intensity measured with the 2-D detector in a reasonable period of the limited beam time becomes strong enough to yield sufficient statistics.

The 2-D data from the water sample should be reduced and averaged as described in Section 2.3.6.1. The mean value for the intensity of the water sample, $I_w(q)$, should be determined over the q-range where the intensity is approximately flat on the 1-D plot. This should include almost the entirely measured q-range except for the smallest q values, where some small-angle scattering is usually seen, and perhaps the last few data points in large q, where the data may be noisy. The expression $I_w(q)$ replaces $I_{\text{std}}(q=0)$ in Eq. (33). Now the cross section of water, $d\sigma_w(q=0)/d\Omega$, which replaces $d\sigma_{\text{std}}(q=0)/d\Omega$ in Eq. (33), is given by

$$\frac{d\sigma_w(q)}{d\Omega} = \frac{(1-T_w)G}{4\pi d_w T_w}, \tag{34}$$

where T_w and d_w represent, respectively, the transmission and thickness (in centimeters) of the water sample. The expression G is a wavelength-dependent factor necessary to account for the fact that the scattering from water is not entirely angularly independent owing to inelastic scattering effect. Values of G have been measured for a 1-mm water sample [37], and a formula that can be used to calculate the factor has been proposed by Jacrot [38]. One simple formula is given by the following equation [38]:

$$G^{-1} = 1 - \exp(-0.6\lambda^{0.5}). \tag{35}$$

Note that in Eq. (35) the unit of the wavelength is Å. The value of $d\sigma_w(q = 0)/d\Omega$ calculated from the preceding equations lies between 0.7 to 1, depending upon the wavelength used.

2.3.6.3 Smearing Effect

In the course of the analysis of the neutron scattering data, the scattering profiles are often to be compared in detail with a scattering model or theory. In such a case, it may be necessary to consider the effects of the instrument resolution function on the data. This is because a neutron scattered with scattering vector $\mathbf{q}$ is counted in a detector pixel that corresponds to a nominal scattering vector $\mathbf{q}_0$; hence, the true intensity profile can be distorted or "smeared." The measured intensity, $I_{\text{meas}}(\mathbf{q})$ is given by a convolution of the true scattering intensity of the sample with the resolution function

$$I_{\text{meas}}(\mathbf{q}) = I_{\text{obs}}(\mathbf{q}) * R(\mathbf{q}), \tag{36}$$

where $I_{\text{obs}}(\mathbf{q})$ is the true (desmeared) scattering intensity. The instrument resolution function for a pinhole collimation SANS instrument with an area detector $R(\mathbf{q})$ may be approximated using a Gaussian resolution function. The pinhole collimation gives a symmetrical contribution to the resolution function, whereas the wavelength spread produces an independent smearing in the magnitude of $\mathbf{q}$ only. Detailed mathematical formulas for $R(\mathbf{q})$ can be found in the literature [22, 39, 40]. Figure 2.8 shows a measured (absolute) scattering intensity (open circles) from the microphase-separated lamellar structure of polystyrene-*block*-poly(ethyrenepropyrene). The filled circles demonstrate the "desmeared" intensity. The inset shows the direct beam profile used for the desmearing process. As a consequence of the desmearing, it is clearly seen that the first-ordered scattering maximum gets narrower.

2.4 Theory of Small-Angle Neutron Scattering

2.4.1 *NONINTERACTING SYSTEMS*

Small-angle neutron scattering (SANS) has been developed as a complementary tool for SAXS. Because both techniques share a theoretical background, it is recommended that the reader be familiar with the theories of SAXS. Here, we briefly review the general theories of scattering. For further reading, several books, such as *Small-Angle Scattering of X-rays* by Guinier [41] and *Small-Angle X-ray*

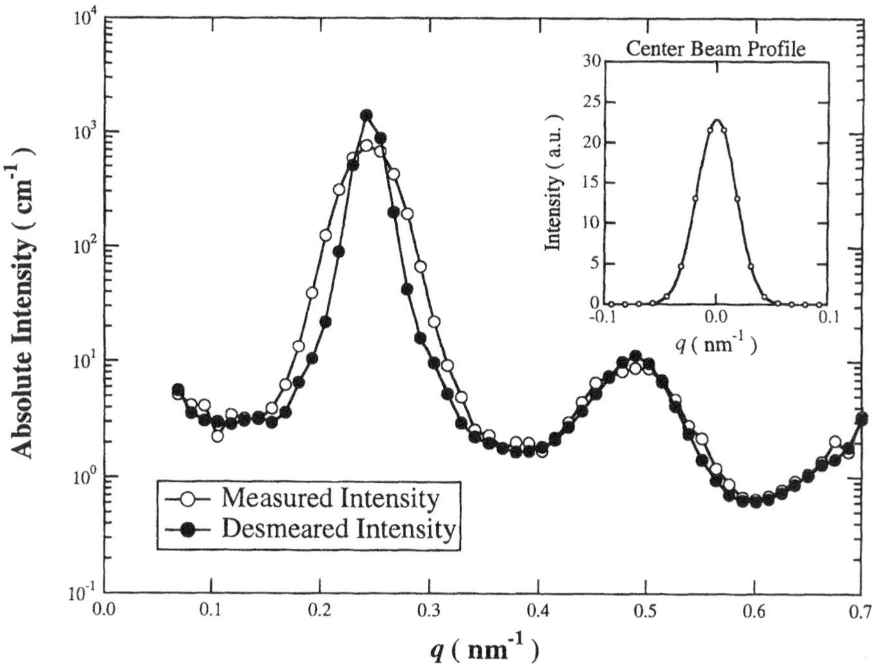

Figure 2.8 The sectorly averaged small-angle neutron scattering from a polystyrene-*block*-poly(ethylene propylene). Comparison between measured (open circles) and desmeared (filled circles) profiles. The inset shows a transmitted neutron beam profile that was used for the correction.

Scattering by Glatter and Kratky [42] are recommended. The book *Polymers and Neutron Scattering* by Higgins and Benoit is also highly recommended [14].

In Section 2.2.2 it is stated that the coherent cross section is obtained by summing the product of the scattering length, where the phase difference is taken into account. Figure 2.9 shows models of (a) one- and (b) two-component systems as well as (c) polymer blend systems in a volume V. Each scatterer can be numbered such as i and j.

2.4.1.1 One-Component System

From Eq. (11), the scattered intensity for a one-component system is written as

$$I(\mathbf{q}) \equiv \left(\frac{d\sigma}{d\Omega}\right)_{\text{coh}} = \frac{b^2}{V}\sum_{i,j}^{N}\langle\exp[-i\mathbf{q}\cdot\mathbf{r}_{ij}]\rangle = \frac{b^2}{V}\sum_{i}^{N}\sum_{j}^{N}\langle\exp[-i\mathbf{q}\cdot\mathbf{r}_{ij}]\rangle \quad (37)$$

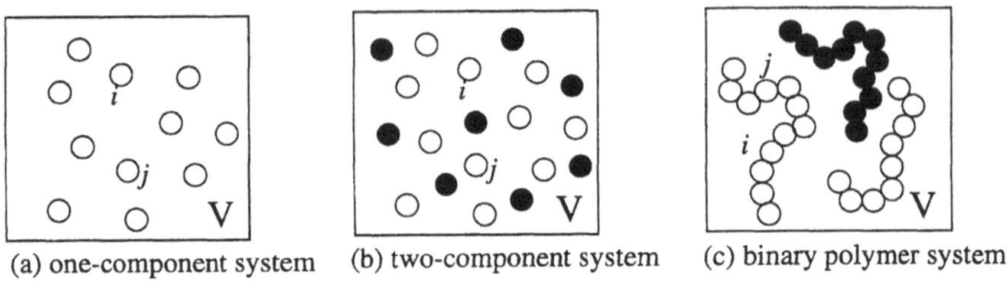

Figure 2.9 Schematic illustration for (a) one-component, (b) two-component, and (c) binary polymer systems.

where $I(\mathbf{q})$ is the absolute scattered intensity with the unit of the inverse of length. The expression $I(\mathbf{q})$ is also called the differential cross section. The vector quantity $\mathbf{r}_{ij}$ is given by $\mathbf{r}_{ij} = \mathbf{r}_i - \mathbf{r}_j$. Note that we treat the molecules as point scatterers. We write $\langle b \rangle^2$ by b^2 without creating confusion because only the coherent scattering is considered, and the relation $\langle b_i b_j \rangle = \langle b_i \rangle \langle b_j \rangle \equiv \langle b \rangle^2$ is assumed for $j \neq i$. (See Section 2.2.2). Now, let us introduce a structure factor $S(\mathbf{q})$ defined by

$$S(\mathbf{q}) = \frac{1}{N} \sum_i^N \sum_j^N \langle \exp\left[-i\mathbf{q} \cdot \mathbf{r}_{ij}\right] \rangle \tag{38}$$

The quantity $S(\mathbf{q})$ does not depend on the kind of scattering method, such as, X rays, neutrons, or light but on the structure of the system itself. Therefore, it is not an exaggeration that our aim in this chapter is to evaluate $S(\mathbf{q})$ for various systems. By using $S(\mathbf{q})$, the SANS intensity $I(\mathbf{q})$ is simply given by

$$I(\mathbf{q}) = \frac{b^2}{v_0} S(\mathbf{q}) \tag{39}$$

where v_0 is the volume of the segment (or the molecule) defined by

$$V = v_0 N \tag{40}$$

In most cases, the scattering elements (e.g., molecules or atoms) are regarded as spreading continuously in a system consisting of N molecules in volume V, and each scattering element has the scattering length b. The scattered intensity is given by

$$I(\mathbf{q}) = \frac{b^2}{v_0} \int \int \langle \Delta n(\mathbf{r}) \Delta n(\mathbf{r}' - \mathbf{r}) \rangle \exp\left[-i\mathbf{q} \cdot (\mathbf{r}' - \mathbf{r})\right] d\mathbf{r}\, d\mathbf{r}' \tag{41}$$

where $\Delta n(\mathbf{r})$ is the fluctuation in the number density of scattering elements at position **r** from the mean value. Thus,

$$\Delta n(\mathbf{r}) = n(\mathbf{r}) - \frac{N}{V} \quad \text{and} \quad \int \Delta n(\mathbf{r})\, d\mathbf{r} = 0 \tag{42}$$

The expression $\langle \Delta n(\mathbf{r}) \Delta n(\mathbf{r}' - \mathbf{r}) \rangle$ means the density correlation of the object, where $\langle \cdots \rangle$ denotes the averaging over the space. We now introduce a correlation function $G(\mathbf{r})$ by

$$G(\mathbf{r}) \equiv \langle \Delta n(\mathbf{r}) \Delta n(\mathbf{r}' - \mathbf{r}) \rangle = \frac{1}{V} \int \Delta n(\mathbf{r}) \Delta n(\mathbf{r}' - \mathbf{r})\, d\mathbf{r}' \tag{43}$$

2.4.1.2 Real Space and Reciprocal Space (q-Space)

Comparing Eqs. (41) and (42), one can understand that Eq. (41) is a Fourier transform of $G(\mathbf{r})$. The scattered intensity $I(\mathbf{q})$ can be obtained by taking an alternative approach. A Fourier transform of $\Delta n(\mathbf{r})$ leads to the scattering amplitude, $f(\mathbf{q})$,

$$f(\mathbf{q}) = \int \Delta n(\mathbf{r}) \exp[-i\mathbf{q} \cdot \mathbf{r}]\, d\mathbf{r} \tag{44}$$

By multiplying its Fourier conjugate, $f^*(\mathbf{q})$ to $f(\mathbf{q})$, we get the scattered intensity

$$I(\mathbf{q}) \equiv \frac{b^2}{v_0} S(\mathbf{q}) = \frac{b^2}{v_0} f(\mathbf{q}) f^*(\mathbf{q}) = \frac{b^2}{v_0} |f(\mathbf{q})|^2 \tag{45}$$

From Eqs. (37), (39), and (43), $S(\mathbf{q})$ is simply given with $G(\mathbf{r})$ by

$$S(\mathbf{q}) = \int G(\mathbf{r}) \exp[-i\mathbf{q} \cdot \mathbf{r}]\, d\mathbf{r} \tag{46}$$

Figure 2.10 shows the relationship among the density fluctuation function $\Delta n(\mathbf{r})$, the correlation function $G(\mathbf{r})$, the scattering amplitude $f(\mathbf{q})$, and the structure factor $S(\mathbf{q})$ as well as the schematic representation of the scattering principle. A monochromated incident neutron beam of wavelength λ is irradiated to the specimen. The incident neutrons are transmitted, scattered, or absorbed by the specimen. Scattered neutrons at positions P and Q interfere with each other because of the phase difference of $\mathbf{r} \cdot \mathbf{q}$. Thus, the scattered intensity becomes a function of $\mathbf{r} \cdot \mathbf{q}$, where $\mathbf{q}$ is the scattering vector, as already defined in Eq. (12). The space described by $\mathbf{q}$ is called the *Fourier space* (the q-space) or the *reciprocal space*. The term "reciprocal" is used because q has the unit of the reciprocal length, i.e., $1/\lambda$. On the other hand, the space defined by $\mathbf{r}$ is called the *real space*. The structure in the real space, as observed through $G(\mathbf{r})$, is related to the structure factor $S(\mathbf{q})$ in the

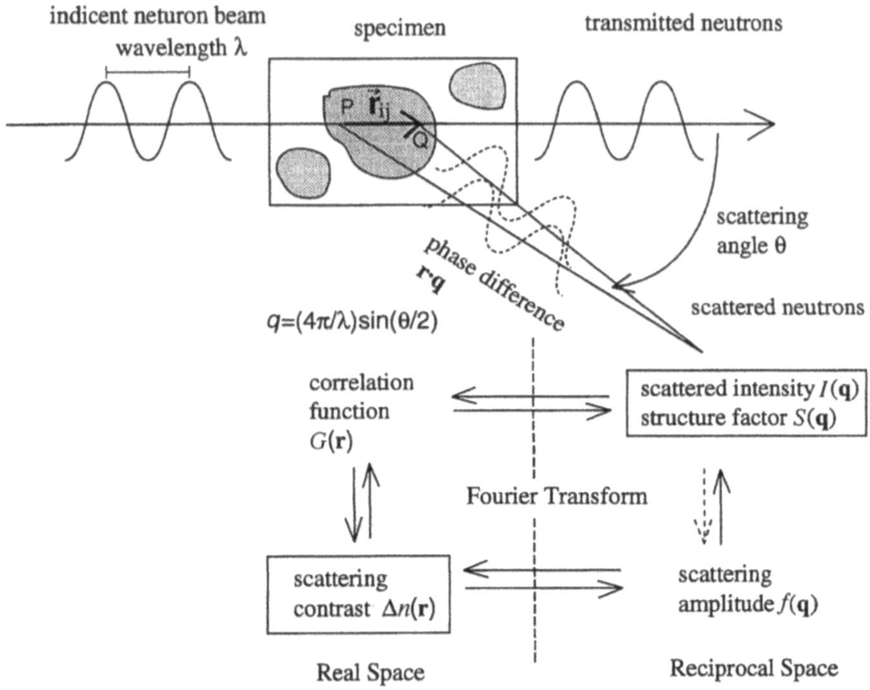

Figure 2.10 Schematic representation of the scattering principle and the relationship among the density fluctuation function $\Delta n(\mathbf{r})$ and structure factor $S(q)$.

q-space by Fourier transformation given by Eq. (46). Therefore, information about large-length scales appears in a low q-region and that of small scales is detected at a large q-region.

The goal of our elastic coherent scattering experiment is to learn the static structures existing in our samples $\Delta n(\mathbf{r})$ by analyzing the scattered intensity $I(\mathbf{q})$ from the specimen. The expression $G(\mathbf{r})$ can be obtained by taking the inverse Fourier transform of $I(\mathbf{q})$ or $S(\mathbf{q})$, which is

$$G(\mathbf{r}) = \frac{1}{(2\pi)^3} \int S(\mathbf{q}) \exp[i\mathbf{q} \cdot \mathbf{r}]\, d\mathbf{q} \tag{47}$$

It is impossible, however, to evaluate $f(\mathbf{q})$ from $I(\mathbf{q})$ because information about the phase in $f(\mathbf{q})$ is missing. Therefore, in many cases, a model describing $\Delta n(\mathbf{r})$ is constructed first. Then, comparison of the experiment with the theory (model) is made in terms of $G(\mathbf{r})$ or $I(\mathbf{q})$. In the following section, we describe some of the common scattering functions calculated on the basis of simple models.

2.4.1.3 Two-Component Systems

Equation (37) is easily extended to the case of a two-component mixture with N_α molecules of scattering length b_α ($\alpha = 1$ or 2), as illustrated in Figure 2.9(b). The scattered intensity in the unit solid angle is given by

$$I(q) = \frac{b_1^2}{V}\sum_i^{N_1}\sum_j^{N_1}\langle \exp[-i\mathbf{q}\cdot\mathbf{r}_{ij}]\rangle + \frac{2b_1b_2}{V}\sum_i^{N_1}\sum_j^{N_2}\langle \exp[-i\mathbf{q}\cdot\mathbf{r}_{ij}]\rangle + \frac{b_2^2}{V}\sum_i^{N_2}\sum_j^{N_2}\langle \exp[-i\mathbf{q}\cdot\mathbf{r}_{ij}]\rangle \tag{48}$$

or

$$\begin{aligned} I(q) &= \frac{b_1^2}{v_0}S_{11}(q) + \frac{2b_1b_2}{v_0}S_{12}(q) + \frac{b_2^2}{v_0}S_{22}(q) \\ &= \frac{1}{v_0}(b_1 \quad b_2)\begin{pmatrix} S_{11} & S_{12} \\ S_{21} & S_{22} \end{pmatrix}\begin{pmatrix} b_1 \\ b_2 \end{pmatrix} \end{aligned} \tag{49}$$

with $S_{12} = S_{21}.S_{\alpha\beta}(q)$ is the partial structure factor defined by

$$\begin{aligned} S_{\alpha\beta}(q) &= \frac{1}{N_1 + N_2}\sum_i^{N_\alpha}\sum_j^{N_\beta}\langle \exp[-i\mathbf{q}\cdot(\mathbf{r}' - \mathbf{r})]\rangle \\ &= \int_V\int_{V'} \exp[-i\mathbf{q}\cdot(\mathbf{r}' - \mathbf{r})]\langle \Delta n_\alpha(\mathbf{r})\Delta n_\beta(\mathbf{r}')\rangle\, d\mathbf{r}\, d\mathbf{r}' \end{aligned}$$

Now let us examine the interrelationship among $S_{\alpha\beta}(q)$. Consider the following identity:

$$\int_V\int_{V'} \exp[-i\mathbf{q}\cdot(\mathbf{r}' - \mathbf{r})]\langle[\Delta n_1(\mathbf{r}) + \Delta n_2(\mathbf{r})]\Delta n_1(\mathbf{r}')\rangle\, d\mathbf{r}\, d\mathbf{r}' = S_{11}(q) + S_{12}(q) \tag{50}$$

In the case of the incompressibility limit, the left-hand side of Eq. (50) becomes zero because

$$\Delta n_1(\mathbf{r}) + \Delta n_2(\mathbf{r}) = 0 \tag{51}$$

Thus,

$$S_{12}(q) = -S_{11}(q) \tag{52}$$

A similar argument leads to

$$S_{11}(q) = S_{22}(q) = -S_{12}(q) = -S_{21}(q) \equiv S(q) \tag{53}$$

Therefore, the scattered intensity function $I(q)$ for an incompressible binary system having different scattering lengths b_1 and b_2 is given by

$$I(q) = \frac{1}{v_0}(b_1 \quad b_2)\begin{pmatrix} S(q) & -S(q) \\ -S(q) & S(q) \end{pmatrix}\begin{pmatrix} b_1 \\ b_2 \end{pmatrix} = \frac{1}{v_0}(b_1 - b_2)^2 S(q) \tag{54}$$

An example of Eq. (54) is given in Section 2.4.1.5, in which the explicit expression for $I(q)$ is given for an isotopic polymer mixture.

The scattered intensity function for a mixture of $(p+1)$ (≥ 3) species can be obtained by generalizing the argument of Eq. (48), which is

$$I(q) = \frac{1}{v_0}(\cdots \quad b_\alpha \quad \cdots)(S_{\alpha\beta}(q))\begin{pmatrix} \cdot\cdot \\ b_\beta \\ \cdot\cdot \end{pmatrix} = \frac{1}{v_0}\sum_{\alpha=0}^{p}\sum_{\beta=0}^{p}(b_\alpha - b_\beta)^2 S_{\alpha\beta}(q) \tag{55}$$

where $0 \leq \alpha \leq p$ and $0 \leq \beta \leq p$. Equation (50) can be rewritten with respect to a reference component. For example, in the case of a polymer solution containing more than two polymer solutes, we denote the solvent as $\alpha = 0$ and rearrange Eq. (55) to [43]

$$I(q) = \frac{1}{v_0}\sum_{\alpha=1}^{p}(b_\alpha - b_0)^2 S_{\alpha\alpha}(q) + \frac{2}{v_0}\sum_{\alpha=1}^{p}\sum_{\beta(>\alpha)}^{p}(b_\alpha - b_0)(b_\beta - b_0)S_{\alpha\beta}(q) \tag{56}$$

Equation (56) means that by adjusting the scattering length of the solvent to be equal to that of either α or β, the structure factor $S_{\alpha\alpha}(q)$ is selectively obtained. It should be noted here that the sign of $(b_\alpha - b_0)$ is not important. In other words, a gray object can be seen with a black or white background. The same rule can be applied to three-constituent polymer mixtures in which one of the polymers is regarded as a "solvent."

2.4.1.4 Intra- and Intermolecular Interference

In the preceding sections, the molecules were treated as point scatterers. Macromolecules, however, cannot be treated in this way. This situation is illustrated in Figure 2.9(c). The typical size of a macromolecule is of the order of a few tens of nanometers, which is much larger than the wavelength of neutrons. It is essential, therefore, to decompose the structure to intramolecular and intermolecular interference functions. Let us start with a one-component macromolecular system of N molecules in the volume V. Each molecule consists of z-segment with the scattering length of b. The segment can be a monomer or a statistical unit of a macromolecule. In this case, the structural factor is given by

$$S(q) = \frac{1}{Nz} \sum_{\alpha}^{N} \sum_{\beta}^{N} \sum_{i}^{z} \sum_{j}^{z} \langle \exp[-i\mathbf{q} \cdot (\mathbf{r}_{\alpha,i} - \mathbf{r}_{\beta,j})] \rangle \tag{57}$$

where $\mathbf{r}_{\alpha,i}$ denotes the position vector for the segment i of chain α. Equation (57) can be decomposed to the intra- and intermolecular interference

$$\begin{aligned} S(q) &= \frac{1}{Nz} \sum_{\alpha}^{N} \sum_{i}^{z} \sum_{j}^{z} \langle \exp[-i\mathbf{q} \cdot (\mathbf{r}_{\alpha,i} - \mathbf{r}_{\alpha,j})] \rangle \\ &\quad + \frac{1}{Nz} \sum_{\alpha}^{N} \sum_{\neq\beta}^{N} \sum_{i}^{z} \sum_{j}^{z} \langle \exp[-i\mathbf{q} \cdot (\mathbf{r}_{\alpha,i} - \mathbf{r}_{\beta,j})] \rangle \\ &= zP(q) + NzQ(q) \end{aligned} \tag{58}$$

where $P(q)$ and $Q(q)$ are, respectively, the normalized intramolecular and intermolecular interference functions given by

$$\begin{aligned} P(q) &= \frac{1}{Nz^2} \sum_{\alpha}^{N} \sum_{i}^{z} \sum_{j}^{z} \langle \exp[-i\mathbf{q} \cdot (\mathbf{r}_{\alpha,i} - \mathbf{r}_{\alpha,j})] \rangle \\ &= \frac{1}{z^2} \sum_{i}^{z} \sum_{j}^{z} \langle \exp(-i\mathbf{q} \cdot \mathbf{r}_{ij}) \rangle \end{aligned} \tag{59}$$

and

$$Q(q) = \frac{1}{z^2} \sum_{i}^{z} \sum_{j}^{z} \langle \exp[-i\mathbf{q} \cdot (\mathbf{r}_{\alpha,i} - \mathbf{r}_{\beta,j})] \rangle \qquad (\alpha \neq \beta) \tag{60}$$

2.4.1.5 Binary Mixture of Identical Isotopic Polymers

In the case of two identical polymers in a molten state, one deuterated (D) and the other protonated (H), we know that the scattered intensity function is simply given by Eq. (49) or more explicitly by

$$I(q) = \frac{1}{v_0}(b_D - b_H)^2 S_{DD}(q) = \frac{1}{v_0}(b_D - b_H)^2 S_{HH}(q) = -\frac{1}{v_0}(b_D - b_H)^2 S_{DH}(q) \tag{61}$$

where b_H and b_D are respectively the scattering lengths of the protonated and deuterated segments. In terms of $P(q)$ and $Q(q)$, on the other hand, $S_{DD}(q)$ and $S_{HH}(q)$ can be written as

$$S_{DD}(q) = \phi z P(q) + \phi^2 NzQ(q) \tag{62}$$

$$S_{HH}(q) = (1 - \phi)zP(q) + (1 - \phi)^2 NzQ(q) \tag{63}$$

where ϕ is the fraction of the deuterated component. By substituting $\phi = 1$ in Eq. (62), one immediately gets

$$Q(q) = -\frac{1}{N}P(q) \tag{64}$$

because $S_{\mathrm{DD}}(q) = 0$ for a one-component system (i.e., $\phi = 1$). Substitution of Eq. (64) into Eqs. (62) and (61) leads to an important expression of the scattered intensity function for a molten polymer mixture of two identical polymers, which is [44]

$$I(q) = \frac{(b_{\mathrm{D}} - b_{\mathrm{H}})^2}{v_0}\phi(1-\phi)zP(q) = \frac{(b_{\mathrm{D}} - b_{\mathrm{H}})^2 N_{\mathrm{A}}}{\hat{v}_0}\phi(1-\phi)zP(q) \tag{65}$$

Here we introduce the molar volume of the segment $\hat{v}_0$, i.e., $\hat{v}_0 = N_{\mathrm{A}}v_0$, for convenience, where N_{A} is the Avogadro's number. Equation (65) has three important features of the scattered intensity function for a molten isotopic polymer mixture composed of two identical polymers: (1) The expression $I(q)$ is described only by the intramolecular function $P(q)$. (2) $I(q)$ is proportional to the product of the fraction $\phi(1-\phi)$; therefore, $I(q)$ can be optimized by setting $\phi = 0.5$ (i.e., a 50/50 mixture). (3) $I(q)$ is proportional to the degree of polymerization z.

It may be tutorial to show an example. Consider a 50/50 mixture of deuterated polystyrene (DPS) and ordinary (hydrogeneous) polystyrene (HPS) having the same degree of polymerization ($z = 1000$). The scattering lengths of the monomers of DPS and HPS are $b_{\mathrm{DPS}} = 106.63 \times 10^{-13}$ cm and $b_{\mathrm{HPS}} = 23.34 \times 10^{-13}$ cm. Thus, by knowing $P(0) = 1$, the scattered intensity at $q = 0$ is given by

$$\begin{aligned} I(q=0) &= \{(106.63 \times 10^{-13}\,[\mathrm{cm}]) - (23.34 \times 10^{-13}\,[\mathrm{cm}])\}^2 \\ &\quad \times 0.5^2 \times \frac{6.022 \times 10^{23}}{100\,[\mathrm{cm}^3]} \times 1000 = 104.4\,[\mathrm{cm}^{-1}] \end{aligned} \tag{66}$$

where we assume that the molar volume of polystyrene monomer is the same for DPS and HPS, $\hat{v}_0 \cong 100\,\mathrm{cm}^3$. This example provides another important fact: The neutron scattered intensity, in the absolute intensity scale, is usually expressed in the unit of cm^{-1}. Applications of Eq. (65) will be given in Section 2.5.1.

2.4.1.6 Radius of Gyration and Law of Guinier

We now examine the average $\langle .. \rangle$. It denotes taking the average over all orientations and conformations. Figure 2.11 shows the relationship between $\mathbf{q}$ and $\mathbf{r}$, where α is the polar angle between the two vectors. For an isotropic system, averaging over

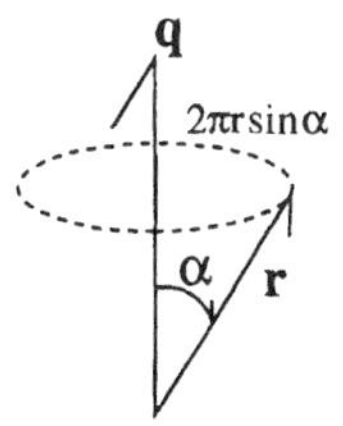

Figure 2.11 The scattering vector **q** and **r**.

orientations can be achieved by

$$\begin{aligned}\langle \exp[-i\mathbf{q}\cdot\mathbf{r}]\rangle_{\text{orient}} &= \frac{\displaystyle\int_{\alpha=0}^{\pi} \exp[-i\mathbf{q}\cdot\mathbf{r}]\,d\mathbf{r}}{\displaystyle\int_{\alpha=0}^{\pi} d\mathbf{r}} \\ &= \frac{\displaystyle\int_{\alpha=0}^{\pi} \exp[-iqr\cos\alpha]2\pi r\sin\alpha\,dr\,d\alpha}{\displaystyle\int_{\alpha=0}^{\pi} 2\pi r\,dr} \\ &= \frac{\sin(qr)}{qr} \end{aligned} \tag{67}$$

Thus,

$$P(q) = \frac{1}{z^2}\sum_{i=1}^{z}\sum_{j=1}^{z}\left\langle \frac{\sin(qr_{ij})}{qr_{ij}} \right\rangle \tag{68}$$

Here the average means only on the distance r_{ij}. Recalling the Taylor expansion of $\sin x/x$,

$$\frac{\sin x}{x} = \sum_{k=1}^{\infty} \frac{(-1)^{k-1}x^{2(k-1)}}{(2k-1)!} = 1 - \frac{x^2}{3!} + \frac{x^4}{5!} - \frac{x^6}{7!} + \cdots \tag{69}$$

one obtains

$$P(q) = 1 - \frac{q^2}{6z^2}\sum_{i=1}^{z}\sum_{j=1}^{z}\langle r_{ij}^2\rangle + \frac{q^4}{5!z^2}\sum_{i=1}^{z}\sum_{j=1}^{z}\langle r_{ij}^4\rangle - \cdots \tag{70}$$

The average $\langle r_{\text{ij}}^2\rangle$ is related to the *radius of gyration* R_{g}, one of the most important parameters describing the size of the object. The square of the radius of gyration is given by

$$\langle R_{\text{g}}^2\rangle = \frac{1}{z}\sum_{i=1}^{z}\langle(\mathbf{r}_i - \mathbf{r}_{\text{G}})^2\rangle \cong R_{\text{g}}^2 \tag{71}$$

where $\mathbf{r}_G$ is the position vector of the center of the gravity of the object. We simply write $\langle R_g^2 \rangle$ by R_g^2 hereafter unless the kind of average has to be specified. The expression R_g^2 is also given by

$$R_g^2 = \frac{1}{2z^2} \sum_{i=1}^{z} \sum_{j=1}^{z} \langle r_{ij}^2 \rangle \tag{72}$$

where the identity

$$r_{ij}^2 = (\mathbf{r}_i - \mathbf{r}_j)^2 = \mathbf{r}_i^2 + \mathbf{r}_j^2 - 2\mathbf{r}_i \cdot \mathbf{r}_j \tag{73}$$

was used. By substituting Eq. (72) into Eq. (70), one obtains

$$P(q) = 1 - \frac{q^2}{3} R_g^2 + \cdots \tag{74}$$

or

$$P(q) \cong \exp\left[-\frac{R_g^2 q^2}{3} \right] \qquad (\text{for } q R_g \ll 1) \tag{75}$$

Equation (75) is called the law of Guinier [41].

The square of the radius of gyration R_g^2 for a given shape of objects can be easily obtained by the following equation:

$$R_g^2 = \frac{\int_V (\mathbf{r} - \mathbf{r}_G)^2 \, d\mathbf{r}}{\int_V d\mathbf{r}} \tag{76}$$

For example,

$$R_g^2 = 2 \int_0^{L/2} \mathbf{r}^2 \, d\mathbf{r} \Big/ 2 \int_0^{L/2} d\mathbf{r} = L^2/12 \qquad (\text{a rod of length } L) \tag{77a}$$

$$R_g^2 = \int_0^{R} 4\pi \mathbf{r}^2 \mathbf{r}^2 \, d\mathbf{r} \Big/ \int_0^{R} 4\pi \mathbf{r}^2 \, d\mathbf{r} = 3R^2/5 \qquad (\text{a sphere of radius } R) \tag{77b}$$

A similar calculation leads to

$$R_g^2 = R^2/2 \qquad (\text{a disk of radius } R) \tag{77c}$$

$$R_g^2 = \frac{R^2}{2} + \frac{L^2}{12} L^2 \qquad (\text{a disk of radius } R \text{ and length of } L) \tag{77d}$$

$$R_g^2 = \frac{a^2 + b^2 + c^2}{5} \qquad (\text{an ellipsoid of half-axes of } a, b, \text{ and } c) \tag{77e}$$

If the shape of the scattering object is known and the object is dispersed in a medium with a low concentration, R_g^2 can be evaluated by employing the law of Guinier. For a dilute-and-dispersed system, we can neglect the intermolecular interference term $Q(q)$, and the scattered intensity is simply given by

$$I(q) \cong \frac{b^2}{V_0}\phi z P(q) \cong \frac{b^2}{V_0}\phi z \exp\left[-\frac{R_g^2 q^2}{3}\right] \tag{78}$$

where ϕ is the volume fraction of the scattering object. Therefore, by plotting the logarithm of $I(q)$ as a function of q^2, R_g^2 is easily evaluated from the slope at low q region. This plot is called the *Guinier plot.* Note that the intercept is proportional to the product of the number of segments consisting of each molecule z and the volume fraction ϕ. Applications of Eq. (78) will be given in Section 2.5.1.

2.4.1.7 Form Factors [41]

The intramolecular interference function $P(q)$ can be calculated for some limited cases. Particularly for particles with a well-defined shape, $P(q)$ is given analytically to some extent. Because $P(q)$ represents the characteristic feature of the shape of the particle, it is also called a *form factor.* We list here exact expressions of the form factors for particles that take all orientations with equal probability.

1. *A sphere of radius R*

$$P(q) \equiv \phi^2(q) = 3\left[\frac{\sin qR - qR\cos qR}{q^3R^3}\right]^2 = \frac{9\pi}{2}\left[\frac{J_{3/2}(qR)}{(qR)^{3/2}}\right]^2 \tag{79}$$

   where $J_{3/2}(x)$ is the Bessel function of the order of 3/2.

2. *An ellipsoid of revolution of axes 2a, 2a, and* $2\mu a$ [45]

$$P(q) = \int_{\alpha=0}^{\pi/2} \Phi^2\left(qa\sqrt{\cos^2\alpha + \mu^2\sin^2\alpha}\right)\cos\alpha\, d\alpha \tag{80}$$

   where α is the angle between the principal axis of the ellipsoid of revolution and the scattering vector.

3. *A cylinder of revolution of diameter* $2R$ *and height* $2H$ [46]

$$P(q) = \int_{\alpha=0}^{\pi/2} \frac{\sin^2(qH\cos\alpha)}{(qH\cos\alpha)^2}\,\frac{4[J_1(qR\sin\alpha)]^2}{(qR\sin\alpha)^2}\sin\alpha\, d\alpha \tag{81}$$

   where $J_1(x)$ is the Bessel function of order 1, and α is the angle between the preferential axis of the rod and the scattering vector.

4. *A rod of infinitesimal transverse dimensions and length* $2H$

$$P(q) = \frac{Si(qH)}{qH} - \frac{\sin^2(qH)}{(qH)^2} \tag{82}$$

where $Si(x)$ is the sine integral defined by

$$Si(x) = \int_0^x \frac{\sin t}{t}\, dt \tag{83}$$

5. *A disk of infinitesimal thickness and diameter* $2R$ [47]

$$P(q) = \frac{2}{q^2R^2}\left[1 - \frac{1}{qR}J_1(2qR)\right] \tag{84}$$

2.4.1.8 $P(q)$ for Gaussian Chains

For a noninteracting polymer chain without excluded volume interaction (i.e., the so-called 'Gaussian chain'), the monomer distribution is given by

$$W(r_{ij}) = \left(\frac{3}{2\pi|i-j|a^2}\right)^{3/2} \exp\left[-\frac{3r_{ij}^2}{2|i-j|a^2}\right] \quad (\text{for large } |i-j|) \tag{85}$$

where $|i - j|$ is the number of monomers between $\mathbf{r}_i$ and $\mathbf{r}_j$. Because of the Gaussian distribution of r_{ij} for large $|i - j|$, the following relation is obtained

$$\langle \exp(i\mathbf{q}\cdot\mathbf{r}_{ij})\rangle = \exp\left[-\frac{|i-j|}{6}a^2q^2\right] \tag{86}$$

Thus,

$$P(q) \equiv g_D(q,z) = \frac{1}{z^2}\sum_i^z\sum_j^z \exp\left[-\frac{|i-j|}{6}a^2q^2\right] = \frac{2}{u^2}[e^{-u} - 1 + u] \tag{87}$$

and

$$u \equiv \frac{a^2z}{6}q^2 = R_g^2q^2 \tag{88}$$

$g_D(q, z)$ in Eq. (87) is often called the *Debye function* and describes the structure factor of a single Gaussian chain [48]. The limiting forms of $P(q)$ are

$$g_D(q,z) \to 1 - \frac{R_g^2}{3}q^2 + O(q^4) \qquad (q \to 0) \tag{89}$$

and

$$g_D(q, z) \rightarrow \frac{12}{a^2 z q^2} = \frac{2}{R_g^2 q^2} \qquad (q \rightarrow \infty) \tag{90}$$

Examination of the single chain function also provides some information about the tacticity of polymer chains. Examples related to this aspect will be given in Section 2.5.1.1 and in Figure 2.16.

2.4.1.9 Asymptotic Behavior of Scattering

In small-angle scattering, asymptotic behavior of the scattered intensity function provides useful information about the geometry of the scattering object. Let us assume here a dilute-and-disperse system. In this case, we do not worry about interparticle interference, i.e., $Q(q) = 0$. Thus, the structure factor of the system $S(q)$ is given by

$$S(q) = Nz^2 P(q) \sim NM^2 P(q) \sim V\phi M P(q) \equiv V\phi M P(x) \tag{91}$$

where ϕ and M are the volume fraction ($\phi \sim NM/V$) and the mass ($M \sim z$) of the scattering object, respectively, and $x(\equiv qR_g)$ is a dimensionless variable. Note that $P(x)$ in Eq. (91) is quite general and represents a particle structure factor. Because the characteristic size of the system (e.g., the radius of gyration of chain molecules) is given by R_g, $P(x)$ is written as

$$P(x) = P(qR_g) \rightarrow x^{-\gamma} \qquad (\text{for large } qR_g \text{ or } x) \tag{92}$$

where γ is the scattering exponent. Now we write the scaling form of the radius of gyration by

$$R_g \sim M^{\alpha} \tag{93}$$

We know the value of α for several systems such as

$$\alpha = \begin{cases} 1/2 & \text{(Gaussian chain)} \\ 3/5 & \text{(SAW chain)} \\ 1 & \text{(rod)} \end{cases} \tag{94}$$

where SAW chains indicate a self-avoiding chain having excluded volume. Note that at a large x (i.e., $q \gg 1/R_g$), $S(q)$ becomes independent of the molar mass M and R_g. This is because our ruler $S(q)$ is now adjusted to a local structure of the object and is too small to measure its global size. Hence, we obtain

$$S(q) \sim V\phi M(qM^{\alpha})^{-\gamma} = V\phi q^{-\gamma} M^{1-\alpha\gamma} \sim M^0 q^{-\gamma} \tag{95}$$

Thus, $\alpha = 1/\gamma$ is obtained and hence $S(q) \sim q^{-1/\alpha}$, i.e.,

$$S(q) \sim \begin{cases} q^{-2} & \text{(Gaussian chain)} \\ q^{-5/3} & \text{(SAW chain)} \\ q^{-1} & \text{(rod)} \end{cases} \tag{96}$$

In the case of two-phase systems in which each phase is homogeneous, the scattering function becomes sensitive to the interface structure at large $q(\gg 1/R_g)$. For example, an assembly of lamellae with an infinitesimally thin thickness and with interface area of A are regarded as two-dimensional objects. The structure factor is proportional to the total area of the lamellae A_T. Because the molar mass M is now in proportion to A for a two-dimensional object and $A_T = NA$, the following equation is given:

$$\begin{aligned} S(q) &\sim NM^2P(qA^{1/2}) \sim A_T A(qA^{1/2})^{-\gamma} \\ &= A_T q^{-\gamma} A^{1-\gamma/2} \sim A^0 A_T q^{-\gamma} \qquad \text{(for large } q\text{)} \end{aligned} \tag{97}$$

Here we assumed that $S(q)$ becomes independent of A for large q. Hence, $\gamma = 2$ is obtained and

$$S(q) \approx A_T q^{-2} \quad \text{(surface scattering from assembly of two-dimensional objects)} \tag{98}$$

Similarly for a three-dimensional object, where the mass $M \sim A^{3/2}$ and $A_T = NA$, the structure factor has the following form:

$$S(q) \sim A_T A^2 P(qA^{1/2}) = A_T A^2 (qA^{1/2})^{-\gamma} = A_T q^{-\gamma} A^{2-2/\gamma} = A^0 A_T q^{-\gamma} \tag{99}$$

Thus, $\gamma = 4$ is given and

$$S(q) \sim A_T q^{-4} \quad \text{(surface scattering from assembly of three-dimensional objects)} \tag{100}$$

This relation, i.e., the q^{-4} power dependence of the structure factor (or the scattered intensity), is called *Porod's law* [49, 50]. The exact form of the Porod equation is given by

$$I(q) = \frac{L}{v_0}(b_1 - b_2)^2 \frac{2\pi A_T}{q^4} \tag{101}$$

where L is a proportional constant having the unit of length. A modification regarding the contrast factor was made here to suit SANS because the original equation was derived for SAXS. Asymptotic behavior for objects having a large aspect ratio and a preferential orientation is highly dependent on their orientation distribution, which is discussed by Shibayama *et al.* [51].

Deviations from Porod's law were theoretically investigated by Ruland [52] and Vonk [53] to take account of the presence of finite interface thickness. Interface curvatures were addressed by Tomita [54, 55]. The principles derived have been applied to determine the interface thickness and interface curvatures (area-averaged scattering mean curvature) [56, 57].

2.4.1.10 Fractal Objects [58]

Small-angle scattering is a powerful technique to study materials with random structure such as fractal objects. A pioneering work to describe fractals or fractal geometry has been published by Mandelbrot [59]. Here we discuss the outline of the scattering law from fractal objects. Fractal objects are those without characteristic length. The shape of the fractal objects has *dilution symmetry* (or *self-similarity*). In other words, the shape looks similar in spite of zooming out or zooming in the view. There are two types of fractals; mass fractals and surface fractals. We introduce the fractal dimension D for the mass fractal and D_s for the surface fractal. These are different from the space dimension d [59].

Consider the topologically $(d-1)$-dimensional surface. The fractal dimension of the surface is given by

$$A \sim R^{D_s} \tag{102}$$

A is the $(d-1)$-dimensional Euclidean measure of the surface area. The lower limit of D_s is $(d-1)$. This is the case of a perfectly smooth minimal surface like a soap bubble. The upper limit of D_s is d. This can be achieved in a highly convoluted surface that fills d-dimensional space. Therefore,

$$d - 1 < D_s < d \tag{103}$$

Percolation clusters, gels, and porous media can be described by the mass fractal where the following relation is held:

$$M \sim R^{\mathrm{D}} \tag{104}$$

The scattering intensity (or the structure factor) becomes extensive in the total surface area A_{T} for surface fractals and in the total mass M for mass fractals. Hence,

$$I(q) \sim A_{\mathrm{T}} \sim R^{D_s} \quad \text{(surface fractal)}, \tag{105}$$

$$I(q) \sim M \sim R^{\mathrm{D}} \quad \text{(mass fractal)}. \tag{106}$$

On the other hand, the asymptotic behavior of the scattered intensity has the following scaling form:

$$I(q) \sim M^2 P(qR) \sim M^2 (qR)^{-\gamma} \tag{107}$$

Thus, by substituting Eq. (106) into (107) and by rearranging, one obtains

$$I(q) \sim M^2 (qR)^{-\gamma} \sim R^{2\mathrm{D}} (qR)^{-\gamma} = q^{-\gamma} R^{2\mathrm{D}-\gamma} \sim R^{\mathrm{D}} \tag{108}$$

This gives $\gamma = D$, hence

$$I(q) \sim \frac{R^{\mathrm{D}}}{q^D} \sim q^{-D} \tag{109}$$

for mass fractals. Similarly, for surface fractals,

$$I(q) \sim M^2 (qR)^{-\gamma} \sim R^{2d} (qR)^{-\gamma} = q^{-\gamma} R^{2d-\gamma} \sim R^{D_s} \tag{110}$$

Then $\gamma = 2d - D_s$, giving rise to

$$I(q) \sim \frac{R^{D_s}}{q^{2d-D_s}} \sim q^{-(2d-D_s)} \tag{111}$$

Note that Eq. (111) is reduced to the Porod equation by letting $d = 3$ (three dimensions) and $D_s = 2$ (smooth surface), i.e.,

$$I(q) \sim q^{-(6-2)} \sim q^{-4} \tag{112}$$

In the preceding discussion, we assumed that the masses are uniform in the system. However, in many cases the masses of the cluster are not uniform but are distributed in size. If the distribution is narrow (the limit of weak polydispersity), it will not change the asymptotic behavior $I(q) \sim q^{-D}$. Contrary to this, clusters having a power-law-type polydispersity (the limit of strong polydispersity) are often found in random structures (e.g., percolation models and sols near the gelation threshold). In this case, the number distribution of clusters of mass M is given by

$$N(M) = M^{-\tau} h(M/M_z) \tag{113}$$

where τ is the polydispersity exponent and M_z is the z-averaged mass and is given by

$$M_z \equiv \frac{\sum M^3 N(M)}{\sum M^2 N(M)} \tag{114}$$

The expression $h(M/M_z)$ is a scaling function for the polydispersity, which decays rapidly for $M > M_z$ [60]. The value of τ is $\tau = 2.2$ for the three-dimensional percolation model and $\tau = 2.5$ for the Bethe lattice. The polydispersity averaged

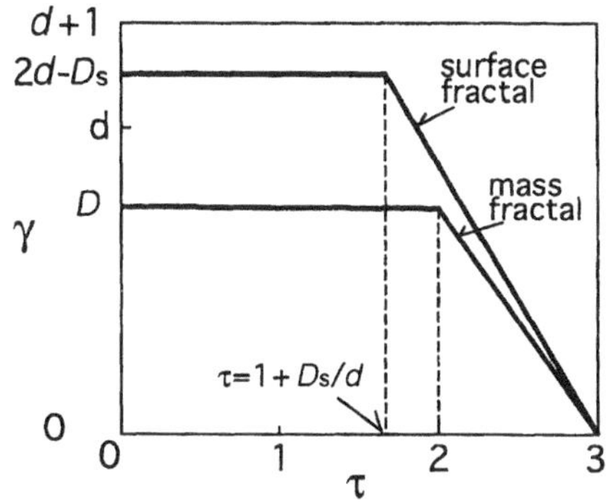

Figure 2.12 Polydispersity exponent τ dependence on the scattering exponent γ for fractal objects.

scattered intensity $I(q)$ is given by

$$I(q) = \frac{\int N(M) I_M(q)\, dM}{\int N(M)\, dM} \tag{115}$$

A scaling argument with respect to R, asymptotic behavior of $I(q)$ for polydisperse mass and surface fractals is given by

$$I(q) \sim \begin{cases} q^{-D(3-\tau)} & (\tau > 2) \\ q^{-D} & (\tau < 2) \end{cases} \quad \text{(mass fractal)} \tag{116}$$

$$I(q) \sim \begin{cases} q^{-d(3-\tau)} & (\tau > 1 + D_s/d) \\ q^{-(2d-D_s)} & (\tau < 1 + D_s/d) \end{cases} \quad \text{(surface fractal)} \tag{117}$$

The relation between the scattering exponent $\gamma(\equiv -[\log I(q)/\log q])$ and τ is shown in Figure 2.12.

2.4.2 MULTICOMPONENT SYSTEMS

In the previous section, we ignored the interaction among different scattering particles (or the interaction among polymer chains). In this section, we deal with interacting (multicomponent) systems in which molecular interactions are important and formulate the structure factor for interacting systems such as polymer solutions, polymer blends, and block copolymer melts.

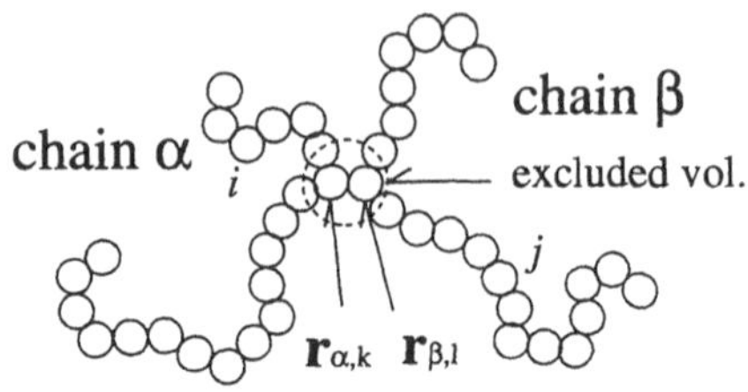

Figure 2.13 Schematic representation of polymer chains in solution. The interaction is transmitted only by way of contact, i.e., through the ith segmant of chain α and jth chain of chain β in the figure. The same place cannot be occupied by any pair of segments ("the excluded volume effect").

2.4.2.1 Polymer Solutions

As has been already discussed in Section 2.4.1.4, the structure factor is decomposed to the intra- and interchain interference functions (see Eq. (56)). Here, the excluded volume interaction becomes significant. Let us derive the scattering function for a polymer solution in terms of the single contact theory for the excluded volume effect [61]. Consider the interaction of the ith segment of chain α and the jth segment of chain β, as shown in Figure 2.13. The interaction is assumed to be transmitted only by way of the contact at $\mathbf{r}_{\alpha,k} = \mathbf{r}_{\beta,l}$. In this case, the average $\langle \exp[-i\mathbf{q}\cdot(\mathbf{r}_{\alpha,i} - \mathbf{r}_{\beta,j})]\rangle$ is written as

$$\begin{aligned}
&\langle \exp[-i\mathbf{q}\cdot(\mathbf{r}_{\alpha,i} - \mathbf{r}_{\beta,j})]\rangle \\
&\quad = \langle \exp[-i\mathbf{q}\cdot(\mathbf{r}_{\alpha,i} - \mathbf{r}_{\alpha,k})]\exp[-i\mathbf{q}\cdot(\mathbf{r}_{\alpha,k} - \mathbf{r}_{\beta,l})]\exp[-i\mathbf{q}\cdot(\mathbf{r}_{\beta,l} - \mathbf{r}_{\beta,j})]\rangle \\
&\quad = \langle \exp[-i\mathbf{q}\cdot(\mathbf{r}_{\alpha,i} - \mathbf{r}_{\alpha,k})]\rangle\langle \exp[-i\mathbf{q}\cdot(\mathbf{r}_{\alpha,k} - \mathbf{r}_{\beta,l})]\rangle\langle \exp[-i\mathbf{q}\cdot(\mathbf{r}_{\beta,l} - \mathbf{r}_{\beta,j})]\rangle
\end{aligned} \tag{118}$$

The first and third averages on the right-hand side of Eq. (118) denote the intrachain interference. Thus,

$$\begin{aligned}
&\frac{\phi}{Nz}\sum_{\alpha}^{N}\sum_{i}^{z}\sum_{k}^{z}\langle \exp[-i\mathbf{q}\cdot(\mathbf{r}_{\alpha,i} - \mathbf{r}_{\alpha,k})]\rangle \\
&\quad = \frac{\phi}{Nz}\sum_{\beta}^{N}\sum_{l}^{z}\sum_{j}^{z}\langle \exp[-i\mathbf{q}\cdot(\mathbf{r}_{\beta,l} - \mathbf{r}_{\beta,j})]\rangle = \phi z P(q)
\end{aligned} \tag{119}$$

Note that the polymer volume fraction ϕ has to be taken into account, i.e., $\phi = v_0 zN/V$. The second average is forbidden because of the excluded volume. In other words, any pair of segments cannot occupy the same space. The volume of this space is called the excluded volume v_{ex}. The excluded volume is given by

$$\langle \exp[-i\mathbf{q}\cdot(\mathbf{r}_{\alpha,k}-\mathbf{r}_{\beta,l})]\rangle = \frac{1}{V}\int [1-g(r)]\exp[-i\mathbf{q}\cdot(\mathbf{r}_{\alpha,k}-\mathbf{r}_{\beta,l})]\cdot 4\pi r^2\,dr$$
$$= \frac{v_{ex}}{V} \tag{120}$$

where $g(r)$ is the radial distribution function of the polymer chain. Because of this negative (or destructive) contribution, the interchain interference function is given from Eqs. (118) to (120) by

$$\frac{\phi}{Nz}\sum_{\alpha}^{N}\sum_{\beta}^{N}\sum_{i}^{z}\sum_{j}^{z}\langle \exp[-i\mathbf{q}\cdot(\mathbf{r}_{\alpha,i}-\mathbf{r}_{\beta,j})]\rangle$$
$$= \frac{\phi}{Nz}N^2z^2\frac{v_{ex}}{V}[zP(q)]^2 = \frac{v_{ex}}{v_0}[\phi zP(q)]^2 \tag{121}$$

This contribution has to be subtracted from the intramolecular interference function in Eq. (58) because it is forbidden. Hence,

$$S(q) = \phi zP(q) - \frac{v_{ex}}{v_0}[\phi zP(q)]^2 \tag{122}$$

Similarly, this type of consideration about contact can be extended to multiple contacts, such as three-body contacts, four-body contacts and so on. Hence, the following equation is finally obtained:

$$S(q) = \phi zP(q) - (v_{ex}/v_0)[\phi zP(q)]^2 + (v_{ex}/v_0)^2$$
$$\times [\phi zP(q)]^3 - (v_{ex}/v_0)^3[\phi zP(q)]^4 + \cdots$$
$$= \frac{\phi zP(q)}{1+(v_{ex}/v_0)\phi zP(q)} \tag{123}$$

Here, we assumed $|(v_{ex}/v_0)\phi zP(q)| \ll 1$. Therefore, the scattered intensity function $I(q)$ is given

$$\frac{(b_1v_0/v_1 - b_0)^2}{I(q)v_0} = \frac{1}{S(q)} = \frac{1}{\phi zP(q)} + \frac{v_{ex}}{v_0} \tag{124}$$

Here, the difference in the monomer (or segment) volume (v_1) and the reference (i.e., solvent) volume (v_0) has been taken into account. Equation (124) has been derived here by using the single-contact theory [61]. The same equation has been derived by many workers from different approaches [44].

The mass concentration of the polymer is given by

$$c = \frac{m\phi}{v_0N_A} = \frac{m\phi}{\hat{v}_0} \tag{125}$$

where m is the molecular mass of the monomer unit of the polymer. By substituting Eq. (125) into (124), one gets

$$\frac{(b_1 v_0/v_1 - b_0)^2}{I(q)v_0} = \frac{1}{S(q)} = \frac{m}{\phi \hat{v}_0 z P(q)} + \frac{v_{ex}}{v_0} = \frac{m}{cN_A v_0 z P(q)} + \frac{v_{ex}}{v_0} \tag{126}$$

By multiplying $v_0 c N_A/m^2$ on the both sides of Eq. (126), one gets

$$\frac{(b_1 v_0/v_1 - b_0)^2 N_A c}{I(q)m^2} = \frac{1}{zmP(q)} + \frac{N_A v_{ex}}{m^2}c = \frac{1}{MP(q)} + 2A_2 c \tag{127}$$

where M is the molecular mass of the polymer and A_2 is the second virial coefficient, which is given by

$$A_2 = \frac{N_A v_{ex}}{2m^2} = \frac{N_A a^3}{2m^2}(1 - 2\chi) \tag{128}$$

The χ is a very important parameter describing the affinity of the polymer and the solvent, which is called the Flory–Huggins interaction parameter. When $\chi = 1/2$, A_2 becomes zero, and the system recovers the unperturbed state. In most cases, however, χ is greater than $1/2$ (a poor solvent case) or smaller than $1/2$ (a good solvent case), resulting in a perturbation of the chain conformation in the solvent to shrink ($\chi > 1/2$) or to expand ($\chi < 1/2$). Note that Eq. (127) leads to the well-known *Zimm equation*,

$$\frac{(b_1 v_0/v_1 - b_0)^2 N_A c}{I(q)m^2} = \frac{1}{M}\left[1 + \frac{1}{3}R_g^2 q^2 + \cdots\right] + 2A_2 c \tag{129}$$

An example of Zimm analysis will be given in Section 2.5.1.

2.4.2.2 Binary Mixture of Isotopic Polymers Having Different Degrees of Polymerization

Polymer chains in a miscible binary polymer mixture are slightly perturbed due to the presence of the other component. The structure factor for miscible polymer blends has been successfully calculated on the basis of the random-phase approximation (RPA) [13]. The structure factor for unperturbed polymer chains is given by

$$S^0_{\alpha\alpha}(q) = \phi_\alpha z_\alpha P(q) \tag{130}$$

The superscript ^{0} indicates the unperturbed state. In this case, $P(q)$ is now equal to the Debye function for the polymer chain, i.e., $g_D(q, z_\alpha)$, which is written as

$$P(q, z_\alpha) = g_D(q, z_\alpha) \equiv \frac{2}{u_\alpha^2}[\exp(-u_\alpha) - 1 + u_\alpha] \tag{131}$$

Hence,

$$S^0_{\alpha\alpha}(q) = \phi_\alpha z_\alpha g_D(q, z_\alpha) \tag{132}$$

It is obvious that there is no α–β correlation if the constituent chains are unperturbed. Hence,

$$S^0_{\alpha\beta}(q) = 0 \qquad (\text{for } \alpha \neq \beta) \tag{133}$$

The main conclusion of the RPA for binary polymer blends of 1 and 2 is the following [13]:

$$\frac{1}{S(q)} = \frac{1}{S^0_{11}(q)} + \frac{1}{S^0_{22}(q)} - 2\chi \tag{134}$$

Equation (134) can be written in a more explicit form,

$$\frac{1}{S(q)} = \frac{1}{\phi_1 z_1 g_D(q, z_1)} + \frac{1}{\phi_2 z_2 g_D(q, z_2)} - 2\chi \tag{135}$$

or

$$\frac{(b_1 - b_2)^2}{v_0} \cdot \frac{1}{I(q)} = \frac{1}{S(q)} = \frac{1}{\phi_1 z_1 g_D(q, z_1)} + \frac{1}{\phi_2 z_2 g_D(q, z_2)} - 2\chi \tag{136}$$

Equation (136) is called the *de Gennes scattering function* for polymer blends [13]. Note that for $q \to 0$ (i.e., the thermodynamic limit), Eq. (135) becomes

$$\frac{1}{S(q = 0)} = \frac{1}{\phi_A z_A} + \frac{1}{\phi_B z_B} - 2\chi = 2(\chi_s - \chi) \tag{137}$$

where χ_s is the χ parameter at the spinodal, given by

$$\chi_s = \frac{1}{2}\left(\frac{1}{\phi_A z_A} + \frac{1}{\phi_B z_B}\right) \tag{138}$$

In general, the constituent polymers differ in the segmental volume v_i. In such a case, Eq. (136) is modified to [62]

$$\frac{(b_1/\hat{v}_1 - b_2/\hat{v}_2)^2 N_A}{I(q)} = \frac{1}{\hat{v}_1 \phi_1 z_1 g_D(q, z_1)} + \frac{1}{\hat{v}_2 \phi_2 z_2 g_D(q, z_2)} - \frac{2\chi}{\hat{v}_0} \tag{139}$$

where $\hat{v}_\alpha$ is the segmental molar volume of component α, i.e., $\hat{v}_\alpha = N_A v_\alpha$. In many cases, $\hat{v}_0$ is defined as the geometric average of $\hat{v}_1$ and $\hat{v}_2$, i.e., $\hat{v}_0 = \sqrt{\hat{v}_1 \hat{v}_2}$ [62]. Note that Eq. (139) is reduced to Eq. (126) by substituting $z_2 = 1$ (solvent) and $\hat{v}_2 = \hat{v}_0 = N_A v_0$. See Section 2.5.2 for applications.

2.4.2.3 Block Copolymer Melt

Diblock copolymers A-*b*-B are intramolecular multicomponent polymeric systems in which the constituent polymer chains A and B are connected at their ends with a chemical junction. Although graft copolymers, multiblock copolymers, and ladder or star block copolymers are also classified as intramolecular multicomponent polymers, we deal with the structure factor of diblock copolymers in melt developed by Leibler [63]. Here, we consider a single-phase state of block copolymer melts at thermal equilibrium. Even in a single-phase state, block copolymers have a nonvanishing correlation that gives rise to a scattering as discussed in the following:

In the case of diblock copolymers, there is nonzero α–β correlation because α and β chains are connected to each other. Hence, Eq. (134) is generalized to the case of $S^0_{\alpha\beta} \neq 0$, which is [63]

$$\frac{1}{S(q)} = \frac{S^0_{11}(q) + 2S^0_{12}(q) + S^0_{22}(q)}{S^0_{11}(q)S^0_{22}(q)} - 2\chi \tag{140}$$

Note that Eq. (140) can certainly be reduced to Eq. (134) when $S^0_{\alpha\beta} \neq 0$. For a diblock copolymer system consisting of block chain 1 of fz segments and block chain 2 of $(1-f)z$ segments, the structure factor for component 1 and 2 is given by

$$\begin{aligned} S^0_{11}(q) &= zg_{\mathrm{B}}(f, u) \\ S^0_{22}(q) &= zg_{\mathrm{B}}(1-f, u) \\ S^0_{12}(q) &= \frac{1}{2}z[g_{\mathrm{B}}(1, u) - g_{\mathrm{B}}(f, u) - g_{\mathrm{B}}(1-f, u)] \end{aligned} \tag{141}$$

where

$$g_{\mathrm{B}}(f, u) = \frac{2}{u^2}(e^{-fu} + fu - 1) \tag{142}$$

and

$$u \equiv \frac{a^2 z}{6}q^2 = R_g q^2 \tag{143}$$

Note an equivalence of the segment length of the block polymer chain, i.e., $a_1 = a_2 \equiv a$ and z is the total number of segments in a chain, is assumed. By writing Eq. (140) more explicitly, one obtains

$$\frac{1}{S(q)} = \frac{g_{\mathrm{B}}(1, u)}{z\left\{g_{\mathrm{B}}(f, u)g_{\mathrm{B}}(1-f, u) - \frac{1}{4}[g_{\mathrm{B}}(1, u) - g_{\mathrm{B}}(f, u) - g_{\mathrm{B}}(1-f, u)]^2\right\}} - 2\chi \tag{144}$$

and

$$I(q) = (b_1/v_1 - b_2/v_2)^2 v_0 S(q) \tag{145}$$

It should be noted that effects of thermal fluctuations are also very important to study the interaction parameter and the order–disorder transition. This aspect was extensively discussed by Fredrickson and Helfand [64]. Applications of SANS for characterizing the nature of phase transition in block copolymers will be discussed later in Section 2.5.3.2.

2.4.2.4 Multicomponent Mixture (Contrast Variation)

As has been already discussed in Section 2.4.1.3, the scattering intensity function for a multicomponent system is given by Eq. (54). The range of the scattering length, b_0 can be varied widely from a fully hydrogenous solvent to a fully deuterated solvent, such as H_2O to D_2O or C_6H_6 to C_6D_6. The scattering lengths of most of solute polymers are located between the two extremes. Therefore, this equation states that one can easily obtain the structure factor of the target part of the system $S_{\alpha\alpha}(q)$ by tuning b_0 (see Eq. (56)). This method is called the *contrast variation method.* In the case of $(p+1)$-component systems with $(p+1) \geq 3$ (three components or above), $I(q)$ in Eq. (54) consists of p of $S_{\alpha\alpha}(q)\,(0 \leq \alpha \leq p)$ and $(p+1)p/2$ of $S_{\alpha\beta}(q)\,(\alpha \neq \beta, 0 \leq \alpha \leq p, 0 \leq \beta \leq p)$. Therefore, several numbers of measurements are required by varying the contrast in order to extract $S_{\alpha\alpha}(q)$.

2.4.3 *EFFECT OF POLYDISPERSITY*

Most synthetic polymers are polydisperse and characterized by several averages of degree of polymerization (or molecular weight). In scattering experiments, three averages, i.e., the number, the weight, and the z-average degree of polymerization are important and are defined by

$$\begin{aligned}
\langle z\rangle_{\mathrm{n}} &= \frac{\sum_i N_i z_i}{\sum_i N_i} = \frac{\sum_i c_i}{\sum_i c_i/z_i} \quad \text{(number average)} \\
\langle z\rangle_{\mathrm{w}} &= \frac{\sum_i N_i z_i^2}{\sum_i N_i z_i} = \frac{\sum_i c_i z_i}{\sum_i c_i} \quad \text{(weight average)} \qquad (146) \\
\langle z\rangle_{\mathrm{z}} &= \frac{\sum_i N_i z_i^3}{\sum_i N_i z_i^2} = \frac{\sum_i c_i z_i^2}{\sum_i c_i z_i} \quad \text{(z-average)}
\end{aligned}$$

where N_i, c_i, and z_i are the numbers and mass concentration of the polymer chain having the degree of polymerization z_i, respectively. The ratio, $\langle z\rangle_\mathrm{w}/\langle z\rangle_\mathrm{n}$ is often called the polydispersity index.

For polydisperse polymeric systems, scattering functions have to be corrected by taking account of the polydispersity. The following Schultz–Zimm exponential distribution function is commonly used to describe the number fraction distribution of chains with the degree of polymerization z:

$$n(z) = \frac{\lambda_\mathrm{s}^{h_\mathrm{s}}}{\Gamma(h_\mathrm{s})} z^{h_\mathrm{s}-1} \exp(-\lambda_\mathrm{s} z) \tag{147}$$

where $\Gamma(h)$ is the gamma function h_s and λ_s are parameters characterizing polydispersity. The averages are simply written by using this continuous expression as follows:

$$\langle z\rangle_\mathrm{n} = h_\mathrm{s}/\lambda_\mathrm{s}, \quad \langle z\rangle_\mathrm{w} = (h_\mathrm{s}+1)/\lambda_\mathrm{s}, \quad \text{and } \langle z\rangle_\mathrm{z} = (h_\mathrm{s}+2)/\lambda_\mathrm{s} \tag{148}$$

Thus,

$$\frac{\langle z\rangle_\mathrm{w}}{\langle z\rangle_\mathrm{n}} = \frac{h_\mathrm{s}+1}{h_\mathrm{s}}, \qquad \frac{\langle z\rangle_\mathrm{z}}{\langle z\rangle_\mathrm{w}} = \frac{h_\mathrm{s}+2}{h_\mathrm{s}+1} \tag{149}$$

and

$$\lambda_\mathrm{s} = 1/(\langle z\rangle_\mathrm{w} - \langle z\rangle_\mathrm{n}) \tag{150}$$

For the most probable distribution, $\langle z\rangle_\mathrm{w}/\langle z\rangle_\mathrm{n}$ is given to be 2. This corresponds to $h_\mathrm{s} = 1$. The weight fraction distribution is calculated with $n(z)$, which is given by

$$w(z) = \frac{zn(z)}{\int zn(z)\,dz} = \frac{\lambda_\mathrm{s}^{h_\mathrm{s}+1}}{\Gamma(h_\mathrm{s}+1)} z^\mathrm{h} \exp(-\lambda_\mathrm{s} z) \tag{151}$$

The scattered intensity from a dilute polymer solution is proportional to the product of the degree of polymerization z and the Debye function, as already discussed in Eq. (78) together with Eq. (65). Thus, for a polydisperse system, $I(q)$ is given by

$$I(q) \sim \langle zP(q)\rangle_\mathrm{w} \tag{152}$$

It is useful to calculate Eq. (152) for the Debye function, i.e., $P(q) = g_\mathrm{D}(q, z)$,

$$\langle zg_\mathrm{D}(q, z)\rangle_\mathrm{w} = \int_0^\infty zg_\mathrm{D}(q, z) w(z)\,dz = \langle z\rangle_\mathrm{n} \langle g_\mathrm{D}(q, z)\rangle_\mathrm{w} \tag{153}$$

In Eq. (153), $\langle g_D(q, z)\rangle_w$ is the weight-average Debye function given by [65]

$$\langle g_D(q, z)\rangle_w \equiv \frac{2}{\langle u\rangle_n^2}\left[\left(\frac{h_s}{h_s + \langle u\rangle_n}\right)^{h_s} - 1 + \langle u\rangle_n\right] \tag{154}$$

where $\langle u\rangle$ is given by

$$\langle u\rangle_n \equiv \langle R_g\rangle_n q^2 = \frac{a^2\langle z\rangle_n}{6} q^2 \tag{155}$$

In the case of polymer blends, the de Gennes scattering function for monodisperse binary blends, Eq. (136), is modified to

$$\frac{(b_1 - b_2)^2 N_A}{v_0} \cdot \frac{1}{I(q)} = \frac{1}{\phi_1\langle z_1 g_D(q, z_1)\rangle_w} + \frac{1}{\phi_2\langle z_2 g_D(q, z_2)\rangle_w} - 2\chi \tag{156}$$

Note that

$$\langle z g_D(q, z)\rangle_w \to \langle z\rangle_w\left[1 - \frac{1}{3}\frac{\langle R_g\rangle_z}{\langle z\rangle_w} q^2\right] \qquad (q \to 0) \tag{157}$$

where

$$\langle R_g\rangle_z \equiv \frac{1}{6} a^2\langle z\rangle_z \tag{158}$$

Therefore, for small q region, Eq. (156) is rewritten by

$$\frac{(b_1 - b_2)^2}{v_0} \cdot \frac{1}{I(q)} = \frac{1}{\phi_1\langle z_1\rangle_w} + \frac{1}{\phi_2\langle z_2\rangle_w} - 2\chi + \frac{1}{18\phi_1\phi_2}\left[\frac{\langle z_1\rangle_z}{\phi_1\langle z_1\rangle_w} + \frac{\langle z_2\rangle_z}{\phi_2\langle z_2\rangle_w}\right] a^2 q^2 \tag{159}$$

which is first derived by Joanny [66]. The polydispersity effect for block copolymers have been also discussed by Leibler and Benoit [67] and others [68].

2.4.4 *EFFECT OF DEUTERATION*

Classical thermodynamics assumes that isotopic mixtures are an example of ideal solutions [69]. However, nontrivial isotope effects have been found, particularly in polymer solutions and polymer blends. For example, deuterated polyethylene crystallizes 7 °C lower than ordinary polyethylene [9], and the lower critical solution temperature of polystyrene/poly(vinyl methyl ether) blends shifts upward by about 40 °C owing to the deuteration of polystyrene [70]. Deuteration effects are of particular significance in the cases of crystallization and phase separation in which polymer effects (e.g., the connectivity effect) become significant. Because potential energies between molecules or atoms are isotope independent (the

Born–Oppenheimer approximation), these isotope effects must result from difference in the masses of isotopes (i.e., the nuclear mass effects) [71]. This results in vapor pressure isotope effect (VPIE) and molar volume isotope effect (MVIE). Buckingham and Hentschel discussed the isotope effects on the basis of MVIE [72]. Bates and Wignall studied isotope mixtures of poly(butadiene) mixtures and of polystyrene and observed a phase separation behavior in the isotopic polymer mixtures [73]. Finite values of χ parameters were obtained by using the de Gennes scattering function [13]. The Buckingham and Hentschel theory was corrected by Singh and van Hook [74]. They took into account the difference in the CH and CD vibrations to the free energy. According to their theory, though the difference is small, it is cumulative with respect to the number of atoms substituted per monomeric unit r and the degree of polymerization z. The critical degree of polymerization for phase separation in an isotope mixture z_c, is given by

$$z_c = \frac{4}{r\Gamma\left(\frac{\Delta v}{v}\right)u'\left(1-\sqrt{\frac{\mu'}{\mu}}\right)} \tag{160}$$

where Γ is the Gruneisen parameter for the effective frequency, $(\Delta v/v)$ is the molar volume isotope effect (MVIE) (i.e., the molar volume change by deuterium substitution), μ' and μ are the reduced masses of the CH and CD oscillators, respectively (i.e., $\mu'/\mu = (2^{-1} + 12^{-1})/(1^{-1} + 12^{-1})$), and $u' = hv/kT$. For example, for poly(protio/deuteriobutadiene), $r = 6$, $\Delta v/v = (4 \pm 1) \times 10^{-3}$ (= 0.035 (benzene)), and $u' = (1.44 \times 3000)/T = 14.4$ at 300 K; therefore, z_c is estimated to be $(1.3 \pm 0.3) \times 10^3$, which is in good agreement with the experimental result, $z_c = (2.3 \pm 0.5) \times 10^3$ [74]. A similar number is also predicted for polystyrene isotope mixtures and is in fairly good accordance with the experimental results obtained [75]. It should be noted here that the deuteration effect is cumulative with the degree of polymerization. Thus, the isotope effect becomes more significant for polymer mixtures with higher molecular weights.

2.5 Experimental Studies

In this section, some of experimental studies will be demonstrated to enable the reader to feel more familiar with SANS techniques and to demonstrate how much SANS has been contributing to polymer science.

2.5.1 *CHAIN DIMENSION*

2.5.1.1 Amorphous Polymers

The first application of SANS was an examination of Flory's prediction [8]. Do amorphous polymer chains in bulk state have a conformation identical to that in the Θ condition? This prediction is based on complete cancellation of intra- and interchain interactions, which has been known to be achieved in a Θ solvent. Benoit *et al.* [5] studied the size of deuterated polystyrene (DPS) in carbon disulfide (CS_2) and in hydrogenous polystyrene (HPS) as a function of concentration. Figure 2.14(a) and (b) show Zimm plots for these systems (a) in CS_2 and (b) in HPS. In both cases, the Zimm analysis, from which the radius of gyration R_g, the weight average molecular weight M_W, and the second virial coefficient A_2, were evaluated (see Eq. (131)) seems to be successful. Note that the slopes of the extrapolated line at $q = 0$ are positive for the case of CS_2 and zero for the case of HPS. Because the slope corresponds to A_2, it is clear from this experiment that CS_2 is a good solvent and HPS is a Θ solvent and DPS, respectively.

The molecular weight dependence of R_g was also obtained for DPS with the molecular weight range between 2.10×10^3 and 3.25×10^5, which is shown in Figure 2.15. A power law relationship between R_g and the weight average molecular weight M_W was obtained in both cases. The same results have been obtained for poly(methyl methacrylate) (PMMA) by Kirste *et al.* [76] and for poly(dimethyl siloxane) (PDMS) [77], molten polyethylene (PE) [78], etc. If the ratio of R_g^2/M_W is independent of M_W, the polymer chains are in an unperturbed state. This relation was indeed obtained for many polymers. The ratios of R_g^2/M_W are shown in Table 2.3 for several molten or glassy polymers [79]. This clearly shows that many of the amorphous polymers obey the relation

$$R_g = K M_W^{1/2} \tag{161a}$$

where K is the constant ($\equiv (R_g^2/M_W)^{1/2}$). Note that the value K is identical to that obtained in a Θ solvent within an experimental error. Hence, Flory's prediction has been proved.

The agreement between the theoretical prediction and the experiments on the relationship between R_g and M_W does not necessarily mean that the local conformation of polymer chains is the same as that for Gaussian chains. Experiments to examine the local structure were carried out by O'Reilly *et al.* for isotactic- (i-), syndiotactic- (s-), and atactic- (a-) -poly(methyl methacrylate) (PMMA) [76].

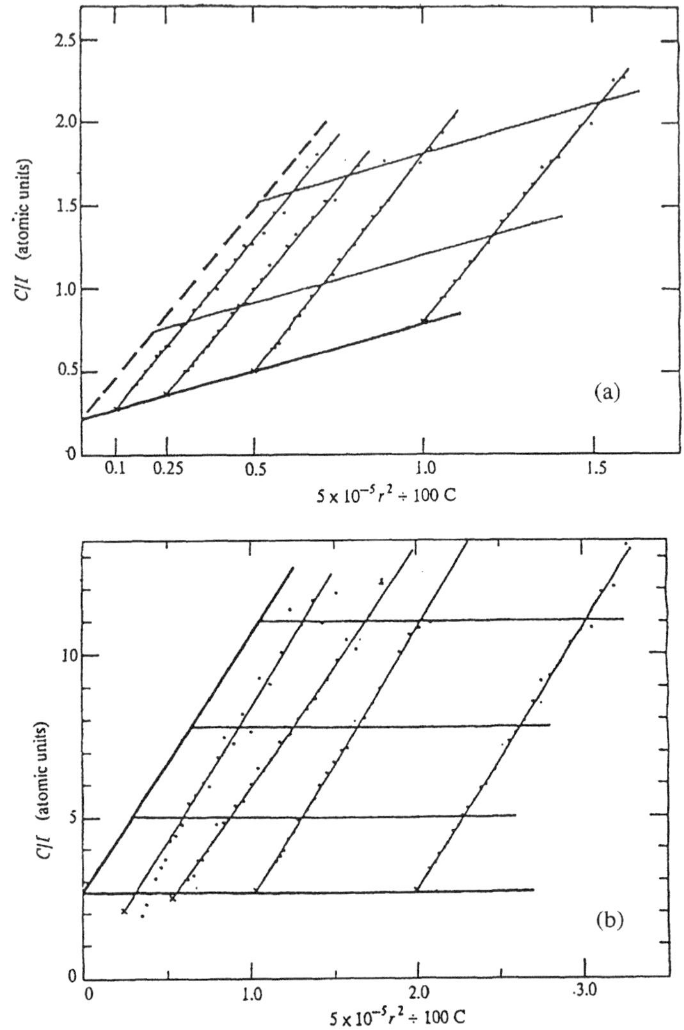

Figure 2.14 Zimm plots for DPS of $M_W = 32{,}500$ in (a) carbon disulfide and in (b) HPS matrix. The r is proportional to the scattering angle θ, $r = L\theta$, where L is the sample-to-detector distance and c is the concentration of DPS (g cm^{-3}). Reprinted from H. Benoit *et al.* (1973) with permission from the author.

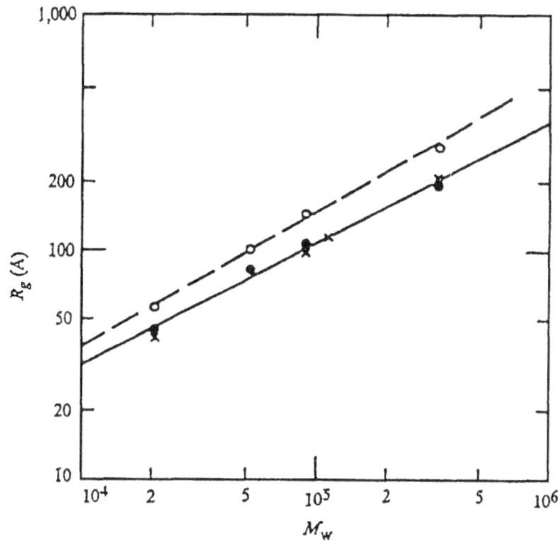

Figure 2.15 Molecular weight dependence of R_g's of DPS in the bulk (open circle), in cyclohexane at 36 °C (solid circle), and in CS_2. Reprinted from H. Benoit *et al.* (1973) with permission from the author and the publisher, Macmillan Magazine Ltd.

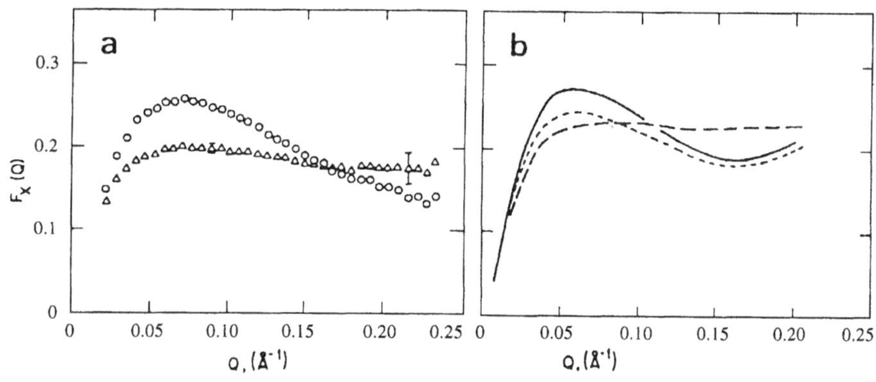

Figure 2.16 Kratky plots for the comparison of intensity data at intermediate q values for PMMA. (a) Observed for syndiotactic (open circles) and for isotactic (triangles) PMMAs, and (b) calculations on the basis of isotactic (dashed curve), atactic (dotted), and syndiotactic (solid curve) polymer chains. Reprinted from J. M. O'Reilly *et al.* (1985) with permission from the author and the publisher, American Chemical Society.

Table 2.3 Molecular dimensions in bulk polymer samples. Reprinted from J. M. Schultz in "Diffraction for Materials Scientists" (1982) with permission from the author.

		$(R_g^2/M_W)^{1/2}\ \frac{\text{Å}\cdot\text{mol}^{1/2}}{\text{g}^{1/2}}$			
Polymer	***State of Bulk***	***SANS Bulk***	***Light-Scattering θ-Solvent***	***SAXS***	***References***
Polystyrene	Glass	0.275	0.275	0.27	a,i
Polystyrene	Glass	0.28	0.275	—	b
Polyethylene	Melt	0.46	0.45	—	c
Polyethylene	Melt	0.45	0.45	—	d
Poly(methyl methacrylate)	Glass	0.31	0.30	—	e
Poly(ethylene oxide)	Melt	0.45	—	—	f
Poly(vinyl chloride)	Glass	0.30	0.37	—	g
Polycarbonate	Glass	0.457	—	—	h

References: [a]J. P. Cotton, D. Decker, H. Benoit, B. Farnoux, J. Higgins, G. Jannink, R. Ober, C. Picot, and J. desCloizeau: *Macromolecules* **7:** 863 (1974). [b]G. D. Wignall, D. G. Ballard, and J, Schelten, *Eur. Polym. J.* **10:** 861 (1974). [c]J. Schelte, D. G. H. Ballard, G. Wignall, G. Longman, and W. Schmatz, *Polymer* **17:** 751 (1976). [d]D. Lieser, E. W. Fischer, and K. Id, *J. Polym. Sci. Polym. Lett. Ed.* **13:** 39 (1975). [e]R. G. Kirste, W. A. Kruse, and K. Ibel, *Polymer* **16:** 120 (1975). [f]G. Alle, *Proc. R. Soc. Lond., Ser. A* **351:** 381 (1976). [g]P. Herchenroeder and M. Dettenmaier, unpublished (1977). [h]D. G. H. Ballard, A. N. Burgess, P. Cheshire, E. W. Janke, A. Nevin, and J. Schelten, *Polymer* **22:** 1353 (1981). [i]H. Hayashi, F. Hamada, and A. Nakajima, *Macromolecules* **9:** 543 (1976). Reproduced from Schultz, ref. [19].

The details of the difference in the structure were detected by so-called Kratky plots, i.e., $q^2 I(q)$ versus q. The i-PMMA shows a Debye-like scattering function, whereas both s- and a-PMMA had a maximum in $q^2 I(q)$. Figure 2.16(a) shows the Kratky plots for s-PMMA and i-PMMA. These results were compared favorably with rotational isomeric calculations by Yoon *et al.* [80, 81], as shown in Figure 2.16(b).

2.5.1.2 Crystallizable Polymers

Compared with amorphous polymers, determining the radius of gyration of labeled chains in the crystalline state has faced several problems, namely, (1) segregation of the labeled species from the matrix polymers due to the difference in the melting temperatures, (2) void scattering arising from contraction during the solidification process, and (3) mismatch in the molecular weights between the labeled and matrix polymer chains. For example, the melting temperature of deuterated polyethylene (DPE) is ≈129 °C, whereas that of hydrogenous polyethylene (HPE) is ≈135 °C [9]. Because of this difference, segregation occurs during the cooling process from the melt, which leads to enormous scattering at low angles. Schelten *et al.* extensively discussed and solved these problems and proposed a method to circumvent them [9]. These problems are particularly severe in the case of a slowly crystallized polyethylene. Thus, most of the studies on PE have been conducted for rapidly crystallized samples by a rapid quenching. Figure 2.17 shows scattering patterns from an HPE/DPE mixuture with 95.5/0.5 wt%/wt%

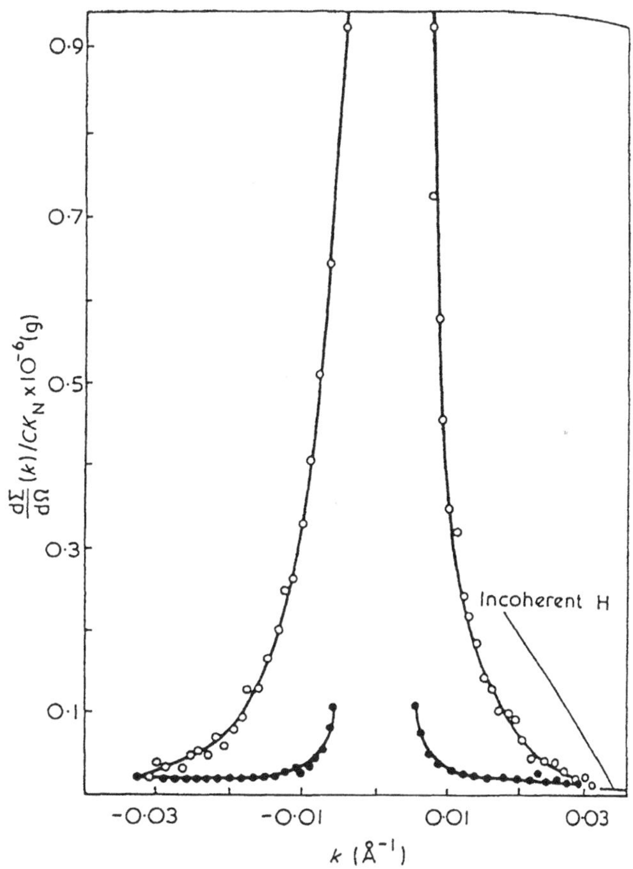

Figure 2.17 Scattering patterns from HPE matrix with 0.5% DPE molecules (open circles) and from blank with HPE matrix (solid circles). Reprinted from J. Scheleten *et al.* (1974) with permission from the author and the publisher, Elsevier Science.

and from pure HPE. These samples were prepared by dissolving HPE (and DPE) in boiling xylene, precipitating into excess methanol, and pressing into bubble-free films of approximately 1-mm thick. The scattering from the blank was mainly due to incoherent scattering of protons, and the upturn at low angles was due to void scattering [9]. By labeling with 0.5% DPE molecules, the scattering was increased dramatically, from which the R_g was successfully obtained. Scheleten *et al.* noted that a measurement for an inversely labeled sample (i.e., a small portion of HPE in DPE) was impossible due to strong scattering from the crystalline DPE matrix. This occurred because the scattering length of HPE is nearly equal to zero as the result of cancellation of the positive (-C-) and negative ($-H_2-$) scattering lengths. Such cancellation does not occur for DPE because both -C- and $-D_2-$ are positive (see Table 2.1).

Similar to amorphous polymers, R_g values have been measured as a function of molecular weight M_W for quenched melts of polyethylene [9], polypropylene (PP) [82], *i*-polystyrene (*i*-PS) [83], hydrogenated polybutadiene (equivalent to PE) [84], and poly(ethylene oxide) [85]. Generally, the same relation as Eq. (161) seems to be obtained for crystalline polymers. Figure 2.18 shows the M_W dependence of R_g for PP prepared by quenching, annealing, and isothermal crystallization [82]. Unlike PE, no cluster formation was observed in PP. It was found that different thermal treatments (i.e., quenching, annealing, or isothermal crystallization) do not have significant effects on R_g, whereas the long spacing in the crystalline state differs from 120 to 250 Å, depending on thermal treatment. In addition R_g was

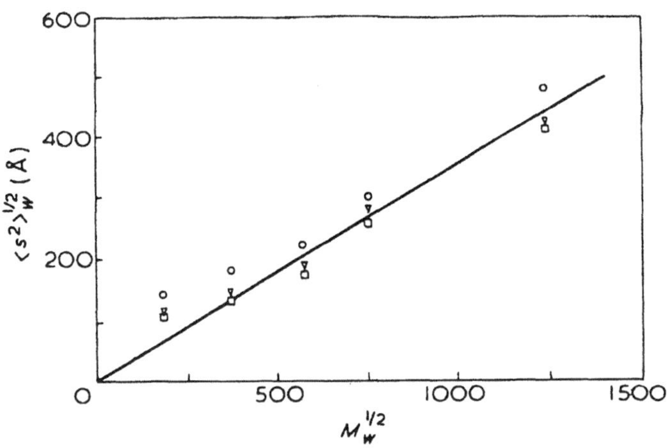

Figure 2.18 Molecular weight dependence of radius of gyration for quenched (squares), annealed (triangles), and isothermally crystallized polypropylene (circles). Reprinted from D. G. H. Ballard *et al.* (1978) with permission from the publisher, Elsevier Science.

Table 2.4 Comparison of molecular dimension in molten and crystallized polymers. Reprinted from G. D. Wignall *et al.* (1982) with permission from the author and the publisher, John Wiley & Sons, Inc.

Polymer	*Method of Crystallization*	$R_g/M_W^{1/2}$ *Melt*	[Å/g/mol)$^{1/2}$] *Crystallized*
Polyethylene	Rapidly quenched from melt	0.46	0.46
Polypropylene	Rapidly quenched	0.35	0.34
	Isothermally crystallized at 139 °C	0.35	0.38
	Rapidly quenched from melt and subsequently annealed at 137 °C	0.35	0.36
Polyethylene oxide	Slowly cooled	0.42	0.52
Isotactic polystyrene	Crystallized at 140 °C (5 h)	0.26–0.28[a]	0.24–0.27
	Crystallized at 140 °C (5 h) then 180 °C (50 min)		0.26
	Crystallized at 200 °C (1 h)	0.22[a]	0.24–0.29

[a]Dimensions in the melt were not available. The values quoted are for amorphous samples (with atactic polystyrene matrices) annealed in the same way as the crystalline material (with isotactic polystyrene matrices). Reproduced from Wignall *et al.*, ref. [86].

found to be proportional to $M_W^{1/2}$ except in the low-M_W region. Table 2.4 shows the relation of $R_g/M_W^{1/2}$ for typical crystalline polymers [86]. This clearly indicates that R_g does not change noticeably in response to crystallization.

Solution-grown polymer crystals have a much smaller R_g compared with those crystallized from their melt. Sadler and coworkers [87] studied molecular trajectory in polyethylene crystallized from xylene solution. It was found that in solution-grown crystals the chains form sheets consistent with adjacent-fold reentry [87]. This is why the molecular dimension increases much less steeply than with $R_g \sim M_W^{1/2}$ (e.g., $R_g \sim M_W^{1/10}$), as shown in Figure 2.19 [88]. Analysis of the anisotropy of polyethylene single crystals was also conducted by SANS, from which the projections of the R_g to the in-plane and parallel to the crystal lamellar normal (R_{xy} and R_z, respectively) were obtained [89], where

$$R_{xy} \equiv \left(\frac{R_x^2}{2} + \frac{R_y^2}{2} \right)^{1/2} \tag{161b}$$

Table 2.5 shows the results of R_{xy} and R_z as well as the lamellar thickness d for polyethylene of $M_W = 90 \times 10^3$. The value of R_z for solution-grown crystals (upper) is significantly less than R_{xy}, whereas that for the other (heat treated; lower) is greater. The former finding indicates that each molecule is in one lamella (R_z

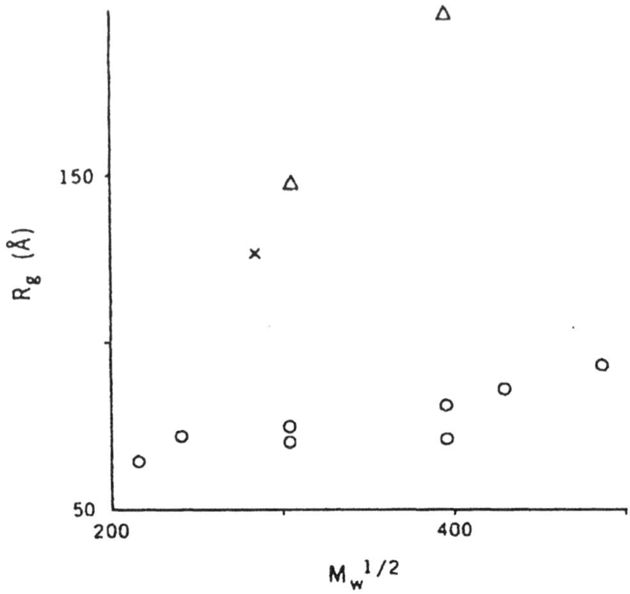

Figure 2.19 Radius of gyration for solution-grown (circles) PE, PE crystallized from its melt (triangles), and molten PE (cross) as a function of the square root of molecular weight. Reprinted from D. M. Sadler *et al.* (1979) with permission from the author.

predicted as 30 Å). The increase in R_z with the heat treating is explained as a result of lamella thickening. More detailed discussion is given by Sadler [12]. These examples eloquently demonstrate the usefulness of the SANS method for polymer sciences.

Table 2.5 Values derived for the average in-plane dimensions (R_{xy}) and dimensions along the lamella normals (R_z). Reprinted from D. N. Sadler (1983) with permission from the publisher, International Union of Crystallography.

	Lamellar Thickness (Å)	R_{xy}(Å) $= \left(\frac{R_x^2}{2} + \frac{R_y^2}{2}\right)^{1/2}$	R_z(Å)	*Uncertainty in* R_z/R_{xy} *(%)*
Crystals grown at 343 K	105	53.5	25.8	17
Crystals annealed at 396 K	210	55.6	81.3	3

Reproduced from Sadler, ref. [89].

2.5.1.3 Polymer Chains in a Confined Space

As demonstrated in the previous subsection, most flexible polymer chains behave as Gaussian chains in bulk as well as in a Θ solvent. This is a consequence of random-walk statistics. The random-walk statistics for polymer chains are usually considered without any spatial restriction. Now we discuss the case of block copolymers in which the constituent block polymer chains obey random-walk statistics in a confined space (i.e., in the microdomain). This case is extensively discussed by Helfand and coworkers [90–92]. It is now well known that the domain spacing D scales with molecular weight with a power of $2/3$ in a strong segregation regime, i.e.,

$$D \sim M^{2/3} \tag{162}$$

This relation has been experimentally verified by SAXS [93]. The same conclusion was also obtained by SANS [94].

The SANS technique is useful to investigate the anisotropy of block chains in their microdomains. By conducting a SANS experiment on a block copolymer system in which small amounts of labeled block copolymers are added to an ordinary block copolymer matrix, one can obtain information about the anisotropy of the R_g of the labeled block chains in the confined microdomains. However, because the difference in the scattering lengths of the constituent block chains is large enough to give rise to so-called domain scattering, it is necessary to subtract this contribution to obtain single-chain scattering. For this purpose, the contrast-matching technique, as discussed in Section 2.4.2, was employed to suppress the domain scattering and to extract the chain scattering [95, 96]. Hadziioannou *et al.* [97] measured the R_g of block chains in poly(styrene-*block*-isoprene) (SI block copolymers) parallel to the lamellar interface and found that block chains are rather contracted. Hasegawa *et al.* [95] reported that block chains are extended along the lamellar normal. The study of anisotropy in block copolymers was extended by Matsushita and coworkers to poly(styrene-*block*-2-vinyl pyridine) (SV) [96] and poly(isoprene-*block*-styrene-2-vinylpyridine) (IPS) [98].

However, even in the case in which the contrast-matching method gives rise to zero contrast between each microdomain, one should note that there is still a remaining contribution of domain scattering that causes intrinsically incomplete contrast matching in the microdomain systems. This is simply because deuterated and undeuterated block chains mixed for the contrast matching give rise to the scattering from the concentration fluctuations of deuterated and undeuterated block chains bounded in the microdomains under considerations. This problem was elaborated by Koizumi *et al.* [99].

2.5.2 MISCIBILITY OF POLYMER BLENDS

Often, the thermodynamics of polymer mixtures is discussed in the context of Flory's interaction parameter χ. Among the many ways to examine the thermodynamics of polymer mixtures, the SANS technique is one of the most powerful methods to examine the miscibility of polymer blends because it offers direct ways to measure the χ parameter. In principle, other scattering methods such as light scattering (LS and SAXS) works in exactly the same way as SANS, however, SANS is a much more powerful tool to measure χ. This is primarily because precise measurement of very weak scattering of polymer blends in the single-phase state is difficult in LS and SAXS: Dusts and impurities are on the same order of the wavelength of light; hence, they scatter light and dominate the scattering of interest. Thorough purification of the sample and sample container is required for accurate measurements. On the other hand, the spatial scale measurable in SANS is much smaller (i.e., the q-range is larger than that of LS), and existence of those impurities would not interfere the scattering. In addition, as mentioned earlier in Section 2.3.1, the deuterium labeling of one of the constituents significantly enhances the scattering contrast for SANS, which makes accurate measurements of the single-phase scattering in SANS even easier. At the top of these experimental advantages of SANS, the theoretical development of the RPA theory and formulation of the RPA theory for multicomponent (polymeric) systems even accelerates the use of SANS.

Isotope mixtures are assumed to form ideal solutions (see Sections 2.4.4 and 2.5.2.1). This assumption has been taken for granted and has become almost "common sense." As explored by Bates and his coworkers [73, 100], an isotope mixture of deuterated polybutadiene (dPB) and polybutadiene (PB) exhibits an upper critical solution temperature (UCST) type of phase diagram. Figure 2.20 shows the SANS profiles at three different temperatures. The scattering intensity at low q increases as the temperature decreases. The solid lines in the figure show the best fit of the theoretical RPA equation (see also Sections 2.4.4 and Eq. 138) to experimental data, from which the χ parameters at corresponding temperatures were obtained. The χ parameters were positive. Bates *et al.* explained this phenomenon by considering a small difference in segmental volume that induces a small-segment polarizability difference between the constituents, as predicted by Buckingham and Hentschel [72]. This is a typical example demonstrating that even the small interaction between the segment strongly affects the phase behavior of high polymers, because the segmental interaction is multiplied by a number of segmental units in a chain ("polymer effect"). Besides the dPB/PB mixture, the isotope effect was experimentally observed in other similar kinds of systems [101, 65].

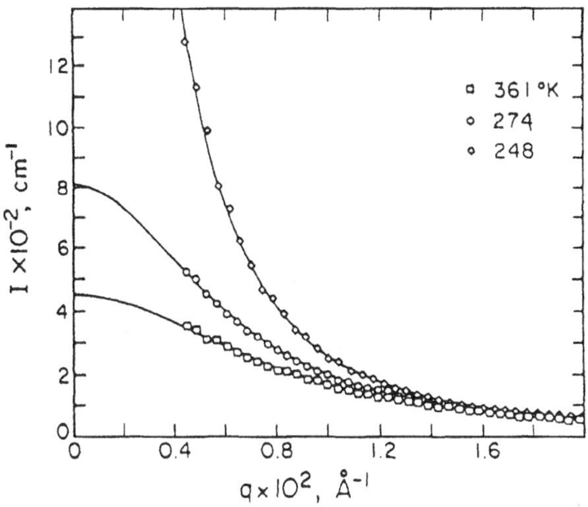

Figure 2.20 Small-angle neutron scattering from a mixture of protonated and perdeuterated 1,4-polybutadiene at the critical composition. The solid curves represent the predicted scattering function (Eq. (136)) by adjusting the interaction parameter χ. Reprinted from Bates *et al.* (1985) with permission from the author and the publisher, American Physical Society.

Sakurai *et al.* also studied the dPB/PB blends from another point of view. They noted that the miscibility of the dPB/PB blends was very sensitive to their geometric isomers (often called "microstructure"), i.e., how the deuterated butadiene or butadiene monomers were chemically connected into chains during the polymerization [65]. Sakurai *et al.* reported negative values of χ in their SANS experiments [65], whereas Bates *et al.* obtained only positive χ [73]. This apparent discrepancy between the results of Bates *et al.* and those of Sakurai *et al.* has been interpreted in terms of "random copolymer effects," namely, dPB and PB are considered to be random copolymers of different amounts of 1,2 and 1,4 geometric isomers. According to the theory, "fundamental" interaction parameters describe bare interaction between each geometric isomer. The fundamental interaction parameters between the geometric isomers were estimated to be positive, which is consistent with the results of Bates *et al.* However, this implies that the net interaction between dPB and PB can be greatly affected by the amounts of the isomers. Moreover, Jinnai *et al.* later demonstrated that the dPB/PB blends showed not only a UCST phase diagram but also a lower critical solution temperature (LCST) type of phase diagram (see Figure 2.21), depending on their microstructures [102]. Similar microstructure effects were discussed in relation to the blends of dPB and

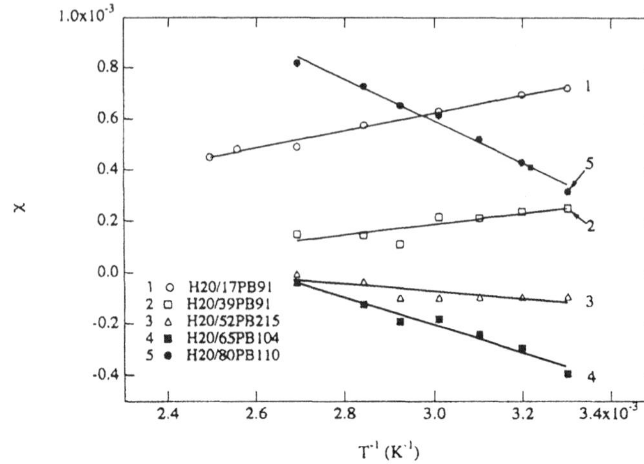

Figure 2.21 Plots of binary interaction parameters χ against reciprocal absolute temperature T^{-1} for five different deuterated (H_2O) and protonated polybutadiene blends. The sample code (e.g., 17PB91) comprises the vinyl content in percentage (first number 17 in this case), polybutadiene (PB), and the weight-averaged molecular weight in units of 1000 (the second number 91 in this case). The vinyl content of H_2O was 16%. Reprinted from H. Jinnai *et al.* (1992) with permission from the publisher, American Chemical Society.

polyisoprene (PI) [103, 104], blends of PB and polystyrene-*random*-polybutadiene (SBR) [105], and polyolefines (hydrogenated (saturated) polydiene) [106]. Along the same line, tacticity effects on the interaction parameter for PS–PVME have been examined [107].

In most of these studies to determine the miscibility of polymer blends, analysis largely depended on the RPA theory and the classical Flory–Huggins (FH) theory for a free-energy expression. In the framework of the FH theory, the interaction parameter is purely energetic in origin, i.e., $\chi \approx 1/T$ (T is an absolute temperature). However, many experimental results clearly show that χ is expressed as $\chi \approx A + B/T$. In other words, χ contains not only the energetic term but also the entropic term independent of temperature. Moreover, as described in the next paragraph, it has been found that χ also depends on composition. Therefore, reasonable modifications to the FH theory are required for more accurate evaluation of χ.

Composition dependence of the interaction parameter $\chi(\phi)$ has also been exclusively studied by SANS [62, 108–111, 101, 65]. Here ϕ denotes the volume fraction of one of the constituent polymers in binary mixtures. Some mixtures

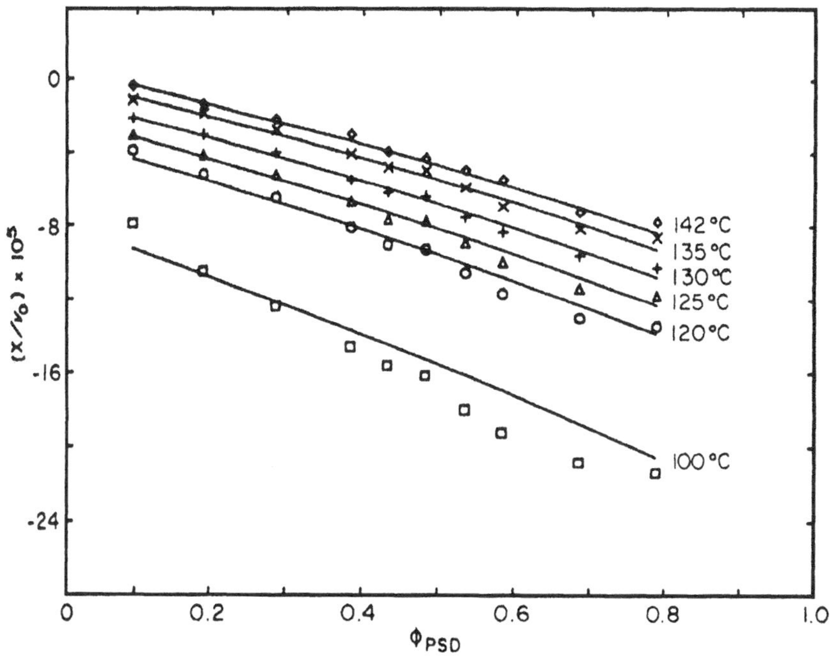

Figure 2.22 Interaction parameter χ/v_0 (v_0 is the reference volume) of a mixture of deuterated polystyrene and poly(vinyl methyl ether) plotted against ϕ at various temperatures. The solid lines are a nonlinear lease-square fit to the data. Reprinted from C. C. Han *et al.* (1988) with permission from the author and the publisher, Elsevier Science.

show a linear dependence, whereas the others have a parabolic dependence in $\chi(\phi)$. Figure 2.22 shows a mixture of such composition dependence in deuterated polystyrene (dPS) and poly(vinyl methyl ether) (PVME) [108]. The linear χ dependence of the blends that show an LCST phase behavior (e.g., dPS–PVME blend) may be understood on the basis of equation-of-state theories in which a large asymmetry in the free volumes is taken into account [112]. An extended version of the lattice theory, a lattice cluster theory, developed by Dudowicz and Freed [113] reproduced the experimentally obtained composition dependence. The parabolic-type composition dependence of χ was reported in some isotopic mixtures [109, 111]. The parabolic dependence with $\partial^2\chi(\phi)/\partial\phi^2 > 0$ has been observed in isotopic mixtures of poly(ethylethylene) (PEE) and poly(vinylethylene) (PVE), which showed a good agreement with a theory developed by Muthukumar [114]. Beside the Muthukumar's theory, the compressible RPA type of theory, such as the one developed by Bidkar and Sanchez [115], or the PRISM (Polymer Reference

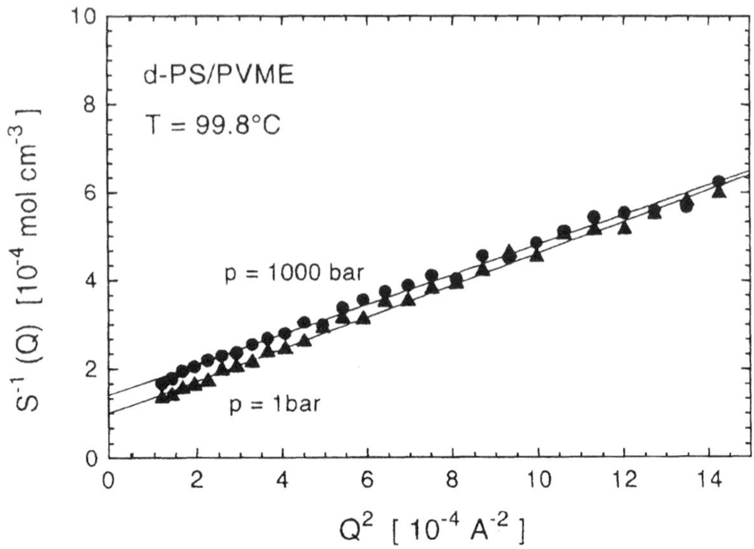

Figure 2.23 Typical Zimm plots of small-angle neutron scattering data of deuterated polystyrene and poly(vinyl methyl ether) blend: pressure (p) $=$ 1 bar and p $=$ 1000 bar and temperature 99.8 °C. Q is the wave number. Reprinted from S. Janssen *et al.* (1993) with permission from the author and the publisher, American Chemical Society.

Interaction Site Model) theory by Schwaeizer and Curro [116] may account for the parabolic composition dependence.

Besides the temperature and composition dependences, pressure is another important thermodynamic variable. More studies have recently been directed at examining the pressure dependence of χ [117, 118]. Figure 2.23 shows a Zimm plot of the dPS–PVME blend. The forward scattering, i.e., scattering intensity, at $q = 0$ for $p = 1000$ (p is a pressure) bar is larger than that for $p = 1$ bar, demonstrating that the miscibility increases as p increases. Hammouda and Bauer found that the enthalpic part, as well as the entropic part in the interaction parameter, showed pressure dependence. The compressible RPA theory of Bidkar and Sanchez [115] was used to analyze their dPS–PVME system in which Hammouda and Benmouna found that the free volume fraction decreases with pressure and increases with temperature. However, the pressure dependence on miscibility is relatively a new area that should be explored in the future.

Apart from the preceding interaction parameter determination in various binary (polymer) systems as a function of temperature, composition, and pressure, many studies have dealt with the critical behavior of binary mixtures. The main issue in these studies is to discuss the universality class of the polymer mixture and the extent to which the mean field approximation is correct. In the vicinity of the critical point, crossover from the meanfield to three-dimensional Ising behavior was observed in many systems [119–123]. The crossover from the mean field to 3-D Ising behavior is characterized by the Ginzburg number G_i, which quantitatively defines where the mean field behavior can be expected [13, 124]. Bates *et al.* first proposed the functional form of G_i that simply depends on the molecular weight of polymers, i.e. $G_i \approx z^{-1}$ [121]. Later, however, it was shown experimentally and theoretically [125] that G_i depends on the segment structure [126] and pressure [127]. An alternative definition of G_i was proposed by Hair *et al.* [122]. A realistic crossover function to describe temperature variation of zero-angle scattering or the correlation length was also introduced [116, 128].

2.5.3 PHASE TRANSITION

2.5.3.1 Polymer Blends

In the past decades, several studies have been carried out to examine the time evolution of self-assembling structures during phase separation, particularly during spinodal decomposition (SD) [129–132, 56]. Dynamics and pattern formation in systems undergoing SD are the subject of many theoretical [133–136], experimental [137, 138], and computational [139–144] investigations as a fascinating example of nonlinear, nonequilibrium phenomena. An advantage of using polymers comes from their unique features of a very large microscopic length and a very long characteristic time [137]. These features facilitate *in situ* and real-time observation of the self-assembling process with better accuracy compared with small molecules or atomic systems such as simple fluid mixtures [145, 146], metallic alloys [147], inorganic glasses [148], etc.

Time-resolved light scattering (LS) is commonly used to investigate the self-assembling process of SD. The results have shown that the coarsening process of polymer mixtures with a critical composition and with nearly symmetric polymers (having nearly equal molecular weights and viscosities) can be classified into at least three stages: (1) early stage, (2) intermediate stage, and (3) late stage [56]. In the early stage of SD, it is said that the time evolution of concentration fluctuations is well described by the linearized theory formulated by Cahn [149, 150] that predicts the time evolution of the scattering maximum and its intensity. However, LS has no power to explore the region of high q where the maximum is expected to appear in the very beginning of SD, except for the very shallow quench case in which the scattering maximum could be inside the q range of LS. Thus, even though it is important to actually see the growth of the scattering maximum to critically test the validity of Chan's theory or Cahn–Hilliard–Cook (CHC) theory [151], one can only see the small q edge of the scattering maximum in LS. Note here that the CHC theory includes the effect of the thermal noise, whereas the CH theory does not. Because the q range covered by SANS is higher than that of light scattering, it is an ideal tool to study the early stage of SD. In addition, a polymer blend is a very good system in the sense that it undergoes very slow SD, the typical characteristic time t_c of which is of the order of 1000 s. Several time-resolved SANS (or SAXS) measurements have been carried out to investigate the very beginning of SD and to examine whether the CHC theory is capable of accurately predicting the early-stage SD [152–154, 101, 155–157]. We note that the effect of thermal fluctuation becomes increasingly important with increasing q.

Figure 2.24 shows one of the examples of such time-resolved measurements of a dPB–PB mixture [157]. Jinnai *et al.* concluded that the CHC theory can satisfactorily describe the early-stage SD in the deep-quench region. This is in accordance with the results by Okada and Han [158] in which they concluded that the effect of thermal concentration fluctuations is considerable even at the small q covered by their LS experiments at very shallow quenches. Jinnai *et al.* also examined the effect of initial thermal concentrations, which are the thermal fluctuations before the system was quenched into the spinodal region, to see how the initial thermal concentration fluctuations in the single phase is transformed into the concentration fluctuations. Although the effect of the initial concentration fluctuations on SD is not incorporated in the CHC theory, this fluctuations experimentally influenced SD in such a way that the time-evolution of SANS profile strongly depended on the initial temperature in the single phase while the phase separation temperature was kept constant.

Some studies have addressed the wave-number dependence of the Onsager kinetic coefficient [155, 157]. The coefficient is taken to be a wave–number-independent quantity in the small q range covered by LS, whereas it becomes significantly dependent on q in the high q range. The measured Onsager coefficient $\Lambda(q)$ was compared with theories predicted by Pincus [159] and Binder [124]. It seems that the Pincus-type q dependence fits slightly better than the other, although R_g may have to be adjusted to reproduce data for both theories [160]. The R_g values thus estimated are generally greater than R_g calculated from molecular weight. The discrepancy was accounted for in terms of a viscoelastic effect on $\Lambda(q)$ [161, 162].

Small-angle neutron scattering has been used to study the intermediate and late stages of SD of a polymer mixture [163, 164]. Jinnai *et al.* [163] showed that there

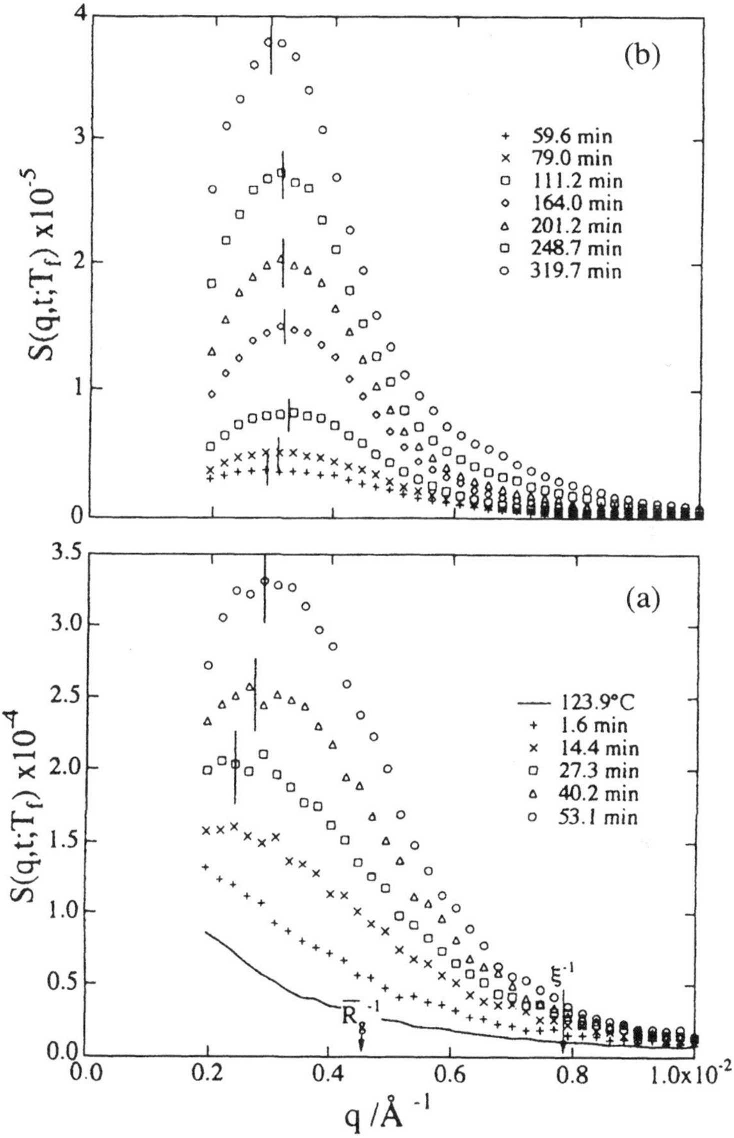

Figure 2.24 Time evolution of small-angle neutron scattering profiles for a critical deuterated and protonated polybutadiene mixture after a temperature drop (T-drop) from 123.9 °C (in the single-phase state) to −7.5 °C (inside the spinodal phase boundary). (a) The early stage SD; (b) the intermidiate stage SD. Vertical bars display the peak positions at different times. Reprinted from H. Jinnai *et al.* (1993) with permission from the publisher, American Institute of Physics.

is q region in the time-dependent scattering profiles where q^{-2} was observed, and the absolute intensity levels of the q^{-2} region decreased with increasing time until these levels got a time-independent constant level. According to the RPA theory, the intensity level of the q^{-2} intensity is a function of the composition of the two components within coexisting phases and R_g. From Eq. (159), the scattering due to thermally induced local composition fluctuations occurring at a length scale much shorter than R_g or at q much larger than R_g^{-1}, $I_T^e(q)$, is given by

$$I_T^e(q) \equiv \lim_{q\to\infty}[I(q)/\{(b_1 - b_2)^2 N_A/\nu_0\}]$$
$$= [12\phi_1(1-\phi_1)/\langle a^2/\nu\rangle]q^{-2} \tag{163}$$

where

$$\langle a^2/\nu\rangle \equiv (a_1^2/\nu_1)(1-\phi_1) + (a_2^2/\nu_2)\phi_1 \quad \text{and} \tag{164}$$

a_1 and a_2 are, respectively, segment lengths of species 1 and 2. Therefore, the net scattering $I_T(q)$ from two coexisting domains having volume fractions of X_1 and $X_2(X_1 + X_2 = 1)$ and compositions (volume fractions of 1 in each domains) equal to ϕ_1^1 and ϕ_1^2 is given by

$$I_T(q) = X_1 I_{T1}(q) + X_2 I_{T2}(q), \tag{165}$$

where the scattering due to thermally induced composition fluctuations from each domain, (i.e., the internal scattering) can be calculated from Eq. (163) as

$$I_{TK}(q) = \{12\phi_1^K[1-\phi_1^K]/\langle a^2/\nu\rangle\}q^{-2} \qquad (K = 1 \text{ or } 2) \tag{166}$$

Thus, SANS offers direct determination of composition difference between the two coexisting phase-separated domains $\Delta\Phi(t)$. Here $\Delta\Phi(t) \equiv |\phi_1^1 - \phi_1^2|$. Figure 2.25 demonstrates the time dependence of $\Delta\Phi(t)$. The expression t_{cr} represents the crossover time from the intermediate to late-stage SD because $\Delta\Phi(t)$ gets constant (i.e., the amplitude of the concentration fluctuations reach equilibrium value at t_{cr}). It is also found that the time evolution of the difference of $\Delta\Phi(t)$ from the equilibrium value $\Delta\Phi_e$ (i.e., $\delta\Delta\Phi(t) \equiv \Delta\Phi_e - \Delta\Phi(t)$) obeys a simple linear relaxation equation $-d\delta\Delta\Phi(t)/dt \cong (1/\tau)\delta\Delta\Phi(t)$.

2.5.3.2 Block Copolymer

The order–disorder transition of the block copolymers is a thermodynamic transition controlled by counterbalancing two physical factors: the energetics and the entropy. This is a common feature of all the phase transitions in atomic or small

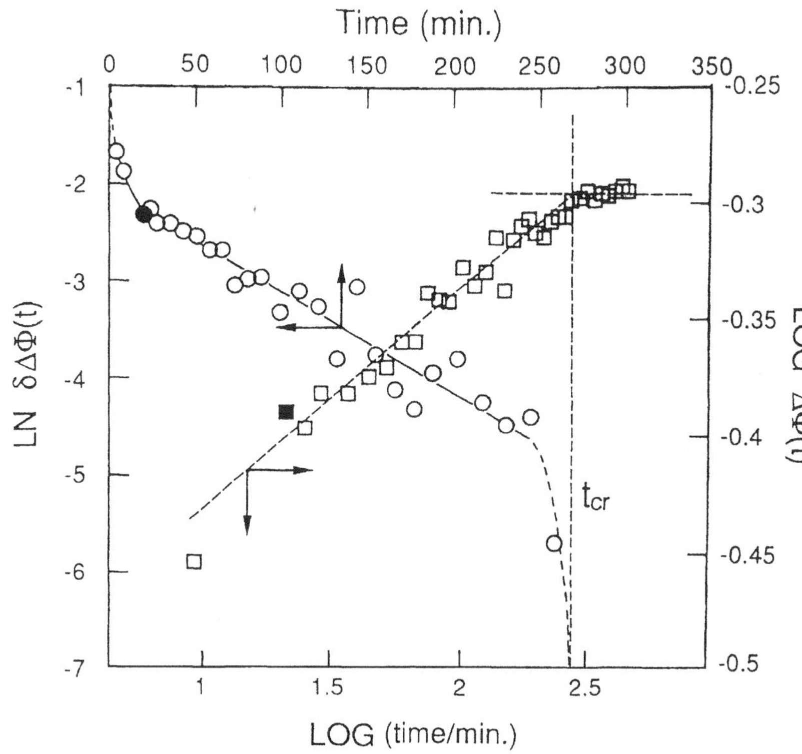

Figure 2.25 Time change of the composition difference of the two coexisting phases $\Delta\Phi(t)$ and that of $\delta\Delta\Phi(t)$, the difference of $\Delta\Phi(t)$ from the equilibrium value $\Delta\phi_e$, for a deuterated polybutadiene and polyisoprene mixture. Reprinted from H. Jinnai *et al.* (1991) with permission from the publisher, American Chemical Society.

molecular systems, such as metallic alloys, lattice gas, and ferromagnetic materials [165].

Figure 2.26 illustrates schematically the order–disorder transition in A–B diblock copolymers. In the ordered state A and B segments segregate themselves in A and B microdomains. However, because of molecular connectivity between the A and B block chains, the segregation would not result in macroscopic phase separation into two coexisting macroscopic liquid phases, as in the mixture of A and B, but would rather result in a regular periodic microdomain with a long-range order, as schematically shown in Figure 2.26(a) (e.g., alternating lamellar microdomains [166]). On the other hand, in the disordered state, A and B polymers are molecularly mixed, as depicted in Figure 2.26(b).

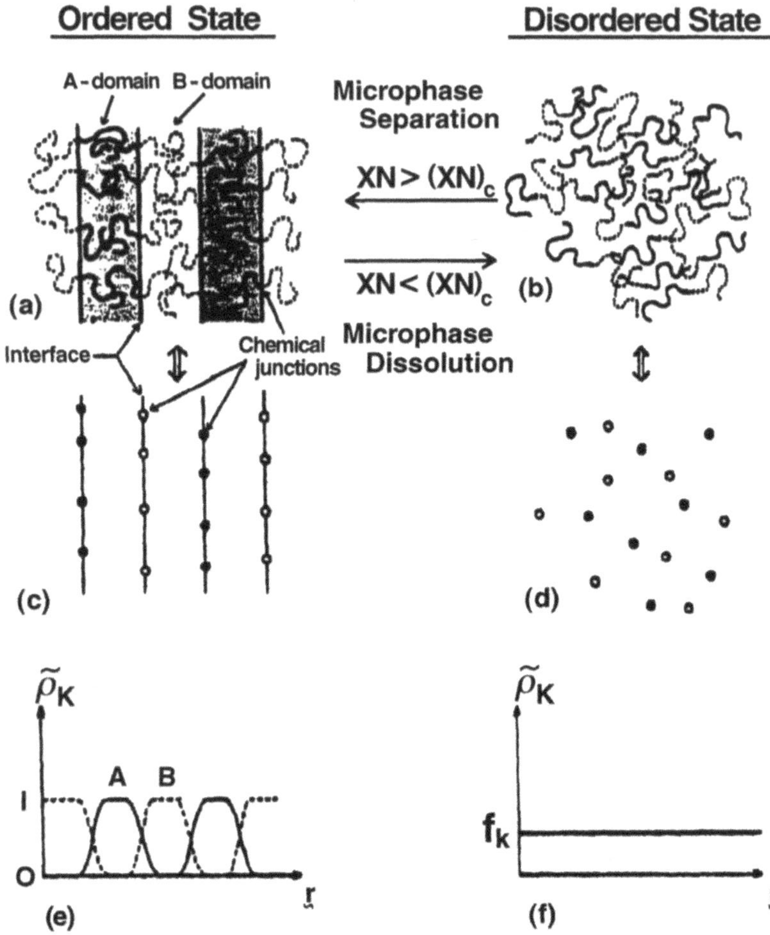

Figure 2.26 Schematic representation of the order–disorder transition of A–B diblock copolymer. (a) and (b) are molecular packing, and (c) and (d) are the spatial distributions of the chemical junctions in the ordered and disordered states, respectively.

The characteristics of the ordered state are symbolically depicted in Figure 2.26(c) where the chemical junctions between A and B block chains are confined to the narrow interfacial region drawn by a series of vertical straight lines. This region has a characteristic interface thickness of about 20 Å for the polystyrene-*block*-polyisoprene (S-I) block copolymers [167, 168, 169]. Generally, the average end-to-end vectors should orient normal to the interface. The solid and open circles differentiate the orientations of the end-to-end vectors of the block chains. The solid circles represent the junctions with their end-to-end vectors of A segments oriented right and, conversely, the open circles represent junctions with the A segments oriented left. Thus, the block copolymers in the ordered state have the liquid crystalline characteristics, although the block copolymer molecules themselves are the flexible chains without any mesogenic groups. On the other hand, in the disordered state, the chemical junctions are randomly distributed in space, and the orientations of the end-to-end vectors for A and B block chains are also random, as schematically illustrated in Figure 2.26(d). Figure 2.26(e) and 2.26(f) represent time-averaged reduced segmental density profiles $\tilde{\rho}_K$ ($K =$ A or B) in the ordered and disordered state, respectively. The expression $\tilde{\rho}_K$ is defined as ρ_K/ρ_{KO}, where ρ_{KO} is the time and space-averaged segmental density of pure K-homopolymer. It should be noted that the systems having the spherical and cylindrical microdomains in the ordered state have essentially identical characteristics to those described above. The same characteristics are seen for the recently found ordered bicontinuous phase [170, 171, 172] as well.

The term order–disorder transition (ODT) has the same meaning as the microphase separation and dissolution conventionally used in the block copolymer literature or the microphase separation transition (MST) designated by Leibler [63]. In case of A–B diblock copolymers, the conventional definition of MST may be based on the following concept: The number of components n is equal to 2, the number of independent variables f is also equal to 2 (e.g., temperature T and pressure P). Then, in a phase-separated state, we have the number of phases a equal to 2 according to the Gibb's phase rule $f = n + 2 - a$. The two phases are the A-rich and B-rich microdomains. This picture holds if our length scale of observation is smaller than R_g or smaller than the size of each domain D_A or D_B. However, for a large length scale, much larger than the domain repeat distance D of the microdomains, we have the homogeneous structure that has a locally periodic variation of compositions inherent to the segregated A and B microdomains or one of which has a thermally induced composition variation inherent to a single-phase state. In other words, in this case our system has $n = 1$, because A and B are linked, and $f = 2$; hence, $a = 1$. Thus, we have physically a homogeneous state or phase with a compositional order (hence designated ordered state or ordered phase) in the segregation regime shown in the left half of Figure 2.26 or with a compositional disorder (hence designated disordered state or disordered phase) in the single-phase regime shown in the right half of Figure 2.26. To our knowledge the term order–disorder transition appeared first in [173] in the block copolymer literature.

Meier [174] and Helfand [90, 175, 91, 92] are the pioneers who developed theories of the microphase transition. Leibler [63] established a firm theoretical basis for the MST of block copolymers on the basis of the Landau-type mean-field theory. His work stimulated experimental investigations of MST or ODT [176, 173, 177, 178, 179, 180, 181, 182, 183]. The experimental results obtained in turn provoked further theoretical explorations. The fact that the block copolymer belongs to a new universality class other than the kinetic Ising universality class, [165, 184] has also attracted theoretical interests.

The transition had been thought to be a gradual transition, and the experimental data obtained by SAXS or SANS methods had been analyzed [176, 173, 177, 178, 179, 180, 181, 182, 183] on the basis of Leibler's Landau-type mean-field theory. The Leibler's theory [63] predicts the following: (1) The SAXS or SANS profile of block copolymer A–B shows a scattering maximum even in the disordered (single-phase) state, and the maximum intensity $I_{\rm m}$ and the wave number $q_{\rm m}$ at the maximum further satisfy the following points: (1) inverse of the maximum intensity $I_{\rm m}^{-1}$ linearly decreases with inverse of absolute temperature T^{1}, (2) The characteristic wavelength D of the composition (or concentration) fluctuations as estimated from $q_{\rm m}$

$$D = 2\pi/q_{\rm m} \tag{167}$$

has the same temperature dependence as that of the unperturbed chain dimension $R_{\rm g0}(T)$. Most of the experimental results were shown to obey this prediction approximately [176, 173, 177, 178, 179, 180, 181, 182, 183], and the deviations of $I_{\rm m}^{-1}$ and D from the predictions were taken to be the onset of ODT or MST [176, 173, 177, 180, 181, 182, 183], as schematically shown in Figure 2.27(a). Here the ODT temperature thus determined will be defined as the mean-field ODT temperature $T_{\rm ODT,MF}$. The system is thought to be in the ordered state (microphase-separated state) at $T < T_{\rm ODT,MF}$ and in the disordered state at $T > T_{\rm ODT,MF}$. In Figure 2.27(a), $D_{\rm r}(T)$ shows the temperature dependence of D as predicted by the mean-field theory

$$D_{\rm r}(T) \equiv D(T_{\rm r})R_{\rm g0}(T)/R_{\rm g0}(T) \tag{168}$$

where $T_{\rm r}$ is a given reference temperature in the disordered state. At $T < T_{\rm ODT}$, $D_{\rm r}(T)$ with $1/T$ deviates from Eq. (167) and has a greater temperature dependence.

However, a more elaborate theory that takes into account the effect of thermal noise (the Brazovskii effect) [185] on the ODT has been proposed by Fredrickson and Helfand [64], a weekly first-order nature of the transition has been illuminated, and the temperature dependence of $I_{\rm m}$ and $q_{\rm m}$ (or D) has been reexamined more carefully. The experimental results obtained by Bates *et al.* [186] for poly(ethyrene propylene)-*block*-poly(ethylethylene) (PEP–PEE) showed the following deviations from the mean-field prediction [63]: (1) the plot $I_{\rm m}^{-1}$

versus T^{-1} in the disordered state is not linear but has a curvature, and (2) shows a small discontinuous change at ODT temperature, as schematically shown in Figure 2.27(b). We shall define this temperature as T_{ODT}. Because the discontinuity of I_m^{-1} at T_{ODT} is small, it might be overlooked with the scattering data alone. However, the discontinuity was supported by T_{ODT} separately determined from the low-frequency rheological behavior as a function of temperature. The experimental discontinuity was further confirmed by the Fredrickson–Helfand theory. Here we should note a surprising report that D had the same temperature dependence as $R_{g0}(T)$ below and above T_{ODT} in the temperature range covered. This point has never been confirmed for S–I and polystyrene-*block*-polybutadiene (S–B) [187].

For S–I and S–B, this discontinuous change in I_m^{-1} at T_{ODT} has been overlooked for a long time and confirmed only recently [188–191]. In particular, high-temperature resolution SAXS experiments [190] unequivocally identified T_{ODT} with accuracy less than 1 °C, though this temperature T_{ODT} generally depends on the heating or cooling rate, as is typical of the first-order phase transition.

Further elaborate experimental works [187] reveal a unified view, as shown in Figure 2.27(c) on the temperature dependence of I_m^{-1} and D across the ODT temperature: We can "operationally" define a crossover temperature T_{MF} such that, at $T > T_{MF}$ in the disordered state, the system approximately obeys the mean-field behavior, as found in the earlier works (Figure 2.27(a)); namely, $T_{ODT,MF}$ in Figure 2.27(a) corresponds to T_{MF} defined in this way but not to T_{ODT} itself; at $T_{ODT} < T < T_{MF}$, the system is still in the disordered state, but the thermal noise effects strongly affect the concentration fluctuations in the disordered state so that the deviations (in I_m^{-1} and D) from the mean-field behavior become significant (Brazovskii's regime); at the ODT temperature, discontinuous changes in I_m^{-1} occur, as shown in Figure 2.27(b), and the system is in the ordered state at $T < T_{ODT}$.

It should be noted that in the Brazovskii regime, I_m^{-1} (solid line) is much larger than that predicted by the mean-field theory (broken line). This indicates that the scattered intensity I_m and hence the concentration fluctuations in real systems are much suppressed compared with those predicted by the mean-field theory. The T_{ODT} in Figure 2.27(c) corresponds to the T_{ODT} in Figure 2.27(b). It should be noted that D has a temperature dependence much greater than that of $D_r(T)$ or $R_{g0}(T)$ in contrast to the case of Figure 2.27(b), although D shows little change at T_{ODT}, as in the case of Figure 2.27(b). It is expected that the discontinuity and the temperature gap $\Delta T \equiv T_{MF} - T_{ODT}$ decrease with increasing z, degree of polymerization of the block copolymer [187, 64]. It is needless to say that, from the theoretical point of view, there is no temperature T_{MF} because the mean-field behavior is achieved only asymptotically. However, T_{MF} was operationally defined

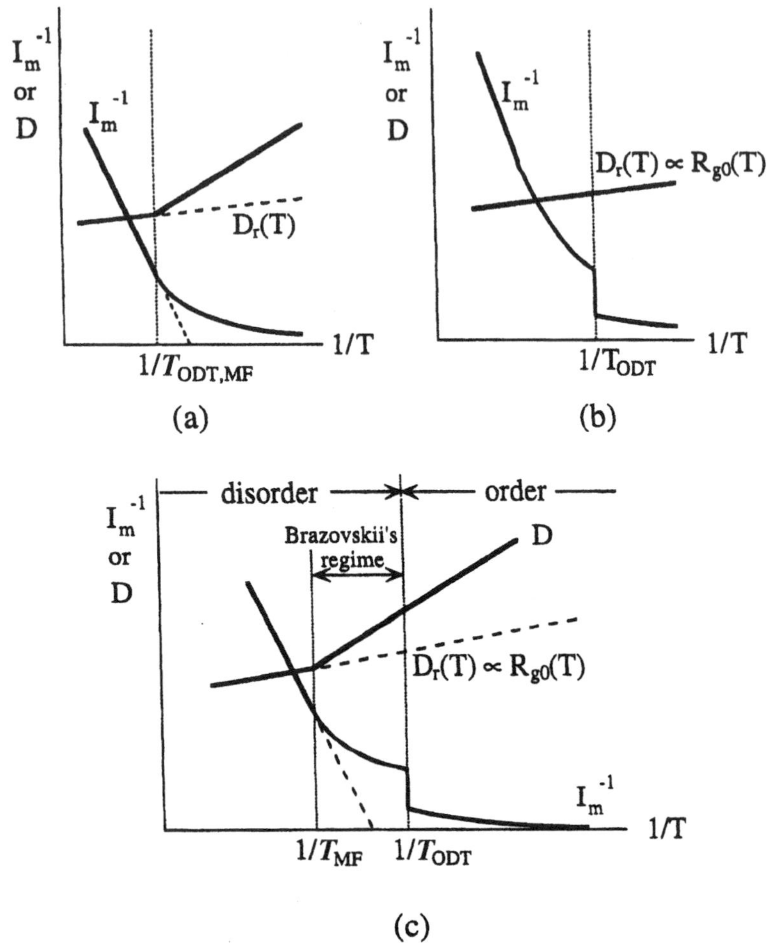

Figure 2.27 Temperature dependence of characteristic scattering parameters, i.e., the inverse maximum intensity I_m^{-1} and the characteristic length D determined from the scattering vector $\mathbf{q}_m$ at the first-order scattering maximum for A–B or an A–B–A diblock copolymer having UCST-type phase behavior: (a) an old view based upon mean-field picture, (b) a new view, and (c) a unified view. The value of D changes with q/T more smoothly across $1/T_{ODT,MF}$ or $1/T_{MF}$ than the trend schematically shown in part (a) or (c).

in such a way that above T_{MF} we can practically analyze our experimental data with good accuracy based on the mean-field theory.

The ordering process below ODT or MST (i.e., whether it involves spinodal decomposition (SD) or nucleation growth (NG) and how it is different from the conventional SD or NG found in mixtures of homopolymers A and B) is by and large left unexplored and deserves future work [192, 193]. If the system has both OOT and ODT, we should be able to distinguish the two transitions. The clear distinction between them is not easy and also deserves future work. All these problems offer important research items for better understanding of the morphology of pure block copolymers. The dynamics of OOT and ODT should be a key physical factor for the morphology control of block copolymers.

2.5.4 POLYMER NETWORKS AND GELS

By cross-linking bulk polymers or polymer solutions, an increase in the scattered intensity is commonly observed in the q-range of $q < 1/\xi$, where ξ is the correlation length [194]. This excess scattering is related to reorganization of the polymer structure and is called "cross-link(ing) inhomogeneity (or heterogeneity)" [195]. The appearance of cross-link inhomogeneity is due to a freezing-in of the topological structure of network heterogeneities by a heterogeneous cross-linking reaction. On the other hand, network polymers also have thermal fluctuations if the temperature of the network is above its glass transition temperature (i.e., rubbers) or the network is immersed in a solvent (i.e., gels). Therefore, those network polymers have both static inhomogeneities and thermal density (or concentration) fluctuations. Because both of them contribute to scattering, the scattering function for network polymers consists of the two contributions. Thermal fluctuations are well treated in the context of statistical thermodynamics and are similar to the concentration fluctuations in polymer solutions. However, formation of the static inhomogeneities is very sophisticated and is dependent on the method and the condition of network formation. Figure 2.28 shows log–log plots of SANS intensity curves for (a) polystyrene (PS) solutions, (b) end-linked PS networks prepared from the same PS precursors, and (c) statistically cross-linked PS networks, respectively. The solvent was toluene, and all of the polymer volume fractions were the same to be $\phi = 0.056$. It is clear from Figure 2.28 that, by introducing cross-links, the scattered intensity increases at low q's, whereas at high q's the intensity is the same irrespective of the difference in the chain architecture. In the high-q region, the asymptotic behavior of $S(q)$ is scaled as $S(q) \approx q^{-5/3}$, which is predicted for self-avoiding chains in Section 2.4.1.9.

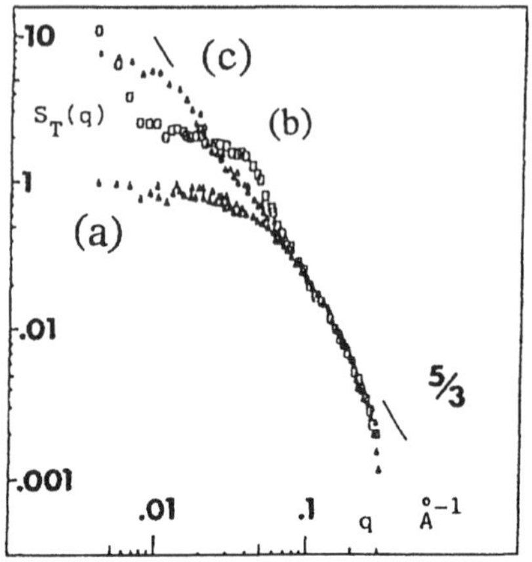

Figure 2.28 Log–log plot of the SANS intensities scattered by (a) polystyrene (PS) polymer solutions, (b) end-linked PS networks prepared from the same PS precursors, and (c) statistically cross-linked PS networks, respectively. The volume fractions of PS were the same in the three systems, i.e., $\Phi = 0.056$. The solvent was deuterated toluene [212]. Reprinted from J. Bastide *et al.* (1989) with permission from the publisher, Springer–Verlag GmbH&Co. KG.

Apart from the asymptotic behavior of the scattered intensity, no explicit expression of the structure factor for a polymer network has been proposed until recently, and only phenomenological scattering functions have been employed to understand the structure of the polymer network. The history of the scattering studies of polymer networks has been well described in reviews [194] and in books [196]. The readers who are interested in the recent progress of the scattering studies of polymer gels may consult the review by Shibayama, who discusses how to decompose the static inhomogeneity from the thermal fluctuations [197].

Regarding experimental aspects, unlike other polymeric systems, there are several labeling techniques for polymer networks and gels. Figure 2.29 schematically shows the labeling methods for gels: (a) the labeled cross-links, (b) partly labeled networks, (c) labeled path, and (d) completely labeled networks. The labeled cross-link method is useful to study the distribution of cross-links. Both of the partly labeled networks and the labeled path are used to study changes of polymer chain conformation by stretching, swelling, or both. The difference lies in the methods of preparation. A partly labeled network is prepared by end-cross-linking of linear polymer chains having functional groups at their ends, some of which are labeled (deuterated). In the case of the labeled path, an isotope mixture of linear polymer

Labeling methods

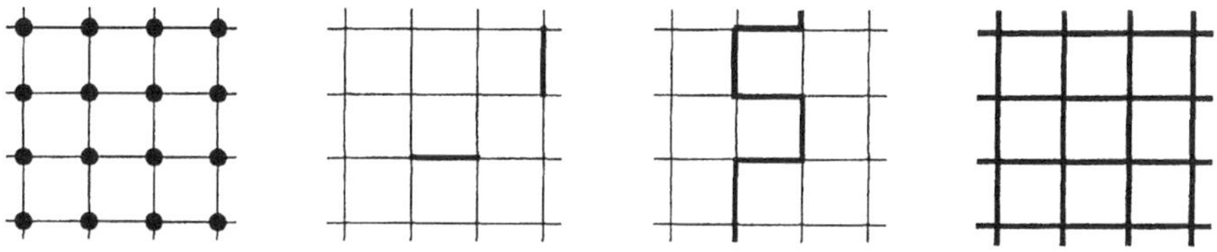

Labeling cross-links (a) | Partly labeled networks (b) | Labeled path (c) | Completely labeled networks (d)

Figure 2.29 Schematic representation of the labeling methods.

chains is randomly cross-linked chemically or by radiation. The completely labeled network method is widely used for investigating polymer gels in which the neutron contrast can easily be introduced by using a deuterated solvent.

2.5.4.1 Spatial Inhomogeneities and Thermal Fluctuations

Historically, the structure factor $S(q)$ or the scattered intensity $I(q)$ for polymer networks has often been approximated to be a sum of the two contributions of the scattered intensity from a corresponding polymer solution, $I_{\text{soln}}(q)$ and that of the excess scattering, $I_{\text{ex}}(q)$:

$$I(q) = I_{\text{soln}}(q) + I_{\text{ex}}(q) \tag{169}$$

For swollen gels, the polymer concentration lies typically in the semidilute regime and, in the long-wavelength limit, $I_{\text{soln}}(q)$ is given by a Lorentz function [13]

$$I_{\text{soln}}(q) = \frac{I_{\text{soln}}(0)}{1 + \xi^2 q^2} \tag{170}$$

where ξ is the thermal correlation length (or blob size). On the other hand, several functional forms of $I_{\text{ex}}(q)$ have been proposed in the literature [194]. For example, (1) another Lorentz function [194]

$$I_{\text{ex}}(q) = \frac{I_{\text{ex}}(0)}{1 + \xi_{\text{G}}^2 q^2} \qquad (\xi_{\text{G}} \gg \xi) \tag{171}$$

(2) a stretched exponential function [198–201]

$$I_{\text{ex}}(q) = I_{\text{ex}}(0) \exp[-(q\,\Xi)^{\alpha}], \quad \text{and} \tag{172}$$

(3) a Debye–Bueche function [202]

$$I_{\text{ex}}(q) = \frac{I_{\text{ex}}(0)}{(1+\beta^2 q^2)^2} \tag{173}$$

where, ξ_{G}, Ξ, and β are some characteristic length scales in the gel, and α is an exponent in the range of 0.7 to 2 [200]. These functions are introduced to describe additional fluctuations, solidlike inhomogeneities, or both. Equation (171) assumes that the gel has another correlation length, $\xi_{\text{G}}(\gg \xi)$ in addition to ξ [194]. Equation (171) is an extended form of the Guinier equation ($\alpha = 2$) [200, 203] for which noninteracting domains of higher or lower monomer densities are assumed to be randomly distributed in the network. For example, Geissler *et al.* examined the value of α and concluded $\alpha = 2$ for end-linked poly(dimethylsiloxane) swollen in toluene [200] and 0.7 poly(vinylacetate) gels [204]. Equation (173) is the so-called Debye–Bueche function [202], which represents a two-phase structure with a sharp boundary. These empirical equations have been used to represent the structure factor of swollen gels.

In some cases, however, $I(q)$ can be better fitted with a power–low-type function with a single correlation length and an exponent. This may be the case of a gel near the gelation threshold in which a self-similar fractal nature can be observed for $q > 1/\xi$, where ξ can be read as a physical cutoff of the fractal structure. In such a case, the following equation may be employed to describe the gel [205]:

$$I(q) = \frac{I(0)}{\left\{1 + \dfrac{(D_{\text{f}}+1)}{3}\xi^2 q^2\right\}^{D_{\text{f}}/2}} \qquad \text{for } \xi q < 1 \tag{174}$$

$$I(q) \approx I(0)\left[\frac{(D_{\text{f}}+1)}{3}\right]^{D_{\text{f}}/2} \xi^{-D_{\text{f}}} q^{-D_{\text{f}}} \qquad \text{for } \xi q \gg 1 \tag{175}$$

where D_{f} is the fractal dimension. Equation (174) is a generalized Zimm function for fractal objects. This equation was used for physical gels of poly(vinyl alcohol)-borate ion complex gels.

The most relevant theory of the structure factor for polymeric gels currently available may be the one proposed by Panyukov and Rabin (the PR theory) [206, 207]. This is the theory of instantaneously cross-linked networks of Gaussian chains with excluded volume. The structure factor consists of a contribution from thermal fluctuations $G(q)$ and static density inhomogeneities $C(q)$:

$$S(q) \equiv \overline{\langle \rho(q)\rho(-q)\rangle} = G(q) + C(q) \tag{176}$$

$$G(q) \equiv \langle \rho(q)\rho(-q)\rangle = \langle \rho^{\mathrm{th}}(q)\rho^{\mathrm{th}}(-q)\rangle = \frac{a^{-3}\phi z g(q)}{1 + wg(q)} \tag{177}$$

$$C(q) \equiv \overline{\rho^{\mathrm{eq}}(q)\rho^{\mathrm{eq}}(-q)} = \frac{a^{-3}\phi z}{[1 + wg(q)]^2(1 + \hat{q}^2)^2} \times \left[6 + \frac{9}{w_0 - 1 + \left\{\left(\hat{q}_{\mathrm{p}}^2\lambda^2 + \hat{q}_t^2/\lambda^2\right)\right\}(\phi_0/\phi)^{5/12}/2}\right] \tag{178}$$

where a, z, ϕ, and ϕ_0 denote the segment length, the average degree of polymerization between cross-linking points, the polymer volume fractions at observation and at preparation, respectively, and λ is the draw ratio. There are two types of averages here: The structure (ensemble) average denoted by $\overline{}$ and the thermal average given by $\langle\ \rangle$. The expressions $r^{\mathrm{eq}}(q)$, $r^{\mathrm{th}}(q)$, and $\rho(q) \equiv \rho^{\mathrm{eq}}(q) + \rho^{\mathrm{th}}(q)$ are the Fourier components of the equilibrium density variations that minimize the free energy of the gel, of the thermal density fluctuations, and of the total density variations, respectively. The expression $\hat{q}$ is defined as the dimensionless wave vector by,

$$\hat{q} = az^{1/2}q \tag{179}$$

and $\hat{q}_{\mathrm{p}}$ and $\hat{q}_t$ are the projections of the wave vector, parallel to and normal to the stretching direction, respectively. Equation (176) looks similar to Eq. (169) in which the contributions from liquidlike thermal fluctuations and solidlike inhomogeneity are separated. However, it should be noted that both terms on the right-hand side of Eq. (176) are coupled with the function $g(q)$, which is given by

$$g(q) = \frac{1}{\hat{q}^2/2 + (4\hat{q}^2)^{-1} + 1} + \frac{2\hat{q}^2(\phi/\phi_0)^{12/5}}{(1 + \hat{q}^2)^2\left(\hat{q}_{\mathrm{p}}^2\lambda^2 + \hat{q}_t^2/\lambda^2\right)} \tag{180}$$

The dimensionless quantities w and w_0 represent the effective second virial coefficients in the final state and in the state of preparation, respectively, and are given by,

$$w = (1 - 2\chi + \phi)\phi z \tag{181}$$

$$w_0 = \phi_0^{5/4} z \tag{182}$$

where χ is the Flory interaction parameter at the state of measurement. It is assumed that the gel was prepared under good solvent conditions. Thus, $S(q)$ is expressed as a function of two sets of parameters in the preparation state (w_0, f_0) and those under conditions of observation (w, f). Although Eqs. (176)–(178) predict the structure factor of gels for a variety of situations, the validity of the PR theory has to be verified experimentally.

2.5.4.2 Swelling and Extension

To emphasize the cross-link inhomogeneity, it is now realized that deforming the polymer network by swelling or stretching is helpful. This is because cross-linking itself does not seriously change the spatial distribution of the concentration of the polymer segment. Therefore, the structure factor does not change noticeably by introduction of cross-links and remains similar to the one from the corresponding polymer solutions. The drastic change of the structure factor results from subsequent change in the concentration or deformation. Let us show an example of visualization of the static inhomogeneities with the aid of light scattering.

Uniaxially stretched gels have a unique SANS intensity pattern called a "butterfly pattern." Figure 2.30 shows a typical butterfly pattern from a poly(dimethylsiloxane) (PDMS) gel [208]. The SANS contour pattern is elongated to the stretching direction, which is opposite to the direction predicted by theories taking into account only the thermal fluctuations. This contradiction is solved by plugging the static inhomogeneities into the total scattering. The static inhomogeneities can be visualized by irradiating laser light and observing the scattered intensity as a function of the sample position. By doing this, so-called speckles appear if the system has static inhomogeneities. This is due to the high coherency of the laser light and frozen inhomogeneities of the refractive index in the polymer network. Figure 2.31

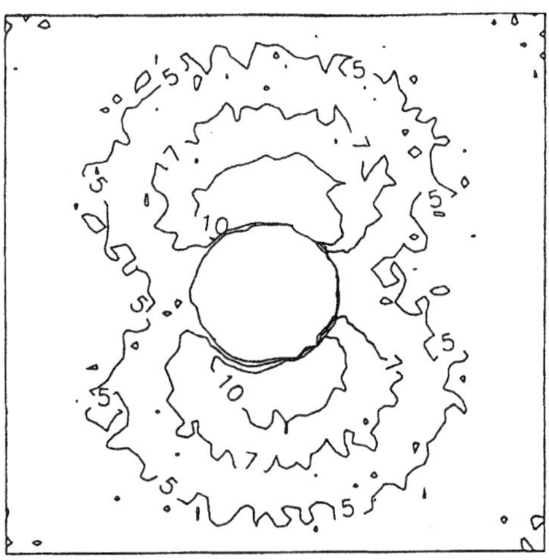

Figure 2.30 Experimental isointensity curves for a PDMS gel deformed uniaxially (in the vertical direction) by a factor of 1.58. The range of the scattering vector is $0.006 < q < 0.019$ Å^{-1} [208]. Reprinted from C. Rouf *et al.* (1994) with permission from the publisher, American Physical Society.

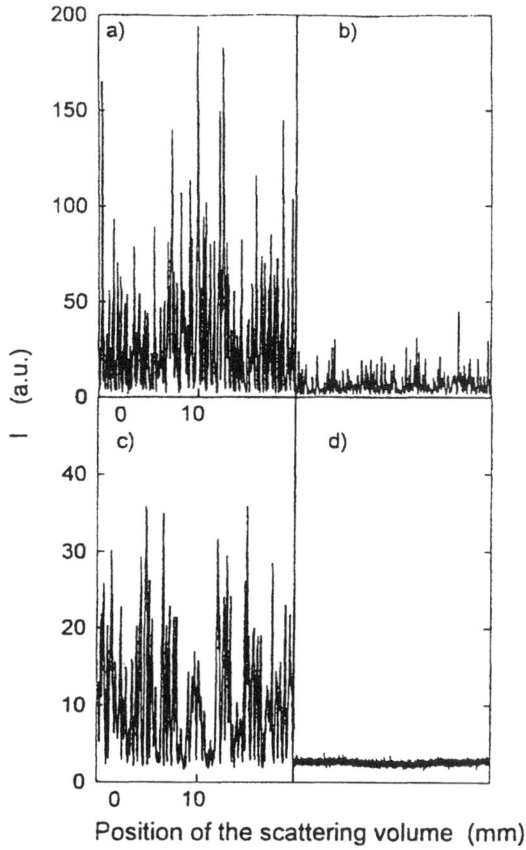

Figure 2.31 Scattering intensities as a function of the position of the scattering volume for $q = 2.73 \times 10^{-3}$ Å^{-1}. (a) Deformed gel with **q** parallel to the elongation axis; (b) deformed gel with **q** perpendicular to the elongation axis, (c) isotropic gel of the same concentration, and (d) semidilute solution with the same concentration. Scattering intensities were derived from ref. [208]. Reprinted from C. Rouf *et al.* (1994) with permission from the author and the publisher, American Physical Society.

shows speckle patterns of PDMS networks swollen in toluene. These are speckle patterns from gels (a) stretched along and (b) perpendicular to the stretching direction, respectively; (c) isotropic gel without deformation; and (d) the speckle pattern from a semidilute polymer solution with the same polymer concentration. From Figure 2.31, it is clear that static inhomogeneities (indicated by the height of speckles) appear by cross-linking [(d) → (c)], increase in the stretching direction [(c) → (a)], and are suppressed in the perpendicular direction [(c) → (b)]. This is a vivid evidence of the origin of the butterfly pattern, which is due to the static (quenched) inhomogeneities. Theoretical predictions by taking account of the static inhomogeneities have been proposed by Onuki [209] and Panyukov and Rabin [206, 207].

2.5.4.3 Charged Gels

If a polymer network carries unmobilized charges, the structure factor has to be modified by taking account of the Coulombic interaction. Particularly, if the solvent is a poor solvent for the major component of the polymer network, a strong scattering maximum has been observed. Figure 2.32 shows the scattered intensity functions $I(q)$ for poly(N-isopropylacrylamide-*co*-acrlyic acid) (NIPA/AAc) gels and the results of curve fitting with the Rabin–Panyukov theory (solid curves) [210]. The polymer volume fraction was $f = 0.0376$, and the ionization degree was estimated to be about 4.57 mol%. As shown in Figure 2.32, $I(q)$ increases with increasing temperature. The Rabin–Panyukov (RP) theory [211] is a modified theory of their theory for uncharged gels. They extended their theory for uncharged gels to charged gels by introducing q-dependent excluded volume parameters $w_0(q)$ and $w(q)$.

$$w(q) = (1 - 2\chi + \phi)\phi z + \frac{\hat{l}_B f^2 \phi z^2}{\hat{q} + \hat{l}_B f \phi z} \tag{183}$$

$$w_0(q) = \phi_0^{5/4} z + \frac{\hat{l}_B f_0^2 \phi_0^{5/4} z^2}{\hat{q}^2 (\phi_0/\phi)^{2/3} + \hat{l}_B f_0 \phi_0^{5/4} z} \tag{184}$$

where $\hat{l}_B$ is the dimensionless Bjerrum length defined by

$$\hat{l}_B = 4\pi l_B / a \tag{185}$$

and l_B is the Bjerrum length ($\approx$7 Å for aqueous solutions at 25 °C). The expressions f and f_0 are, respectively, the degree of ionization at observation and at preparation. Though the RP theory (Eqs. (176–178) and (181–183)) looks complicated, there are only three fitting parameters (i.e., χ, z, and the prefactor of the intensity scaling). The fitted χ values are tabulated in the inset. By considering that the number of fitting parameters is only two, the fitting seems to be quite satisfactory. However, it is unnecessary to mention that justification of the theory requires further experimental investigations.

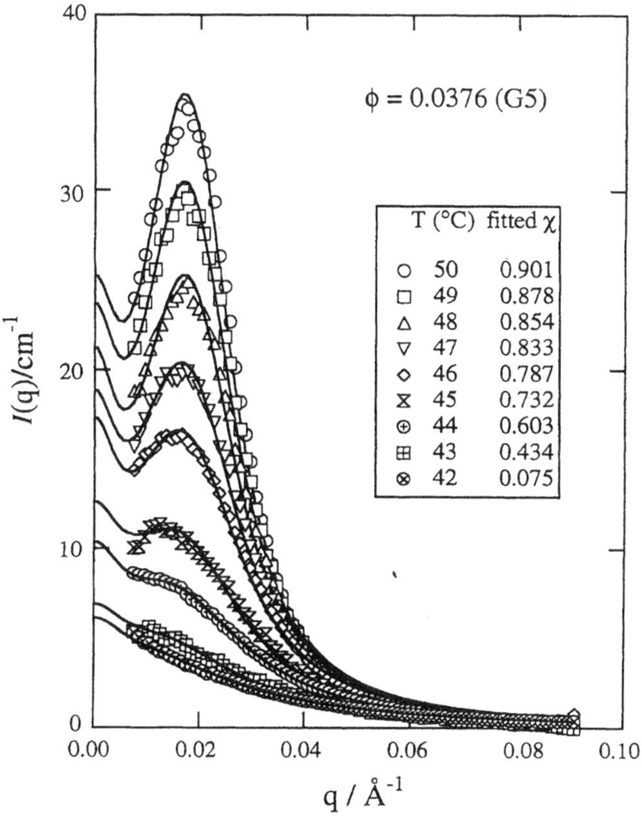

Figure 2.32 Scattered intensity functions for weakly charged NIPA/AAc gels at different temperatures. The solid curves denote the results of curve fitting with the Rabin–Panyukov theory. Reprinted from M. Shibayama *et al.* (1998) with permission from the publisher, American Chemical Society.

References

1. R. Pecora, R. F. Reinisch Ed., Plenum, New York, 1970.
2. R. Pecora, *Discuss. Faraday Soc.* **49:** 222 (1970).
3. R. Pecora, *Nature Phys. Sci.* **231:** 73 (1971).
4. B. Zimm, *J. Chem. Phys.* **16:** 1093 (1948).
5. H. Benoit, D. Decker, and J. S. Higgins, C. Picot, J. P. Cotton, B. Farnoux, G. Jannink, and R. Oben, *Nature Phys. Sci.* **24:** 13 (1973).
6. J. Kumaki, *Macromolecules* **21:** 749 (1988).
7. J. Kumaki, Y. Nishikawa, and T. Hashimoto, *J. Am. Chem. Soc.* **118:** 3321 (1996).

8. P. J. Flory, *J. Chem. Phys.* **17:** 303 (1949).
9. J. Scheleten, G. D. Wignall, and D. G. H. Ballard, *Polymer* **15:** 682 (1974).
10. L. H. Sperling, *Polym. Eng. Sci.* **24:** 1 (1984).
11. F. Boué, *Adv. Polym. Sci.* **82:** 49 (1987).
12. D. M. Sadler, in *Comprehensive Polymer Science*, C. Booth and C. Price Eds., Vol. 2, Pergamon Press, Oxford, 1989.
13. P. G. de Gennes, *Scaling Concepts in Polymer Physics*, Cornell University Press, Ithaca, 1979.
14. J. S. Higgins and H. C. Benoit, *Polymers and Neutron Scattering*, Clarendon Press, Oxford, 1994.
15. D. R. Lide, Handbook of Chemistry and Physics, CRC Press, New York, 1995.
16. G. E. Bacon, *Application of Neutron Diffraction in Chemistry*, Pergamon Press, New York, 1963.
17. G. E. Bacon, Neutron diffraction Vol. 1, Clarendon, Oxford, 1975.
18. M. Born and E. Wolf, *Principles of Optics*, Pergamon Press, New York, 1974.
19. J. M. Schultz, *Diffraction for Materials Scientists*, Prentice-Hall, Inc., Englewood Cliffs, New Jersey, 1982.
20. V. F. Sears, *Neutron News* **3:** 26 (1992).
21. H. Maier-Leibnitz and T. Springer, *Reactor Science and Technology* **7:** 217 (1963).
22. C. J. Glinka, J. M. Rowe, and J. G. LaRock, *J. Appl. Cryst.* **19:** 427–439 (1986).
23. C. J. Glinka, J. G. Barker, B. Hammouda, S. Kruger, J. J. Moyer, and N. J. Orts, *J. Appl. Cryst.* **31:** 430 (1998).
24. L. Rosta, *Physica B* **156&157:** 615 (1989).
25. H. Friedrich, V. Wagner, and P. Wille, *Physica B* **156&157:** 547 (1989).
26. B. Hammouda, *Nucl. Instrum. Methods Phys. Res. A* **321:** 275–283 (1992).
27. J. R. D. Copley, *Nucl. Instrum Methods Phys. Res. A* **332:** 511–520 (1993).
28. K. Ibel, *J. Appl. Cryst.* **9:** 296–309 (1976).
29. R. P. May, *J. Appl. Cryst.* **27:** 298–301 (1994).
30. A. N. Falcao, J. S. Pedersen, and K. Mortensen, *J. Appl. Cryst.* **27:** 330–337 (1994).
31. C. J. Borkowski and M. K. Kopp, *J. Appl. Cryst.* **11:** 430 (1978).
32. T. Hashimoto, S. Suehiro, M. Shibayama, K. Saijo, and H. Kawai, *Polym. J.* **13:** 501–516 (1981).
33. Y. Amemiya, K. Wakabayshi, H. Tanaka, Y. Ueno, and J. Miyahara, *J. Sci.* **237:** 164 (1987).
34. T. Hashimoto, S. Okamoto, K. Saijo, K. Kimishima, and T. Kume, *Acta Polymer* **46:** 463–470 (1995).
35. G. D. Wignall and F. S. Bates, *J. Appl. Cryst.* **20:** 28–40 (1987).
36. C. J. Glinka and S. Kruger, NBS Memorandum for SANS Users, Dec. 29, 1986 (1986).
37. R. P. May, K. Ibel, and J. Haas, *J. Appl. Cryst.* **15:** 15 (1982).
38. B. Jacrot, *Reports on Progress in Physics* **39:** 911 (1976).
39. J. S. Pedersen, D. Posselt, and K. Mortensen, *J. Appl. Cryst.* **23:** 321–333 (1990).
40. J. G. Barker and J. S. Pedersen, *J. Appl. Cryst.* **28:** 105–114 (1995).

41. A. Guinier and G. Fournet, *Small-Angle Scattering of X-rays*, John Wiley & Sons, Inc., New York, 1955.
42. O. Glatter and O. Kratky, *Small-Angle X-rays Scattering*, Academic Press, New York, 1982.
43. H. Benoit, J. Koberstein, and L. Leibler, *Die Makromol. Chem. Suppl.* **4:** 85 (1981).
44. M. Daoud, J. P. Cotton, B. Farnoux, G. Jannink, G. Sarma, and H. Benoit, *Macromolecules* **8:** 804 (1975).
45. A. Guinier, *Ann. Phys.* **12:** 161 (1939).
46. G. Fournet, *Bull. soc. franc. mineral. et. crist.* **74:** 39 (1951).
47. O. Kratky and G. Porod, *J. Colloid Sci.* **4:** 35 (1949).
48. P. Debye, *J. Phys. Colloid Chem.* **51:** 18 (1947).
49. G. Porod, *Kolloid Z.* **124:** 83 (1951).
50. G. Porod, *Kolloid Z.* **125:** 51 (1952).
51. M. Shibayama, S. Nomura, T. Hashimoto, and E. L. Thomas, *J. Appl. Phys.* **66:** 4188 (1989).
52. W. Ruland, *J. Appl. Cryst.* **4:** 70 (1971).
53. C. G. Vonk, *J. Appl. Cryst.* **6:** 81 (1973).
54. H. Tomita, *Prog. Theor. Phys. Prog. Lett.* **72:** 656 (1984).
55. H. Tomita, *Prog. Theor. Phys.* **72:** 482 (1986).
56. T. Hashimoto, M. Takenaka, and H. Jinnai, *J. Appl. Cryst.* **24:** 457 (1991).
57. M. Takenaka and T. Hashimoto, *J. Chem. Phys.* **96:** 6177 (1992).
58. J. E. Martin, *J. Appl. Cryst.* **19:** 25 (1986).
59. B. B. Mandelbrot, *The Fractal Geometry of Nature*, Freeman, San Francisco, 1982.
60. D. Stauffer, *Phys. Rep.* **54:** 1 (1979).
61. H. Benoit and M. Benmouna, *Polymer* **25:** 1059 (1984).
62. M. Shibayama, H. Yang, R. S. Stein, and C. C. Han, *Macromolecules* **18:** 2179 (1985).
63. L. Leibler, *Macromolecules* **13:** 1602 (1980).
64. G. H. Fredrickson and E. Helfand, *J. Chem. Phys.* **87:** 697 (1987).
65. S. Sakurai, H. Hasegawa, T. Hashimoto, I. G. Hargis, S. L. Aggarwal, and C. C. Han, *Macromolecules* **23:** 451 (1990).
66. J. F. C. Joanny and R. Hebd, *Seances Acad. Sci.* **286B:** 89 (1978).
67. L. Leibler and H. Benoit, *Polymer* **22:** 195 (1981).
68. R. W. Richards and J. L. Thomason, *Macromolecules* **16:** 982 (1983).
69. E. A. Guggenheim, *Thermodynamics*, North Holland Pub., Amsterdam, 1986.
70. H. Yang, G. Hadziioannou, and R. S. Stein, *J. Polym. Sci., Polym. Phys. Ed.* **21:** 159 (1983).
71. G. Jansco, L. P. N. Rebeto, and W. A. van Hook, *Chem. Soc. Rev.* **23:** 257 (1994).
72. A. D. Buckingham and H. G. E. Hentschel, *J. Polym. Sci., Polym. Phys. Ed.* **18:** 853 (1980).
73. F. S. Bates, G. D. Wignall, and W. C. Koehler, *Phys. Rev. Lett.* **55:** 2425–2428 (1985).
74. R. R. Singh and W. A. Van Hook, *Macromolecules* **20:** 1855 (1987).

75. F. S. Bates and G. D. Wignall, *Phys. Rev. Lett.* **57:** 1429 (1986).
76. R. G. Kirste, W. A. Kruse, and J. Schelten, *Makromol. Chem.* **162:** 299 (1973).
77. R. G. Kirste and B. R. Lehnen, *Makromol. Chem.* **177:** 1137 (1976).
78. G. Lieser, F. W. Fischer, and K. Ibel, *J. Polym. Sci., Polym. Lett. Ed.* **13:** 39 (1975).
79. J. S. Higgins and R. S. Stein, *J. Appl. Cryst.* **11:** 346 (1978).
80. D. Y. Yoon and P. J. Flory, *Polymer* **16:** 645 (1975).
81. D. Y. Yoon and P. J. Flory, *Macromolecules* **9:** 299 (1976).
82. D. G. H. Ballard, G. W. Longman, T. L. Crowley, A. Cunningham, and J. Scheleten, *Polymer* **19:** 379 (1978).
83. J. M. Guenet, *Macromolecules* **13:** 387 (1980).
84. B. Crist, W. W. Graessley, and G. D. Wignall, *Polymer* **23:** 1561 (1982).
85. E. W. Fischer, M. Stamm, M. Dettenmaier, and P. Herschenraeder, *Polym. Prepr. Am. Chem. Soc. Div. Polym. Chem.* **20:** 219 (1979).
86. G. D. Wignall, L. Mandelkern, C. Edwards, *J. Polym. Sci., Polym. Phys. Ed.* **20:** 245 (1982).
87. D. M. Sadler and A. Keller, *Macromolecules* **10:** 1128 (1977).
88. D. M. Sadler and A. Keller, *Science* **203:** 263 (1979).
89. D. M. Sadler, *J. Appl. Cryst.* **16:** 519 (1983).
90. E. Helfand and Z. R. Wasserman, *Macromolecules* **9:** 879 (1976).
91. E. Helfand and Z. R. Wasserman, *Macromolecules* **13:** 994 (1980).
92. E. Helfand and Z. R. Wasserman, Developments in Block Copolymers, I. Goodman Ed., Applied Science Publishers, 1984.
93. T. Hashimoto, M. Shibayama, and H. Kawai, *Macromolecules* **13:** 1237 (1980).
94. R. W. Richards and J. L. Thomason, *Macromolecules* **16:** 982 (1980).
95. H. Hasegawa, T. Hashimoto, H. Kawai, T. P. Lodye, E. J. Amis, C. J. Glinka, and C. C. Han, *Macromolecules* **18:** 67 (1985).
96. Y. Matsushita, Y. Nakao, R. Saguchi, K. Mori, H. Choshi, Y. Muroga, I. Noda, and M. Nagasawa, *Macromolecules* **21:** 1802 (1988).
97. G. Hadziioannou, L. H. Wang, R. S. Stein, and R. S. Porter, *Macromolecules* **15:** 880 (1982).
98. Y. Mogi, K. Mori, H. Kotsuji, Y. Mutsushita, and I. Noda, *Macromolecules* **26:** 5169 (1993).
99. S. Koizumi and T. Hashimoto, *Physica B* **213&214:** 703 (1995).
100. F. S. Bates, S. B. Dierker, and G. D. Wignall, *Macromolecules* **19:** 1938 (1986).
101. D. Schwahn, J. Hahn, J. Streib, and T. Springer, *J. Chem. Phys.* **93:** 8383 (1990).
102. H. Jinnai, H. Hasegawa, T. Hashimoto, and C. C. Han, *Macromolecules* **25:** 8154 (1992).
103. S. Sakurai, H. Jinnai, H. Hasegawa, T. Hashimoto, and C. C. Han, *Macromolecules* **24:** 4839 (1991).
104. R. N. Thudium and C. C. Han, *Macromolecules* **29:** 2413 (1996).
105. S. Sakurai, T. Izumitani, H. Hasegawa, T. Hashimoto, and C. C. Han, *Macromolecules* **24:** 4844 (1991).

106. N. P. Balsara, L. J. Fetters, N. Hadjichristidis, P. Lohse, C. C. Han, W. W. Graessley, and R. Krishnamoorti, *Macromolecules* **25:** 6137 (1992).
107. G. Beaucage and R. S. Stein, *Macromolecules* **26:** 1617 (1993).
108. C. C. Han, B. J. Bauer, J. C. Clark, Y. Muroga, Y. Matsushita, M. Okada, Q. Tran-Cong, T. Chang, and I. C. Sanchez, *Polymer* **29:** 2002 (1988).
109. F. S. Bates, G. D. Muthukumar, G. D. Wignall, and L. J. Fetters, *J. Chem. Phys.* **89:** 535 (1988).
110. H. Hasegawa, S. Sakurai, M. Takenaka, T. Hashimoto, and C. C. Han, *Macromolecules* **24:** 1813 (1991).
111. R. Krishnamoorti, W. W. Graessley, N. P. Balsara, and D. Lohse, *J. Chem. Phys.* **100:** 3894 (1994).
112. I. C. Sanchez and C. G. Panayiotou, *Macromolecules* **11:** 1145 (1978).
113. J. Dudowicz and K. F. Freed, *Macromolecules* **24:** 5112 (1991).
114. M. Muthukumar, *J. Chem. Phys.* **85:** 4722 (1986).
115. U. R. Bidkar and I. C. Sanchez, *Macromolecules* **28:** 958 (1991).
116. K. S. Schweizer and J. G. Curro, *Adv. Polym. Sci.* **116:** 319 (1994).
117. S. Janssen, D. Schwahn, and K. Mortensen, *Macromolecules* **26:** 5587 (1993).
118. B. Hammouda and B. J. Bauer, *Macromolecules* **28:** 4505 (1995).
119. D. Schwahn, K. Mortensen, and Y. Yee-Madeira, *Phys. Rev. Lett.* **58:** 1544 (1987).
120. D. Schwahn, K. Mortensen, T. Springer, M. Yee-Madeira, and R. Thomas, *J. Chem. Phys.* **87:** 6078 (1987).
121. F. S. Bates, J. H. Rosedale, P. Stepanek, T. P. Lodge, P. Wilzius, G. H. Fredrickson, and R. P. Hjelm, *Phys. Rev. Lett.* **65:** 1893 (1990).
122. D. W. Hair, E. K. Hobbie, A. I. Nakatani, and C. C. Han, *J. Chem. Phys.* **96:** 9133 (1992).
123. S. Janssen, D. Schwahn, and T. Springer, *Phys. Rev. Lett.* **68:** 3180–3183 (1992).
124. K. Binder, *J. Chem. Phys.* **79:** 6387 (1983).
125. J. Dudowicz, M. Lifschitz, K. F. Freed, and J. F. Douglas, *J. Chem. Phys.* **99:** 4804 (1993).
126. D. Schwahn, H. Takeno, L. Willner, H. Hasegawa, H. Jinnai, T. Hashimoto, and M. Imai, *Phys. Rev. Lett.* **73:** 3427 (1994).
127. D. Schwahn, T. Schmackers, and K. Mortensen, *Phys. Rev. E* **52:** R1288 (1995).
128. E. K. Hobbie, L. Reed, C. C. Haung, and C. C. Han, *Phys. Rev. E* **48:** 1579 (1993).
129. H. L. Snyder and P. Meakin, *J. Chem. Phys.* **79:** 5588 (1983).
130. T. Hashimoto, M. Itakura, and N. Shimizu, *J. Chem. Phys.* **85:** 6773 (1986).
131. T. Kyu and M. Saldanha, *Macromolecules* **21:** 1021 (1988).
132. F. S. Bates and R. Wiltzius, *J. Chem. Phys.* **91:** 3258 (1989).
133. J. D. Gunton, M. San Miguel, P. S. Sahni, C. Domb, and J. L. Lebowitz, Eds., Vol. 8, p. 269, Academic Press, New York, 1983.
134. K. Kawasaki, *Progr. Theor. Phys.* **57:** 826 (1977).
135. K. Kawasaki and T. Ohta, *Progr. Theor. Phys.* **59:** 362 (1978).

136. K. Binder, in *Phase Transformation in Materials*, P. Haasen, Ed., Vol. 5, p. 40, VCH, Weinheim, Germany, 1990.
137. T. Hashimoto, in *Structure and Properties of Polymers*, E. L. Thomas, Ed., Vol. 12, p. 251, VCH, Weinheim, Germany, 1993.
138. T. Nose, *Phase Transitions* **8:** 245 (1987).
139. J. S. Langer, M. Bar-on, and H. D. Miller, *Phys. Rev. A* **11:** 1417 (1975).
140. A. Chakrabarti, A. Toral, J. D. Gunton, and M. Muthukumar, *Phys. Rev. Lett.* **63:** 2072 (1989).
141. A. Chakrabarti, A. Toral, J. D. Gunton, and M. Muthukumar, *J. Chem. Phys.* **92:** 6899 (1990).
142. T. Koga and K. Kawasaki, *Phys. Rev. A* **44:** R817 (1991).
143. Y. Oono and S. Puri, *Phys. Rev. A* **38:** 434 (1988).
144. A. Sinozaki and Y. Oono, *Phys. Rev. Lett.* **66:** 173 (1991).
145. Y. C. Chou and W. I. Goldburg, *Phys. Rev. A* **20:** 2105 (1979).
146. N. C. Wong and C. M. Knobler, *J. Chem. Phys.* **69:** 725 (1978).
147. S. Komura, K. Osamura, H. Fujii, and T. Tanaka, *Phys. Rev. B* **31:** 1278 (1985).
148. A. F. Craievich, J. M. Sanchez, and C. E. Williams, *Phys. Rev. B* **34:** 2762 (1986).
149. J. W. Cahn, *Acta Metall.* **9:** 1961 (1961).
150. J. W. Cahn, *J. Chem. Phys.* **42:** 93 (1965).
151. H. E. Cook, *Acta Metall.* **18:** 297 (1970).
152. J. S. Higgins, H. Fruitwala, and P. E. Tomlins, *Br. Polym. J.* **21:** 247 (1989).
153. J. S. Higgins, H. Fruitwala, and P. E. Tomlins, *Macromolecules* **22:** 3674 (1989).
154. W. Pyckhout, D. Schwahn, I. Sosnowska, T. Springer, and H. Yee-Madeira, *Physica B* **156–157:** 402 (1989).
155. D. Schwahn, S. Janssen, and T. Springer, *J. Chem. Phys.* **97:** 8775 (1992).
156. M. Motowoka, H. Jinnai, T. Hashimoto, Y. Qiu, and C. C. Han, *J. Chem. Phys.* **99:** 2095 (1993).
157. H. Jinnai, H. Hawegawa, T. Hashimoto, and C. C. Han, *J. Chem. Phys.* **99:** 4845 (1993).
158. M. Okada and C. C. Han, *J. Chem. Phys.* **85:** 5317 (1986).
159. P. Pincus, *J. Chem. Phys.* **75:** 1996 (1981).
160. H. Jinnai, H. Hawegawa, T. Hashimoto, and C. C. Han, *J. Chem. Phys.* **99:** 8154 (1993).
161. K. Kawasaki and T. Koga, *Physica A* **201:** 115 (1993).
162. A. Onuki, *Non-Cryst. Solids* **172–174:** 1151 (1994).
163. H. Jinnai, H. Hasegawa, T. Hashimoto, and C. C. Han, *Macromolecules* **24:** 282 (1991).
164. T. Hashimoto, H. Jinnai, H. Hasegawa, and C. C. Han, *Physica A* **204:** 261–276 (1994).
165. S. K. Ma, *Modern Theory of Critical Phenomena*, Benjamin/Cummings Pub. Co., London, 1976.
166. T. Hashimoto, K. Kowsaka, M. Shibayama, and H. Kawai, *Macromolecules* **19:** 754 (1986).
167. T. Hashimoto, *Thermoplastic Elastomers*, Hanser, Munich, 1996.

168. M. Shibayama and T. Hashimoto, *Macromolecules* **19:** 140 (1986).
169. F. S. Bates, C. V. Berney, and R. E. Cohen, *Macromolecules* **16:** 1101 (1983).
170. E. L. Thomas, D. B. Alward, D. J. Kinning, D. C. Martin, D. J. Handlin, Jr., and L. J. Fetters, *Macromolecules* **19:** 2197 (1986).
171. H. Hasegawa, H. Tanaka, K. Yamasaki, and T. Hashimoto, *Macromolecules* **20:** 1651 (1987).
172. D. A. Hajduk, P. E. Harper, S. M. Gruner, C. C. Honeker, G. Kim, E. L. Thomas, and L. J. Fetters, *Macromolecules* **27:** 4063 (1994).
173. T. Hashimoto, M. Shibayama, H. Kawai, H. Watanabe, and T. Kotaka, *Macromolecules* **16:** 361 (1983).
174. D. J. Meier, *J. Polym. Sci. C* **26:** 81 (1969).
175. E. Helfand and Z. R. Wasserman, *Macromolecules* **11:** 960 (1978).
176. R. J. Roe, M. Fishkis, and J. C. Chang, *Macromolecules* **14:** 1091 (1981).
177. W. C. Zin and R. J. Roe, *Macromolecules* **17:** 183 (1984).
178. K. Mori, H. Hasegawa, and T. Hashimoto, *Polym. J.* **17:** 799 (1985).
179. F. S. Bates and M. A. Hartney, *Macromolecules* **18:** 2478 (1985).
180. F. S. Bates, *Macromolecules* **20:** 2221 (1987).
181. T. Hashimoto, Y. Ijichi, and L. J. Fetters, *J. Chem. Phys.* **89:** 2463 (1988).
182. Y. Ijichi, T. Hashimoto, and L. J. Fetters, *Macromolecules* **22:** 2817 (1989).
183. J. N. Owens, J. T. Gancarz, J. T. Koberstein, and T. P. Russel, *Macromolecules* **22:** 3380 (1989).
184. H. E. Stanley, *Phase Transition and Critical Phenomena*, Oxford Univ. Press, Oxford, 1971.
185. A. Brazovskii, *Sov. Phys. JETP* **41:** 85 (1975).
186. F. S. Bates, J. H. Rosedale, and G. H. Fredrickson, *J. Chem. Phys.* **92:** 6225 (1990).
187. N. Sakamoto and T. Hashimoto, *Macromolecules* **28:** 6825 (1995).
188. B. Stün, R. Mutter, and T. Albrecht, *Europhys. Lett.* **18:** 427 (1992).
189. T. Wolff, C. Burger, and W. Ruland, *Macromolecules* **26:** 1707 (1993).
190. T. Hashimoto, T. Ogawa, and C. D. Han, *J. Phys. Soc. Jpn.* **63:** 2206 (1994).
191. G. Floudas, T. Pakula, E. W. Fischer, N. Hadjuichristidis, and S. Pispas, *Acta Polymer* **45:** 176 (1994).
192. T. Hashimoto and N. Sakamoto, *Macromolecules* **28:** 4779 (1995).
193. T. Hashimoto, N. Sakamoto, and T. Koga, *Phys. Rev. E* **54:** 5832 (1996).
194. S. Candau, J. Bastide, and M. Delsanti, *Adv. Polym. Sci.* **44:** 27 (1982).
195. R. S. Stein, *J. Polym. Sci. B* **7:** 657 (1969).
196. J. Bastide and S. J. Candau, in *The Physical Properties of Polymer Gels*, J. P. Cohen Addad, Ed., Vol. Chapt. 9, p. 143, John Wiley, New York, 1996.
197. M. Shibayama, *Macromol. Chem. Phys.* **199:** 1 (1998).
198. S. Mallam, F. Horkay, A. M. Hecht, and E. Geissler, *Macromolecules* **22:** 3356 (1989).
199. S. Mallam, A. M. Hecht, and E. Geissler, *J. Chem. Phys.* **91:** 6447 (1989).
200. S. Mallam, F. Horkay, A. M. Hecht, A. R. Rennie, and E. Geissler, *Macromolecules* **24:** 543 (1991).

201. F. Horkay, A. M. Hecht, and E. Geissler, *Macromolecules* **27:** 1795 (1994).
202. W. Wu, M. Shibayama, S. Roy, H. Kurakawa, L. D. Coyen, S. Nomura, and R. S. Stein, *Macromolecules* **23:** 2245 (1990).
203. Y. Cohen, O. Ramon, I. J. Kopelman, and S. Mizrahi, *J. Polym. Sci., Polym. Phys. Ed.* **30:** 1055 (1992).
204. F. Horkay, A. M. Hecht, S. Mallam, E. Geissler, and A. R. Rennie, *Macromolecules* **24:** 2896 (1991).
205. M. Shibayama, H. Kurokawa, S. Nomura, M. Muthukumar, R. S. Stein, and S. Roy, *Polymer* **33:** 2883 (1992).
206. S. Panyukov and Y. Rabin, *Macromolecules* **29:** 7960 (1996).
207. S. Panyukov and Y. Rabin, *Physics Report* **269:** 1 (1996).
208. C. Rouf, J. Bastide, J. M. Pujol, F. Schosseler, and J. P. Munch, *Phys. Rev. Lett.* **73:** 830 (1994).
209. A. Onuki, *J. Phys. II France* **2:** 45 (1992).
210. M. Shibayama, K. Kawakubo, and T. Norisuye, *Macromolecules* **31:** 1608 (1998).
211. Y. Rabin and S. Panyukov, *Macromolecules* **30:** 301 (1997).
212. J. Bastide, F. Boue, and M. Buzier, in *Molecular Basis of Polymer Networks*, A. Baumgartner and C. E. Picot, Eds., p. 48, Springer-Verlag, Berlin, 1989.

Chapter 3 | Fluorescence Spectroscopy

H. Itagaki

Department of Chemistry
Faculty of Education
Shizuoka University
Ohya, Shizuoka, Japan

3.1 Introduction

The application of fluorescence spectroscopy is growing remarkably as a powerful and effective tool to study the physical and chemical behavior of macromolecules. This chapter deals with fluorescence measurements in polymer systems. The main advantage of spectrofluorometry is its high sensitivity. The amount of precious sample used for a measurement can be kept very low because traces of fluorescent species can be detected quantitatively. Ultraviolet–visible (Uv–vis) absorption is also known to have a high sensitivity and can, for instance, effectively be employed to monitor chemical species in blood in medical tests; however, the sensitivity of fluorometry would be 10 to 100 times as much as that of uv–vis absorption spectroscopy. This high sensitivity makes fluorescence one of the best available methods for trace analysis.

The probe method allows us to get information from a system in which probe molecules are mixed. It is similar to sending a spy to a hostile country. If many spies have to be sent to get information, the existence of spies could be noticed by the country. The information obtained by a spy could include the information influenced by another spy itself. Because the fluorescence probe methods do not require a great number of probe molecules, the system is not distorted too much by the addition of the probes, and thus we can acquire the inherent properties of the system.

Simply speaking, fluorescence is a photon released from an excited state of a molecule. The excited state can readily be induced by the irradiation of light to a normal molecule in the ground state, but because the duration time of the excited state is usually very short, it goes back to the ground state by releasing the excitation energy as heat, fluorescence, or both. What we are interested in is how a molecule behaves during its excited state; for example, if it dies normally or associates with another molecule to form another excited species and so on. Thus, we can label a chromophore that is not a special one in a polymer only by the

EXPERIMENTAL METHODS
IN POLYMER SCIENCE

ISBN 0-12-683265-X/$30.00.

irradiation of light and can observe its behavior, especially in the case of inherently fluorescent polymers such as polystyrene and poly(ethylene terephthalate).

The second advantage of fluorescence measurements is that they are able to make use of equipment to measure directly the time change of photophysical processes that take place in the time range longer than 100 femtoseconds (1 fs = 10^{-15} s). As will be described in detail later, fluorescence lifetime is usually 10^{-12} to 10^{-6} s, and thus we can directly observe the dynamic process of polymer molecules by measuring the transient fluorescence. It is also a benefit that commercially available spectrophotometers are good and not-too-expensive instruments with which to measure steady-state and transient fluorescence.

The third advantage to mention is a variety of phenomena relating to fluorescence to be observed because the fluorescence process is influenced very much by the microenvironment. If two chromophores are within 0.4 nm, they probably form a complex either in the ground state or in the excited state: The fluorescence of the complex should be different from that of the isolated chromophores. Thus, we can infer both the distance and the interaction among chromophores from the fluorescence measurements. By means of fluorescence measurements, we can get information on interactions over a wide range of radii from a few to 10 nm, and information about a wide range of times longer than 10^{-13} s. Because a molecule usually employed for the fluorescence measurements is quite stable in the dark, we can observe the fluorescence of a sample even 10 years after the preparation of the sample solution, meaning that we can monitor a process with a long relaxation time such as those processes that occur in polymer solids or gels.

Another advantage of fluorescent probe techniques is the possibility of obtaining different information from a single sample prepared for measurement. For example, a dansyl moiety is often employed as a probe to monitor polarity within a certain environment. Moreover, the measurement of fluorescence anisotropy of the dansyl chromophore will provide information on its mobility in the same sample. An excited state of a molecule is not formed unless it is irradiated. Thus, even if two or more dyes are dispersed in a polymer system, one kind of dye can be excited by light of a certain wavelength, whereas other dyes are not excited. This demonstrates that one sample can be used for several different experiments by changing the excitation and emission wavelengths.

Thus far we have enumerated the merits of fluorescence measurements and the fluorescent probe method: high sensitivity, imperviousness to influence by changing circumstances, nondestructiveness, accuracy of transient measurements ($>10^{-13}$ s), ability to yield valuable information on interactions and motion, versatility in obtaining several different pieces of information from a dye, and relatively low cost. In particular, the fluorescent probe method is more effective for obtaining information on the microenvironment at the molecular level near a probe molecule and for monitoring dynamic changes taking place over pico- (10^{-12}) and nanosecond (10^{-9}) ranges.

This chapter is intended as a guide for beginners in carrying out research by means of fluorescence measurements. I assume that the reader has experience in studies of polymer science but that he or she is applying fluorescence spectrometry for the first time. Accordingly, I prefer the explanation of the practical use of the techniques over the detailed description of the principles.

The chapter consists of six major sections. In Section 3.2, the main photophysical phenomena are introduced in brief. Here I try to emphasize what parameters need to be kept in mind throughout the fluorescence measurement experiments. In particular, the principles of a variety of fluorescence phenomena are introduced one by one in Section 3.2.3. Understanding each type of fluorescence behavior is of help for the application of fluorescent probes in polymer systems. Section 3.3 summarizes detailed descriptions of the experiments. Section 3.4 deals with fluorescence as a probe and thus describes the same phenomena as in Section 3.3 but from another point of view. A few examples of the practical uses of fluorescence spectroscopy are described in detail in Section 3.5. On the basis of the understanding of fluorescence spectroscopy and fluorescent probe methods acquired from this chapter, the reader is referred to reviews in the literature that deal with certain topics in more detail.

3.2 Introduction to Fluorescence Processes

3.2.1 GENERAL PHOTOPHYSICAL PROCESSES

Here we shall mainly deal with the photophysical processes that occur in organic compounds because most of the fluorescent probes employed for study in polymer science are condensed aromatic hydrocarbons and organic dye molecules.

First, we will explain the principle of spectroscopy. Every electron of a molecule is in a molecular orbital for which the energy is determined. One orbital can be occupied by two electrons, provided they have different spin vectors. Let us consider the case of benzene (C_6H_6) as an example. Because carbon has 6 electrons and hydrogen has 1, a benzene molecule will have a total of 42 electrons. Therefore, 1 benzene molecule has 21 orbitals occupied with these 42 electrons, but it also has many vacant orbitals with energy levels higher than those of the occupied orbitals. A situation in which only orbitals with the lowest energy levels are occupied within the molecule is called the electronic ground state. An electron that can be activated by light (a photon) is one staying in the orbital whose energy is highest, i.e., within the highest occupied molecular orbital (HOMO), because this electron exists at the surface of the molecule. When a molecule is irradiated by a photon, the transition of an electron can take place from the ground state to a vacant orbital with higher energy. This new state is then called an excited state.

The total energy (E_t) of a molecule is usually considered to be the sum of three components: the electronic energy (E_e), the vibrational energy (E_v), and the rotational energy (E_r) as a first approximation:

$$E_t = E_e + E_v + E_r \tag{1}$$

In order to form an excited state of a molecule, a photon should be available whose energy is equal to the difference between that of the excited state and the ground state. The energy of a photon is expressed as

$$E = h\nu = hc/\lambda \tag{2}$$

where h is Planck's constant ($= 6.626 \times 10^{-34}$ J·s), c is the velocity of light (2.998×10^8 m/s), ν is frequency (s^{-1}), and λ is wavelength (m). The energy for transitions between rotational states can be provided by photons whose wavelength is in the far-infrared region but for vibrational transitions comes from photons in the near-infrared region. Transitions between electronic states only occur with visible and ultraviolet light.

A photophysical process is a physical process resulting from the electronic excitation of a molecule or system of molecules by light. Figure 3.1 shows the typical photophysical phenomena of an isolated molecule or group. Here singlet state S indicates that there is no net electronic spin associated with the state: The total number of spins up is exactly same as that of spins down. The triplet state T indicates a threefold degeneracy of the state owing to the three possible alignments of two unpaired spins: If we set up a magnetic field to a molecule in a triplet state, three different states in energy come up. The subscript (0, 1, 2, ...) shows the energy ranking of the state. In typical aromatic compounds, the electronic ground state is S_0.

The main photophysical processes can be classified into three types: (1) absorption, (2) radiative deactivation (luminescence), and (3) nonradiative deactivation. Absorption corresponds to the transition from a lower energy level to an upper level, whereas the two deactivation processes correspond to the transition from an upper level to a lower level. During the transitions, the nuclear configuration coordinate of a molecule usually remains the same as that of the initial state. This is known as the Franck–Condon principle.

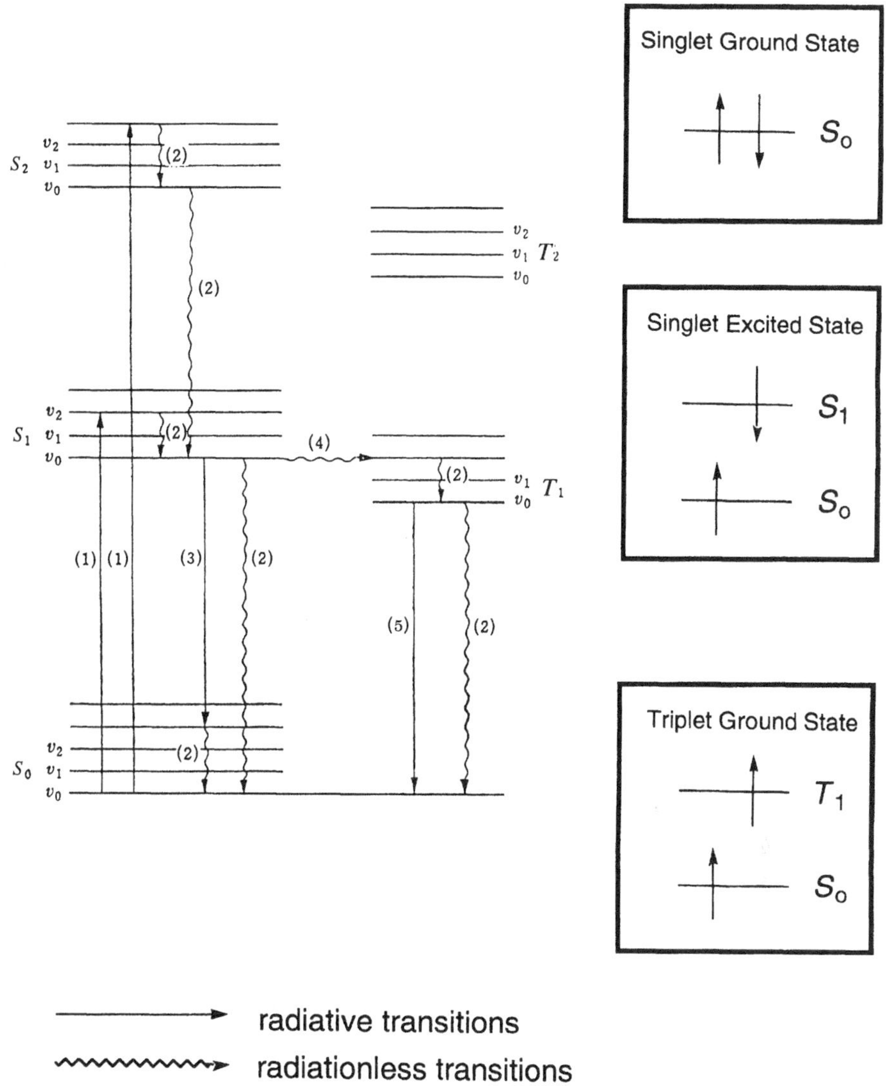

Figure 3.1 A Jablonski diagram showing the main photophysical processes that generally takes place in aromatic compounds: (1) absorption, (2) internal conversion, (3) fluorescence, (4) intersystem crossing, (5) phosphorescence.

Direct excitation processes from S_0 to a triplet excited state (T_n) can be neglected due to the spin-forbidden rule. The triplet excited state is typically generated by an intersystem crossing process by way of the singlet excited state (S_m), which is produced by the absorption from S_0 to S_m. When an electron goes down from an upper energy level to a lower one, energy is released as either a photon (radiative deactivation) or as heat (nonradiative deactivation). Luminescence released accompanying the transition either from S_m to S_n or from T_m to T_n is called fluorescence, whereas the luminescence accompanying transitions between S_m and T_n is called phosphorescence. However, in most cases, fluorescence arises from the transition between S_1 and S_0, whereas phosphorescence comes from the transition between T_1 and S_0.

Because each electronic state possesses multiple vibrational levels, both the transition from the ground state to an excited state and the reverse transition occur at various vibrational levels. A state involving electronic and vibrational levels is referred to as a vibronic state, and a transition between two such states is a vibronic transition. Thus, the uv–vis absorption spectrum of a molecule is given by observing the vibronic transitions.

Figure 3.2 shows (1) that the absorption and fluorescence spectra have fine structures reflecting vibrational levels and (2) that there are mirror symmetry relations between the absorption and fluorescence spectra, which are commonly observed. Before irradiation with light, almost all of the molecules in a sample are at their zero-point vibrational level of the electronic ground state ($S_0{:}v_0$) (the fraction of molecules in higher energy levels can be neglected). Thus, the fine structures in the uv–vis absorption spectrum correspond to transitions from $S_0{:}v_0$ to $S_m{:}v_n$. On the other hand, fine structures in the fluorescence spectrum usually correspond to transitions from $S_1{:}v_0$ to $S_0{:}v_n$. Even if a molecule or a group is excited to a higher electronic excited state S_m or a higher vibrational level of S_1 ($S_1{:}v_n$), the electron finally settles down in $S_1{:}v_0$ by releasing the excitation energy as heat because the nonradiative deactivation process between these levels takes place very rapidly. Because the nuclear configurations of S_0 and S_1 are sufficiently similar that the vibrational modes are the same, it is quite common that the transition probability from $S_1{:}v_0$ to $S_0{:}v_n$ (fluorescence) agrees well with that from $S_0{:}v_0$ to $S_1{:}v_n$ (absorption). The vibronic transitions from $S_1{:}v_0$ to $S_0{:}v_0$ (fluorescence) and from $S_0{:}v_0$ to $S_1{:}v_0$ (absorption) are referred to as the 0–0 transition.

If the conversion of spin (i.e., intersystem crossing) occurs during the electronic excited singlet state, an excited triplet state is produced. Similarly, as in the singlet state, an electron then rapidly goes down to the lowest possible vibrational level of T_1 ($T_1{:}v_0$) and settles for a relatively long time before releasing the excitation energy

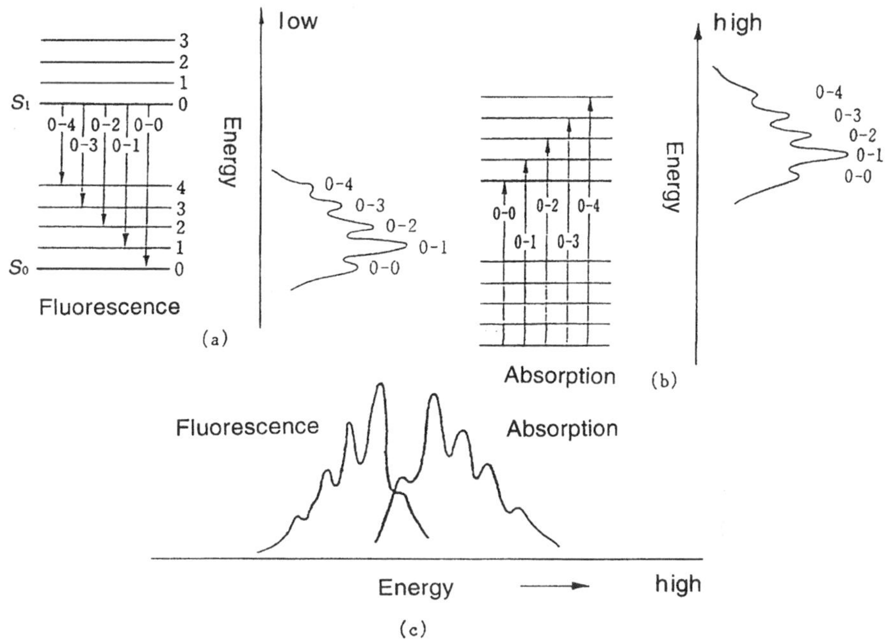

Figure 3.2 (a) Relationship of fine structure of fluorescence spectrum with vibrational levels. (b) Relationship of fine structure of absorption spectrum to vibrational levels. (c) Mirror symmetry relation between fluorescence spectrum and absorption spectrum.

because the transition from $T_1{:}v_0$ to $S_0{:}v_m$ is spin-forbidden. Roughly speaking, fluoresence lifetimes of aromatic compounds are 1 ps to 1 μs (1 $\mu\text{s} = 10^{-6}$ s), whereas phosphorescence lifetimes are from 1 μs to 10 s. In particular, the triplet excited state is usually so long-lived that it is subject to nonradiative deactivation by collision with oxygen or solvent molecules. Hence, phosphorescence can usually be neglected in fluid solution at room temperature.

In summary, absorption of a photon is the first step in a process that produces an electronically excited state. After the rapid radiationless deactivation, the electron settles in the lowest vibrational level of S_1 ($S_1{:}v_0$) for a while and then goes back either (1) to $S_0{:}v_0$ by releasing heat or (2) directly to a vibrational level of S_0 ($S_0{:}v_m$) by emitting fluorescence and back to $S_0{:}v_0$ by releasing heat ($v_m{-}v_0$).

For more detail, the reader is referred to those textbooks and review papers that describe general photophysical processes [1–11] and polymer photophysics, including the fluorescent probe method [12–16].

3.2.2 *WHAT IS NORMAL FLUORESCENCE BEHAVIOR?*

Now we can imagine how to obtain fluorescence of a chemical species because we have learned the general photophysical processes: First of all, we should measure the uv–vis absorption spectrum of the compound to determine the wavelength necessary to excite this compound. If a molecule absorbs a photon at such a wavelength it will reach the electronically excited state, after which we will be able to monitor the fluorescence of this molecule. To observe strong fluorescence, we should excite this compound at the peak wavelength of its absorption spectrum. However, we may not always be able to observe measurable fluorescence because the efficiency of absorbance and fluorescence depends on each chemical species. Some species are inherently nonfluorescent. In this section we will demonstrate typical fluorescence behavior, and we will discuss how to treat it quantitatively.

3.2.2.1 Information to Be Obtained from the uv–vis Absorption Spectrum

First, let us consider what information we should get from the uv–vis absorption spectrum. As shown already in Figure 3.2, the peak wavelength corresponds to the difference in energy between $S_0{:}v_0$ and $S_n{:}v_m$. The absorption intensity is based on the probability of such a transition. If a monochromatic (wavelength: λ nm) beam of light of intensity I_0 is pointed at a 1-cm-thick specimen containing a solution of c mol/liter and I is the intensity of the light that can transmit through this specimen, then the absorbance A at this wavelength is expressed according to the well known Lambert–Beer law's as

$$A = \mathrm{OD} = \log_{10}(I_0/I) = \varepsilon c l \tag{3}$$

where ε is the molar extinction coefficient, which has an inherent value for each chemical species and changes with wavelength. Here I for intensity is a measure proportional to the number of photons at wavelength λ. The absorbance is also known as the optical density OD. The Lambert–Beer's law holds at relatively low concentrations for which the OD is below 1.5.

The important information obtained from the uv–vis absorption spectrum is (1) the peak wavelength (λ_{max}) and (2) the molar extinction coefficient at this wavelength (ε_{max}). When we change the solvent, λ_{max} may shift to the blue or to the red, and ε_{max} may increase or decrease. In some cases, new absorption bands come out. The observation of such behavior gives precious information on the excited singlet electronic states. Uv–vis absorption spectroscopy is also a useful technique for determining the concentration of a sample for which the ε is already known. It is a general method to quantify the concentration of chemical species, and, as such, it is often applied in medical tests of blood and urine and for the assessment of environmental problems such as those that occur with water pollution. Because an excitation wavelength (λ) using a usual laser is a definite value, we have to determine the molar extinction coefficient (ε_λ) at such a wavelength to prepare a solution with an adequate concentration for the measurements of fluorescence.

3.2.2.2 Information to Be Obtained from the Steady-State and the Transient Fluorescence Spectra

For a given sample, the absorbance at a given wavelength must be true no matter who measures the absorption spectrum and irrespective of the spectrophotometer used. However, the fluorescence intensity at a given wavelength depends on a spectrofluorometer: the fluorescence intensity is not a universal property but rather is arbitrary. Usually the value of the fluorescence intensity I_F at a wavelength is expressed as

$$I_F = A_0 \Phi_f (1 - 10^{-\varepsilon cl}) \tag{4a}$$

where A_0 is a machine constant and Φ_f is the fluorescence quantum yield, which is a universal value. In the case of absorption spectra, the absorbance is proportional to concentration in the range of OD < 1.5. In the case of fluorescence spectra, the fluorescence intensity is in proportion to the concentration in the range of OD < 0.07, because $(1 - 10^{-\varepsilon cl})$ can be approximated as εcl at these low concentrations. Then,

$$I_F = A_0 \Phi_f \varepsilon cl \tag{4b}$$

The expression Φ_f is the fraction of photons released through fluorescence per one excited molecule; thus, the unit value means that each excited molecule definitely releases one photon through fluorescence. Figure 3.3 shows the kinetic scheme of unimolecular processes in a fluorescent molecule. Because, in general, no phosphorescence is detectable at room temperature in the case of aromatic compounds, the nonradiative deactivation process ought to include all the processes that possibly take place after the intersystem crossing. Thus, the rate equation is described as

$$d[\mathrm{M}^*]/dt = I_0 - (k_f + k_i)[\mathrm{M}^*] \tag{5}$$

A usual fluorescence measurement is performed under photostationary conditions by using a spectrofluorometer whose light source is often a xenon lamp. Under this condition, $d[\mathrm{M}^*]/dt$ should be zero. Therefore $[\mathrm{M}^*] = I_0/(k_f + k_i)$.

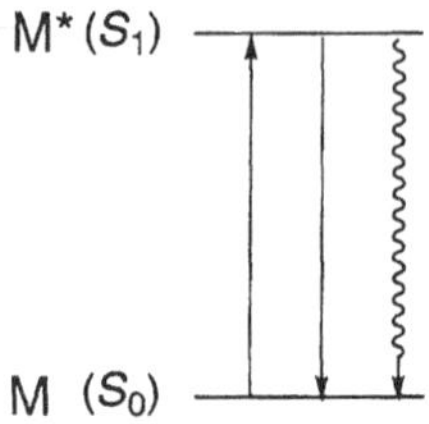

$$M \xrightarrow{I_0} M^* \quad \text{(absorption)}$$

$$M^* \xrightarrow{k_f} M \quad \text{(fluorescence)}$$

$$M^* \xrightarrow{k_i} M \quad \text{(heat) (nonradiative deactivation)}$$

The rate equation is

$$\frac{d[M^*]}{dt} = I_0 - (k_f + k_i)\,[M^*]$$

Figure 3.3 Kinetic scheme relating to typical fluorescence: unimolecular process with one excited state when one photon is given to one molecule.

Accordingly, the fluorescence intensity ($k_f[M^*]$) is expressed as $k_f I_0/(k_f + k_i)$. Because Φ_f is the value for the situation in which one photon is given to a molecule ($I_0 = 1$),

$$\Phi_f = k_f/(k_f + k_i) \tag{6}$$

If the fluorescence measurement is performed under transient conditions by means of pulse excitation such as with a pulsed laser or with a nanosecond flash lamp, we can observe the transient behavior. Rate equation, Eq. 5, is also valid under the transient condition, where I_0 can be treated as a δ-function. The solution is then expressed as a function of time t : $[M^*](t) = M_0 \exp[-(k_f + k_i)t]$, where M_0 is the concentration of M^* at $t = 0$. Thus, the time dependence of the fluorescence at a wavelength observed, $I_F(t)$, is expressed as

$$I_F(t) = k_f M_0 A_1 \exp[-(k_f + k_i)t] = F_0 \exp[-(k_f + k_i)t] \tag{7}$$

where A_1 and F_0 are constants. In general, normal fluorescence decays according to a single exponential function. The reciprocal of $(k_f + k_i)$ is a time constant measured in seconds, and it is defined as the fluorescence lifetime τ_f by

$$\tau_f = 1/(k_f + k_i) \tag{8}$$

Let us summarize the information obtained from the photostationary and transient fluorescence measurements. Figure 3.4(a) shows the fluorescence spectrum of 1-pyrenebutanoic acid (PB) in tetrahydrofuran (THF) at room temperature.

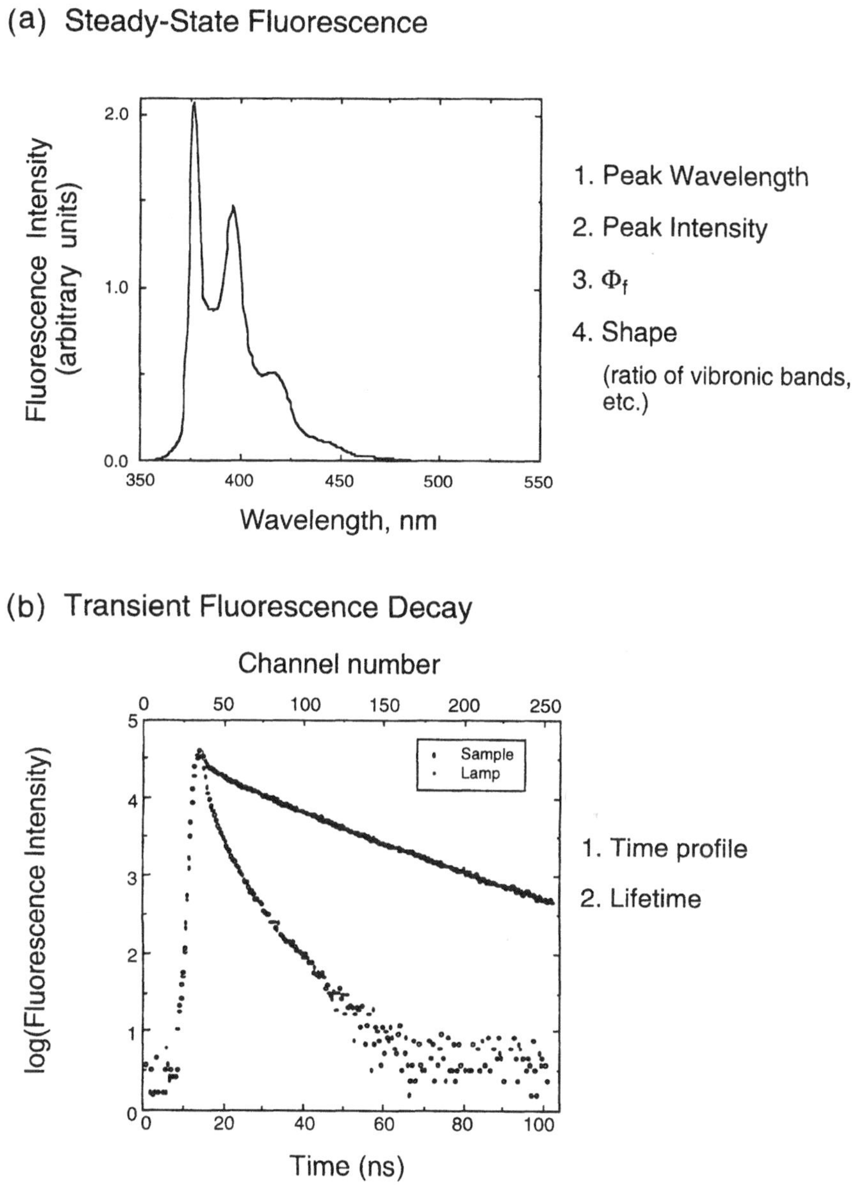

Figure 3.4 (a) Fluorescence spectrum of 1-pyrenebutanoic acid in degassed THF at room temperature (excitation wavelength: 344 nm). (b) Decay curve of fluorescence at 380 nm of 1-pyrenebutanoic acid in degassed THF at room temperature (excitation and emission light is separated through interference filters of 340 and 383 nm, respectively).

Important are (1) the wavelengths of several fluorescence peaks and (2) their intensities or Φ_f. A microenvironmental change can induce (1) a shift in the peak wavelengths, (2) a change in intensities of some vibronic peaks, and (3) a change in their ratios. A change in temperature, solvent, pH, and concentration could all influence the fluorescence behavior of a chemical species.

Figure 3.4(b) shows the time profile of PB fluorescence in degassed THF. The fluorescence at 380 nm is best fitted to a single exponential function with a time constant of 210 ns. The information we have to obtain from the fluorescence decay curves is (1) the decay pattern and (2) the decay parameters. As described in the following paragraphs, the lifetime will be shortened if other deactivation processes additionally take place from S_1. If new excited states are produced in a system, the fluorescence should follow a different function such as $\sum_{i=1}^{n} A_i \exp(-t/\tau_i)$. It is also possible to monitor the transient whole fluorescence spectrum, i.e., time-resolved fluorescence spectrum, by using a photodiode as a detector. Time-resolved spectra are very useful to grasp the total dynamic change such as when one excited component turns to another excited one. However, they are of no use if, as usual, only one fluorescent component is present.

Table 3.1 summarizes several values showing photophysical properties of some aromatic compounds. The introduction of some functional groups to an aromatic chromophore induces a shift of peak wavelengths and a change in their intensities; however, usually the total shape of the fluorescence spectrum is not changed dramatically.

3.2.2.3 The Fluorescence Depolarization Method

I would now like to introduce the fluorescence depolarization method. Figure 3.5 shows the principle of the measurements. An isotropic fluorescent sample situated at the origin is irradiated from the minus part of the x-axis, and its fluorescence is detected in the direction of the plus part of the y-axis. Because fluorescence is radiated to all directions, the fluorescence intensity S_{pol} proportional to the total fluorescence can be defined as follows:

$$S_{pol} = I_x + I_y + I_z \tag{9}$$

where I_x, I_y, and I_z are the intensities of the polarized fluorescence in the direction of x-, y-, and z-axes, respectively. If the sample is excited by polarized light in the z-axis direction, the radiated fluorescence will then be distributed symmetrically around the z-axis; then, $I_x = I_y$. Therefore,

$$S_{pol} = 2I_x + I_z \tag{10}$$

Table 3.1 Spectroscopic properties of some aromatic hydrocarbons [170].

			Absorption Properties			***Fluorescence Properties***	
Sample	***Solvent***	***State***	***0–0 transition (nm)***	***Peak (nm)***	ε_{max}	Φ_f	τ_f ***(ns)***
Benzene	n	S_1	260	254	250	0.053–0.07	26–32
		S_2		204	8,800		
		S_3		184	68,000		
	p	S_1	260			0.042	31
Toluene	n	S_1	269	262	260	0.14–0.20	34–38
		S_2		208	7,900		
		S_3		189	55,000		
	p						34
o-Xylene	n	S_1	271	263	260	0.16–0.19	32
		S_2		210	8,900		
		S_3		191	56,000		
Naphthalene	n	S_1	311	301	270	0.19–0.23	96–120
		S_2	286	275	5,600		
		S_3	221	221	117,000		
	p	S_1	311			0.12–0.21	52–118
1-Methylnaphthalene	n					0.18–0.25	67–83
	p	S_1	317	312	340	0.12	42
		S_2	291	282	6,000		
		S_3	224	224	60,000		
2-Methylnaphthalene	n					0.24–0.32	59
	p	S_1	319	319	530	0.16	47
		S_2		275	4,800		
		S_3	224	224	81,000		
Anthracene	n	S_1	375	375	8,500	0.24–0.36	3.9–5.8
		S_2	Not observed				
		S_3	252	252	220,000		
	p	S_1	375			0.22–0.30	4.5–6.0
Phenanthrene	n	S_1	346	330	250	0.11–0.16	56–60
		S_2	292	292	14,800		
		S_3		251	67,000		
	p	S_1	345			0.10–0.24	61–63
Pyrene	n	S_1	372	351	510	0.36–0.58	450
		S_2	334	334	55,000		
		S_3	272	272	54,000		
	p	S_1	372			0.53	475

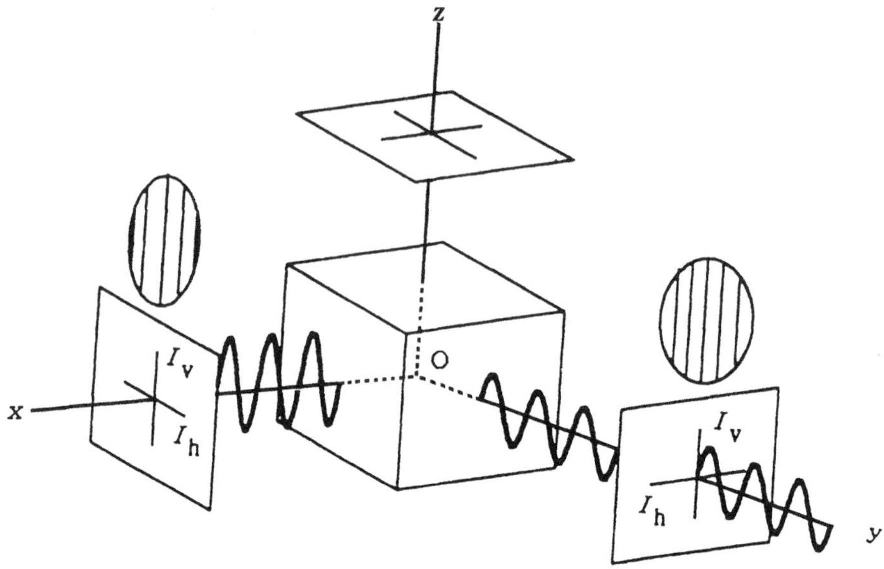

Figure 3.5 Principle of fluorescence depolarization method.

The fluorescence detected in the plus y-axis direction only consists of I_x and I_z when a polarizer is set. Thus, under the conditions described, we can determine the intensity proportional to the total fluorescence by measuring the intensity of the fluorescence in only one direction. The degree of polarization can be defined as

$$D_{pol} = I_z - I_x \tag{11}$$

The ratio of D_{pol} to S_{pol} can be a measure of the fluorescence polarization and is referred to as the anisotropy of the fluorescence.

$$r_{pol} = \frac{I_z - I_x}{I_z + 2I_x} \tag{12a}$$

The equation by which fluorescence anisotropy is obtained experimentally is Eq. (12b) in Section 3.3.2.5. In general, possible causes for fluorescence depolarization are (1) the inherent value itself is nearly zero (see Eq. (13): $\cos^2\omega = 1/3$), (2) molecular motion (rotational diffusion), (3) transportation of excitation energy (Section 3.2.3.3.2), (4) reabsorption, and (5) multiple scattering. However, the main causes must be (2) and (3). Thus, when a chromophore is excited by polarized light, the emission of the chromophore will be observed to be polarized if (1) the molecular motion of the chromophore is slow enough and (2) energy transfer and energy migration do not take place. Each fluorescent molecule has an inherent value of anisotropy, which can be determined by measurements using a plastic solution in which molecules are randomly distributed and isolated from one another. When ω is defined to be the angle between the transition moments of absorption and fluorescence, the inherent value of fluorescence anisotropy, $r_{\text{pol},0}$, is

$$r_{\text{pol},0} = 0.4 \times \frac{3\cos^2\omega - 1}{2} \tag{13}$$

Therefore, the values of $r_{\text{pol},0}$ are limited to be $-0.2 \leq r_{\text{pol},0} \leq 0.4$.

The measurements of the emission anisotropy and their comparison with $r_{\text{pol},0}$ give information on molecular motions and energy transport in the very system we would like to observe.

Here I briefly introduce the typical way to analyze the fluorescence anisotropy. It is general to assume that (1) a fluorescent probe is in an isotropic medium and (2) the rotational Brownian motion of a fluorescent molecule can be described by using the second moment of the orientational autocorrelation function $M(t)$ defined by

$$M(t) = \left\langle \frac{3\cos^2\theta(t) - 1}{2} \right\rangle \tag{14}$$

where $\theta(t)$ is the angle through which the emission transition moment of the fluorescent probe rotates from time zero to time t. The brackets indicate an average over all fluorescent molecules. Here the transient fluorescence anisotropy at time t can be expressed as follows:

$$r_{\text{pol}}(t) = r_{\text{pol},0} M(t) \tag{15}$$

The stationary fluorescence anisotropy r_{pol} is expressed as [17, 18]

$$r_{\text{pol}} = \frac{r_{\text{pol},0}}{\tau} \int_0^\infty M(t) \exp(-t/\tau)\, dt \tag{16}$$

where τ is the fluorescence lifetime.

According to the model for Brownian rotational motion of a spherical body, $M(t)$ can be described as

$$M(t) = \exp(-3t/\rho) \tag{17}$$

where ρ is the relaxation time of rotation. Usually ρ relates to the fluorescence rotational correlation time θ as

$$\theta = \rho/3 \tag{18}$$

and is also expressed as follows:

$$\theta = D_r/6 \tag{19}$$

$$= \frac{V\eta}{k_B T} \tag{20}$$

where D_r is a rotational diffusion coefficient, k_B is the Boltzmann constant, η is the viscosity of the solvent or the local viscosity, T is the absolute temperature, and V is the molecular volume. According to the preceding model,

$$r_{pol}(t) = r_{pol,0} \exp(-t/\theta) \tag{21}$$

$$\frac{1}{r_{pol}} = \frac{1}{r_{pol,0}}\left(1 + \frac{\tau}{\theta}\right)$$

$$= \frac{1}{r_{pol,0}}\left(1 + \frac{k_B T \tau}{\eta V}\right) \tag{22}$$

Equation (22) is known as the Perrin–Weber equation [19, 20]. Thus, a plot of $1/r_{pol}$ versus T/η would give a straight line, and a value of V can be calculated if τ is also determined. Usually the value of V is not identical to the real volume, because V is a hydrodynamic volume. For example, in the case of spherical proteins in water, the values obtained by the fluorescence anisotropy method are nearly equal to the volume of the protein surrounded by one layer of water molecules.

Several theoretical descriptions of $M(t)$ and $r_{pol}(t)$ are known. For example, if a fluorescent probe molecule is attached to a large macromolecule whose shape is far from that of a spherical body, or whose motion is not isotropic, or both, it is quite obvious that the Perrin–Weber equation is not valid at all. For example, Valeur *et al.* [21, 22] proposed Eq. (23) by assuming that the polymer chain performs three bond-jump motions on a diamond lattice.

$$r_{pol}(t) = r_{pol,0} \exp[(1/\tau_1 - 1/\tau_2)t]\mathrm{erf}\sqrt{t/\tau_1} \tag{23}$$

For details of the theoretical model, the reader is referred to the reviews in the literature [16, 23–25].

3.2.3 INTRODUCTION TO FLUORESCENCE BEHAVIOR USEFUL FOR THE MOLECULAR PROBE METHOD

I have explained the general fluorescence process of a molecule or a group isolated from other molecules and groups. In this section, I will introduce the more specific

Table 3.2 Main fluorescence behavior competent as probes.

Process		Main Information Possible to Obtain from Probes
Unimolecular process less dependent on circumstances		Concentration, tracer
Unimolecular process subject to influence from circumstances	Vibronic transitions (pyrenyl group)	Polarity
	Twisted intramolecular charge transfer (TICT)	Viscosity, polarity, molecular motion, free volume
	Molecular rotor	Viscosity, free volume
	N-Arylaminonaphthalene-sulfonate	Viscosity, polarity
	Dansyl group	Viscosity, polarity
Bimolecular process	Ground-state dimer	Orientation, concentration
	Charge transfer complex	Orientation, concentration, polarity
	Excimer	Orientation, concentration, molecular motion, free volume
	Exciplex	Orientation, concentration, polarity, molecular motion, free volume, distance
	Delayed fluorescence	Molecular motion, distance, concentration
	Singlet energy transfer	Distance, molecular motion, concentration
Fluorescence anisotropy		Molecular motion, free volume, viscosity, orientation

and more complicated processes relating to fluorescence. These processes supply more effective probing techniques for obtaining information on the microenvironment.

The processes introduced in this section are summarized in Table 3.2 along with the information necessary for using (A) unimolecular processes and (B) bimolecular processes as probes. In the case of (A), I will explain several instances of a fluorescent molecule being easily influenced by its surroundings. Understanding the reasons for its fluorescence being dependent on the microenvironment would be helpful for those readers wanting to employ these fluorescent molecules as probes.

The bimolecular processes (B) are roughly divided into two cases: (C) association (formation of a complex) and (D) transportation of excited energy. In the case of (C), two molecules have to contact each other. On the other hand, it is useful to consider the distance among two molecules in the case of (D).

3.2.3.1 Unimolecular Process Less Dependent on the Microenvironment

It is quite general to say that fluorescence properties change with the solvent more or less because of a change in the polarity, viscosity, or both, of the microenvironment. Figure 3.6 shows the polarity dependence of the fluorescence peak wavelength of dibenzyl terephthalate, a monomer model for poly(ethylene terephthalate) (PET) [26]. This is an example of fluorescence changing in response to a change in environment. Especially the fluorescence of a molecule or a group in a nonpolar solvent such as cyclohexane is generally blue-shifted and has clear, fine vibronic structures compared with those in polar solvents.

The molecules in the following sections are more subject to environmental circumstances, and the reasons are so interesting that we can make use of them.

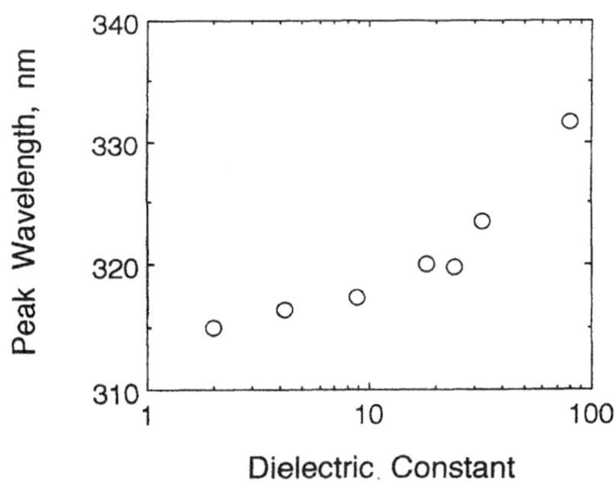

Figure 3.6 Polarity dependence of fluorescence peak wavelength of dibenzyl terephthalate, a model compound of PET, in various solvents. The solvents used are cyclohexane (dielectric constant = 2.0), diethyl ether (4.2), dichloromethane (8.9), 2-propanol (18.3), ethanol (24.3), methanol (32.6), and water (80.2), excitation wavelength is 260 nm. This figure was derived from ref. [26].

3.2.3.2 Unimolecular Process Subject to Influence from the Microenvironment

Here I introduce five case studies: (1) the pyrenyl group, (2) twisted intramolecular charge transfer (TICT), (3) the molecular rotor, (4) *N*-arylaminonaphthalenesulfonate, and (5) the dansyl group.

1. **Pyrenyl group** Figure 3.7 shows the solvent dependence of the pyrene fluorescence spectrum. Pyrene is one of the few condensed aromatic hydrocarbons that shows significant fine structure due to vibronic transitions in its monomer fluorescence spectrum in liquid solution. Pay attention to the first (peak I) and the third (peak III) vibronic peaks in Figure 3.7. The intensity ratio of peak III/peak I decreases very much with an increase in polarity of solvent. Thus, this ratio can be an effective tool as a probe to determine the polarity (i.e., bulk solvent dielectric constant) around a pyrene molecule.

   The change of the ratio is mainly due to the remarkable enhancement of peak I, the 0–0 vibronic band, with an increase in polarity of solvent, whereas peak III (0–737 cm^{-1} band) shows a minimal intensity variation with the polarity of the solvent. This is because the vibronic bands in pyrene fluorescence consist of allowed transitions (b_{1g} vibrations) and forbidden ones (a_g vibrations). The allowed transitions, such as peak III, are not influenced by polarity; however, the forbidden transition, including the 0–0 transition (peak I), readily takes place when one increases the polarity around a pyrene molecule. As described below, pyrene is often employed as a fluorescent probe because it shows strong excimer fluorescence in an area in which its local concentration is high.
2. **Twisted intramolecular charge transfer (TICT)** A series of derivatives of 4-(*N*,*N*-dimethylamino)benzonitrile or 4-(*N*,*N*-dimethylamino)benzoate exhibit characteristic fluorescence spectra in polar solvents, showing two emission bands corresponding to two different excited states (i.e., a* and b* states). According to Grabowski *et al.* [28, 29], this phenomenon is explained by the schematic diagram shown in Figure 3.8. The dimethylamino group can be an electron donor in polar solvents. This group rotates around the amino–phenyl bond from the initial planar structure b* to a* structure in the excited state. In the a* state, a charge separation between donor and acceptor moieties takes place, as shown in Figure 3.8. Thus, in polar solvents, the twisted and polar a* band appears to the red relative to the

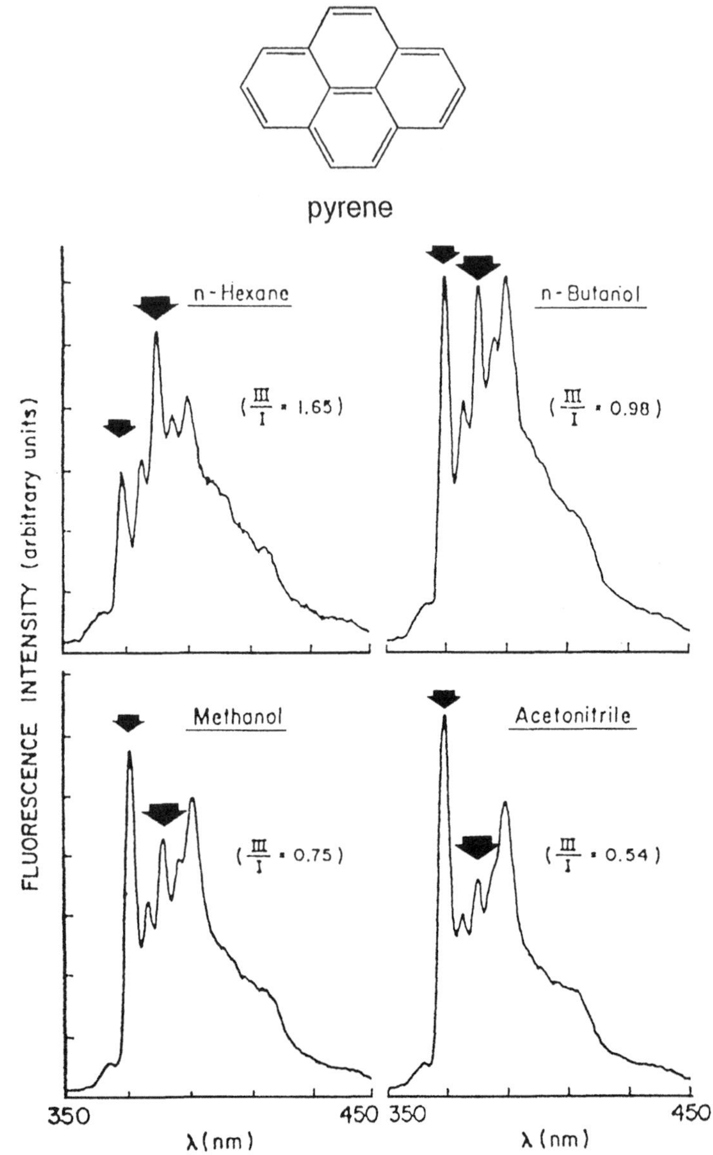

Figure 3.7 Solvent dependence of vibronic band intensities in pyrene fluorescence: [pyrene] = 2 μmol/liter (excitation wavelength is 310 nm) from ref. [27]. Small and large arrows correspond to peak I and peak III, respectively.

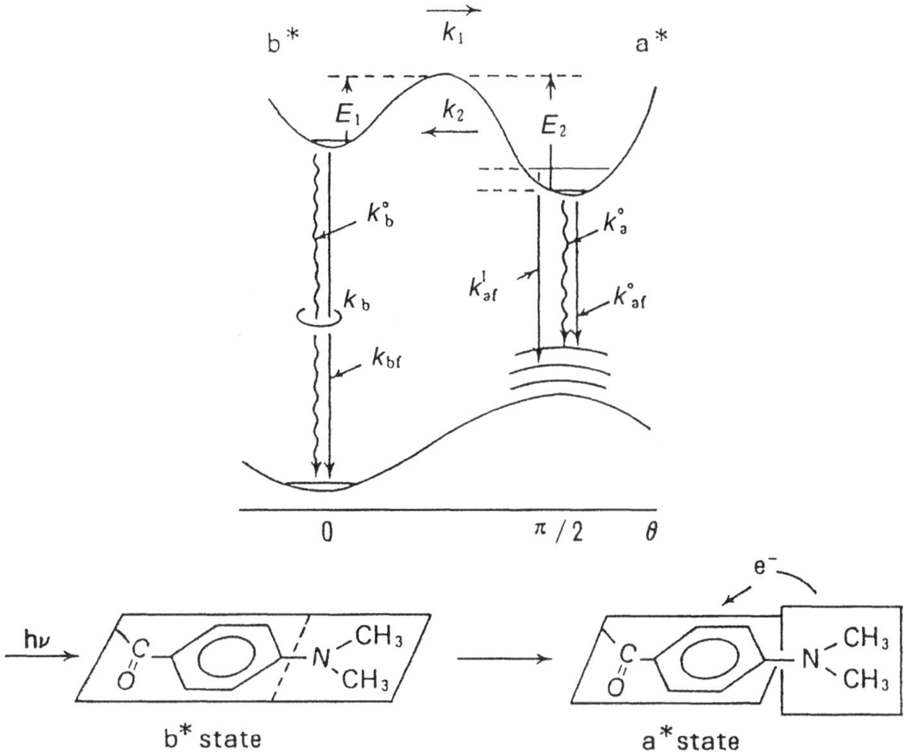

Figure 3.8 Schematic diagram showing twisted intramolecular charge transfer (TICT) proposed by Grabowski *et al.* in refs. [28, 29]. The θ represents the twist angle between the donor and acceptor planes.

planar noncharge-transfer b* band. Nonpolar and viscous environments would prevent the formation of the a* state. Thus, the TICT compound is anticipated to act as a probe to detect microscopic polarity and viscosity. For example, Hayashi *et al.* attached 4-(N,N-dimethylamino)benzoate groups onto poly(methyl methacrylate) and observed the effect of polymer concentration on local viscosity and microscopic polarity [30].

3. **Molecular rotor** The molecular rotors are similar to TICT; however, their emissions do not consist of dual components. Only the intensity of the fluorescence changes with polarity and viscosity. Figure 3.9 shows the structure of a series of molecular rotors, [p-(N,N'-dialkylamino)-benzylidene]malononitriles (1–3) along with schematic diagrams of the rotors. The internal rotation of the NR_2 group (route I) and of two plane parts (route II) urges the nonradiative deactivation. Thus, the fluorescence intensity of compound 1 is consistently lower than that of

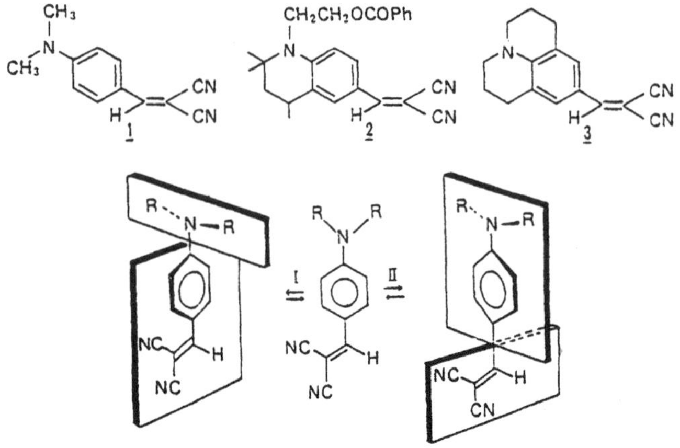

Figure 3.9 Structure of molecular rotors and schematic diagram showing the internal molecular rotation of the NR_2 group (route I) and of the donor and acceptor moiety (route II). The diagram is taken from ref. [31].

compounds 2 and 3 in fluid solution because compound 1 has two routes to radiationless deactivation. Moreover, if the viscosity of the fluid solution is raised, the fluorescence intensities of 1 and 3 increase remarkably, as shown in Figure 3.10 because routes I and II prevent free internal rotation [32]. Thus, the molecular rotors are important probes for measuring especially the local viscosity and free volume around them.

4. ***N*-Arylaminonaphthalenesulfonate** The application of fluorescent probes to biochemical research began with the introduction of 1-anilinonaphthalene-8-sulfonate (ANS) by Weber and Laurence in 1954 [33]. The derivatives of arylaminonaphthalenesulfonates have the interesting property of fluorescing weakly in water and strongly in nonpolar organic solvents or when bound to proteins. In fact, the peak intensity of ANS fluorescence increases by 200 times with a decrease in polarity of solvent, whereas the peak wavelength is shifted to the blue by 60 nm. These drastic changes make ANS one of the most famous fluorescent probes to detect hydrophobic environments. In addition to ANS, 2-*N*-*p*-toluidinylnaphthalene-6-sulfonate (TNS) [34] and *N*-methyl-*N*-2-anilino-naphthalene-6-sulfonate (MANS) [35] are often employed.

The mechanism for the dependency of *N*-arylaminonaphthalene-sulfonate's fluorescence on microenvironment is shown in Figure 3.11 together with the polarity dependence of fluorescence of 2-*N*-phenylamino-6-naphthalenesulfonate (PNS) [36]. In the ground

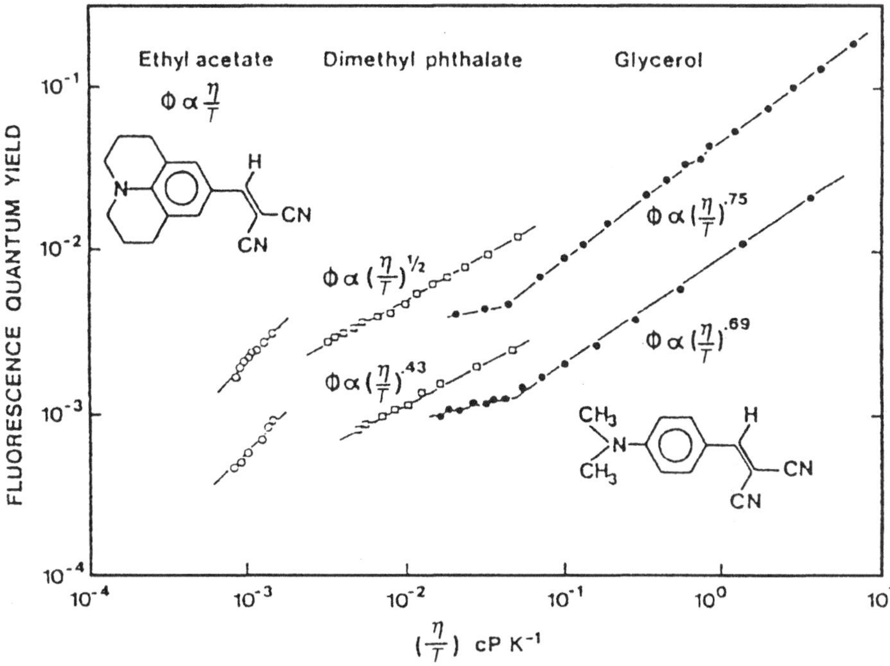

Figure 3.10 Dependence of the fluorescence quantum yield of 1 and 3 (in Figure 9) on η/T in (○) ethyl acetate, (□) dimethyl phthalate, and (●) glycerol as derived from ref. [32].

state of PNS, the N-aryl group and the naphthyl moiety are not in the same plane but are situated in a twisted configuration. This nonplanar configuration is excited to the $S^*_{1,\mathrm{np}}$ state, which can change to become a donor-acceptor complex $S^*_{1,\mathrm{ct}}$ by intramolecular electron transfer. The fluorescence from $S^*_{1,\mathrm{ct}}$ is shifted to the red due to its resonance. Polar solvents enhance the formation of $S^*_{1,\mathrm{ct}}$ by accelerating the electron transfer process; however, nonpolar, or viscous media, or both, prevent electron transfer from $S^*_{1,\mathrm{np}}$ to $S^*_{1,\mathrm{ct}}$, meaning that a strong fluorescence shifted to the blue will be observed. In polar and nonviscous solvents, most excited states are $S^*_{1,\mathrm{ct}}$, which can fluoresce but more readily deactivates through nonradiative processes because of rapid electron transfer to form the $S_{0,\mathrm{np}}$. In polar and viscous solvents such as ethylene glycol, a strong fluorescence from $S^*_{1,\mathrm{ct}}$ can be observed. In summary, both polarity and viscosity influence the fluorescence behavior of ANS derivatives, which consists of two electron transfer processes:
(1) $S^*_{1,\mathrm{np}} \rightarrow S^*_{1,\mathrm{ct}}$ and (2) $S^*_{1,\mathrm{ct}} \rightarrow S_{0,\mathrm{np}}$.

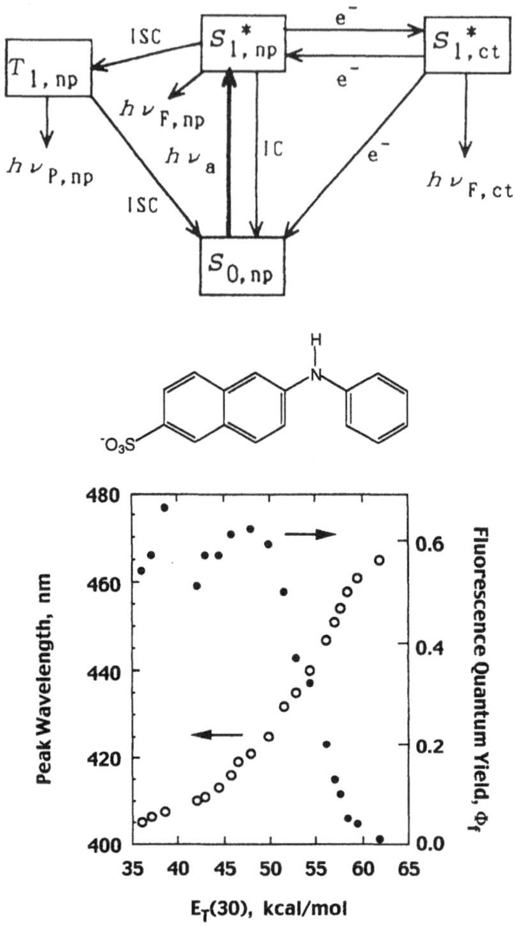

Figure 3.11 A plot of fluorescence peak wavelength (○) and fluorescence quantum yield (●) versus the solvent polarity parameter, E_T (30) (from ref. [37]) along with mechanism for the fluorescence of 2-*N*-phenylamino-6-naphthalenesulfonate (from ref. [36]).

5. **Dansyl group [1-(dimethylamino)-5-naphthalenesulfonic acid derivatives]** With an increase in polarity around a dansyl group, the fluorescence peak shifts to the red along with a decrease in intensity. For example, the wavelength of the fluorescence peak of dansylhydrazine is 466 nm in cyclohexane and 541 nm in methanol. Both the fluorescence behavior and the structure are similar to those of ANS derivatives; however, the interpretations valid for the ANS system are not adequate to

explain the dansyl system. In fact, the effects are observed only in the case of dansyl probes for which (1) the rate constant of fluorescence k_f decreases as the solvent polarity increases and (2) the half-width of the fluorescence spectral band is greater in polar solvents than in nonpolar solvents [38]. Li *et al.* explained the polarity dependence of dansyl fluorescence by using the concept of inversion of the excited states shown in Figure 3.12 [38].

Platt [39] has termed the two lowest $\pi \rightarrow \pi^*$ transitions for polyacenes as ^1L_a and ^1L_b. For naphthalene, the ^1L_b state is lower in energy than ^1L_a by about 3400 cm^{-1} and is long-axis polarized, whereas the ^1L_a state is short-axis polarized. Because of the difference in electron distribution in the two excited states, a substituent at the 1-position of naphthalene will interact more effectively with the ^1L_a excited state than with the ^1L_b state; conversely, 2-position substituents will interact more strongly with the ^1L_b state. In aromatic systems, intramolecular charge transfer transitions result from mixing of the $\pi \rightarrow \pi^*$ transition of the aromatic ring with the lone-pair transition from the amino group ($n \rightarrow \pi^*$). Such a mixed transition will occur in 1-aminonaphthalenes involving the ^1L_a transition of naphthalene and the nonbonding electron pair of the amino group. The ^1L_b band of 1-aminonaphthalene has much less charge transfer character and remains essentially a $\pi \rightarrow \pi^*$ transition, meaning that this band is insensitive to solvent polarity. Conversely, in 2-aminonaphthalene, the ^1L_b band will acquire charge transfer character, whereas the ^1L_a band will remain largely a $\pi \rightarrow \pi^*$ transition. Thus, the solvent effects on the 1- and 2-aminonaphthalene result in the conversion of the energy levels shown in Figure 3.12.

It is self-evident that the cause for the sensitivity of the ^1L_a band of 1-aminonaphthalene to polarity is the same as that for the ANS system. A change in the configuration of the amino group from tetrahedral (sp^3) in S_0 to trigonal (sp^2) in S_1 induces the increase in dipole moment upon excitation [40].

3.2.3.3 Bimolecular Processes

As described previously, the bimolecular processes are divided into two cases: (C) association (formation of a complex) and (D) transportation of excited energy. The bimolecular processes described here include the case in which two moieties or two groups belong to a same macromolecule.

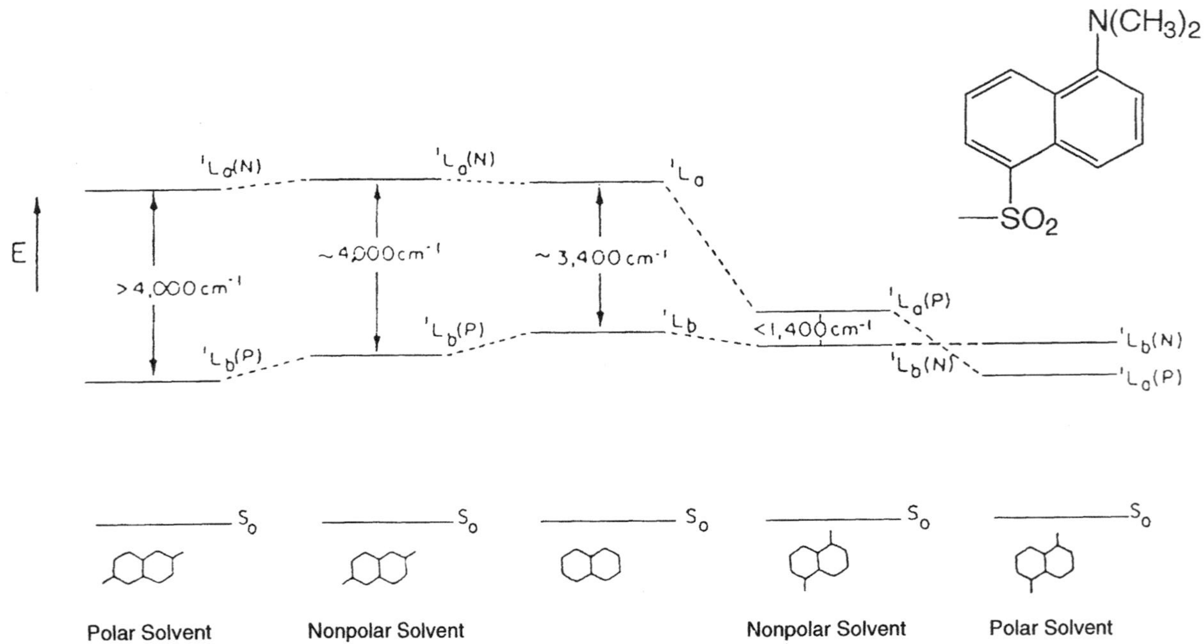

Figure 3.12 Solvent effects on the ^1L_a and ^1L_b transition energies for 1- and 2-aminonaphthalenes from ref. [38]: P = polar state; N = nonpolar state.

3.2.3.3.1 Association of Two Molecules

Two factors are used to classify the association processes of two molecules relating to fluorescence process: (1) whether each molecule is in the excited state (S_1) or in the ground state (S_0), and (2) whether two molecules are the same species or not.

3.2.3.3.1.1 Two Molecules in the Ground state If two separate molecules are of the same chemical species, the complex formed is referred to as a ground-state dimer. If two molecules are of different species, the complex formed is described as a charge-transfer (CT) complex or an electron donor–acceptor complex. Because two chromophores interact with each other in their ground states, the complex formed should behave like a new chemical species; that is to say, both absorption and fluorescence spectra of a ground-state dimer and a CT complex are different from those of their constituent molecules. Generally, the peak wavelengths of absorption and fluorescence spectra are shifted to the red when a ground-state dimer or a CT complex is formed.

Aromatic hydrocarbon molecules themselves do not normally associate in the ground state; however, once they are introduced into a main chain or a side chain of a macromolecule, the fluorescence of their ground-state dimer can be observed in the aggregated or solid state. The formation of a ground-state dimer is due to the resonance of a π-electron among two of the same species; thus, two aromatic groups are assumed to be in parallel configuration within a distance of, at most, 0.35 nm. Figure 3.13 shows the fluorescence spectra of the ground-state dimer of PET solids along with the spectra of a monomeric phenylene moiety in fluid solution and solids. Some chromophores in a macromolecule show the fluorescence of the ground-state dimer in the solid state even if the concentrated fluid solution of their monomer model compounds do not show any at all.

The formation of a CT complex is often observed between an electron-rich aromatic compound and a compound that can accept an electron.

$$\mathrm{D^+ + A^- \rightarrow (D^+ - A^-)}\ \text{(a CT complex)} \tag{24}$$

Because Mulliken's CT theory holds [41, 42], the peak wavelengths of absorption and fluorescence shift to the red when the ionization potential of donors $\mathrm{IP_D}$ and the electron affinity of acceptors $\mathrm{E_a}$ increases. Namely,

$$h\nu_{\mathrm{CT}} = \mathrm{IP_D} - \mathrm{E_a} + C \tag{25}$$

where $h\nu_{\mathrm{CT}}$ is the absorption energy in the CT band (see Eq. (2)) and C is a constant. The characteristic features in fluorescence of a CT complex are (1) the fluorescence peaks appear in a considerably longer wavelength region than those

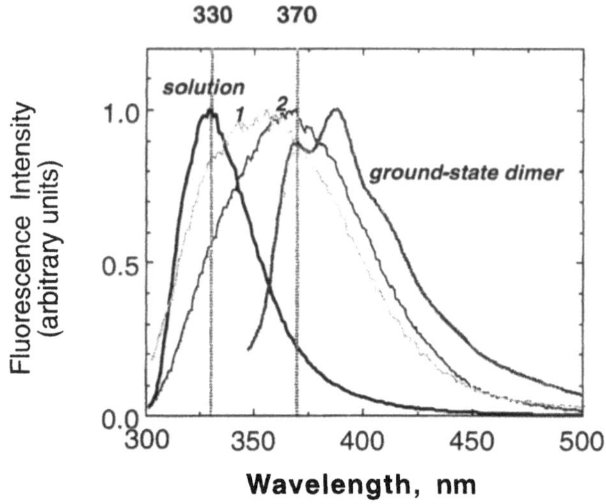

Figure 3.13 Fluorescence spectra of solid poly(ethylene terephthalate) (PET) with a spectrum of a dilute PET solution of chloroform and trifluoroacetic acid (9:1 v/v). 1 and 2 are PET solids prepared by a spin-casting method cast at 160 °C (1) and 40 °C (2). These three spectra were obtained by the excitation at 290 nm. The fluorescence of the PET ground-state dimer was obtained by exciting commercial PET film used for a bottle at 336 nm.

of the constituent molecules, (2) the fluorescence is broad and without structure, (3) the fluorescence intensities as a whole are weak, and (4) the difference between the wave numbers of absorption and fluorescence, the so-called Stokes shift, is largely due to the large dipole moment of the excited CT complex state. Usually, lifetimes of CT complexes are from 10^{-9}–10^{-8} s.

In both cases, the observation of the association in the ground state means that the two molecules are in contact with each other in the solid state or aggregated state, or both, such as in a gel.

3.2.3.3.1.2 One Molecule in the Ground State Associates with Another in the Excited State If two molecules are of the same chemical species, the complex formed is referred to as an excimer (excited dimer). If two molecules are of different species, the complex formed is described as an exciplex (excited complex). Both of them are formed in an excited state, and thus their properties are summarized as follows: (1) both the absorption spectrum and the fluorescence excitation spectrum are definitely identical with the spectra of their constituent molecules because there is no interaction between two molecules in the ground state, (2) the

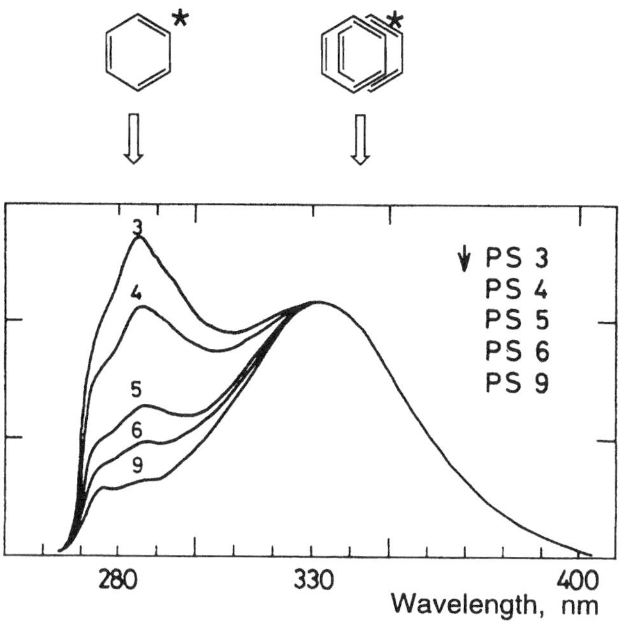

Figure 3.14 Fluorescence spectra of atactic oligostyrenes, PS*n* (*n*-degree of polymerization) with symmetric propyl ends in degassed cyclohexane solution at 25 °C. The spectra are normalized at the excimer peak and have been taken from ref. [43].

fluorescence spectrum of an excimer or an exciplex differs from the spectra of their constituent molecules, and (3) an excimer or an exciplex turns into two molecules in the ground state after the deactivation of excited energy by either fluorescence or heat.

1. **Excimer** By analogy with a ground-state dimer, two aromatic groups in an excimer state are in parallel configuration within a distance of 0.35 nm. Figure 3.14 demonstrates the fluorescence spectra of oligostyrenes in cyclohexane [43]. The peaks near 285 and 330 nm have been assigned to a phenyl monomer singlet and excimer emission, respectively. The excimer formed in oligostyrenes is among two adjacent side-chain phenyl moieties. An excimer is a good probe for getting information on local concentration, mobility, and orientation because two chromophores must encounter, and take a parallel configuration, within its excited lifetime.

   Broad fluorescence without structure due to excimers with wholly overlapping aromatic rings usually appears in the longer-wavelength region from 5000 to 6000 cm^{-1} from the monomer fluorescence. However, in the case of large condensed aromatic hydrocarbons, a partially overlapping structure of rings sometimes demonstrates another structureless fluorescence whose peak is situated in the middle of monomer and normal excimer fluorescence. It is known as a second excimer or a partially overlapped excimer; we have thus shown the kinetics of two different excimers in dimer model [44, 45] and in polymer systems [46]. The discrimination of two excimers would yield the information on very local areas such as whether there exists free volume for aromatic rings to rotate freely.

The kinetic scheme of a system in which only normal intramolecular excimer fluorescence is observable is shown as Figure 3.15. The relations corresponding to Eqs. (5–8) are obtained as follows: First, the rate equation is described as Eq. 26 in which two excited states exist: i.e., the monomer singlet excited state and the excimer state.

$$\frac{d[\mathrm{M}^*]}{dt} = I_0 - (k_{\mathrm{FM}} + k_{\mathrm{IM}} + k_{\mathrm{EM}})[\mathrm{M}^*] + k_{\mathrm{ME}}[\mathrm{E}^*]$$
$$\frac{d[\mathrm{E}^*]}{dt} = k_{\mathrm{EM}}[\mathrm{M}^*] - (k_{\mathrm{FE}} + k_{\mathrm{IE}} + k_{\mathrm{ME}})[\mathrm{E}^*] \tag{26}$$

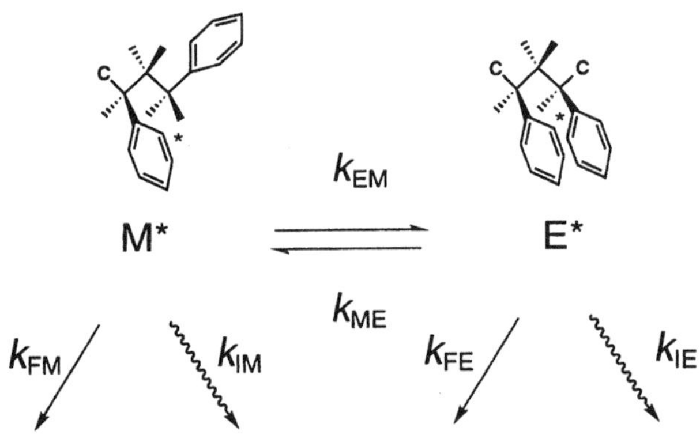

Figure 3.15 Kinetic scheme for intramolecular excimer formation (M* and E* are monomer singlet and excimer, respectively, C means the main chain; k_{FM} and k_{FE} are the rate constants for the radiative deactivation from excited monomer and excimer state, respectively; k_{IM} and k_{IE} are those for the non-radiative deactivation, and k_{EM} and k_{ME} are those for excimer formation and dissociation, respectively). The figure is a meso diad of polystyrene for which the conformations of M* and E* correspond to g^+t/tg^- and tt.

Under photostationary conditions ($d[\mathrm{M}^*]/dt = d[\mathrm{E}^*]/dt = 0$), the ratio of excimer fluorescence intensity I_E to monomer fluorescence intensity I_M is expressed as

$$\frac{I_\mathrm{E}}{I_\mathrm{M}} \propto \frac{k_\mathrm{EM} k_\mathrm{FE}}{k_\mathrm{FM}(k_\mathrm{FE} + k_\mathrm{IE} + k_\mathrm{ME})} \tag{27}$$

Generally speaking, excimer dissociation can be neglected at low temperatures; in the case of polystyrene, the excimer does not dissociate very much at room temperature. Therefore, if k_ME can be neglected in Eq. (27), the rate constant for excimer formation is proportional to the ratio $I_\mathrm{E}/I_\mathrm{M}$.

Under transient conditions, I_0 can be approximated with a δ-function. Because Eq. (26) consists of simultaneous differential equations, the time dependence of monomer fluorescence $I_\mathrm{M}(t)$ and excimer fluorescence $I_\mathrm{E}(t)$ is calculated as follows under the condition where $I_\mathrm{M}(0) = 1$, $I_\mathrm{E}(0) = 0$, and k_ME are not neglected.

$$\begin{aligned} I_\mathrm{M}(t) &= \frac{1}{(\lambda_2 - \lambda_1)}[(\lambda_2 - X)\exp(-\lambda_1 t) + (X - \lambda_1)\exp(-\lambda_2 t)] \\ I_\mathrm{E}(t) &= \frac{k_\mathrm{FE} k_\mathrm{EM}}{k_\mathrm{FM}(\lambda_2 - \lambda_1)}[\exp(-\lambda_1 t) - \exp(-\lambda_2 t)] \end{aligned} \tag{28}$$

where

$$\lambda_{2,1} = \frac{1}{2}[X + Y \pm \{(Y - X)^2 + 4k_\mathrm{EM}k_\mathrm{ME}\}^{1/2}] \tag{29}$$

$$X = k_\mathrm{FM} + k_\mathrm{IM} + k_\mathrm{EM}$$

$$Y = k_\mathrm{FE} + k_\mathrm{IE} + k_\mathrm{ME}$$

If k_ME can be neglected,

$$\begin{aligned} I_\mathrm{M}(t) &= \exp[-(k_\mathrm{FM} + k_\mathrm{IM} + k_\mathrm{EM})t] \\ I_\mathrm{E}(t) &\propto (\exp[-(k_\mathrm{FE} + k_\mathrm{IE})t] - \exp[-(k_\mathrm{FM} + k_\mathrm{IM} + k_\mathrm{EM})t]) \end{aligned} \tag{30}$$

In this case, the monomer fluorescence decays single-exponentially with a time constant whose reciprocal is the sum of k_FM, k_IM, and k_EM. $k_\mathrm{FM} + k_\mathrm{IM}$ can be substituted by the experimental value of the same chromophore (or a monomer model) in the case in which the excimer is not formed, thus, each k_EM can be determined by measuring fluorescence decay curves. This method is effective to monitor the dynamic motion of two molecules (or groups) that encounter.

2. **Exciplex** The behavior of exciplexes is similar to that of excimers except that the two chemical species are different. An exciplex is a CT complex formed only in the excited state, and thus its structure is not so strict as that of an excimer. The typical exciplex is formed between aromatic hydrocarbons and amines, or between aromatic hydrocarbons and cyano compounds: An exciplex has an polar structure whose electron distribution is so localized that it is subject to the influence of polarity around it. Therefore, the fluorescence spectrum of an exciplex is remarkably shifted to the red with an increase in polarity of the solvent. Note that the polarity should greatly influence the fluorescence intensity and the fluorescence lifetime of the exciplex.

   As a fluorescence probe, an excimer has much merit because the steric structure of an exciplex is loose. However, there is one interesting property that only an exciplex has. In some cases, an exciplex consists of three species and is then known as a triple exciplex. In a toluene solution of naphthalene (donor) and *p*-dicyanobenzene (acceptor), the fluorescence whose peak is at 400 nm is first observed, corresponding to a 1:1 exciplex. If the concentration of naphthalene is increased after the formation of this exciplex, the fluorescence intensity of the 400-nm band decreases, and a new band near 480 nm appears. This is due to the formation of 2:1 exciplexes, i.e., $(DD^+A^-)^*$ [47]. A triple exciplex is sometimes formed between an excited state of a CT complex, $(D^+A^-)^*$ and a donor molecule; for example, an exciplex of benzene with a complex between 1,2,4,5-tetracyanobenzene (TCNB) and benzene [48].

3.2.3.3.1.3 Two Molecules in the Excited State Strictly speaking, this case does not result in the formation of an associated complex, but the interaction between the two identical molecules is necessary to give rise to this phenomenon. Consequently, the encounter of two molecules, each in an excited triplet state, induces the T–T annihilation process, yielding one molecules in the excited singlet state (S_1) and the other in the ground state (S_0). Figure 3.16 shows the schematic diagram of the T–T annihilation process. The molecule in S_1 releases its excitation energy by either fluorescence or heat. In this case, the shape of the fluorescence spectrum obtained is perfectly identical with that normally measured under photostationary conditions. However, the time profile is completely different from the normal fluorescence because this fluorescence is by way of the triplet excited state

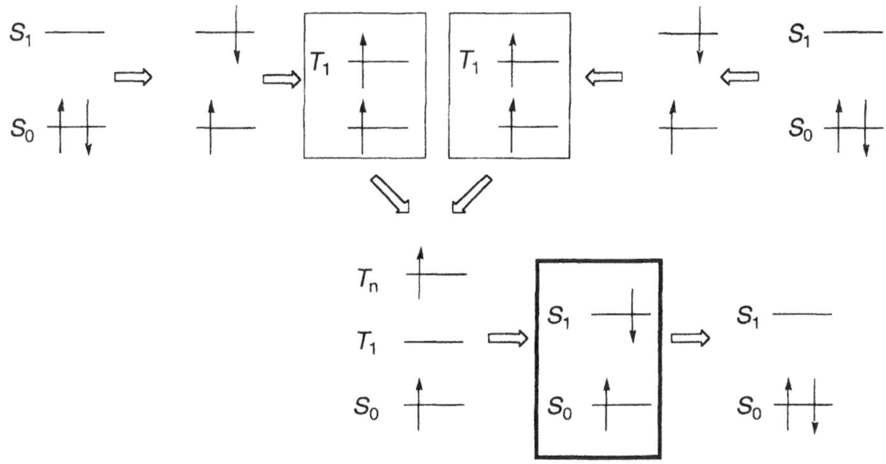

$$T_1^* + T_1^* \longrightarrow S_1^* + S_0 \xrightarrow{\text{Fluorescence}} 2\,S_0$$

Figure 3.16 The schematic diagram of the $T-T$ annihilation process to yield a P(pyrene)-type delayed fluorescence.

whose lifetime is remarkably long compared with the singlet excited state. This fluorescence is called the delayed fluorescence.

As a matter of fact, there are two types of delayed fluorescence: one is the P(pyrene)-type and the other is the E(eosin)-type. The E-type delayed fluorescence corresponds to the case in which the S_1 state is produced again by thermal activation of the excited triplet state (T_1) during its excitation lifetime to a vibrational level isoenergetic with S_1. The very delayed fluorescence described in this section is the P-type, which is anticipated to be a good probe for monitoring two encountering molecules. There are successful papers by Ushiki *et al.* [49–52] studying the behavior of polymer chains by means of the T–T annihilation process of anthryl groups.

3.2.3.3.2 Transportation of Excited Energy

3.2.3.3.2.1 Introduction to Energy Transfer and Migration Here I present the case in which only excitation energy is transported. Thus far I have described the bimolecular processes that influence the shapes of absorption and fluorescence spectra; however, each chromophore excited by the transportation of excitation energy demonstrates its own inherent photophysical process as well as when excited directly. The transportation process is shown as

$$A^* + B \rightarrow A + B^* \tag{31}$$

where the donor A and acceptor B are molecules or groups. The process is divided into two cases, namely radiative and nonradiative transportation. The latter is more important because of its dependence on distance between a donor and an acceptor. The former is due to the reabsorption by the acceptor of the photon emitted by the donor; however, this process would influence the spectrum when the concentration of acceptors is very low. Thus, the reader should always check whether or not radiative energy transportation occurs in his or her experimental system.

If A and B are the same chemical species, this transportation of the electronic excitation energy is described as energy migration or excitation migration. If A and B are different chemical species, the transport is described as energy transfer or excitation transfer.

Quenching is the term describing the decrease or disappearance of A* fluorescence in Eq. (31), and sensitization is a term describing the increase of B* fluorescence. If observation of only the quenching of A* fluorescence is desired, one should use the nonfluorescent reagent B as a quencher. If one would like to observe only sensitization of B* fluorescence, nonfluorescent reagent A should be used as a sensitizer.

3.2.3.3.2.2 Mechanism of Energy Transfer Energy transportation usually takes place when the excitation energy level of a donor is higher than, or isoenergetic with, that of an acceptor during the lifetime of the donor. Because excitation transportation depends on the coupling strength of the donor and the acceptor, the mechanism without any molecular motion is theoretically developed and established for three cases: (1) the exciton band theory, (2) the Förster mechanism (long-range, singlet), and (3) the Dexter mechanism (short-range, triplet).

Case (1) takes place in aromatic molecular crystals in which the molecules are oriented together by weak van der Waals forces. Thus, their coupling strength is so strong that an excited band orbital is formed among molecules. Consequently, the energy migration takes place as a coherent migration of the exciton wave through the crystal when the exciton–phonon scattering is sufficiently small [53, 54]. In typical polymer systems, the regular orientation, such as it occurs in aromatic crystals, is not expected, and the coupling strength should be weak. Therefore, the excitation energy is considered to be localized in a molecule or a group and to be transferred to another molecule or group by a one-step hopping process such as (2) and (3).

Energy transport due to coulombic resonance interactions (mechanism 2) was developed by Förster [55, 56] and Galanin [57]: it accurately describes the singlet energy transfer and migration processes. The theory is based on the assumption that the leading term in the interaction is the dipole transition moments between a donor and an acceptor. The rate constant for energy transfer k_{DA} is expressed as a function of the distance between the donor and the acceptor r. Here R_0 is a critical radius: when $r = R_0$, the value of k_{DA} is identical with that of the reciprocal of donor fluorescence lifetime τ_0 in the absence of acceptors.

$$k_{DA} = \frac{(9000 \ln 10)\kappa^2 \Phi_D}{128\pi^5 n^4 N_A r^6 \tau_0} \int_0^\infty F_D(\tilde{\nu})\varepsilon_A(\tilde{\nu})/\tilde{\nu}^4 \, d\tilde{\nu} \tag{32}$$

$$= \frac{1}{\tau_0}\left(\frac{R_0}{r}\right)^6 \tag{33}$$

where ε_A is the extinction coefficient of the acceptor, F_D is the fluorescence intensity of the donor at a wave number $\tilde{\nu}$; satisfying $\int_0^\infty F_D \, d\tilde{\nu} = 1$; Φ_D is the fluorescence quantum yield of the donor in the absence of the acceptor; n is the refractive index; N_A is Avogadro's number; and κ^2 is a molecular orientation factor. Figure 3.17 shows the relationship between the vectors of donor fluorescence

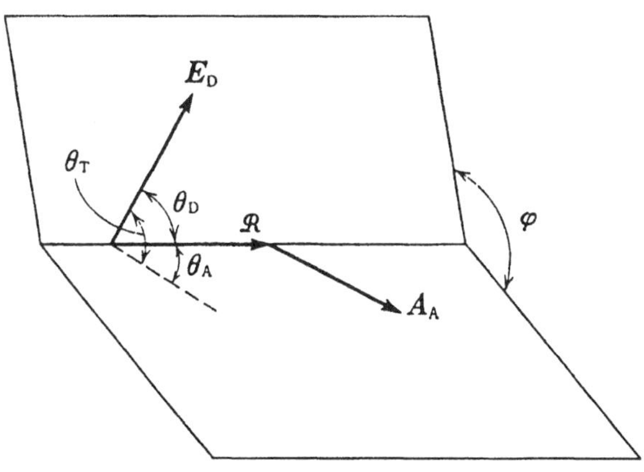

Figure 3.17 Relationship between the vectors of donor fluorescence, E_D, and acceptor absorption, A_A.

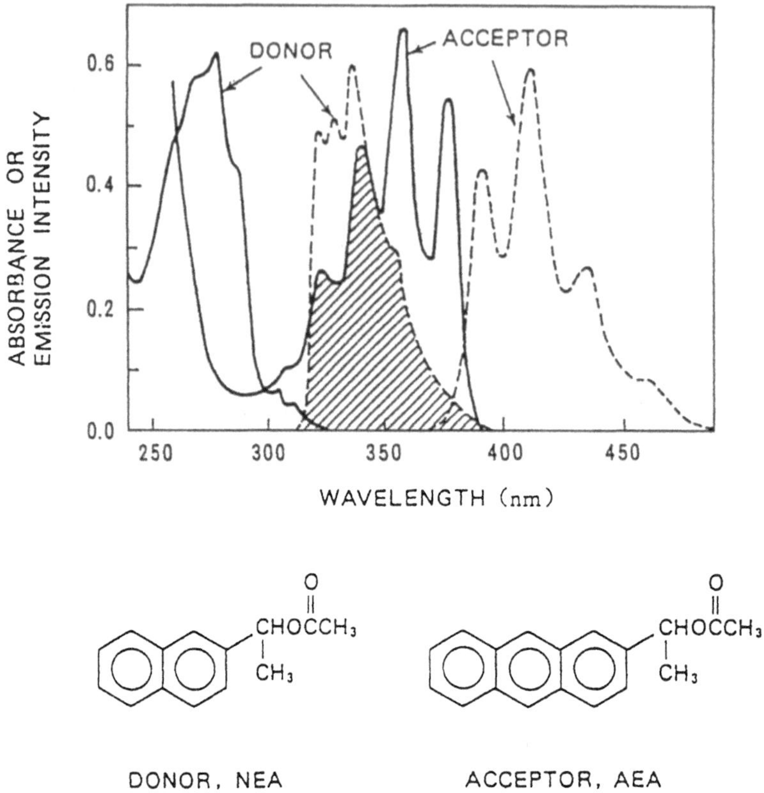

Figure 3.18 Absorption (—) and fluorescence (----) spectra of NEA (naphthyl moiety) and AEA (anthryl moiety) from ref. [59]. The overlap integral is crosshatched. The efficiency of Förster-type energy transfer depends on this overlap.

and acceptor absorption. Here κ^2 is expressed as follows:

$$\begin{aligned}\kappa^2 &= (\cos\theta_T - 3\cos\theta_D\cos\theta_A)^2\\ &= (\sin\theta_D\sin\theta_A\cos\varphi - 2\cos\theta_D\cos\theta_A)^2 \end{aligned}\qquad(34)$$

If the transition dipoles are averaged over a random distribution of orientations, κ^2 takes the value of 2/3.

In the case of singlet energy transfer and migration, the effective distance between a donor and an acceptor would be 1 to 10 nm. Figure 3.18 shows the absorption and fluorescence spectra of naphthyl (donor) and anthryl (acceptor) moieties. The value of k_{DA} is strongly dependent on the overlap between the fluorescence spectrum of the donor and the absorption spectrum of the acceptor.

The value of R_0 is calculated to be 2 nm in this case. Morawetz *et al.* [58, 59] prepared two polymers with either NEA or AEA and studied the compatibility of two polymers by observing the energy transfer efficiency from the naphthyl to the anthryl moiety. The fluorescence intensity of the anthryl group increases, when two polymers interpenetrate each other, because the distance between the naphthyl and anthryl moieties, i.e., r in Eq. (33), is shortened.

The transfer quantum efficiency Φ_{DA} from the donor to the acceptor is a function of the concentration of acceptor around a donor molecule, when all the molecules are fixed in an inert solid film, and is defined by

$$\Phi_{DA} = \sqrt{\pi}(C_A/C_0)\exp(C_A/C_0)^2\left[1 - \frac{2}{\sqrt{\pi}}\int_0^{C_A/C_0}\exp(-x^2)\,dx\right] \qquad (35)$$

where

$$C_0 = \frac{3000}{2\pi^{1.5}N_A R_0^3} \qquad (36)$$

C_A is the acceptor concentration and C_0 is a critical acceptor concentration (mol/liter). The values of R_0 and C_0 are listed in a book by Berlman [60].

One-step energy transport due to electron-exchange interactions is theorized by Dexter [61]. This mechanism is valid for describing transfer and migration of triplet excitation energy. The rate constant for energy transfer k_{DA} is expressed as follows:

$$k_{DA} = \left(\frac{2\pi}{h}\right)Z^2\int_0^{\infty} F_D(\nu)\varepsilon_A(\nu)\,d\nu \qquad (37)$$

$$= \frac{1}{\tau_0}\exp\left[\frac{2R_0}{L}\left(1 - \frac{r}{R_0}\right)\right] \qquad (38)$$

where

$$Z^2 = K^2\exp(-2r/L) \qquad (39)$$

in which Z^2 is a parameter that cannot be directly related to optical quantities. Dexter derived Eq. (38) by approximating Z^2 to be expressed as Eq. (39) where L is a positive constant called the "effective average Bohr radius" and K is a constant with the dimension of energy. Other parameters have the same meaning as in Eqs. (32) and (33). The effective distance between a donor and an acceptor would be 0.4 to 1.5 nm.

3.2.3.3.2.3 Kinetics of Singlet–Singlet Energy Transfer As described previously, we can determine the value of the average distance between a donor and an acceptor by measuring k_{DA} and Φ_{DA} in a solid film without molecular motion. However, in the case of a fluid solution, we have to consider the influence of molecular motion such as translational diffusion against energy transfer and migration. Here I explain the singlet–singlet energy transfer process in which we take into account molecular diffusion. With regard to the details of energy transfer and migration, including triplet state, the reader is referred to the literature [10].

North and Treadaway [62] defined a relative "molecular energy migration" distance R such that

$$R = \sqrt{2\tau_0(D + \Lambda)} \tag{40}$$

where D is the relative molecular diffusion coefficient of a donor and an acceptor and Λ is the donor excitation migration coefficient. D and Λ are measured in cm^2/s. If R is large, only the relative diffusion of molecules determines the overall transfer rate. On the contrary, if R is nearly zero, energy transfer follows the Förster mechanism itself. Thus, kinetics of singlet–singlet energy transfer are divided into three cases: (1) $R \ll R_0$ (Förster kinetics), (2) $R_0 \ll R$ (Stern–Volmer kinetics [63]), and (3) $R \approx R_0$ (intermediate condition, Yokota–Tanimoto kinetics [64]).

1. **Förster kinetics** [55, 56] In the case of (1), Eqs. (32, 33, 35, and 36) are valid. The time profile of the donor fluorescence after a δ-pulse excitation is written in the following form:

$$I_D(t) = I_0 \exp[-t/\tau_0 - 2(C_A/C_0)\sqrt{t/\tau_0}] \tag{41}$$

   where I_0 is a constant.

2. **Stern–Volmer kinetics** [63] In the case of (2), the excitation mobility is so high that the transfer rate becomes time-independent. Thus, the time profile of the donor fluorescence is expressed as a single exponential function:

$$I_D(t) = I_0 \exp[-t(1/\tau_0 + k_q C_A)] \tag{42}$$

   where k_q is the rate constant for quenching. Using photostationary measurements, the efficiency of energy transfer is obtained by the following equation:

$$\frac{F_0}{F} = 1 + K_Q C_A = 1 + k_q \tau_0 C_A \tag{43}$$

   where F_0 and F are the fluorescence intensities in the absence and presence of acceptor A, respectively. The expression K_Q is described the

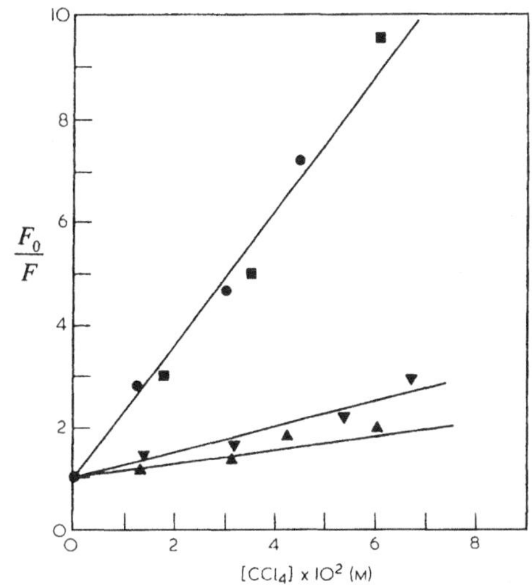

Figure 3.19 A Stern–Volmer plot from ref. [65] in dichloroethane solutions (donor: ethyl benzene (EB): ●, dibenzyl (DB): ■, head-to-head polystyrene: ▼, and head-to-tail polystyrene: ▲, acceptor (quencher): carbon tetrachloride).

Stern–Volmer constant and is given by the product of k_q and the donor fluorescence lifetime τ_0 without acceptors.

Figure 3.19 shows an example of a Stern–Volmer plot. Note that there is no large difference for the experimental results between EB and DB because their photophysical properties such as fluorescence and absorption never influence the efficiency of the Stern–Volmer type energy transfer. Usually, the value of k_q is compared with the rate constant for diffusion k_{diff} (measured in $M^{-1}s^{-1}$) derived from the Einstein–Smoluchowski diffusion theory

$$k_q = p \cdot k_{diff} \tag{44}$$

where p is the probability of fluorescence quenching.

In the case of small molecules,

$$k_{diff} = \frac{4\pi L N_A D}{1000}\left(1 + \frac{L}{\sqrt{D\tau_0}}\right) \tag{45}$$

where L is the distance between the reaction centers, or the sum of the van der Waals radii of a donor and an acceptor. In the case of the quenching of EB by carbon tetrachloride shown in Figure 3.19, k_{diff} is calculated to be $1.25 \times 10^9\ \mathrm{M^{-1}\,s^{-1}}$ by using $D = 1.53 \times 10^{-5}\ \mathrm{cm^2\,s^{-1}}$, $L = 5.0 \times 10^{-8}$ cm [66]. Thus, the efficiency of energy transfer p is 0.79.

In the case of a polymer with side-chain chromophores in which singlet energy migration can take place,

$$k_{\mathrm{diff}} = \frac{4\pi L N_{\mathrm{A}}(D + \Lambda)}{1000}\left(1 + \frac{L}{\sqrt{\tau_0(D + \Lambda)}}\right) \tag{46}$$

It is general to adopt two assumptions [67]: that (a) the diffusion of the polymer molecule is negligible in comparison with that of a small molecule quencher, and (b) the collisional radii of a donor to obtain L are not that of the entire macromolecule but of an excited side-chain chromophore.

3. **Yokota–Tanimoto kinetics** [64] In the case of (3) (intermediate kinetics), several models are published. Here we introduce the Yokota–Tanimoto kinetics [64] developed from Förster's theory. The equation was solved by using the Padé approximation. The time profile of the donor fluorescence after a δ-pulse excitation is written in the following form:

$$I_{\mathrm{D}}(t) = I_0 \exp[-t/\tau_0 - 2B(C_{\mathrm{A}}/C_0)\sqrt{t/\tau_0}] \tag{47}$$

where I_0 is a constant and B is expressed as Eq. (47).

$$B = \left[\frac{1 + 10.87X + 15.50X^2}{1 + 8.743X}\right]^{3/4} \tag{48}$$

with

$$X = D\alpha^{-1/3}t^{2/3} \tag{49}$$

and

$$\alpha = R_0^6/\tau_0 \tag{50}$$

Notice that Eq. (47) reduces to Förster kinetics (Eq. (41)) when $B = 1$, which is the case when molecular diffusion is negligible ($D \approx 0$). The functional form is a little complicated to handle; however, the numerical calculation [68] enables us to analyze fluorescence yields and decay profiles by means of the Yokota–Tanimoto kinetics.

3.3 Introduction to Fluorescence Measurements

Fluorescence measurements guarantee high sensitivity, which also means that the contamination of impurities may give rise to wrong results. One always has to make sure that sufficiently clean cells and vessels are employed to prepare solutions in order to measure the photostationary and transient fluorescence spectra of a sample correctly. One also has to measure the uv–vis absorption spectra correctly before the fluorescence measurement is done. Thus, we will explain the method to measure absorption first.

3.3.1 THE MEASUREMENT OF THE UV–VIS ABSORPTION SPECTRUM

A standard cell used for the uv–vis absorption experiment is a rectangular cuvette having two parallel transparent sides and a 1 cm × 1 cm square as the base. In the case where wavelengths to observe absorption are shorter than 330 nm, the material of cells should be quartz; otherwise, a glass cell is sufficient. It is recommended that the solvents employed for the measurements have no absorption bands in the wavelength region where the sample has absorption. Even if a solvent compound should not have any inherent absorption, there is a possibility that the very solvent used is contaminated; for example, fluorescent antioxidant reagent is mixed on purpose in some solvents such as tetrahydrofuran. For measurements of absorption and fluorescence spectra it is recommended that commercial solvents for spectroscopic uses or those distilled freshly be used.

Before the fluorescence measurements, one should always measure the absorption spectrum of the solution of the same solvent and determine the molar extinction coefficient at the peak wavelength. The concentration should be low enough to give an OD below 1.5. A situation for which the OD reaches 2 means that the number of photons that can transmit through the cell is only one hundredth of all the photons given. One should not determine the value of ε by measuring only one solution whose OD is nearly 2 at this wavelength. In the case in which one has to measure a solution of high concentration, a cell with a short optical path length should be used (1 or 5 mm).

As described earlier, there are some possibilities in a polymer system of forming a CT complex with a solvent or a ground-state dimer in an aggregated state. Thus, it is strongly recommended that one measure the absorption spectra of (1) a monomer model for a polymer having side-chain chromophores and (2) a dilute solution of a chromophore as an isolated model against the aggregated state together with the sample under the same condition for fluorescence measurements.

3.3.2 THE MEASUREMENT OF THE PHOTOSTATIONARY FLUORESCENCE SPECTRUM

3.3.2.1 Instrument

Figure 3.20 shows a fundamental construction of a typical commercial spectrofluorometer whose detector is situated in the perpendicular direction against the beam of the excitation light. A xenon lamp is well used as an excitation source because it emits photons of a wide range of wavelengths. The light of the xenon lamp is separated through a monochromator into various wavelengths. If a narrower bandwidth of the monochromator is taken, the accuracy of the excitation wavelength increases, but the intensity of the light is weakened very much. Another monochromator is set to resolve fluorescence in front of a detector, i.e., a photomultiplier. Compare Figure 3.4(a) with Figure 3.7; both fluorescence spectra are of a pyrenyl moiety. The bandwidth used for the separation of the fluorescence in Figure 3.4(a) is 5 nm, whereas that in Figure 3.7 might be narrower than 1 nm. To measure fluorescence spectra of a compound having sharp, fine structures due to vibronic transitions, the bandwidth of the monochromator should be adjusted to be narrower.

Wavelengths of both excitation and emission can be scanned in the ultraviolet (> 200 nm) and visible regions. Usually, the sensitivity of the photomultiplier and

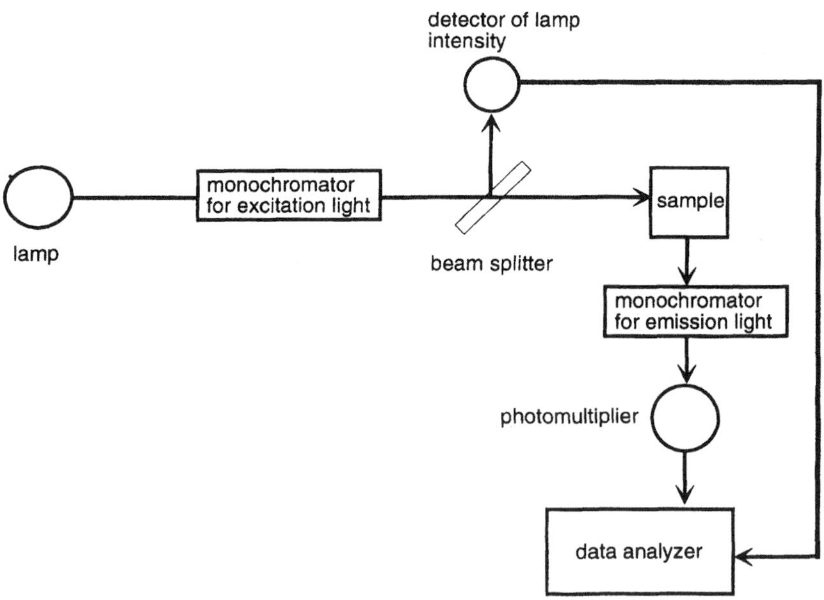

Figure 3.20 Setup of a typical spectrofluorometer.

the intensity of light depend on wavelength; thus, a correction of the fluorescence spectrum is necessary. It is generally advised to calculate the correction function for fluorescence up to 600 nm by making use of rhodamine B in ethylene glycol. The measurements for red or infrared fluorescence require a photomultiplier sensitive to light of a longer wavelength.

Recently, it has become possible to measure the fluorescence of a sample that cannot be set in a spectrofluorometer; for example, the fluorescence from a Langmuir–Blodgett (LB) membrane. The fluorescence of a sample, which is excited through optical fibers by the xenon lamp, is guided through other optical fibers to a detector. Another remarkable recent improvement is the shortening of the scanning rate of the wavelength (for example 500 nm/s); it is now possible to measure an entire fluorescence spectrum within 1 s under the photostationary condition. Some spectrometers and spectrofluorometers adopt the combination of a photodiode with a grating to take entire spectra in a moment.

3.3.2.2 Measurement of Fluid Solution

3.3.2.2.1 Cell

A standard cell used for fluorescence measurements is a rectangular cuvette having four transparent sides and a 1 cm × 1 cm square as the base. In the case where wavelengths to observe absorption and fluorescence are shorter than 330 nm, the material of cells should be quartz: otherwise, a glass cell is sufficient.

3.3.2.2.2 Solvent

The solvents employed for the fluorescence measurements should have no absorption and no fluorescence bands in the wavelength region for which the spectrum of a sample is observed. Recommended are the solvents distilled freshly or commercial solvents specifically made for absorption and fluorescence measurements.

3.3.2.2.3 Contamination

Because the fluorescence measurement is 10 to 100 times as sensitive as the absorption measurement, attention to contamination of impurities is very important. Let us examine a case in which 1 mol% of anthracene is contaminated by 99 mol% benzene. At 250 nm, the molar extinction coefficients of benzene and anthracene in nonpolar solvent are about 250 and 220,000, respectively, whereas the quantum efficiencies of fluorescence are about 0.1 and 0.3. Accordingly, when the mixture is excited at 250 nm, the fluorescence intensity of anthracene is about 26 times as much as that of benzene, although the fraction of anthracene is just 1 mol% (see Eq. (4b)).

The conditions to ascertain whether the fluorescence spectrum obtained is of a compound to be measured are (1) the absorption spectrum is appropriate in comparison to that of parent aromatic compounds such as naphthalene and phenanthrene, (2) the fluorescence excitation spectrum (Section 3.3.2.4) agrees well with the absorption spectrum, and (3) the relation of the mirror image holds between the fluorescence and absorption spectra. Condition (2), in particular, must be ascertained.

The existence of oxygen in a solution is also referred to as another problem of contamination. Oxygen acts as a quencher mainly followed by the Stern–Volmer kinetics (Section 3.2.3.3.2.3). In comparison with an experiment without oxygen, fluorescence intensity decreases and lifetime is shortened, as shown in Eqs. (42) and (43). In particular, the intensity of fluorescence whose lifetime is long decreases dramatically owing to this collisional quenching. The removal of oxygen is not always necessary when the system to study is under aerated conditions; however, one should keep in mind that oxygen quenches fluorescence. When it is necessary to measure an inherent value of fluorescence quantum yield, sample solutions should be degassed by freeze–pump–thaw cycles under high vacuum.

3.3.2.2.4 Scattering Light

Figure 3.21(a) shows a general emission spectrum obtained by exciting only solvent without any fluorescence. The main emitting components are (1) Rayleigh scattering light, whose wavelength is identical with that of the excitation light, and (2) Raman scattering light. The wavelength of the latter, λ_{Raman} (nm), is dependent on that of excitation light, λ_{Ex} (nm); i.e, the value of $10^7(\lambda_{\mathrm{Ex}}^{-1} - \lambda_{\mathrm{Raman}}^{-1})$ (cm^{-1}) corresponds to the vibrational energy of a solvent molecule. Usually the C–H stretching vibration is strong, and thus the difference in energy is around 3000 cm^{-1}. Notice that Raman scattering light shifts when the excitation wavelength is changed, whereas fluorescence does not shift at all. As shown in Figure 3.21(a), light with wavelengths two times as long as wavelengths of real emission appears in the longer-wavelength region owing to light separation using a monochromator. Figure 3.21(b) shows a general emission excitation spectrum (Section 3.3.2.4) obtained by exciting only the solvent without any fluorescence. Notice that peaks appear corresponding to Rayleigh and Raman scattering light.

3.3.2.2.5 Concentration

Fluorescence intensity is measured in arbitrary units according to Eq. (4a), which holds under the condition in which the optical density is less than 0.5. Under the

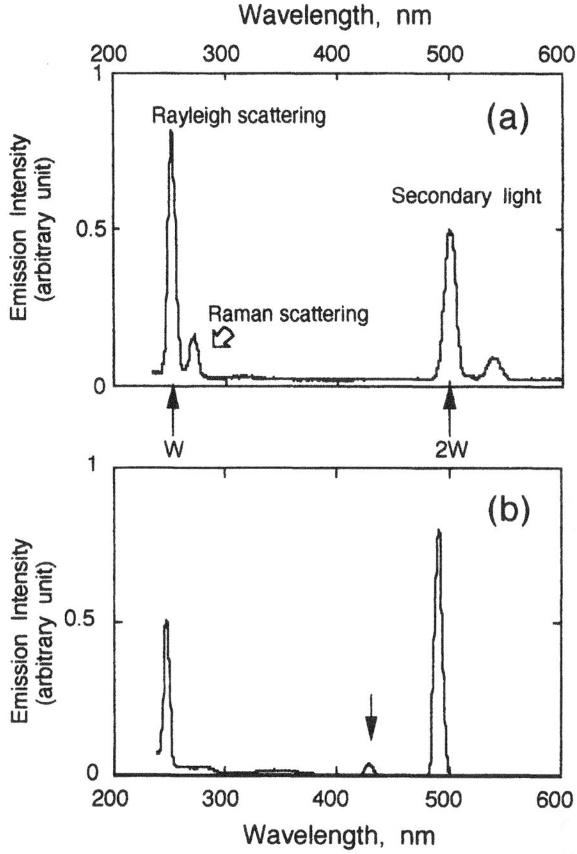

Figure 3.21 (a) Emission spectrum of solvent without any fluorescence in the case of excitation at 250 nm. The emission peaks shift according to the wavelength of the excitation. (b) Emission excitation spectrum of solvent without any fluorescence for 490 nm. When the solvent is excited at the wavelength pointed at by the arrow, Raman scattering appears at 490 nm. The peak at 245 nm is due to the secondary light of 490 nm.

condition in which the optical density is less than 0.07, fluorescence intensity is proportional to concentration, and thus it is quite convenient to compare several fluorescence spectra with one another. When one would like to determine only the shape of the fluorescence spectrum, it is not necessary to dilute a solution so much—especially in the case where a quantum efficiency of fluorescence is so low that the fluorescence is not clearly distinguished from Raman scattering. However, too concentrated a solution decreases the fluorescence intensity, as shown in Figure 3.22(a). Further increases in concentration induce change in the shape

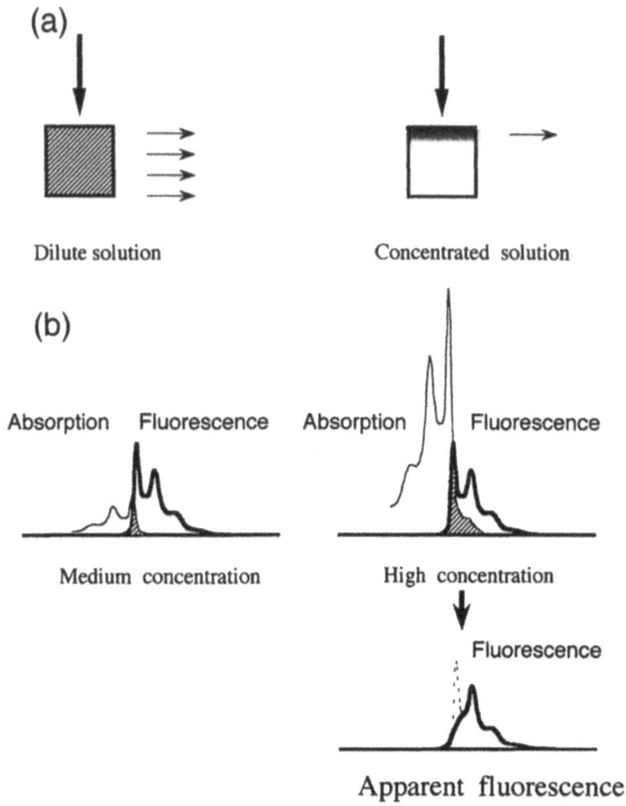

Figure 3.22 Cause of decrease in intensity and distortion in shape of fluorescence spectrum when the concentration is too high.

of the fluorescence spectrum because the fluorescence at shorter wavelengths is absorbed by other molecules of the same species (Figure 3.22(b)). The way to measure fluorescence for a concentrated solution is described in the following section.

3.3.2.2.6 Determination of Fluorescence Quantum Efficiency

Generally speaking, what is discussed about the fluorescence behavior is not a fluorescence quantum yield, Φ_f (Eq. (4)) but rather relative intensity measured in arbitrary units. However, Φ_f is still one of the important inherent values to characterize fluorescence, and it is especially necessary to determine Φ_f in the study of energy transfer. The value of fluorescence quantum yield of a sample Φ_f is obtained by comparing the fluorescence intensity of a sample with that of a standard reagent whose fluorescence quantum yield Φ_0 is already determined.

$$\Phi_f = \Phi_0 \cdot \left(\frac{\int F_f(\tilde{v})\, d\tilde{v}}{\int F_0(\tilde{v})\, d\tilde{v}} \right) \cdot \left(\frac{A_0}{A_f} \right) \cdot \left(\frac{I_{ex,0}}{I_{ex,f}} \right) \cdot \left(\frac{n_f^2}{n_0^2} \right) \tag{51}$$

where the subscripts 0 and f show a standard reagent and a sample whose Φ_f is unknown, respectively, the integration means the total area of the corrected fluorescence spectrum, A is the absorbance at the excitation wavelength, I_{ex} is the intensity of excitation light (photons/s), and n is the refractive index obtained by averaging over the wavelengths of the entire fluorescence spectrum. The ratio of the refractive index is necessary for only the case in which different solvents are used. If both fluorescence measurements are engaged for the same solvent, the same excitation wavelength, and the same absorbance, the Φ_f is determined only by the ratio of fluorescence areas.

3.3.2.3 Measurement of Polymer Solids and Concentrated Solution

Figure 3.23 shows the way to set a sample in a spectrofluorometer. In the case of the measurements for concentrated solutions of dyes or films, the sample should be set at 45° to the exciting beam (Figure 3.23(b)). To irradiate the surface of a cell containing a concentrated solution, we usually place a cell with an optical path length of 1 mm on the hand-made holder.

Plastic films cast on disks and samples bound by two plates such as liquid crystal and membrane are recommended to be set as in Figure 3.23(c). If they are set as in

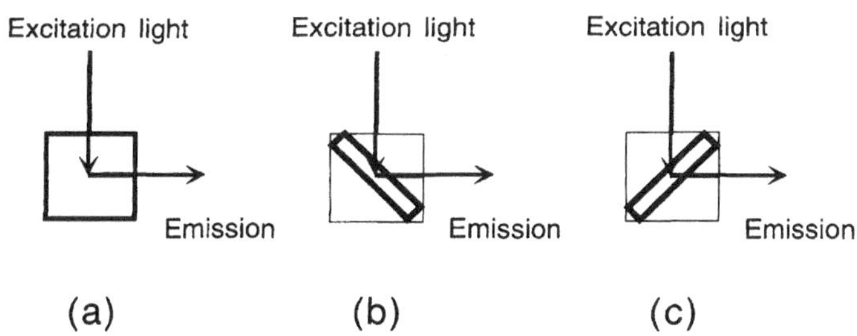

Figure 3.23 Position of a sample in a spectrofluorometer for fluorescence measurements. (a) Typical measurement for dilute solution in a standard cell. (b) Measurement for bulk film or for concentrated solution in a microcell with an optical path length of 1 mm, etc. (c) Measurement for film cast on a disk or for a sample such as gels, liquid crystal, membrane, etc., bound by two plates or disks.

Figure 3.23(b), the lights reflected on the front surface and back surface interfere with one another, giving many regular peaks at the same interval in the emission spectrum.

When the wavelength of the absorption and fluorescence to observe is shorter than 330 nm, the material of disks and plates should be quartz; otherwise, glass is acceptable. One of the convenient ways of preparing plastic films is by a spin-casting method: the films for the fluorescence measurements can be prepared on quartz or glass disks by using a spin-casting method from a polymer solution. The films are left on the disks for ease of handling during subsequent fluorescence measurements.

3.3.2.4 Fluorescence Excitation Spectrum

Fluorescence measurements are engaged by scanning wavelengths of an emission detector, whereas the excitation wavelength is fixed (Figure 3.20). The fluorescence excitation spectrum is obtained by scanning wavelengths of an excitation light while a wavelength to monitor with the emission detector is fixed (λ_{em}). It clearly shows which molecule absorbs the photon and finally releases the emission at λ_{em}. If there should be only one component, shapes of fluorescence excitation spectra at any emission wavelength coincide with one another and perfectly agree with its absorption spectrum. The excitation at a wavelength at which the absorbance is large should give rise to many excited states, yielding strong fluorescence.

Figure 3.24 shows an example of fluorescence and excitation spectra. The fluorescence peaks of 9-methylcarbazole (MCz) in THF appear at 350 and 380 nm, whereas those of 1-naphthylmethyl-L-glutamate in THF appear around 330 nm. The THF solution of the carbazolyl group attached to one end of poly(γ-1-naphthyl-L-glutamate)(PNLG–Cz, line 1) shows nearly same shape of fluorescence as MCz, whereas the excitation spectra (line 2) for the fluorescence of the carbazolyl moiety of PNLG–Cz differ from the absorption spectrum of MCz (line 3); rather, they almost coincide with the absorption spectra of PNLG–Cz in THF (line 4). This occurs because the excitation energy absorbed by naphthyl moieties of PNLG–Cz transfers to the carbazolyl group at the end of PNLG–Cz [169].

3.3.2.5 Measurement of Fluorescence Depolarization

It is usual to employ polarized filters to measure fluorescence anisotropy (Eq. (12)). In Figure 3.5, v and h refer to the case in which light is polarized in the direction of

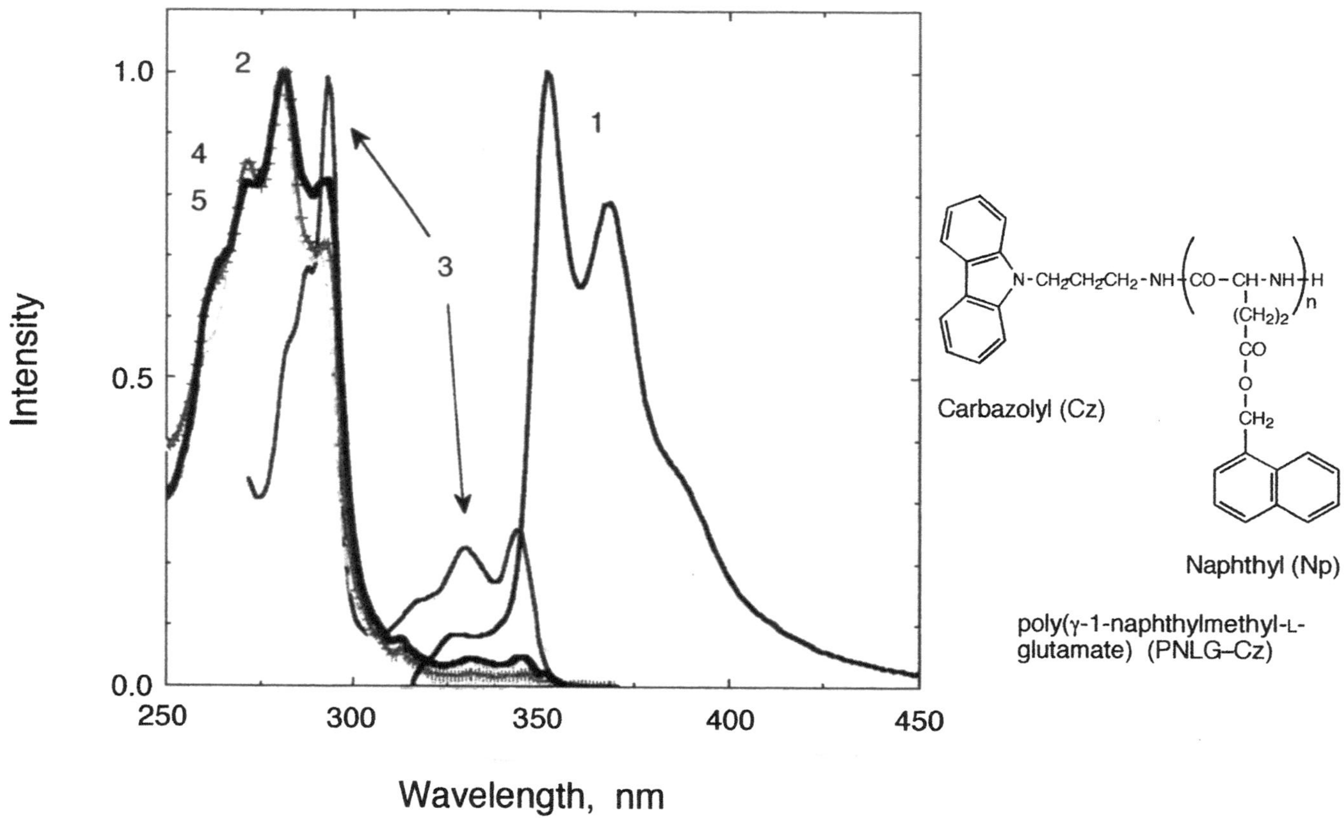

Figure 3.24 Spectra of poly(γ-1-naphthylmethyl-L-glutamate) in THF: (1) fluorescence spectrum (excitation at 282 nm); (2) (solid line), excitation spectrum for fluorescence at 352 nm, which is fluorescence of Cz; (3) absorption spectrum of a monomer model of Cz in PNLG; (4) (+++), absorption spectrum of PNLG-Cz; (5) (broken line) absorption spectrum of a monomer model of Np in PNLG.

the z-axis (i.e., vertical direction (90°)) and the case in which light is polarized in the direction of the x- or y-axis (i.e., horizontal direction (0°)), respectively. Here I_{pq} (p, q = h, v) is defined as the fluorescence intensity obtained by using a p-type polarizer for the excitation light and a q-type polarizer for the emission light. In order to correct the anisotropy induced by the sensitivity of a spectrofluorometer, the machine constant G should be taken into account: $G = \frac{I_{hv}}{I_{hh}}$

$$r_{pol} = \frac{I_{vv} - GI_{vh}}{I_{vv} + 2GI_{vh}} \tag{12b}$$

Values of anisotropy depend on excitation wavelength. Before the measurements of fluorescence anisotropy of a probe molecule in a system of interest are made, an inherent value of anisotropy, $r_{pol,0}$, should be determined by measurements using a plastic solution in which a small amount of probe molecules is randomly distributed. In fluid solution, the value of r_{pol} is generally observed to be zero owing to rapid molecular motion.

The value of r_{pol} is determined by averaging the values obtained for a stronger fluorescence region. In the case of the measurements by means of an automatic polarizer, the value of r_{pol} is determined by averaging the values for a few minutes.

3.3.2.6 Measurement of Temperature Dependence of Fluorescence

The temperature change of fluorescence may be able to give precious information on the kinetics of a system. Because the fluorescence intensity or Φ_f is dependent on temperature, the spectrum obtained in summer may sometimes not agree with that measured in winter, and temperature considerations are critical. Thus, a spectrofluorometer should be located in an air-conditioned room and a cell holder should be kept at a constant temperature controlled by using a circulating system of water or ethylene glycol. Usually, cell temperatures are kept at 20 or 25 °C at room temperature.

The large change of sample temperature can be controlled by using a cryostat with a digital temperature controller. Temperature regulation by a commercial cryostat is easily better than ±0.1 K. Typically, the temperature the controller demonstrates is not that of the sample but of a position of the temperature sensor. Thus, independent temperature measurement on a sample may be necessary by means of a second thermocouple and potentiometer. In my laboratory, the samples are kept at each set temperature, and spectra are run repeatedly for quite a long time even after perfect duplication is obtained.

3.3.3 MEASUREMENT OF TRANSIENT FLUORESCENCE SPECTRUM

3.3.3.1 Introduction to Transient Fluorescence Spectrum

The measurements of transient photophysical processes give good accuracy in time and so can be used as one of the best methods to observe rapid dynamic processes directly. We can choose any time region down to 100 fs. As for transient fluorescence measurements, picosecond and nanosecond flashphotolysis systems are available as a commercial product and are even easy to handle for a beginner in transient fluorometry.

As described earlier, if photons are given to a sample by a δ-function, all the molecules in excited states are formed simultaneously at $t = 0$. If we observe an emission at one wavelength, it gives (1) a fluorescence decay curve that it would decay according to its own function, such as a single exponential, or according to slightly more complicated functions such as Eqs. (41) and (47) [55, 64, 186]. On the other hand, if we observe the whole spectrum wavelength region, it gives (2) a time-resolved fluorescence spectrum. It is also possible to detect (3) transient fluorescence anisotropy.

The important information obtained from measurements of transient fluorescence is the shape of the transient fluorescence spectrum (peak wavelength and its intensity at a given time) and its function of time. The parameters observed should be reproduced by assuming a kinetic scheme appropriate for the system. Thus, unless the kinetic scheme accurately describes the experimental system, we cannot get any information on polymer properties from the fluorescence of probe molecules or groups. Generally speaking, the probing method using transient fluorometry is strongly related to the understanding of photophysics itself. In other words, a trial to get information of polymer behavior may yield a new photophysical process. I will briefly explain the method to determine the kinetics before introducing the equipment.

3.3.3.2 Determination of Kinetics

Figure 3.25 shows a flow chart for determining kinetics. Usually a photostationary measurement including the excitation spectrum and the absorption spectrum indicates how many excited states the system has. Even if there is only one emitting species, the decay is not always of a single exponential form because there are several possibilities such as that some excimers and exciplexes that can dissociate are nonfluorescent and that the fluorescence is quenched by nonfluorescent species, and so on.

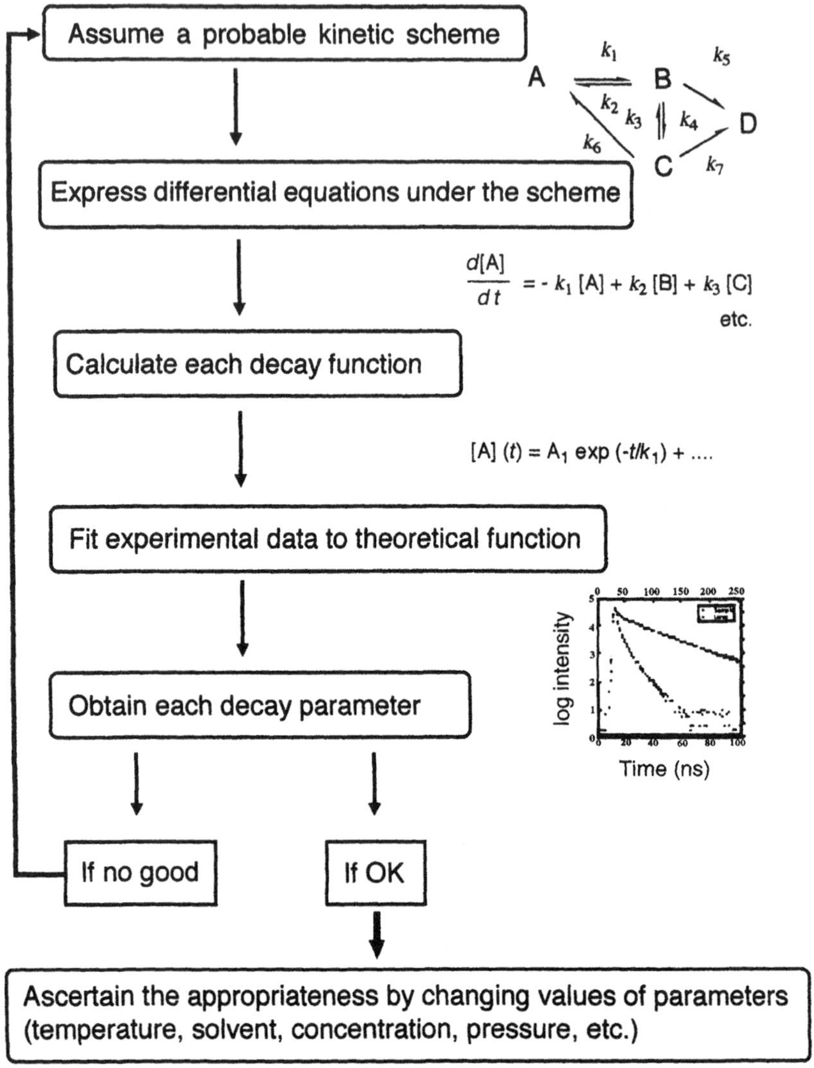

Figure 3.25 Flow chart of the process for determining kinetics.

The decay function can be calculated theoretically by solving the differential rate equations developed if one assumes all the processes possible take place. Here some rate constants may be time-dependent. Next, the experimental decay curves are fitted to the theoretical functions. When (1) the theoretical function never reproduces the experimental data and (2) the time constants obtained are far from the values assumed, even if the fitting is perfect, we have to improve the kinetic scheme.

In the case in which the kinetic scheme describes the experimental fluorescence decay, it is preferable to change a parameter such as temperature, concentration, solvent, pressure, or species of quencher; then, the appropriateness of the kinetic scheme will be proved by the coincidence of the experimental decay curves with the theoretical ones. Before we can get some values of rate constants by fitting the experimental data to theoretical functions deduced by assuming the appropriate kinetic scheme, we cannot start with analyzing the meaning of the rate constant values. Please notice that rate constants obtained sometimes have nothing to do with the phenomenon that we would like to investigate because of the large difference between the observation time (fluorescence lifetime) and the time required for the phenomenon occurring (for example, with the rate constant for very slow diffusion). Even if the time change observed by transient fluorescence is accurate, we always have to prove how the transient fluorescence is related to the physical and chemical changes that we would like to investigate.

3.3.3.3 Instrument

The main setup to monitor transient fluorescence is similar to that for photostationary fluorescence (Figure 3.20) except (1) the excitation light is to be a pulse of very short duration, (2) the starting time to detect the signal fluorescence is to be synchronized with the discharge of the light pulse, and (3) the detector should have a rapid time response.

The main excitation sources are a pulsed laser and a nanosecond flash lamp. When a pulsed laser is employed, we can obtain an entire decay curve or a time-resolved spectrum at a given time by just one shot of the laser because a great number of photons are given to a sample at once. It is needless to say that the integrated repetition of measurements must increase accuracy and precision of the data. A pulse width of light is described by the full width at half maximum. The pulse width of a pulsed laser is so short that the initial process can be monitored accurately just after the irradiation. However, there are two main disadvantages with the use of a pulsed laser for one-shot detection. First, the excitation wavelength usually cannot be changed; for example a Nd:YAG (yttrium aluminum garnet) laser can only generate a pulse of 1.06-μm light. A solution of a dye or a harmonic generator such as a KDP crystal is employed to change the excitation wavelength of a pulsed laser; in this way, the fourth harmonic wavelength of the Nd:YAG laser (265 nm) can be used. However, this method is not convenient and it still does not allow one to change the wavelength at random. A second disadvantage is that the irradiation of strong laser beams sometimes induces biphotonic or multiphotonic processes that can distort the photophysical processes. Biphotonic processes are negligible for the photostationary measurements with a typical spectrofluorometer, and thus a fluorescence spectrum may be contradictory when its transient data are obtained by using a strong pulsed laser.

A nanosecond flash lamp is employed for measurements with the time-correlated single photon counting method. In this method, the experimental conditions should be adjusted to detect, at most, one emitting photon for one pulse of irradiation. Although the frequency of the flash irradiation is quite large, it takes a long time to accumulate enough emitting photons to determine a decay function accurately. However, fluorescence decay curves can be obtained more accurately because multiphotonic processes will never take place. One usually accumulates up to 60,000 photon signals for a single decay curve. Any excitation wavelength can be chosen in either the visible or near-ultraviolet light range. The pulse width of a flash lamp using H_2 gas is, at the shortest, 2 ns. Single photon counting methods can be set up with a picosecond or nanosecond pulsed lasers as the excitation source.

The emission wavelength can be resolved by using a grating monochromator for strong light or an interference filter for weak light. The transient emission is usually measured with (1) a photomultiplier, (2) a photodiode-alley, or (3) a streak camera. The signals detected are normally stored in a sampling oscilloscope or in a multichannel digital memory: the emission signals monitored by the gated streak camera followed by a silicon-intensified target tube are directly transferred to a personal computer. Basically, the data storage has a limited number of channels (256, 512, etc.), and thus the collection time per channel is changeable. Consequently, several experimental decay curves are usually measured for one sample by changing the time per channel in order to review an entire time profile of fluorescence. When a grating is set in front of a photodiode-alley, a time-resolved fluorescence spectrum is measurable for one shot of an excitation pulse; an alley usually consists of 256 or 512 elements that can detect different light separated through a grating.

3.3.3.4 Concentration

It is self-evident that measurements of absorption and fluorescence of a sample in the same solvent are necessary before one measures the transient fluorescence. High concentrations of the sample in solution would be better if no spectral changes appear in comparison with a spectrum having a low optical density. Thus, the absorbance at the excitation wavelength should preferably be around 1. We can measure transient fluorescence spectra of polymer solids even with a time-correlated single-photon counting technique; however, it is difficult to obtain fluorescence intensity and remove strong scattering light. Here, a pulsed laser is preferable as an excitation light source.

3.3.3.5 Calculation

To obtain a fluorescence decay curve of a sample, we usually measure several fluorescence decay curves of the same sample over different time ranges. The time range for the measurements is recommended to be three to five times as long as the lifetime of a sample. For example, because the lifetime of 1-pyrenebutanoic acid in degassed THF is 210 ns, the time range to determine it should be 512 to 1024 ns in the full time scale (Figure 3.4(b)). The observation of transient fluorescence in a shorter time range is also necessary to determine whether it also has a short lifetime component or not. When the response time of the measurement system including a pulse width of an excitation source is fast enough compared with a lifetime, an experimental fluorescence decay can be calculated by using a nonlinear least-squares method without any deconvolution. In such a case one would need to exclude several initial channels in which signals are distorted by the instrumental response and by the scattering of light. On the other hand, in the situation in which the lifetime of a sample is comparable to the response time of the system, one would have to calculate most of the fluorescence decay, including the initial time region, to determine the short time constant by using the deconvolution method. In this situation, the excitation pulse cannot be considered as a δ-function. If $I(t)$ is the theoretical decay function of fluorescence obtained by a δ-function lamp excitation and $g(t)$ is the response function of the equipment, the apparent fluorescence decay $F(t)$ is described as follows:

$$F(t) = \int_0^t g(T)I(t-T)\,dT \tag{52}$$

In Figure 3.4(b), the fluorescence curve of a sample is $F(t)$, and the lamp curve is $g(t)$. Our aim is to determine $I(t)$. Equation (52) is called the integration of convolution. A theoretical function is fitted to an experimental decay using a nonlinear least-squares curve-fitting approach incorporating the Levenberg–Marquardt algorithm [69]. In this process, a fitting function with varying preexponential factors and lifetimes is adjusted in successive iteration after giving an appropriate initial value for each parameter. The process continues until the best fit is obtained between the calculated and actual decay curves. The calculation method of the deconvolution is explained in detail in the literature [69–71]. A book by O'Connor and Phillips [71] is one of the best to describe the single-photon counting method and also provides a model program for the computer calculations.

We usually use three methods to assess the validity of the trial fitting function: (1) One can graph the weighted residual R_i against channel i,

$$R_i = \frac{F_i - I_i}{\sqrt{F_i}} \tag{53}$$

where F_i and I_i are the values of experimental data and trial decay function corresponding to the time channel i. (2) The autocorrelation function of R_i, and (3) the Durbin–Watson factor DW [72, 73].

$$DW = \frac{\sum_{i=2}^{N}(R_i - R_{i-1})^2}{\sum_{i=1}^{N} R_i^2} \tag{54}$$

where N is the number of experimental points and DW is a value to express the autocorrelation function of R_i. For the best fitting, the value of DW should approach 2.0.

Our calculation method is as follows when decay functions are the sum of exponentials: $F(t) = \sum_{i=1}^{m} A_i \exp(-t/\tau_i)$, $\sum_{i=1}^{m} A_i = 1$ $(m = 1, 2, 3)$. First we analyze the fluorescence decay curves whose full time scale is long enough to calculate the longest lifetime and determine it by fitting the data to a single exponential. After determining the longest time constants, we try to fit experimental data in the shorter time range to the sum of two exponentials by using the deconvolution method. If values of time constants are calculated and appear to be nearly the same, the decay function should be a single exponential. Otherwise, there should be more than two components. Here, we usually calculate the same data repeatedly by changing the fitting regions (channels). To determine shorter time constants, it is better to measure fluorescence decays in the shorter time range. The problem is the determination of preexponential factors because the fraction of the shortest lifetime should be exaggerated in the data measured in the shorter time range. We usually determine each fraction of components having different time constants by analyzing the decay data in the middle time range. It is not easy to determine the complicated function such as the sum of four exponentials or a relationship having a time-dependent term such as $t^{1/2}$. The appropriateness of a fitting can also be ascertained by comparing the integration of decay functions calculated with steady-state fluorescence intensity as follow:

$$I_{\mathrm{F}} = \int_0^{\infty} \left[\sum_{i=1}^{m} A_i \exp(-t/\tau_i) \right] dt \tag{55}$$

3.3.4 *MEASUREMENT OF TRANSIENT FLUORESCENCE ANISOTROPY*

Recent developments in the fluorescence polarization decay technique are outstanding. The technique has come to be used to study dynamics of polymer local motion; however, the problem for a beginner is that the most appropriate theoretical model is not established yet to describe the decay functions $r_{\mathrm{pol}}(t)$, as described in Section 3.2.2.3.

However, there is another approach to employ a correlation time τ_{c} [74] rather than function form of $r_{\mathrm{pol}}(t)$ itself.

$$\tau_{\mathrm{c}} = r_{\mathrm{pol},0}^{-1} \int_0^{\infty} r_{\mathrm{pol}}(t)\, dt \tag{56}$$

In this case, decay curves are first fitted to the correlation functions such as the Hall–Helfand (HH) model (Eq. (57)) [75] or the generalized diffusion and loss (GDL) model (Eq. (58)) [76], and τ_{c} is calculated by Eq. (56).

$$r_{\mathrm{pol}}(t) = r_{\mathrm{pol},0}(t) \exp(-t/\tau_1) \exp(-t/\tau_2) I_0(t/\tau_1) \tag{57}$$

$$r_{\mathrm{pol}}(t) = r_{\mathrm{pol},0}(t) \exp(-t/\tau_1) \exp(-t/\tau_2) [I_0(t/\tau_1) + I_1(t/\tau_1)] \tag{58}$$

Ediger *et al.* discuss the dynamics of polymer chains in dilute solution by using τ_{c} obtained from the transient fluorescence anisotropy measurements of the anthryl group attached to polystyrene [77] and polyisoprene [78, 79].

The experimental method is the same as that described in Section 3.3.2.5. Here, $I_{\mathrm{pq}}(t)$ (p, q = h, v) is defined as the fluorescence intensity at a time t obtained by using a p-type polarizer for the excitation light and a q-type polarizer for the emission light. The meanings of h and v are identical as those described in Section 3.3.2.5. In transient measurements, a probe molecule or a group is excited by a pulse of vertically polarized light. The emission light is then recorded as a function of time in both parallel ($I_{\mathrm{vh}}(t)$) and perpendicular ($I_{\mathrm{vv}}(t)$) polarization. In order to correct the anisotropy induced by the sensitivity of the equipment, the machine constant G should be taken into account: $G = \frac{I_{\mathrm{hv}}(t)}{I_{\mathrm{hh}}(t)}$. If possible, a photostationary lamp should be used to determine G to shorten the measurement time. Notice that the same sample should be measured under the same conditions to determine G. The emission anisotropy is described as

$$r_{pol}(t) = \frac{I_{vv}(t) - GI_{vh}(t)}{L[I_{vv}(t) + 2GI_{vh}(t)]} \tag{59}$$

where L is a factor to correct the influence of light scattering and is independent of time. Usually L can be 1; however, L has to be determined when the sample has turbidity. L can be calculated by a comparison with the anisotropy value r_{pol} obtained under photostationary conditions in dilute solution by

$$L = \frac{r_{pol}}{\int_{\tau_1}^{\tau_2} r_{pol}(t)\,dt} \tag{60}$$

where τ_1 and τ_2 are arbitrary time constants. Equation (60) can not be used if L is less than 0.9.

3.4 Application of the Fluorescent Probe Method to Polymer Science

3.4.1 APPLICATION OF THE FLUORESCENT PROBE

I have introduced the principle of fluorescence behavior together with the experimental technique. Now I would like to summarize the fluorescence probe method. Because the method gives a wide variety of information such as concentration, viscosity, polarity, pH, volume, distance, time, and so on, what we have to do is to choose an appropriate fluorescent probe.

The form of a fluorescence probe in a polymer system can be classified into three categories: (1) internal probe, i.e., a sample being fluorescent by itself; (2) fluorescent groups introduced at the end or in the middle of a polymer molecule by covalent bonds; and (3) fluorescent molecules doped or added to a polymer system.

In the case of (1), one can study the physical properties of polymers having fluorescent aromatic groups in the main chain or side chain easily by their fluorescence measurements. Categories (2) and (3) are described as external probe methods. In the case of (2) there is an example such as cytochrome type c whose porphyrin moiety is nonfluorescent owing to the formation of a complex with iron but which becomes fluorescent after the removal of iron [80, 81].

Many fluorescent probes relating to the biosciences or to polymer chemistry are developed and commercially available. Some easily react only with a specific group such as an amino group or readily penetrate into a hydrophobic part of a protein.

The fluorescent properties of a probe molecule or group are summarized into two categories: (1) those dependent on the microenvironment around the probe (Section 3.2.3.2) and (2) those not dependent so much on the microenvironment around the probe (Section 3.2.3.1). The latter can also be effective for the detection and the determination of a small amount of a particular compound. If the fluorescence quantum yield of a probe is independent of the microenvironment, its fluorescence intensity should be proportional to the concentration. Moreover, the fluorescence depolarization method can be applied to a system when the fluorescence property of a probe does not change in the system.

It is difficult to summarize the use of fluorescent probes in few words. For example, the transient fluorescence anisotropy of a chromophore with a large molecular volume would yield information on the dynamic processes of its rotational motion. If this molecule is added into a fluid solution of a polymer and the change in anisotropy is measurable with the change in polymer concentration, the molecule can be described to work as a fluorescence probe monitoring the viscosity. Moreover, if the change in viscosity corresponds to the gelation, the change can be a good probe to detect the transition point of gelation. Table 3.3 summarizes several examples of fluorescence probes employed for studies in polymer science.

3.4.2 PROBLEMS WITH THE FLUORESCENT PROBE METHOD

The fluorescent probe method is very useful for obtaining information on microenvironments nondestructively and easily. The merits of this method are enumerated in Section 3.4.1. However, there are five serious points to be aware of with regard to the application of any fluorescent probe method to a polymer system.

First, we always have to make clear whether the change in fluorescence observed is due only to the photophysical change without reflecting its microenvironment or the change induced by its microenvironment. Careful analysis is necessary, especially in the case where a fluorescent probe is first applied to a new system. For example, if one finds a dramatic change in fluorescence with a change in the temperature of a thermotropic liquid crystal system, one must determine whether this change reflects only temperature dependence of the fluorescence or is induced by the change of orientation accompanied by the liquid crystallization process.

Second, the probe molecules should be uniformly distributed in the system to be monitored, if one intends to monitor an average phenomenon in this system. This is a common problem with all probe methods. Because of the high sensitivity of

Table 3.3 Some examples of fluorescence probes employed for polymer science studies.

System	*Information to Monitor*	*Fluorescence Behavior Used as Probes*	*Examples of References*
Dilute fluid solution	Local motion of polymer chain	Excimer	91, 92
	Segment motion of polymer chain	Singlet energy transfer	68
		Fluorescence anisotropy	21, 79
	End-to-end distance	Excimer	112, 173
		Singlet energy transfer	166, 167
		Delayed fluorescence	49–52
Semidilute fluid solution	Interpenetration	Singlet energy transfer	174–176
Liquid crystal	Aggregation, interaction	Fluorescence change	139
		CT complex	138
		Fluorescence anisotropy	141, 142
	Microenvironment	Dansyl group	140
Micelle	Aggegation number	Singlet energy transfer	117, 118
		Excimer	118, 183
LB films	Orientation	Singlet energy transfer	113, 114
	Diffusion constant	Singlet energy transfer	115
Gel	Molecular motion	Depolarization	123, 126, 129
		Excimer	107, 108, 128
	Microenvironment	Dansyl	126, 127
		Pyrenyl	128
	Solvent behavior	Depolarization	109
	Aggregation, interaction	Ground-state dimer	135, 136
Solid	Crystallinity, orientation	Fluorescence change	26
		Excimer	144
		CT complex	147–150
	T_g, T_β, T_γ	Excimer	177, 178
	Recrystallization	Fluorescence change	145
	Compatibility	Singlet energy transfer	58, 179, 180
		Excimer	181, 182
	Free volume	Depolarization	99, 162
		Molecular rotor	32, 160
		Excimer	156, 159, 172
	Molecular motion	Depolarization	76, 162
	Diffusion constant (interpenetration)	Singlet energy transfer	155
	(phase separation)	Excimer	184, 185

fluorescence, the change that takes place in a quite local area is often exaggerated so much that the phenomenon is misunderstood in the entire system.

Third, some fluorescence probes can be subject to photodegradation under the experimental conditions. It is quite natural that the photochemical reaction takes place during the fluorescence measurements because the sample is successively irradiated by light. The decrease in fluorescence intensity may not be due to energy transfer but rather to the photoreaction.

The fourth point is that the method is never effective to determine exact chemical structure, although changes in structure can be monitored qualitatively and quantitatively.

Lastly, the fluorescent probe method is not very efficient for observing macroscopic change. Because the method is most effective in monitoring microenvironmental and rapid change, we sometimes find out that there is a large gap with the macroscopic information during the fluorescence measurements. We have to combine several measurements with fluorometry to understand certain phenomena clearly.

3.5 The Use of Fluorescence Measurements in Polymer Science

The aim of this chapter is not to review extensively all the results obtained by using fluorescence measurements in polymer science but rather to serve merely as an introduction for those who want to use this technique for the study of polymer science. Thus, I will pick up several examples from the area relating to fluorescence in polymer science and I will describe them in this section.

It is possible to take two approaches to review the studies using fluorescence as a tool. One is categorized by the phenomenon of observing and another is by the probing method. Thus far, I have explained fluorescence behavior that can be applied for a polymer system as a probe in Section 3.2 and summarized the probing method in Table 3.2. Here I introduce some papers from the standpoint of polymer science. They are summarized in Table 3.3.

I design two subsections: one concerns fluorescence measurements in a polystyrene system, and the other is a review of papers dealing with different phenomena in polymer systems. The first subsection is an example of the measurements of an inherently fluorescent polymer together with basic properties of polymer photophysics. This subsection includes our papers dealing with thermoreversible physical gelation by using polystyrene fluorescence as an internal probe. The second subsection deals with papers studying Langmuir–Blodgett (LB) films, micelles, gels, liquid crystals, and polymer solids by means of fluorescence measurements. The object of this section is to deepen the understanding of the readers.

3.5.1 *FLUORESCENCE BEHAVIOR OF POLYSTYRENE*

This section describes the behavior of polystyrene fluorescence and the behavior of polystyrene molecules monitored by fluorescence measurements. A phenyl moiety is the smallest aromatic chromophore able to fluoresce. Because polystyrene is one of the most general macromolecules, its physical behavior has been investigated extensively. Accordingly, polystyrene is a good polymer to use for assigning a photophysical phenomenon to a particular physical behavior in a polymer molecule, solution, or solid.

First, I will introduce our studies concerning the fluorescence behavior of polystyrene in fluid and solid solutions to provide a typical example of fluorescence measurements of polymers having side-chain chromophores. The studies will also demonstrate that excimer fluorescence, as it occurs in polystyrene, can be a powerful tool for monitoring molecular motions and ordered structures of polymer molecules.

3.5.1.1 Conformational Change of Polystyrene in Fluid Solution

Many authors have reported intramolecular excimer formation in dilute solutions of vinyl polymers containing aromatic pendant groups. Figure 3.14 shows the fluorescence spectra of oligostyrenes in degassed cyclohexane. The peaks near 285 and 330 nm have been assigned to phenyl monomer singlet and excimer emission, respectively. Intramolecular excimers in such polymer systems are formed between adjacent side-chain chromophores [82–84], and their formation requires two important processes: one is the formation of excimer conformations (Figure 3.26: tt in meso diad; g^-t/tg^- in racemo diad) by local motion, and the other is the influx of excitation energy into the preformed excimer conformation by singlet energy migration [85–87]. In particular, the former process suggests that the transient measurement of intramolecular excimer formation can be the direct measurement of dynamic local motion of macromolecules.

One of the dramatic features of intramolecular excimers in aromatic vinyl polymers is their molecular weight dependence, which was first reported for poly(1-vinylnaphthalene) by Nishijima *et al.* [88] and systematically shown for polystyrene by Ishii *et al.* [89]. Each chromophore attached to a polymer molecule is assumed to be in the same environment whatever the molecular weight. Thus, the

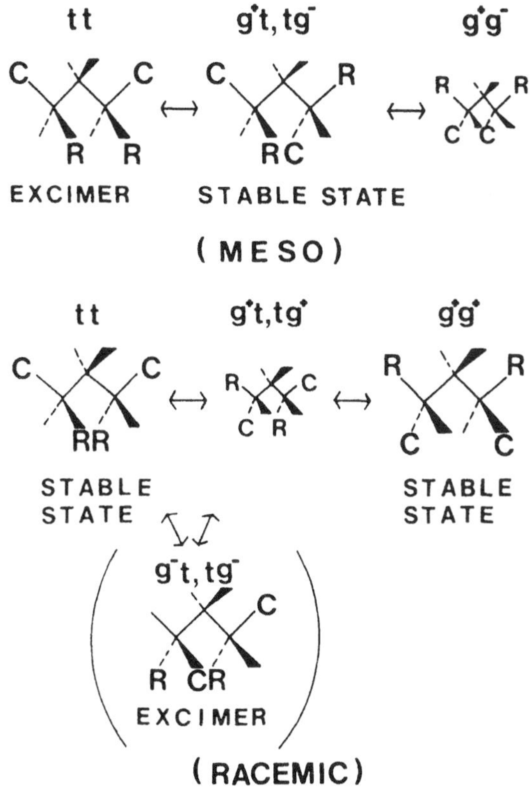

Figure 3.26 Main conformational changes for meso (top) and racemic (bottom) diads of a vinyl polymer with side-chain aromatic groups (C-main chain; R-aromatic group).

variation of molecular weight can provide information on photophysical processes without introducing any other new parameters. By observing k_{EM} in Figure 3.15 for oligostyrenes with degrees of polymerization (DP) ranging from 2 to 13, we showed that the formation of the excimer conformation is the rate-determining process of excimer formation in the low-molecular-weight region [43]. The studies on the intramolecular excimer formation process of dimeric diastereomers with symmetric alkyl end groups are quite important for understanding that of vinyl polymers with aromatic pendant groups [11], although there is still a big gap between polymers and their dimer models.

To elucidate the local motion of polystyrene, we measured the dynamic process of intramolecular excimer formation in diastereoisomeric oligostyrenes, as shown in Table 3.4 [90–93]. Table 3.4 shows the decay parameters obtained for diastereoisomeric oligostyrenes in degassed cyclohexane. Monomer fluorescence

Table 3.4 Decay parameters of monomer fluorescence of diastereoisomeric PS2, PS3, and PS4 in degassed cyclohexane at 20 °C. Fluorescence decays owing to equation $I_M = \sum A_i \exp(-t/\tau_i)$, where $\sum A_i = 1$.

Sample	*Ref.*	A_1	τ_1 *(ns)*	A_2	τ_2 *(ns)*	k_{EM}^a *(10^7 s^{-1})*	$1/k_c^b$ *(ns)*
m PS2	90	0.3 ± 0.1	6 ± 1	0.7 ± 0.2	0.5 ± 0.2	13	7.7
r PS2	90	1.0 ± 0.1	11 ± 1	–	–	5.4	19
mm PS3	91	0.3 ± 0.1	4.0 ± 0.5	0.7 ± 0.2	0.5 ± 0.2	21	9.4
mr PS3	91	0.4 ± 0.1	4.0 ± 0.5	0.6 ± 0.2	0.4 ± 0.2	21	9.4
rr PS3	91	1.0 ± 0.1	7.4 ± 0.4	–	–	9.8	20
mmm PS4	92	0.4 ± 0.1	2.0 ± 0.4	0.6 ± 0.1	0.7 ± 0.2	46	6.5
mmr PS4	92	0.4 ± 0.1	2.0 ± 0.4	0.6 ± 0.1	0.7 ± 0.2	46	6.5
mrm PS4	92	0.3 ± 0.1	2.0 ± 0.5	0.7 ± 0.2	0.6 ± 0.2	46	6.5
mrr PS4	92	1.0 ± 0.1	2.0 ± 0.4	–	–	46	6.5
rrr PS4	92	1.0 ± 0.1	8.2 ± 0.5	–	–	8.5	35

[a] $k_{EM} = 1/\tau_1 - 1/27$ ns (cumene).
[b] $k_c = k_{EM}/(n-1)$. (–) Not observable.

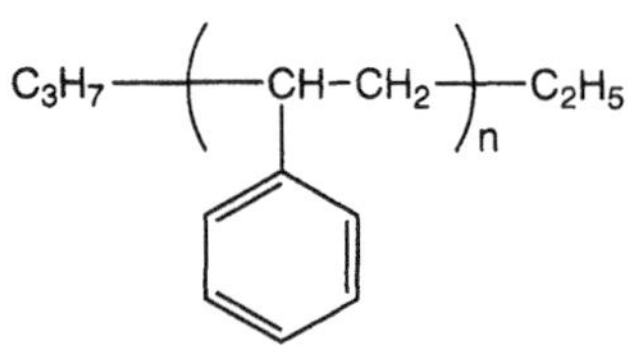

PSn

of all-racemic isomers was found to decay with a single exponential function; the overlap of the excimer fluorescence was taken into account in the fitting calculation.

$$I_M(t) = \exp(-t/\tau_1) \tag{61}$$

The monomer fluorescence of other isomers was found to decay according to the sum of two exponentials,

$$I_M(t) = A_1 \exp(-t/\tau_1) + A_2 \exp(-t/\tau_2) \qquad (A_1 + A_2 = 1) \tag{62}$$

The excimer dissociation is neglected in styrene oligomer and polymer systems because the decay profiles of excimer fluorescence agree very well with one another. Because each integration ($\sum(A_i\tau_i)$) of monomer fluorescence decay corresponds to each peak intensity in the photostationary fluorescence, the analysis must be acceptable.

Since Longworth and Bovey [94], several authors have reported that an excimer is more easily formed in meso than in racemic dimeric isomers. Almost all meso dimers have, at the most, stable ground state conformation (g^+t/tg^-) at the initial stage of excitation (Figure 3.26). Here the excimer conformation (tt) is easily formed from g^+t/tg^- conformation by a pathway with small steric hindrance. On the other hand, in the racemic isomers, the excimer conformers (g^-t/tg^-) cannot be formed from the most stable conformers (tt, g^+g^-) without passing over a higher rotational energy barrier.

The results for our oligostyrenes demonstrate that the efficiencies of excimer formation of all racemic isomers (r PS2, rr PS3, and rrr PS4) were by far lower than those of the isomers with more than one meso diad [91, 92]. Under the assumptions that (1) the decay constant $1/\tau_1$ is expressed by the sum of k_{FM}, k_{IM}, and k_{EM} (see Eq. (30)) and (2) the lifetime of cumene in cyclohexane (27 ns) can be substituted into the sum of k_{FM} and k_{IM}, we have obtained the values of k_{EM} to be $5.4 \times 10^7\ s^{-1}$ (r PS2, 19 ns), $9.8 \times 10^7\ s^{-1}$ (rr PS3, 10 ns), and $8.5 \times 10^7\ s^{-1}$ (rrr PS4, 12 ns). The value of k_{EM} for rr PS3 is about twice as large as that of r PS2, whereas that of rrr PS4 is rather smaller than that of rr PS3. The increase in k_{EM} from r PS2 to rr PS3 appears to be reasonable because the possibility of excimer formation is proportional to the number of pairs capable of attaining the excimer conformer between two adjacent phenyl groups (r PS2, 1:rr PS3, 2). However, the decrease in k_{EM} from rr PS3 to rrr PS4 is in contrast to the change from mm PS3 to mmm PS4. This clearly suggests that in polystyrene it is not easily possible for a racemo diad to take an excimer conformer (g^-t/tg^-) owing to the higher steric hindrance, for the time required for a racemo diad to take an excimer conformer is at least 19 ns. Thus, our results have proved that the excimer is formed mainly as a meso diad, which coincides with the theoretical prediction [187].

In particular, the most important conclusion deduced from the results with PS3 and PS4 is that the phenyl groups do not rotate independently around the carbon–carbon bonds of the backbone chain, but rather their rotation involves a cooperative motion [91, 92], such as a crankshaft transition [95–98] or a three-bond jump motion [21], in the backbone chain. As described earlier, the excimer is formed in a meso diad; thus, the excimer formation in mr PS3 and mrr PS4 would give information on the dynamic motion of the phenyl group in a meso structure of the trimer and tetramer, respectively. If side-chain groups could rotate around each carbon–carbon bond of the main chain independently, the efficiency of excimer formation in mm PS3 would be expected to be two times as high as that in mr PS3, and that in mmm PS4 to be three times as high as that in mrr PS4. However, the experimental results show that the longer time constants (τ_1) are identical among the trimers (mm and mr PS3) or the tetramers (mmm, mmr, mrm, and mrr PS4). Because efficiency of excimer formation depends not on the number of meso diads but on the number of phenyl groups (or the number of bonds) indicates that conformational changes take place by a cooperative motion, for the frequency of a cooperative motion in a molecule should be proportional to the number of carbon–carbon bonds in the backbone chain.

Under the same assumption applied to the calculation for all racemic isomers, values of k_{EM} of the isomers having at least one meso diad are calculated to be $1.3 \times 10^8\ s^{-1}$ (PS2, 7.7 ns), $2.1 \times 10^8\ s^{-1}$ (PS3, 4.8 ns), and $4.6 \times 10^8\ s^{-1}$ (PS4, 2.2 ns). Because the number of pairs capable of attaining the excimer conformer of two adjacent phenyl groups is $(n - 1)$ for an n-mer (PS2, 1; PS3, 2; PS4, 3), we calculated by dividing each k_{EM} by $(n - 1)$ the values of rate constants for conformational change [$(g^+t/tg^-) \rightarrow tt$] for a meso diad by way of a cooperative motion: $1.3 \times 10^8\ s^{-1}$ (PS2, 7.7 ns), $1.1 \times 10^8\ s^{-1}$ (PS3, 9.1 ns), and $1.5 \times 10^8\ s^{-1}$ (PS4, 6.7 ns). Similarly, the rate constants for conformational change [$(tt, g^+g^-) \rightarrow (g^-t/tg^-)$] for a racemo diad by way of a cooperative motion were $5.4 \times 10^7\ s^{-1}$ (PS2, 19 ns) and $4.9 \times 10^7\ s^{-1}$ (PS3, 20 ns). This coincidence itself suggests the presence of a cooperative motion in the main chain.

On the other hand, another relaxation time (τ_2, 0.5 ns) is observable in isomers having more than one meso diad. Because the values of τ_2 are identical with one another, the time constant τ_2 is related to another cooperative motion to induce conformational changes. We assume [91] that this is due to a conformational change [$(g^+t/tg^-) \rightarrow tt$] in a meso diad by means of jump motion of a few main chain bonds such as a three-bond jump motion [21].

3.5.1.2 Physical Gelation of Stereoregular Polystyrene

We have clarified the formation process of intramolecular excimers in polystyrene systems and obtained information on the dynamic process of local motion. Then, we tried to observe the thermoreversible gelation process of isotactic polystyrene (iPS) in decalin by monitoring polystyrene fluorescence.

More than 60 reports have been published for iPS thermoreversible gel as a typical example of the physical gel [100] since Girolamo *et al.* [101] first demonstrated that quenching of a solution of iPS leads to a solution of a gel. Several authors have presented the hypothesis about the structure of iPS in decalin. Their common conclusion is that iPS in a gel form has a helical structure, although the type of the helix was in the air. Among the possibilities are (1) a 12_1 helical form having extended tt chain conformation [102, 103], and (2) a near-3_1 (Guenet's ladderlike model) [104, 105] or 3_1 helical form (coiled-coil molecular model) [106] consisting of g^+t or tg^- conformers. With regard to (2), both models are not conflicting with each other.

Figure 3.27 [108] shows the temperature dependence of fluorescence spectra of iPS in THF (A), which does not form gel, and iPS in *cis*-decalin (B), which

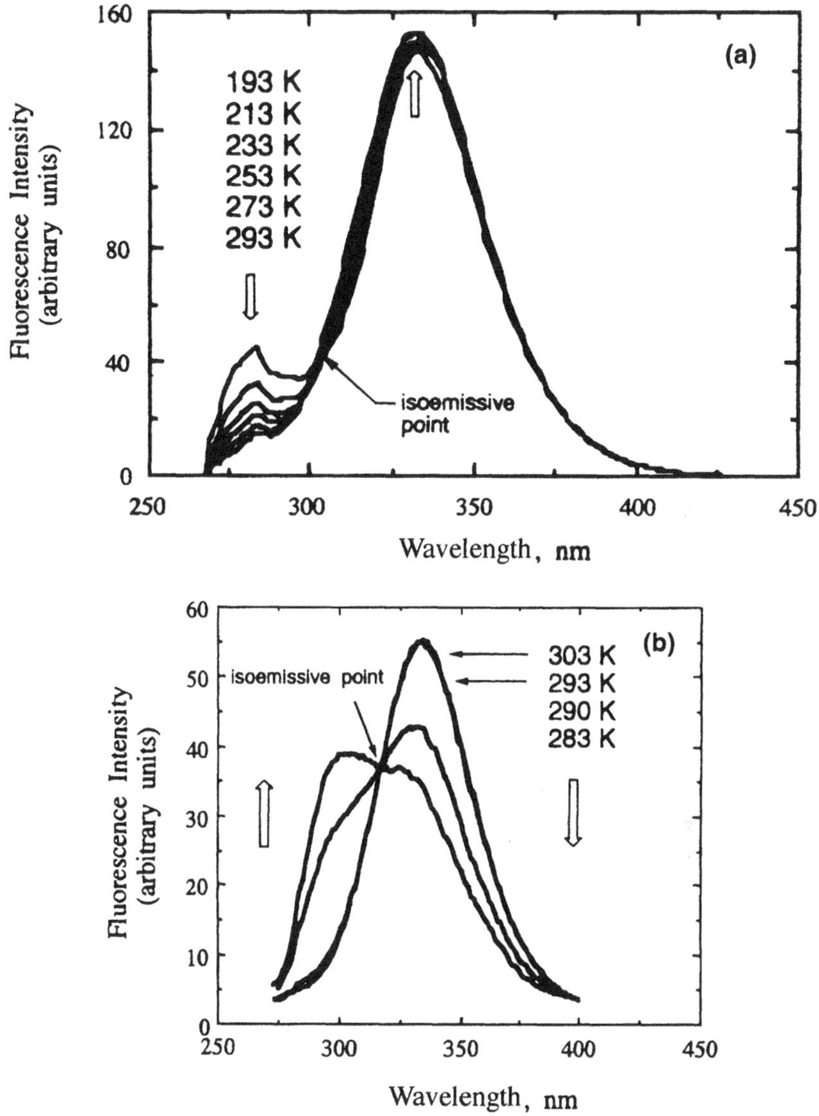

Figure 3.27 Temperature dependence of fluorescence spectra of (a) dilute isotactic polystyrene (iPS) in THF between 193 and 293 K and (b) 1.4% iPS in aerated *cis*-decalin (gel) in the cooling process from 303 to 283 K (the excitation wavelength is 257 nm) from ref. [108].

forms gel. The former is measured in a quartz cell with an optical path length of 1 cm set as Figure 3.23(a), whereas the latter is measured in a quartz cell with an optical path length of 1 mm set as Figure 3.23(b). The fluid solution of iPS shows very strong fluorescence of an excimer because the excimer is formed in the meso diad as described above: iPS has a stereoregular structure with all meso diads. However, once the gelation takes place, the formation of excimer, i.e., the local motion of side chains, is restricted so much that a strong monomer fluorescence can be observed. We can present two explanations for why the formation of the helix structure restricts the local motion. First, it requires higher energy to change conformations when an oriented structure having a helical rod is formed because the total energy of the helical form is thought to be low. Second, the conformational change takes place by way of cooperative motion of the polymer chain as described earlier [91, 92]. The formation of a helical form would restrain a diad, which is both in a helix and near the helix, from taking part in a cooperative motion of the chain.

Generally speaking, although the fluorescence probe method gives detailed information on the microscopic environment around a probe, it does not yield macroscopic information such as structure. However, in our case, fluorescence data have played an important role in determining the structure of the helix.

If a side-chain phenyl group in the 12_1 helical form rotated freely about the C_α–C_{para} axis, a strong excimer emission would be observed, because almost all diads are in tt (excimer) conformers in the case of 12_1 helical form, and the free reorientation motion of phenyl rings in the tt conformer definitely forms an excimer with small energy. On the contrary, in the case of a 3_1 or near-3_1 helical form, the excimer cannot be formed without the conformational change from g^+t or tg^- to tt: it requires large energy. Our fluorescent data clearly show that the excimer formation was restricted very much. Thus, we concluded that the 12_1 helix is not likely in the nascent gel and a near-3_1 helical form is more probable [108].

As a matter of fact, it took more than 2 weeks to obtain Figure 3.27(b). In the cooling process down to 293 K, the equilibrium is attained at least within several hours after setting a new temperature of the solution. However, the fluorescence spectra at 290 K changed with time for 2 weeks. Figure 3.28(a) shows the time dependence of the excimer on the monomer fluorescence intensity ratio I_E/I_M. We judged that the I_E/I_M reached the equilibrium value $(I_E/I_M)_{t=\infty}$, after 300 hours because the value of I_E/I_M did not change for 2 days. Figure 3.28(b) shows the logarithm plot of $(I_E/I_M - (I_E/I_M)_{t=\infty})$ against time. This transient curve is found to be the best fit for the sum of two exponentials with time constants of 11 and 92 hours [108]. It is clear that this fluorescence change is not due to a photochemical process because Figure 3.27(b) shows the isoemissive point. Figure 3.28 suggests that some physical properties are different between the nascent gel and the gel stored for some weeks.

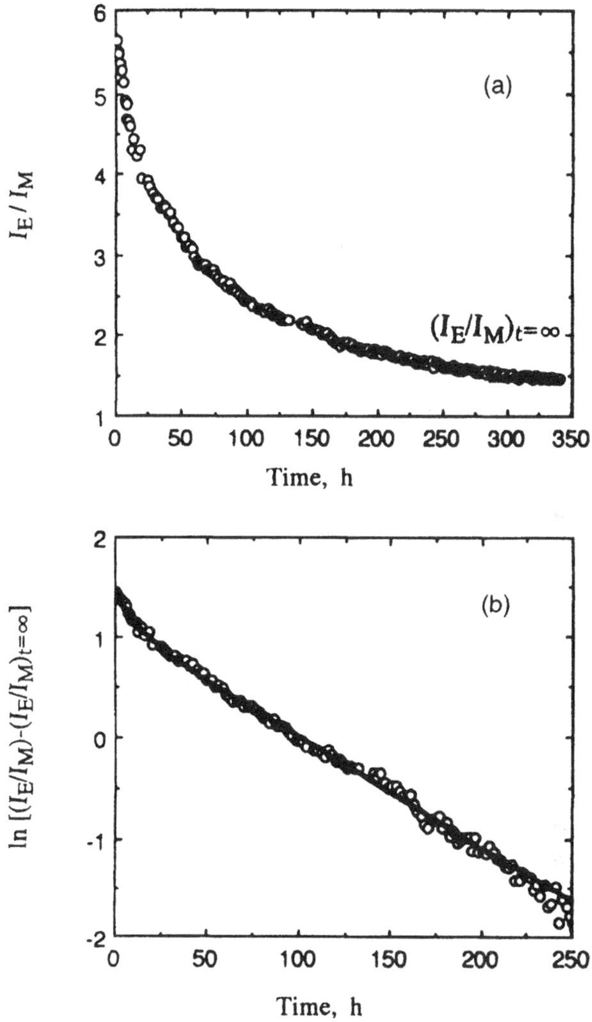

Figure 3.28 Time dependence of the excimer-to-monomer fluorescence intensity ratio I_E/I_M, for 1.4% iPS in aerated *cis*-decalin (gels) at 290 K (excitation wavelength is 259 nm). Variation of I_E/I_M (a) and $\ln[(I_E/I_M) - (I_E/I_M)_{t=\infty}]$ (b) as a function of time; $(I_E/I_M)_{t=\infty}$ is the equilibrium value. The solid line in b shows a model curve of the sum of two exponentials $([(I_E/I_M)(t) - (I_E/I_M)_{t=\infty}]/[(I_E/I_M)_{t=0} - (I_E/I_M)_{t=\infty}] = 0.68 \exp(-t/92\,\mathrm{h}) + 0.32 \exp(-t/11\,\mathrm{h}))$ [108].

The time profile of iPS in *cis*-decalin was found to be different from that in *trans*- and mixed decalin: a dual exponential function in *cis*-decalin versus a single exponential in *trans*- and mixed decalin. Guenet determined the number of solvent molecules adsorbed per iPS monomer unit to be 1.89 for *cis*-decalin and 1.15 for *trans*-decalin by using differential scanning calorimetry [104] to present the proof of his ladderlike model in which iPS and decalin form a polymer–solvent compound. Thus, the difference of time profiles between *trans*- and *cis*-decalin is supposed to be induced by their disparity in molecular size [108]. It is reasonable to assume that the *cis*-decalin gel has a longer time profile because smaller *cis*-decalin molecules are more mobile in a limited free volume in a molecular compound and the total amount of *cis*-decalin that is intercalated into iPS helices is larger than that of *trans*-decalin.

Figure 3.29 shows the temperature dependence of I_E/I_M of the 1.4% iPS in *cis*-decalin, which is obtained by continuous measurements for a month from 373 to 223 K and again from 223 to 373 K. All the I_E/I_M values are of the equilibrium

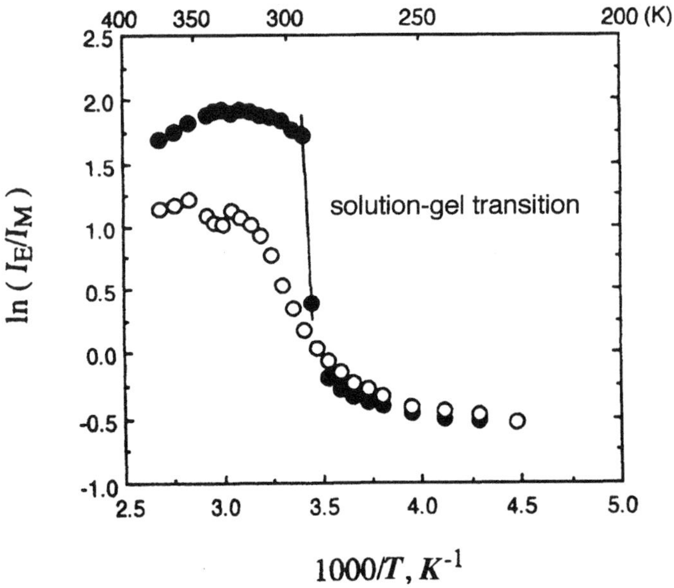

Figure 3.29 Temperature dependence of the excimer-to-monomer fluorescence intensity ratio I_E/I_M for 1.4% iPS in aerated *cis*-decalin (gels); (●) cooling from 373 to 223 K; (○) heating from 223 to 373 K. All the I_E/I_M values are the equilibrium ones. The solution-to-gel transition is observed by cooling from 293 to 290 K.

state. During the cooling process from 373 to 293 K and the heating process from 333 to 373 K, iPS–decalin is of the solution form. The continuous increase in I_E/I_M in the heating process to 328 K indicates that the local motion to form excimer conformation (tt) is released gradually, although it is still of gel form.

Figure 3.29 demonstrates that (1) a hysteresis loop is observed in the iPS solution–gel transition, (2) the solution-to-gel transition against temperature is observable as a dramatic change of I_E/I_M at 290 K, (3) the gel-to-solution transition is not clearly observable, and (4) after a cycle of cooling and heating of the iPS in *cis*-decalin, the value of I_E/I_M at 373 K is lower than the value of I_E/I_M that was first measured at 373 K before cooling. Observation (4) is due to the association of helical rods followed by the formation of a node, which appears to be a part of the crystal structure. A helical rod of an isolated iPS molecule goes back to a random coil at temperatures higher than the temperature at which the helix-to-coil transition takes place. However, if some helical rods encounter each other and line up to become a node by interactions such as van der Waals forces, this association consequently becomes a sort of microcrystal. The energy required for breaking up a node into some single helical rods should be higher than that for losing a helical polymer chain to a random-coil form.

Consequently, the physical gelation of iPS–decalin systems is shown to take place by the formation of nodes that are a sort of microcrystal made up of the oriented helical rods and that can be a cross-linking point of the gel network. Consequently, conditions for the formation of helical rods appear to be favorable near nodes [108].

So far, I have explained our studies that employed the inherent fluorescence of polystyrene. Next, I show our results by doping some chromophores into the gel system, because certain aspects of the motion and orientation of the solvent decalin in such gels is still poorly understood. Figure 3.30 shows the possible location of solvent molecules within the gel. Region I is the area where solvent molecules gather, and Region II is the area where helical rods of iPS associate. We can assume another interface area among iPS chains and decalin molecules; however, it is difficult to distinguish this area from Region I. We have been interested in investigating if the solvent decalin can exist in Region II and if the solvent decalin can move around when in Region I and II or when in an interface region.

To answer these questions, we doped naphthalene (NP), 1-methylnaphthalene (MN), 1,5-dimethylnaphthalene (DMN), and anthracene (AT) into iPS–decalin gels uniformly and measured the iPS concentration dependence of their fluorescence anisotropy [109]. Probe molecules larger than NP in molecular size were found to be mobile (i.e., their values of anisotropy were zero) in the iPS gel, indicating that solvent molecules are mobile in the area where solvent gathers (Region I)

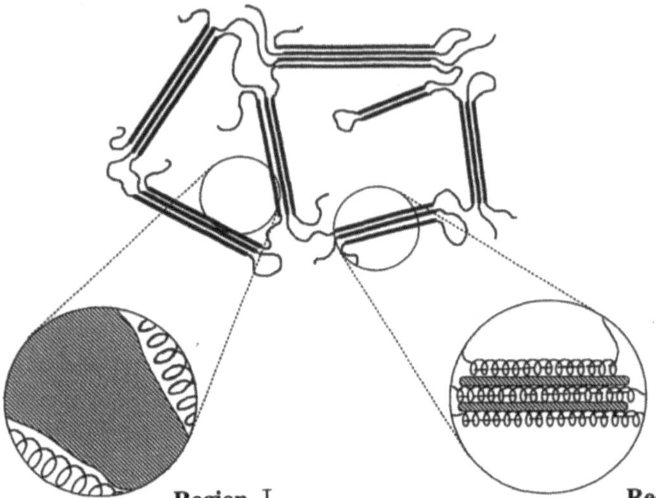

Figure 3.30 Sketch of iPS chains in the gel form. The hatched areas represent the space in which solvent molecules can be located.

and that motion of solvents near the interface with iPS is not limited so much. However, NP, the smallest probe molecule considered, showed higher values of fluorescence anisotropy in the iPS gel, and this effect increased with an increase in concentration of iPS. From these results we conclude that a free volume exists, into which NP molecules can break, among iPS helical rods in an area where the iPS chains associate (Region II). This means that the possibility is also very high for decalin molecules to be incorporated into a polymer–solvent compound claimed by Guenet [104, 105] and that their motion is limited because *cis*- and *trans*-decalin molecules are smaller than naphthalene molecules.

This study is an example of the use of fluorescence anisotropy for probing free volume in polymer gels and polymer solids.

3.5.1.3 Segmental Motion of Polystyrene [68]

What follows is an example making use of energy transfer to obtain a diffusion constant. The Yokota–Tanimoto kinetics (Section 3.2.3.3.2.3) give us the information on diffusion, as described in Eqs. (47–50). We employed seven anionically polymerized polystyrenes having an anthryl group at one end (PSA) and anthryl groups at both ends (APSA) with a variety of molecular weights and measured the energy transfer efficiency from the side-chain phenyl groups to the anthryl group at an end by monitoring the quenching yield of the phenyl excimer fluorescence. In general, the quantum yield of the donor fluorescence Φ_D is calculated from Eq. (47) as:

$$\Phi_D(t) = \frac{\int_0^{\infty} I_0 \exp[-t/\tau_0 - 2B(C_A/C_0)\sqrt{t/\tau_0}]\,dt}{\int_0^{\infty} I_0 \exp[-t/\tau_0]\,dt}$$

$$= \tau_0 \int_0^{\infty} \exp[-t/\tau_0 - 2B(C_A/C_0)\sqrt{t/\tau_0}]\,dt \qquad (63)$$

In this case, τ_0 is of polystyrene excimer (20 ns). First, we calculated C_A/C_0 (Förster kinetics) or a set of C_A/C_0 and D (Yokota–Tanimoto kinetics) to reproduce the experimental value (Φ_D) of PSA and APSA. The Durbin–Watson factors [72, 73] are used for the assessment of the kinetics. Table 3.5 shows the experimental results together with calculated values. Each calculation depends on the model of the kinetics, and thus the assessment is carried out for the whole fitting: The Yokota–Tanimoto kinetics has reproduced our experimental data more exactly. Finally we obtained the diffusion constant of a segment having some decades of monomer units near the end group to be $(2.1 \pm 0.2) \times 10^{-6}$ cm^2/s. This value is reasonable in comparison with the diffusion constants obtained from other methods such as the self-diffusion constant of low-molecular-weight polystyrene measured by means of dynamic light scattering [110] and the pulsed-field-gradient NMR method [111] and the diffusion constants calculated from the cyclization behavior of a polystyrene chain [112].

3.5.2 *THE FLUORESCENT PROBE METHOD EMPLOYED IN POLYMER SYSTEMS*

3.5.2.1 Mono- and Multilayer Films Prepared by the Langmuir–Blodgett Technique

The Langmuir–Blodgett (LB) technique enables one to prepare monolayered and multilayered films with a thickness of nanometers and with an oriented structure at the molecular level. The fluorescence probe method has been demonstrated to be the most effective to assess the structure of such layers.

Yamamoto *et al.* [113, 114] investigated the layered structure of LB films of poly(vinyl octal) (PVO) by time-resolved fluorometry of the energy transfer between layers containing phenanthrene (P: energy donor) and anthracene (A: acceptor) moieties. The PVO containing P and A were synthesized by the acetalization

Table 3.5 Quenching yield of excimer fluorescence by terminal anthryl groups (Φ_D) and comparison of calculated values of Φ_D with experimental values from ref. [68].

Polymer	Φ_D	*Calculated Values Using Förster Kinetics*	*Calculated Values Using Yokota–Tanimoto Kinetics*
PS25A	0.60	0.55	0.60
PS42A	0.68	0.68	0.68
PS70A	0.74	0.78	0.76
PS400A	0.94	0.96	0.94
PS800A	1.0	0.98	0.97
APS135A	0.75	0.77	0.74
APS775A	0.95	0.95	0.94
DW		1.10	1.72

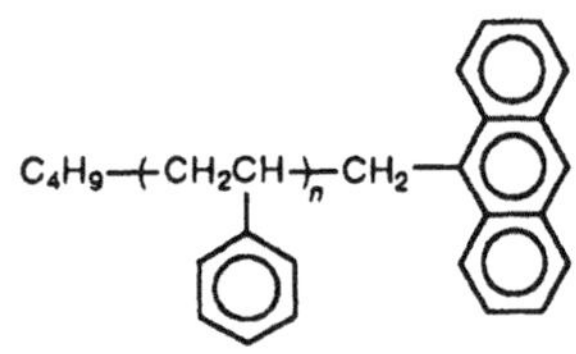

PSnA

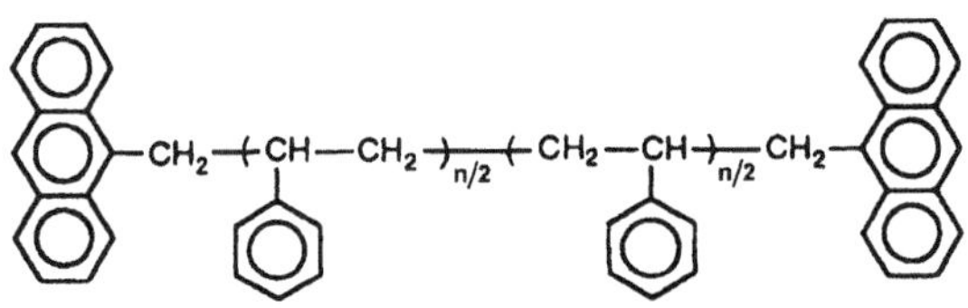

APSnA

of poly(vinyl alcohol) with octanal and chromophoric aldehyde. The polymer was spread onto pure water and transferred by dipping a quartz plate vertically. The sample (P(n)A), which consisted of two layers of polymer P, n layers of PVO (n = 0–8), and two layers of polymer A, was prepared for fluorescence measurements. Figure 3.31 shows the fluorescence spectra of P(n)A films: the fluorescence of P decreases with a decrease in the number of the spacing layer (n) because of singlet energy transfer. The energy transfer efficiency for each P(n)A was larger than the theoretical value predicted by the Förster mechanism (Section 3.2.3.3.2.2), indicating that the segmental relaxation of the polymer chain makes the chromophores move away from the original plane on which they were deposited.

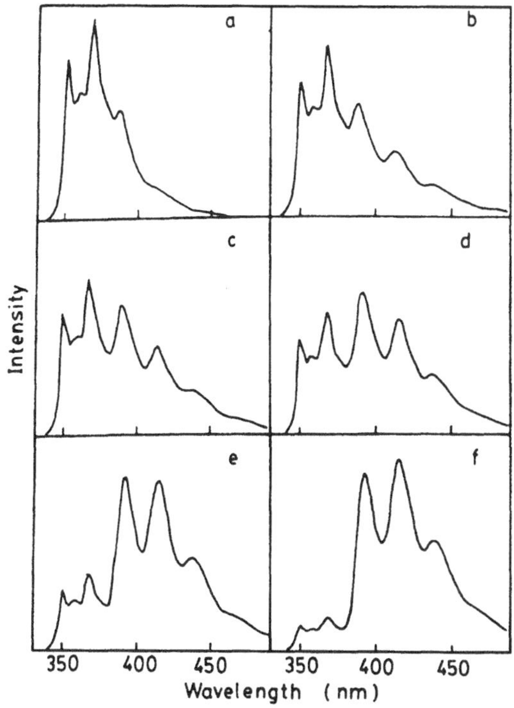

Figure 3.31 Fluorescence spectra of P(*n*)A films: (a) P, (b) P8A, (c) P6A, (d) P4A, (e) P2A, and (f) P0A (the excitation wavelength is 298 nm with a bandwidth of 3 nm) from ref. [113].

To estimate the degree of disordering, the fluorescence decay curve of P was analyzed by a theoretical model in which the chromophores the distributed according to a Gaussian function in the direction of plane normal [113]. The calculation by means of the Monte Carlo method enabled the quantitative evaluation of the partly disordered structure of the LB films; the standard deviation for the P(4)A, P(6)A, and P(8)A films was calculated to be 1.35, 1.70, and 2.25 nm, respectively.

In addition to some initial distortion of the layer structure, appreciable increase of transfer efficiency occured during a few days after the deposition. This aging effect shows that segmental diffusion takes place in the LB films. Then, Yamamoto *et al.* [115] tried to monitor the polymer diffusion by applying the same method to a poly(vinyl pentanal acetal) LB film system labeled by P and A. The thermal relaxation clearly takes place at temperatures higher than the glass transition temperature. The theoretical treatment shown above, which is based on the Förster kinetics by assuming Gaussian distribution of chromophores in the direction normal to the layer plane, has yielded a diffusion constant of polymer segments on the order of 10^{-16} to 10^{-18} cm^2/s and an apparent activation energy of 150 kJ/mol. The values are similar to those reported for the diffusion phenomena of bulk polymers [116]. The polymer chains in the LB films could behave like ones in a polymer bulk, although this point must be clarified in detail.

3.5.2.2 Micelles

One of the methods to determine micelle aggregation numbers, N_{agg}, is established by using fluorescence measurements. Here, I briefly present only the equations to derive N_{agg} by using a singlet energy transfer method. Readers are invited to refer to some papers concerning the method [117, 118].

The method is based on the quenching of a fluorescent probe molecule in a micelle in which mobile quenchers are dispersed. The model is valid under certain conditions: (1) the micelles are monodisperse in size, (2) the probes and quenchers remain associated with the host micelles for longer time compared with the fluorescence lifetime of the probes, (3) emission of the probe is observed only from micelles with probe but without quenchers, and (4) the quenchers do not interact with one another and the probes and quenchers follow a Poisson distribution law among the micelles. The following combination of fluorescence probe and quencher is often employed: ruthenium tris(bipyridyl) complex cation, dichloride (probe) and 9-methylanthracene (quencher) [119]; 1-ethylpyrene (probe) and N-cetylpyridinium chloride (quencher) [120]; 1-methylpyrene (probe) and m-dicyanobenzene [121].

3.5.2.2.1 Steady-State Fluorescence Quenching Methods

The simplest method is given by Turro and Yekta [119]. The relative intensity of a probe is related to the quencher concentration $[Q]$ and the micelle concentration $[M]$ by the relationship

$$I_F = I_F^0 \exp\left(-\frac{[Q]}{[M]}\right) \tag{64}$$

where I_F^0 is the fluorescence intensity in the absence of quencher, and I_F is the fluorescence intensity in the presence of quencher. The micelle concentration is given by

$$[M] = (S - \text{cmc})/N_{agg} \tag{65}$$

where S is the total surfactant concentration and cmc is the critical micelle concentration. From Eqs. (64) and (65),

$$\ln\left(\frac{I_F^0}{I_F}\right) = \frac{N_{agg}[Q]}{S - \text{cmc}} \tag{66}$$

Thus, a plot of $\ln(I_F^0/I_F)$ versus $[Q]$ allows N_{agg} to be determined. The method is quite accurate up to an aggregation number of about 120 [119]. Beyond this value, the condition (3) is no longer valid, and N_{agg} is underestimated.

3.5.2.2.2 Transient Fluorescence Quenching Methods

A general formula for the fluorescence decay of a probe in a micellar solution containing mobile quenchers was given as Eq. (67) by Tachiya [122]:

$$I(t) = I(0)\exp[-A_1 t - A_2(1 - \exp(-A_3 t))] \tag{67}$$

Here, some rate constants are defined as follows: k_f is the rate constant for fluorescence in a micelle without quenchers, k_{qm} for intramicellar quenching when only one quencher is present in the micelle, k_m^- for a quencher leaving the micelle, k_m^+ for a quencher entering the micelle, and equilibrium constant $k = k_m^+/k_m^-$. In this case, the quenching of a fluorescent probe is expressed as $nk_{qm}[P_n^*]$ in a micelle P_n that consists of a probe and n quenchers. Here A_1, A_2, and A_3 can be expressed as follows on the basis of the derivation by De Schryver *et al.* [121]:

$$\begin{aligned} A_1 &= k_f + \frac{k_{qm}[Q]k_m^+}{(1 + K[M])(k_m^- + k_{qm})} \\ &= k_f + S_1[Q] \end{aligned} \tag{68}$$

$$\begin{aligned} A_2 &= \frac{Kk_{qm}^2[Q]}{(1 + K[M])(k_m^- + k_{qm})^2} \\ &= S_2[Q] \end{aligned} \tag{69}$$

$$A_3 = k_m^- + k_{qm} \tag{70}$$

where

$$S_1^{-1} = \frac{(1 + K[M])A_3}{k_m^+ k_{qm}} \tag{71}$$

$$S_2^{-1} = \frac{A_3^2}{Kk_{qm}^2} + \frac{A_3^2}{k_{qm}^2}[M] \tag{72}$$

A plot of A_2 versus $[Q]$ yields a straight line and gives a value of S_2 (Eq. (69)). A plot of A_1 versus $[Q]$ yields a slope S_1 and an intercept that is identical with

k_f (Eq. (68)). Because $[M]$ is expressed as Eq. (65), a plot of S_1^{-1} and S_2^{-1} versus $(S - \text{cmc})$ gives a linear relationship to determine A_3. Thus, k_m^- can be determined by Eq. (73):

$$k_m^- = \frac{(S_1/S_2)A_3}{S_1/S_2 + A_3} \tag{73}$$

and k_{qm} can be obtained from $(A_3 - k_m^-)$ (Eq. (70)). If B and C are defined as follows, they can be described as Eqs. (74) and (75) by substituting Eqs. (71) and (72) into S_1 and S_2, respectively and using Eq. (65): $B = \frac{A_3}{k_{qm}} S_1$, $C = \frac{A_3^2}{k_{qm}^2} S_2$.

$$B^{-1} = \frac{1}{k_m^+} + \frac{S - \text{cmc}}{k_m^- N_{agg}} \tag{74}$$

$$C^{-1} = \frac{1}{K} + \frac{S - \text{cmc}}{N_{agg}} \tag{75}$$

Thus, a plot of B^{-1} versus $(S - \text{cmc})$ gives a slope equal to $(k_m^- N_{agg})^{-1}$ and an intercept equal to $1/k_m^+$ (Eq. (74)), whereas a plot of C^{-1} versus $(S - \text{cmc})$ gives a slope equal to N_{agg}^{-1} and an intercept equal to $1/K$ (Eq. (75)). The calculation described above gives all the parameters.

3.5.2.3 Gels

It was Horie *et al.* [123] who first applied fluorescence probe methods to polymer gel systems, and many papers have been published dealing with gelation monitored by means of fluorescence spectroscopy.

3.5.2.3.1 Gel Phase Transition Revealed by Means of Fluorescent Probes

Since the volume phase transitions (VPT) of some organic polymer hydrogels such as polyacrylamide gel were discovered [124, 125], this phenomenon has attracted much attention because of its scientific interest and technological significance. In order to monitor the microenvironment of the gel, Hu *et al.* introduced a dansyl group into poly(acrylamide) (PAAm) gel ([dansyl unit]/[total unit] = 0.13 mol%) and observed the fluorescence change during a pH-induced [126] or a solvent-induced [127] VPT. Here I introduce the successful experiments [127] of VPT induced by the composition change in an acetone–water mixed solvent as an example. Irrespective of the presence or absence of dansyl groups attached to the side group of PAAm gels, the PAAm gel swells about 35 times in volume between acetone contents from 10 to 59.5% and abruptly shrinks at 60% acetone content (Figure 3.32(a)).

As described in Section 3.2.3.2, the dansyl fluorescence intensity increases sharply, and its peak shifts to shorter wavelengths, with a change of the microenvironments to a less polar medium than water. Figure 3.32(b) shows the peak wavelength shift with the change in acetone content. As a reference, the fluorescence behavior is also shown for the dansyl group attached to a linear PAAm ([dansyl unit]/[total unit] = 0.13 mol%). The sharp change in peak wavelength to the red and sharp return to the blue at 60% acetone content would be caused by the PAAm network fluctuation near the discontinuous volume phase transition (DVPT) because the twisting motion of the dimethylamino group was enhanced owing to the rapid fluctuation of water rearrangement.

The dansyl fluorescence of PAAm gels and linear PAAm decays according to the sum of two exponentials with time constants of 4–12 and 11–33 ns (acetone content: 10 to 90%). The authors claim that the shorter-lifetime component corresponds to the TICT state (Section 3.2.3.2) in which the dimethylamino group and naphthyl group in the dansyl probe are in a nonplanar excited state in a hydrophilic microenvironment. On the other hand, the longer-lifetime component corresponds to the dansyl probe with a coplanar excited state in a hydrophobic microenvironment. The shorter time constant shows a discontinuous change near the VPT of the gel. It is due to the rapid twisting motion of the dimethylamino group around the naphthyl group caused by the network fluctuation.

These dynamic fluctuation of the gel network are also supported by the change in the rotational diffusion coefficient D_r determined by the fluorescence anisotropy measurements (Eq. (76), see Eqs. (19–22) in Section 3.2.2.3).

$$\frac{r_{\text{pol},0}}{r_{\text{pol}}} = 1 + 6D_r\tau \tag{76}$$

Figure 3.32(c) shows the acetone content change for two values of D_r calculated for each time constant of two decay components. The values become extremely large near the DVPT. The rotational diffusion coefficient of the side group of the network is supposed to become infinite owing to the rapid microscopic fluctuation of the network and solvent in spite of the collective diffusion coefficient of the network's becoming zero in macroscopic DVPT. The authors claim that the rapid microscopic transition between the solidlike network aggregate and the liquidlike swollen state is induced by the density fluctuation and nonequilibrium situation near DVPT.

The method can be summarized as follows:

1. Preparation of a sample polymer having a dansyl group. A fluorescent probe should be located in a spot sensitive to the change in microenvironment.

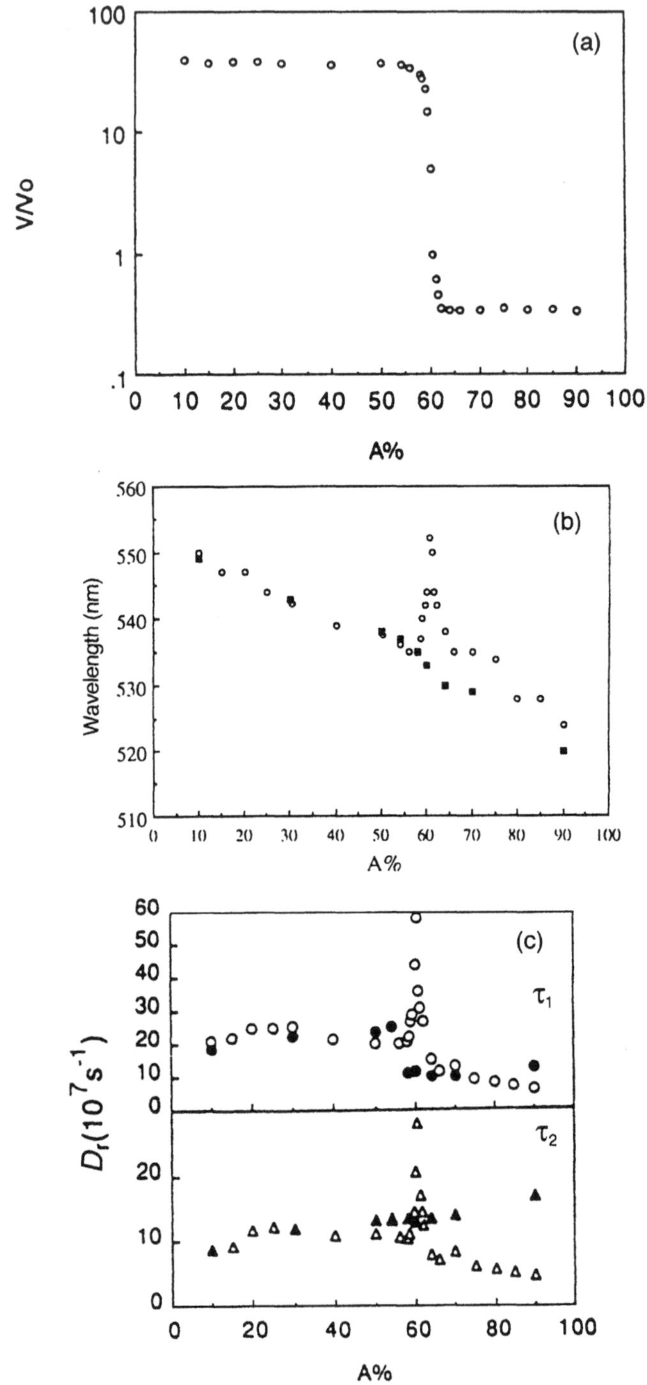
(a)
V/Vo
A%
(b)
Wavelength (nm)
A%
(c)
τ1
τ2
Dr(10^7 s^-1)
A%

2. Steady-state fluorescence measurements. The fluorescence intensity and the peak wavelength shifts with the change in polarity, viscosity, and mobility around the probe.
3. Fluorescence decay measurements. The time constants and the preexponential term show the change in microenvironments.
4. Transient and steady-state fluorescence anisotropy measurements. By using Eq. (76), D_r, can be determined.

Pyrene is another effective and popular fluorescent probe employed to obtain microscopic information in polymer systems. Hu *et al.* [128] studied the same DVPT of PAAm gels by using fluorescence behavior of pyrenyl group, i.e., (1) excimer fluorescence, (2) change in ratio of vibrational peaks (peak I/peak III), and (3) fluorescence anisotropy. Horie *et al.* has applied the fluorescent probe method as described previously to other hydrogels [129], showing VPT and random coil-to-helix transitions or aggregation of polyamides [130].

3.5.2.3.2 Physical Gelation of Cellulose Derivatives

The characteristic features of cellulose molecules such as the rigidity of the chain and the insolubility in many solvents are attributed to three hydroxyl groups (2-, 3- and 6-position) in an anhydroglucose unit of cellulose forming both inter- and intramolecular hydrogen bonds. The substitution of the hydroxyl groups by functional groups is known to induce gelation [131] or liquid crystallization [132–134] where the intermolecular hydrogen bonds may play an important role. Thus, we aimed at elucidating the function of each hydroxyl group of cellulose in the aggregated states such as gel [135, 136]. What follows is an example of the fluorescent probe method using a phenyl moiety. Phenyl groups are relatively easily introduced into a polymer sample, for polymers or monomers having phenyl moieties are sometimes available commercially and phenyl groups have the advantage of being small.

Figure 3.32 Dependence of some properties of PAAm from ref. [127] on acetone volume content in an acetone–water mixed solvent: (a) degree of equilibrium swelling of PAAm gel labeled with a dansyl group; (b) fluorescence peak wavelength of the dansyl probe attached to PAAm gel (○) and linear PAAm (■); (c) rotational diffusion coefficient, D_r, of the dansyl probe attached to PAAm gel (○, △) and linear PAAm (●, ▲) obtained from lifetime τ_1 (○, ●), lifetime τ_2 (△, ▲), and $r_0 = 0.325$.

Table 3.6 Structure of the cellulose derivatives used for measurements.

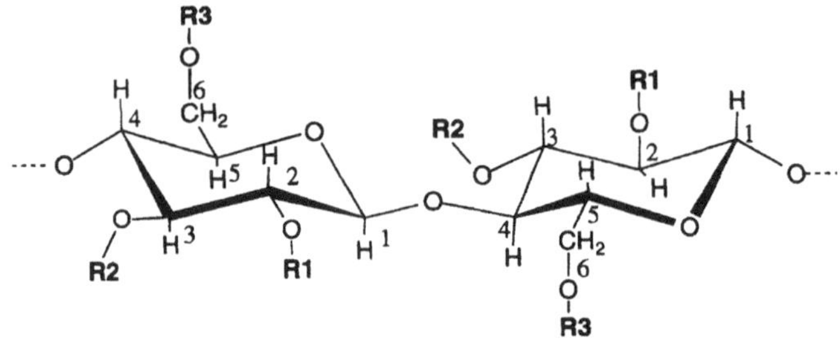

Sample		*R1*	*R2*	*R3*
Cellulose		H	H	H
6-*O*-benzylcellulose	(23O6B)	H	H	Benzyl
2,3-di-*O*-methyl-6-*O*-benzylcellulose	(23M6B)	Methyl	Methyl	Benzyl
2,3-di-*O*-benzylcellulose	(23B6O)	Benzyl	Benzyl	H
2,3-di-*O*-benzyl-6-*O*-methylcellulose	(23B6M)	Benzyl	Benzyl	Methyl
2,3,6-tri-*O*-benzylcellulose	(236B)	Benzyl	Benzyl	Benzyl

Five cellulose derivatives whose hydroxyl groups are regioselectively substituted by benzyl ether and methyl ether groups were prepared (Table 3.6) [136]. The gelation did not take place in THF solutions of the cellulose derivatives whose hydroxyl group at the 6-position was substituted to methyl ether or benzyl ether groups; however, only the cellulose derivative having the 6-position hydroxyl group, i.e., 2,3-di-*O*-benzylcellulose (23B6O), was found to form gels. The samples other than 23B6O showed only the usual fluorescence of benzyl group over the temperature region between 200 and 310 K.

Figure 3.33(a) shows the temperature dependence (from 300 to 200 K) of the fluorescence spectra of 2.3% 23B6O in THF. The spectra are found to change, apparently, depending on temperature. In order to clarify the temperature dependence of 23B6O fluorescence, we compared the shape of the fluorescence spectra normalized at each peak. The change for the cooling process can be divided into

Figure 3.33 Temperature dependence of fluorescence spectra of 2.3% 23B6O in aerated THF in the cooling process from 300 to 200 K. (a) 1, 305 K; 2, 295 K; 3, 283 K; 4, 275 K; 5, 270 K; 6, 250 K; 7, 230 K; 8, 210 K (the excitation wavelength is 257 nm). The spectra normalized at the peak are also shown separately (b) 305–270 K; (c) 270–210 K. The spectra are from ref. [136].

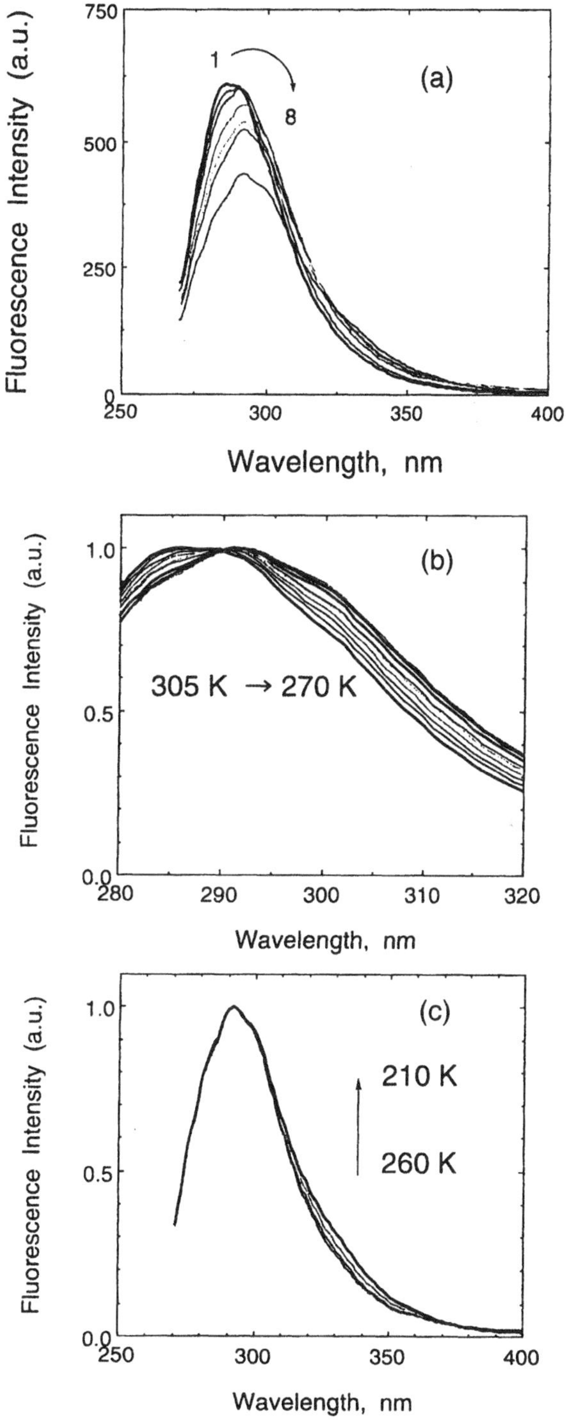
750
500
250
0
250
300
350
400
1
8
(a)
Fluorescence Intensity (a.u.)
Wavelength, nm
1.0
0.5
0.0
280
290
300
310
320
(b)
305 K → 270 K
Fluorescence Intensity (a.u.)
Wavelength, nm
1.0
0.5
0.0
250
300
350
400
(c)
210 K
260 K
Fluorescence Intensity (a.u.)
Wavelength, nm

two temperature regions, i.e., (I) 305 to 270 K (Figure 3.33(b)) and (II) 270 to 210 K (Figure 3.33(c)). The change in region (I) was due to the red shift of the fluorescence, whereas the change in region (II) was due to the formation of excimers. When the temperature of 23B6O in THF was raised to 300 K, the spectral shape that was shifted to the red at low temperatures returned to the shape before cooling. The excitation spectra for 320-nm fluorescence of 23B6O showed the appearance of a new peak near 280 nm, which is concluded to be the absorption peak of the ground-state dimer formed intermolecularly among benzyl groups, meaning that a hydrophobic interaction exists among them. Notice that the existence of a ground-state dimer formed intermolecularly is a decisive proof of the existence of the hydrophobic interaction.

Thus, our fluorescent probe method could elucidate the gelation process with a change of temperature in terms of the molecular association involving a hydrogen bond. In conclusion, (1) the main cause for the gel formation in our system is the hydrogen bonds by means of the 6-position hydroxyl group and (2) the hydrophobic interaction among benzyl groups also keeps 23B6O molecules associated with one another as well as the hydrogen bonds after aggregation [136].

3.5.2.4 Liquid Crystals

The formation process and microscopic structures such as molecular orientation in liquid crystals is an important problem to clarify. Generally speaking, the intermolecular interaction of macromolecules can be monitored by excimer or exciplex formation, energy transfer, and the formation of a CT complex. Here I introduce the application of fluorescent probes to the studies on thermotropic and lyotropic liquid crystallization.

Figure 3.34(a) shows the proposed structure of B–C16, synthesized by the condensation of 1,4-dihexadecylesters of pyromellitic acid with 4,4′-biphenol based on X-ray observations [137].

B–C16

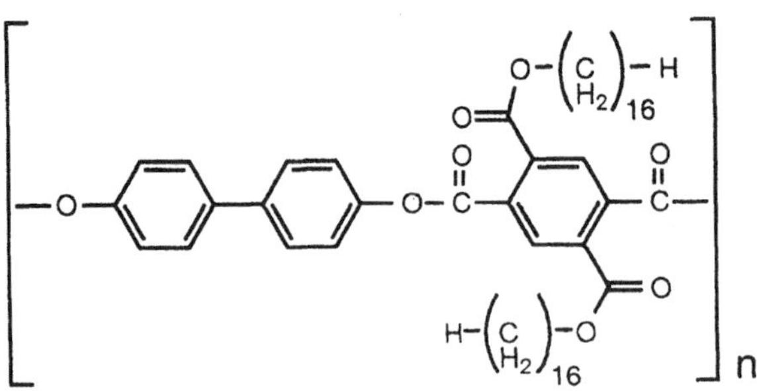

(a) (b)

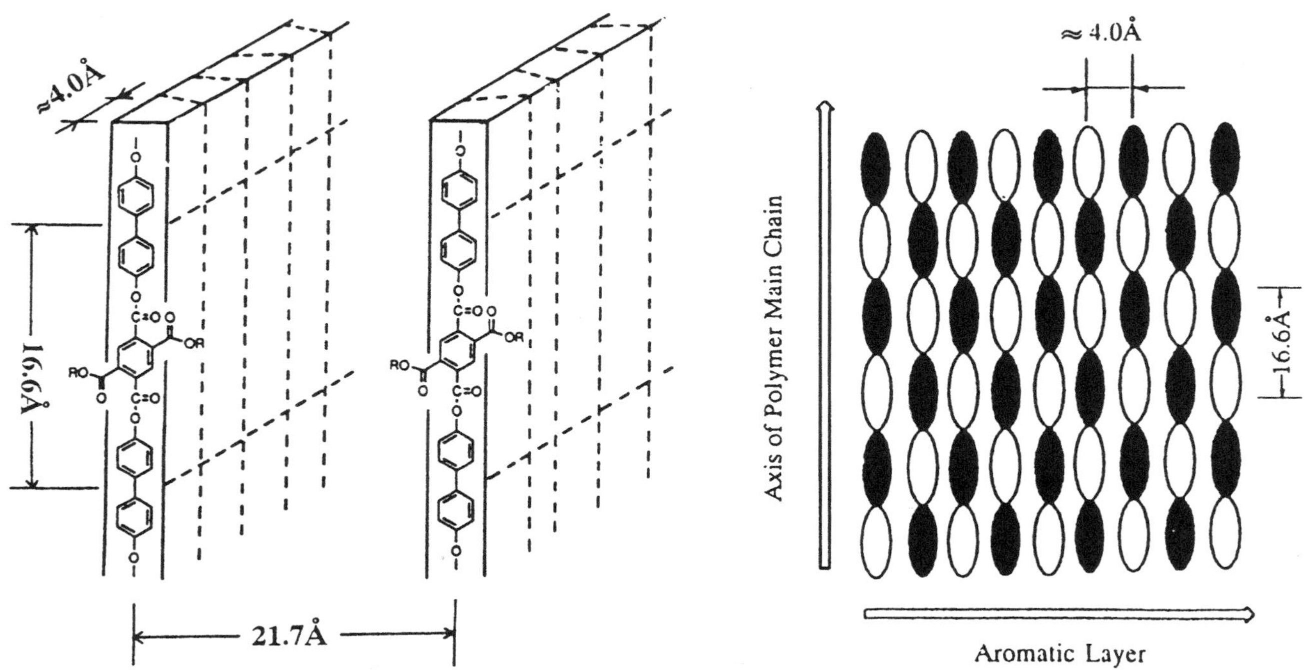

Figure 3.34 (a) An illustration of the proposed structure for the layered mesophase of B–C16 observed in the lower temperature region based on X-ray observations from ref. [137]. In this layered structure, the aromatic main chains in the extended form are packed into a layer, and the molten alkyl side chains occupy the space between the layers. (b) The illustration for ideal packing structure of aromatic main chains within a layer, as proposed according to the fluorescence measurements from ref. [138] in which the biphenyl moieties (open ellipses) lie adjacent to the pyromellitic ester groups (closed ellipses).

Before the fluorescence measurements, the position between main chains had not been determined. Sone *et al.* [138] found out the fluorescence and absorption of a CT complex formed between biphenyl (donor) and pyromellitic ester (acceptor) chromophores by the measurements of the concentrated solutions of B–C16. The equimolar mixture of 4,4′-diacetoxy biphenyl and 1,4-dihexadecyl pyromellitic acid exhibits the same fluorescence as B–C16; thus, the existence of a CT complex is perfectly proved. Therefore, the fluorescence results dictate that the main chains within a layer be laterally packed in such a way that the biphenyl moieties of one polymer chain are adjacent to the pyromellitic ester moieties of the neighboring chain. The formation of a CT complex is also supported by the fact that the distance between the main chains in the individual layers is about 0.4 nm. In conclusion, the ideal packing structure is determined as Figure 3.34(b) by these fluorescence measurements.

Fluorescence is an effective and powerful tool for providing molecular-level information on a liquid crystal polymer. Figure 3.35 shows the fluorescence and its excitation spectra of PB–10, a thermotropic liquid–crystalline polyester [139]. The PB–10 shows two enantiotropic transitions: one is the crystal–smectic H transition at 207 °C and the other is the smectic H-isotropic phase transition at 257 °C.

PB–10

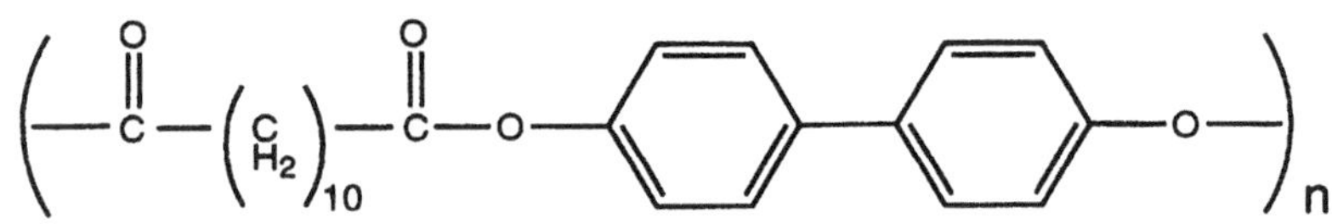

The fluorescence behavior in Figure 3.35 is divided into three classes according to excitation wavelength: monomer fluorescence of a biphenyl group excited at 260–300 nm (emission peak at 343 nm), fluorescence from intermolecular ground-state complexes excited at 320–360 nm, and fluorescence excited at 380–400 nm. The fluorescence peak wavelength excited at 320 nm shifts from 365 to 429 nm during heating. This is due to the change of molecular interactions between mesogenic moieties from the interaction between two fully overlapping biphenyl groups to that between a biphenyl and an ester moiety. Figure 3.35 also illustrates the probable molecular arrangements of mesogenic moieties.

In regard to lyotropic liquid–crystalline polymers, just as with thermotropic ones, fluorescence measurements are effectively employed; i.e., cellulose derivatives and aromatic polyamides. We introduced a dansyl group (i.e. dansylhydrazine) into one end of 2,3,6-tri-*O*-ethylcellulose (EC) by using a small amount

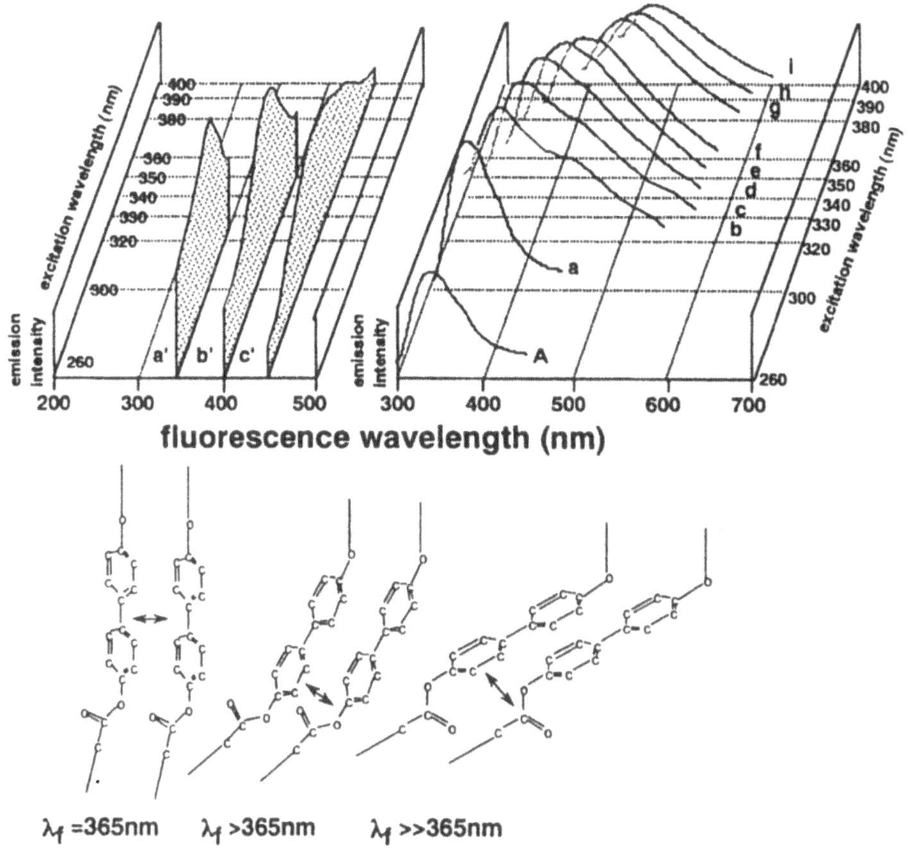

Figure 3.35 Fluorescence (right) and its excitation spectra (left) for PB–10 cooled from 270 °C and measured at 26 °C. A and a–i are the fluorescence spectra excited at 260, 300, 320, 330, 340, 350, 360, 380, 390, and 400 nm, respectively; a′-c′ are the fluorescence excitation spectra monitored at 343, 400, and 450 nm, respectively. The spectra are from ref. [139].

of terminal glucose unit, being a reduced form, and observed the development process of the lyotropic liquid crystalline formation in dichloromethane (DCM). In the dilute DCM solution of ethylcellulose derivatives (EC–D), the fluorescence peak of the dansyl probe appears at 520 nm. The liquid crystallization takes place at concentrations higher than 34%. The change of dansyl fluorescence was observed with a change in EC–D concentration. The peak wavelength changes slightly to the blue with an increase in EC–D concentration during the liquid crystalline state. This indicates that lyotropic liquid crystallization takes place gradually and that the increase in polymer concentration directly induces the removal of solvent DCM molecules from the surroundings of a polymer molecule together with changes to the orientation of polymer molecules [140].

With regard to the lyotropic liquid crystal transition of polyamides, Ushiki *et al.* [141] studied the transition phenomenon using the fluorescence polarization method for poly(γ-benzyl-L-glutamate) (PBLG) having rhodamine B at an end. Values of r_{pol} became zero due to the aggregation in a liquid crystalline region. This behavior is demonstrated to be an indication of transition of solution to the liquid crystalline state [142].

3.5.2.5 Polymer Solids

Concerning the use of the luminescent probe to study polymer solids, please refer to our review paper [16]. Here I present just a few examples from a large number of studies.

3.5.2.5.1 Fluorescence Behavior of Fluorescent Material Polymers

The luminescent probe method has been widely known to be useful as a tool to investigate microstructures and motions of polymer molecules. However, I have come to realize through an extensive literature survey that the characterization of fluorescence properties of luminescent polymers has focused on very few polymers such as polystyrene (Section 3.5.1) and polyvinylcarbazole only. Thus, I started with the investigation of the luminescence of the material polymers. Here I introduce the fluorescence behavior of poly(ethylene terephthalate) (PET) [26] along with bisphenol A polycarbonate (PC) [143, 144], both of which have fluorescent phenylene chromophore in their main chain.

The fluorescence properties of bisphenol A polycarbonate (PC) solids were studied in detail with a change in the casting and annealing temperatures between 40 and 180 °C. The pure fluorescence spectrum of PC solids was first presented as a result of establishing the way to measure the fluorescence. The existence of the excimer formed between intermolecular phenyl groups in the PC main chain was ascertained, showing some main chains located closely within a distance of 0.35 nm of each other in the amorphous area of PC solids. However, the amount of excimer formation is not large owing to the large steric hindrance [144].

The PET solid is soluble in chloroform–trifluoroacetic acid (9:1 v/v) abbreviated as C–TFA. Figure 3.13 displays the fluorescence spectrum of PET film prepared by a spin-casting method together with that of a C–TFA liquid solution. The results show that the fluorescence peak wavelength of PET solids (350–370 nm) is by far longer than that of PET in fluid solution (330 nm). The fluorescence excitation curves of PET solids for 350 to 370 nm coincide with the absorption spectrum of PET solution; thus, they are not due to the formation of a CT complex. Then we tried to examine polarity effects on the fluorescence spectrum of a monomer model compound using various solvents. Dibenzyl terephthalate (DBT) is employed as a monomer model to reflect the surroundings when phenylene moieties are abundant. The fluorescence spectra of DBT in various solvents show that the peak wavelength is red-shifted with an increase in polarity of the solvent (Figure 3.6): the distribution of electron density of the terephthaloyl group is found to be sensitive to polarity. In conclusion, the fluorescence of PET films at longer wavelengths is from the phenylene group and is induced by interactions involving some contributions of charge transfer stabilization among phenylene groups, or phenylene and carbonyl groups, or all three [26].

We prepared many PET films with various crystallinities and measured their infrared spectra (ir), differential scanning calorimetry (dsc), and densities to determine weight percentage crystallinity as well as fluorescence spectra. In conclusion, the fluorescence whose peak is near 330 nm corresponds to that of the phenylene groups in the crystalline region, whereas that at longer wavelengths corresponds to the main-chain phenylene in the amorphous region [26, 145].

In the crystalline region, every phenylene moiety is regularly isolated from others, and thus no interaction among phenylene moieties occurs; the shortest distance between phenylenes and between phenylene and carbonyl is still longer than 0.35 nm. On the other hand, some phenylene moieties can contact other phenylene or carbonyl moieties within the distance of 0.35 nm in the amorphous region where some PET chains are closer than in the crystal region; as a matter of course, some PET chains are farther away from one another.

Thus the ratio of the fluorescence intensity at 330 nm I_{330}, to that at 370 nm I_{370} can be an indication of the crystallinity of PET films [145]. Then we tried to observe the dynamic process of PET recrystallization by monitoring the fluorescence change of amorphous PET solids. Figure 3.36 shows the temperature dependencies of I_{330}/I_{370}, each of which is obtained by averaging several data and determined as an equilibrium value; the value of I_{330}/I_{370} at temperatures higher than 410 K is difficult to determine owing to an increase in radiationless deactivation. The I_{330}/I_{370} does not change very much up to 340 K, which is T_g. However, it increases slightly between 340 and 370 K and dramatically above 380 K. The differential scanning calorimetry (DSC) measurements for the PET films showed that the recrystallization process took place at temperatures from 380 to 410 K; thus, the change shown in Figure 3.36 completely corresponds to the change in crystallinity.

Next we annealed an amorphous PET film at 370 K for a day and heated it up to 390 K to monitor the recrystallization process. Figure 3.37(a) shows the time change of I_{330}/I_{370} of two films at 390 K; the reproducibility was found to be very

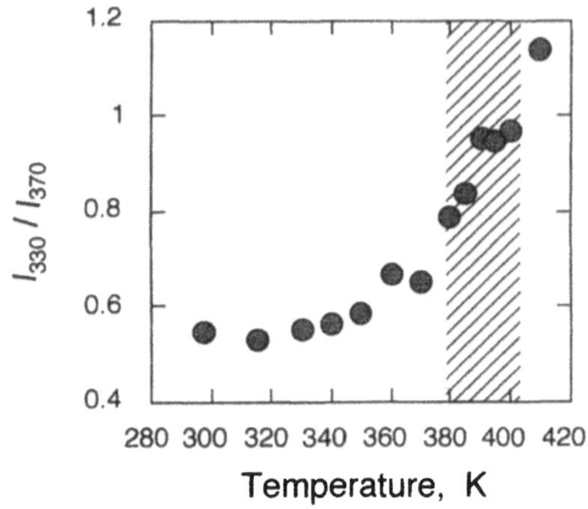

Figure 3.36 Annealing temperature dependence of I_{330}/I_{370} of amorphous PET films (crystallinity is $11 \pm 2\%$). Each point is an average value of some experimental values. Shaded parts indicate temperature range at which exothermicity was observed for differential scanning calorimetry measurements corresponding to the recrystallization process.

good. The I_{330}/I_{370} kept constant after 150 min, and thus the equilibrium value $(I_{330}/I_{370})_\infty$ was determined by averaging. $[(I_{330}/I_{370})_\infty - (I_{330}/I_{370})]$ was found to be best fitted to a single exponential with a time constant of 33 ± 5 min (Figure 3.37(b)). This time constant shows the reciprocal of the rate constant for recrystallization k_{RC} at 390 K. We obtained the values of k_{RC} in the temperature range at which the recrystallization process takes place. A plot of $\ln(k_{RC})$ versus $1/T$ gives a straight line with a value of 120 ± 10 kJ/mol as an apparent activation energy of recrystallization. This value is so large for one terephthaloyl group that more than a few terephthaloyl units would take part in the recrystallization process [146].

3.5.2.5.2 CT Complex of Aromatic Polyimides

Aromatic polyimides have outstanding thermal stability and excellent mechanical properties. These properties are correlated with both the rigidity of the main chain and strong intermolecular interaction caused by the stacking of aromatic rings. Aromatic moieties can be electron donors, whereas imide groups can be electron acceptors. Thus, it is reasonable to assume the formation of a CT complex in polyimide solids. Hasegawa *et al.* [147–150] systematically have measured fluorescence of commercial polyimides and have studied the aggregation state of polyimides. In order to examine the fluorescence behavior in accordance with

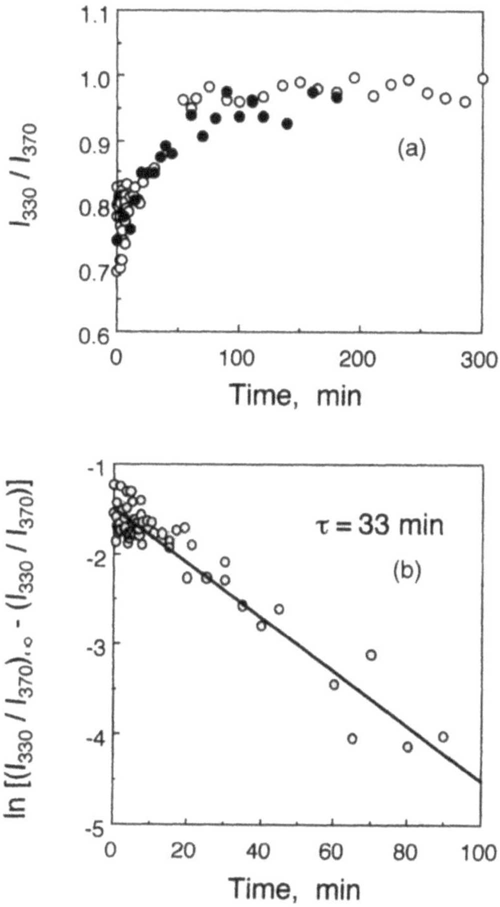

Figure 3.37 Time dependence of I_{330}/I_{370} for amorphous PET films just after the film temperature reached 390 K from 370 K. Variation of I_{330}/I_{370} (a) and ln $[(I_{330}/I_{370})_{t=\infty} - (I_{330}/I_{370})]$ (b) as a function of time; $(I_{330}/I_{370})_{t=\infty}$ is the equilibrium value. In a, ○ and ● show different experimental data by using other PET films prepared in the same way. The solid line in b shows a model curve of a single exponential with time constant of 33 min.

Mulliken's CT theory (Section 3.2.3.3.1.1) [41, 42], they measured two series of fluorescence measurements in which the diamine components were (1) *p*-phenylene diamine (PDA) and (2) oxydianiline (ODA). A weak CT complex with a fixed electron donor showed the following relationship (see Eq. (25)):

$$h\nu_{CT} = -E_a + C \tag{77}$$

where $h\nu_{CT}$ is the absorption energy in the CT band, E_a is the electron affinity of the acceptors, and C is a constant of a series of the components used. The plot of E_a of dianhydrides corresponding to imide group versus wave number of fluorescence peak corresponding to $h\nu_{CT}$ is shown in Figure 3.38. The linearity of the two plots verifies strongly that the fluorescence is due to the CT complex.

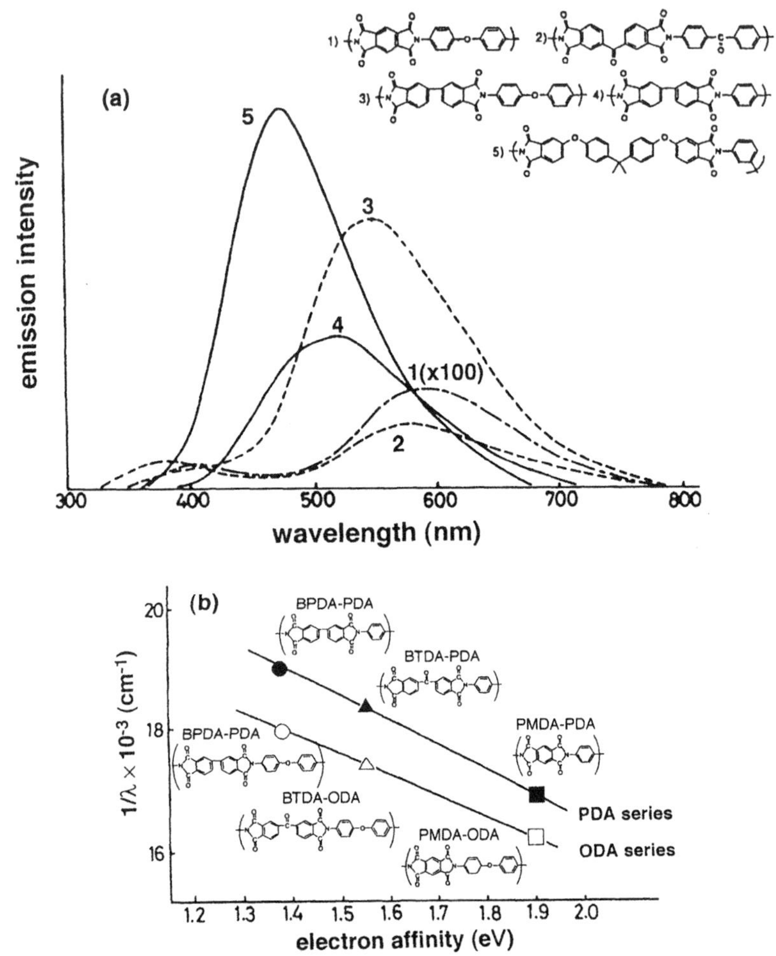

Figure 3.38 (a) Fluorescence spectra excited at 300 nm for various polyimides. (b) The plots of electron affinities of dianhydride components versus wave numbers at the fluorescence peak. The diamine components were *p*-phenylene diamine (PDA) and oxydianiline (ODA). BPDA: biphenyltetracarboxylic dianhydride; BTDA: benzophenonetetracarboxylic dianhydride; BMDA: pyromellitic dianhydride. The information has been taken from ref. [147].

Once the fluorescence behavior is clearly understood, it can be employed as a probe. Miscibility of fluorescent polyimide with nonfluorescent polyimide blends can be monitored by the change in CT fluorescence intensity [151]; the smaller intensity shows the better miscibility. The fluorescence results were supported by the results of dynamic mechanical analysis, which also clearly demonstrated that the miscibility of two polyimides blends depends strongly on casting temperature: The CT fluorescence intensity of the blends decreases dramatically with an increase in casting temperature.

3.5.2.5.3 Interdiffusion in Films Prepared from Latex Particles

Films prepared from aqueous dispersions of latex particles are widely used as coatings. When freshly applied, the films formed are mechanically weak; however, their strength will increase when the films are dried at temperatures higher than T_g. This growth in strength is due to the interdiffusion of polymer molecules across an interface of the latex and to the formation of entanglements on both sides of the original boundary [152, 153]. Winnik *et al.* have studied the interdiffusion of polymer in melt-pressed films of latex by means of the energy transfer method [154, 155]. Here we introduce a part of this work as an example.

Wang *et al.* [154] prepared poly(methyl methacrylate) (PMMA) latex with 1 mol% phenanthrene and 1 mol% anthracene moieties. Both lattices have particle diameters of 127 and 120 nm. Singlet energy transfer measurements were carried out on films composed of a 1:1 mixture of these particles as a function of annealing time and oven temperature. The fluorescence decay curves of donor phenanthrene were fitted to the expression

$$I_D(t) = B_1 \exp[-t/\tau_D - P\sqrt{t/\tau_D}] + B_2 \exp(-t/\tau_D) \tag{78}$$

The B_1 term describes the contribution to the decay profile of the phenanthrene label, which can undergo energy transfer owing to diffusive mixing with the anthracene-labeled polymer, and the B_2 term describes the decay of the phenanthrene label that remains unmixed at the time the sample is removed from the oven. P is proportional to the concentration of anthracene label in the mixing region. The apparent volume fraction of mixing is given by the expression

$$f'_m(t) = B_1/(B_1 + B_2) \tag{79}$$

These apparent values must be correlated first for the amount of interparticle energy transfer $f_m(0)$ taking place in the nascent film and for the possibility that, at long times, $f'_{\mathrm{m}}(\infty)$ is not equal to unity. Thus,

$$f_{\mathrm{m}}(t) = \frac{f'_{\mathrm{m}}(t) - f'_{\mathrm{m}}(0)}{f'_{\mathrm{m}}(\infty) - f'_{\mathrm{m}}(0)} \tag{80}$$

By integrating $I_{\mathrm{D}}(t)$ to determine the area under the normalized decay curve, one obtains the efficiency of energy transfer without making specific assumptions about the shape of the $I_{\mathrm{D}}(t)$. Here $f_{\mathrm{m}}(t)$ can be calculated by Eq. (81)

$$f_{\mathrm{m}}(t) = \frac{\mathrm{Area}(t) - \mathrm{Area}(0)}{\mathrm{Area}(\infty) - \mathrm{Area}(0)} \tag{81}$$

and $f'_{\mathrm{m}}(0)$ values can be calculated by comparing the integrated area in the nascent film with that of a film composed exclusively of the phenanthrene-labeled PMMA. In the case of Wang *et al.*'s experiments, $f'_{\mathrm{m}}(0)$ values were commonly on the order of 0.15, whereas $f'_{\mathrm{m}}(\infty)$ values were very close to unity. Here the volume fraction of mixing, $f_{\mathrm{m}}(t)$, can be monitored at each annealing temperature.

The diffusion coefficients of polymer molecules were also calculated from the mixing data by using Fick's law of diffusion. The values in the case of PMMA are 10^{-15} cm^2/s at 149 °C to 10^{-18} cm^2/s at 120 °C, corresponding to an apparent activation energy of 420 kJ/mol.

3.5.2.5.4 Free Volume in Polymer Solids

The basic way employed so far to determine amounts of free volume and its distribution in polymer solids is to dope small molecules with different sizes into the solids and to monitor their photophysical properties sensitive to microenvironments. The method is based on the assumption that the molecules doped can be quiescent only when free space exists for them to enter in the polymer solids; otherwise, a sort of phase separation takes place between the polymer and the additives. Figure 3.39 shows the concept of a fluorescence probe to monitor the free volume. Basically, the probe used is a molecule either whose fluorescence behavior is sensitive to its own molecular motion or whose photoreaction requires molecular motion. Finally, we can estimate a size of free volume in polymer solids to monitor whether free space exists for a probe capable of rotating or changing its conformation.

The next photoproperties sensitive to the molecular motion of a dye have been employed: (1) intramolecular excimer formation of dimer compounds [156–158] such as dicarbazolylpentane [159], (2) fluorescence of molecular rotors such as [*p*-(*N*,*N*-dialkylamino)benzylidene]malononitriles [31, 32, 160, 161], and (3) fluorescence depolarization of rigid probes of different sizes [99, 162]. On the

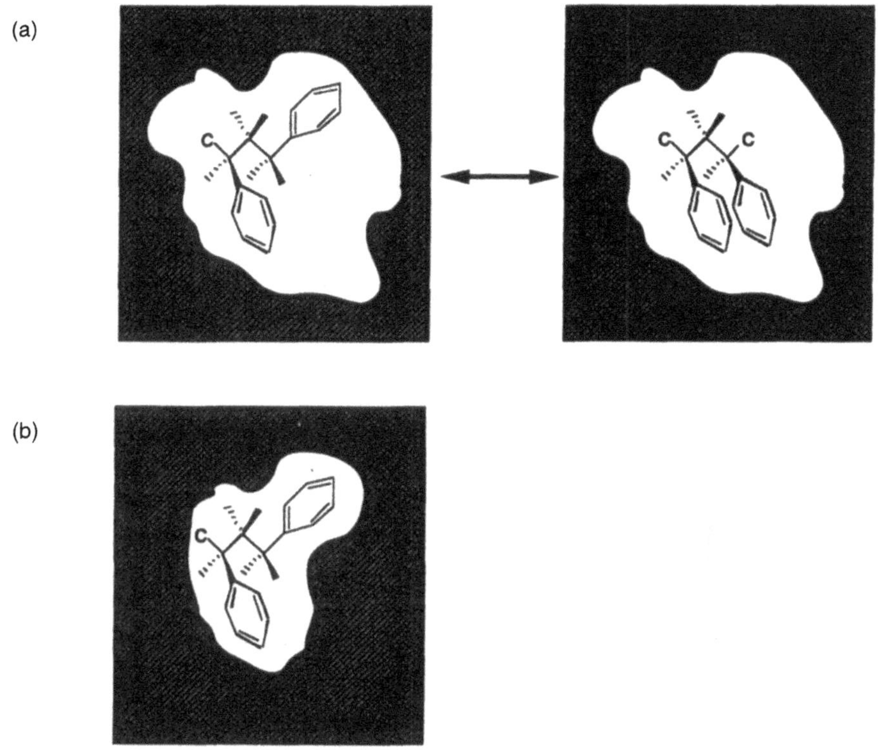

Figure 3.39 Sketch of an example of a fluorescent probe to monitor free space volume in polymer solids. A probe can change its conformation in the case of (a), whereas the motion of a probe is restricted in the case of (b). In this example, excimer fluorescence can be monitored only in the case of (a), where its fluorescence is depolarized.

other hand, since photochromic labels such as azobenzene compounds need some extra volume to isomerize (*trans*→*cis*), photoisomerization reaction monitored by ultraviolet absorption give information on free space in a polymer matrix [163–165].

The preceding photophysical methods successfully present information on free volume in a polymer matrix. However, one of the problems is that the size of molecules doped into polymer solids cannot be adjusted systematically over a wide range, although Jarry *et al.* [99, 162] used three serial diphenyl-*n*-ene compounds ($n = 2$–4). Therefore, we employed serial *n*-alkylbenzenes (C_n: C_6H_5–C_nH_{2n+1} ($n = 4$–16)) as probe molecules to estimate sizes of free space in an amorphous area of poly(methyl methacrylate) (PMMA) films systematically.

Let us explain our method [172]. In general, an aromatic compound in dilute solution shows a usual fluorescence spectrum while it displays excimer fluorescence at high concentrations. Thus, it would be possible to evaluate local concentrations of aromatic compounds by monitoring the fluorescence yields of excimer. When probe molecules compatibile with a polymer matrix are doped into polymer solids, a molecule whose size is smaller than the average free volume in a polymer matrix can remain in a free space area, and it will be isolated from other probe molecules. However, a larger molecule cannot penetrate into a space with average free volume and finally gathers together by way of a kind of microscopic phase separation. This local concentration can be evaluated by measuring excimer fluorescence yields, if a dopant is an aromatic compound.

Figure 3.40 displays the relationship between the excimer formation efficiency (modified value of I_E/I_M), f_E, of C_n in PMMA and the fraction of phenyl groups of each C_n (phenyl amount/PMMA (g/gPMMA)). This weight dependence of excimer formation of C_n is summarized as follows: (1) C_n with $n \leq 5$ does not exhibit excimer fluorescence at all even at quite high concentrations, (2) C_n with $n \geq 6$ shows excimer emission at higher concentrations and f_E increases with an increase in the amount of C_n added into PMMA as far as spectra are measurable, (3) in the case of $6 \leq n \leq 11$, C_n with larger n displays stronger excimer fluorescence if the amounts of the phenyl moieties are the same, (4) in the case of $n \geq 12$, the spectra of C_n agree well with one another if the amounts are same.

It is self-evident that this experimental condition induces the plasticization of PMMA; however, the data clearly show the inherent properties referring to the free volume of PMMA solids. The calculation results by means of semiempirical molecular orbital methods suggest that the location of benzene is very close to the PMMA main chain because of the stabilization due to the interaction with carbonyl groups of PMMA. Thus, the C_n molecules that are in the free-volume area in PMMA can be defined as the molecules whose phenyl ring is partially or wholly surrounded by the PMMA chain, resulting in a situation in which intermolecular excimer formation is prevented.

In conclusion, all the C_n with $n \leq 5$ added into PMMA are estimated to be in the free volume of PMMA and isolated from each other because C_n with $n \leq 5$ does not exhibit any excimer fluorescence. In the case of C_n with $n \geq 6$, the volumes of some conformations of C_n are larger than the average inherent free volume, and thus their excimer formation is not restricted. All the C_n with $n \geq 12$ are estimated to be out of the area covered with PMMA chains, and they show the same intensities of excimer fluorescence. In summary, the method clarifies that (1) the average free volume of PMMA is larger than the volume that all the conformers of C_5 occupy in three-dimensional space and that (2) the free volume is

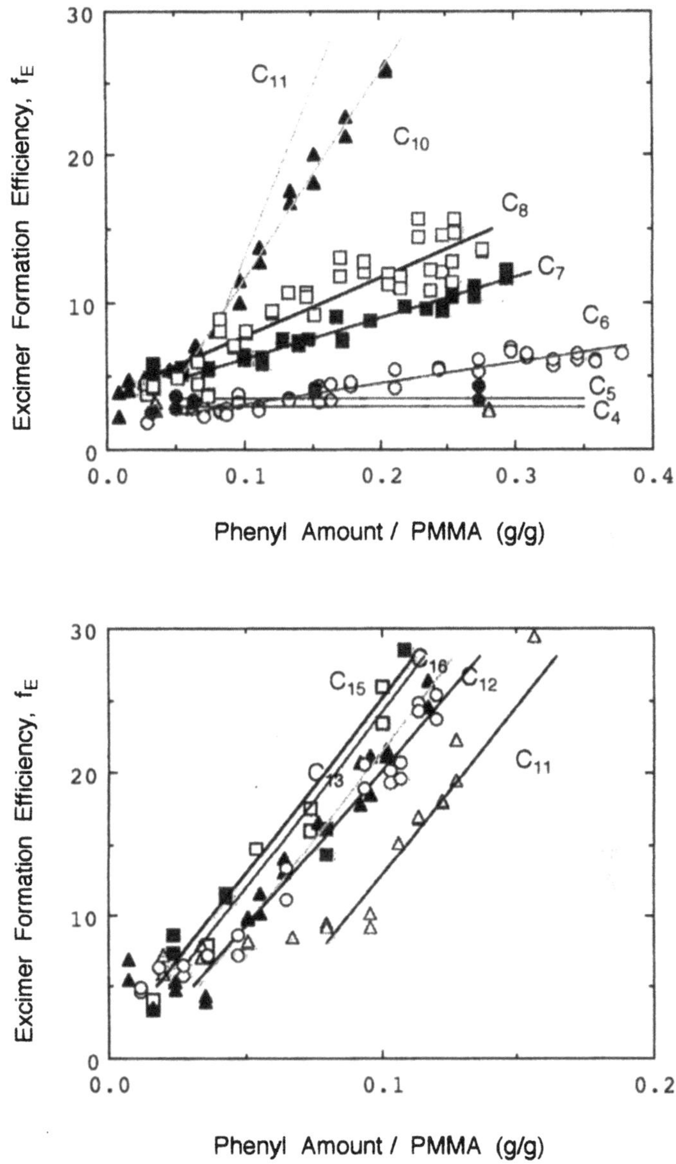

Figure 3.40 Concentration dependence of excimer formation efficiency f_E at 25 °C of smaller C_n (top) and larger C_n (bottom) in dried PMMA; (top): C_4 (●), C_5 (△), C_6 (○), C_7 (■), C_8 (□) and C_{10} (▲) with a line C_{11} of as a reference; (bottom): C_{11} (△), C_{12} (○), C_{13} (■), C_{15} (□), and C_{16} (▲).

smaller than the volume that almost all the conformers of C_{12} occupy in three-dimensional space.

Under the steady-state approximation, the ratio of excimer to monomer intensity is described as the product of K_1 and $[^1M]$, concentration of C_n. Thus,

$$f_E = K_1[^1M] \tag{82}$$

where

$$K_1 = \frac{k_{FE}k_{EM}}{k_{FM}(k_{FE} + k_{IE} + k_{ME})}. \tag{83}$$

Consequently, f_E should be proportional to $[^1M]$ because K_1 is considered to be constant. Figure 3.40 shows that f_E is in proportion to phenyl concentration of C_n in PMMA and proves that the conventional scheme is valid for our C_n system. The slopes of straight lines were calculated by a least-squares method and compared with one another. Here we define P_{out} as the percentage of C_n outside the inherent free-space area of PMMA solids that can form excimer. P_{out} is calculated from the slope of a linear plot in Figure 3.40:

$$P_{out}(\%) = \frac{100 \times (\text{slope of } C_n)}{(\text{average slope of } C_n \text{with } n > 11)} \tag{84}$$

Here the percentage of C_n in the free-volume area is defined as P_{free}:

$$P_{free}(\%) = 100 - P_{out} \tag{85}$$

The values of P_{free} and P_{out} are summarized in Table 3.7. They show a kind of distribution of free-volume sizes in PMMA solids because the average volume of C_n is determined.

3.6 Concluding Remarks

Recently quite a few studies on macromolecules by means of the luminescent probe method have clarified microscopic information such as interaction and orientation of molecules or information on the dynamic behavior of polymer systems. The numbers of papers dealing with fluorescence spectroscopy is increasing remarkably year by year.

The main part of polymer photophysics seems to me to have been established in the 1980s. I think that Section 3.5.1.1 describes one part of history that polymer photophysics has established because polystyrene is the most basic fluorescent polymer. Up to the middle of the 1980s, fluorescence measurements were not always effectively employed in polymer science compared with bioscience partly

Table 3.7 The slope of a plot of f_E versus phenyl concentration of C_n and the obtained values of P_{out} and P_{free}.

C_n	*Concentration Range of Data Used for Calculation (g/g PMMA)*	*Number of Data Used for Calculation*	*Slope of Linear Plot* [a]	P_{out} *(%)*	P_{free} *(%)*
C_6	0.029–0.518	41	14	6	94
C_7	0.074–0.293	31	28	12	88
C_8	0.066–0.275	38	40	17	83
C_{10}	0.079–0.210	14	140	58	42
C_{11}	0.080–0.201	9	230	96	4
C_{12}	0.036–0.120	16	220	92	8
C_{13}	0.023–0.132	9	240	100	0
C_{15}	0.016–0.100	9	240	100	0
C_{16}	0.035–0.117	17	250	100	0

[a] The average value of the slope of C_n ($n = 13$–16) is 240.

because of the unclearness of polymer photophysics and partly because of the immaturity of instruments. However, the present situation is completely different, for many papers have presented results that would never have been obtained without using fluorescence spectroscopy.

The instruments to monitor emission have been developed greatly, and the cost is decreasing. Nowadays, even polymer scientists who are not familiar with photophysics can easily operate a picosecond-pulsed photolysis apparatus. In this sense, fluorescence spectroscopy becomes a convenient method to monitor microscopic change in polymer science.

However, I would like to repeat the problem (Section 3.4.2) of the application of the fluorescent probe method at the end of this chapter. Needless to say, we always have to check the same system by other macroscopic measurements such as X-ray diffraction, light scattering, nuclear magnetic resonance, IR, DSC, or mechanical tests such as tensile stress and so on. If the results obtained from fluorescence measurements coincide with those obtained from other macroscopic measurements, we can ascertain that the application is successful. In the case in which the results obtained from fluorescence measurements differ from those obtained from other macroscopic measurements or are not compared with other measurements, we have to be careful. The change in fluorescence might be induced by only photophysical processes instead of being influenced by a change in microenvironment. Thus, the understanding of the mechanism of fluorescence behavior or photophysical properties is still most important.

In such a case, I recommend two methods. One is to apply another fluorescent probe method to the same polymer system; for example, from the nonradiative energy transfer method to excimer. It is a good example that Hu *et al.* applied dansyl group [127] and pyrenyl group [128] to the PAAm gels (Section 3.5.2.3.1). Another approach is to simulate fluorescence behavior by using computer calculation; for example, a Monte Carlo method or a molecular dynamics method. As an example, we refer to the works of Liu *et al.* [166, 167] to measure the root-mean-square end-to-end distance of polymer chains by using energy transfer between two different chromophores attached both ends of the polymer. They succeeded in presenting proof of computer simulation [168].

Now I would say the time has come that one can design the application of adequate fluorescent probes corresponding to the information that one would like to get. I hope that this chapter will be of help for readers who for the first time are trying to employ a fluorescent probe method.

Acknowledgment

I wish to express my sincere gratitude to Dr. Simon Hesp of Queen's University in Canada for his useful suggestion for this manuscript.

References

1. R. S. Becker, *Theory and Interpretation of Fluorescence and Phosphorescence*, Wiley, New York, 1969.
2. J. B. Birks, *Photophysics of Aromatic Molecules*, Wiley-Interscience, New York, 1970.
3. J. B. Birks, Ed., *Organic Molecular Photophysics*, Wiley, New York, 1973.
4. N. S. Allen, Ed., *Developments in Polymer Photochemistry 1*, Appl. Sci., Lond., 1980.
5. S. N. Semerak and C. W. Frank, *Adv. Polym. Sci.* **54:** 31 (1983).
6. J. E. Guillet, *Polymer Photophysics and Photochemistry*, Cambridge University Press, Cambridge, 1985.
7. D. Phillips, Ed., *Polymer Photophysics*, Chapman and Hall, London, 1985.
8. N. J. Turro, *Modern Molecular Photochemistry*, University Science, Mill Valley, California, 1991.
9. W. Klöpffer, *Spectroscopy Lett.* **11:** 863 (1978).
10. H. Itagaki and I. Mita, in *Degradation and Stabilization of Polymers II*, H. H. G. Jellinek, Ed., Elsevier, Amsterdam, 1989.
11. F. C. De Schryver, P. Collart, J. Vandendriessche, R. Goedeweeck, A. M. Swinnen, and M. Van der Auweraer, *Acconts Chem. Res.* **20:** 159 (1987).

12. I. Mita and K. Horie, in *Degradation and Stabilization of Polymers*, H. H. G. Jellinek, Ed., Elsevier, Amsterdam, 1983.
13. K. Horie and I. Mita, *Adv. Polym. Sci.* **89:** 77 (1989).
14. C. E. Hoyle and J. M. Torkelson, Eds., *Photophysics of Polymer Systems*, ACS Symp. Series, ACS, Washington, 1987.
15. M. A. Winnik, Ed., *Photophysical and Photochemical Tools in Polymer Science (Conformation, Dynamics, Morphology)*, NATO ASI Series, Reidel, Dordrecht, The Netherlands, 1986.
16. H. Itagaki, K. Horie and I. Mita, *Prog. Polym. Sci.* **15:** 361 (1990).
17. F. Perrin, *J. Phys. Radium* **7:** 390 (1926).
18. F. Perrin, *Ann. Phys.* **12:** 169 (1929).
19. G. Weber, *Biochem. J.* **51:** 145 (1952).
20. G. Weber, *Adv. Protein Chem.* **8:** 415 (1953).
21. B. Valeur, J. P. Jarry, F. Geny, and L. Monnerie, *J. Polym. Sci., Polym. Phys.* **13:** 667 (1975).
22. B. Valeur, J. P. Jarry, F. Geny, and L. Monnerie, *J. Polym. Sci., Polym. Phys.* **13:** 675 (1975).
23. T. Tao, *Biopolymers* **8:** 609 (1969).
24. R. Rigler and M. Ehrenberg, *Q. Rev. Biophys.* **6:** 139 (1973).
25. J. L. Vivoy and L. Monnerie, *Adv. Polym. Sci.* **67:** 100 (1985).
26. H. Itagaki, Y. Inagaki, and N. Kobayashi, *Polymer* **37:** 3553 (1996).
27. K. Kalyanasundaram and J. K. Thomas, *J. Am. Chem. Soc.* **99:** 2039 (1977).
28. K. Rotkiewicz, K. H. Grellmann, and Z. R. Grabowski, *Chem. Phys. Lett.* **19:** 315 (1973).
29. Z. R. Grabowski, K. Rotkiewicz, A. Siemiarczuk, D. J. Cowley, and W. Baumann, *Nouv. J. Chim.* **3:** 443 (1979).
30. R. Hayashi, S. Tazuke, and C. W. Frank, *Macromolecules* **20:** 983 (1987).
31. R. O. Loutfy, *Macromolecules* **14:** 270 (1981).
32. R. O. Loutfy and B. A. Arnold, *J. Phys. Chem.* **86:** 4205 (1982).
33. G. Weber and D. J. R. Laurence, *Biochem. J.* **56:** xxxi (1954).
34. W. O. McClure and G. M. Edelman, *Biochem.* **5:** 1908 (1966).
35. R. P. Cory, R. R. Becker, R. Rosenbluth, and I. Isenberg, *J. Am. Chem. Soc.* **90:** 1643 (1968).
36. E. M. Kosower, H. Dodiuk, K. Tanizawa, M. Ottolenghi, and N. Orbach, *J. Am. Chem. Soc.* **97:** 2167 (1975).
37. K. Dimroth, C. Reichert, T. Siermann, and F. Bohlmann, *Justus Liebigs Ann. Chem.* **661:** 1 (1963).
38. Y. H. Li, L. M. Chan, L. Tyer, R. T. Moody, C. M. Himel, and D. M. Hercules, *J. Am. Chem. Soc.* **97:** 3118 (1975).
39. J. R. Platt, *J. Chem. Phys.* **17:** 484 (1949).
40. N. Mataga, *Bull. Chem. Soc. Jpn.* **36:** 654 (1963).
41. R. S. Mulliken, *J. Am. Chem. Soc.* **74:** 811 (1952).

42. R. S. Mulliken, *J. Phys. Chem.* **56:** 801 (1952).
43. H. Itagaki, K. Horie, I. Mita, M. Washio, S. Tagawa, and Y. Tabata, *J. Chem. Phys.* **79:** 3996 (1983).
44. H. Itagaki, N. Obukata, A. Okamoto, K. Horie, and I. Mita, *Chem. Phys. Lett.* **78:** 143 (1981).
45. H. Itagaki, N. Obukata, A. Okamoto, K. Horie, and I. Mita, *J. Am. Chem. Soc.* **104:** 4469 (1982).
46. H. Itagaki, A. Okamoto, K. Horie, and I. Mita, *Eur. Polym. J.* **18:** 885 (1982).
47. H. Beens and A. Weller, *Chem. Phys. Lett.* **2:** 82 (1968).
48. N. Tsujino, H. Masuhara, and N. Mataga, *Chem. Phys. Lett.* **21:** 301 (1973).
49. H. Ushiki, K. Horie, A. Okamoto, and I. Mita, *Polym. J.* **11:** 691 (1979).
50. H. Ushiki, K. Horie, A. Okamoto, and I. Mita, *Polym. J.* **12:** 35 (1980).
51. H. Ushiki, K. Horie, A. Okamoto, and I. Mita, *Polym. J.* **13:** 191 (1981).
52. K. Horie, W. Schnabel, I. Mita, and H. Ushiki, *Macromolecules* **14:** 1422 (1981).
53. A. S. Davydov, *Zh. Eks. Theor. Fiz.* **18:** 210 (1948).
54. A. S. Davydov, *Theory of Molecular Excitons*, M. Kasha and M. Oppenheimer, Jr. Trans., McGraw–Hill, New York, 1962.
55. Th. Förster, *Naturwiss.* **33:** 166 (1946).
56. Th. Förster, *Disc. Faraday Soc.* **27:** 7 (1959).
57. M. D. Galanin, *Zh. Eksperim. Soviet Phys., JETP* **1:** 317 (1955).
58. H. Morawetz and F. Amrani, *Macromolecules* **11:** 281 (1978).
59. F. Amrani, J. M. Hung, and H. Morawetz, *Macromolecules* **13:** 649 (1980).
60. I. B. Berlman, *Energy Transfer Parameters of Aromatic Compounds*, Academic Press, New York, 1973.
61. D. L. Dexter, *J. Chem. Phys.* **21:** 836 (1953).
62. A. M. North and M. F. Treadaway, *Eur. Polym. J.* **9:** 609 (1973).
63. O. Stern and M. Volmer, *Physik. Z.* **20:** 83 (1919).
64. M. Yokota and O. Tanimoto, *J. Phys. Soc. Jpn.* **22:** 779 (1967).
65. B. R. M. Kyle and T. Kilp, *Polymer* **25:** 989 (1984).
66. F. Heisel and G. Laustriat, *J. Chim. Phys.* **66:** 1881 (1969).
67. M. Heskins and J. E. Guillet, *Macromolecules* **3:** 231 (1970).
68. H. Itagaki, *Macromolecules* **24:** 6531 (1991).
69. J. R. Lakowicz, *Principles of Fluorescence Spectroscopy*, p. 280, Plenum Press, New York, 1983.
70. J. N. Demas, *Excited State Lifetime Measurements*, Academic Press, New York, 1983.
71. D. V. O'Connor and D. Phillips, *Time-Correlated Single Photon Counting*, Academic Press, Orlando, Florida, 1984.
72. J. Durbin and G. S. Watson, *Biometrika* **37:** 409 (1950).
73. J. Durbin and G. S. Watson, *Biometrika* **38:** 159 (1951).
74. R. Kubo, M. Toda, and H. Hashitsume, *Statistical Physics II: Nonequilibrium Statistical Mechanics*, p. 42, Springer-Verlag, Berlin, 1985.

75. C. K. Hall and E. J. Helfand, *J. Chem. Phys.* **77:** 3275 (1982).
76. J. L. Viovy, L. Monnerie, and J. C. Brochon, *Macromolecules* **16:** 1845 (1983).
77. D. A. Waldow, M. D. Ediger, Y. Yamaguchi, Y. Matsushita, and I. Noda, *Macromolecules* **24:** 3147 (1991).
78. D. A. Waldow, B. S. Johnson, P. D. Hyde, M. D. Ediger, T. Kitano, and K. Ito, *Macromolecules* **22:** 1345 (1989).
79. D. B. Adolf, M. D. Ediger, T. Kitano, and K. Ito, *Macromolecules* **25:** 867 (1992).
80. M. Abe, H. Inoue, T. Nagamura, T. Yagi, and H. Itagaki, *J. Photopolym. Sci. Tech.* **10:** 271 (1997).
81. M. Abe, T. Yagi, H. Itagaki, and T. Nagamura, *J. Photochem. Photobiol. A*, **120:** 125 (1999).
82. R. B. Fox, T. R. Price, R. F. Cozzens, and W. H. Echols, *Macromolecules* **7:** 937 (1974).
83. W. E. Lindsell, F. C. Robertson, and I. Soutar, *Eur. Polym. J.* **17:** 203 (1981).
84. J. M. Torkelson, S. Lipsky, and M. Tirrell, *Macromolecules* **14:** 1601 (1981).
85. R. B. Fox, T. R. Price, R. F. Cozzens, and J. R. McDonald, *J. Chem. Phys.* **57:** 534 (1972).
86. L. A. Harrah, *J. Chem. Phys.* **56:** 385 (1972).
87. C. W. Frank and L. A. Harrah, *J. Chem. Phys.* **61:** 1526 (1974).
88. Y. Nishijima, K. Mitani, S. Katayama, and M. Yamamoto, *Rep. Prog. Polym. Phys. Jpn.* **13:** 421 (1970).
89. T. Ishii, T. Handa, and S. Matsunaga, *Macromolecules* **11:** 40 (1978).
90. H. Itagaki, K. Horie, I. Mita, M. Washio, S. Tagawa, Y. Tabata, H. Sato, and Y. Tanaka, *Macromolecules* **20:** 2774 (1987).
91. H. Itagaki, K. Horie, I. Mita, M. Washio, S. Tagawa, Y. Tabata, H. Sato, and Y. Tanaka, *Macromolecules* **22:** 2520 (1989).
92. H. Itagaki, K. Horie, I. Mita, M. Washio, S. Tagawa, and Y. Tabata, *Radiat. Phys. Chem.* **34:** 597 (1989).
93. H. Itagaki, K. Horie, I. Mita, M. Washio, S. Tagawa, Y. Tabata, H. Sato, and Y. Tanaka, *Macromolecules* **23:** 1686 (1990).
94. J. W. Longworth and F. A. Bovey, *Biopolymers* **4:** 1115 (1966).
95. T. F. Schatzki, *J. Polym. Sci.* **57:** 496 (1962).
96. R. F. Boyer, *Rubber Chem. Technol.* **34:** 1303 (1963).
97. W. Pechhold, S. Blasenbrey, and S. Woerner, *Kolloid Z. Z. Polym.* **189:** 14 (1963).
98. E. Helfand, *J. Chem. Phys.* **54:** 4651 (1971).
99. J. P. Jarry and L. Monnerie, *Macromolecules* **12:** 927 (1979).
100. J. M. Guenet, *Thermoreversible Gelation of Polymers and Biopolymers*, Academic Press, London, 1992.
101. M. Girolamo, A. Keller, K. Miyasaka, and N. Overbergh, *J. Polym. Sci., Polym. Phys. Ed.* **14:** 39 (1976).
102. R. P. Sundararajan, *Macromolecules* **12:** 575 (1979).
103. E. D. T. Atkins, A. Keller, J. S. Shapiro, and P. J. Lemstra, *Polymer* **22:** 1161 (1981).
104. J. M. Guenet, *Macromolecules* **19:** 1961 (1986).

105. M. Klein, J. M. Guenet, A. Brulet, and F. Boue, *Polymer* **32:** 1943 (1991).
106. Y. Chatani and N. Nakamura, *Polymer* **34:** 1644 (1993).
107. H. Itagaki and I. Takahashi, *Chem. Phys. Lett.* **205:** 446 (1993).
108. H. Itagaki and I. Takahashi, *Macromolecules* **28:** 5477 (1995).
109. H. Itagaki and Y. Nakatani, *Macromolecules* **30:** 7793 (1997).
110. K. Huber, S. Bantle, and P. Lutz, *Macromolecules* **18:** 1461 (1985).
111. D. N. Pinder, *Macromolecules* **23:** 1724 (1990).
112. A. E. C. Redpath and M. A. Winnik, *Ann. N. Y. Acad. Sci.* **366:** 75 (1981).
113. S. Ohmori, S. Ito, and M. Yamamoto, *Macromolecules* **24:** 2377 (1991).
114. T. Ueno, S. Ito, S. Ohmori, Y. Onogi, and M. Yamamoto, *Macromolecules* **25:** 7150 (1992).
115. T. Hayashi, T. Okuyama, S. Ito, and M. Yamamoto, *Macromolecules* **27:** 2270 (1994).
116. C. Zhao, Y. Wang, Z. Hruska, and M. A. Winnik, *Macromolecules* **23:** 4082 (1990).
117. M. Almgren and J.-E. Löfroth, *J. Chem. Phys.* **76:** 1 (1982).
118. F. Grieser and C. J. Drummond, *J. Phys. Chem.* **92:** 5580 (1988).
119. N. J. Turro and A. Yekta, *J. Am. Chem. Soc.* **100:** 5951 (1978).
120. H. Faes, F. C. De Schryver, A. Sein, K. Bijma, J. Kevelam, and J. B. F. N. Engberts, *Macromolecules* **29:** 3875 (1996).
121. Y. Croonen, E. Geladé, M. Van der Zegel, M. Van der Auweraer, H. Vandendriessche, F. C. De Schryver, and M. Almgren, *J. Phys. Chem.* **87:** 1426 (1983).
122. M. Tachiya, *Chem. Phys. Lett.* **37:** 289 (1975).
123. K. Horie, I. Mita, J. Kawabata, S. Nakahama, A. Hirao, and N. Yamazaki, *Polym. J.* **12:** 319 (1980).
124. T. Tanaka, *Phys. Rev. Lett.* **40:** 820 (1978).
125. T. Tanaka, D. Fillmore, S. T. Sun, I. Nishio, G. Swislow, and A. Shah, *Phys. Rev. Lett.* **45:** 1636 (1980).
126. Y. Hu, K. Horie, and H. Ushiki, *Macromolecules* **25:** 6040 (1992).
127. Y. Hu, K. Horie, H. Ushiki, F. Tsunomori, and T. Yamashita, *Macromolecules* **25:** 7324 (1992).
128. Y. Hu, K. Horie, H. Ushiki, T. Yamashita, and F. Tsunomori, *Macromolecules* **26:** 1761 (1993).
129. M. Asano, F. M. Winnik, T. Yamashita, and K. Horie, *Macromolecules* **28:** 5861 (1995).
130. T. Torii, T. Yamashita, and K. Horie, *Biopolymers* **34:** 101 (1994).
131. A. H. Clark and S. B. Ross-Murphy, *Adv. Polym. Sci.* **83:** 129 (1987).
132. D. G. Gray, *J. Appl. Polym. Sci. Appl. Polym. Symp.* **37:** 179 (1983).
133. D. G. Gray, *Faraday Discuss Chem. Soc.* **79:** 257 (1985).
134. T. Yamagishi, T. Fukuda, T. Miyamoto, and J. Watanabe, *Polym. Bull.* **20:** 373 (1988).
135. H. Itagaki, I. Takahashi, M. Natsume, and T. Kondo, *Polym. Bull.* **32:** 77 (1994).
136. H. Itagaki, M. Tokai, and T. Kondo, *Polymer* **38:** 4201 (1997).
137. B. R. Harkness and J. Watanabe, *Macromolecules* **24:** 6759 (1991).

138. M. Sone, B. R. Harkness, J. Watanabe, T. Yamashita, T. Torii, and K. Horie, *Polym. J.* **25:** 997 (1993).
139. H. W. Huang, K. Horie, T. Yamashita, S. Machida, M. Sone, M. Tokita, J. Watanabe, and Y. Maeda, *Macromolecules* **29:** 3485 (1996).
140. T. Kondo and H. Itagaki, 207th ACS National Meeting, CELL 65 (1994).
141. H. Ushiki, F. Tanaka, and I. Mita, *Polym. J.* **22:** 827 (1986).
142. T. Torii, T. Yamashita, H. Ushiki, and K. Horie, *Eur. Polym. J.* **29:** 465 (1993).
143. H. Itagaki, *J. Photopolym. Sci. Tech.* **6:** 101 (1993).
144. H. Itagaki and Y. Umeda, *Polymer* **36:** 29 (1995).
145. H. Itagaki, *J. Luminescence* **72–74:** 435 (1997).
146. H. Itagaki and S. Kato, *Polymer* **40:** 3501 (1999).
147. M. Hasegawa, I. Mita, M. Kochi, and R. Yokota, *J. Polym. Sci., Part C: Polym. Lett.* **27:** 263 (1989).
148. M. Hasegawa, M. Kochi, I. Mita, and R. Yokota, *Eur. Polym. J.* **25:** 349 (1989).
149. M. Hasegawa, H. Arai, I. Mita, and R. Yokota, *Polym. J.* **22:** 875 (1990).
150. M. Hasegawa, Y. Shindo, T. Sugimura, S. Ohshima, K. Horie, M. Kochi, R. Yokota, and I. Mita, *J. Polym. Sci., Part B: Polym. Phys.* **31:** 1617 (1993).
151. M. Hasegawa, I. Mita, M. Kochi, and R. Yokota, *Polymer* **32:** 3225 (1991).
152. K. Hahn, G. Ley, H. Schuller, and R. Oberthur, *Colloid. Polym. Sci.* **264:** 1092 (1986).
153. Y. Wang and M. A. Winnik, *J. Phys. Chem.* **97:** 2507 (1993).
154. Y. Wang and M. A. Winnik, *Macromolecules* **26:** 3147 (1993).
155. J. Feng, M. A. Winnik, R. R. Shivers, and B. Clubb, *Macromolecules* **28:** 7671 (1995).
156. E. Pajot-Augy, L. Bokobza, L. Monnerie, A. Castellan, H. Bouas-Laurent, and C. Millet, *Polymer* **24:** 117 (1983).
157. L. Bokobza, E. Pajot-Augy, L. Monnerie, A. Castellan, and H. Bouas-Laurent, *Polym. Photochem.* **5:** 191 (1984).
158. E. Pajot-Augy, L. Bokobza, L. Monnerie, A. Castellan, and H. Bouas-Laurent, *Macromolecules* **17:** 1490 (1984).
159. C. P. V. Cang, L. Bokobza, L. Monnerie, J. Vandendriessche, and F. C. De Schryver, *Polym. Comm.* **27:** 89 (1986).
160. R. O. Loutfy, *Macromolecules* **16:** 678 (1983).
161. K. Y. Law and R. O. Loutfy, *Macromolecules* **14:** 587 (1981).
162. J. P. Quelel, J. P. Jarry, and L. Monnerie, *Polymer* **27:** 1228 (1986).
163. Ref. 13, the references therein (p. 106).
164. W. C. Yu and C. S. P. Sung, *Macromolecules* **21:** 365 (1988).
165. I. Mita, K. Horie, and K. Hirao, *Macromolecules* **22:** 558 (1989).
166. G. Liu and J. E. Guillet, *Macromolecules* **23:** 1388 (1990).
167. G. Liu, J. E. Guillet, E. T. B. Al-Takrity, A. D. Jenkins, and D. R. M. Walton, *Macromolecules* **23:** 1393 (1990).
168. G. Liu, *Macromolecules* **26:** 1144 (1993).
169. H. Itagaki and H. Nagura, *J. Photopolym. Sci. Tech.* **8:** 107 (1995).
170. The values are from the two books: ref. 2, pp. 70 and 120 and ref. 171.

171. S. L. Murov, *Handbook of Photochemistry*, Dekker, New York, 1973.
172. H. Itagaki, A. Ochiai, and T. Nakamori, submitted to *Macromolecules*.
173. T. Kanaya, K. Goshiki, M. Yamamoto, and Y. Nishijima, *J. Am. Chem. Soc.* **104:** 3580 (1982).
174. I. Nagata and H. Morawetz, *Macromolecules* **14:** 87 (1978).
175. L. P. Chang and H. Morawetz, *Macromolecules* **20:** 428 (1987).
176. J. M. Torkelson, M. Tirrell, and C. W. Frank, *Macromolecules* **17:** 1505 (1984).
177. C. W. Frank, *Macromolecules* **8:** 305 (1975).
178. H. Itagaki, K. Horie, and I. Mita, *Eur. Polym. J.* **19:** 1201 (1983).
179. G. H. Fredrickson and E. Helfand, *Macromolecules* **19:** 2601 (1986).
180. J. M. Ubrich, F. B. C. Larbi, J. L. Halar, L. Monnerie, B. J. Baner, and C. C. Han, *Macromolecules* **19:** 810 (1986).
181. C. W. Frank and M. A. Gashgari, *Macromolecules* **12:** 163 (1979).
182. S. N. Semerak and C. W. Frank, *Macromolecules* **17:** 1148 (1984).
183. P. Lianos and R. Zana, *J. Phys. Chem.* **84:** 3339 (1980).
184. R. Gelles and C. W. Frank, *Macromolecules* **16:** 1448 (1983).
185. F. J. Tsai and M. Torkelson, *Macromolecules* **21:** 1026 (1988).
186. H. Itagaki, K. Horie, and I. Mita, *Macromolecules* **16:** 1395 (1983).
187. S. Bloński and K. Sienicki, *Macromolecules* **19:** 2936 (1986).

Chapter 4 | NMR Spectroscopy in Polymer Science

I. Ando, M. Kobayashi, M. Kanekiyo, S. Kuroki, and S. Ando

Department of Polymer Chemistry
Tokyo Institute of Technology
Ookayama, Meguro-ku
Tokyo, Japan

S. Matsukawa

Department of Food Science
Tokyo University of Fisheries
Konan, Minato-ku
Tokyo, Japan

H. Kurosu

Department of Textile and Aparrel Science
Nara Women's University
Kita-uoya, Nishimachi
Nara, Japan

H. Yasunaga

Department of Chemistry and Materials Technology
Kyoto Institute of Technology
Goshokaido-machi, Matsugasaki
Kyoto, Japan

S. Amiya

Analytical Research Center
Kuraray Co., Ltd.
Kurashiki, Okayama, Japan

4.1 Introduction

Since Purcell *et al.* [1] and Bloch *et al.* [2] independently discovered nuclear magnetic resonance (NMR) in 1945, NMR spectroscopy has greatly advanced with the extensive developments of the basic NMR theory [3], the computer, and electronics. Consequently NMR has become an important methodology for studying polymer science, materials science, chemistry, biochemistry, physics, medicine, etc. [4]. In particular, the quality and quantity of information about

ISBN 0-12-683265-X/$30.00.

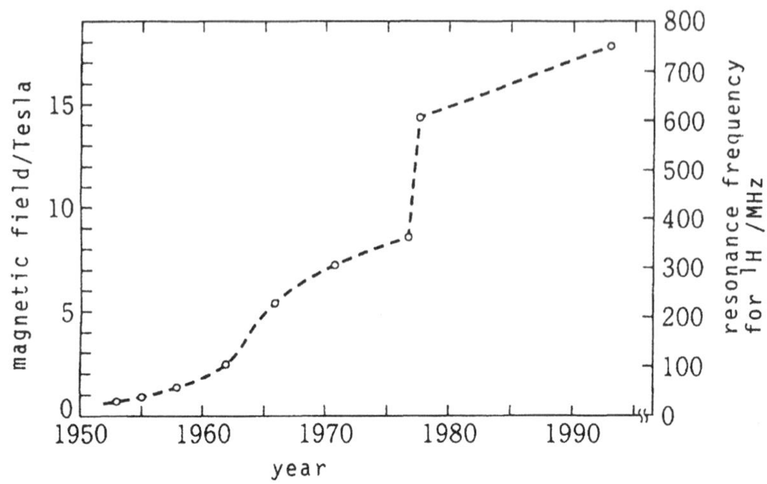

Figure 4.1 Plot of the strength of the NMR magnet developed against year.

the structure and dynamics of polymers in the solid, liquid crystalline, gel, and solution states as obtained by NMR have become very great [5].

The NMR techniques that have been applied to polymers are pulse NMR, solution-state NMR, solid-state NMR and NMR imaging methods [5, 6]. At present a high-field magnet with a superconducting coil is mainly used in these methods. Increasing the field strength of the NMR magnet leads to the enhancement of signal sensitivity and resolution. Therefore, the field strength of the magnet has been a measure of the advance of NMR spectroscopy. It is informative to plot the strength of the NMR magnet against years, as shown in Figure 4.1.

One can see that the field strength of the NMR magnet has been increased every year. Since the discovery of NMR, NMR magnets have been used in the following order: the permanent magnet, electromagnet, and superconducting coil magnet. The transitional increase of the field strength in 1967 was due to the appearance of the superconducting coil magnet. At present, the superconducting coil magnet is mainly used, and the high-field magnet generating 18.6 T (tesla) (corresponding to 800 MHz for ^{1}H) is also commercially available.

On the other hand, the pulse technique, which is another important method, as well as the magnet, has been considerably advanced on the basis of NMR theory [7]. Observation by NMR started with the continuous wave (CW) method in which an external alternating electromagnetic field is continuously applied to a sample under consideration in a static magnetic field. Later, the Fourier transform (FT) method involving the use of pulse technique and data processing with a computer was developed by Ernst *et al.* [8]. The advantages of the pulse technique led to the development of two-dimensional (2-D) NMR and multidimensional NMR and then to the separation of individual interactions such as chemical shift, spin–spin, dipole–dipole, quadrupole, etc. (3i, 14). The conventional spectrum with frequency as the horizontal axis and intensity as the vertical axis is called the one-dimensional (1-D) spectrum. The spectrum representing the contour map with the frequency for the horizontal and vertical axes is called the two-dimensional spectrum. Such a 2-D NMR technique has led to large increase of the application of NMR to polymer science. For example, the speed and precision of the structural analysis of polymers have been considerably advanced. In the past, the analysis of three-dimensional structure of proteins had been carried out using the X-ray diffraction method. However, at present such analysis is performed using the distance geometry method based on 2-D or multidimensional NMR [14]. Such examples show that NMR spectroscopy is an indispensable method for structural analysis of polymers.

Most polymers that form materials are almost always used as solids. A structural and dynamic characterization of the polymers is necessary to understand the relations between properties and structure and to design new polymer materials. The X-ray diffraction method has contributed to the structural determination of polymers with a crystalline region. However, the crystallinity of polymers is generally low. The chain segments in the noncrystalline region sometimes undergo molecular motion, and thus the X-ray diffraction method does not give structural information in this situation. On the other hand, solid-state NMR gives useful structural and dynamic information for the noncrystalline region in addition to the crystalline region. Further, solid-state NMR has provided characteristic information that cannot be obtained by other spectroscopic methods.

The spatial distribution of nuclear spins in a polymer sample in the mobile or immobile state can be determined by NMR imaging [9], which has made it possible to study the structure and dynamics of polymer samples macroscopically on the basis of microscopic, spatial, and dynamics information obtained the NMR imaging pattern.

Against such a background, NMR spectroscopy and its applications to structural and dynamic characterization of polymers will be described in this chapter. Many excellent books and periodical monographs on fundamental and advanced NMR spectroscopies have been published.

4.2 Overall Survey of NMR [3]

NMR-observable nuclei must have the magnetic moment, that is, spin number I, of an integer or half-integer. For example, hydrogen has some isotopic elements such as ^1H, ^2H, and ^3H, that have spin number 1/2, 1, and 1/2, respectively, and so ^1H (proton), ^2H (deuteron), and ^3H (tritium) are NMR-observable nuclei. Carbon, a very popular element in polymers, has isotopic elements such as ^{12}C and ^{13}C. The spin number of ^{12}C is zero, and that of ^{13}C is 1/2. Thus, the former isotopic element is unobservable, but the latter observable. Most of elements in the periodical table are observable. This means that NMR is useful in various fields.

4.2.1 RESONANCE PHENOMENON

In the explanation of the principle of resonance, we will be concerned with nuclei of $I = 1/2$ such as ^1H, ^{13}C, ^{15}N, ^{19}F, etc., which are popular in polymers. If a nucleus with $I = 1/2$ is placed in a magnetic field, it takes two energy states by quantization, as indicated in Figure 4.2. This energy difference ΔE is proportional to the strength of applied magnetic field H_0. Such a splitting is called the Zeeman splitting. The energy difference is expressed by

$$\Delta E = \mu H_0 / I = 2\mu H_0 \tag{1}$$

where μ is the nuclear magnetic moment and is given by $\gamma h I/2\pi$. The γ denotes the magnetogyric ratio because it is defined in terms of the ratio of the magnetic moment to the angular momentum as $2\pi\mu/Ih$. The magnitude of γ depends on the property of the nucleus, as indicated in Table 4.1. As seen from Figure 4.1, it is clear that transition between the two states can be induced by the use of appropriate electromagnetic radiation. Such a transition is governed by the Bohr condition, and energy is transferred when ΔE is just equal to $h\nu$, where ν is the frequency of the radiation in the radio-frequency (rf) region. Thus, we have the relation for the transition of a nucleus as

$$\nu = \gamma H_0 / 2\pi \tag{2a}$$

$$\omega = 2\pi\nu = \gamma H_0 \tag{2b}$$

where ω is the angular frequency. This relation means that the frequency of NMR is determined by the magnitude of γ and H_0, as shown in Table 4.1, and if the strength of the applied magnetic field is increased, the frequency of NMR will be proportionally increased.

4.2.2 NMR SIGNAL INTENSITY

The NMR signal intensity is closely related to the population difference in nuclear spins between the two states, as shown in Figure 4.2. The relation in the populations

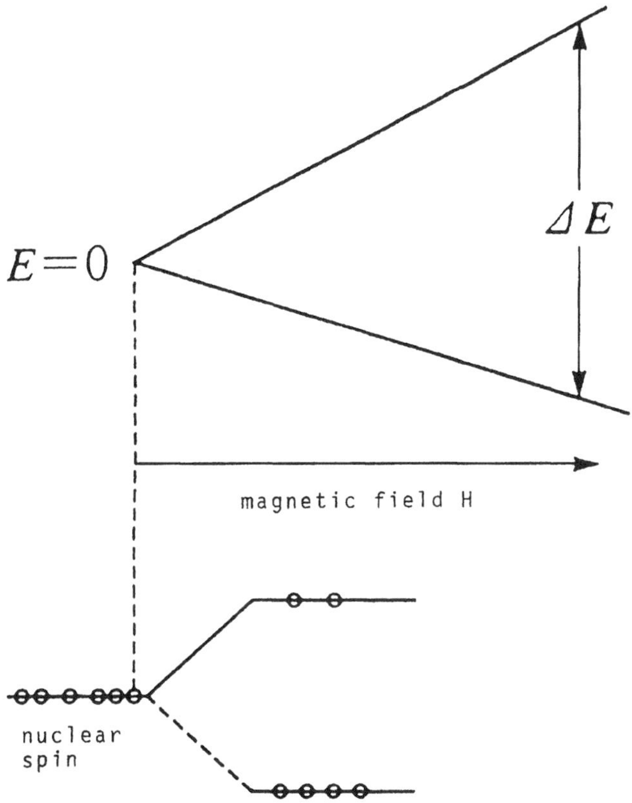

Figure 4.2 A diagram of the Zeeman splitting of a proton nucleus ($I = 1/2$) in an applied magnetic field.

of nuclear spin assemblies (e.g., for $I = 1/2$) in the two energy states is expressed by the Boltzmann distribution as

$$N_u/N_d = \exp(-\Delta E/kT) \tag{3}$$

where N_u and N_d are the number of nuclear spins in the higher and lower energy states, respectively, k the Boltzmann constant, and T the absolute temperature. From this relation, it is seen that the population in the lower energy state is higher than that in the higher energy state. To increase the population difference, temperature must be as low as possible. Thus, the decrease in temperature leads to an enhancement in NMR signal intensity. The signal-intensity relation will be derived as follows.

The nuclear spin system may absorb energy from rf radiation with some factors such as the population difference between the two energy states, the probability

Table 4.1 Characteristic parameters for selected NMR-observable nuclei.

Isotope	Spin	Natural Abundance (%)	Magnetogyric Ratio $\gamma/10^7$ rad $T^{-1}s^{-1}$	Sensitivity: Relative[a]	Sensitivity: Absolute[b]	NMR Frequency (MHz) at Some Magnetic Fields (T): 1.8790	2.3483	4.6975	5.8719	7.0463	9.3950	11.7440	14.0926	17.6157
1 H	1/2	99.98	26.7519	1.00	1.00	80.000	100.000	200.000	250.000	300.000	400.000	500.000	600.000	750.000
2 H	1	1.5×10^{-2}	4.1066	9.65×10^{-3}	1.45×10^{-6}	12.280	15.351	30.701	38.376	46.051	61.402	76.753	92.102	115.128
3 H	1/2	0	28.535	1.21	0	85.331	106.663	213.327	266.658	319.990	426.664	533.317	639.980	799.974
6 Li	1	7.42	3.9371	8.50×10^{-3}	6.31×10^{-4}	11.772	14.716	29.431	36.789	44.146	58.862	73.578	88.292	110.367
7 Li	3/2	92.58	10.3975	0.29	0.27	31.091	38.863	77.727	97.158	116.590	155.454	194.317	233.180	291.474
10 B	3	19.58	2.8746	1.99×10^{-2}	1.39×10^{-3}	8.5972	10.746	21.493	26.886	32.239	42.986	53.732	64.478	80.598
11 B	3/2	80.42	8.5843	0.17	0.13	25.667	32.084	64.167	80.209	96.251	128.335	160.419	192.502	240.627
13 C	1/2	1.108	6.7283	1.59×10^{-2}	1.76×10^{-4}	20.115	25.144	50.288	62.860	75.432	100.577	125.721	150.864	188.580
14 N	1	99.63	1.9338	1.01×10^{-3}	1.01×10^{-3}	5.779	7.224	14.447	18.059	21.671	28.894	36.118	43.342	54.177
15 N	1/2	0.37	−2.712	1.04×10^{-3}	3.85×10^{-6}	8.1063	10.133	20.265	25.332	30.398	40.531	50.664	60.796	75.996
17 O	5/2	3.7×10^{-2}	−3.6279	2.91×10^{-2}	1.08×10^{-5}	10.845	13.557	27.113	33.892	40.670	54.227	67.784	81.340	101.676
19 F	1/2	100	25.181	0.83	0.83	75.261	94.077	188.154	235.192	282.231	376.308	470.385	564.462	705.576
23 Na	3/2	100	7.0801	9.25×10^{-2}	9.25×10^{-2}	21.161	26.451	52.902	66.128	79.353	105.805	132.256	158.706	198.384
25 Mg	5/2	10.13	−1.639	2.67×10^{-3}	2.71×10^{-4}	4.895	6.1195	12.238	15.298	18.358	24.477	30.597	36.716	45.894
27 Al	5/2	100	6.9760	0.21	0.21	20.846	26.057	52.114	65.143	78.172	104.229	130.287	156.344	195.429
29 Si	1/2	4.7	−5.3188	7.84×10^{-3}	3.69×10^{-4}	15.892	19.865	39.730	49.662	59.595	79.460	99.325	119.190	148.986
31 P	1/2	100	10.841	6.63×10^{-2}	6.63×10^{-2}	32.385	40.481	80.961	101.202	121.442	161.923	202.404	242.884	303.606

[a] At constant field for equal numbers of nuclei.

[b] Product of relativity and natural abundance.

per unit time per spin of a transition, etc. By considering such estimations, the observed NMR signal intensity S in a steady-state experiment is given by

$$S = \gamma^4 H_0^2 H_1 / T \tag{4}$$

This relation indicates that the signal intensity is proportional to H_0^2 and $1/T$. The increase in applied magnetic field very effectively enhances the signal intensity. The nuclear spins with large γ originally have high sensitivity.

On the other hand, the signal-to-noise ratio (SNR) is improved by using the NMR signal accumulation with a computer. The number of accumulations n leads to an increase n times the signal intensity and leads to an increase $n^{\frac{1}{2}}$. Thus, if the number of accumulations is n, the obtainable SNR is improved by $n^{\frac{1}{2}}$ times. At present, very high stability is afforded by the superconducting magnet, and thus we can have a long accumulation time for the NMR measurement. For example, in the solid-state ^{17}O NMR measurement on 10% enriched polypeptides, the number of accumulations becomes more than 100,000 to obtain a reasonable SNR, and this corresponds to an accumulation time of more than 24 h.

4.3 NMR Parameters

NMR parameters such as chemical shift, spin–spin coupling, dipolar interaction, relaxation times, nuclear Overhauser effects, etc., provide useful information on the structure and dynamics of polymers. The deep understanding of these NMR parameters leads to structural and dynamic analyses with high decision.

4.3.1 CHEMICAL SHIFT

When a sample to be measured is placed in a magnetic field, nuclei experience the more reduced strength of magnetic field H' compared with the strength of applied magnetic field H_0 owing to the existence of electrons around a nucleus. Such shielding depends on the spatial distribution of electrons, as expressed by σ. Thus, we have the following equation.

$$H' = (1 - \sigma)H_0 \tag{5}$$

It is not easy to determine the absolute magnitude of the shielding constants in the NMR experiment. However, usually we need to know the absolute magnitudes, and so it is very convenient to use the difference in the resonance frequency between any specified nucleus (ν_{sample}) and the standard nucleus (ν_{standard}). It is called the chemical shift δ and is expressed by

$$\delta = 10^6 \, (\nu_{\text{sample}} - \nu_{\text{standard}}). \tag{6}$$

Correspondingly, if the shielding constants are used, δ is expressed by

$$\delta = 10^6 \, (\sigma_{\text{sample}} - \sigma_{\text{standard}}). \tag{7}$$

The standard substance must be used, depending on the nucleus to be considered, as follows:

Tetramethylsilane (TMS)	for ^{1}H, ^{13}C, and ^{29}Si
Glycine	for ^{15}N
H_2O	for ^{17}O
CCl_3F	for ^{19}F

The standard chemical compound is used either internally or externally to the sample. In the internal standard case, a small amount of the chemical compound is added to the liquid and solution samples. On the other hand, in the external standard case, the chemical compound is placed in a capillary tube. In these experiments, if we want to determine the exact chemical shift of the sample, the chemical shift correction must be made. In the internal standard case, there is the solvent effect in the observed chemical shift to be induced by intermolecular interactions, and in the external standard case, the bulk susceptibility correction must be made, which arises from the difference in the bulk susceptibility between the sample and standard chemical compound.

NMR chemical shift is originally the second-rank tensor $\boldsymbol{\sigma}$, which comes from the distortion from the spherical distribution of electrons around the nucleus as expressed by the second-rank tensor

$$\boldsymbol{\sigma} = \begin{pmatrix} \sigma_{xx} & \sigma_{xy} & \sigma_{xz} \\ \sigma_{yx} & \sigma_{yy} & \sigma_{yz} \\ \sigma_{zx} & \sigma_{zy} & \sigma_{zz} \end{pmatrix} \tag{8}$$

This tensor can be diagonalized as follows:

$$\boldsymbol{\sigma} = \begin{pmatrix} \sigma_{11} & 0 & 0 \\ 0 & \sigma_{22} & 0 \\ 0 & 0 & \sigma_{33} \end{pmatrix} \tag{9}$$

where σ_{11}, σ_{22}, and σ_{33} are the principal values. In the solid state these principal values are observed, and in the liquid or solution the average of the three principal values $(\sigma_{11} + \sigma_{22} + \sigma_{33})/3$ is observed. Typical examples for ^{13}C spectra of solid

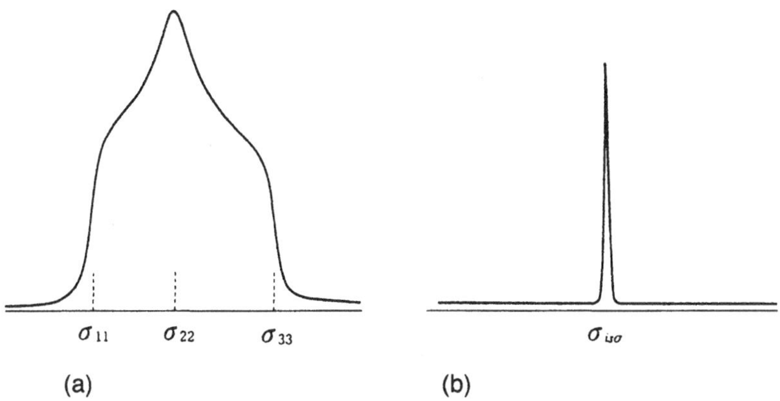

Figure 4.3 A typical example of ^{13}C spectra of solid polyethylene with (a) the static CP method and (b) the CP/MAS method.

polyethylene with the chemical shift tensor and isotropic chemical shift are shown in Figure 4.3. The former was performed using the ^{13}C CP (cross-polarization) NMR method and the latter by using the ^{13}C CP/MAS (cross-polarization/magic angle spinning) NMR method, which will described below.

Such a chemical shift originates from the electronic state associated with the structure. Thus, the chemical shift behavior of various polymers in the solid or solution states should be understood through changes in the electronic state of the polymers, depending on structural changes. This will be described in the following section.

4.3.2 SPIN–SPIN COUPLING

A small interaction between nuclear spins through electrons induces a splitting of the signal in a spectrum in addition to the chemical shift displacement of signals. The magnitude of such a splitting is affected by the primary and secondary structures. This interaction is called the spin–spin coupling *J* and sometimes the indirect or scalar coupling (3d). It is distinguished from the direct dipolar interaction. The couplings of two nuclei with one bond, two bonds, three bonds, and so on between them are called one-bond, two-bond, and three-bond spin–spin coupling, respectively.

The magnitude of three-bond spin–spin coupling depends on the conformation, described below [10]. For this, three-bond spin–spin coupling is used to determine the conformation of polymers. In general, the magnitude of spin–spin coupling between two nuclei is decreased with the number of bonds between them. This has great significance when determining the chemical structure. The 2-D NMR techniques automatically determine the chemical structure when obtaining the connectivity of chemical bonds through the observation of the connectivity of the spin–spin coupling.

4.3.3 DIPOLAR COUPLING

A nuclear spin acts as a small magnet and so generates a local magnetic field. Two nuclei interact with each other in a direct through-space. It is called the dipolar coupling [3]. For example, when two nuclei interact with the distance r between them and with the angle θ between the vector $\mathbf{r}$ and the direction of applied magnetic field H, the local magnetic field H_{loc} arising from the dipolar interaction H is expressed as

$$H_{\text{loc}} = \gamma I h(3\cos^2\theta - 1)/2\pi r^3 \tag{10}$$

Such an interaction sometimes leads to the broadening of the NMR signal. If we want to obtain a spectrum with very sharp peaks, this interaction must be removed. However, if we can obtain the magnitude of the dipolar coupling from a spectrum, we can determine the distance between two interacting nuclei. In the ^{13}C–^{13}C case, the determinable distance is below 5 Å. This limitation comes from the magnitude of the magnetogyric ratio for the nuclei.

4.3.4 RELAXATION TIMES

Dipolar interactions between nuclear spins in polymers predominantly govern relaxation times. There are two kinds of popular relaxation times: spin–lattice relaxation time T_1 and spin–spin relaxation time T_2. According to the Bloembergen–Purcell–Pound (BPP) theory [11], T_1 and T_2 are expressed by

$$1/T_1 = (3/8\pi^2)\gamma^4 h^2 I(I+1)[J_1(\omega_I) + J_2(2\omega_I)] \tag{11}$$

$$1/T_2 = (1/4\pi^2)\gamma^4 h^2[(3/8)J_0(0) + (15/4)J_1(\omega_I) + (3/8)J_2(2\omega_I)] \tag{12}$$

where

$$J_0(\omega) = (24/16r^6)\left[\tau_c/\left(1+\omega^2\tau_c{}^2\right)\right] \tag{13a}$$

$$J_1(\omega) = (4/16r^6)\left[\tau_c/\left(1+\omega^2\tau_c{}^2\right)\right] \tag{13b}$$

$$J_2(\omega) = (16/15r^6)\left[\tau_c/\left(1+\omega^2\tau_c{}^2\right)\right] \tag{13c}$$

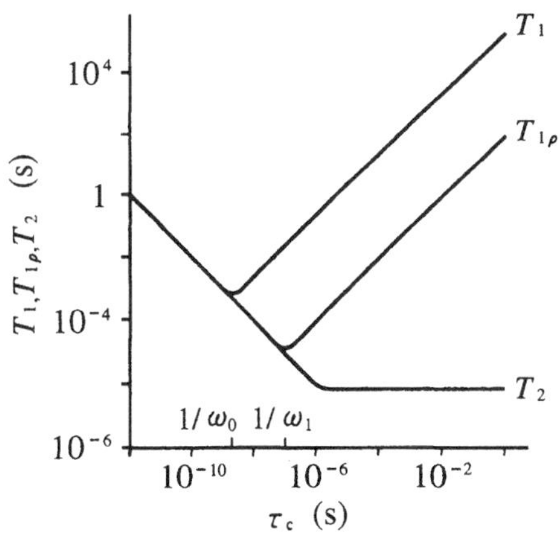

Figure 4.4 Plots of T_1 and T_2 against the correlation time τ_c obtained by the BPP theory.

The plots of T_1 and T_2 are shown against the correlation time in molecular motion τ_c as indicated in Figure 4.4. In the T_1 curve at the minimum T_1, $\tau_c = 1/\omega$, where ω is the resonance frequency. From the observation of the minimum T_1 we can determine the absolute correlation time in molecular motion. On the other hand, T_2 decreases with an increase in τ_c. From these T_1 and T_2 behaviors, we can discuss dynamics of polymers in the solid, liquid crystalline, and solution states.

From the rotating frame experiments, we can obtain $T_{1\rho}$ as expressed by [12]

$$1/T_{1\rho} \propto H_d^2 \tau_c / 1 + 4\omega_1^2 \tau_c^2 \tag{14}$$

From this equation, $T_{1\rho}$ provides information about the molecular motion in the Kilohertz region.

4.3.5 *NUCLEAR OVERHAUSER EFFECT (NOE)*

In a system having dipolar interactions between nuclear spins that neighbor each other, decoupling is accompanied by intensity changes in the signal. Such intensity changes depend on the distance r between interaction spins, which is called the nuclear Overhauser effect (NOE) [13]. The NOE is proportional to the inverse of r^{-3} and so is used to obtain information on conformations and configurations. Using the 2-D NMR with NOE information, one can determine the three-dimensional structure of proteins as well as the X-ray diffraction cases.

4.4 NMR Chemical Shift and Structure

In the liquid and solution states, NMR chemical shifts of polymers are often the averaged values for all of the possible conformations because of rapid interconversion by rotations about bonds. On the other hand, in the solid state, chemical shifts are often characteristic of specific conformations because of strongly restricted rotation about the bonds. The chemical shift is affected by a change of the electronic structure arising from structural changes. In the solid state the components of the full chemical shift tensor can be determined. The complete chemical shift tensor provides information on the local symmetry of the electron cloud around the nucleus and so yields much more detailed knowledge of the electronic structure of the polymer compared with the average chemical shift associated with the structure [15].

4.4.1 *THE ORIGIN OF CHEMICAL SHIFT*

The chemical shift of an atom depends upon its electronic and molecular environments [142]. Note that the chemical shift relative to a standard reference is expressed by δ, and the chemical shielding is expressed by σ. The chemical shielding σ for atom A can be estimated by a sum of the following terms:

$$\sigma_{\mathrm{A}} = \sigma^{\mathrm{d}} + \sigma^{\mathrm{p}} + \sigma' \tag{15}$$

where σ^{d} is the diamagnetic term, σ^{p} is the paramagnetic term, and σ' is another term that comes from the magnetic anisotropy, polar, and ring current effects. For nuclei with 2p electrons such as ^{13}C, ^{15}N, etc., the relative chemical shift is predominantly governed by the paramagnetic term, and for the ^{1}H nucleus by the first and third terms in Eq. [15].

The paramagnetic term is expressed as a function of excitation energy, bond order, and electron density according to the sum-over-states (SOS) method in the simple form as follows:

$$\sigma^{\mathrm{p}} = -C \sum \langle r^{-3} \rangle_{2\mathrm{p}} (E_{\mathrm{m}} - E_{\mathrm{n}})^{-1} Q \tag{16}$$

where $E_{\mathrm{m}} - E_{\mathrm{n}}$ is the singlet–singlet excitation energy from the nth occupied to the mth unoccupied orbitals, and Q is a factor including the bond order and electron density. The quantity $\langle r^{-3} \rangle_{2\mathrm{p}}$ is the spatial dimensions for a 2p electron, and C is the coefficient incorporating universal constants. This term is calculated by semiempirical or *ab initio* molecular orbital (MO) methods. The former has some features that give the substantial aspects of the chemical shift behavior associated with the spatial structure, the electronic structure, or both. The diamagnetic term is estimated from the calculated electron density. Using these procedures, one can calculate the chemical shielding σ_i of the model compound with any specified conformation. From these results, for the chemical shift calculations of nuclei with 2p electrons, it is seen that it is very important to estimate the paramagnetic term exactly.

4.4.2 MEDIUM EFFECTS ON CHEMICAL SHIFTS

Most MO calculations of nuclear shielding relate to the case of a molecule in a vacuum. For nuclei forming the molecular skeleton, such as ^{13}C, and nuclei with small shielding ranges, such as ^{1}H, this may not be an unreasonable approximation. This is particularly true if comparison of the theoretical results is to be made with experimental data taken on a molecule dissolved in an inert solvent.

More reactive atoms, especially those with lone-pair electrons, such as ^{14}N, ^{15}N, ^{17}O, and ^{19}F, are very likely to have their nuclear shieldings influenced by interactions with solvent molecules. Such interactions may be specific (e.g., hydrogen bonding), nonspecific (e.g., polarizability–polarity), or perhaps a combination of specific and nonspecific solute–solvent interactions. An empirical procedure has been developed for quantitatively unraveling the contributions made to the shielding of solute nuclei by specific and nonspecific interactions.

Nonspecific solute–solvent effects on nuclear shielding may be described by MO calculations that include the influence of solvent polarity–polarizability by means of the solvent dielectric constant ε. This approach is epitomized by the use of the solvaton model with a semiempirical MO basis set [16–23]. This method has successfully accounted for the observed variations of ^{13}C and ^{15}N shielding for a number of solute molecules in solvents with a variety of ε values. The solvaton model has been demonstrated to predict the sign and magnitude of the solute shielding variation correctly as the value of ε of the medium is changed. This approach has led to a good understanding of the various solute–solvent interactions with the ^{13}C chemical shifts of polymers such as poly(vinyl chloride) associated with the stereochemical configurations [17]. *Ab initio* MO calculations of the effect of the solvent on the ^{29}Si shielding of some solvated molecules have been reported [22]. The extended basis set calculations employed the individual gauge for different localised orbitals (IGLO) nuclear shielding procedure coupled with a continuum solvent model in which the solute was placed inside an appropriately dimensioned cavity within a polarizable continuum with a given value of ε [23]. The calculated variation in the silicon nuclear shielding, as a function of ε, is found to be in satisfactory agreement with experimental result. In general, the variation in the silicon shieldings of the solvated molecules is small but not negligible.

Specific solute–solvent interactions, such as hydrogen bonding or protonation, may be included in the calculation of the shielding of solute nuclei by a supermolecule approach. The appropriate structure of the solute–solvent supermolecule may be obtained by the use of molecular mechanics simulations. At the semi-empirical MO level this approach has been successfully used to describe the effects of hydrogen bonding on the nuclear shielding of small molecules. *Ab initio* MO calculations using the gauge-invariant atomic orbital (GIAO) procedure has been applied in a study of the effects of hydrogen bonding on the ^{1}H, ^{13}C, and ^{17}O nuclear shieldings. Calculations using a 6–31G** basis set show that hydrogen bonding can result in a change in ^{17}O nuclear shielding by up to about 70 ppm, whereas a few parts per million change at most is predicted for the ^{1}H and ^{13}C shielding upon hydrogen bond formation.

4.4.3 *CHEMICAL SHIFT AND CONFORMATION OF POLYMER CHAINS IN SOLUTION OR THE AMORPHOUS STATE*

A polymer chain can assume an enormous number of conformations because of the various possibilities of rotation around the chain bonds owing to molecular motion [24]. In solution or the amorphous state, if molecular motion in the polymer chains is extremely slow on the NMR time scale, the spectrum represents the superposition of the spectra for the various conformations. However, if the rotation around the chain bonds is very fast on the NMR time scale, the experimentally observable chemical shift for nucleus A is given as (15)

$$\langle \delta_{\mathrm{A}} \rangle = \sum_{i=1}^{n} P_i \delta_i \tag{17}$$

The numerical indices refer to the preferred conformations, and P_i and δ_i are the probability of occurrence and the chemical shift of the preferred conformation i, respectively. This indicates that the chemical shift of a given nucleus can be obtained by a combination of quantum chemical and statistical mechanical methods, as reviewed by Ando *et al.* [15, 25–29].

4.4.4 *CHEMICAL SHIFT AND STRUCTURE OF POLYMER CHAINS IN THE SOLID STATE*

Sometimes the estimation of the electronic structures of polymer chains necessitates including long-range interactions and intermolecular interactions in the chemical shift calculations. To do so, it is necessary to use a sophisticated theoretical method that can take account of the characteristics of polymers. In this context, the tight-binding molecular orbital (TB MO) theory from the field of solid-state physics is used in the same sense in which it is employed in the linear combination of atomic orbitals (LCAO) approximation in molecular quantum chemistry to describe the electronic structures of infinite polymers with a periodical structure [30–34]. In a polymer chain with linearly bonded monomer units, the potential energy of an electron varies periodically along the chain. In such a system, the wave function $\Psi(k)$ for electrons at a position r can be obtained from Bloch's theorem as follows [35]:

$$\Psi(k) = N^{-1/2} \sum \exp(ikj) C_{\nu n}(k)_{\nu} \phi(r - j\mathbf{a}) \tag{18}$$

where k is the wave number, n is the band index, ν is an orbital index in the jth cell, $\mathbf{a}$ is the unit vector of translational symmetry, N is the total number of cells, and 1 is the number of atomic orbitals in the cell. The term $\phi(k)$ represents the υth atomic orbital in the jth cell and $C_{\nu n}(k)$ is the expansion coefficient. The formulas needed to calculate the shielding of polymers using the tight-binding MO theory incorporating the SOS method have been derived as a function of k as follows:

$$\sigma^{\mathrm{d}}(k) = C \sum P(k) \langle \phi_{\nu} | r^{-1} | \phi_{\nu'} \rangle \tag{19}$$

$$\sigma^{\mathrm{p}}(k) = -C' \sum_{m>n} \langle r^{-3} \rangle_{2p} (E_m - E_n)^{-1} Q \tag{20}$$

where C and C' are constants and $P(k)$ is the bond order as a function of k. To compare the calculated and the experimental values of the nuclear shielding, it is necessary to average the calculated data over k within the first Brillouin zone as given by

$$\sigma = \sum_{k} D(k) \sigma(k) \tag{21}$$

where $D(k)$ is the density of states, namely, the number of states per unit amount of energy. With this methodology, elucidation of the conformation-dependent chemical shift of various polymer chains such as polyethylene, polyacetylene, polypyrrole, polyoxymethylene, and polyoxyethylene in the solid state has been performed systematically [30–34].

The TB MO calculation of the ^{15}N chemical shift of polypyrrole in the solid state allows useful information to be extracted from the observed spectra, namely, that the two peaks obtained are correctly assigned to the quinoid and aromatic structures [36]. (The quinoid structure is closely related to the electric conductivity.) A decrease in the band gap leads to a high frequency. These results from conducting polymers demonstrate that the chemical shift behavior provides information about the band gap, which, in turn, is a measure of the electrical conductivity. It can be said that TB MO calculations offer useful perspectives in interpreting the results of

NMR nuclear shieldings in polymers, both in terms of the structure in the solid state and in understanding the effect of intermolecular interactions on nuclear shieldings.

4.5 Basic NMR Techniques

In this section, we introduce some basic NMR techniques that play an important role for characterizing the structure and dynamics of polymers in the solid, solution, and gel states. These techniques include pulse NMR, solution NMR, and solid-state NMR methods as well as the spatial distance method and NMR imaging, as described below.

4.5.1 *PULSE NMR METHOD*

In the pulse NMR experiment, high-power rf pulses are applied to a sample for rotating the equilibrium longitudinal magnetization, and, subsequently, the free-induction decay (FID) of the transverse component is detected as a time-dependent function. The Fourier transform converts FID into the corresponding frequency spectrum, of which the chemical shift gives information about chemical structures. On the other hand, the FID contains information about molecular dynamics, and so the experiment with some rf pulses gives T_1, T_2, and diffusion coefficient D. The T_1 and T_2 are closely related to the correlation time in molecular motion by the BPP theory [11] described earlier when they are governed by dipolar interactions, and the D, as obtained by the pulse NMR method, is a parameter for a measure of the velocity of the translational motion. Therefore, pulse NMR is a very useful method for the elucidation of molecular dynamics in polymer systems.

4.5.2 *EXCITATION BY PULSES*

In the pulse NMR experiments for weakly interacting spins, such as nuclei in solution, the motion of the ensemble of spins is described by using the macroscopic magnetization vector in a classical manner, although a single spin is considered to obey quantum mechanics. This classical formalism is helpful to understand the pulse NMR experiments, and the behavior of the magnetization $\mathbf{M}$ in the magnetic field $\mathbf{H}$ during experiments is the described by the Bloch equation as follows [37]:

$$d\mathbf{M}/dt = \gamma\mathbf{M} \times \mathbf{H} - (M'_x\mathbf{i}' + M'_y\mathbf{j}')/T_2 - (M'_z - M_0)\mathbf{k}'/T_1 \qquad (22)$$

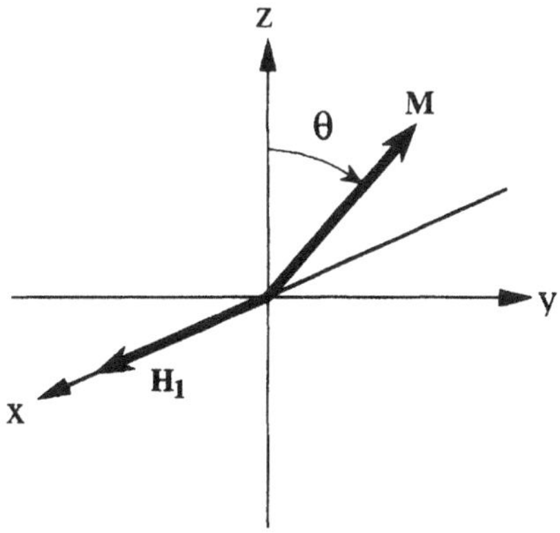

Figure 4.5 Rotation of the magnetization vector **M** around the x-axis in the y–z plane in the rotating flame of the Larmor frequency ω_0 for the static field $\mathbf{H}_0$ by the rf pulse of ω_0, which induces the magnetic field $\mathbf{H}_1$ in the rotating flame.

where $\mathbf{i}'$, $\mathbf{j}'$, and $\mathbf{k}'$ are the unit vectors in the laboratory frame. The M_x', M_y', and M_z' components are along $\mathbf{i}'$, $\mathbf{j}'$, and $\mathbf{k}'$, respectively, and M_0 is the equilibrium value of M_z'. In Bloch equations, T_1 and T_2 are used as parameters for the decay of longitudinal magnetization and transverse magnetization, respectively. The application of an rf pulse of the same frequency with the Larmor frequency ω_0 for the static field $\mathbf{H}_0$ causes the rf magnetic field $\mathbf{H}_1$ on the x-axis in the rotating frame of ω_0. Then, **M** is rotated around the x-axis into the y–z plane (Figure 4.5). For the rf pulse much shorter than T_1 and T_2, the time rate of change of **M** in the rotating flame of ω_0 is given by

$$d\mathbf{M}/dt = \gamma \mathbf{M} \times \mathbf{H}_1 \tag{23}$$

and the angle of the rotation θ is given by

$$\theta = \gamma H_1 t_p \tag{24}$$

where t_p is the pulse width, which is the length of time for the application of the rf pulse. In many pulse NMR experiments, $\pi/2$ and π pulses, for which $\theta = 90°$ and $180°$, respectively, are frequently used and made by the corresponding t_p for the H_1.

4.5.3 *T_1 MEASUREMENTS*

Spin–lattice relaxation is a process through which the equilibrium population in spin states is reached after a perturbation for the population through the energy transfer from an excited spin to the lattice, the surrounding thermal reservoir, which is in equilibrium. On the basis of the phenomenological description (Eq. (22)), the z component of the magnetization M_z obeys the equation

$$dM_z/dt = -(M_z - M_0)/T_1 \qquad (25)$$

where M_0 is the equilibrium value of the magnetization. The inversion recovery (IR) method is frequently used to measure the T_1 value. Figure 4.6 shows the pulse sequence and magnetization trajectories; π pulse is used as a first pulse to invert the magnetization vector lying along the z-axis to that lying along the $-z$-axis (a),

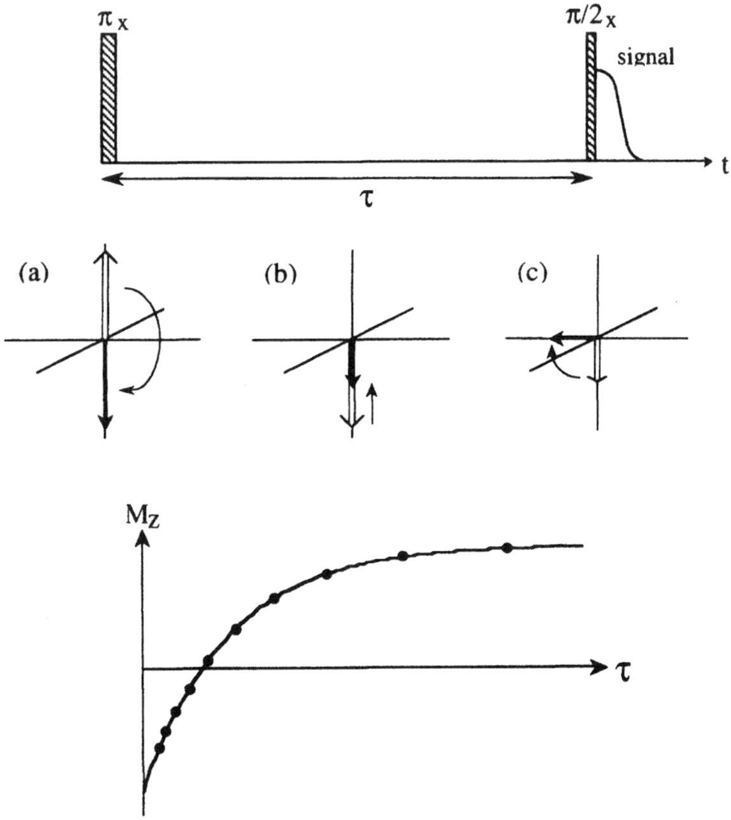

Figure 4.6 Pulse sequence and magnetization trajectories for the inversion recovery method: (a) the π pulse inverts the magnetization vector lying along the z-axis to that lying along the $-z$-axis, (b) the inverted magnetization reverts to the equilibrium condition, and (c) the $\pi/2$ pulse is used as a second pulse after delay time τ to inspect the z component of the magnetization M_z.

the inverted magnetization reverts to the equilibrium condition (b), $\pi/2$ pulse is used as a second pulse after delay time t to inspect the z component of the magnetization M_z, which follows Eq. (4) (c). When $t = 0$, $M_z = -M_0$; therefore, Eq. (25) gives

$$M_z = M_0(1 - 2\exp(-t/T_1)). \tag{26}$$

From a series of measurements varying t, the plot of M_z against t, such as that shown in Figure 4.6, is obtained and the T_1 value is given by a least-squares fitting to the experimental data by using Eq. (26). Before the first $\pi/2$ pulse, every experiment has to be done after enough retention time, at least five times T_1, to reach the equilibrium condition.

4.5.4 T_2 MEASUREMENTS

Spin–spin relaxation is a process of decay in the net transverse magnetization owing to the loss of phase coherence by the mutual exchange of spin energies after a pulse to make the transverse magnetization. The phase coherence also decreases by the effect of magnetic field inhomogeneity, and so this effect should be removed for the exact T_2 measurement. In the pulse NMR experiments with the spin echo sequence, the phase coherence is refocused at some time after rf pulses. This phenomenon is described by the classical formalism using the macroscopic magnetization vector.

4.5.4.1 Spin Echo Method

After the rf pulse is switched off, the time rate of change of the net transverse magnetization $M_+ = M_x + iM_y$, where M_x and M_y are the x and y components of $\mathbf{M}$, respectively, is derived from Eq. (22) as follows:

$$\begin{aligned} dM_+/dt &= -i\gamma H_0 M_+ - M_+/T_2 \\ &= i\omega_0 M_+ - M_+/T_2 \end{aligned} \tag{27}$$

where H_0 is the z component of $\mathbf{H}_0$. The time rate of change of the net transverse magnetization in the rotating frame M_+ introduced by $M_+ = M_+ \exp(i\omega_0 t)$ is written as

$$dM_+/dt = -M_+/T_2 \tag{28}$$

In Eq. (28), the T_2 emerges as a time constant for the decay where the perturbed M_+ returns to the equilibrium state; $M_+ = 0$. When there is an imhomogeneity of the static magnetic field, the z component of the total external magnetic field H_z at a point r is expressed by using field gradient $\mathbf{G}$ in the first approximation:

$$H_z = H_0 + \mathbf{G} \cdot \mathbf{r}. \tag{29}$$

Then, $\mathbf{M}$ in Eq. (27) is rewritten as a time- and space-dependent function:

$$\begin{aligned} \partial M_+/\partial t &= -i\gamma(H_0 + \mathbf{G} \cdot \mathbf{r})M_+ - M_+/T_2 \\ &= i\omega_0 M_+ - i\gamma(\mathbf{G} \cdot \mathbf{r})M_+ - M_+/T_2 \end{aligned} \tag{30}$$

and M_+ is given by

$$\begin{aligned} \partial M_+/\partial t &= -i\gamma(\mathbf{G} \cdot \mathbf{r})M_+ - M_+/T_2 \\ &= -M_+/[1/T_2 + i\gamma(\mathbf{G} \cdot \mathbf{r})]^{-1}. \end{aligned} \tag{31}$$

In Eq. (31), the $T_2^* = [1/T_2 + i\gamma(\mathbf{G} \cdot \mathbf{r})]^{-1}$ emerges as a time constant for the decay under the existence of an imhomogeneity of the static magnetic field. The FID after a pulse decays with the time constant of T_2^*.

With $\Psi(\mathbf{r}, t)$ introduced by $M_+ = \Psi(\mathbf{r}, t)\exp(-t/T_2)$, Eq. (31) is rewritten as

$$\partial\Psi(\mathbf{r}, t)/\partial t = -i\gamma(\mathbf{G} \cdot \mathbf{r})\Psi(\mathbf{r}, t). \tag{32}$$

This indicates that spin vectors precise in the rotating flame with the individual rates of $-\gamma(\mathbf{G} \cdot \mathbf{r})$ corresponding to the locations and that the macroscopic magnetic vector dephases, which causes a decay of the net transverse magnetization.

In the spin echo experiments [38], this rotation is canceled at the echo point (Figure 4.7); the individual $\Psi(\mathbf{r}, t)$ induced by the $\pi/2_x$ pulse (the $\pi/2$ pulse along the x-axis) indicated as spin vectors in Figure 4.7, precesses to become the angle from the x-axis $\alpha = -\gamma(\mathbf{G} \cdot \mathbf{r})\tau$ during the τ period (a); the π_x pulse (the π pulse along the x-axis) flips spin vectors to become $\alpha = \pi - \{-\gamma(\mathbf{G} \cdot \mathbf{r})\tau\}$ (b); and the individual spin vectors precess $-\gamma(\mathbf{G} \cdot \mathbf{r})\tau$ during the next τ period and are aligned on the $-y$-axis (c). The solution for $\Psi(\mathbf{r}, t)$ after the π pulse is given by

$$\Psi(\mathbf{r}, t) = -A(t)\exp[-i\gamma(\mathbf{G} \cdot \mathbf{r})(t - 2\tau)]. \tag{33}$$

where $A(t)$ is a space-independent function. Equation (33) shows that $\Psi(\mathbf{r}, t)$ for the individual spins in the hole of the sample is identical with $A(t)$ at $t = 2\tau$; therefore, the net transverse magnetization becomes a maximum. Similarly, in the spin echo experiment with the $\pi/2_x$ pulse and π_y pulse (the π pulse along y-axis), the spin vectors in the hole of the sample are aligned on the y-axis, and $\Psi(\mathbf{r}, t)$ is given by the expression with the opposite sign of the right-hand side of Eq. (33).

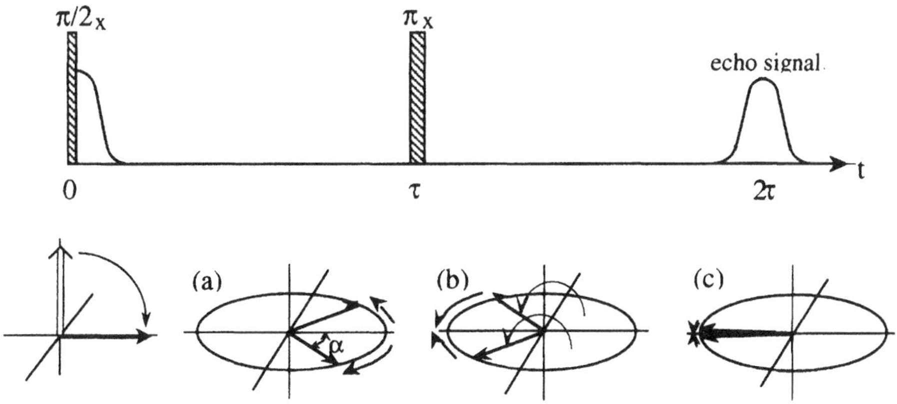

Figure 4.7 Pulse sequence and magnetization trajectories for the spin echo experiment: (a) the individual spin vectors induced by the $\pi/2_x$ pulse precess to be the angle from the x-axis $\alpha = -\gamma(\mathbf{G}\cdot\mathbf{r})\tau$ during the τ period corresponding to the locations, (b) the π_x pulse flips spin vectors to be $\alpha = \pi - \{-\gamma(\mathbf{G}\cdot\mathbf{r})\tau\}$, (c) the individual spin vectors precess $-\gamma(\mathbf{G}\cdot\mathbf{r})\tau$ during the next τ period and are aligned on the $-y$-axis.

4.5.4.2 Effect of Diffusion on Spin Echo

When a nucleus in a minute area at **r** diffuses to another area following Fick's law, the decay of $\Psi(\mathbf{r}, t)$ by the diffusion is given by

$$\partial\Psi(\mathbf{r}, t)/\partial t = D\nabla^2\Psi(\mathbf{r}, t) \tag{34}$$

where D is the diffusion coefficient. The nuclear mean square displacement in the z direction $\langle z^2\rangle$ is given by

$$\langle z^2\rangle = 2DT_{\mathrm{d}} \tag{35}$$

where T_{d} is the time allowing diffusion. With the effect of the diffusion, taken into account Eq. (32) is rewritten as

$$d\Psi(\mathbf{r}, t)/dt = -i\gamma(\mathbf{G}\cdot\mathbf{r})\Psi(\mathbf{r}, t) + D\nabla^2\Psi(\mathbf{r}, t). \tag{36}$$

From Eq. (33), the diffusion term is given as

$$D\nabla^2\Psi(\mathbf{r}, t) = -D\gamma^2G^2(t - 2\tau)^2\Psi(\mathbf{r}, t). \tag{37}$$

On the basis of Eqs. (33) and (37), Eq. (36) is rewritten as

$$dA(t)/dt = -D\gamma^2G^2(t - 2\tau)^2A(t). \tag{38}$$

From the result of the integration of Eq. (38) from $t = 0$ to $t = \tau$ and from $t = \tau$ to $t = 2\tau$, $A(2\tau)$ is given by

$$A(2\tau) = -A(0)\exp(-2/3D\gamma^2G^2\tau^3). \tag{39}$$

Therefore, $M_+(2\tau)$, the net transverse magnetization at $t = 2\tau$, is given by Eq. (39) as

$$M_+(2\tau) = -M_+(0)\exp(-2\tau/T_2 - 2/3D\gamma^2G^2\tau^3). \tag{40}$$

Equation (40) shows that the spin echo intensity decays by the T_2 relaxation process and also by the effect of the diffusion, which becomes greatly pronounced in large τ measurements.

4.5.4.3 Carr–Purcell–Meiboom–Gill (CPMG) Method

In the Carr–Purcell–Meiboom–Gill (CPMG) method [39, 40], a $\pi/2_x$ pulse and a series of π_y pulses are used to form an echo train (Figure 4.8); the individual spin vectors induced by the $\pi/2_x$ pulse precess to become $\alpha = -\gamma(\mathbf{G}\cdot\mathbf{r})\tau$ during the τ period (a), flip to become $\alpha = -[-\gamma(\mathbf{G}\cdot\mathbf{r})\tau]$ after the π_y pulse (b), precess to become $-\gamma(\mathbf{G}\cdot\mathbf{r})\tau$ during the next τ period and align on the y-axis to make an echo (c), and flip at $t = 2n\tau$ and align on the y-axis at $t = (2n+1)\tau$ in a similar manner. As seen in Figure 4.8, the angle error of the π pulse is compensated every two π pulses by using the π pulse along the y-axis. The solution for $\Psi(\mathbf{r}, t)$ between $t = (2n-1)\tau$ and $t = (2n+1)\tau$ is given by

$$\Psi(\mathbf{r}, t) = A(t)\exp[-i\gamma(\mathbf{G}\cdot\mathbf{r})(t - 2n\tau)]. \tag{41}$$

Considering the diffusion term $D\nabla^2\Psi(\mathbf{r}, t) = -D\gamma^2G^2(t-2n\tau)^2\Psi(\mathbf{r}, t)$, we have

$$dA(t)/dt = -D\gamma^2G^2(t-2n\tau)^2A(t). \tag{42}$$

The integration from $t = 0$ to $t = \tau$ from $t = (2m-1)\tau$ to $t = (2m+1)\tau$, where $m = 1, 2, \ldots, n-1$, and from $t = (2n-1)\tau$ to 2τ gives

$$A(2n\tau) = A(0)\exp(-2/3D\gamma^2G^2\tau^3 n). \tag{43}$$

and $M_+(t)$ at $t = 2n\tau$ is given by

$$\begin{aligned} M_+(t) &= M_+(0)\exp(-2n\tau/T_2 - 2/3D\gamma^2G^2\tau^3 n) \\ &= M_+(0)\exp(-t/T_2 - tD\gamma^2G^2\tau^2/3). \end{aligned} \tag{44}$$

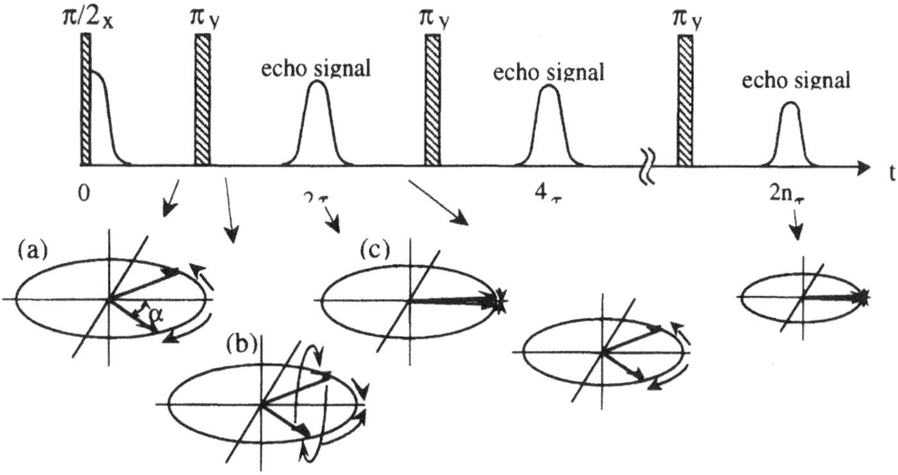

Figure 4.8 Pulse sequence and magnetization trajectories for the Carr–Purcell–Meiboom–Gill method: (a) the individual spin vectors induced by the $\pi/2_x$ pulse precess to be $\alpha = -\gamma(\mathbf{G}\cdot\mathbf{r})\tau$ during the τ period, (b) flip to be $\alpha = -\{-\gamma(\mathbf{G}\cdot\mathbf{r})\tau\}$ after the π_y pulse, and (c) precess $-\gamma(\mathbf{G}\cdot\mathbf{r})\tau$ during the next τ period and aligns on y axis to make an echo. The spin vectors flip at $t = 2n\tau$ and align on the y-axis at $t = (2n+1)\tau$ in a similar manner. The angle error of the π pulse is compensated every two π pulses by using the π pulse along the y-axis.

As indicated in Eq. (44), it is possible to make the effect of diffusion negligible by employing a small τ value with large n in the CPMG experiment, and thus we have

$$M_+(t) = M_+(0)\exp(-t/T_2). \tag{45}$$

Under the condition that the effect of diffusion is not negligible, the FID in the CPMG experiment decays single exponentially with the time constant of $T_2' = \{1/T_2 - D\gamma^2G^2\tau^2/3\}^{-1}$, whereas the FID in spin echo decays with the stronger time dependence, which is indicated in Eq. (40), rather than with exponential decay.

4.5.5 *DIFFUSION-COEFFICIENT MEASUREMENTS*

It is possible to measure the diffusion coefficient D using the CPMG experiments by varying τ, as seen from Eq. (44). However, the large magnitude of the constant gradient, which is required for the measurement of a small diffusion coefficient, makes it impossible to carry out the resonance condition over all the sample. In the pulsed-gradient spin-echo (PGSE) experiment, pulsed field gradients are applied

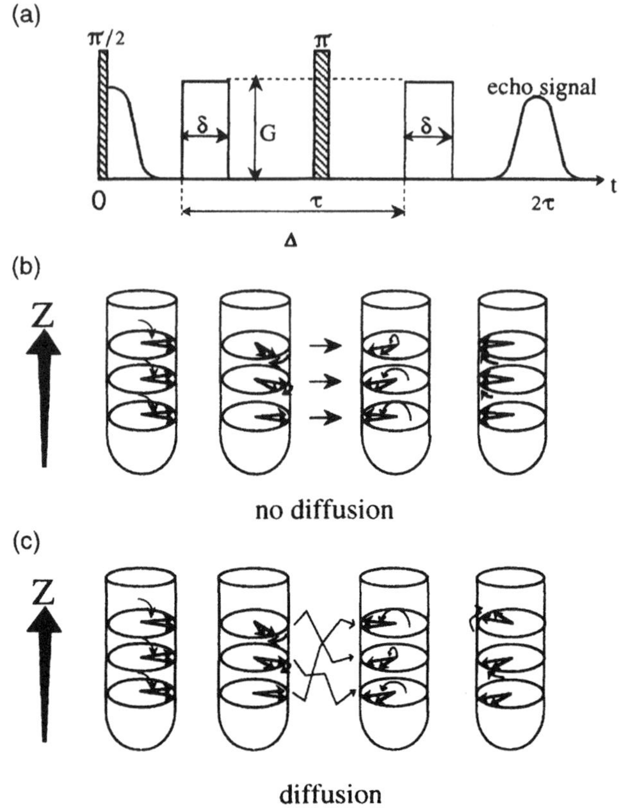

Figure 4.9 Pulse sequence and schematic representation of magnetization trajectories in each location in the sample for the pulsed-gradient spin-echo experiment. (b) The magnetization rephases perfectly at the echo point $t = 2\tau$ if there was no diffusion of spins. (c) The rephasing is not perfect under the existence of spin diffusions.

between the two rf pulses and between the first rf pulse and the spin echo (Figure 4.9) [41, 42]. When the time-dependent field gradient $\mathbf{G}(t')$ is used, $\Psi(\mathbf{r}, t)$ in the period from the $\pi/2_x$ pulse to the π_y pulse is given by Eq. (41) as

$$\Psi(\mathbf{r}, t) = A(t)\exp\left[-i\gamma\mathbf{r}\cdot\int_0^t \mathbf{G}(t')\,dt'\right]. \tag{46}$$

and $\Psi(\mathbf{r}, t)$ after π_y pulse at $t = \tau$ is given by

$$\Psi(\mathbf{r}, t) = A(t)\exp\left\{-i\gamma\mathbf{r}\cdot\left[\int_0^t \mathbf{G}(t')\,dt' - 2\int_0^\tau \mathbf{G}(t')\,dt'\right]\right\}. \tag{47}$$

It is seen in Eq. (47) that the effect of the π pulse is to set back the phase of $\Psi(\mathbf{r}, t)$ by twice the phase shift until the π pulse. Equations (46) and (47) are combined by using the Heaviside step function $H(t)$ and inserted into Eq. (46), and we have (42)

$$dA(t)/dt = -D\gamma^2 \left[\int_0^t \mathbf{G}(t')\,dt' - 2H(t-\tau) \int_0^\tau \mathbf{G}(t')\,dt' \right]^2 A(t). \tag{48}$$

In the pulse sequence shown in Figure 4.9(a), where the effect of the background gradient is negligible, the integration of Eq. (48) from $t = 0$ to $t = 2\tau$ gives

$$M_+(2\tau, G\delta)/M_+(2\tau, 0) = A(2\tau)/A(0) = \exp[-D\gamma^2 G^2 \delta^2 (\Delta - \delta/3)] \tag{49}$$

where $M_+(2\tau, G\delta)$ and $M_+(2\tau, 0)$ are echo signal intensities at $t = 2\tau$ with and without the magnetic field gradient pulse being the strength G and the length δ, respectively, and Δ the gradient pulse interval. The run of events for the decay of $A(t)$ by the diffusion is schematically represented by using the magnetization vectors in Figure 4.9. The magnetization rephases perfectly at the echo point $t = 2\tau$ if there were no diffusion of spins (Figure 4.9(b)); on the other hand, the rephasing is not perfect under the existence of the diffusion of spins (Figure 4.9(c)). A series of PGSE experiments varying δ is commonly employed for the determination of the D value using Eq. (49).

4.5.6 *ANALYSES FOR T_1, T_2 AND D MEASUREMENTS*

When the spin has a single component motion, the magnetization M_z in the inversion recovery (IR) method and the echo intensity in the CPMG method are given by Eq. (26) and Eq. (45), respectively. For the spin with more than one component motion, when spins of individual components do not change the state of motion during the measurement, the intensity of the magnetization is given as a superposition of individual components, and thus the intensity is given by

$$M_z(t) = M_0 \left[1 - 2 \sum_{i=1}^{n} f_i \exp(-t/T_{1,i}) \right] \tag{50}$$

for $M_z(t)$ in the IR method, and

$$M_+(t) = M_+(0) \sum_{i=1}^{n} f_i \exp(-t/T_{2,i}) \tag{51}$$

for $M_+(t)$ in the CPMG method, where f_i is the fractional number of the spin of ith component and $\sum_{i=1}^{n} f_i = 1$. When the motion of spin has a continuous distribution and the fraction is expressed by $f(1/T_1)$ and $f(1/T_2)$ for $1/T_1$ and

$1/T_2$, respectively, the intensity is given by

$$[M_0 - M_z(t)]/2M_0 = \int_0^\infty f(1/T_1)\exp(-t/T_1)\,d(1/T_1) \tag{52}$$

for $M_z(t)$ in the IR method, and

$$M_+(t)/M_+(0) = \int_0^\infty f(1/T_2)\exp(-t/T_2)\,d(1/T_2) \tag{53}$$

for $M_+(t)$ in the CPMG method. The functions for decay curves expressed on the right-hand sides of Eqs. (52) and (53) are the Laplace transforms of the distribution functions $f(1/T_1)$ and $f(1/T_2)$, respectively. Therefore, the decay curves experimentally obtained can be inverted to the distribution function of the relaxation times by the inverse Laplace transform using numerical methods. When the distribution can be regarded as Gaussian, the decays are given by

$$\{M_0 - M_z(t)\}/2M_0 = \exp(-\overline{1/T}_1 t + M_{T1}t^2/2) \tag{54}$$

for the $M_z(t)$ in the IR method, and

$$M_+(t)/M_+(0) = \exp(-\overline{1/T}_2 t + M_{T2}t^2/2) \tag{55}$$

for $M_+(t)$ in the CPMG method, where $\overline{1/T}_1$ and $\overline{1/T}_2$ are the average values for $1/T_1$ and $1/T_2$ given by

$$\overline{1/T}_{1/2} = \int_0^\infty 1/T_{1/2} f(1/T_{1/2})\,d(1/T_{1/2}) \tag{56}$$

respectively, and M_{T1} and M_{T2} are the variances for the distribution of $1/T_1$ and $1/T_2$ given by

$$M_{T1/2} = \int_0^\infty (1/T_{1/2} - \overline{1/T}_{1/2})^2 f(1/T_{1/2})\,d(1/T_{1/2}) \tag{57}$$

In the PGSE method for the spin with the multicomponent translational motion, the decay of the echo intensity is given as a superposition of individual components in which the decay owing to the T_2 relaxation is taken into account, and thus the intensity is given by

$$\begin{aligned} M_+(2\tau, K) &= M_+(0,0)\sum_{i=1}^{n} f_i \exp[-2\tau/T_{2,i} - KD_i(\Delta - \delta/3)] \\ &= M_+(2\tau, 0)\sum_{i=1}^{n} f_i^*(2\tau)\exp[-KD_i(\Delta - \delta/3)] \end{aligned} \tag{58}$$

where $K=\gamma^2G^2\delta^2$ and $f_i^*(2\tau)=f_i\exp(-2\tau/T_{2,i})/\sum_{j=1}^{n} f_i\exp(-2\tau/T_{2,j})$. When the translational motion of spin has a continuous distribution and the fraction is expressed by $f(D)$, the experimental decay is expressed by

$$M_+(2\tau, K)/M_+(2\tau, 0) = \int_0^\infty f(D)\exp[-K(\Delta - \delta/3)D]\,dD. \qquad (59)$$

Therefore, the distribution function for the diffusion coefficient $f(D)$ can be obtained by the inverse Laplace transform of the experimental decay curves in the same manner as those for relaxation times.

When the spin changes the state of motion during the measurement, the observed signal decay depends on the residence life time of the spin in each of the states. For the change of motion attended by the environmental change, such as chemical exchanges, the decay of the net transverse magnetization is accelerated by the additional dephasing process, which arises from the difference of the Larmor precession frequencies in each state $\Delta\omega$ and becomes significant when $\Delta\omega$ is comparable with the rate constant of the chemical exchange. For the change of motion without the change of resonant frequency, the time rate of the change of the magnetization in each state Ψ_i can be described as follows:

$$\begin{aligned} d\Psi_1/dt &= -R_1\Psi_1 - C_1\Psi_1 + p_{21}C_2\Psi_2 + \cdots + p_{n1}C_n\Psi_n \\ d\Psi_2/dt &= -R_2\Psi_2 + p_{12}C_1\Psi_1 - C_2\Psi_2 + \cdots + p_{n2}C_n\Psi_n \\ &\vdots \\ d\Psi_n/dt &= -R_n\Psi_n + p_{1n}C_1\Psi_1 + p_{2n}C_2\Psi_2 + \cdots - C_n\Psi_n \end{aligned} \qquad (60)$$

where $R_i = 1/T_{2,i}$ in the CPMG experiment, $R_i = \gamma^2G^2\delta^2D_i + 1/T_{2,i}$ in the PGSE experiment, $R_i = 1/T_{1,i}$ in the IR experiment, $C_i = 1/\tau_i$, where τ_i is the mean residence life time in the ith state, and p_{ij} is the conditional probability for a spin to change the state from the ith state to the jth state. For the motional change between two sites, Eq. (60) gives

$$\Psi(t) = \Psi_1(t) + \Psi_2(t) = f_1'\exp(-R_1't) + f_2'\exp(-R_2't) \qquad (61)$$

where

$$\begin{aligned} 2R_1' &= R_1 + C_1 + R_2 + C_2 - B \\ 2R_2' &= R_1 + C_1 + R_2 + C_2 + B \\ f_1' &= (R_2' - R_{av})/B \\ f_2' &= (R_{av} - R_1')/B \end{aligned} \qquad (62)$$

with $B = [(R_1 + C_1 - R_2 - C_2)^2 + 4C_1C_2]^{0.5}$ and $R_{av} = f_1R_1 + f_2R_2$. For the diffusion measurement by the PGSE method, Eq. (61) is available when $2\tau H \Delta - \delta/3$, which means that the time period for the T_2 relaxation is identical with that for the diffusion. In most PGSE methods, $2\tau - \Delta - \delta/3$, and the experimental decay $\Psi(2\tau, K)/\Psi(2\tau, 0) = [\Psi_1(2\tau, K) + \Psi_2(2\tau, K)]/[\Psi_1(2\tau, 0) + \Psi_2(2\tau, 0)]$ is numerically calculated by using Eq. (60).

4.6 Solution NMR Method

Solution NMR spectroscopy is now a main tool for analyzing organic compounds. This technique has been widely used to elucidate the polymer structure, polymer dynamics, and the polymerization mechanism since the early 1960s. Especially in the last 10 years, NMR spectroscopy has dramatically been improved in sensitivity and the chemical shift separations. Therefore, we can obtain the most detailed information concerning the trace components in a polymer chain and the microstructures of polymers. For example, if the analysis of end groups of polymer or irregular structures could be performed, we would be able to elucidate the mechanism of polymerization, determine the absolute molecular weight of the polymer, and have some information on improving its heat resistance. Determining the fine microstructures such as high-order tacticities or the sequence distribution of the copolymer makes it possible to improve the mechanical properties. The relationships between the microstructure and the property have provided some valuable information on the development of advanced polymers. NMR methods are extremely useful for designing high-performance polymers. Thus, we must pay sufficient attention to NMR experiments to obtain an NMR spectrum of high quality.

This section is an overview of FT–NMR spectroscopy, sample preparation, system adjustment, and data processing. Basic information needed for one-dimensional NMR measurement is also described. Finally, basic information about two-dimensional NMR measurement principles is provided.

For a more thorough introductory explanation, refer to references [3, 37a, and 43].

4.6.1 OUTLINE OF FT–NMR SPECTROSCOPY

A block diagram of an FT–NMR spectrometer is shown in Figure 4.10. An alternating-current signal generated from an rf oscillator is shaped into a combination of desired pulses. The pulses are amplified by a power amplifier and sent to a sample probe. A sample is positioned at the most homogeneous part of the magnetic field in a superconducting magnet. A strong rf pulse is applied to the sample, and a weak rf signal is output. A duplexer separates the directions of the rf signals. The weak rf signals are enhanced by a preamplifier and sent to a receiver.

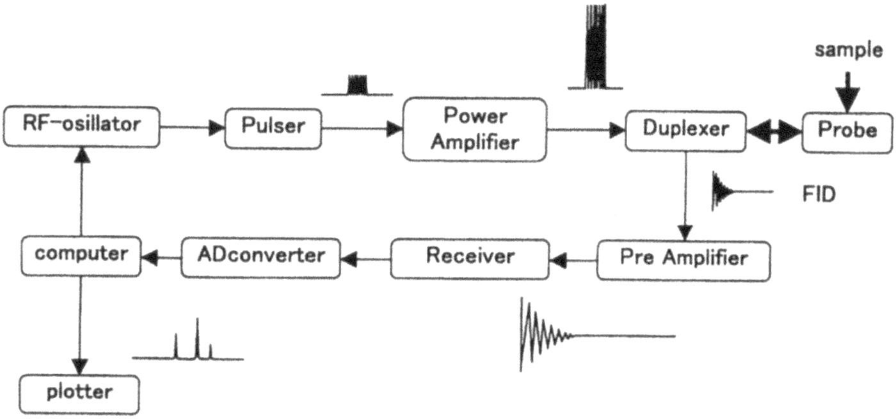

Figure 4.10 A block diagram of a pulsed FT–NMR spectrometer.

The signals are passed through the audio filter to eliminate useless frequency components and amplified by the amplifier built into the receiver to appropriate intensities. This signal is called free-induction decay (FID). The FIDs are sent to an analog-to-digital (AD) converter. The AD converter converts these signals from an analog representation to a digital one. These digital signals are finally stored in the computer. Fourier transformation of the time domain signal results in the usual frequency-domain spectrum.

NMR signals for nuclei with low natural abundance like ^{13}C, ^{15}N, and ^{29}Si are generally very weak. Therefore, many signals are accumulated before a spectrum with adequate SNR can be obtained. This is the so-called "accumulation" in which the signal intensity increases in proportion to the accumulation times. The SNR of a spectrum improves with the square root of the accumulation times. The repetition time between pulses must be long enough for a T_1 relaxation process to be completed.

4.6.2 SAMPLE PREPARATIONS

In solution NMR, a target polymer is dissolved in a suitable deuterated solvent, and a standard substance is added to the polymer solution as a reference of the chemical shift. To obtain a spectrum of high quality, it is necessary to be sufficiently careful in selecting the solvent and a standard substance as well as in handling the sample itself; moreover, several kinds of contamination in a polymer must be completely eliminated. This section deals with the methods of sample preparation.

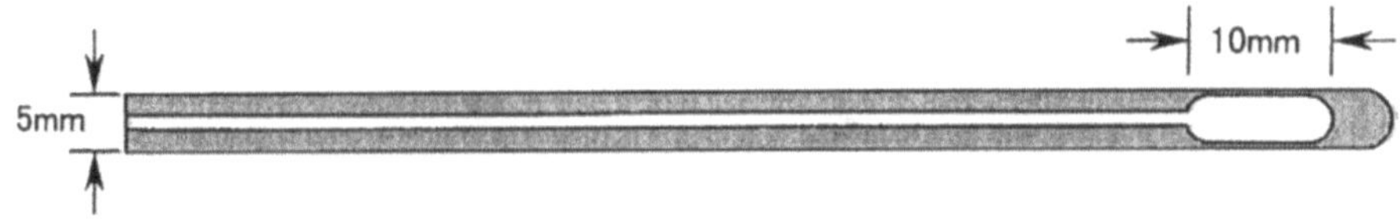

Figure 4.11 A microcell tube for an NMR experiment.

4.6.2.1 Sample Tube

As a sample tube, a glass tube with an outer diameter of 5 mm (for ^{1}H nuclei) or 10 mm (for ^{13}C, ^{19}F, etc.) is usually used. In the presence of a very small amount of a sample, it is convenient to use a microcell tube, as shown in Figure 4.11. The special sample tube characteristics required are as follows: high dimension precision, thinness and uniformity in the thickness of the tube, and a uniform rate of spinning. The selection of a cap covering the sample tube is also important. Because the role of the cap is to prevent the solvent used from volatilizing in the solution, it should be fitted to the tube tightly.

4.6.2.2 Selection of a Solvent

For the stability of a high homogenous magnetic field in a magnet, an NMR internal lock system has been adopted. As the lock signal, usually a ^{2}H signal of a deuterated compound is used. When selecting the solvent, pay attention to the following points:

1. Select a solvent that dissolves the sample well.
2. Ensure that the solvent signal does not overlap with the signal of the target sample.
3. When measuring the sample at high temperature, use a solvent whose boiling point is as high as possible and exercise caution not to make the measuring temperature higher than the solvent's boiling point.
4. To enhance separation of peaks, use a solvent whose viscosity is as low as possible.

Refer to the list of frequently used solvents in Table 4.2. A commercial deuterated solvent sometimes contains a trace amount of H_2O, D_2O, additives, decomposed materials, and so on. When observing ^{1}H nuclei of a very small amount of sample, it is necessary to use a solvent whose degree of deuteration is high. It will be useful to pay special attention to sample preparation and to absorb water by

Table 4.2 ^{1}H and ^{13}C–NMR chemical shifts of typical solvents.

Solvent	*Melting Point/ Boiling Point (°C)*	*^{1}H-shift (ppm)*	*^{13}C-shift (ppm)*	*$^{n}J_{^{13}C-^{2}H}$ (Hz)*
Acetic acid-d_4	17/118	2.035	20.03	20
		(11.569)	(178.40)	
Acetone-d_6	−94/57	2.051	29.83	20
			(206.08)	0.9
Acetonitrile-d_4	−45/82	1.938	118.28	21
Benzene-d_6	5/80	7.186	128.05	24
Chloroform-d	−64/62	7.258	77.01	32
Cyclohexane-d_{12}	6/81	1.383	26.40	19
Carbon tetrachloride		—	95.93	
Deutrium oxide	3.8/101.4	4.755	—	
Dimethyl formamide-d_7	−61/153	2.744	30.11	21
		2.915	35.21	21
		8.022	162.70	21
Dimethyl sulfoxide-d_6	18/189	2.500	39.50	21
p-Dioxane-d_8		3.529	66.50	22
Ethyl alcohol-d_6	−130/79	1.105	17.23	19
		3.563	56.85	22
Methyl alcohol-d_4	−98/65	3.305	49.10	21.5
Nitrobenzene-d_5		o 7.511	o 123.50	26
		m 7.678	m 134.80	25
		p 8.121	p 129.46	24.5
Nitromethane	−29/101	4.335	62.85	22
Pyridine-d_5	−42/116	α 8.727	α 149.89	27.5
		β 7.207	β 123.55	25
		γ 7.580	γ 135.51	24.5
Tetrahydrofuran-d_8	−109/66	1.724	25.35	20.5
		3.575	67.40	22
Toluene-d_8	−95/111	2.088	20.43	19
		o 7.095	o 128.90	23
		m 6.975	m 128.00	24
		p 7.011	p 125.15	24
			C_1 137.50	
2,2,2-Trifluororethyl alcohol-d_3	−44/75	CH_2 3.900	CH_2 61.50	22
				C–F 36
		CF_3 5.135	CF_3 126.15	C–F 277

using an activated molecular sheave. And it is very important to be careful not to allow impurities to mix with the solvent when handling it. Because deuterated chloroform frequently decomposes to produce a small amount of hydrochloric acid, it is not appropriate for a sample that is unstable under acidic conditions. Trifluoroacetic acid (TFA), which is used as a solvent of polyester, is reacted with a hydroxyl group in the polymer to produce esterification, and TFA sometimes decomposes some kinds of aliphatic polyesters to monomers. Therefore, when using TFAs, exercise caution. Instead of TFA, hexafluoroisopropanol (HFIP) is a good solvent for polyesters, polyamides, and poly(vinyl alcohol). If deuterated dimethylsulfoxide (DMSO–d_6) is used, an active proton signal of NH or OH groups in polyamides, polyurethanes, or poly(vinyl alcohol) can be observed.

In the case of polyolefines such as polyethylene or polypropylene, the mixture of *o*-dichlorobenzene (or trichlorobenzene) and deuterated benzene (4/1 v/v) can be used, and the dissolution of the polymer should be performed at a condition of high temperature (more than 100 °C). Solvents purchased in a lot may be contaminated by water in the air or by repetitive preparation techniques during a long period of use. To avoid this, it is useful to subdivide the solvent into small ampoules immediately after opening the package. These small ampoules should be kept in a desiccator or a refrigerator.

4.6.2.3 Reference Substances

The standard position of the chemical shift for ^{1}H or ^{13}C nuclei is determined to be zero ppm for a proton or carbon peak of tetramethylsilane (TMS). Therefore, TMS as a reference substance is added to the solvent of the sample. However, because TMS is volatile (boiling temperature $T_b = 27\,°C$), hexamethyldisilane (HMDS, $T_b = 100\,°C$) or octamethyl-cyclotetrasiloxane (OTS, $T_b = 175\,°C$) is frequently used during the high-temperature measurement. In the case of using heavy water as a solvent, sodium trimethylsilylpropionate (TSP) or sodium dimethylsilylpentane sulfonate (DSS) is suitable. When no reference substance is added, the solvent signal is used as the standard, and chemical shift values are converted to the value from TMS. Because the chemical shifts of the solvent's signals are varied owing to the measurement condition, attention should be paid to the temperature or polymer concentration. When a reference substance is not dissolved in the solvent, the reference substance put in a capillary tube is placed at the center of the sample.

4.6.2.4 Sample Preparations

A polymer material contains varying amounts of several kinds of contamination such as impurities and dust. To get a spectrum of high quality, it is very important to avoid this contamination from the polymer solution by a glass filter. A very simple method of filtration is to use absorbent cotton, which is filled in the Pasteur pipette. When dissolving a crystalline polymer, one must spend enough time to get a uniform solution by using a block heater. To enhance separation of peaks for ^{1}H observation, it is necessary to reduce the viscosity of polymer solution as much as possible. However, in ^{13}C observation, to increase the sensitivity as much sample as possible should be used. To obtain the best spectrum, therefore, it is necessary to adjust the sample amount according the observed nucleus. The best SNR and resolution of the spectrum can be obtained at a proper height of the solvent, which depends on an NMR machine or a sample tube. For measurements at high temperatures over the boiling point of the solvent, observe relaxation times precisely, for it is necessary to remove oxygen gas dissolved in the solution. Usually, it is removed by repeated freezing, melting, and evacuation using a vacuum system. A simplified method of substituting inert gas (N_2, He, etc.) for oxygen is also frequently used.

4.6.3 SYSTEM ADJUSTMENTS

4.6.3.1 Probe Tuning

Initially, it is very important to adjust probe tuning to obtain a sufficiently good spectrum. The probe tuning should be adjusted to increase the irradiation efficiency of rf pulses to the samples. As a normal practice, adjust the probe tuning whenever the probe is changed. And even if the same probe is used, readjust the probe tuning when the solvent, amount of solvent, or measurement temperature is changed. If the probe condition is out of tune, an increased pulse width and a decrease of the resolution and of the SNR of a spectrum would occur. Especially for ^{13}C nuclei having low sensitivity, adjustment of the probe tuning should be performed carefully.

4.6.3.2 Resolution Adjustments

High-resolution NMR measurement requires magnetic field homogeneity of the order of 10^{-10}, which cannot be obtained by the superconducting magnet itself. To obtain such a high homogeneity, coils for correcting distortion of the magnetic field, called shim coils, are used. The process of correcting deviation of the magnetic field from its ideal homogeneous state by changing the electric current running through the shim coils is called shimming. Generally, because the high-order shim adjustments have been made at the time of installation, it is necessary to adjust only axial ($Z1$, $Z2$, $Z3$, $Z4$) and radial shim (X, Y) as a lower-order shim. Whether the shim condition are good or not can be judged by observing the intensity of the lock signal. The stronger the lock signal is, the higher the homogeneity of the resulting magnetic field. When adjusting the axial shims, always spin the sample tube. First, optimize $Z1$ and $Z2$ to maximize the lock signal. Then, adjust the radial shim after the spinning of the tube stops. The same procedure for adjusting the shim should be performed repeatedly until the sensitivity of the lock signals is maximized. Instead of the lock signal, the FID signal, which has only a singlet and a narrow natural line width, can be used for adjusting the shims. Adjust the shims to trail the FID signal, as long as possible.

4.6.4 MEASUREMENT CONDITIONS

4.6.4.1 Measurement Temperature

Generally, the NMR signals of a polymer sample are relatively broad compared with those of a low-molecular compound. This phenomenon is due to the low mobility of a local molecular motion in the polymer. When measuring a polymer, it is sometimes necessary to increase temperature properly to enhance the separation of peaks and to improve the SNR in a spectrum. It is possible to increase temperature up to the boiling point of the solvent used. If the temperature is too high, a part of the polymer, the solvent, or both may be sometimes decompose and change the color of the solution; therefore, pay special attention to this consideration. During low-temperature measurement, the sample tube may be broken by the expansion of the frozen solution. Be careful to keep the tube above the freezing point of the solvent.

4.6.4.2 Spectral Width

Because the observation range of signals depends on the species of the target polymer, the observation range of the known sample is generally set so that the target peaks enter 70% of the whole observation range. In the case of unknown samples, the observation range is set wider than that of a typical sample. For example, when ^{1}H nuclei are observed, the range should be set to 20 ppm (10 K for 500 MHz); for ^{13}C nuclei, the range is set to 250 ppm. After the measurement is performed simply to check the range of each signal, reset the spectral width. When doing so, it is important to check the observing frequency (carrier frequency). The carrier frequency should be set at the center of the whole signals, and then the spectral width (observation range) can be set. If the carrier frequency is shifted from the center or a spectral width is too narrow, ghost peaks may appear owing to the many foldings of peaks from outside the spectral width.

The NMR resolution depends on the spectral width and data points. If the spectral width is narrow and the number of data points is large, the resolution is enhanced. However, when the spectral width is too narrow, the peaks observed at both ends of the observed range decrease owing to the wide-range filter. In 2-D measurements, because the spectral width or data points need to be restricted given the memory limit of the computer, it is necessary to set the spectral width as narrow as possible and the data points as low as possible. In such a case, the measurement is not performed quantitatively.

4.6.4.3 Data Points

The term *data points* denotes the number of sampling points. Therefore, with an increase of data points, the spectral resolution is enhanced and a precise spectrum can be obtained. The digital resolution is obtained from a spectral width (*SW*) and data points as follows:

$$\text{digital resolution (Hz)} = [2\text{x}SW(\text{Hz})/(\text{data points})]$$

In ^{1}H–NMR measurement, data points should be relatively large to yield a high resolution. Because more than 32 K is set to data points and 10 K to spectral width, the digital resolution is finally 0.6 Hz, which is enough to obtain a good spectrum. On the other hand, as the data points increase, the SNR is decreased because a noise component after the FID signals is incorporated. Much accumulation should be performed to enhance the SNR. Therefore, after the content of a sample is fully examined, data points should be carefully set.

4.6.4.4 Pulse Width

At first, an rf pulse is applied to the sample coil to generate the magnetization of the sample. The rf pulse is called a pulse width. Usually, the $\pi/2$ pulse, as the excitation pulse, is typically at 5–10 μs. The pulse width enables excitation of a certain frequency range. Therefore, when the output from the oscillator deteriorates, the range of excitation is narrowed. It is sometimes necessary to check the width of the $\pi/2$ pulse. Because a $\pi/2$ pulse is selected as an excitation pulse, the SNR of a spectrum obtained by a single scan is at a maximum. When executing accumulation using the $\pi/2$ pulse, the repetition time t between pulses should be sufficiently long ($t > 5T_1$). Because the accumulation is performed after each magnetization is thoroughly recovered to its original position, we can get a spectrum with the same SNR as when a single-scan spectrum is simply added. In this case, however, total measurement time is lengthened unnecessarily. In practice, the value selected for a pulse width should be somewhat less than $\pi/2$. When using this pulse, the SNR in the spectrum per one pulse decreases. On the other hand, the time required for the recovery of T_1 shortens, and thus the accumulation with a short repetition time can be performed. Finally, a spectrum with a better SNR can be obtained in the same measurement time.

For ^{13}C observation, the repetition time should be short and the number of accumulations should be increased. In the case of a polymer sample, it is better to select a 45° or 30° pulse as an excited pulse and 1–2 s as the repetition time. However, if a quantitative measurement is performed, the repetition time should be set longer. The precise value of the $\pi/2$ pulse depends on the observation nuclei, target sample, and solvent used. In Particular, the value should be determined for T_1 measurement, INEPT, DEPT, 2-D NMR measurements, and so on.

4.6.5 DATA PROCESSING

Data obtained by an FT–NMR spectrometer are represented by a digitized signal called FID. Generally, FID has the maximum intensity at $t = 0$, and the oscillations gradually decay with time.

Although a spectrum can be obtained simply through a Fourier transformation of FID, a much better spectrum can be obtained by executing Fourier transformation after some appropriate processing of FID data.

A flow of the 1-D data processing is shown below.

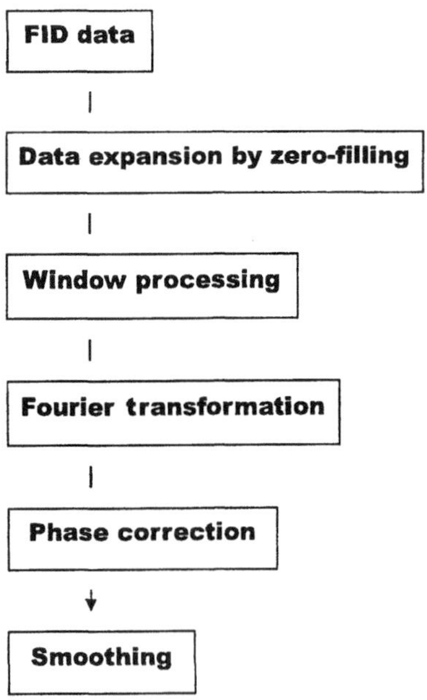

4.6.5.1 Zero-Filling

When inputting an FID signal into a computer, a sufficient number of data points cannot be used because of the restricted measurement time or memory capacity. If the number of data points of the spectrum is sufficient, peaks that were originally separated may look as if they were only a single peak. In such cases, apparent resolution can be enhanced by expanding the number of data points by means of zero-filling.

The spectrum was obtained by Fourier transformation the FID after expansion its time-axis data points four times by means of zero-filling. When data points on the time axis are expanded by zero-filling, the data points of the spectrum are interpolated and thus the apparent resolution is enhanced.

4.6.5.2 Window Data Processing

Time-domain window processing entails multiplying the FID by appropriate functions before executing Fourier transformation. The functions that the FID is multiplied by are called *window functions*. When window processing is performed, the appearance of the spectrum changes. By selecting appropriate window functions according to purpose, analysis of the spectrum can be facilitated. Typical examples of window-processing purposes and functions prepared for them are presented in the following paragraphs.

4.6.5.3 Elimination of Ripples

According to the measurement conditions, sometimes, the data acquisition time becomes shorter than the FID signal decay time. If such an FID signal is Fourier-transformed as it is, ripples are generated at the skirt of a peak. If one applies an appropriate window function, these ripples can be eliminated. To eliminate ripples, a function smoothing the end point of the FID is used. The function usually used for this purpose is a "trapezoidal function." By applying a trapezoidal function to the FID, one can eliminate the ripples; however, the line width of a peak will be broadened.

4.6.5.4 Enhancement of SNR

A window function with the same shape as that of the decay curve of the FID is called a matched filter function. When the matched filter function is used as the window function, the SNR of the spectrum is maximized. However, the line width of peaks is broadened, and resolution of the spectrum deteriorates. In many cases, the decay curve of FID is expressed by

$$F(t) = \exp(-\pi BF\, t) \tag{63}$$

In this case, selecting an exponential function expressed by Eq. (63) and setting an appropriate *BF* (broadening factor) maximizes the SNR of the spectrum. In the present case, the window function in which $BF = 1/(\pi T_2)$ is the matched filter function. The process of practically selecting the *BF* to maximize SNR is as follows: At first, perform Fourier transformation the FID without multiplying it by any window function. Expand the vicinity of a peak that can be used as a standard. Use the half-height width of that peak as a broadening factor. In practice, it is better to select a broadening factor smaller than that mentioned earlier because applying such a window function causes the resolution of the spectrum to deteriorate.

4.6.5.5 Enhancement of Resolution

By selecting the exponential window function and using a negative value for *BF*, one can enhance the resolution of the spectrum. However, because this window function emphasizes noises in the tail of the FID signal, the SNR of the spectrum deteriorates markedly. However, by using a function that makes the tail of the FID smoothly approach zero, the resolution of the spectrum can be enhanced without sacrificing the SNR so much.

The window function frequently used to enhance the resolution of the spectrum is the Lorentz–Gauss transformation function expressed by Eq. (64)

$$f(t) = \exp(\pi BF\, t)x\, \exp(-(\pi G t)^2/\log 2) \tag{64}$$

When $BF = 1/(\pi T_2)$, the line shape of a peak is transformed from a Lorentzian to Gaussian having a half-height width of G. The skirt of a Gaussian peak is narrower than that of a Lorentzian peak, and separation of peaks is better by that degree in the former. Also, by adjusting the Gaussian coefficient G, the resolution of the spectrum can be enhanced. However, if the resolution is enhanced over the natural line width, the SNR dramatically deteriorates.

4.6.5.6 Phase Correction

After Fourier transformation of the FID as it is, an operation called phase correction is necessary. If an FID undergoes Fourier transformation, a peak shape, is obtained.

The peak shape of the real part is called the absorption waveform and that of the imaginary part the dispersion waveform. The real part in the out-phase is a mixture of the absorption and dispersion components. This is because the phase of the observed peak is shifted from the phase of the actual peak. The shift of the phase depends on the measurement conditions, conditions of the spectrometer, and the difference of the frequency from the center frequency. The operation to correct this shift and to obtain a peak shape, is called phase correction. The phase shift of NMR data can be performed by the phase correction P_0 and P_1. P_0 is the zeroth-order phase, which operates all the peaks equally, and P_1 is the first-order phase, which corrects the remaining phase shift of the first-order formula of resonance frequency. Usually, we have to adjust P_0 to correct the phase of the maximum peak and then adjust P_1 to correct the phase of the total spectrum. After enlarging the spectrum as much as possible, we should readjust P_0, P_1, or both slightly to correct the phase of the whole peaks.

4.6.6 *FUNDAMENTAL PRINCIPLES AND ANALYSIS OF 1-D NMR SPECTRA*

4.6.6.1 Proton

In a polymer solution, a dipolar–dipolar interaction or quadrupole interaction, which affects the line width of a peak very greatly disappears owing to the rapid local motion of polymer chain segments. Therefore, a high-resolution NMR spectrum can be obtained in a solution. In the spectrum, only Zeeman interaction due to chemical shift and spin–spin interaction remains.

When isolated nuclei having magnetic moment $\mu z = \gamma h I/2\pi$ is placed in a static magnetic field H_0, their energy is given by

$$E = -\gamma h I H_0/2\pi \tag{65}$$

where γ is called the magnetogyric ratio, h is Planck's constant, and I is the spin quantum number. When Iz is the component of the nuclei spin I in the H_0 direction, Eq. (65) is rewritten as

$$E = -\gamma h I z H_0/2\pi \tag{66}$$

The Hamiltonian, which shows total energy of all spin systems, is defined as follows:

$$H = -h/2\pi \Sigma \gamma_i I z(i) H_0(1 - \sigma_i) + h \Sigma J_{ij} I(i) I(j) \tag{67}$$

where γ_i is the magnetrogyric ratio of i nuclei, σ_i is a shielding coefficient of i nuclei, and J_{ij} is a coupling constant between nuclei i and j. The eigenvalue of the

Hamiltonian is calculated, and then the energy levels can be determined. When all nuclei have spin $I = 1/2$, the quantum number may have only the value $+1/2$ and $-1/2$, and thus the system is either in the state α or β. The spin functions for the individual nuclei are assumed to be independent of one another; therefore, the spin functions for n nuclei with individual spins α or β can be written as

$$|s\rangle = \alpha(1)\alpha(2)\beta(3)\cdots\beta(N) \tag{68}$$

For $I = 1/2$, there are 2^N product functions. These basic product functions are the eigenfunctions of the first term (Zeeman interaction) in Eq. (68) and are not the eigenfunction of the second term (J interaction). Therefore, we can calculate the matrix elements in Eq. (68) using the basic product functions. The diagonal matrix can be calculated from the first term, and the off-diagonal matrix with the nonzero off-diagonal element can be obtained from the second term. The eigenvalues E can also be calculated by solving a secular equation composed of $2^N \times 2^N$ determinants. The secular equation can be written as

$$|H_{ss'} - E\delta'| = 0 \tag{69}$$

where $H_{ss'}$ is the ss' matrix element of the Hamiltonian and δ' is the Kronecker delta. The eigenvalues of total Iz are

$$Fz = \Sigma Iz(i) \tag{70}$$

The determinant can be decomposed into several product determinants of lower order. As H in Eq. (67) commutes with Fz, the matrix elements of $H_{ss'}$ are zero between the different conditions of the eigenvalues of Fz. Because the number of the different eigenvalues of Fz is $(N + 1)$, the determinants in Eq. (69) are the products of $(N + 1)$ submatrices of lower order.

In an AB system ($N = 2$), the four basic product functions are used as a basis of wave functions for the molecule. The Hamiltonian matrix decomposes into three submatrices whose dimensions are $1 \times 1, 2 \times 2$, and 1×1 corresponding to $Fz = 1$, 0, and -1. The matrix element for $Fz = 1(|\alpha\alpha\rangle)$ is

$$\begin{aligned}\langle\alpha\alpha|H|\alpha\alpha\rangle &= -\gamma H_0(1/2)(2 - \sigma_A - \sigma_B) + hJ/4 \\ &= -(h/2)(\nu_A + \nu_B) + hJ/4\end{aligned} \tag{71}$$

where $\nu_A = \gamma H_0(1 - \sigma_A)/2\pi$, $\nu_B = \gamma H_0(1 - \sigma_B)/2\pi$.

For $Fz = -1(|\beta\beta\rangle)$,

$$\begin{aligned}\langle\beta\beta|H|\beta\beta\rangle &= \cdot\gamma H_0(1/2)(2 - \sigma_A - \sigma_B) + hJ/4 \\ &= (h/2)(\nu_A + \nu_B) + hJ/4\end{aligned} \tag{72}$$

and for $Fz = 0(|\alpha\beta\rangle$ and $|\beta\alpha\rangle)$, the following four matrix elements are obtained:

$$
\begin{aligned}
\langle\alpha\beta|H|\alpha\beta\rangle &= -(h/2)(\nu_A - \nu_B) - hJ/4 \\
\langle\beta\alpha|H|\beta\alpha\rangle &= (h/2)(\nu_A - \nu_B) - hJ/4 \\
\langle\alpha\beta|H|\beta\alpha\rangle &= \langle\beta\alpha|H|\alpha\beta\rangle = hJ/2
\end{aligned}
\tag{73}
$$

Solutions obtained from the 1×1 matrices are

$$
\begin{aligned}
E_1 &= -(h/2)(\nu_A + \nu_B) + hJ/4 \\
E_4 &= (h/2)(\nu_A + \nu_B) + hJ/4
\end{aligned}
\tag{74}
$$

and solutions of the 2 × 2 matrix give the two roots

$$
\begin{aligned}
E_2 &= (h/2)[\{(\nu_A - \nu_B)^2 + J^2\}^{1/2} - J/2] \\
E_3 &= -(h/2)[\{(\nu_A - \nu_B)^2 + J^2\}^{1/2} + J/2]
\end{aligned}
\tag{75}
$$

Although $|1\rangle$ and $|4\rangle$ stationary state correspond to $\alpha\alpha\rangle$ and $\beta\beta\rangle$, respectively, the $|2\rangle$ and $|3\rangle$ state are mixed with $|\alpha\beta\rangle$ and $|\beta\alpha\rangle$. Namely,

$$
\begin{aligned}
|2\rangle &= (\cos\theta|\alpha\beta\rangle + \sin\theta|\beta\alpha\rangle) \\
|3\rangle &= (-\sin\theta|\alpha\beta\rangle + \cos\theta|\beta\alpha\rangle)
\end{aligned}
\tag{76}
$$

where $C\cos 2\theta = (\nu_B - \nu_A)/2$, $C\sin 2\theta = J/2$ and $C = [(\nu_A - \nu_B)^2 + J^2]^{1/2}/2$.

The energy level is shown in Figure 4.12. The intensity of a transition between states p and q is proportional to the square, as follows:

$$|\langle p|Fx|q\rangle|^2 \tag{77}$$

where p and q are the stationary state wavefunctions and Fx is the total spin of the x-component.

$$Fx = \Sigma Ix(i) \tag{78}$$

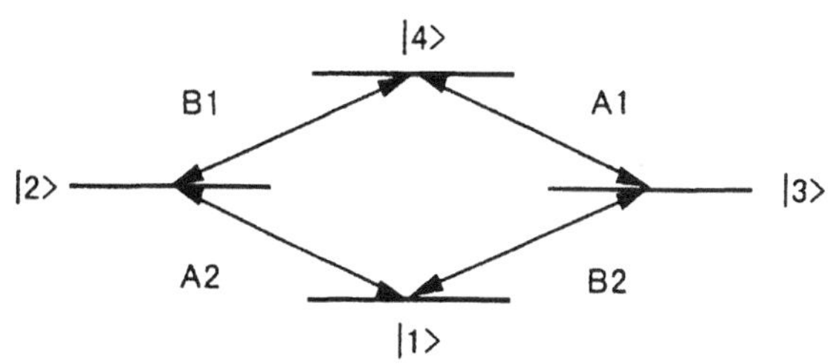

Figure 4.12 Energy levels of AB system.

Table 4.3 Transition energies and relative intensities for the AB system.

Transition	*Transition Energy (Hz)*	*Relative Intensity*
$1 \longleftrightarrow 2$(A2)	$\{(\nu_A - \nu_B)^2 + J^2\}^{1/2}/2 - J/2 + (\nu_A - \nu_B)/2$	$1 + J/\{(\nu_A - \nu_B)^2 + J^2\}^{1/2}$
$1 \longleftrightarrow 3$(B2)	$-\{(\nu_A - \nu_B)^2 + J^2\}^{1/2}/2 - J/2 + (\nu_A + \nu_B)/2$	$1 - J/\{(\nu_A - \nu_B)^2 + J^2\}^{1/2}$
$2 \longleftrightarrow 4$(B1)	$-\{(\nu_A - \nu_B)^2 + J^2\}^{1/2}/2 + J/2 + (\nu_A + \nu_B)/2$	$1 + J/\{(\nu_A - \nu_B)^2 + J^2\}^{1/2}$
$3 \longleftrightarrow 4$(A1)	$\{(\nu_A - \nu_B)^2 + J^2\}^{1/2}/2 + J/2 + (\nu_A + \nu_B)/2$	$1 - J/\{(\nu_A - \nu_B)^2 + J^2\}^{1/2}$

The matrix elements of $Ix(i)$ are zero unless α and β differ only by -1 in the values with $\Delta Fz = -1$; thus, transitions take place only between energy levels with $\Delta Fz = -1$. The selection rules for transitions between the levels of α and β are defined by

$$\Delta Fz = \pm 1 \tag{79}$$

From the rules, only the transitions between 1⇔2, 1⇔3, 2⇔4, and 3⇔4 are allowed. The transition energies and relative intensities for the AB system are shown in Table 4.3. Theoretical spectra for the AB spin system are illustrated in Figure 4.13 for several values of $J/\nu_0\delta$. When the coupling constant J is zero, two absorption bands of equal intensity are observed. For very small values of $J(J/\nu_0\delta = 0.02)$, two doublets are observed, each doublet having components of equal intensity; this type of spectrum is described as AX. As the coupling constant is increased with respect to the chemical shift difference, two asymmetric doublets characteristic of an AB system are observed. The central components increase in intensity at the expense of the outer components as $J/\nu_0\delta$ increases. At very large values of this ratio, the central bands coalesce and the outer peaks disappear.

The spin systems, more so than the AB system, could be analyzed in principle by solving the secular equations of higher order in the same manner. For further details, refer to [44–46]. The general appearance of the spectrum can be estimated if the chemical structure of a molecule can be determined. From the experimental rules so far, the methyl protons appear at about 1 ppm (with TMS as a standard reference), and aromatic protons also appear near 7 ppm. The J coupling between nonequivalent protons on the same carbon, which is called geminal coupling, has constants generally in the range of -12 to -16 Hz, and the coupling constants for CH–CH separated by three chemical bonds, $^3J_{HH}$, exhibit a range of 5–8 Hz. Vicinal couplings are markedly dependent on geometry. Karplus [10, 47] has given a valence-bond interpretation and concludes that vicinal coupling of protons on

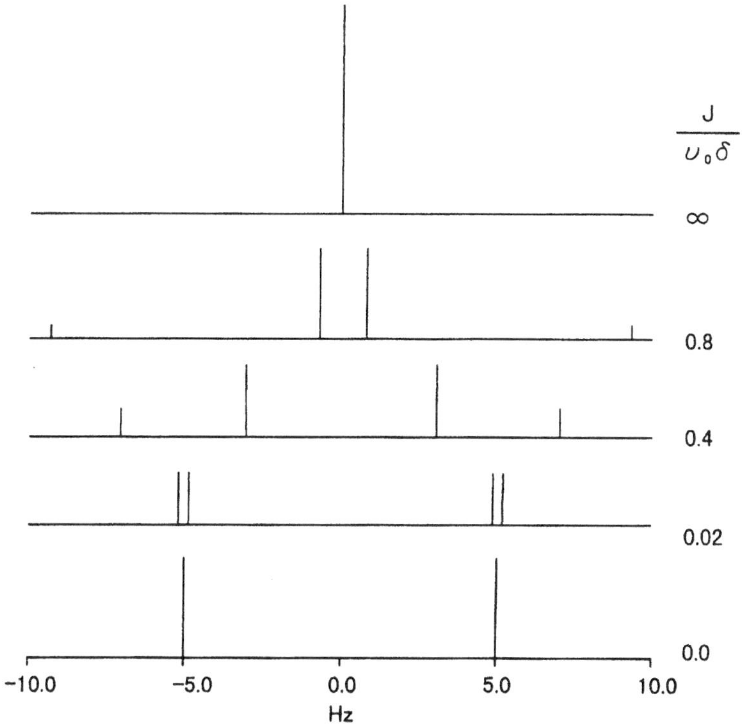

Figure 4.13 Theoretical AB-type spectra for two nuclei of spin 1/2 having $\nu_0\delta = 10\,\text{Hz}$.

adjacent sp^3 carbon atoms should depend on ϕ, the dihedral angle, as follows:

$$^3J_{HH} = A\cos^2\phi + 0.28, \qquad A = 8.5(0^\circ \le \phi \le 90^\circ), \quad 9.5(90^\circ < \phi \le 180^\circ) \tag{80}$$

A plot of these functions shows that J should be 1.8 Hz when $\phi = 60^\circ$ and 9.2 Hz when $\phi = 180^\circ$, as shown in Figure 4.14. Proton–proton couplings through more than four bonds, which is called long-range coupling, are nearly zero. However, the $^4J_{HH}$ coupling constant separated by four bonds, which has a stereo structure resembling an uppercase W, is also about 1 Hz. This is the so-called W rule [48].

4.6.6.2 Double Resonance

The purpose of a double resonance experiment is to simplify a complex spectrum by removing the spin–spin coupling between nuclei or groups of nuclei. Peaks that are spin–spin coupled with the irradiated peak are decoupled. The splitting of

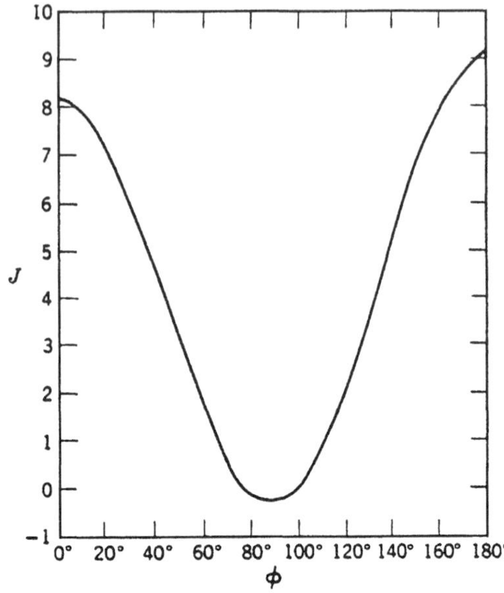

Figure 4.14 The Karplus function expressing the magnitude of vicinal proton–proton J coupling as a function of dihedral angle θ in the H—C—C—H bond system.

peaks caused by the spin–spin coupling with the irradiated peak is eliminated by decoupling. With this, the peaks that are spin–spin coupled with each other can be identified.

4.6.6.3 Differential NOE Measurement

When two different protons are presented within about 5 Å spatially, even if they are not bonded through a carbon atom directly, the proton peak that is spin coupled with the irradiated peak is enhanced by decoupling. This phenomenon is called the nuclear overhauser effect (NOE) [49]. The NOE does not arise from J-coupling but rather from direct dipole–dipole interaction. Therefore, the NOE depends on the internuclear distance. For protons, it can be shown that maximum enhancement is 50%.

4.6.6.4 ^{13}C–NMR

^{13}C–NMR is a powerful method for analyzing the fine structure of polymers [50, 51]. The ^{13}C resonances of organic compounds are found over a chemical shift range of 300 ppm compared with under 20 ppm for protons. Moreover, the lines of

^{13}C nuclei are narrower than those of ^1H nuclei. Thus, it is not difficult in ^{13}C–NMR to be able to identify individual resonances for each carbon in an organic compound. Additional advantages of ^{13}C–NMR relate to the direct observation of molecular backbones, the direct observation of carbon-containing functional groups with no attached proton (e.g., C=O and CN), and the direct observation of carbon reaction sites of interest. In the ^{13}C-experiment, however, much measurement time will be required to obtain a spectrum signal with the same SNR as that of ^1H nuclei, and it is necessary to pay attention to the quantitative analysis.

Listed below are a variety of ^{13}C–NMR measurement methods.

1. Broad-band decoupling [52] (Figure 4.15(a))
   Because the ^{13}C natural abundance is low enough to make ^{13}C–^{13}C spin–spin coupling interaction unlikely, the ^{13}C-spectrum peaks are split owing to the ^{13}C–^1H J_{CH} coupling. The methyl groups (CH_3) are split to a quartet signal, the methylene groups (CH_2) to a triplet signal, the methine groups (CH) to a doublet signal, and the quaternary carbon having no directly bonded proton to a singlet. Therefore, we can assign nonprotonated carbons CH, CH_2, and CH_3 carbons observed as singlets, doublets, triplet, and quartets, respectively. A ^{13}C–NMR spectrum is so complicated because of the spilttings that proton decoupling is usually performed to obtain a simple spectrum without the splittings caused by the J_{CH} coupling.
2. NOE-Eliminated ^1H complete decoupling [53] (Figure 4.15(b))
   This measurement allows the number of carbons to be accurately measured. The pulse repetition time should be set to a value that is 5 to 10 times as large as the ^{13}C's longest relaxation time T among the peaks to be measured. If proper conditions are selected, the intensity ratio of each ^{13}C signal is proportional to the number of carbons.
3. J_{CH} Measurement using NOE [53] (Figure 4.15(c))
   The spin–spin coupling constant J_{CH} between the carbon and proton is measured in this mode. The J_{CH} is restored and measured by stopping irradiation at the ^1H channel during data acquisition. Irradiating the ^1H channel during the pulse waiting time gives rise to NOE, resulting in an improved SNR. Information on the direct coupling with proton J_{CH} and the long-rang coupling constant $^nJ_{CH}$ can be obtained by comparing the NOE spectrum with a completely decoupled ^{13}C spectrum.
4. Selective decoupling (Figure 4.15(d))
   The signals from ^{13}C nuclei directly bonded with a specific ^1H nucleus are identified by this method. A signal from ^{13}C nuclei that are

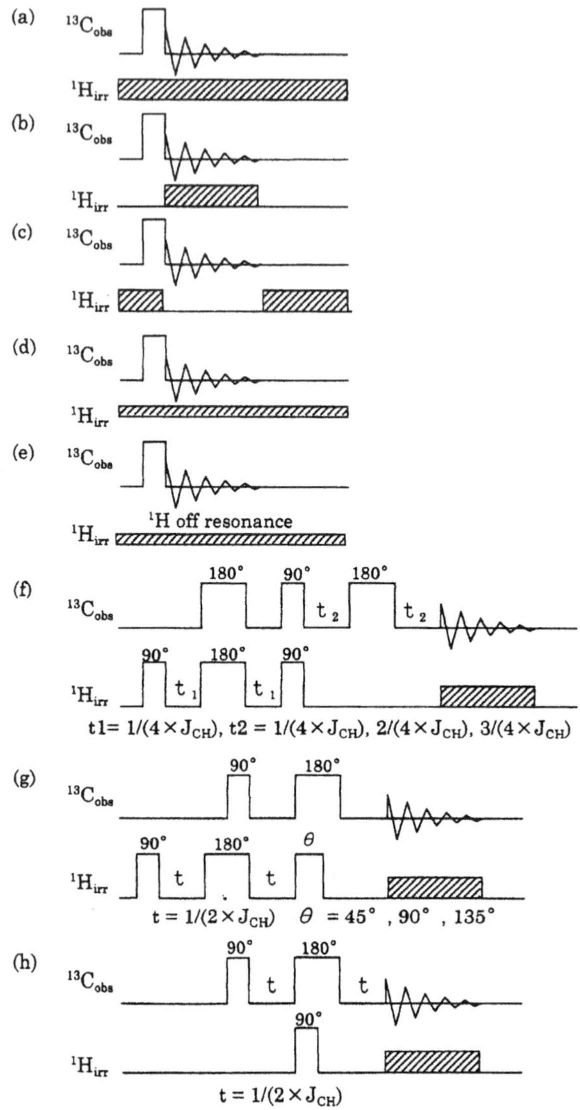

(a) Noise decoupling
(b) NOE eliminated decoupling
(c) J_{CH} measurement using NOE
(d) Selective decoupling
(e) Off resonance
(f) INEPT
(g) DEPT
(h) quaternary carbon measurement

Figure 4.15 Various measurement modes in a ^{13}C experiment.

directly bonded with a selectively decoupled ^1H nucleus appears as a singlet.

5. Off-resonance (Figure 4.15(e))

   This measurement mode allows the number of protons directly bonded with each carbon to be measured by performing ^1H incomplete decoupling. The methyl group gives rise to a quartet signal, the methylene group to a triplet, the methine group to a doublet, and the quaternary carbon to a singlet.

6. INEPT: Insensitive nuclei enhanced by polarization transfer [54] (Figure 4.15(f))

   The number of protons directly bonded with each carbon is measured. This method, compared with the off-resonance method, greatly reduces measurement time owing to high sensitivity by population transfer from ^1H to ^{13}C and as a function of pulse repetition with relaxation time T_1 of ^1H. Patterns with respect to the multiplicity of resonance signals and features of this measurement method are given below.

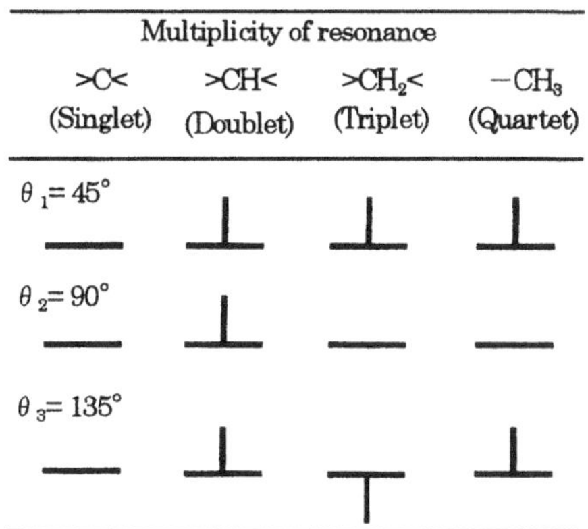

7. DEPT: Distortionless enhancement by polarization transfer [55] (Figure 4.15(g))

   The number of protons directly bonded with each carbon is measured. This method is also capable of compiling spectra effectively using addition and subtraction, for the phase deviation is less than that in the INEPT mode; however, signals of quaternary carbon cannot be observed in this mode.

8. Quaternary carbon measurement (Figure 4.15(h))

   This method allows only quaternary carbon to be detected. Namely, it allows signals from ^{13}C nuclei directly bonded with ^1H to be eliminated by using spin–spin coupling with ^1H.

4.6.6.5 2-D NMR Measurements [14, 56, 57]

A 1-D NMR spectrum can usually be obtained from Fourier transformation of the 1-D FID data. In 2-D measurements, an additional axis is introduced. This axis is called the evolution axis. The evolution axis is defined by the time intervals (evolution time) between particular pulses in the pulse sequence. In a single scan of accumulation, data at a point on the evolution axis are measured. Every time the evolution time is changed, data at another point on the evolution axis are measured. Two-dimensional FID data are a series of FIDs measured by changing the evolution time with a fixed increment and arranged in the order of measurements. A general method of time division for obtaining 2-D FIDs is as follows:

Preparation	Evolution	Mixing	Detection
	t_1	t_m	t_2

The nuclear spins are permitted to equilibrate with their surroundings via spin–lattice relaxation during the preparation period. After a 90° pulse is applied, the transverse magnetization is evolved during t_1. Interactions are given between the specific components of the evolved transverse magnetization. The design of the mixing time represents features of many 2-D NMR methods. The FID acquired in the detection period is one whose intensity or phase is modulated during the evolution period t_1.

On the basis of the obtained results, solution 2-D NMR can roughly be divided into the following classifications:

4.6.6.5.1 Shift Correlation Spectroscopy

Both the ω_1 and ω_2 axes represent the chemical shift. A cross peak appearing in the 2-D spectrum indicates that there is a correlation between the two nuclei assigned to this peak. The meaning of the correlation differs according to the design of the mixing period.

1. COSY: Correlation spectroscopy [58, 59] (pulse sequence shown in Figure 4.16(a))

   Correlation signals between directly J-coupled peaks are observed. In this way, the connectivity between ^{1}H nuclei can be clarified. By adjusting parameters, small long-range coupling can also be observed.
2. NOESY: nuclear Overhauser effect spectroscopy [60] (Figure 4.16(b))

   Correlation signals due to NOE and chemical exchange are observed. This method is useful for three-dimensional molecular structural analysis

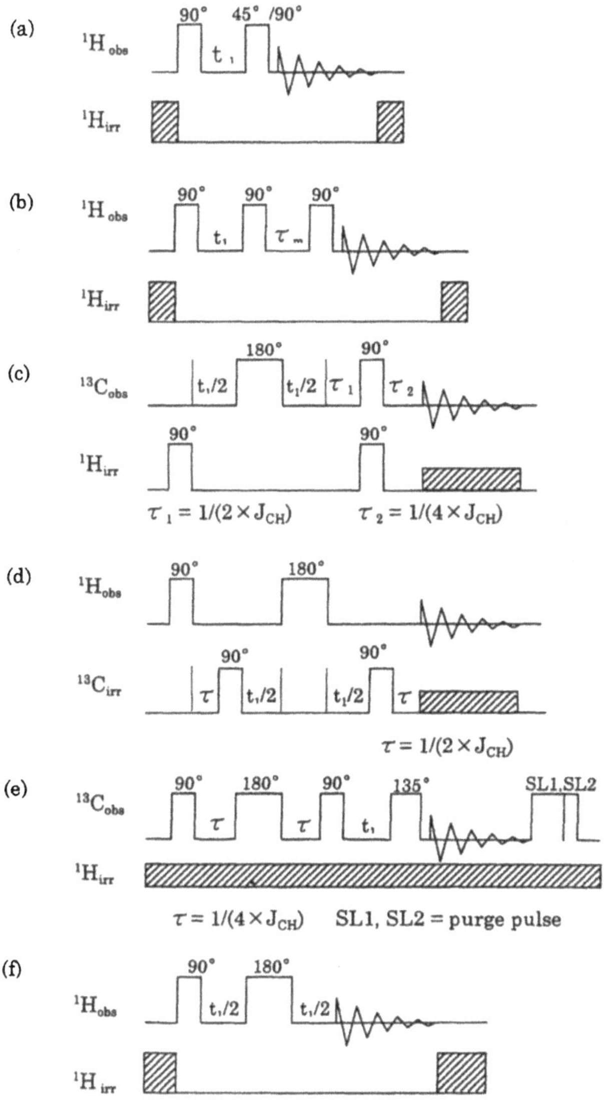

(a) COSY
(b) NOESY
(c) C—H COSY
(d) HMQC
(e) INADEQUATE
(f) J-resolved

Figure 4.16 Various types of pulse sequences using a 2-D experiment.

because nuclei existing spatially close to each other can be identified by observing NOE.

3. C–H COSY: ^{13}C-detected heteronuclear shift correlation [61] (Figure 4.16(c))
   Correlation signals between directly bonded ^{1}H and ^{13}C nuclei are observed.
4. HMQC: ^{1}H-detected heteronuclear shift correlation [62] (Figure 4.16(d))
   Correlation signals between directly coupled ^{1}H and ^{13}C nuclei are observed. The SNR is enhanced by observing ^{1}H compared with the case of the ^{13}C-detected hetronuclear shift correlation method.

4.6.6.5.2 Multiple Quantum Coherence Correlation Spectroscopy

The ω_1 axis represents multiquantum coherence. Multiquantum coherence is observed as a modulation component given to observable single-quantum coherence.

1. INADEQUATE: ^{13}C-detected double quantum coherence shift correlation [63] (Figure 4.16(e))
   Connectivities between ^{13}C nuclei are observed. Although measurement is difficult to perform because the sensitivity is low, once the data are obtained, information on the carbon skeleton can be obtained directly.
2. J-resolved spectroscopy [64] (Figure 4.16(f)): This method separates the spin–spin coupling and the chemical shift. The ω_2 axis represents the chemical shift, and the ω_1 axis the spin–spin coupling constant (J-constant). The cross section taken in parallel to the ω_1 axis at the top position of a peak appearing on the ω_2 axis represents the size of the spin–spin coupling affecting that peak.

4.7 Solid-State NMR Method

Line widths of NMR signals in solid polymers are generally broadened more than 20 kHz by several interactions such as homonuclear and heteronuclear dipolar interactions, quadrupolar interaction, and chemical shielding anisotropy. The broadenings can be classified into two groups, namely, inhomogeneous and homogeneous. Figure 4.17 shows schematic representations of the origins of the two broadenings. A variety of nuclei or their pairs that take different configurations or directions

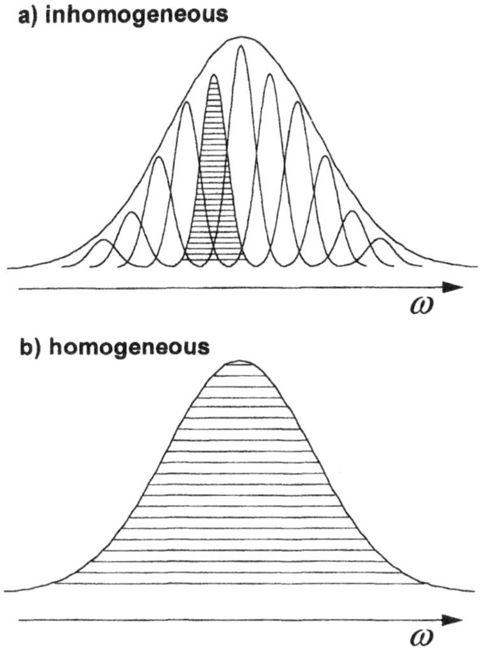

Figure 4.17 Schematic representations of (a) inhomogeneous and (b) homogeneous broadening.

with respect to the magnetic field give a wide distribution of the positions of NMR peaks. This corresponds to the inhomogeneous broadening that can be sharpened or manipulated using appropriate pulse techniques for respective interactions. On the other hand, the homogeneous broadening that originates from a very short T_2 makes it difficult to sharpen line shapes by pulse techniques. This line reacts to all pulses within the line width as an indivisible entity. Most of the efforts for obtaining high-resolution solid-state NMR spectra for solid polymers have been devoted to developing new techniques to manipulate the interactions that give inhomogeneous broadening.

4.7.1 NMR INTERACTIONS BETWEEN NUCLEAR SPINS

The total Hamiltonian H of nuclear spins in solids can be expressed as (3b, 3e, 3f, 3g)

$$H = H_Z + H_D + H_{CS} + H_Q \qquad (81)$$

where H_Z is the Zeeman interaction, H_D is the dipolar interaction, H_{CS} is the chemical shielding anisotropy, and H_Q is quadrupolar interaction. The average magnitudes of H_D, H_{CS}, and H_Q interactions are on the orders of 2×10^4, 10–10^4, and 10^5–10^6 Hz, respectively. In this section, for convenience in understanding the principle of solid-state NMR, the nature of these interactions is described.

4.7.1.1 Zeeman Interaction

The Zeeman interaction H_Z, which is the strongest interaction among those appearing in Eq. (81) is expressed as

$$H_Z = -\gamma \hbar H_0 \mathbf{I} \tag{82}$$

where H_0 the external field and $I = (I_x, I_y, I_z)$ is the spin. This Hamilton has eigenvalues of

$$E_m = -\gamma H_0 m, \qquad m = I, I-1, \ldots, -I \tag{83}$$

which indicates that the energy levels of the nuclear spin are split into $2I + 1$ levels, and energy absorption, that is, nuclear magnetic resonance, will occur between levels under the selection rule of $\Delta m = \pm 1$. The energy difference between all adjacent levels is the same with a unique frequency of $\nu = (\gamma/2\pi)H_0$, which indicates that this interaction gives only one sharp peak in the spectrum. This situation corresponds to an isolated bare nucleus located in a magnetic field. Because the resonance frequency ν is determined by the γ of the nucleus, the resonant frequency is varied depending on its γ. Because NMR spectra are illustrated by units of parts per million (ppm) of H_0 from a certain resonant peak of a standard sample, absolute values of H_Z have no important meaning. The difference in ν caused by the shielding effects of surrounding electrons, i.e., chemical shift, is described.

4.7.1.2 Dipolar Interaction

A nucleus having a nonzero spin number has a dipolar moment μ, and it produces a local magnetic field around the nucleus. The classical interaction energy E between two magnetic moments μ_1 and μ_2 is

$$E = \frac{\mu_1 \cdot \mu_2}{r^3} - \frac{3(\mu_1 \cdot r)(\mu_2 \cdot r)}{r^5} \tag{84}$$

where r is the internuclear distance. This divides a single peak resulting from Zeeman interactions into two split peaks. A Hamiltonian can be formulated by treating μ_i as operators $\mu_i = \gamma h I_i$. The general dipolar Hamiltonian for N spins becomes

$$H_{\rm D} = \frac{1}{2}\sum_{j=1}^{N}\sum_{k=1}^{N}\left[\frac{\mu_j \cdot \mu_k}{\mathbf{r}_{jk}^3} - \frac{3(\mu_j \cdot \mathbf{r}_{jk})(\mu_k \cdot \mathbf{r}_{jk})}{\mathbf{r}_{jk}^5}\right] \tag{85}$$

where $\mathbf{r}_{jk} = \mathbf{r}_j - \mathbf{r}_k$. Equation (85) is sometimes written in tensor notation as

$$H_{\rm D} = I_j \cdot {\rm D} \cdot I_k$$

where D the dipolar coupling tensor. Because D is traceless, this interaction does not affect the average resonant position of the NMR resonance. This Hamiltonian can be transformed into spherical coordinates (r, θ, ϕ), where z is chosen as the polarization axis:

$$H_{\rm D} = \frac{\gamma_i \gamma_k \hbar^2}{r^3}[A + B + C + D + E + F] \tag{86}$$

where

$$A = (1 - 3\cos^2\theta) I_{zj} I_{zk}$$

$$B = -1/4(1 - 3\cos^2\theta)(I_j^+ I_k^- + I_j^- I_k^+)$$

$$C = -3/2 \sin\theta \cos\theta \exp(-i\phi)(I_j^+ I_{zk} + I_{zj} I_k^+)$$

$$D = -3/2 \sin\theta \cos\theta \exp(+i\phi)(I_j^- I_{zk} + I_{zj} I_k^-)$$

$$E = -3/4 \sin^2\theta \exp(-i2\phi)(I_j^+ I_k^+)$$

$$F = -3/4 \sin^2\theta \exp(+i2\phi)(I_j^- I_k^-)$$

$I_+ = I_x + i I_y$ and $I_- = I_x - i I_y$ are raising and lowering operators that increase or decrease the spin state of the nuclei. The term A corresponds to the classical dipolar interaction, and the term B, the socalled flip-flop term, enables nuclei to exchange their spin states simultaneously. Note that the spin-exchange process usually occurs only among homonuclear spins. However, the term B plays an important role in generating cross-polarization between heteronuclei. Other terms are closely related to spin relaxation and the NOE. As shown in the preceding equations, the splitting of the line depends considerably on r and the angle between the external magnetic field and the internuclear vector θ, indicating that the dipolar interaction includes information of orientation and distance of internuclear vectors of a pair of interacting nuclei. When a single crystal sample and its exact crystal structure are available, both can be determined from a successive sample rotation in an NMR probe using a goniometer. In contrast, a randomly oriented or powder

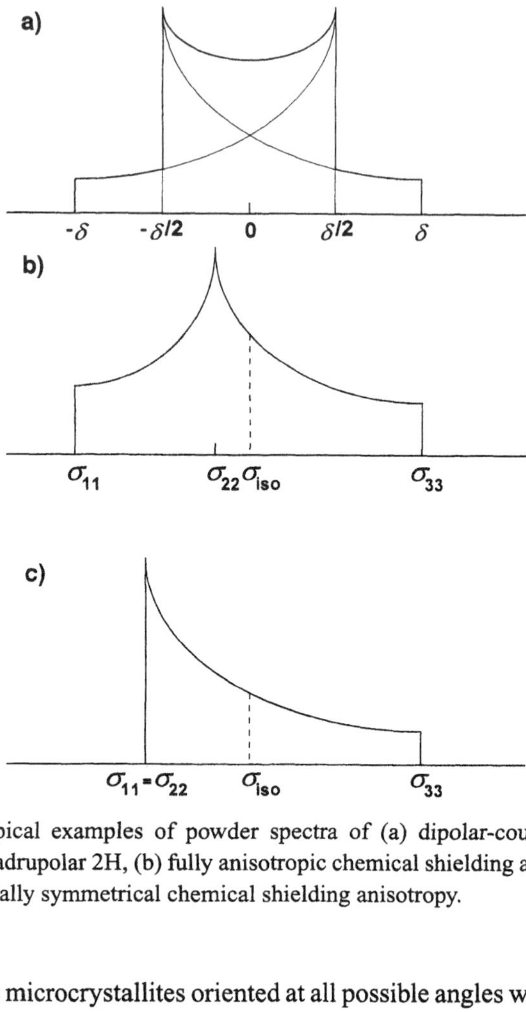

Figure 4.18 Typical examples of powder spectra of (a) dipolar-coupled spin pair or quadrupolar 2H, (b) fully anisotropic chemical shielding anisotropy, and (C) axially symmetrical chemical shielding anisotropy.

sample contains microcrystallites oriented at all possible angles with respect to the applied magnetic field. In this case, I spin is surrounded by randomly distributed I and S spins. The NMR spectrum of such sample is characteristically broadened, which is referred to as a powder spectrum. Powder samples and polycrystalline samples having a dipolar interaction give powder spectra, as shown Figure 4.18(a). The line shape of powder spectra gives information of internuclear distance r.

Molecular motion or fast sample rotation affects the geometrical factors in Eq. (86). In liquids and solutions, rapid isotropic tumbling causes the averaging of these factors. It is easy to show that the average of the terms A to F is zero because $\langle 3\cos^2\theta - 1\rangle = 0$ and $\langle \exp(\pm i\phi)\rangle = \langle \exp(\pm 2i\phi)\rangle = 0$. Thus, the isotropic tumbling makes all the terms zero, and this means that dipolar interactions do not affect the chemical shifts and intensities of solution NMR. The considerably broadened spectra observed for organic polymers are composed of overlapped dipolar powder spectra.

4.7.1.3 Chemical Shielding Anisotropy

The resonant frequency of a particular nucleus in a molecule depends on the type of nucleus (e.g., ^{1}H and ^{13}C) and the electronic environment of the nucleus. The surrounding electrons cause diamagnetic and paramagnetic current, and this shields or deshields the nucleus from the external field. This is the cause of chemical shift. Because the spatial distribution of electrons is generally anisotropic around molecules, the induced diamagnetic and paramagnetic current considerably depends on the direction of the molecule relative to the magnetic field. The Hamiltonian of chemical shielding anisotropy is expressed as

$$\begin{aligned} H_{\mathrm{CS}} &= \hbar \sum_i \gamma_i \mathbf{H}_0 \cdot \boldsymbol{\sigma}_i \cdot \boldsymbol{I}_i \\ &= \hbar \sum_i \gamma_i (H_{0x}, H_{0y}, H_{0z}) \begin{pmatrix} \sigma_{xx} & \sigma_{xy} & \sigma_{xz} \\ \sigma_{yx} & \sigma_{yy} & \sigma_{yz} \\ \sigma_{zx} & \sigma_{zy} & \sigma_{zz} \end{pmatrix} \begin{pmatrix} I_{0x} \\ I_{0y} \\ I_{0z} \end{pmatrix} \end{aligned} \tag{87}$$

where $\boldsymbol{\sigma}$ is the chemical shielding tensor. Unlike the dipolar interaction tensor, this $\boldsymbol{\sigma}$ is not traceless; hence, this interaction displaces the average resonant position of the NMR resonance.

The second-rank tensor $\boldsymbol{\sigma}$ can be diagonalized along the principal axis system in the respective nucleus, as represented by Eq. (87). Thus, the observed chemical shielding (chemical shift) can be expressed as [65]

$$\begin{aligned} \sigma_{zz} &= \lambda_1\sigma_{11} + \lambda_2\sigma_{22} + \lambda_3\sigma_{33} \\ \lambda_1 &= \cos\alpha \sin\beta \\ \lambda_2 &= \sin\alpha \sin\beta \\ \lambda_3 &= \cos\beta \end{aligned} \tag{88}$$

where α and β are Euler angles that relates the principal axis system and the experimental axis system. The principal elements are taken in order of $\sigma_{11} < \sigma_{22} < \sigma_{33}$. The principal values of σ and their directions can be determined from the angular dependence of chemical shifts of single crystals.

On the other hand, powder samples and polycrystalline samples having a chemical shielding anisotropy give powder spectra as shown in Figure 4.18(b) and (c), which consist of the overlaps of the respective peaks with all different α and β.

The three principal values of σ can easily be obtained from the spectra; however, no information about α and β can be deduced. A typical line width of the powder pattern for a ^{13}C nucleus is 50–200 ppm. The orientation of a polymer chain can be inferred from the symmetry of the powder pattern, which corresponds to axially symmetrical tensors. The single-valued chemical shifts observed in the solution state are the isotropic averages of anisotropic shielding resulting from rapid molecular orientation.

4.7.1.4 Quadrupolar Interaction

Nuclei with $I > 1/2$ possess a quadrupolar moment Q subject to an electric field gradient tensor:

$$V^{j}_{\alpha\beta} = \frac{\partial E_\alpha}{\partial r_\beta} = \frac{\partial^2 \Phi}{\partial r_\alpha \partial r_\beta} \qquad (\alpha \text{ or } \beta = x, y, z) \tag{89}$$

where Φ is the electric potential. This electric field gradient produces an interaction that can be described by the following Hamiltonian:

$$H_{\mathrm{Q}} = \sum_i \frac{eQ_i}{6I_i(2I_i - 1)} \mathbf{I}_i \cdot \mathbf{V}_i \cdot \mathbf{I}_i \tag{90}$$

where e is the elemental charge and Q_i and I_i are the nuclear quadrupolar moment and the nuclear spin quantum number of the ith nucleus, respectively. It should be noted that this interaction is entirely intramolecular (Eq. (90)). Because V is traceless, this interaction does not affect the average resonant position of the NMR resonance. Powder samples and polycrystalline samples having a quadrupolar interaction ($I > 1/2$) give, in principle, the same powder spectra as that of the dipolar coupled spin pair (Figure 4.18(a)).

4.7.2 METHODS FOR OBTAINING HIGH-RESOLUTION NMR SPECTRA

To achieve high-resolution solid-state spectra of nonabundant nuclei with $I = 1/2$ like ^{13}C, ^{15}N, and ^{29}Si, the following problems have to be resolved:

1. Broadening due to heteronuclear dipolar interaction with protons.
2. Broadening due to chemical shielding anisotropy.
3. Low sensitivity due to the low natural abundance and long spinlattice relaxation.

On the other hand, the broadening due to homonuclear dipolar interaction should first be eliminated to obtain high-resolution solid-state spectra of abundant nuclei with $I = 1/2$ like ^{1}H and ^{19}F in solids. In addition, the heteronuclear dipolar interaction between abundant spins (e.g., ^{1}H–^{19}F) is to be eliminated for fluoropolymers containing hydrogen. In this section, several important methods to solve the problems are summarized.

4.7.2.1 Magic Angle Spinning (MAS) [65]

The dipolar interaction, which is a main cause of the broadening for solids, involves the geometric factor $(3\cos^2\theta - 1)$ in its Hamiltonian. The average of $(3\cos^2\theta - 1)$ caused by coherent rotation of a solid sample at β about an axis to B_0 can be expressed as

$$\langle 3\cos^2\theta - 1\rangle = 1/2(3\cos^2\beta - 1)(3\cos^2\gamma - 1) \tag{91}$$

where γ is the angle between the rotation axis and internuclear vector $\boldsymbol{r}$. If β is set equal to 54.7°, $\langle 3\cos^2\beta - 1\rangle = 0$ is attained because $\cos^2\beta = 1/3$. This situation is called magic angle spinning (MAS), and 54.7° is called the magic angle. Both the heteronuclear and homonuclear dipolar broadenings are averaged to zero by MAS when the spin speed is larger than the static linewidth.

The Hamiltonian of chemical shielding anisotropy H_{CS} is also affected by MAS. Magic angle spinning at an angular velocity of ω introduces time dependence into the factors of Eq. (88) as

$$\lambda_i = \cos\beta\cos\chi_i + \sin\beta\sin\chi_i\cos(\omega t + \psi_i) \tag{92}$$

where χ_i is the angle between the rotation axis and a principal axis of the shielding tensor $\boldsymbol{\sigma}$ at $t = 0$. The time average of observed chemical shielding σ_{zz} can be expressed as [65]

$$\sigma_{zz} = 1/2\sin^2\beta\bar{\sigma} + 1/2(3\cos^2\beta - 1)\sum\sigma_{\mathrm{ii}}\cos 2\chi \tag{93}$$

Just as in the case of the dipolar interaction, the second term $\langle 3\cos^2\beta - 1\rangle$ becomes zero when the β is set equal to 54.7°. When combined with the relation $\langle\sin 2\beta\rangle = 2/3$, the high-speed MAS gives the isotropically averaged chemical shielding for solids as

$$\sigma_{zz} = 1/3\,Tr(\sigma) = 1/3(\sigma_{11} + \sigma_{22} + \sigma_{33}) \tag{94}$$

This value is the same as that in liquids and solutions.

To attain an appropriate MAS condition, a sample is filled in a cylindrical rotor and spun at 3 to 20 kHz by compressed air. When the spinning speed is faster than the static line width expressed in hertz, high-resolution spectra are obtained. Unless the condition is attained, spinning side bands (SSB) are observed on the both sides of the true peak. The intervals between the real peak and the SSBs are equal to the spinning speed. Hence, the SSBs can be distinguished from the real peak by varying the spinning speed. However, when several SSBs appear in a spectrum, they make spectra more complicated. The total suppression of sidebands (TOSS) pulse sequence solved this problem (Figure 4.19).

4.7.2.2 High-Power Decoupling

The homo- and heteronuclear dipolar interactions can be eliminated by sufficiently fast MAS technique, as described in the preceding paragraphs. However, it is practically difficult to rotate samples at the spinning speed of several tens of Kilohertz for eliminating dipolar interactions. In the case of dilute spins that frequently appear in polymeric systems (e.g., ^{13}C, ^{15}N, and ^{29}Si), dipolar interaction among the dilute spins can be ignored. Hence, the interactions between the dilute spins and abundant hydrogens (or fluorines) should be considered. The broadening due to surrounding protons can also be eliminated by high-power rf irradiation at the resonant frequencies of the protons, which corresponds to the averaging of the spin state I in Eq. (68). This technique is sometimes called dipolar decoupling (DD). The rf power should be strong enough to suppress all the dipolar interactions (at least 10 gauss is needed for proton decoupling). However, this can be attained more easily when compared with the high-speed MAS. When the heteronuclear dipolar interaction is eliminated by DD, and chemical shielding anisotropy is eliminated by MAS, high-resolution solid-state NMR spectra are obtained.

4.7.3 CROSS POLARIZATION (CP) AND RELATED PULSE SEQUENCE

The remaining problem for obtaining high-resolution solid-state spectra of nonabundant nuclei with $I = 1/2$ is the low sensitivity due to their low natural abundance and long spin–lattice relaxation time T_1. Naturally abundant nuclei, on the contrary, have relatively larger magnetization and short T_1 values. The cross-polarization (CP) technique transfers the magnetization from the abundant nuclei (^1H, ^{19}F) to the nonabundant nuclei (^{13}C, ^{15}N, and ^{29}Si) using the cross-relaxation mechanism [66]. A typical pulse sequence of $^1H \rightarrow {}^{13}C$ CP is shown in Figure 4.20(a). After a short 90° pulse along the x-axis of ^1H, the rf phase is shifted by 90° and continued irradiation along the y-axis. The magnetization rotates around

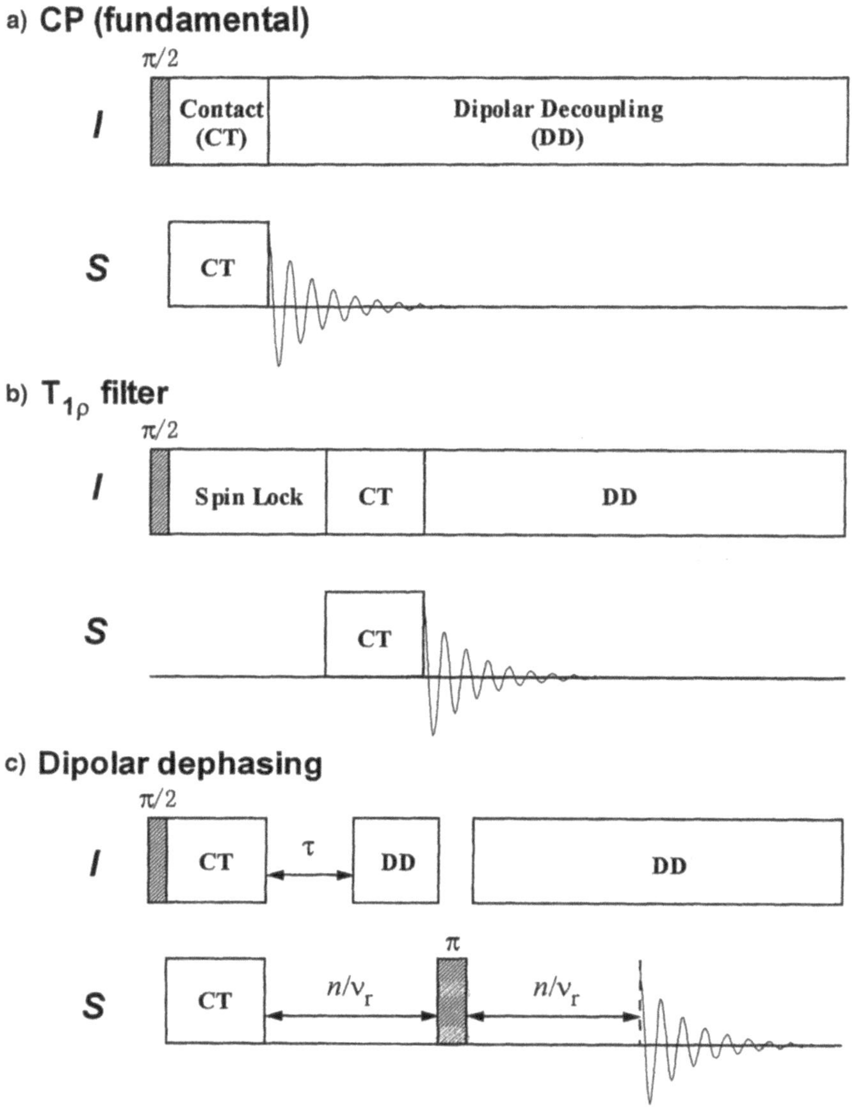

Figure 4.19 Representative pulse sequences of high-resolution solid-state NMR.

the y-axis because the rf field H_1 is larger than the local field based on dipolar interaction and the effect of the external magnetic field is eliminated in the rotating frame. This is the so-called spin-lock.

The mechanism of CP can be explained using the concept of spin temperature. According to Curie's law, the equilibrium magnetization of spin I in the magnetic

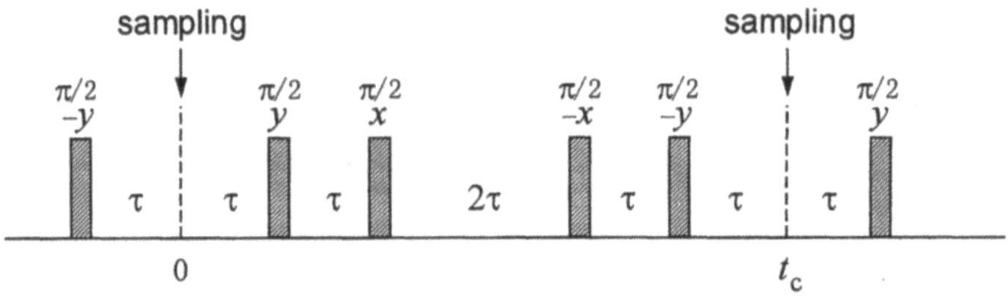

Figure 4.20 WAHUHA 4-pulse sequence for suppressing homonuclear dipolar interaction.

field H_0 can be expressed as

$$M_z^I = \frac{N_I \gamma_I^2 \hbar^2 I(I+1)}{3k} \frac{H_0}{T_{\mathrm{L}}^I} \tag{95}$$

where T_{L} is the lattice temperature and N_I is the number of spin I. Because the magnetization after a 90° pulse and a subsequent spin lock pulse (M_R^I) is equal to M_z^I, the spin temperature drastically changes in the rotating frame. From Eq. (95), the spin temperature in the rotating frame is expressed as

$$T_R^I = T_{\mathrm{L}}^I \frac{H_1^I}{H_0}. \tag{96}$$

When typical values of $H_0 = 10^4$ gauss and $H_1 = 10$ gauss are inserted, the spin temperature in the rotating frame is lower than that in the experimental frame by three orders of magnitude. This considerable cooling is carried out, not by an actual lowering of the temperature, but by a lowering of the effective field seen by protons. The cross-polarization process occurs when S spin is irradiated with an rf field H_1^S of sufficient the strength that the so-called Hartmann–Hahn condition is satisfied.

$$\gamma_I H_1^I = \omega_I = \omega_S = \gamma_S H_1^S \tag{97}$$

This indicates that the z-components of the I and S spins have same time dependency, which enables magnetization exchange between heteronuclear spins based on the flip-flop term of dipolar interaction (Eq. (85)). Unless this condition is attained, magnetization exchange occurs among abundant homonuclei (e.g., ^1H–^1H and ^{19}F–^{19}F). Because the I spins are cold and the S spins are hot, there will be calorimetric effect and the energy exchange will proceed. In other words, the magnetization of S spin generated by external magnetic field

$$M_z^S = \frac{N_S \gamma_S^2 \hbar^2 S(S+1)}{3k} \frac{H_0}{T_L^S} \tag{98}$$

is changed by a contact to I spins, and the spin temperatures of both nuclei become the same ($T_R^I = T_S^I$). When the I spin is ^{1}H and S spin is ^{13}C, ^{15}N, or ^{29}Si, the M_R^I is not affected by the CP because the number of the I spin is much larger than that of the S spin ($N_I \gg N_S$). In contrast, the S spin obtains magnetization as expressed by

$$M_R^S = M_R^I \frac{N_S \gamma_S}{N_I \gamma_I} \tag{99}$$

which is larger than M_z^S by a factor of (γ_I/γ_s). This indicates that the CP of $^1\mathrm{H} \rightarrow {}^{13}\mathrm{C}, {}^1\mathrm{H} \rightarrow {}^{15}\mathrm{N}$, and $^1\mathrm{H} \rightarrow {}^{29}\mathrm{Si}$ can give 4.0, 9.9, and 5.2 times larger signals, respectively, than those of direct polarization measurements using a 90° pulse. Another important advantage of CP is the reduction of repetition time in measurement. Because of the low natural abundance of S spins, the spin-lattice relaxation time T_1^S in solids is on the orders of minutes. However, a short repetition time can be adopted in the CP experiments because the spin–lattice relaxation time of the I spin T_1^I, which is usually less than 1 s for polymers, is much shorter than T_1^S.

The typical time dependence of generated magnetization of S spins during CP is inserted in Figure 4.19(a). The M_R^S increases with a time constant of cross polarization T_{IS} and then relaxes to a lattice with the same time constant of spin-locked proton $T_{1\rho}^I$. For obtaining a large signal enhancement expressed by Eq. (101), T_{IS} should be much smaller than $T_{1\rho}^I$. This $T_{1\rho}^I$-dependence of signal intensity can be used to distinguish two phases with different values of $T_{1\rho}^I$, i.e., amorphous and crystalline phases, with a pulse sequence called the $T_{1\rho}$ filter (Figure 4.19(b)). During the spin-locking period of I spin prior to CP, the I spin magnetization of the phases with a shorter $T_{1\rho}^I$, which is likely due to amorphous phases, decays faster than the other. Hence, by choosing an appropriate spin-locking time, only signals originating from the phases with a longer $T_{1\rho}^I$, which is likely due to crystalline phases, are intensified. The pulse sequence of Figure 4.20(c) serves for selectively observing S spin signals with small heteronuclear dipolar couplings. By switching off the high power decoupling of I spin for 30–150 μs, the magnetization of S spin directly bonded to I spin is dephased by their I–S dipolar coupling. The chemical shift evolution of the S spin not directly bonded to the I spin during the gate is refocused by a 180° pulse to generate a spin echo.

4.7.4 MULTIPULSE EXPERIMENTS

For abundant nuclei such as ^{1}H and ^{19}F, neither high-power decoupling nor cross polarization can be used for obtaining high-resolution spectra in solids. Both techniques can only be applied to heteronuclear, dipolar coupled systems. Although high-speed MAS can eliminate homonuclear dipolar interactions, sample spinning at several tens of kilohertz is impractical. The terms A and B in Eq. (86) cause spectral broadening in homonuclear, dipolar coupled systems, and they can be rewritten as

$$A + B = \frac{(3\cos^2\theta - 1)}{2}(3\mathbf{I}_i\mathbf{I}_j - I_{iz}I_{jz}) \tag{100}$$

In liquids or solution and under MAS condition, the time dependence of the first term makes its average zero $\langle 3\cos^2\theta - 1\rangle = 0$. Another averaging can be attained by introducing the time dependence into the latter term of Eq. (100) by specifically designed cycles of several pulses, possibly separated by windows without irradiation. They are repeated many times with stroboscopic observation of the magnetization in certain windows. This corresponds to the rotation in the spin states, and this technique is called multiple pulse operation. Figure 4.20 shows the simplest sequence consisting of a cycle of four pulses known as WAHUHA [67]. The consecutive 90° pulses along $x, -x, y, -y$ axes rotate the polarization vector in order of $\mathbf{I}_x \rightarrow \mathbf{I}_y \rightarrow \mathbf{I}_z \rightarrow \mathbf{I}_x$.

This indicates that the components of the dipolar Hamiltonian are rotated in the order of $H_D^x \rightarrow H_D^z \rightarrow H_D^y \rightarrow H_D^y$ as well. After one cycle of pulse sequence $t_c = 6\tau$, H_D is averaged to zero because each H_D component stays for the same period 2τ in x, y, and z directions. This sequence requires pulses that are very accurate in their timing and phases. Moreover, the cycle time t_c, must be short to fulfill the condition $t_c \ll \pi T_2$. In practice, typical values of t_c for WAHUHA are about 25 μs, and the duration of the 90° pulses should be much shorter than t_c, which means that very high H_1 powers are necessary. This sequence does not average chemical shielding anisotropy to zero but only scales it by a factor of $1/\sqrt{3}$. Hence, the remaining chemical shielding anisotropy can be removed by using MAS. The combination of multiple pulse and MAS yields proton line width of a few tens of hertz. This technique is called combined rotation and multiple pulse spectroscopy or CRAMPS.

4.7.5 RELAXATION TIMES

Mobility of molecules and polymers in solids can be inferred from spin-lattice relaxation time T_1, spin–spin relaxation time T_2, spin-lattice relaxation time in a rotating frame $T_{1\rho}$, and cross-relaxation time T_{IS} [7]. In particular, the efficiency of the CP process is determined by T_{IS}, $T_{1\rho}$, and the CP contact time, t_{cp}. A schematic diagram of the abundant I-spin and dilute S-spin systems in a typical solid is shown in Figure 4.21, and the buildup and decay of carbon magnetization have already been shown in Figure 4.22(a). The cross-polarization rate under spin lock condition ($1/T_{IS}$) is determined by the effective strength of the dipolar interaction when the Hartmann–Hahn condition is fulfilled. The cross-polarization rate also depends on the initial magnetization of I spin; however, in the case of S spin it is ^{13}C, ^{15}N, or ^{29}Si and for I spin is ^{1}H; the I spin reservoir can be considered infinite. Typical T_{IS} values for protonated and nonprotonated carbons are about 0.1 ms and several milliseconds, respectively. After the buildup of the signal, decay of the spin-locked proton magnetization depletes the S signal with a time constant $T_{1\rho}^{I}$. Because $T_{1\rho}^{S}$ is much longer than $T_{1\rho}^{I}$, this is not an effective relaxation pass. When $T_{1\rho}^{I}$ is longer than t_{cp} and T_{CH}, the growth and decay of S spin under long t_{cp} can be expressed as

$$M(t) = \frac{M_0}{1 + \frac{T_{CH}}{T_{1\rho}^{H}}}\left[1 - \exp\left(-\frac{1}{T_{CH}} + \frac{1}{T_{1\rho}^{H}}\right)t\right]\exp\left(-\frac{t}{T_{1\rho}^{H}}\right) \tag{101}$$

$T_{1\rho}^{I}$ and T_{CH} are obtained by observing the growth and decay of S-spin signals using the pulse sequence of Figure 4.22(a) and fitting the data to Eq. (101). More simply, the slopes of the signal grow, and the decay in logarithm scale can be approximated as $-1/T_{CH}$ and $1/T_{1\rho}^{I}$, respectively. The $T_{1\rho}^{S}$ and $T_{1\rho}^{I}$ can be

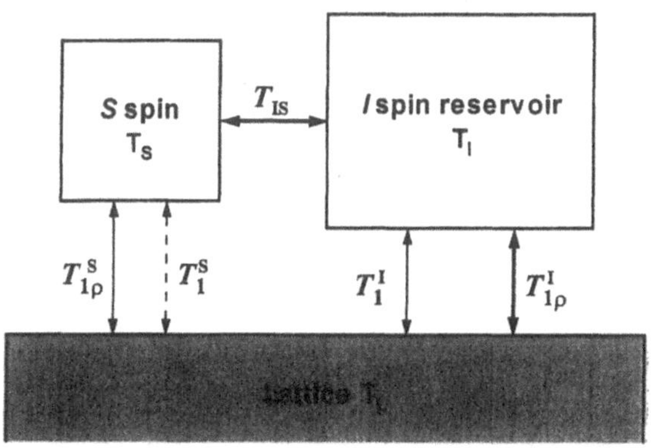

Figure 4.21 Schematic diagram of the abundant I-spin and dilute S-spin systems in a typical solid. The T_S, T_I, and T_L are spin temperatures, and the italic T's indicate various relaxation processes that may occur among I-spin, S-spin, and lattice systems.

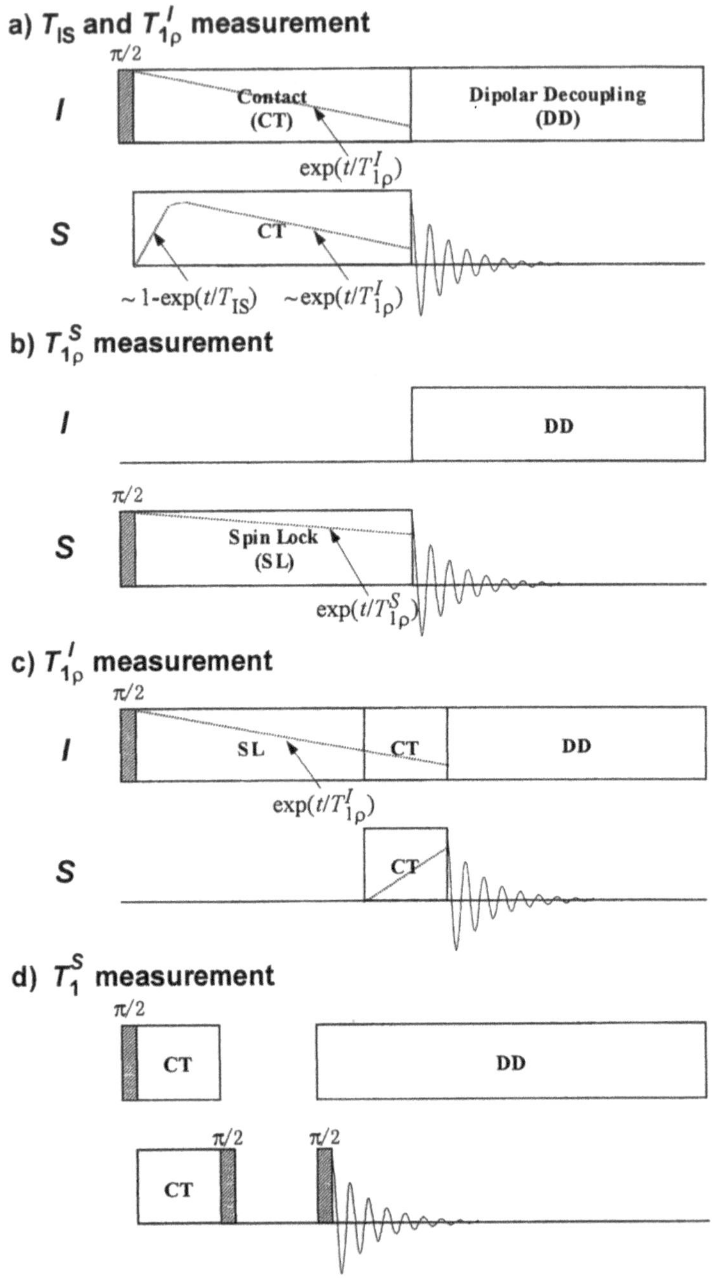

Figure 4.22 Representative pulse sequences for measuring relaxation time constants in solids.

determined independently by using the sequence of Figures 4.22(b) and (c). The pulse sequence, namely Torchia sequence, in Figure 4.22(d) can be used for determining T_1^S. This pulse sequence has two parts. The first part is that the magnetization of ^{13}C along the y'-axis, which arises from the CP method, is returned along the z-axis by the $\pi/2$ pulse at the $-x'$-axis. Then waiting for τ with weak $\pi/2$ pulses to ^{1}H, the $\pi/2$ pulse at the x'-axis is applied, and the FID signal is observed. The signal intensity $M_{z1}(t)$ of this experiment is

$$M_{z1}(t) = [M_{\text{CB}}(0) - M_0]\exp(-\tau/T_1) + M_0 + [MA(t)]_z. \qquad (102)$$

Here $M_{\text{CP}}(0)$ is the intensity only for CP efficiency, M_0 is the intensity at thermal equilibrium, and $[MA(t)]_z$ is the intensity from other forms of efficiency. The second part is that the magnetization of ^{13}C along the y'-axis is returned along $-z$-axis by the $\pi/2$ pulse at the $+x'$-axis, and the FID signal is observed. The signal intensity $M_{z2}(t)$ of this experiment is

$$M_{z2}(t) = [-M_{\text{CP}}(0) - M_0]\exp(-\tau/T_1) + M_0 + [MA(t)]_z. \qquad (103)$$

The difference of these two experiments is $NM_{\text{cp}}(0)\exp(-\tau/T_1)$ at N times (N is even). From the plot of signal intensity for τ, we can obtain the T_1 value. Very long τ is not needed for this experiment.

4.7.6 MULTIDIMENSIONAL NMR

The basic idea of two-dimensional NMR (2-D NMR) was introduced by Jeener in 1971 [68] and developed by Ernst and coworkers [14]. The CP–MAS spectrum, which is usually used for the high-resolution, solid-state, one-dimensional NMR, can be observed by reducing dipolar interaction and chemical shift anisotropy. However, dipolar interaction and chemical shift anisotropy involve much information for the molecular structure and electronic structure. The extension of spectroscopy to two and more dimensions makes it possible to obtain more detailed information about the system under investigation.

The 2-D NMR experiment involves four basic intervals (Figure 4.23).

1. Preparation period: During the preparation period, a nonequilibrium spin system is created by an rf pulse, which will evolve in the subsequent periods. The preparation period is normally of fixed length for all experiments and consists of a recycle delay to allow z magnetization followed by single-pulse excitation or cross polarization from abundant to rare nuclei.

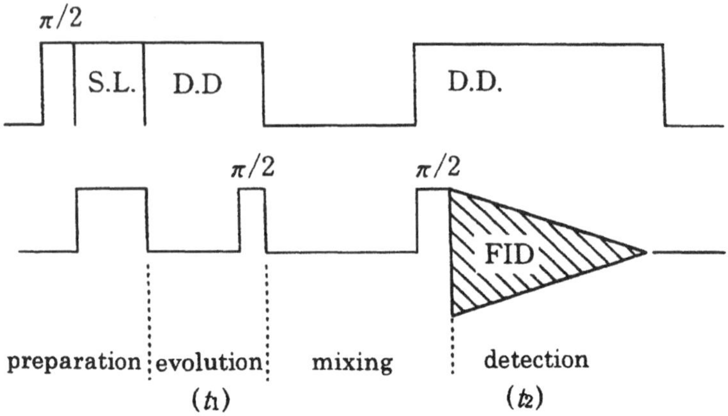

Figure 4.23 Pulse sequence for solid-state 2-D exchange NMR.

2. Evolution period: During the evolution period for time t_1, the changes of the magnetization are detected indirectly as modulations of the signal. This period determines the frequencies in the ω_1-domain.
3. Mixing period: The mixing period, which consists of one or more pulses separated by intervals, allows for transfer of spin information among different sets of spins.
4. Detection period: The transverse magnetization is observed during the detection period t_2. This period determines the frequencies in the ω_2-domain.

This cycle is repeated for various t_1 values to yield a series of FIDs. Two-dimensional spectra can be obtained by the 2-D Fourier transformation of the FIDs. To detect more effects for observation, three or more parameters must be varied independently and Fourier transformation leads to a three- or higher-dimensional spectrum.

Many multidimensional NMR techniques have been developed. Here, several 2-D NMR methods are introduced.

4.7.6.1 2-D Exchange NMR

Off-diagonal peaks are observed if chemical or spin exchange exists in the solution-state 2-D exchange NMR. On the other hand, in the case of rare spin such as ^{13}C, solid-state 2-D exchange NMR can neglect spin diffusion. Therefore, chemical exchange can be observed by solid-state 2-D NMR.

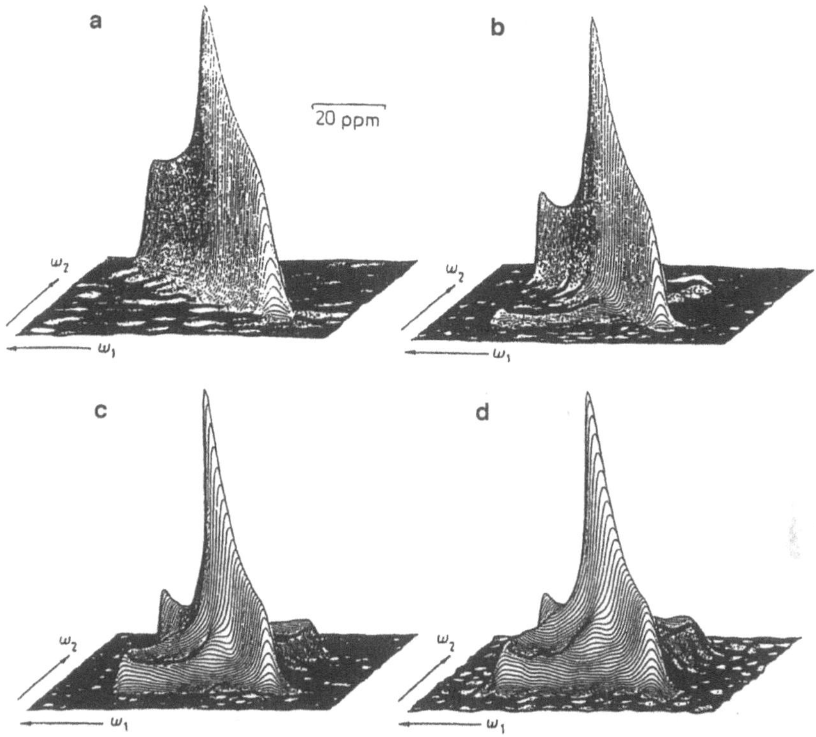

Figure 4.24 ^{13}C solid-state 2-D exchange NMR spectra of polyoxymethylene (operating at 75.47 MHz): (a) $T = 252$ K, $t_m = 1$ s, (b) $T = 330$ K, $t_m = 1$ s, (c) $T = 360$ K, $t_m = 2$ s, (d) $T = 370$ K, $t_m = 4$ s.

Hagemeyer *et al.* observed solid-state 2-D exchange NMR spectra of polyoxymethylene [69]. The pulse sequence is shown in Figure 4.23, and observed variable-temperature, solid-state, 2-D exchange NMR spectra are shown in Figure 4.24. At low temperature ($T = 252$ K), the 2-D exchange NMR spectrum (mixing time (t_m): 1 s) shows that only diagonal peaks are observed and there is no off-diagonal peak. This means that slow motion is not detected in the mixing time of 1 s. At $T = 330$ K, off-diagonal peaks (exchange peaks) appear, and this shows a simple ridge structure proving immediately that it must be due to a jump motion involving mainly two sites. The exchange patterns grow with increasing temperature and mixing time, whereas the intensity of the diagonal spectrum is drastically reduced, as shown in Figures 4.24(c) and (d). A detailed line shape analysis (Figure 4.25) provides information about the geometry and the time-dependent populations of the different sites.

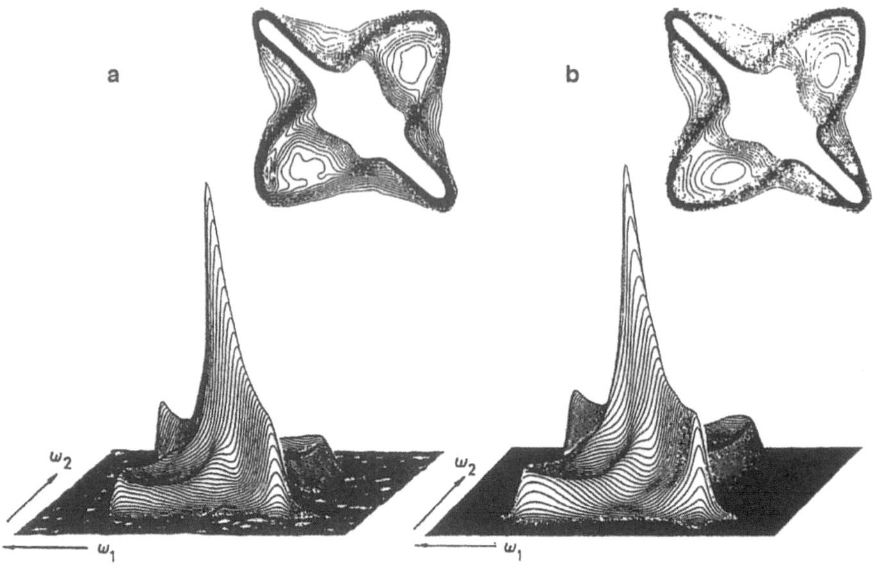

Figure 4.25 ^{13}C solid-state 2-D exchange NMR spectra of polyoxymethylene (operating at 75.47 MHz): (a) $T = 360$ K, $t_m = 2$ s and (b) theoretical spectrum of (a).

4.7.6.2 Dipolar–Chemical Shift Solid-State 2-D NMR

High-power decoupling is applied to eliminate the dipolar field produced by protons to obtain high-resolution solid-state NMR spectra. However, the dipolar interaction depends on the interatom distance and angle, and thus the dipolar interaction has detailed information for the molecular structure. Therefore, in principle, information on the molecular structure can be obtained by the measurement of the dipolar interaction.

The dipolar–chemical-shift, solid-state, 2-D NMR spectrum is measured to separate dipolar interaction and chemical shift into two dimensions (Figure 4.26).

Roberts *et al.* observed the dipolar–chemical-shift, solid-state, 2-D NMR spectrum of acetylvaline by using pulse sequence as shown in Figure 4.27 [72]. The observed spectrum is shown in Figure 4.28. After the cross polarization, the rotational echo produced by the MAS of rare spins refocuses by a 180° pulse by applying ^{1}H homodecoupling (MREV-8) during t_1, where heteronuclear dipolar interaction is reserved. The obtained peak separation Δ_{IS} is given as a function of internuclear vector r_{IS}.

$$\Delta_{IS} = \frac{3}{2\pi}\gamma_I\gamma_S h\left[(3\cos^2\theta_{IS} - 1)/2\gamma_{IS}^3\right] \qquad (104)$$

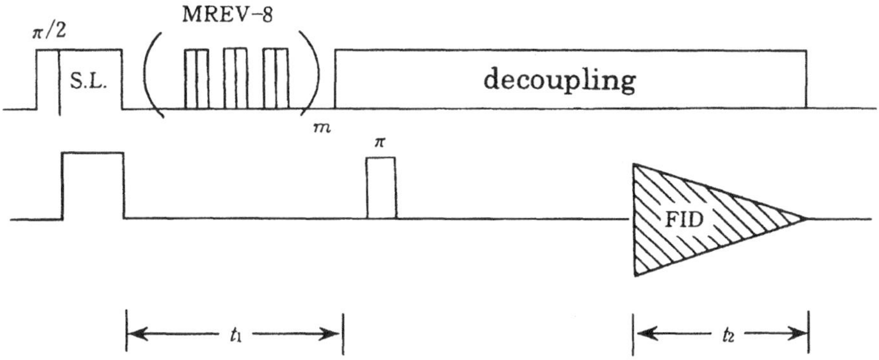

Figure 4.26 Pulse sequence for 2-D dipolar–chemical shift NMR method.

θ can be determined by the intensities and phase of dipolar sidebands, and then internuclear distance can be obtained.

Another example of dipolar–chemical-shift 2-D NMR spectra of $H^{13}COOCH_3$ is shown in Figure 4.29 [70]. The experimental spectrum (left) and simulated spectrum (right) show a good agreement for an orientation of the **CH** vector with respect to the principal axis of the shielding tensor given by $\beta = 85°$, which is the

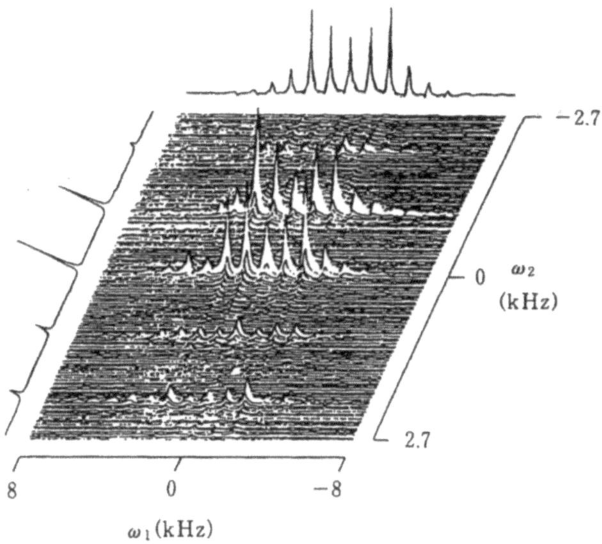

Figure 4.27 Two-dimensional ^{15}N–^{1}H dipolar–chemical shift NMR spectrum of acetylvaline.

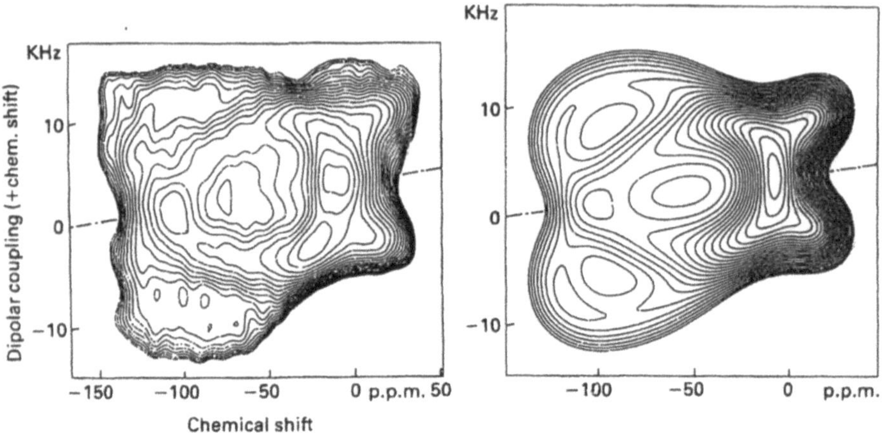

Figure 4.28 Two-dimensional separated local-field spectra of polycrystalline powders. Comparison of an experimental spectrum (a) of ^{13}C-enriched methyl formate ($H^{13}COOCH_3$) with a computer simulation (b) that assumes a relative orientation of the shielding and dipole tensors defined by $\alpha = 33°$ and $\beta = 85°$.

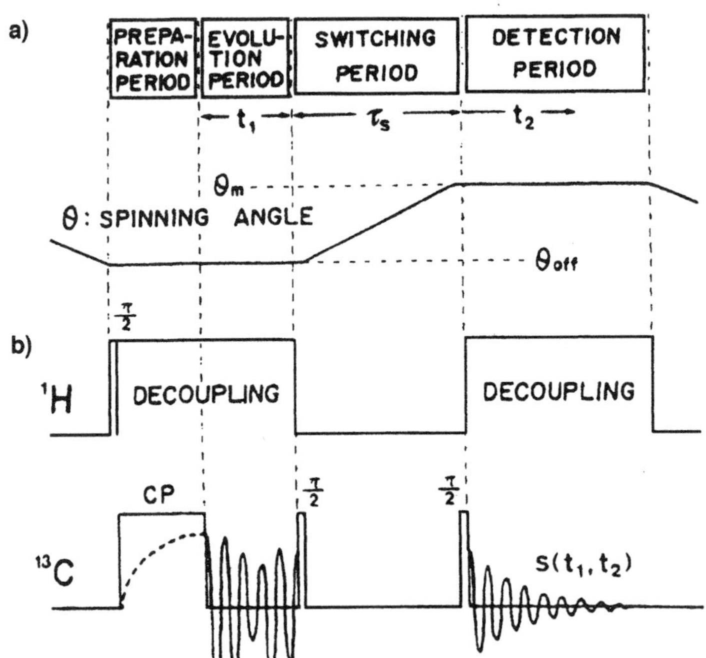

Figure 4.29 (a) Timing blocks of SASS spectroscopy. (b) Pulse sequence for separating ^{13}C chemical-shift powder patterns.

polar angle with respect to σ_{33} and $\sigma = 33°$ (azimuthal angle with respect to σ_{11} in the plane spanned by σ_{11} and σ_{22}).

4.7.6.3 Chemical Shift Anisotropy Separation

The solid-state ^{13}C CP spectrum of powder sample measured without MAS shows a broad-line spectrum, the so-called powder pattern. This is because the chemical shift anisotropy (CSA) is not eliminated, and overlap of powder patterns of chemically nonequivalent nuclei makes it impossible to interpret a spectrum. The high-resolution, solid-state NMR technique eliminates the chemical shift anisotropy by MAS. However, the chemical shift anisotropy itself has variable information for molecular structure. Therefore, it is very useful to obtain chemical shift anisotropy for detailed discussion of molecular structure.

Terao *et al.* proposed the switching angle sample spinning (SASS) technique to separate powder patterns by isotropic ^{13}C chemical shifts [71]. Figure 4.29 shows the timing blocks and pulse sequence of SASS spectroscopy. The angle θ between the sample-sapinning axis and the static magnetic field is switched between an off-magic angle θ_{off} and the on-magic angle θ_{m}. Figure 4.30 shows the ^{13}C CP–MAS spectrum of 3,4-dimethoxybenzaldehyde and the ^{13}C chemical shift powder patterns separated by SASS spectroscopy ($\theta_{\text{off}} = 65.24°$, scaling factor: -0.237). From these experiments, three principal values of chemical shift tensor (σ_{11}, σ_{22} and σ_{33}) can be obtained.

4.8 Spatial Distance Method

The measurement of nuclear magnetic dipole–dipole coupling in NMR spectroscopy provides a means of determining atomic distances. In this section, we will show some NMR techniques to obtain the distance between two atoms using the dipole–dipole interaction. The first technique is for the heteronulcear case such as ^{13}C and ^{15}N, and the other two techniques are for homonuclear cases.

4.8.1 REDOR (ROTATIONAL ECHO DOUBLE RESONANCE)

REDOR (Rotational Echo Double Resonance) was introduced by Gullion and Schaefer [73–75] to determine the dipolar interaction strength between two dilute heteronuclear spins [76]. In this MAS experiment, the heteronuclear interaction is recoupled by the application of two π pulse per rotor period applied to either the

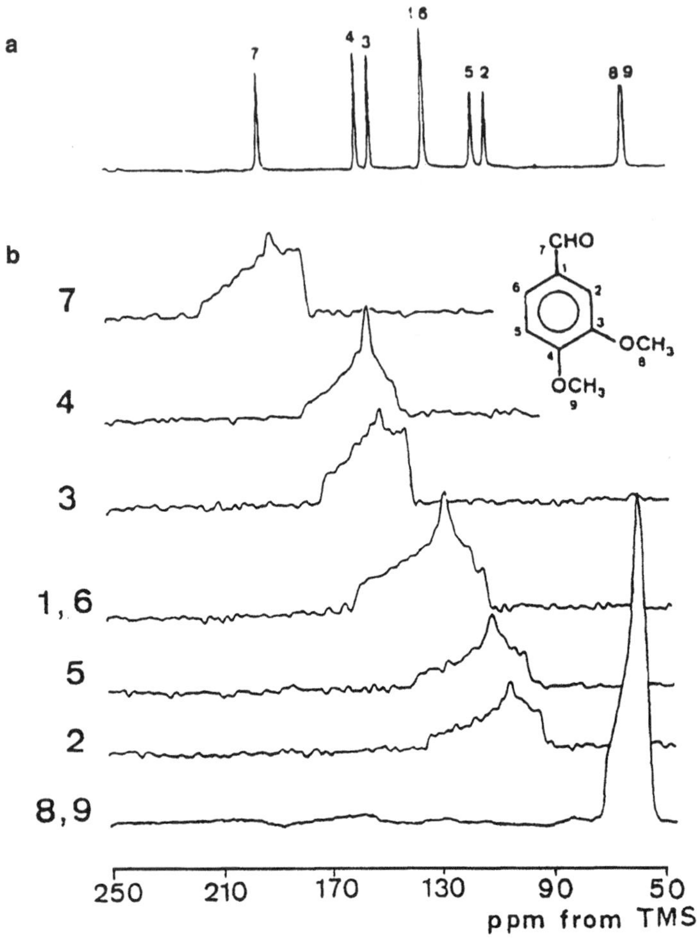

Figure 4.30 (a) ^{13}C CP/MAS spectrum of 3,4-dimethoxybenzaldehyde. (b) ^{13}C chemical-shift powder patterns separated by SASS spectroscopy with a scaling factor of −0.237.

I or *S* spin. REDOR experiments are designed to measure the dipolar interaction between heteronuclear spins other than protons such as ^{13}C and ^{15}N. In these experiments, the detected spin is first excited by the proton cross polarization. Following excitation, the observed transverse magnetization evolves under the influence of the dipolar and *J*-coupling interaction, the chemical shift interactions of two spins, and the π pulses applied to the system. The location of the two π pulses within each rotor period can be varied, resulting in various expressions for the recoupled dipolar interaction. In all REDOR experiments, the combined effect of the two π pulses is to hinder the refocusing of the dipolar interaction under MAS and cause dephasing of the signal. The rate of dephasing reflects the magnitude of the dipole–dipole coupling. The pulses are equally spaced, but the overall phase Ψ of the pulse sequence is left variable. To eliminate the influence of the chemical shift (CS) interaction of the S spin on the overall spin evolution, one π pulse is applied per rotor period to refocus its chemical shift interaction, including the chemical shift anisotropy. In the absence of dipolar dephasing, the result is the formation of a full-spin echo and recovery of the transverse signal after every two-rotor period. Following the REDOR dephasing sequence, the FID of the S spin is acquired and transformed into its Fourier spectrum with varying periods of dephasing $N\tau_R$. This procedure yields the total spectral intensity $S(N\tau_R)$ of the monitored S spin magnetization as a function of the REDOR evolution time $N\tau_R$.

Because the acquisition begins at rotational echo positions and is therefore synchronous with rotor period, an effective Hamiltonian for the REDOR pulse sequence can be constructed that characterizes the evolution of the spin pair. The π pulses are taken to be ideal and instantaneous rotations. The most convenient way to evaluate the effective Hamiltonian is to transform the dipolar coupling into the toggling frame and calculate it using average Hamiltonian theory (AHT). However, it is also possible to apply the Floquet theory to the analysis of dipolar recoupling by REDOR. The dephasing experiment consists of n periods of two rotor cycles each, and thus there are $N = 2n$ rotor periods altogether in the dephasing interval. The first pulse is applied at time τ_P following cross polarization, which corresponds to the phasing angle $\Psi = \omega_R \tau_P$ with respect to the rotor period. The dipolar Hamiltonian assumes the following form in the toggling frame:

$$H_D^{tg}(t) = 2\omega_{IS}^D(t)F(t)I_zS_z, \tag{105}$$

where $F(t)$ accounts for the action of the p pulses in the toggling frame, which is to flip the sign of the dipolar coupling at each point a pulse is applied. A π pulse on either channel switches the sign of the toggling function $F(t)$ but does not modify the form of the spin operators. A similar result applies to the chemical shift interactions. However, in the case of chemical shifts, the π pulses flip only the sign of the CS interaction associated with the spin to which the pulse is applied. Because in all cases the π pulses preserve the form of the spin operators, the dipolar and CS interactions of the two spins commute with one another in the toggling frame in exact analogy with the usual situation in the absence of π pulses. Consequently, the dipolar REDOR Hamiltonian is completely independent of the CS parameters, and only consideration of the heteronuclear interaction is required for its derivation. The commutation of the CS terms with the spin–spin coupling terms, which is retained under π pulse sequences, is also useful in the design of dipolar chemical shift correlation experiments, but it does not apply to more general pulse sequences, such as the continuous rf field employed in rotary resonance experiments. In the latter case, the chemical shift anisotropy (CSA) interaction of the I spin plays a major role in the spin dynamics at relatively low spinning speeds, implying that its parameters must be estimated to obtain the dipolar coupling from the observed line shapes. Similar considerations apply to all of the homonuclear methods.

To calculate the AHT results for REDOR, the function $F(t)$, which defines the toggling frame transformation

$$U_{\mathrm{RF}}^{1}(t)I_zS_zU_{\mathrm{RF}}(t) = F(t)I_zS_z, \tag{106}$$

is written as a Fourier series

$$F(t) = \sum_{n=-\infty}^{\infty} F_n \exp\{in\omega_R t\} \tag{107}$$

When combined with the expansion of $\omega^D(t)$, the Fourier expansions facilitate the evaluation of the average Hamiltonian. Because the Hamiltonian commutes with itself at different times in the toggling frame, the zeroth-order term of the Magnus expansion provides the exact result. Additionally, after a pair of π pulses has been applied to each spin, the toggling frame transformation is cyclic. Consequently, the AHT method can be straightforwardly applied, and the effective Hamiltonian assumes the form

$$\bar{H}_{\mathrm{eff}} = \sum_{n=-2}^{2} 4\omega_n^{IS} F_{-n} I_z S_z = 2\omega_{\mathrm{eff}}^{-D} I_z S_z \tag{108}$$

The part of the average Hamiltonian corresponding to the CS interactions of the spins and their J-coupling vanishes over two rotor cycles. Therefore, the evolution of the transverse magnetization is determined only by the effective dipolar frequency $\omega_{\mathrm{eff}}^{-D}$.

All Fourier coefficients ω_n^{IS} of $\omega_{IS}^D(t)$ are functions of the initial polar angles (θ, ϕ) of the dipolar vector in the rotor frame, and therefore the effective REDOR frequency $\omega_{\mathrm{eff}}^{-D}$ depends on the crystalline orientation. Likewise, the effective coupling depends on the phase angle Ψ associated with the rf pulse sequence and is proportional to the dipolar interaction constant ω_{IS}^D:

$$\omega_{\mathrm{eff}}^{-D} = \frac{2}{\pi}\omega_{IS}^D \sin 2\theta \sin(\phi + \Psi) \tag{109}$$

In the particular case where $\Psi = \pi$, the pulses are applied every $1/2\tau_R$, and the dipolar frequency is given by

$$\omega_{\text{eff}}^{-D} = \frac{-2}{\pi}\omega_{IS}^{D} \sin 2\theta \sin\phi \tag{110}$$

For each crystallite in a powder sample, the dephasing trajectory follows the course

$$S(N\tau_R) = Tr[\exp(-i\bar{H}_{\text{eff}}N\tau_R)S_x \exp(i\bar{H}_{\text{eff}}N\tau_R)S_x] = \cos\left(\omega_{\text{eff}}^{-D}N\tau_R\right) \tag{111}$$

The ensemble-averaged signal $S(N\tau_R)$ arising from all possible crystallite orientations in a powder sample depends explicitly on $\omega_{IS}^{D}N\tau_R$ and can be expressed as a function of this parameter alone by

$$S\left(\omega_{IS}^{D}N\tau_R\right) = \frac{1}{4\pi}\int_0^{2\pi}\int_{-1}^{+1} S(N\tau_R)\, d\cos\theta\, d\phi \tag{112}$$

The inhomogeneous distribution of the effective frequencies leads to the decay of the transverse magnetization. Determination of ω_{IS}^{D} from the observed dephasing of the signal permits evaluation of the nuclear distance. Because of the mutual commutation among the terms of the spin Hamiltonian, prior knowledge or estimation of the CS parameters is not required to extract the dipolar coupling constant from the data.

The signal from a powder does not depend on the phase angle Ψ of the pulse sequence because it serves only to add an additional constant to the phase angle ϕ associated with the dipolar vector of each crystallite. Although the dipolar frequency for any given crystallite is modified by the phase Ψ, the signal from the isotropic distribution of crystallites remains unchanged after averaging over ϕ. In addition, the CS interactions are refocused following every two rotor periods, regardless of the angle Ψ. Typically, the phase Ψ is chosen to be π, and thus the pulses are applied every half rotor cycle and alternate between the two spins.

For the experimental determination of $S(\omega_{IS}^{D}N\tau_R)$, the REDOR signal is monitored in conjunction with the reference signal $S_0(N\tau_R)$ obtained without the application of dipolar dephasing pulses to the I spin. Alternating detection of $S(\omega_{IS}^{D}N\tau_R)$ and $S_0(N\tau_R)$ provides a relative difference function $\Delta S/S$ that is independent of the rate of transverse relaxation:

$$\frac{\Delta S}{S} = \frac{S_0(N\tau_R) - S\left(\omega_{IS}^{D}N\tau_R\right)}{S_0\left(N\tau_R\right)} \tag{113}$$

The time dependence of the REDOR difference function $\Delta S/S$, illustrated in Figure 4.31 [75], is uniquely determined by the pulse sequence and the coupling

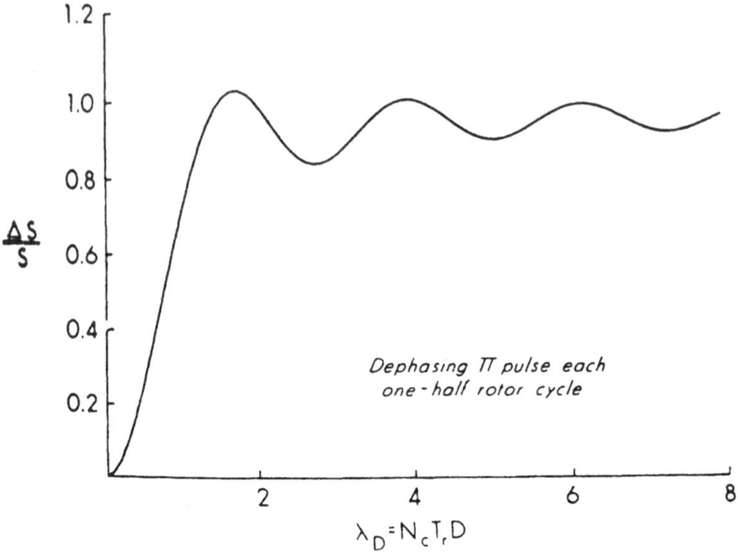

Figure 4.31 Calculated REDOR difference as a function of the parameter $\omega_{IS}^{D} N \tau_R$.

constant ω_{IS}^{D}. Because uncoupling spins contribute to $\Delta S/S$ through the denominator of Eq. (103), their relative abundance must be known to perform accurate distance measurements.

Most REDOR measurements of internuclear distances have been performed on ^{13}C–^{15}N spin pairs, but applications to other spin pairs have been reported. The REDOR method has also been extended to the case of coupling between a dilute spin-half nucleus and ^{2}H [77]. An experiment demonstrating the application of the REDOR technique to alanine is shown in Figure 4.32 [75]. The REDOR difference ΔS removes uncoupled spins from the MAS spectrum and emphasizes the resonances of nuclei experiencing the largest interactions with the nonobserved spins.

4.8.2 ROTATIONAL RESONANCE

Very precise measurements of internuclear distance of ^{13}C–^{13}C pairs in the solid state are achieved by a rotational–resonance (R^2) technique introduced and analyzed by Levitt, *et al.* [78]. The rotational resonance exchange experiment is closely related to rotor-driven spin diffusion: under ^{13}C–^{1}H decoupling, through matching of the rotating frequency with the chemical-shift difference of the species of interest, time independent ^{13}C–^{13}C dipolar couplings result that naturally produce a transfer of magnetization between the carbon spins. In the R^2 experiment, the

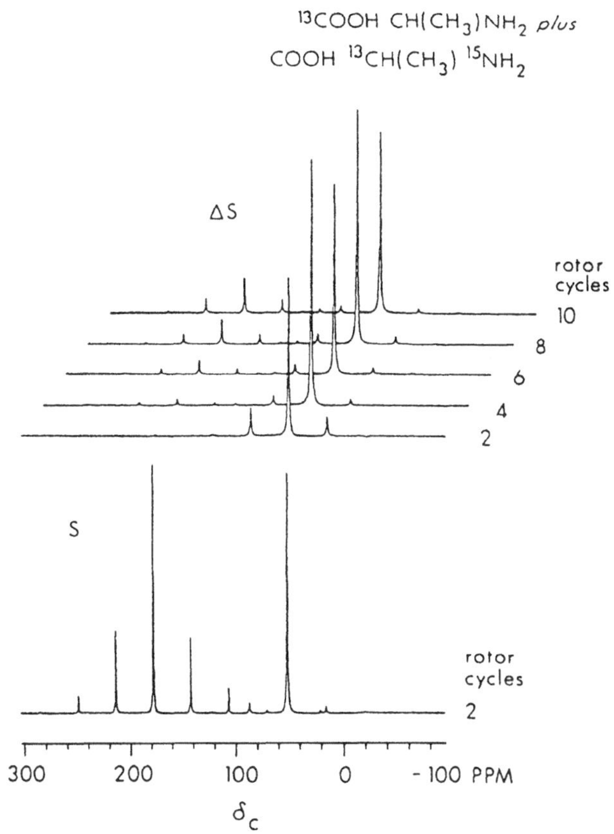

Figure 4.32 ^{13}C REDOR spectra of an equimolar, recrystallized mixture of ^{13}C-, ^{15}N-labeled L-alanines. The full signal *S* is shown at the bottom.

exchange behavior is dominated by the ^{13}C–^{13}C spin pairs, whereas rotor-driven spin diffusion on ^{13}C in natural abundance depends on the wealth of ^{13}C spin configurations resulting from the statistical distribution of ^{13}C nuclei.

Owing to the relative complexity of the expression for effective dipolar frequency, the extraction of the internuclear distances from the rotational–resonance *tm*-dependence of the magnetization difference requires knowledge of the chemical shift tensors involved, including their relative orientation. As an additional complication, instead of the oscillations with the dipolar coupling frequency ω_D, monotonic magnetization transfer functions are often observed experimentally.

There are also other NMR methods for determining internuclear distances in ^{13}C–^{13}C pairs. One example is CPMG (Carr–Purcell–Meiboom–Gill) dipolar

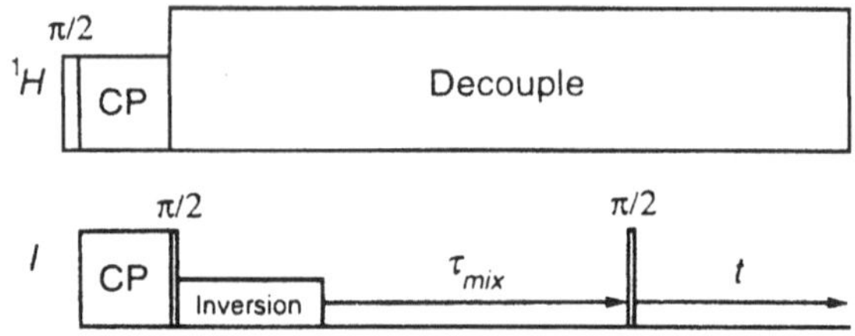

Figure 4.33 Pulse sequence used to prepare difference magnetization between two spins and monitor its time evolution at R^2 condition. Following selective inversion of one of the resonances, exchange occurs at rotational resonance, and the signal is acquired.

spectroscopy in which a train of π ^{13}C pulses, under ^{13}C–^1H decoupling and stroboscopic signal detection, refocuses the chemical shift, whereas the ^{13}C–^{13}C interactions are unaffected. In the homonuclear dipolar Pake pattern that results, the distance of the singularities is a sensitive measure of the internuclear distance without other parameters entering the analysis.

Figure 4.33 demonstrates the pulse sequence used to prepare the difference magnetization between two spins and monitor its time evolution at the R^2 condition [78]. The R^2 data obtained for a double-labeled peptide fragment of the noncrystalline amyloid-forming protein of Alzheimer's disease are shown in Figure 4.34 [79]. The plotted decrease of the magnetization difference as a function of the mixing time after selective inversion of one of the lines is significantly faster than expected for the commonly assumed structure of an ideal antiparallel β sheet. Instead of the expected distance of 4.74 Å, the experimental data yield 4.0 Å, which suggests an unusual peptide conformation that may be linked to the stability of the amyloid deposits.

4.8.3 DIPOLAR RECOVERY AT THE MAGIC ANGLE (DRAMA)

Tycko and Dabbagh [80, 81] have introduced a technique for recovering full continuous powder patterns for the homonuclear dipolar interaction of isolated spin pairs under fast MAS and heteronuclear decoupling. Rotation-synchronized $\pi/2$ pulses are applied during the evolution period, changing the homonuclear dipolar Hamiltonian in the toggling frame in such a way that MAS is prevented from averaging out this dipolar interaction. As shown in Figure 4.35, per rotor cycle, a pair of $\pi/2$ pulses of phases x and $-x$ are applied, spaced by a delay τ, and centered on a time t. They toggle the toggling-frame Hamiltonian between $H_{zz}(t)$ and $H_{yy}(t)$, where the t-dependence is caused by the sample rotation. In the limit

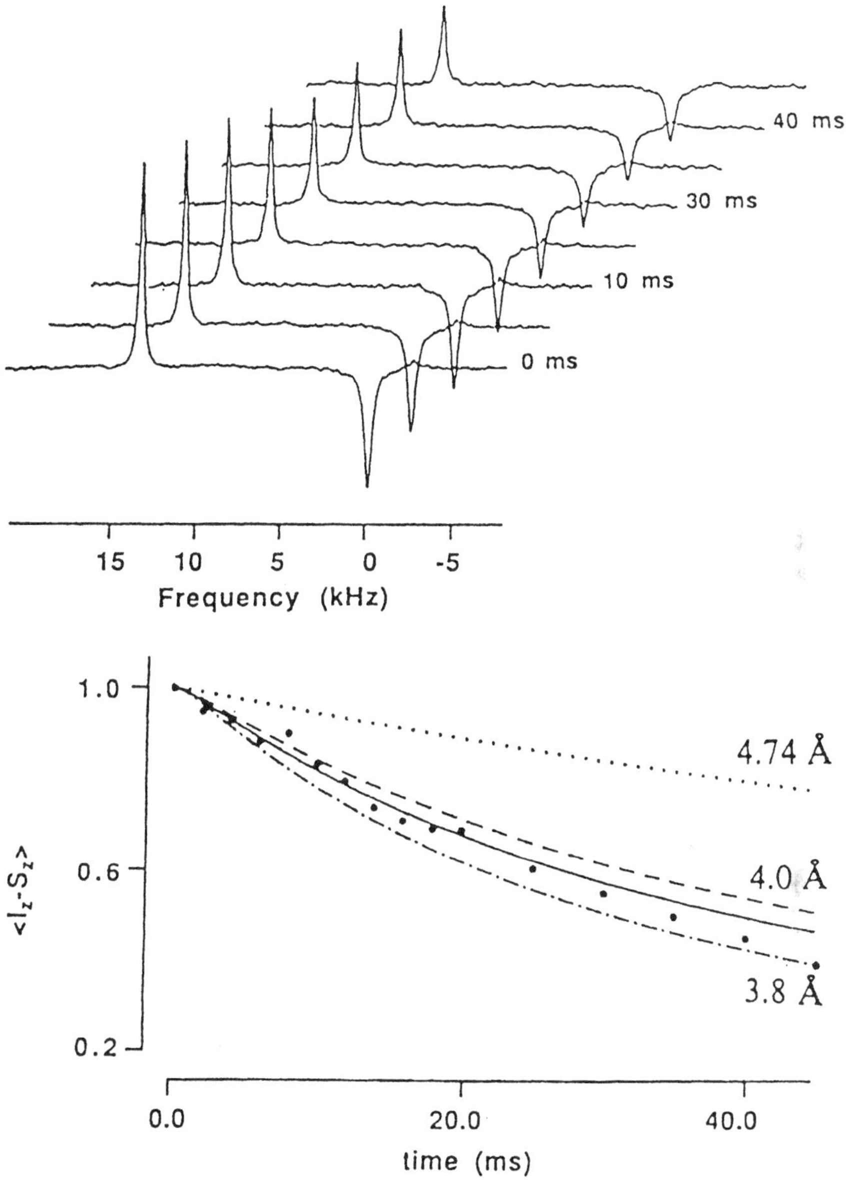

Figure 4.34 R^2 data in the amyloid of a doubly ^{13}C-labeled synthetic fragment of the amyloid-forming protein of Alzheimer's disease (a). Mixing-time dependence, after selective inversion at a rotation frequency, of the resonance of the labeled carbon of a glycine residue and labeled carbonyl in the neighboring glycine, which is separated by four bonds (b).

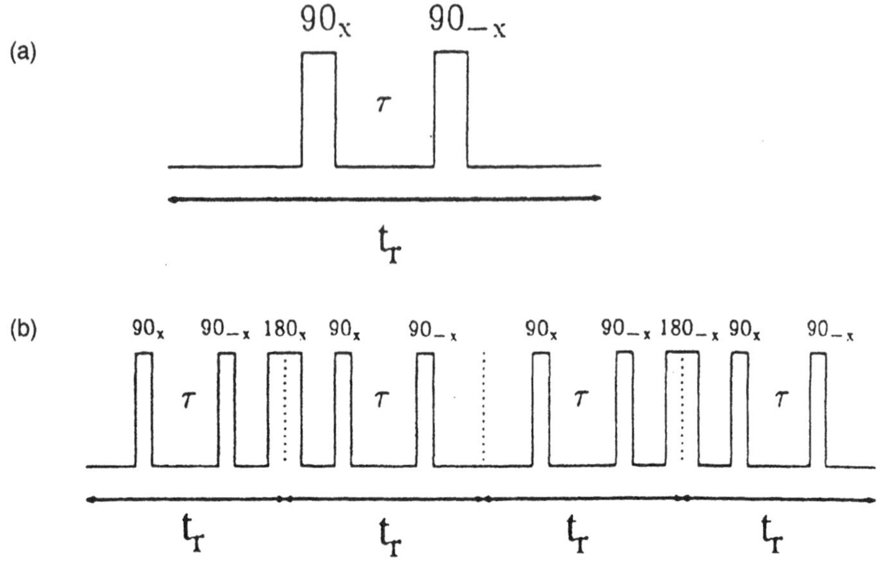

Figure 4.35 Pulse sequence for the DRAMA experiment: (a) Basic cycle with $\pi/2$ pulse within one rotor period. (b) Four rotor periods with the cycle of (a) and π pulses for refocusing the chemical shift.

of $\tau \ll t_r$, the evolution is quasistatic during τ under a toggling-frame propagator $\exp[-i\tau H_{yy}(t)]$. At the end of the rotor period, the gap in the H_{zz} evolution amounts to $\exp\{i\tau H_{zz}(t)\}$. Consequently, the average Hamiltonian propagator per rotor cycle is $\exp\{-i\tau[H_{yy}(t) - H_{zz}(t)]\}$, which results in a quasistatic dipolar pattern. To average out chemical-shift and resonance-offset effects, π ^{13}C pulses are applied at time $n{*}2t_r$. If the time is lengthened, the quasistatic approximation breaks down, but the effective dipolar evolution becomes longer, which is advantageous for measuring small dipolar coupling. The internuclear distance is then determined by comparison of experimental and simulated spectra such as those of Figure 4.36.

4.9 NMR Imaging Method

NMR imaging is a noninvasive and powerful technique based on nuclear magnetic resonance that provides spatial information on chemical and physical properties of a sample as a clear image. This technique gives spatial information on the

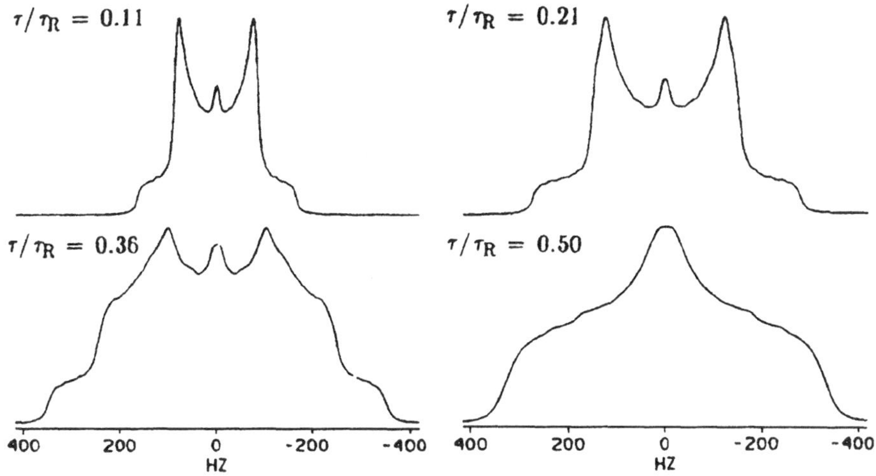

Figure 4.36 Calculated Fourier transforms of DRAMA dephasing trajectories. Dipolar powder spectra obtained with various pulse spacings τ are shown. The coupling constant is 1.5 kHz, the rotor period 399 μs, and the π pulse length is 10 μs.

spin density, the chemical shift, relaxation time, and diffusion and flow rate of a sample. NMR imaging is generally applied to medical diagnoses as a method for obtaining and examining the cross-sectional images of a human body; for example, the ^{1}H nuclei of water in tissues are usually measured because the natural relative abundance and the detection sensitivity of them are higher as compared with that of the other nuclei. Further, the ^{13}C, ^{23}Na, and ^{31}P NMR imaging, and so on, are also studied and used in the appropriate fields.

The advantages of NMR imaging as a characterization technique are as follows:

1. Spatial information on chemical and physical properties of a sample are obtained.
2. The method is noninvasive and no probes are needed.
3. Any sectional images in any direction of a sample are obtained (the section is a flat or a curved surface).
4. Change of a sample with time is measurable.
5. Any kinds of samples, such as those in the liquid, solid, and gel states are measurable.
6. Chemical components contained in a gel sample such as solvents, ions, molecules, and networks are detected separately.

However, the technique has some disadvantages to be overcome such as the following:

1. The spatial resolution is not high enough ($>10\ \mu$m).
2. The duration time for getting an image is not short enough.

Many researchers are studying energetically to break through such difficulties, and various NMR imaging methods have been developed [82, 83].

In contrast to other normal NMR spectroscopies in which a homogeneous magnetic field is employed, a three-dimensional magnetic field gradient is used in NMR imaging. A particular nucleus located at a specific point in three-dimensional space in a sample is placed under a specific magnetic field. The magnetic fields that such nuclei in the sample are subjected to different from each other. Therefore, all the frequencies to which the nuclei resonate are different and correspond to a position in three dimensions. The section of the sample to be considered can be selected freely by controlling the magnetic field gradient of each axis.

The information obtained by normal one-dimensional NMR such as chemical shift, signal intensity, relaxation time, and the self-diffusion constant are also represented by NMR imaging. The important point is that such information is imaged as a spatial distribution. The technique demonstrates its ability when heterogeneous samples and reactions are measured with it.

4.9.1 HOW TO GET SPATIAL IMAGES WITH NMR IMAGING

To study polymer gels, spin-warp imaging [84, 85] has generally been employed, which is based on Fourier imaging [86], and its pulse sequence is the spin-echo pulse sequence of Hahn [38].

4.9.1.1 Fourier Imaging

The data processing for a two-dimensional image is performed by the Fourier imaging method [86]. In Fourier imaging, phase-encoding periods during which the gradient magnetic fields are applied influence the observed signal through the initial phase of the sample's nuclei. The three-dimensional signal of the sample is retrieved by a systematic variation in the duration of the phase-encoding periods. The contributions of the signal volume elements (voxels) from the sample can be separated by a three-dimensional Fourier transformation. The two-dimensional Fourier imaging excitation is performed by a selective rf pulse applied in the presence of a magnetic field gradient. The magnetic field gradient is called a slice selection gradient and is used to select a section. The NMR signals ζ (x-direction) and η (y-direction) obtained under magnetic field gradients G_X (x-direction) and G_Y (y-direction) are represented as function of duration times t_X (x-direction) and t_Y (y-direction) [86] as

$$\zeta = \gamma G_X t_X \tag{114}$$

$$\eta = \gamma G_Y t_Y \tag{115}$$

where γ is the magnetogyric ratio. The η signal $s(t_X)$ can be separated from all contributions by two-dimensional Fourier transformation $F(\zeta, \eta)$ by

$$s(t_X) = F(\gamma G_X t_X, \gamma G_Y t_Y) \tag{116}$$

All the signals obtained with systematically varied G_Y or t_Y on the xy-coordinates are treated as the same and are collected. The two-dimensional image of the sample can be retrieved from the data by the reverse of two-dimensional Fourier transformation. Clear, colored, or black and white images, according to the spin density, can easily be obtained by the Fourier imaging.

4.9.1.2 Spin-Warp Imaging

The spin-warp imaging represents a variant of the Fourier imaging and differs from it by the way the phase-encoding gradient is applied. In the spin-warp imaging, the phase-encoding period is fixed in length, and the amplitude of the applied magnetic field gradient is varied. The constant phase-encoding period gives the same relaxation effects through all experiments, and the resolution in the phase-encoded direction is independent of relaxation. This is a great advantage of spin-warp imaging.

A schematic representation of basic time periods and magnetic field gradients of spin-warp imaging and its pulse sequence is given in Figures 4.37 [88] and Figure 4.38, respectively. The slice selection and excitation period (1) in Figure 4.37 consist of selective excitation of a plane in a sample using a selective $\pi/2$ rf pulse and a slice selection gradient (G_Z). The rf pulse and gradient determine the position of the region to be observed. The slice selection for a sample is performed during the period (1) of selecting a plane section perpendicular to the z-axis. The selective $\pi/2$ rf pulse is a $(\sin x)/x$ form convoluted with a gaussian. In the following period (2), the phase-encoding gradient (G_Y) is applied and varied in intensity systematically from experiment to experiment. The G_Y serves to obtain the differentiation of the volume elements in the y-direction. The G_Y causes the nuclei across the sample to precess at phases that are characteristic of their location

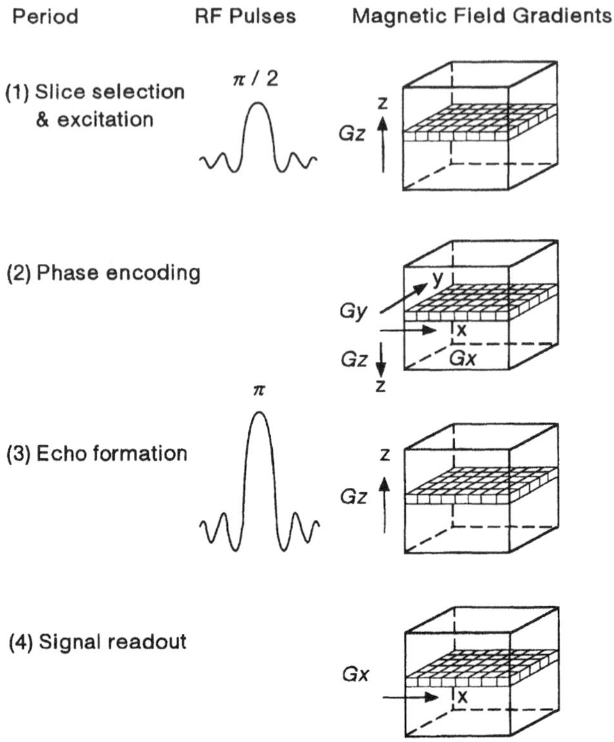

Figure 4.37 Basic time periods in measuring an NMR image.

with respect to the G_Y. At the end of the period (2), the magnetic moments from each voxel in the sample have precessed through an angle related to their position in the gradient. During the period, a readout gradient G_X is applied in the x-direction for advancing the position of the echo along the time axis and away from the time of the pulses, and a z-directional reversed gradient $(-G_Z)$ is also applied to refocus the selectively excited magnetization and to obtain a maximum signal. The echo formation period (3) consists of a selective π rf pulse and G_Z. During the signal readout (acquisition) period (4), the G_x is applied to disperse the voxels in the x-direction, and a spin echo is detected.

4.9.2 EXPERIMENTAL DETAILS

It is very important to obtain maximum image contrast and SNR in the image of the sample. The image contrast is influenced by several NMR parameters.

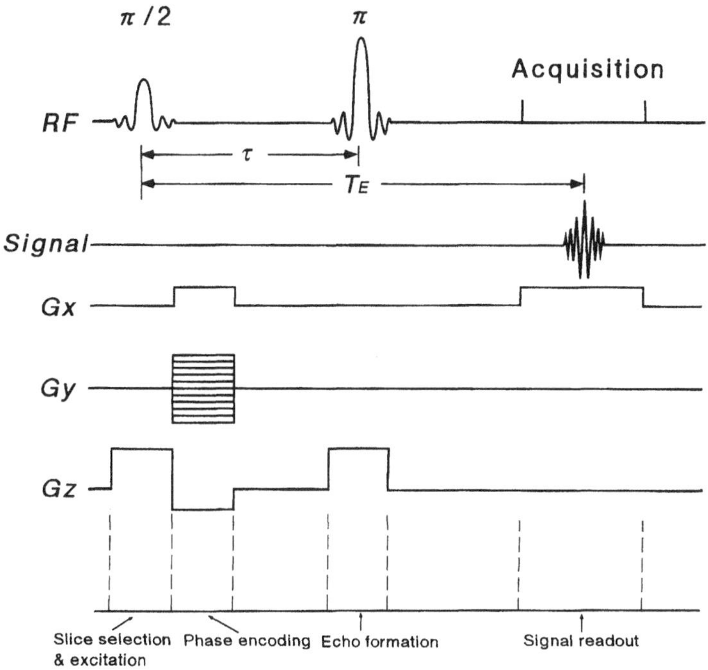

Figure 4.38 Timing diagram for the spin-echo NMR imaging pulse sequence.

^{1}H–NMR images get their contrast from the differences in ^{1}H densities, ^{1}H–NMR relaxation times, and self-diffusion coefficients in various parts of the sample, and the distributions of them are mapped out [88]. The spin-echo method is used combined with spin-warp imaging to obtain ^{1}H T_1 and T_2 images and ^{1}H T_1 and T_2 enhanced images.

4.9.2.1 Spin Density and Relaxation Times

Several NMR imaging techniques are available to measure the distributions of the spin density ρ and relaxation times T_1 and T_2 in a sample. The techniques include the spin-echo method (for ρ, T_1, T_2, and T_1- and T_2-enhanced images), the inversion recovery method (for T_1-enhanced images), and the gradient-echo method (for T_1-enhanced images) [88]. The spin-echo method modified as the spin-warp imaging is most widely used in commercial instruments. The magnetization of the sample M is expressed as the attenuation of the equilibrium magnetization, M_0 by

$$M = M_0[1 - \exp(-T_R/T_1)]\exp(-T_E/T_2), \qquad T_R \gg \tau \qquad (117)$$

where the T_R is the repetition time and the T_E is the echo time. The τ is the time interval between the $\pi/2$ and π rf pulses and $T_E = 2\tau$, as shown in Figure 4.38. The ρ distribution in the sample is measured under long T_R and short T_E. In a multicomponent sample such as a gel sample, an image from the mobile component that has long T_R and short T_E is mainly detected with this method.

The T_1 distribution in the sample is determined by collecting its several images with systematic variation of the T_R with short T_E. The M is expressed as Eq. (118) obtained from Eq. (117):

$$M = M_0[1 - \exp(-T_R/T_1)] \tag{118}$$

The T_1 of a pixel in the obtained image is calculated from the fitting curve for the $T_R - M$ of the pixel. The contrast of the obtained image is based on the distribution of the T_1 in the sample. The T_1-enhanced image is given by the same single measurement with short T_R ($T_R < T_1$) and short T_E ($T_E < T_2$). In the measurement, spins with longer T_1 in the sample are more saturated with short T_R, and signals from spins with shorter T_1 are detected as brighter images.

Similarly, the T_2 distribution in the sample is determined by collecting several images with varying T_E under long T_R. The M is expressed as Eq. (119) obtained from Eq. (117):

$$M = M_0 \exp(-T_E/T_2) \tag{119}$$

The T_2 of a pixel in the obtained image is calculated from the fitting curve for the $T_E - M$ of the pixel. The T_2-enhanced image is obtained by the same single measurement with long T_R ($T_R > T_1$) and long T_E ($T_E \approx T_2$). In this case, the intensity of signals from spins with longer T_2 is higher than that from shorter ones.

4.9.2.2 Diffusion Coefficient

Diffusion coefficients in a sample are measured by NMR spectroscopy using the pulsed-field-gradient spin-echo (PGSE) pulse sequence [41, 42, 89]. The magnetic field gradient is applied in a single direction to encode the location of spins in this method. Common principles using field gradients are employed in the PGSE method and in NMR imaging. Spatial distribution of self-diffusion coefficients of translationally mobile spins in a sample can be determined by PGSE NMR imaging, whereas the spatially diffusion coefficient values of them are obtained by the normal PGSE method. In PGSE NMR imaging, the $\pi/2 - \pi$ rf pulse sequence based on the spin-echo method is used, as shown in Figure 4.39. The first pulsed-field-

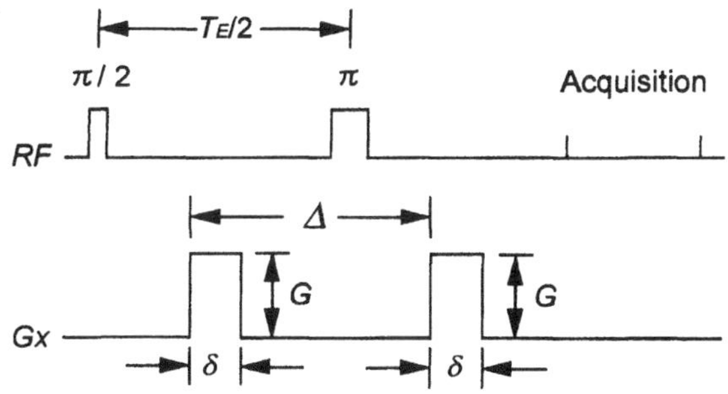

Figure 4.39 Timing diagram for the PGSE NMR imaging pulse sequence for determining self-diffusion coefficients.

gradient (G) causes the spins to dephase at a spatially dependent precession rate, and, after an interval Δ, the second gradient causes the spins to refocus at the same spatially dependent precession rate. When the sample molecules move translationally during Δ, their spins are located under different fields and do not refocus correctly. As the result, the echo signal intensity is attenuated according to Eq. (120).

$$\ln(M/M_0) = -T_E/T_2 - \gamma^2 G^2 D\delta^2(\Delta - \delta/3) \quad (120)$$

where T_E is the echo time, ($T_E/2$ is an interval between the $\pi/2$ and π rf pulse), D is the diffusion coefficient, and δ is the duration time of the gradient G. The M values are measured with changing δ or G, and $\ln(M/M_0)$ values are plotted against $\gamma^2 G^2 \delta^2(\Delta - \delta/3)$. The D is determined as the slope of a plotted line, and the results are mapped out at the selected section in the sample.

In the case of the diffusion of solvents, ions, and molecules contained in a gel sample, their mobility is restrained, and their diffusion coefficients become small [90]. If we want to measure D values of such samples, we should employ large G or long Δ in PGSE NMR imaging. However, making G large causes a disturbance of the magnetic field. The long Δ inevitably requires long T_E, with the consequence that the effect of T_2 relaxation on the echo signal intensity is not negligible. In such a case, as an alternative approach, stimulated echo acquisition mode (STEAM) imaging is used [91, 92]. Figure 4.40 is the STEAM imaging pulse sequence. The second $\pi/2$ rf pulse stores the magnetization of a sample along the z-direction, and the third one recovers the magnetization. The spins diffuse with signal attenuation owing mainly to the T_1 relaxation during the interval between the two rf pulses T_1. We can employ long Δ using long T_I for measuring small D values because the T_1 values of the sample for which mobility is

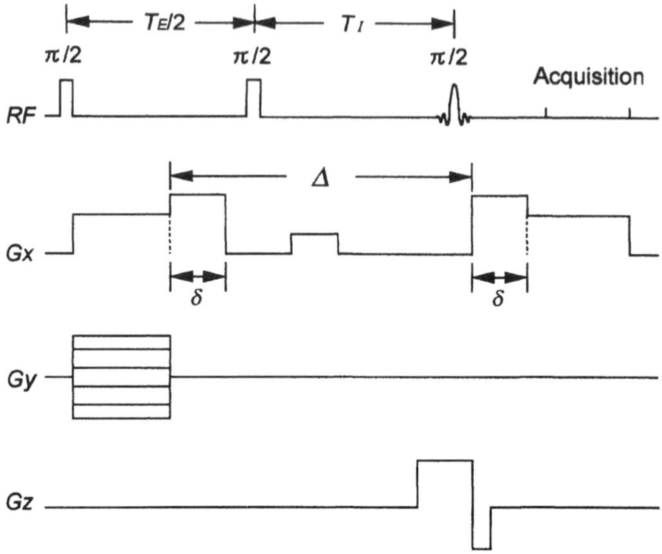

Figure 4.40 Timing diagram for the STEAM NMR imaging pulse sequence for determining self-diffusion coefficients.

restrained are generally longer than the T_2 values. The D is determined according to Eq. (121).

$$\ln(M/M_0) = -T_E/T_2 - T_I/T_1 - \gamma^2 G^2 D\delta^2(\Delta - \delta/3) \tag{121}$$

The result of the distribution of the D's in the sample is mapped out, and the D image is obtained.

4.9.2.3 Hard Materials

Many polymers are used as hard materials. It is important to characterize such hard polymer samples for getting information on their properties. Some of them, like resins, are insoluble in solvent, and we must characterize them in the solid state. When such samples are measured by NMR imaging, special techniques are necessary. Because the T_2 of hard materials is too short (<1 ms) as compared with the T_E, their NMR signals are broad and show low SNR and poor resolution.

Several techniques for NMR imaging such as MAS [93], the magic sandwich echo method [94], and the multiples method [95] are available for working with hard materials. Such techniques are not common methods performed by commercial instruments today. Many efforts are being made to create a useful NMR imaging technique for measuring hard materials [234].

4.9.2.4 NMR Imaging Equipment

The system for NMR imaging experiments consists of the normal NMR spectrometer, data processing system, and field gradient equipment containing gradient generators, controls, amplifiers, and coils. The development of computer technology led to a quick-image process for treating much NMR data.

Careful adjustments of the field gradients and rf pulses are required for obtaining clear and artifactless images. For these, experimental conditions such as gradient and rf pulse strength, duration times of gradients and rf pulses, and repetition time are adjusted by using a standard phantom. Samples are placed in a sealed cell and must be fixed in the center of the probe coils. In the experiments for relaxation times, careful degassing of a sample should be made for getting correct values.

Figure 4.41 is a photograph of a ready-made probe for NMR imaging (JEOL). This is a probe for polymer materials, small animals, insects, and plants. We can apply stimuli (stress, light, and changes in temperature, pressure, pH, and solvent) to samples by using glass or solid plastic equipment. The image of spatial change in chemical and physical properties caused by such stimuli can be measured and obtained.

An active shield gradient coil [96, 97] is recommended for measuring self-diffusion coefficients to prevent eddy currents around gradient magnetic coils from being induced.

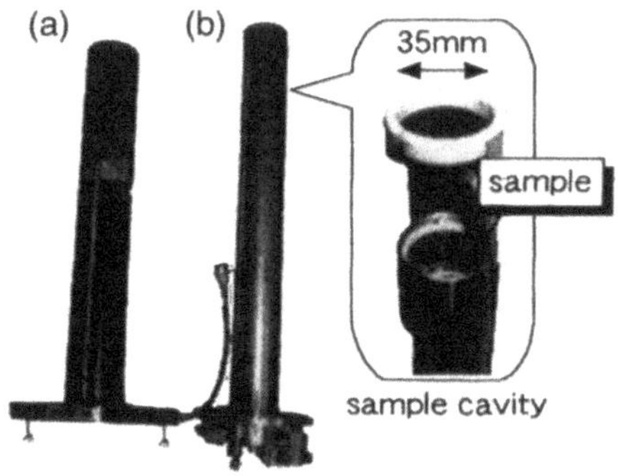

Figure 4.41 Probe for NMR imaging (JEOL NM-GIM 270). (a) field gradient coil part and (b) probe part.

4.10 Applications to Some Polymer Systems

4.10.1 POLYMERS IN SOLUTION

The property–structure relationship of polymers leads us to improve properties of common polymers. The polymer structures have been determined through many kinds of spectroscopy. In particular, solution NMR is now an indispensable tool for analyzing the detailed fine structure of polymers. Of course, we have many different types of polymers. Poly(vinyl alcohol) (PVA) and PVA-related copolymers introduce one of those typical polymers in this section and were selected for analysis because they have many kinds of microstructures such as tacticity, sequence distribution, and several kinds of irregular structures. The microstructures of PVA and poly(vinyl alcohol-acetate) copolymers were studied by 1-D and 2-D NMR techniques. The assignments of all CH proton resonance of PVA relating to 1,2-1,4-glycol and terminal 1,2-glycol linkages were made. The concentrations of 1,2-glycol linkages located at inner and terminal positions of a polymer chain were determined to be about 1.61 and 0.06 mol %. End groups of PVA arising from chain transfers to solvent, monomer, and initiator or from the termination of hydrogen abstraction could readily be determined by using α, α, β-trideuterated PVA. The short chain branches consisting of two monomer units (butyl branches) resulting from intramolecular chain transfer to methine carbons have also been identified. Other structural irregularities, polyene structures arising from heat treatment, were analyzed by 2-D COSY NMR spectroscopy. The sequence distributions of (PVA–Ac) copolymers were studied by ^{13}C–NMR spectroscopy. In the saponification reactions of poly(vinyl acetate), the presence of an irregular bonding could make the rate of saponification slower. The relationship between the sequence distribution and a saponification distribution was also discussed, and the saponification distribution of (PVA–Ac) copolymers was evaluated from a block character η obtained by ^{13}C–NMR spectroscopy. The combination of 1-D and 2-D NMR techniques provides a powerful method for gaining information on the microstructure of these polymers.

It has clearly been established that the structures of polymer chains are an important factor in determining their properties. In obtaining the structure–property relationship of a polymer, the most important work is to collect high-quality information about its structure. Various kinds of microstructural information are present in polymers such as tacticity and sequence distribution for major structures and end groups in addition to irregular bonds and short branching within minor structures. It is therefore very important to analyze these microstructures using advanced quantitative techniques. NMR spectroscopy is a powerful method for characterizing synthetic polymers. High-resolution ^{1}H and ^{13}C–NMR spectroscopy and, more recently, multidimensional homonuclear and heteronuclear correlation spectroscopy have revealed structural features of complex polymers in detail [87, 98–106]. In particular, two-dimensional (2-D) NMR proved to be a useful technique to identify small amounts of irregular structures in synthetic polymers [107–110]. This study focuses on the application of conventional 1-D and 2-D NMR to determination of the microstructures of PVA and also describes the structure–property relationship. PVA is usually obtained by radical polymerization of vinyl acetate followed by saponification of poly(vinyl acetate). Physical and mechanical properties of PVA are influenced by the degree of polymerization as well as the degree of saponification. Moreover, a small amount of irregular structures also affects properties such as durability, stability of the concentrated solution of PVA, and so on. As shown in Figure 4.42, there are many kinds of microstructures in PVA. Each of the items has been studied using conventional 1-D and a variety of 2-D techniques.

PVA samples (Kuraray Co. Ltd.) considered here are commerically available and consist of derivatives of poly(vinyl acetate) (PVAc) synthesized from vinyl acetate (VAc) with α, α'-azobisisobutyronitrile (AIBN) as an initiator in methanol (viscosity-average molecular weight $P_n = 1700$, degree of saponification; $DS = 98.5$ mol% and $P_n = 170$, $DS = 99.9$ mol%). An α, α, β-trideuterated PVA

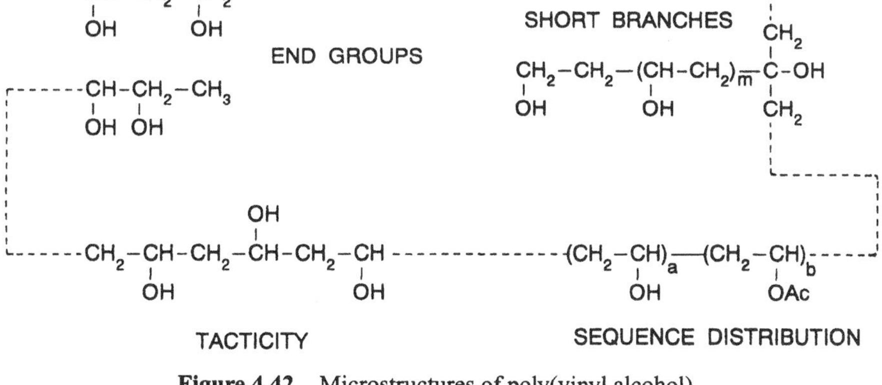

Figure 4.42 Microstructures of poly(vinyl alcohol).

(d_3–PVA) was derived from a sample taken during polymerization of d_3–VAc by using AIBN in methanol at 60 °C ($P_n = 1650$, $DS = 98.0$ mol %) The d_3–VAc was synthesized through the intermediate acetylene–d_2 ($CD \equiv CD$) and acetic acid–d_1 (CH_3COOD). Poly(vinyl alcohol–acetate) copolymers with about the degree of saponification of 90 mol % were prepared in three ways:

1. Reacetylation: PVA was dissolved and acetylated in a mixture of acetic acid and water by HCl as catalyst at 30 °C. The reaction was controlled by the relative amounts of acetic acid and water.
2. Acid hydrolysis: PVAc was dissolved in methanol and hydrolyzed by H_2SO_4 at 63 °C for 2 h. The concentration of H_2SO_4 was almost 0.17 N.
3. Alkali hydrolysis: PVAc was dissolved in methanol and hydrolyzed by NaOH at 40 °C. The degree of saponification was controlled by the molar ratio of PVAc and NaOH (2/100).

The ^{1}H–NMR spectra were obtained at 60 °C with a JEOL JNM–GX500 spectrometer at 500 MHz. The sample concentration was 5% (w/v) in perdeuterated dimethyl sulfoxide (DMSO–d_6), which provided the internal lock signal. Approximately 256–10,000 free induction decays were accumulated using a pulse width of 4 ms, a pulse interval of 5.0 s, and spectral width of 10 kHz.

Homonuclear 2-D correlation (COSY) spectra were acquired according to the method of Nagayama *et al.* [111] and the homonuclear Hartmann–Hahn experiment (HOHAHA) according to Bax and Davis [112]. The 2-D data were zero filled and weighted prior to Fourier transformation as appropriate. A sine-bell filtering function was used in both dimensions.

The ^{13}C–NMR spectra were obtained with the JNM–GX500 at 125 MHz. The sample concentration was 10% (w/v) in DMSO–d_6, D_2O, and a mixture of D_2O and perdeuterated methyl alcohol (CD_3OD). The spectra were recorded at 80 °C using a pulse width of 10 ms and pulse intervals of 3.0 s. The spectra were obtained after accumulation of 5000–20,000 scans. TMS was added as an internal standard to the DMSO–d_6 solution and sodium 3-trimethylsilylpropionate (TSP) to the D_2O or the mixture solution. The concentrated solution (12% by weight) viscosity was determined with a Brookfield viscometer.

4.10.1.1 Tacticity

Various 2-D NMR methods, including ^{1}H J-resolved spectroscopy, ^{1}H–^{1}H COSY spectroscopy, ^{13}C–^{1}H COSY spectroscopy, and INADEQUATE spectroscopy have been applied to determine absolute assignments of ^{1}H and ^{13}C–NMR spectra at

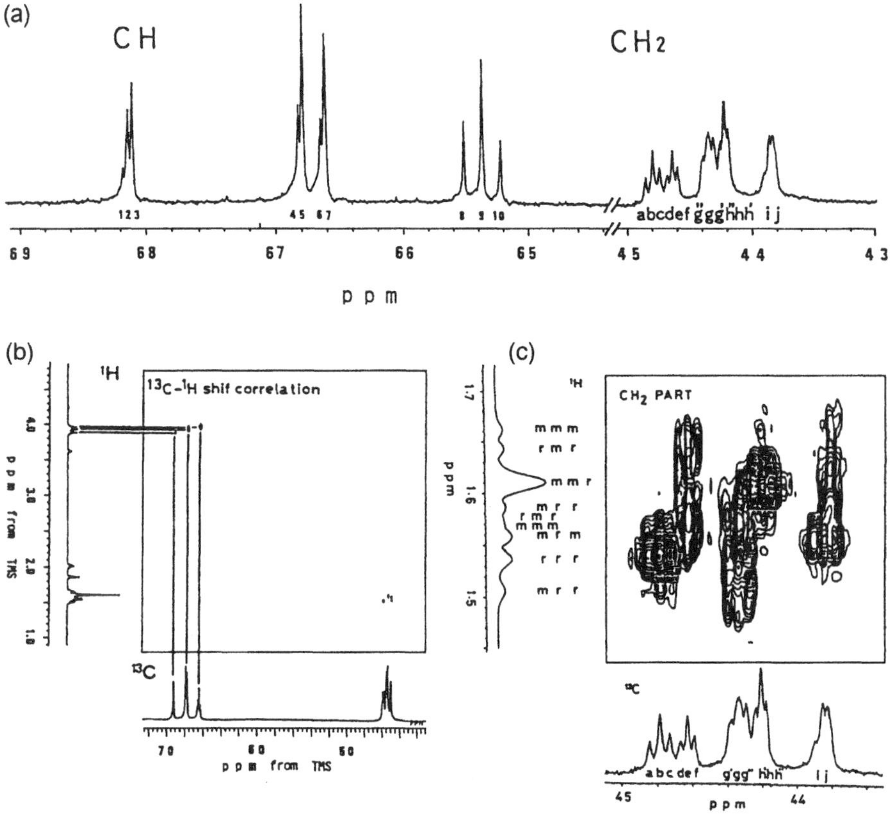

Figure 4.43 (a) ^{13}C–NMR spectrum of poly(vinyl alcohol) in a mixture of H_2O and D_2O (1:1). (b) F1(^1H)-axis broad-band decoupled C–H COSY spectrum. (c) Expanded view of methylene region of poly(vinyl alcohol).

the triad–tetrad and pentad–hexad level shown in Figure 4.43. The results of the assignment of the ^{13}C spectrum of PVA are also shown in Table 4.4 [98, 113].

The configurational assignments of PVA have been almost finished through these 2-D NMR methods.

4.10.1.2 Sequence Distribution

It is very important for understanding structure–property relationships to determine the sequence distributions of vinyl acetate and vinyl alcohol units as well as the degree of saponification. The former, in particular, strongly affects the solution and solid properties (e.g., the viscosity, the surface activity, the melting point

Table 4.4 Assignments of the ^{13}C spectra of PVA.

	Peak No.	*Configuration*	*Chemical Shift (ppm)*
CH	1	*rmmr*	68.18
	2	*mmmr*	68.14
	3	*mmmm*	68.10
	4	*rmrr*	66.83
	5	*mmrr*	66.79
	6	*rmrm*	66.65
	7	*mmrm*	66.61
	8	*rrr*	65.53
	9	*mrrr*	65.37
	10	*mrrm*	65.22
CH_2	a	*mrrrm*	44.85
	b	*mrrrr*	44.79
	c	*rrrrr*	44.74
	d	*mrmrm*	44.68
	e	*mrmrr*	44.64
	f	*rrmrr*	44.59
	g″,g,g′	*mrr*	44.38, 44.34, 44.30
	h″,h,h′	*mmr*	44.24, 44.22, 44.19
	i,j	*mmm,mrm*	43.86

[114] and so on). The determination of the sequence distribution of (PVA–Ac) copolymers using ^1H and ^{13}C–NMR techniques have been reported by several researchers [115–118]. Figure 4.44 shows ^{13}C–NMR spectra in the region of CH_2 carbons in (PVA–Ac) copolymers with various degree of saponification. Resonances (X) at 41–43 ppm were assigned to the methylene carbons of poly(vinyl acetate) (see Figure 4.44(a)). With an increase in the degree of saponification, the peak (X) gradually decreases in intensity and, conversely, the peak (Z) at around 47 ppm increases. After the middle peak (Y) increases slightly, the intensity falls off again. These three peaks have been assigned to the three dyad sequences: acetate–acetate, alcohol–acetate (including acetate–alcohol), and alcohol–alcohol with decreasing field strength. The mole fractions of the dyad sequences will be abbreviated (OAc–OAc), (OAc–OH), and (OH–OH), respectively. Thus, the integrated intensities of the carbon resonance are directly proportional to the number of carbons contributing to those signals. A quantitative description of the sequence distribution and a chemical composition are possible by applying the following equations to the integrated areas of the three methylene carbons

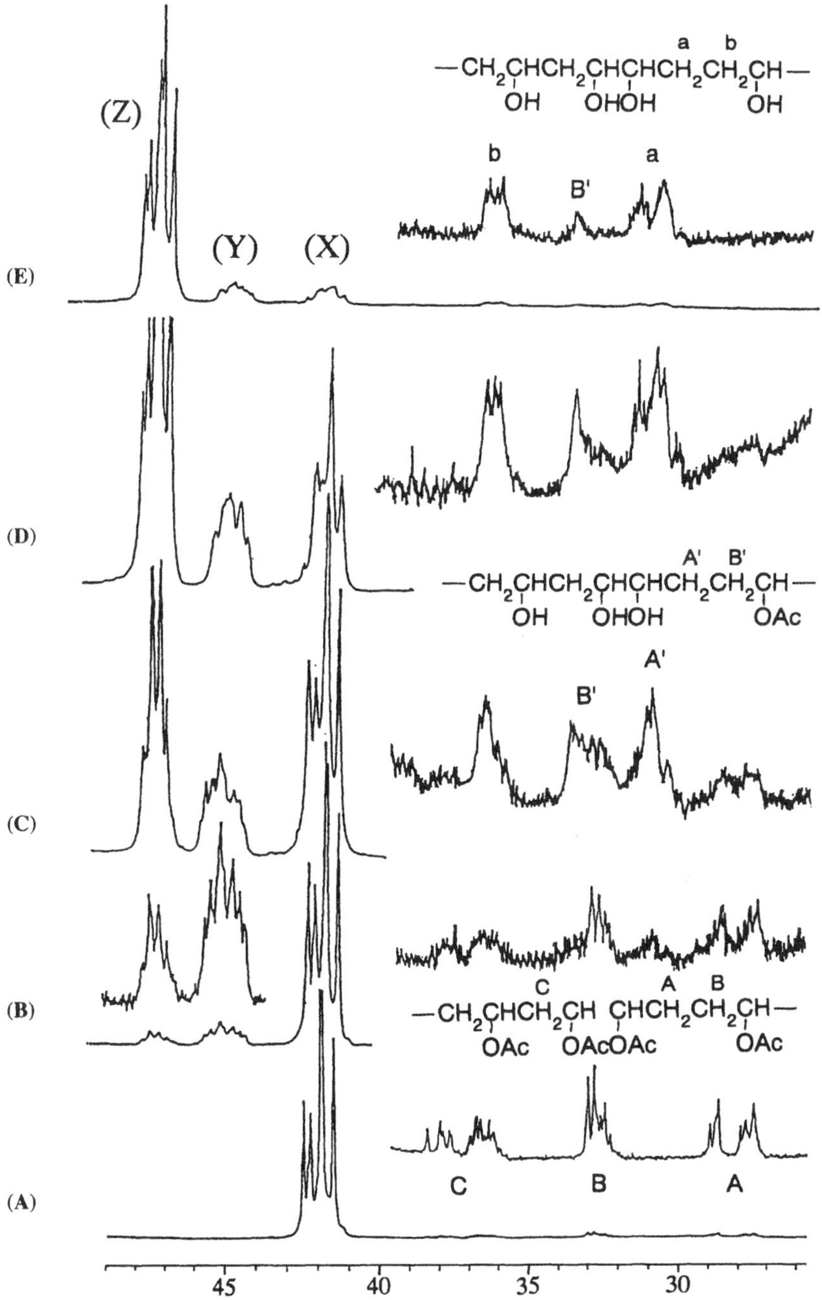

Figure 4.44 ^{13}C–NMR spectra of methylene carbon of (PVA–Ac) copolymers prepared by alkali hydrolysis. Degree of saponification (mol %): (a) 0, (b) 6, (c) 45, (d) 63, and (e) 88.

dyads:

$$(\text{OH}) = (\text{OH–OH}) + (\text{OH–OAc})/2 \qquad (122)$$

$$(\text{OAc}) = (\text{OAc–OAc}) + (\text{OH–OAc})/2 \qquad (123)$$

$$\eta = (\text{OH–OAc})/2 \cdot (\text{OH})(\text{OAc}) \qquad (124)$$

where (OH) is the mole fraction of vinyl alcohol units, (OAc) is the mole fraction of vinyl acetate, units and η is the block character.

The η value is a useful guide to characterize a sequence distribution in binary copolymers; it takes $0 \leq \eta < 1$ for bulkier distributions, $\eta = 1$ for completely random cases, and $1 < \eta \leq 2$ for alternate like cases. The sequence distributions of each sample in Figure 4.44 are bulky with $\eta \leq 0.45$. The samples prepared by reacetylation from PVA are close to random distributions with $\eta = 1.0$. In the sample obtained by acid hydrolysis, the sequence distributions are bulky, but their randomness increases compared with the copolymers prepared by alkali hydrolysis for which η =0.70–0.80.

On the other hand, some small peaks are observed in the region between 27 and 39 ppm in the spectra. As indicated in Figure 4.44, these peaks are identified as the methylene carbons belonging to the 1,2- and 1,4-linkages. Peak assignments have been made by comparison with the spectra of polymers like PVAc, PVA, and (PVA–Ac) copolymers as well as the empirical additivity rules [119, 120]. Although acetyl groups 1, 2, 3 (as seen from Eqs. (122)–(124)) could readily be converted to hydroxyl groups, the acetyl group 4 would not be converted as far as the degree of saponification of 90 mol %.

Consequently, the presence of the irregular bonding could make the rate of the saponification reaction slower. It is generally believed that the sequence distribution of PVA–Ac copolymers becomes random if the rate decreases. In the PVA–Ac copolymer with the irregular bonding of about 2.2 mol %, the block character η is about 0.50, which is larger than that of the commercially available copolymers ($\eta = 0.45$) having the irregular bonding of about 1.7 mol %. The block character increases; namely, the randomness increases with increasing content of the irregular bonding. The mechanism of saponification by the alkali catalyst has been reported [121], that is, that the catalyst is adsorbed on the free hydroxyl groups and that this induces successive hydrolysis in the acetyl groups of the direct neighbors. Therefore, the rate of the reaction depends on the distance between the hydroxyl group and acetyl group. To accelerate the rate of saponification reaction, the structures of the 1,2-sequences are favorable; however, the 1,4-sequences are unfavorable structures. The relationship between the sequence distribution and the distribution of a degree of saponification (or composition distribution) has not been fully elucidated. Because the sequence distribution is defined by the distribution of each monomer in a polymer chain, the distribution of a degree of saponification indicates the distribution of each polymer chain with various degrees of saponification.

The η value is substituted for the sequence distribution, and D_c for the composition distribution. D_c indicates the index of deviation of a compositional uniformity and is defined as follows [122]:

$$D_c = \Sigma |W_i \Delta A_{ci}| \tag{125}$$

Here W_i represents the weight fraction at the fraction number i, and ΔAc_i represents the difference between the mean degree of saponification and the degree of saponification at the fraction number i. D_c is the sum of the absolute value of the product of W_i by ΔA_{ci}. As D_c decreases, the composition distributions become sharp. The D_c values of (PVA–Ac) copolymers prepared in three ways were determined by the method of Spencer [123] as follows:

1. reacetylated copolymer, 0.65–0.75
2. acid-hydrolyzed copolymer, 0.75–0.80
3. alkali-hydrolyzed copolymer, 1.15–1.20

Figure 4.45 shows the relationship between η versus D_c. This reveals the strong correlation between η and D_c. D_c decreases with increasing η; namely, the composition distribution becomes sharp as the randomness of the sequence distribution

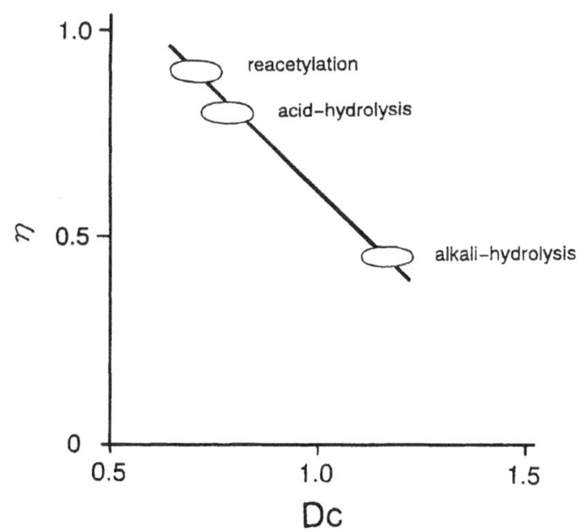

Figure 4.45 Relationship between η and D_c.

increases. These results suggest that the composition distributions of (PVA–Ac) copolymers can be readily evaluated from the η values. Figure 4.46 shows ^{13}C–NMR spectra of methylene carbons in (PVA–Ac) copolymers with and without heat treatment (at 135 °C and for 10 and 12 hr). In the figure, the intensity of (OAc–OAc) lines slightly decreases. The intensity of the (OAc–OH) lines, on the other hand, increases. The η values of the copolymers before and after heat treatment could be determined to be 0.45 and 0.54, respectively. Namely, the randomness of the sequence distribution of copolymers increases with heat treatment. The degree of saponification of each copolymer was nearly equal; therefore, the heat treatment could give rise to the change in the sequence distribution but not to change the degree of saponification. These results reveal that the ester-exchange reactions between acetyl groups and hydroxyl groups occur with both inter- and intra-molecular chains. Heat treatment of (PVA–Ac) copolymers improves the properties—in particular the surface activity [124] and the ability of foam formation to some extent.

4.10.1.3 1,2-Glycol Linkage

1,2-glycol linkages as irregular bondings are present along the PVA chains. The determination of 1,2-glycol linkages is important because the irregular bonding produces serious undesirable effects on the properties of PVA such as thermal deterioration and reduced mechanical properties. Chemical assay methods [125] are usually adopted for 1,2-glycol bonding determinations, but there is some disagreement between methods for determining 1,2-glycol content of low-molecular-weight PVA (e.g., viscometry after 1,2-cleavage with $NaIO_4$ and $NaIO_4$ titrimetry). Therefore, new techniques are continually being developed. ^{1}H–NMR spectra of PVA dissolved in DMSO–d_6 are shown in Figure 4.47 in which hydroxyl proton signals were removed from the observable range by adding a few drops of trifluoroacetic acid to the solution. The expanded peaks that resonated in the chemical shift region of 3.0–3.8 ppm are due to methine protons evaluated against the standard chemical shifts. Triplet resonance at 3.55 ppm has been identified as the methyrol proton ($-CH_2CH_2OH$). The asymmetrical triplet due to unresolved stereo sequence effects at about 1.03 ppm has been assigned to methyl group belonging to terminal 1,2-glycol linkages [126]. In addition to these known peaks, undetermined ones also appeared.

Specific sequence assignments can be made from the COSY spectrum shown in Figure 4.48. In the COSY spectrum, the normal spectrum is contained along the diagonal, and symmetric pairs of cross peaks connecting peaks that correspond to J-coupled protons. The assignment of the peaks in the downfield region from 3.1 to 3.7 ppm is made as follows:

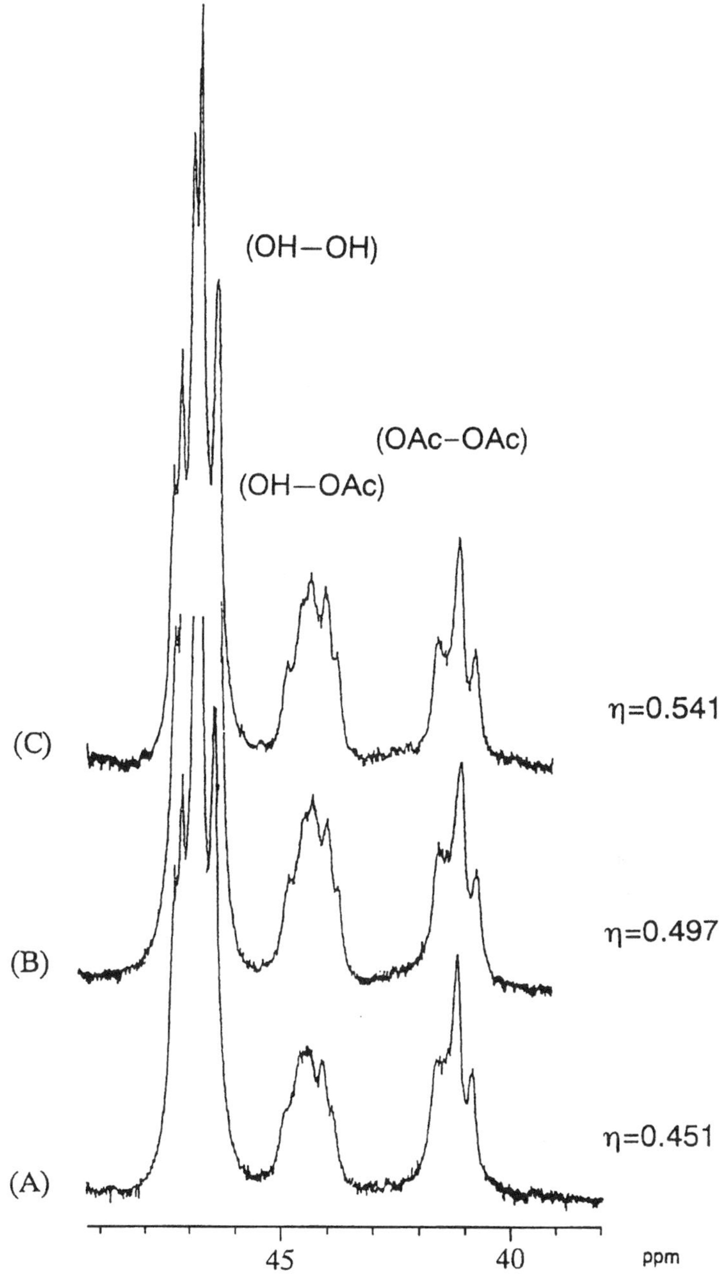

Figure 4.46 ^{13}C–NMR spectra of methylene carbon in (PVA–Ac) copolymers treated with heat: (a) original, (b) at 135 °C for 10 h, and (c) at 135 °C for 12 h.

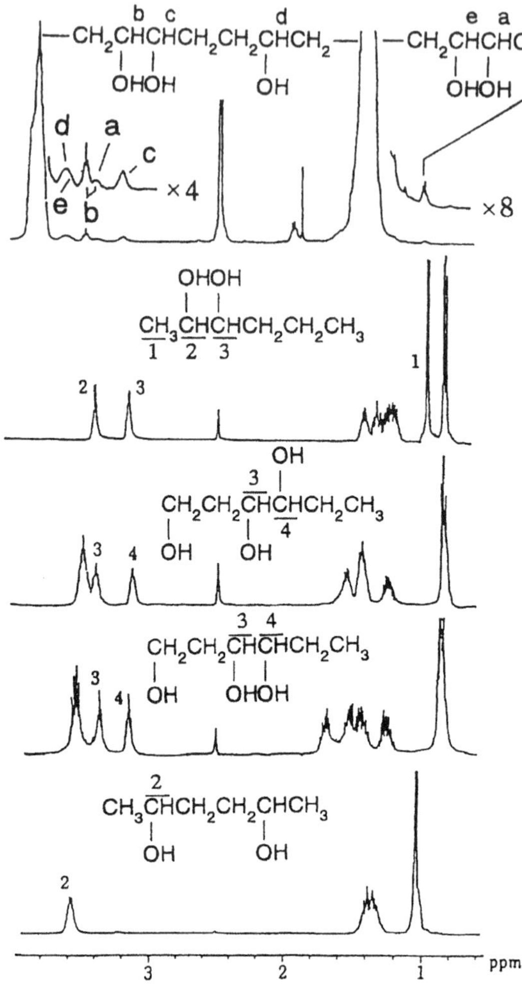

Figure 4.47 ^{1}H–NMR spectra of poly(vinyl alcohol) and its model compounds with irregular bonds in PVA.

The highest field cross peak (cross peak 1) is assigned to CH_3 (1.03 ppm) and the methine proton H_B (3.43 ppm). The methine proton H_B also couples to the methine proton H_C. Therefore, the cross peak comes from HB and H_C coupling. Two cross peaks (cross peak 2, 4) correlated with H_B were observed in the figure.

The assignment of the H_C proton can be made from the HOHAHA spectrum. In the HOHAHA spectrum, the cross peak connecting peaks due to long-range

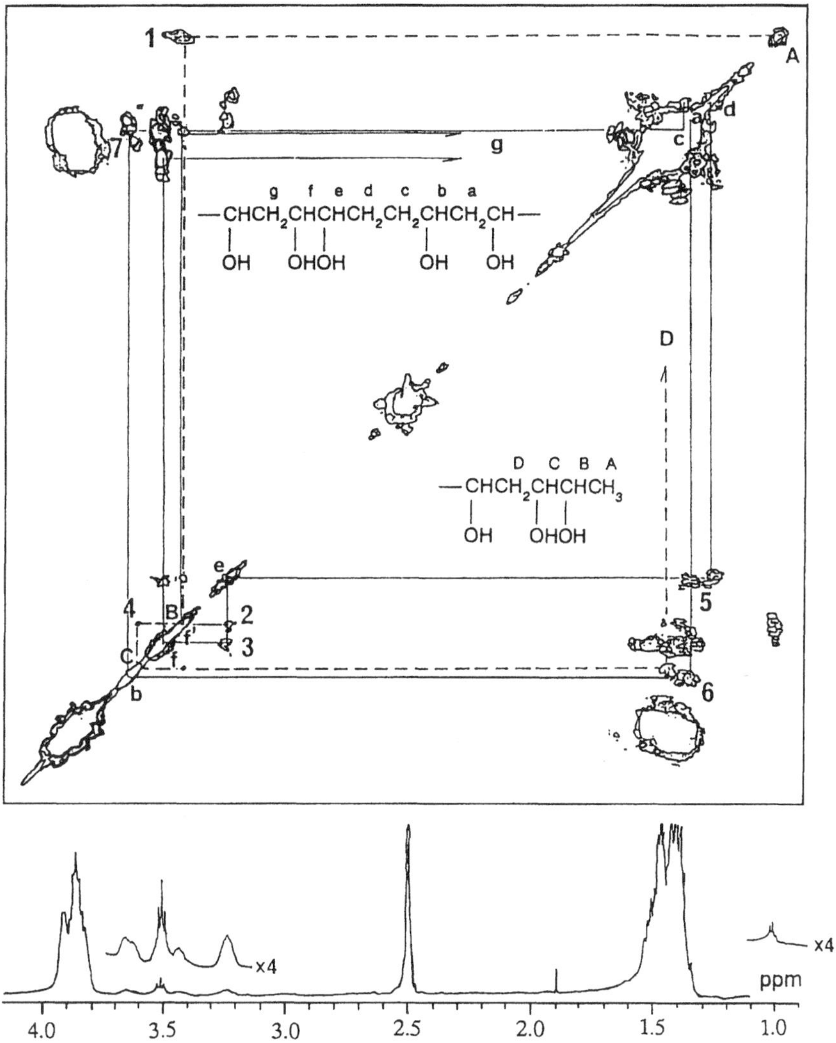

Figure 4.48 COSY spectrum of poly(vinyl alcohol) in DMSO–d_6.

J-coupled protons can be observed. Then, the connectivities of more than three bonds through the long-range coupling can readily be determined (Figure 4.48). The cross peaks due to H_A and H_C (A/C) as well as H_A and H_B (A/B) were observed; however, the cross peak due to H_A and CH(e) (3.23 ppm) was not observed. Consequently, the peak at 3.62 ppm can be assigned to the H_C proton.

The determination of 1,2-glycol linkages located at the inner position in a polymer chain is made as follows: As shown in Figure 4.48, the peak at 3.23 ppm has been assigned to CH(e). The CH_2(d) couples to the CH(e) and CH_2(c). Therefore, the cross peak 5 comes from the CH_2(d) and CH(e) coupling. Then, cross peaks 2 and 3 are due to CH(e) and CH(f) coupling. The CH(f) peak is probably two decomposed peaks (3.43 and 3.52 ppm). This phenomenon may be due to threo and erythro stereo configurational isomers around 1,2-glycol linkages, whereas cross peaks (6) and (7) are due to CH_2(a)–CH(b) and CH(b)–CH_2(c) coupling, respectively, because the CH(b) couples to both the CH_2(a) and CH_2(c). The CH(f) peaks at 3.43 and 3.52 ppm cannot be identified as erythro or threo isomers from the 2-D spectra.

Using the proton chemical shifts of some model compounds (1,3-4(threo)-hexanetriol, 1,3-4(erythro)-hexanetriol), the peak at 3.43 ppm could be assigned to the erythro isomer, and the peak at 3.52 ppm to the threo isomer, as shown in Figure 4.48. Therefore, we could accurately assign all the CH proton resonances of PVA relating to 1,2-1,4-glycol and terminal 1,2-glycol linkages. From the signal intensity measurements, the concentrations of 1,2-glycol linkages located at the terminal and inner positions of a polymer chain were determined to be about 0.06 and 1.61 mol %, respectively.

Figure 4.49(a) shows the ^{13}C–NMR spectrum of PVA in D_2O. The expanded signals have been assigned, as illustrated in the figure, according to the previous data [127]. In Figure 4.49(a), ^{13}C peaks of terminal 1,2-glycol linkages were not observed because of the low concentration. The ^{13}C signal of each carbon of 1,2-1,4-glycol linkages should basically be split into eight peaks arising from the stereochemical configuration of both of the neighboring hydroxyl groups to 1,2-glycol linkages. The 1,2-glycol linkages were cleaved to a long-chain aldehyde by $NaIO_4$. After scission with $NaIO_4$, the aldehyde was reduced with $NaBH_4$, resulting in PVA having both the end groups of methyrols ($-CH_2OH$). Figure 4.49(b) also shows the ^{13}C–NMR spectrum of the PVA in D_2O.

The ^{13}C-peaks arising from 1,2-glycol linkages were completely undetectable. Some new peaks were observed instead. These signals were probably assigned to the methyrol end groups on the basis of the ^{13}C additivity rules and the relative intensity, as shown in the figure.

It is generally believed that normal VAc chain growth occurs in a head-to-head addition but that occasional head-to-head followed by tail-to-tail addition polymerization occurs (Figure 4.50). The growing radical chain mainly undergoes termination by chain transfer in the VAc polymerization. The chain-transfer reaction occurring with the normal polymer chain radical leads to an end group, which is an acetate ester of a primary alcohol. The chain transfer with abnormal polymer chain radicals would lead to a methyl end-group,

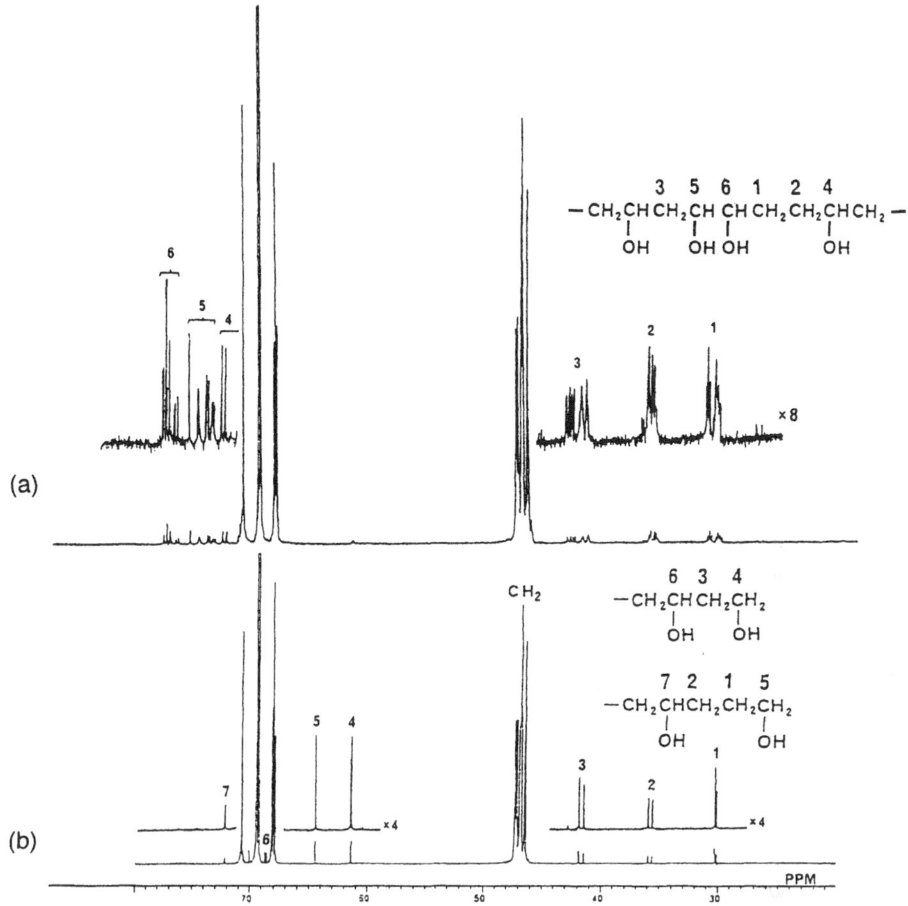

Figure 4.49 ^{13}C–NMR spectra of poly(vinyl alcohol) in D_2O: (a) original and (b) treated with $NaIO_4$ followed by $NaBH_4$.

The transfer constants are defined as

$$C = k_{tr}/k_p \tag{126}$$

$$C' = k'_{tr}/k'_p \tag{127}$$

where k_p and k'_p are the rate constants of normal and abnormal polymer propagation, and k_{tr} and k'_{tr} are also the rate constant of the chain transfer for the normal and abnormal polymer chain radical, respectively. These values indicate the situation in which chain transfer occurs. The C value for VAc or methanol is small (on the order of 10^{-4}), which suggests that the chain-transfer reaction occurs every 10^4 propagation reactions. That is, the degree of polymerization maximizes at about

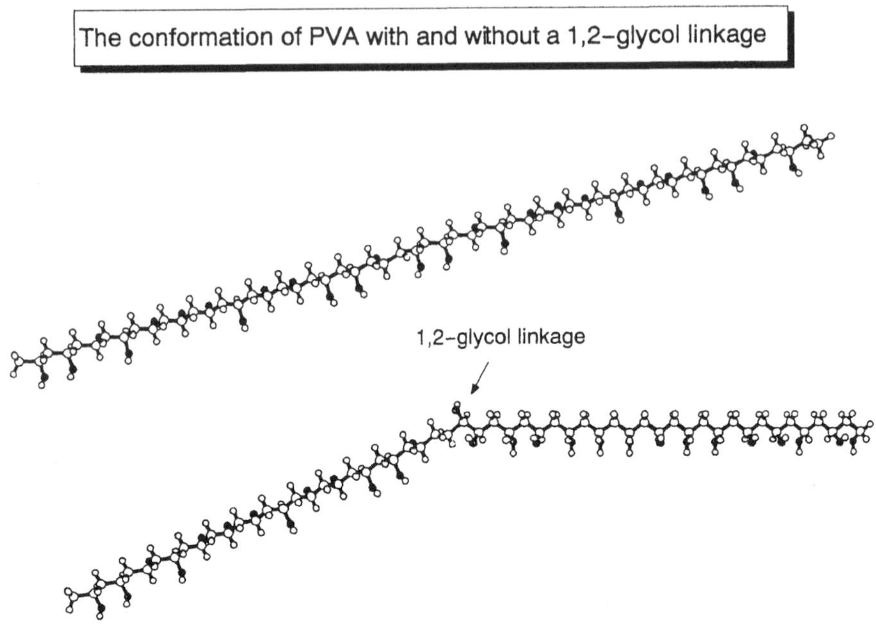

Figure 4.50 Structure of poly(vinyl alcohol) with and without a 1,2-glycol linkage.

10^4, whereas the value of C' can be calculated to be $3 \times 10^{-2}(= 0.06/1.68)$, which is much larger than that of C. Because 0.06 of a –CH(OH)CH(OH)CH_3 group is formed every 1.68 vicinal glycol groups, about 3% of the steps following abnormal addition result in chain transfer. The most important factor affecting the content of 1,2-glycol linkages is the temperature of polymerization, as was first pointed out by Flory and Leutner [128]. With an increase of the temperature, the 1,2-glycol content goes up; as a result, the degree of polymerization decreases. The properties of PVA seem to be affected by 1,2-glycol linkages even if the amount is very small (1–2 mol %). It has been reported that 1,2-glycol linkages are present in the non-crystal region but not in the crystal region [129]. As a result, 1,2-glycol linkages may be cited as a structure-confusing factor. Figure 4.51 shows the change in the viscosity of concentrated aqueous solution (12%) of two PVA samples with the same molecular weights ($P_n = 170$). Although the solution viscosity of commercial PVA, which contains about 1.7 mol% of the 1,2-glycol linkages, is almost stable with the elapse of time, the viscosity of PVA treated with $NaIO_4$ followed by $NaBH_4$, which has no 1,2-glycol linkage, increases gradually. The 1,2-glycol linkage in PVA is one of the most important factors in the stability of concentrated aqueous solutions. This phenomenon indicates that 1,2-glycol linkages may play an important role in interrupting the coagulation of polymer chains due to the

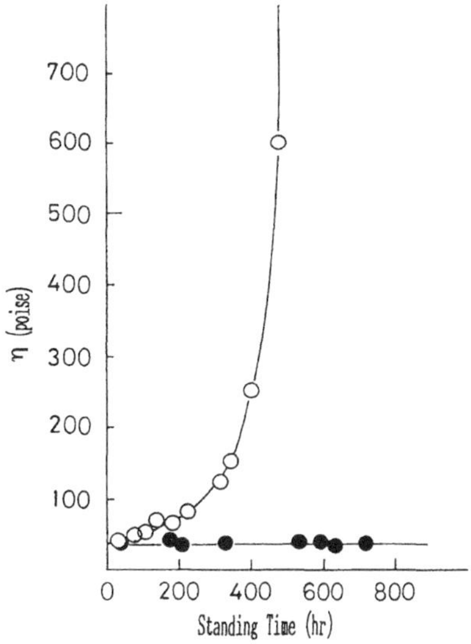

Figure 4.51 The change in the viscosity of aqueous solutions (12%) of two poly(vinyl alcohol) samples containing different amounts of 1,2-glycol linkages: (o) 0 mol % and (•) 1.7 mol % (degree of polymerization = 170 at 20 °C).

interhydrogen bonding. To investigate, we tried to determine the conformation of 1,2-glycol linkages from information about a J-coupling constant between both of the methine protons of 1,2-glycol linkages. The peak of each proton is too broad to determine the J-coupling constant. The broadness of these peaks probably arises from the configurational effect surrounding 1,2-glycol linkages and scalar couplings among neighboring methine and methylene protons. Then, a combination of the conformation obtained from the J-coupling constants of model compounds and the preferred conformations calculated from a molecular mechanics method (MM2) provided the determination of the conformations at 1,2-glycol linkages in PVA. As a result, the preferred conformation of 1,2-glycol linkages is a *gauche* conformer rather than a *trans* conformer. For a 1,3-glycol linkage as a regular structure, the *trans* zigzag conformation was preferable. Therefore, the stereo structure in PVA polymer chains may be bent at the point of 1,2-glycol linkages shown in Figure 4.50.

4.10.1.4 End groups

Chain transfer predominates over termination under the usual conditions of polymerization of vinyl acetate. Consequently, major end groups can arise from chain

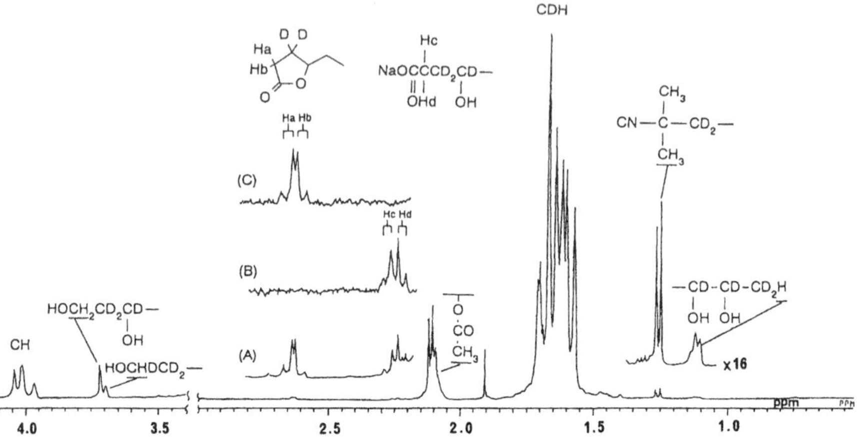

Figure 4.52 ^{1}H–NMR spectra of d_3-poly(vinyl alcohol) in D_2O at 80 °C: (a) normal spectrum, (b) spectrum in the presence of excess sodium hydroxide, and (c) spectrum in the presence of excess hydrochloric acid.

transfer reactions to monomers, solvents, polymers, initiators, and some kinds of additives. Thus, various kinds of end groups are present in PVA. The determination of end groups in PVA reveals the condition of polymerization in details such as temperature of polymerization, concentration of monomer to solvent, and kind of initiator and additives used. Figure 4.52 shows the ^{1}H–NMR spectrum of d_3–PVA in D_2O. In the figure, peaks at 1.50–1.80 ppm and 3.95–4.15 ppm are assigned to residual protonated CHD proton and CH proton of the main chain. The splitting of these peaks is due to only the configurational effect because of the absence of scalar couplings between methylene and methine. Peaks at 2.05–2.15 ppm are due to the CH_3 proton in the unconverted residual acetate. In the polymerization of d_3–vinyl acetate, the termination of hydrogen abstraction, the initiation by AIBN and chain transfer would lead to some kinds of end groups having nondeuterated hydrogen as described in the following paragraph.

The peak at the highest field, 1.13 ppm is assigned to one H of the methyl group in terminal 1,2-glycol linkages. The splitting into two or three peaks may be due to the tactility of the 1,2-glycol sequence and monomer sequence at the chain end. Resonances at 1.25 and 1.27 ppm are assigned to an initiation fragment, a 1-cyano-1-methylethyl group located at the chain end of a polymer molecule, because these peaks were not observed when PVA was prepared without AIBN. The splitting into two peaks of equal intensities is due to the existence of an adjacent asymmetric center in the polymer chain. Peaks at 2.20–2.30 and

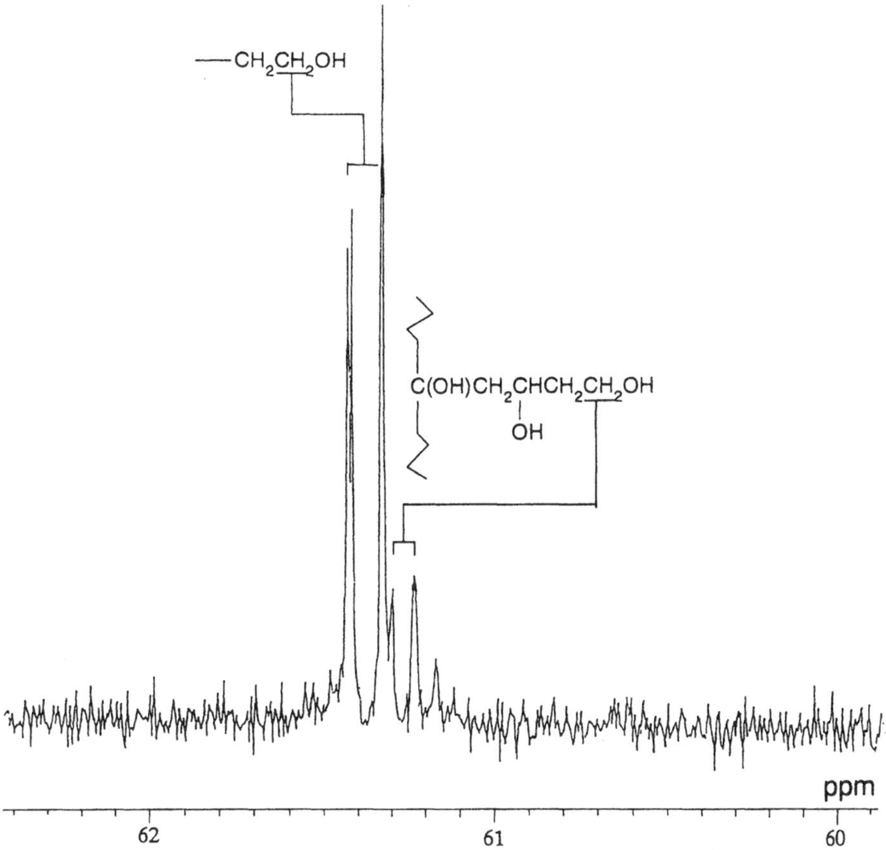

Figure 4.53 Expanded ^{13}C–NMR spectrum of poly(vinyl alcohol) in D_2O at 80 °C showing only terminal CH_2 resonances.

2.55–2.68 ppm, which show a typical AB spin system, are probably assigned to the methylene proton next to a carboxylic acid end group. When NaOH was added into the D_2O solution of PVA, the resonance centered at 2.63 ppm disappeared, and the resonance centered at 2.25 ppm increased in intensity, as shown in Figure 4.53(b). An opposite phenomenon occurred on addition of hydrochloric acid to PVA solution (see Figure 4.53(c)). These results and our previous article [127] indicate that carboxylic acid sodium salts and neighboring hydroxyl groups in PVA undergo an intramolecular reaction to form the γ-lactone structures in the presence of hydrochloric acid and reverse in the presence of sodium hydroxide. Thus, peaks Ha and Hb are assigned to methylene groups in the closed form, and peaks Hc and Hd are also assigned to those in the open form, as noted in the figure.

No carboxylic acid proton was observed in the spectrum of PVA in DMSO–d_6 at 60 °C (not shows here). These spectra revealed a pH-dependent equilibrium.

Among various kinds of end groups formed by the chain transfer reactions to monomer or polymer, one of the end groups was found to be present in the γ-lactone form under the acidic condition, whereas in the carboxylic acid sodium salt form under the alkaline condition no free carboxylic acid group was detected in ordinary PVA. The concentrations of these end groups depend on the degree of saponification of PVA used. With increasing degrees of saponification, the concentration of the open form increases.

Singlet peaks at 3.69 and 3.72 ppm may be assigned to the $-CD_2CDHOH$ and $-CD_2CH_2OH$ groups, respectively. The assignment was on the basis of the relative intensity. The latter group would occur as end groups formed by chain transfer with the methanol solvent, and the former group at 3.69 ppm occurs as end groups formed by the termination of hydrogen abstraction. The structures of these two end groups would be the same in an undeuterated polymer chain; therefore, it is impossible for the commercially available PVA to distinguish between methanol fragment groups and methyrol end groups formed by normal termination.

4.10.1.5 Short Branches

Although several researchers [129–132] indicated that small amounts of short branches were present in PVA, NMR signals of proton or carbon atoms at or near resonance in $-CH_2OH$, peaks at 61.33 and 61.42 ppm are assigned to a methylene carbon of the $-CH_2OH$ group in the main-chain end. These resonances are split into two peaks that are attributable to the configurational effect of near terminal, that is, random dyad tacticity. The peak at 61.42 ppm is further split into two peaks owing to triad tacticity. As seen from Figure 4.53 two weak peaks appear at 61.30 and 61.24 ppm, which are observed at a higher field than those of the main-chain end by about 0.1 ppm. On the basis of a backbiting mechanism proposed by Roedel [133], short branches in PVA may take either of the following model structures designated by B1 or B2, and ^{13}C chemical shifts of these structures are made by the ^{13}C additivity rules.

The structure of B1 corresponds to a monomer unit of vinyl alcohol, whereas the structure B2 is two monomer units. By comparison of observed chemical shifts with those of the model compounds, the signals at 61.30 and 61.24 ppm are assigned to the ultimate carbon atom in the end group in model B2 and two peaks also coming from essentially dyadic tacticity. The content of the PVA was about 0.027 B2 end groups 100 repeat units. When VAc was polymerized in bulk or methanol solution at the range of conversion, 40–90%, the concentrations of short branches in the PVAs were almost the same. The correlation between the concentration and the conversion was not found.

4.10.1.6 Polyene Structure

Treatment of PVA with heat results in a formation of conjugated polyene structures. Consequently, PVA is yellow. The conjugated polyene structure has usually been studied by ultraviolet (UV) spectrum [134]. Two-dimensional NMR experiments could be performed on the sample with thermal treatment to determine the polyene structure in detail. Figure 4.54 shows a COSY spectrum of PVA treated with heat in DMSO–d_6. COSY gives correlation maps showing the connectivity of spin–spin coupled protons. The F2 axis (horizontal) represents the ^{1}H–NMR spectrum of PVA, as shown in Figure 4.54. In the case of triene ($n = 3$, in which n indicates the conjugation sequence of $-(CH{=}CH)_3-$ starting from the H-1 signal at 6.08 ppm), a unique transverse connectivity with H-2 at 7.21 ppm is then possible. Hence H-2 is correlated with H-3. From H-3 on, the same mechanism reveals connectivity with H-4, H-5, and H-6. Figure 4.54 also shows other kinds of connectivity with monoene ($n = 1$) and diene ($n = 2$) structures. From analyzing 2-D COSY spectra, we found that three kinds of conjugated sequences were contained in the polymer. The ^{1}H–NMR spectrum was utilized in the quantitative analysis of PVA treated with heat. The concentration of n-polyenes, $-(CH{=}CH)_n$, was as follows: monoene ($n = 1$), 0.137 mol %; diene ($n = 2$), 0.083 mol %; triene ($n = 3$), less than 0.01 mol %. We concluded that various kinds of microstructures in PVA were determined by 1-D and 2-D NMR spectroscopy. Moreover, the determinations of these microstructures could reveal a detailed property–structure relationship of PVA.

4.10.2 SOLID POLYMERS

Polymer materials are almost always used as solids. It is well known that the properties of polymers depend on their structure and dynamics. For this reason, in order to design new polymer materials it is necessary to understand their property–structure relations more deeply. The X-ray diffraction method has significantly enhanced the structural determination of polymers with high crystallinity. However, most polymers have low crystallinity, and thus structural information about the noncrystalline region, which is the major component, cannot be obtained by X-ray studies. Therefore, the X-ray diffraction method has a limitation for the structural analysis of such systems. Chain segments in the noncrystalline region are sometimes in a mobile state, and thus the X-ray diffraction method provides

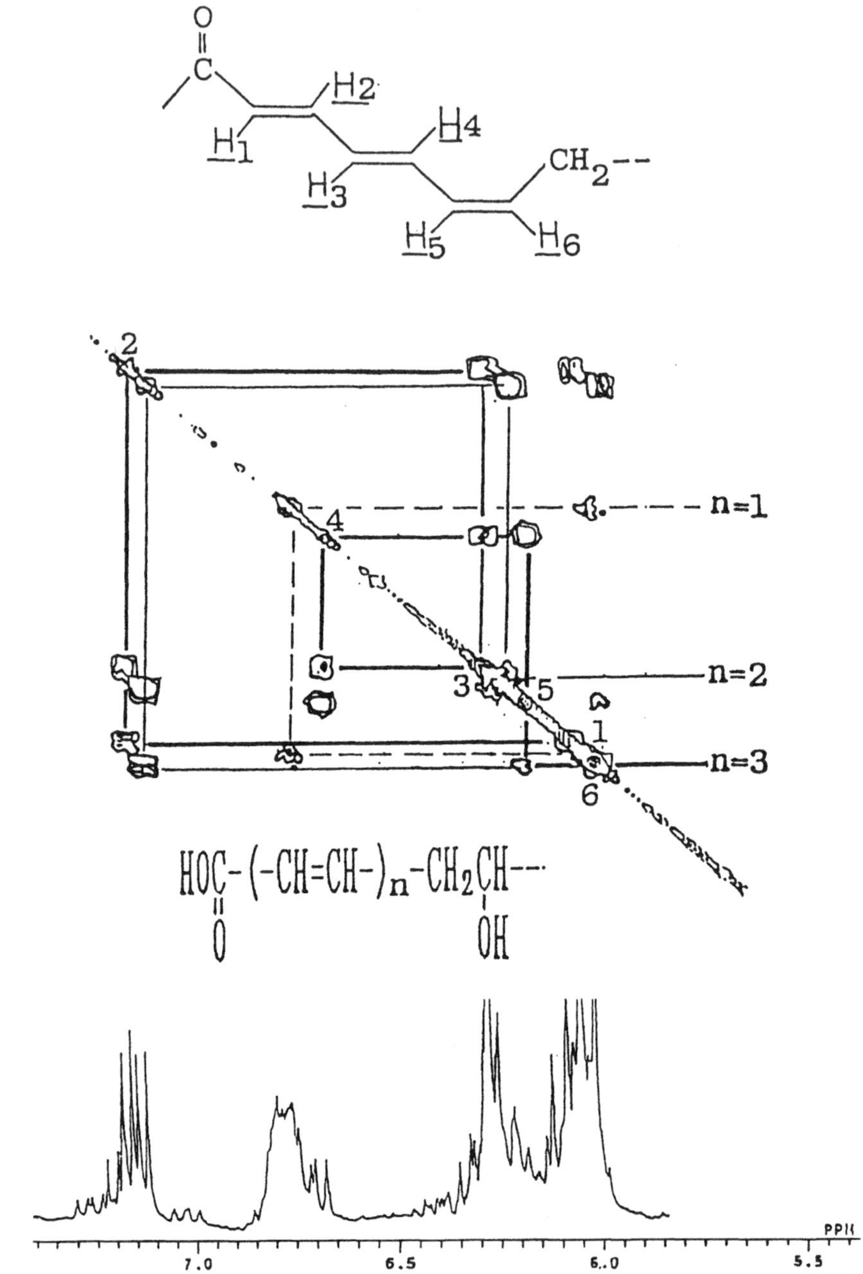

Figure 4.54 COSY spectrum of poly(vinyl alcohol) treated with heat at 150 °C for 14 h under vacuum in DMSO–d_6 at 80 °C.

no structural or dynamic information. On the other hand, the solid-state NMR method provides a source of knowledge on the structure and dynamics of a sample irrespective of whether the region studied is crystalline or noncrystalline.

4.10.2.1 Approaches by Solid-State ^{13}C–NMR

4.10.2.1.1 Synthetic Polymers

4.10.2.1.1.1 Polyethylene Polyethylene (PE) is a typical engineering plastic. The polymer takes a *trans* and two kinds of *gauche* conformations, between which the conformational energy difference is small, and thus PE takes various configurations under appropriate conditions. Polyethylene takes two kinds of crystal forms as determined by several kinds of spectroscopic methods such as NMR, X-ray diffraction, electron diffraction, infrared (IR), and neutron diffraction methods. One of them is the orthorhombic form that occurs under normal conditions and another is the monoclinic form that occurs under high pressure or the drawn condition. In both crystal forms the conformation is the all-*trans* zigzag conformation, but the chain arrangements are different from each other. The all-*trans* zigzag planes in the orthorhombic form are perpendicular to each other, and in the monoclinic form they are parallel to each other. Further, noncrystalline and interfacial phases also exist. Their existence strongly affects physical properties of the polymer. Therefore, it is very important to analyze the detailed structure of PE in the solid state to understand its physical properties.

Figure 4.55 shows ^{13}C CP/MAS NMR spectra of single-crystal PE (SC–PE) [135], melt-quenched PE (MQ–PE) [135], and drawn PE (DR–PE) [136]. Only one sharp peak at 33.0 ppm appears in the SC–PE spectrum. On the other hand, MQ–PE and DR–PE have two and three peaks, respectively. This means that MQ–PE and DR–PE have at least two or three kinds of magnetically inequivalent carbon atoms. Earl and VanderHart have assigned these peaks by T_1 measurements; peak A (31 ppm) comes from the noncrystalline region, which is a mobile state, and peak I (33 ppm) from the crystalline region, which is in an immobile state [137]. Further, the high-frequency peak of DR–PE (35.0 ppm) is assigned to the methylene carbons in the monoclinic crystal region. These assignments were justified by quantum chemical shielding calculations with the tight-binding (TB) sum-over-states (SOS) method [138, 139].

The crystal structure of PE has been proposed from some models such as the sharp-fold, switchboard, and loose-loops models [135]. To obtain detailed information on fold structure, Ando *et al.* have studied SC– and MQ–PE by high-resolution solid-state ^{13}C–NMR [135]. This study reveals that SC– and MQ–PE

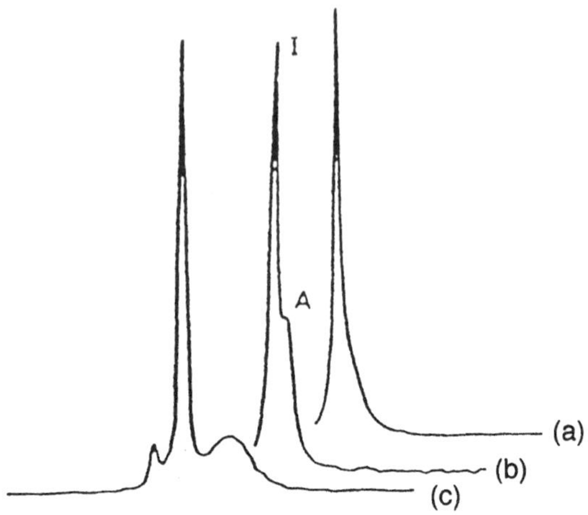

Figure 4.55 ^{13}C CP/MAS NMR spectra of polyethylene: (a) SC–PE, (b) MQ–PE, and (c) DR–PE.

take the adjacent reentry type of macroconformation in addition to loose and long loops in the fold surface, as shown Figure 4.56. The number of carbon atoms in the *trans* zigzag chain from one fold to the next is estimated to be approximately 100, and this leads to an estimate of the stem length of about 125 Å. Its magnitude is consistent with the crystal thickness (120–150 Å) measured directly for PE single crystals by electron microscopy.

The physical properties of polymers in the solid state are strongly affected by temperature. Variable-temperature (VT) NMR techniques should provide useful information on the structural and dynamic aspects of polymers in the solid state [141]. The advantage of VT solid-state NMR techniques is that the temperature dependence of the conformation and molecular motion can be studied by observing the chemical shift. Akiyama *et al.* have investigated the structure of ultrahigh-molecular-weight PE (UHMW–PE) in the solid state by VT ^{13}C CP/MAS NMR [140]. The ^{13}C–NMR spectrum of UHMW–PE at room temperature has two peaks (the low- and high-frequency peaks are designated by A and I, respectively), as shown in Figure 4.57. Because their chemical shifts agree with those for the melt-crystallized PE (MC–PE), peak I and peak A have been assigned to the crystalline region with the *trans* zigzag conformation and to the noncrystalline region, respectively. When the temperature is decreased, peak A shifts to low frequency but peak I shifts to high frequency. The ^{13}C chemical shift behavior for peak A of UHMW–PE is similar to that of the MQ–PE [140]. At −108 °C, the molecular motion is frozen, and the chemical shift for peak A is about 32 ppm. Therefore,

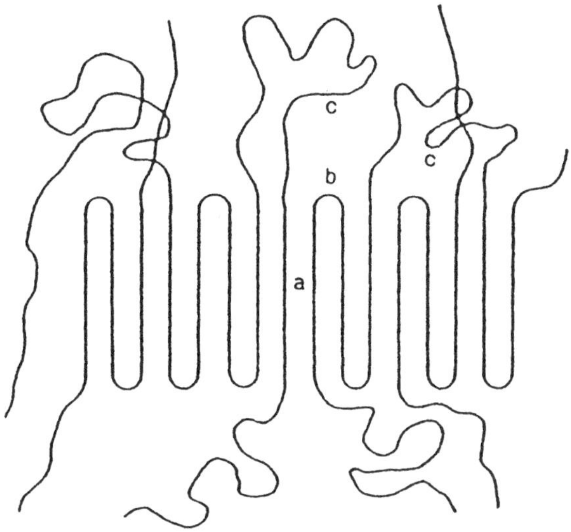

Figure 4.56 A schematic illustration of the folded-chain conformation form melt-quenched polyethylene: (a) *trans* zigzag, (b) sharp fold, and (c) loose and long loop.

peak A comes from the methylene carbons in the *trans* conformation in the noncrystalline region because in the frozen state a methylene carbon in the *gauche* conformation should appear at about 27 ppm by the γ-*gauche* effect. On the other hand, the ^{13}C chemical shift of peak I shifts to low frequency, which is the opposite of the case for the MQ–PE, with decreasing temperature. At −108 °C, the chemical shift is about 32 ppm, and peaks A and I coalesce. It is hard to believe that the cause for the low-frequency shift of peak I is a change of structure in going from the orthorhombic form to a different crystal structure. It is known that the ^{13}C chemical shifts of the methylene carbons in the orthorhombic, triclinic, and monoclinic forms are about 33, 34, and 35 ppm, respectively [140]. From these results, the ^{13}C chemical shift of 32 ppm at −108 °C indicates that the crystalline structure is different from any other kinds of crystal structures found at room temperature. The other possibility for the low-frequency shift is the distortion of the orthorhombic form.

The relaxation parameters T_1, dipolar-dephasing relaxation times T_{DD}, etc., can provide useful information with which to study the dynamics of polymer motion in the solid state and to monitor solid-state polymer phase transitions.

The noncrystalline region of two kinds of PE samples, one a single ^{13}C-labeled solution-crystallized PE (PE–SL) and the other a single ^{13}C-labeled melt-quenched PE (MQ–PE–SL), were studied by VT ^{13}C CP/MAS NMR [143]. The dynamics of

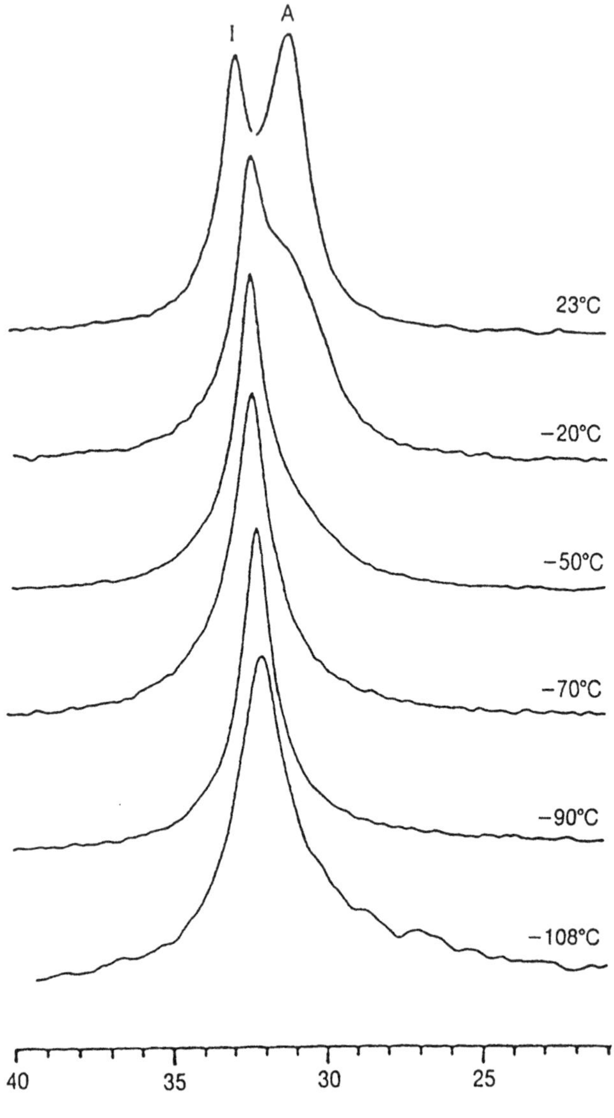

Figure 4.57 ^{13}C CP/MAS NMR spectra of ultrahigh-molecular-weight polyethylene as a function of temperature.

the noncrystalline region were studied by measuring ^{13}C T_1 and T_{DD} over a wide temperature range from -120 to 44 °C. Each of these spectra consists of three peaks corresponding to an orthorhombic crystalline peak at 33.0 ppm, monoclinic crystalline peak at 34.4 ppm (a small shoulder on the left side of the orthorhombic peak), and a noncrystalline peak that appears at 30.8–31.3 ppm. The ^{13}C T_1 data of these samples over a wide range of temperatures obtained using the inversion-recovery method with the pulse saturation transfer (PST) pulse sequence are given. The PST pulse sequence enhances the intensity of mobile methylene carbons. The T_1 values for samples PE–SL and MQ–PE–SL are plotted against the inverse of the absolute temperature ($1/T$) in Figure 4.58(a). It has been suggested previously [144–146] that local molecular motion in the noncrystalline-region PE is independent of the degree of crystallinity, and the presence of higher-order structures or morphologies. However, these suggestions are not supported by the experimental results of PE–SL and MQ–PE–SL. There are two reasons: one is the difference of T_1 values and another is the large difference of T_1 minimum values of the noncrystalline region between PE–SL and MQ–PE–SL, as shown Figure 4.58(a). These facts show that local molecular motions in the noncrystalline regions of sample PE–SL are more constrained than those of MQ–PE–SL. To study whether the dynamics behavior of the two kinds of noncrystalline regions is also different on the T_2 time scale, T_{DD} values of samples of PE–SL and MQ–PE–SL were measured over a wide range of temperature. The relative intensity of the noncrystalline peak obtained from computer simulation was plotted against the delay time τ in Figure 4.62(b). It can clearly be seen that the noncrystalline peak of the sample of PE–SL relaxes more quickly than that of the sample MQ–PE–SL. The T_{DD} usually depends on molecular motion, carbon–proton dipolar interactions, and MAS rate and spin diffusion [147]. Fundamentally, it can be said that the dipolar dephasing time in the noncrystalline region becomes a measure of molecular motion because of the high mobility. Therefore, the longer T_{DD} value of the sample of MA–PE–SL, compared with that of the sample of PE–SL, obviously suggests that the carbon–proton dipolar interaction is partially averaged by molecular motion on the T_2 time scale.

4.10.2.1.1.2 Polyoxymethylene The molecular motion of polyoxymethylene (POM), which is the first member of the polyether series, has been investigated by a 2-D exchange experiment [148]. The observed and calculated 2-D exchange spectra of POM are shown in Figures 4.59(a) and (b), respectively. The spectrum observed at 252 K shows only diagonal signals, and this means that no molecular motion occurred during the mixing time. On the other hand, if molecular motion occurred during the mixing time, the spectrum would show steric effects, as shown

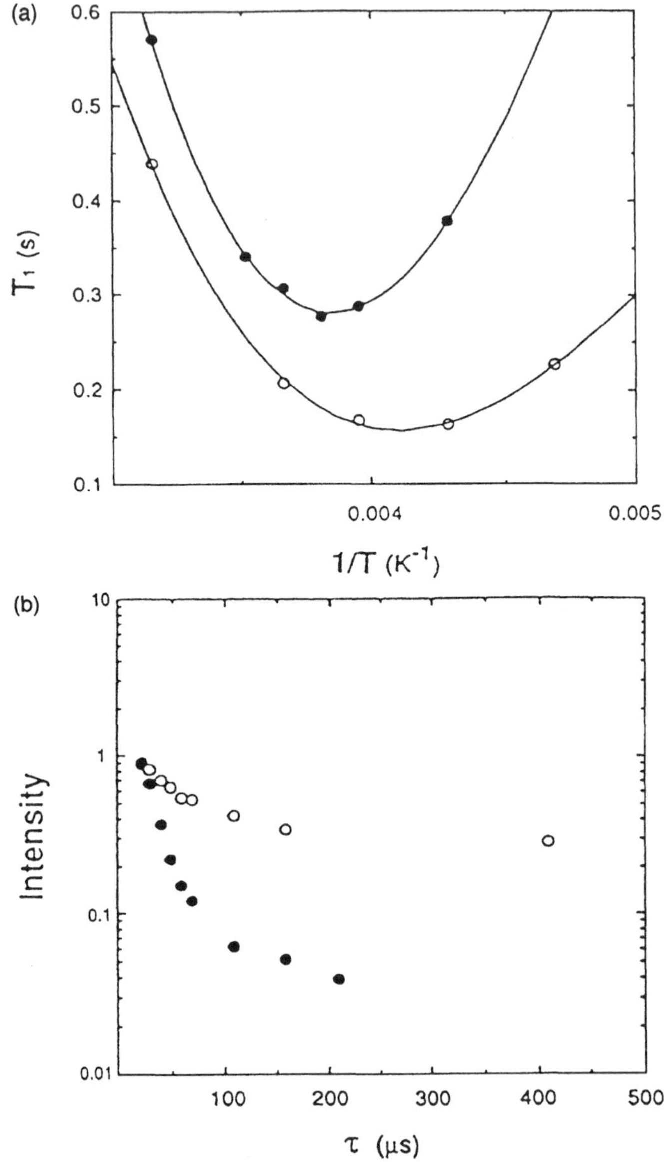

Figure 4.58 (a) Plots of noncrystalline ^{13}C spin-lattice relaxation times T_1 of samples PE–SL (●) and MQ–PE–SL (○) versus the reciprocal of the absolute temperature. (b) Intensity of noncrystalline peaks of sample of PE–SL and MQ–PE–SL versus delay time τ. The peak intensity was obtained from computer simulation of the ^{13}C partially relaxed dipolar-dephasing NMR spectra.

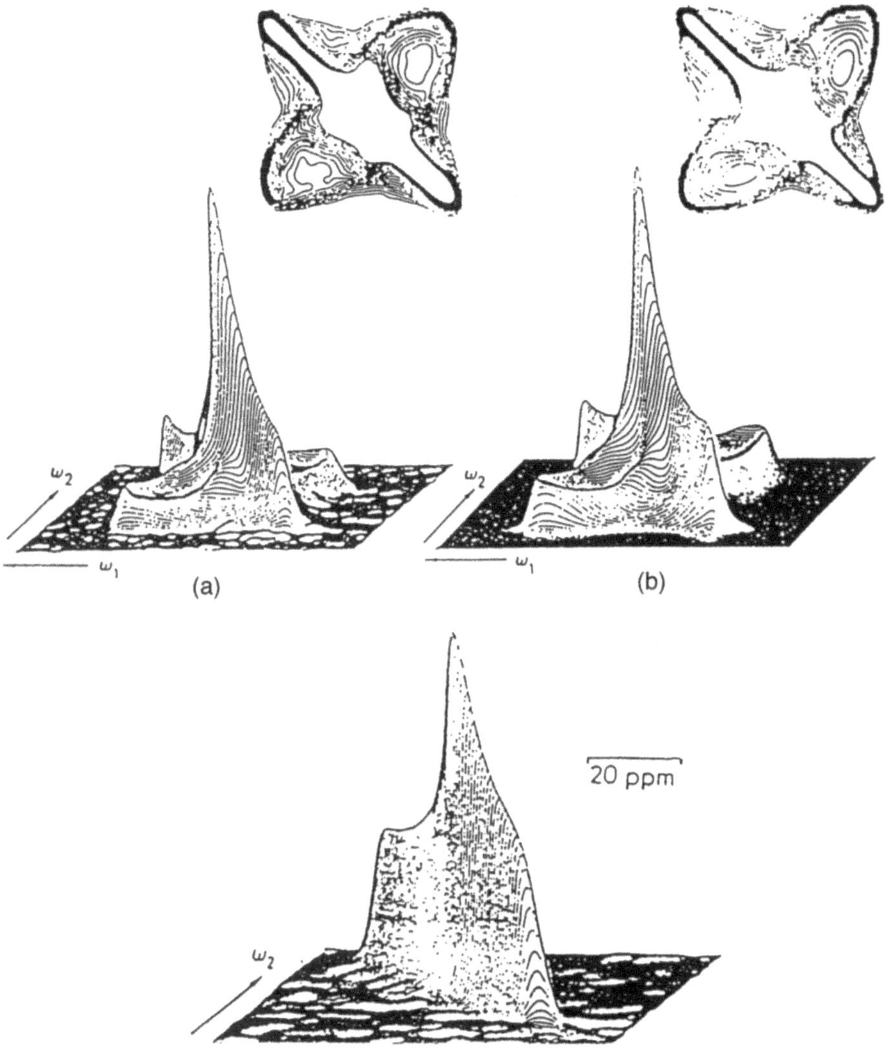

Figure 4.59 Pure absorption-mode 2-D exchange spectra of isotactic POM: (a) temperature $T = 360$ K and mixing time $t_m = 2$ s, (b) theoretical spectrum of (a) with $x = 2\omega$, $t_m = 2.5$ s, and (c) temperature $T = 252$ K and $t_m = 1$ s.

Table 4.5 Solid-state ^{13}C–NMR chemical shifts of nylon 6.

Specimen	*C=O*	*α-CH$_2$*	*β-CH$_2$*	*γ-CH$_2$*	*δCH$_2$*	*ω-CH$_2$*
α-Form	171.1	34.4	24.3	28.1	28.1	41.1
γ-Form	171.0	35.4	28.2	28.2	32.0	37.9
$\Delta\delta c^a$	−0.1	+1.0	+3.9	+0.1	+3.9	−3.2
Solution^b	176.4	34.6	24.4	25.0	27.2	39.2

$^a\Delta\delta c = \delta c(\gamma) - \delta c(\alpha)$.
^{b}in 10% HFIP solution.

in Figure 4.59(c). The calculations were performed by using the model of helical jump motions with a one-dimensional random walk in continuous time assumed—an elementary process of 200° jumps. Thus, at $T = 360$ K the subspectra for the discrete steps of a given CH_2 group of POM experiences, as the helix rotates, can be identified.

4.10.2.1.1.3 Nylons The molecular structure of nylons has been investigated by high-resolution solid-state ^{13}C–NMR and X-ray diffraction [149–152]. Kubo *et al.* have measured the ^{13}C CP/MAS and PST/MAS NMR spectra, which emphasize the carbons with high mobilities, as well as the CP/MAS technique, ofthe single-crystal and melt-quenched nylon 6 [153, 154]. The detailed comparison of these spectra reveal that the noncrystalline peaks of the ω–CH_2 and $(\delta+\gamma)$–CH_2 carbons of the melt-quenched sample are increased compared with those of the crystal sample. These spectra show that the noncrystalline peaks appear at lower frequency than the crystalline peaks. The chemical shift behavior of these peaks can be understood on the basis of the *γ-gauche* effect. The crystalline state has a *trans* zigzag structure, and the noncrystalline carbons are undergoing rapid transitions between the *trans* and *gauche* conformations. Also, peaks that come from some crystallographic forms are observed in some nylons [150].

It is known that nylon takes two types of α- and γ-forms as polymorphs. Weeding *et al.* [155] and Okada *et al.* [156] observed the ^{13}C CP/MAS NMR spectra of nylon 6 samples with the α- and γ-forms. It has been shown that there is a significant chemical shift difference between these forms. It has also been demonstrated that solid-state ^{13}C–NMR provides useful information about the crystal structure of nylons (Table 4.5). This has led to solid-state ^{13}C–NMR studies on nylon 7 [157] and 11 [158] with polymorphs.

4.10.2.1.1.4 Polyimides Polyimides are usually prepared by thermal imidization of poly(amic acid). Water and evaporated solvent bring about voids in the polyimide resins during the reaction. This problem has been solved mainly by the use of acetylene-terminated polyimide prepolymeric resins in which homopolymerization at elevated temperatures and pressures occurs without the formation of volatile products. The mechanism of cure is considered to be aromatization of the acetylenic end groups, but this has been impossible to verify because of the intractability of the cross-linked resin. The ^{13}C CP/MAS NMR spectra of the resin and the cure polyimide polymer revealed changes in the line shape of the aromatic carbon region as well as a diminution of intensity in the acetylene carbon region [159–161]. Reactions involving the terminal acetylenic groups of the polyimide resin are involved in the polymerization process, but less than 30% of these acetylenes can undergo cyclotrimerization or other condensation reactions. The remainder appear to be consumed by addition reactions. To make the reaction clear, Swanson *et al.* have studied the solid-state chemical reactions in the ^{13}C-labeled polyimides at sites that are thought to be involved in the cure reaction and derived the detailed structure of reaction products [162].

Bismaleimides are also used as void-free, high-temperature matrix resins for fiber-reinforced composite materials. The high-strength and high-modulus properties of these resins persist after hot–wet or hot–dry exposures to 250 °C. Bismaleimides by themselves tend to form highly cross-linked networks that produce brittle resins. To alleviate this problem, chain extenders are introduced into the resins to lower the cross-link density and thereby increase the fracture toughness of the cured composite. Fry and Lind have measured the ^{13}C CP/MAS NMR spectra of 1,1-(methylene di-4,1-phenylene)-bismaleimide (BMI) as received, after quenching, and after curing for 4 h at 493 K [163]. The cured BMI spectrum clearly shows a new carbonyl resonance at 175 ppm. A decrease is observed in the intensity of the resonance at 132 ppm, and a new resonance occurs at about 45 ppm. From these results, it was discovered that there are two primary reactions, one is the cross-linking and the other is the chain-extender reaction.

4.10.2.1.1.5 Fluoropolymers Fluoropolymers have excellent properties and have been widely used for various purposes under severe conditions. Although high-resolution solid-state NMR is a powerful tool to elucidate the structures and dynamics of polymers in the solid state as in the aforementioned ^{13}C CP/MAS experiment, this method often fails to provide detailed information regarding structures of fluoropolymers because their spectra give rise to broad lines. This is mainly due to two dominant factors; namely, large dipolar coupling and shielding anisotropy of fluorine atoms. Although the high-speed magic angle spinning (HS MAS) method leads to a significant reduction in such broadening for ^{19}F NMR in the solid state and has been used for characterizing fluoropolymers [164], only a few ^{13}C solid-state NMR experiments have been reported [165, 166]. The simultaneous application of proton and fluorine decoupling drastically reduces line broadening caused by ^{1}H–^{13}C and ^{19}F–^{13}C dipolar interactions. Holstein *et al.* reported that the two kinds of poly(vinylidene fluoride), the nonpolar α-phase and polar β-phase, have been characterized by means of solid-state triple-resonance spectroscopy [166]. There was a small difference (about 4 ppm) in the chemical shift positions for these distinct crystalline modifications.

4.10.2.1.1.6 Blend Polymers One of the recent industrially important organic polymer developments is multicomponent systems such as polymer blends. Their macroscopic properties are determined not only by the molecular properties of the individual polymers but also by the miscibility, morphology, and structure of the interfaces between the polymers. Solid-state NMR is a useful tool for studying the microscopic structure of heterogeneous polymer materials.

The miscibility of PMMA and poly(vinylidene fluoride) (PVDF) has been established by various methods. ^{19}F–^{13}C and ^{19}F–^{1}H CP techniques have been developed [167, 168] and molecular miscibility in blends of PVDF, and isotactic, syndiotactic, and atactic PMMAs were investigated by triple resonance ^{1}H, ^{19}F, and ^{13}C solid-state CP/MAS NMR [169]. The fraction of nonmixed PMMA was determined for these blends, and this fraction was found to be smaller for isotactic than for atactic and syndiotactic PMMAs.

White and Miran reported about polymer blends of poly(vinyl phenol) and poly(methyl acrylatre) using 2-D ^{13}C–^{1}H correlation NMR [170]. From the spectrum, a direct interaction between the hydroxyl hydrogen of poly(vinyl phenol) and the carbonyl carbon of poly(methyl acrylate) can be deduced. Tang *et al.* have discussed the miscibility in annealed 50/50 blends of a random, copolyester Vectra-A containing 73% *p*-hydroxybenzoic acid, 27% hydroxynaphthoic acid, and poly(ethylene terephthalate) (PET) by using a standard 2-D exchange experiment [171]. The small cross peaks between the peak of aliphatic PET and the peak of quaternary Vectra-A carbon diagonal lines indicated intimate mixing.

4.10.2.1.2 Biopolymers

In the solution state, the NMR chemical shifts of biopolymers with internal rotations are often the averaged values for all internal rotations because of rapid interconversion by rotation about peptide bonds. In the solid state, however, chemical shifts are frequently characteristic of specific conformations because of highly restricted rotations about the peptide bonds. The NMR chemical shift is affected by a change in the electronic state arising from the conformational change. NMR chemical shifts in the solid state, therefore, provide useful information about the electronic state and conformation of a biopolymer with fixed structure, especially the membrane polymer [172]. Several research works involving typical synthetic polypeptide and membrane proteins studied by high-resolution ^{13}C–NMR in the solid state are given here.

4.10.2.1.2.1 Synthetic Polypeptides Synthetic polypeptides consist of repeating sequences of certain amino acids, and their primary structures are not as complicated as those in proteins. The polypeptides are very important polymers in polymer science and protein science. The characteristic properties related to their structure which are very different from conventional synthetic polymers, have led to expansion for research fields in polymer science. For example, the concept of the liquid crystal is expanded by revealing a variety of structures and properties of liquid crystals. Further, the polypeptides are sometimes used as biomimic materials. On the other hand, synthetic polypeptides are sometimes used as model biomolecules for proteins because they take the α-helix, β-helix, ω-helix, and so on, under appropriate conditions. From such situations, it can be said that synthetic polypeptides are "interdisciplinary" macromolecules that are very important for research in both polymer and protein science.

As is well known, most of the peptides, polypeptides, and proteins considered here consist of repeating sequences of peptide bonds with 20 different types of substituents at the Cα carbon. The limited conformations such as α-helix, ω-helix, beta-sheet, etc., are taken by a possible set of dihedral angles (φ, Ψ) around N–Cα and Cα–C(=O) bonds. In the solution state, the NMR chemical shift of these biomolecules, with possible rotation around the bonds, becomes the averaged value for all rotations because of rapid rotation about the peptide bonds in the NMR time scale. In the solid state, however, the chemical shift is characteristic of specific conformations because the internal rotation around the peptide bonds is fixed. This shows that the NMR chemical shift can be used for elucidating the conformation of polypeptides and proteins in the solid state. It has been experimentally and theoretically shown that the NMR chemical shift of polypeptides and proteins is a very important NMR parameter for determining the main-chain conformation, and some recent studies on the structural characterization of polypeptides and proteins by using such a methodology have been published [173].

As an example, it has been shown that the ^{13}C chemical shifts of the Cα, Cβ, and amide C=O carbons of poly(L-alanine)[$(\text{Ala})_n$] are related to particular

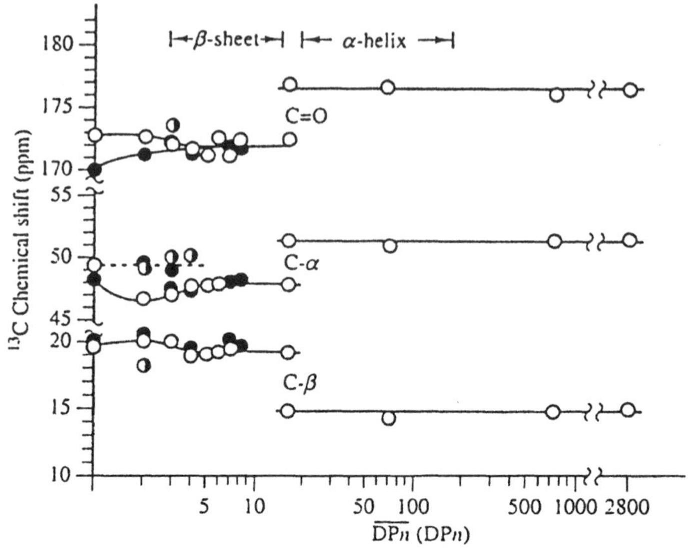

Figure 4.60 Plots of the ^{13}C chemical shifts of $(Ala)_n$ against the number average degree of polymerization (DP*n*).

conformations [174] ^{13}C CP/MAS (cross polarization/magic angle spinning) NMR spectra of solid poly(L-alanine) show the Cα, Cβ, and amide C=O carbon signals to be well resolved between the α-helix. The ^{13}C chemical shifts are plotted against the number-average degree of polymerization in Figure 4.60 [174]. Clearly, ^{13}C chemical shifts of $(Ala)_n (n > 16)$ are unchanged for the peptides of various molecular weights within experimental error and thus can serve to characterize the alpha-helix form. The chemical shifts of the Cα and carbonyl carbons of the α-helix are displaced significantly downfield by 4.2 and 4.6 ppm, respectively, relative to those of the β-sheet form, whereas the chemical shift of the Cβ carbon of the α-helix is displaced upfield by about 5 ppm with respect to that of the beta-sheet. For this reason, the value of the ^{13}C chemical shift can be used to describe the local conformation. In addition, the ^{13}C chemical shifts of randomly coiled $(Ala)_n$ in trifluoroacetic acid solution have values between those of the α-helix and β-forms.

The existence of such characteristic displacements of ^{13}C chemical shifts is not limited to the Ala residue. In Table 4.6 the ^{13}C chemical shift values of various amino acid residues in the α-helix and β-sheet forms are summarized [173]. Here it is seen that the Ca and C=O peaks of the alpha-helix form are all displaced downfield with respect to those of the β-sheet form, which is consistent with the

Table 4.6 ^{13}C chemical shifts of polypeptides characteristic of α-helix, β-sheet, and ω-helix forms.

			^{13}C Chemical Shift			
Sample[a]	*Conformation*[b]	*Carbonyl*[c]	C_α	C_β	*Phenyl*	*Benzyl*
$(Gly^*)_n$	β	168.5^+	43.2			
$(Gly^*)_n$	3_1	171.8^+	43.2			
$(Gly^*)_5$		171.4^+	41.7			
$(Gly)_n$	β	168.4	43.2			
		169.2	44.3			
$(Gly)_n$	3_1	172.3	43.2			
		172.1	42.0			
$(Ala, Gly^*)_n$	α	176.7	52.2	14.6		
		171.7^+				
$(Ala)_n$	α	176.4	52.4	14.9		
		176.8	52.8	15.5		
	β	171.8	48.2	19.9		
		172.2	49.3	20.3		
$(Leu, Gly^*)_n$	α	171.4^+	55.4	39.0		
$(Leu)_n$	α	175.7	55.7	39.5		
	β	170.5	50.5	43.3		
$(Glu(OBzl), Gly^*)_n$	α	172.1^+	56.5	25.3	127.5	65.2
$(Glu(OBzl)_n$	α	175.6	56.4	25.6	~128.5	66.0
		175.4	56.8	25.9		
	β	171.0	51.2	29.0		
		172.2	51.1	29.7		
$(Asp(OBzl), Gly^*)_n$	α	172.0^+	53.2	33.8	127.8	65.9
$(Asp(OBzl), Gly^*)_n$	ω	171.1^+	50.8	33.2	127.7	65.3
$(Asp(OBzl))_n$	α	174.9	53.4	33.8	129.8	65.7
		174.9	53.6	34.2	128.2	66.1
	α_1	171.1	50.9	33.8	129.2	66.1
	ω	171.3	50.5	32.9	129.0	65.3
	β	169.6	49.2	35.1	129.2	65.9
$(Val, Gly^*)_n$	β	168.5^+	58.0	32.0		
$(Val)_n$	α	174.9	65.5	28.7		
	β	171.8	58.4	32.4		
		171.5	58.2	32.4		

[a] Symbol * indicates carbonyl carbon enriched by ^{13}C.
[b] α-helix is right-handed unless otherwise specified.
[c] Symbol + indicates chemical shift of glycine residue.

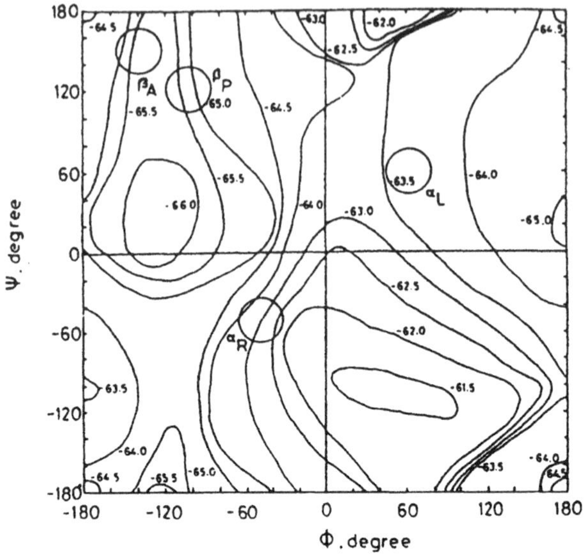

Figure 4.61 The calculated ^{13}C chemical shielding contour map for the C-β carbon of the L-alanine residue in *N*-acetyl-*N'*-methyl-L-alanine amide as the model peptide of L-alanine-containing polypeptides as a function of the dihedral angles (f, y) using the FPT INDO method. The negative sign means deshielding.

data of $(Ala)_n$. Further, it is significant to show the ^{13}C chemical shift values for various conformations of PBLA such as the α_R-helix, α_L-helix, ω_L-helix, and β-sheet forms in Table 4.6, which are taken by appropriate treatments [175]. The absolute ^{13}C chemical shifts of the Cα and Cβ carbons are affected by the chemical structure of the individual amino acid residues and can be used effectively for conformational studies of particular amino acid residues in polypeptides and proteins. On the other hand, the C=O chemical shifts do not seem to be affected by residue structure and can be used for diagnosing the main-chain conformation.

Such sizeable displacements of the ^{13}C chemical shifts can be characterized by variations in the electronic states of the local conformation as defined by the dihedral angles (φ, Ψ) [173, 174, 177]. The calculated contour map for the Cβ carbons of an alanine dipeptide was made using the FPT INDO method [176] within the semiempirical MO framework, as shown in Figure 4.61 [178].

It should be noted that the negative sign of the shielding constant sigma indicates deshielding, and thus shielding variations can be compared with the experimental chemical shift d for which a positive sign denotes deshielding. Such chemical shift maps successfully predict the ^{13}C chemical shifts and conformations of the L-alanine (L-Ala) residues in polypeptides and proteins. For example, the experimental ^{13}C chemical shift for the α-helix form appears at a higher field by about 5.5 ppm than that for the β-sheet form. The calculated map reasonably predicts the experiment.

Most recently, the calculated ^{13}C chemical shift maps of the Cα and Cβ carbons of the L-Ala residue were obtained using the GIAO–CHF method with the *ab initio* 4-31G basis set as shown in Figure 4.62, may be more reliable compared with the case of the semiempirical MO calculation. The chemical shift behavior as shown in these maps is very similar to that obtained by the FPT INDO MO method [181]. On the basis of these calculated maps, the relationship between the experimental ^{13}C chemical shift and the main-chain conformation (Table 4.6) for the series of L-Ala-containing peptides and polypeptides in the solid state was elucidated. As is seen from Figure 4.62(a), the calculated isotropic ^{13}C shielding constants for the Cβ carbon are 186.4 ppm for the β-sheet conformation, 189.4 ppm for the right-handed $\alpha(\alpha_R)$-helix, 189.6 ppm for the left-handed $\alpha(\alpha_L)$-helix, 188.7 ppm for the 31 helix, and 186.5 ppm for the Silk I and (AlaGly(glycine))*n* form I [175, 177]. Here the larger value means higher field shift. The experimental isotropic 13C chemical shifts are 21.0 ppm for the anti-parallel β-sheet, 15.5 ppm for the α_R-helix, 15.9 ppm for the α_L-helix, 18.4 ppm for the 3_1 helix, and 17.6 ppm for the (AlaGly)*n* form II. Here the larger value means lower field shift. It was found that the change of the main-chain dihedral angles dominantes the isotropic chemical shift behavior of the Ala residue Cβ carbon, except for Silk I and (AlaGly)*n* form II. It is predicted that the structure of Silk I and (AlaGly)*n* form II may be a loose, four-fold helical conformation, which was originally proposed in the X-ray diffraction study. From the calculation, the loose helix model could not be accepted either; at this point, however, it is still difficult to determine the main-chain dihedral angles.

As is seen from Figure 4.62(b), the calculated ^{13}C chemical shieldings for the Cα carbon are 160.4 ppm for the β-sheet, 159.6 ppm for the α_R-helix, 159.2 ppm for the α_L-helix, 161.4 ppm for the 3_1-helix, and 157.9 ppm for Silk I and (AlaGly)*n* form II; for these calculated chemical shieldings, the experimental isotropic ^{13}C chemical shifts are 48.7 ppm for the β-sheet, 53.0 ppm for the α_R-helix, 50.1 ppm for the α_L-helix, 49.7 ppm for the 3_1-helix, and 51.5 ppm for (AlaGly)*n* form II. Although it is obvious that there is a main-chain dihedral-angle dependence on the chemical shift for the Cα carbon, it seems more complicated than that for the Cβ carbon because orientation of the chemical shift tensor for the Cα carbon with respect to the molecular fixed frame is varied from one structure with any specific dihedral angles to one another, and also because, in a case in which the L-Ala residue carbonyl or amide group would form the hydrogen bond, the hydrogen bonding structure can also affect the behavior of the ^{13}C chemical shift of the Cα carbon.

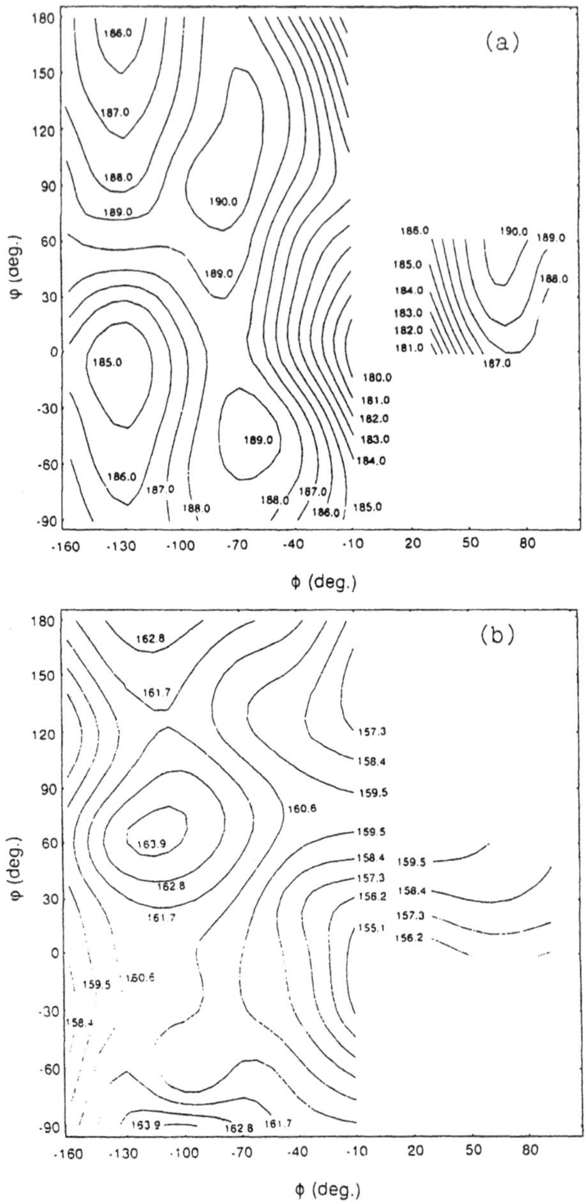

Figure 4.62 The calculated ^{13}C chemical shielding contour maps for the C-β (a) and C-α (b) carbons of the L-alanine residue in N-acetyl-N'-methyl-L-alanine amide as a function of the dihedral angles (φ, Ψ) using the GIAO–CHF with *ab initio* 4-31G basis set. The positive sign means shielding.

With regard to the structure of (AlaGly)*n* form II and Silk I, they require a chemical shift calculation that takes into consideration the hydrogen bonding; the investigation of the chemical shift anisotropy of the carbonyl carbon is needed [178].

The side-chain conformation of poly(L-proline)(PP; (Pro)*n*)) form II and form I in the crystalline state has been studied by a combination of solid-state ^{13}C–NMR [180] and the tight-binding INDO/S sum-over-states theory [177], which is used for calculating chemical shieldings of polymer chains with infinite regular structure [180]. In poly(L-proline) form II, the pyrrolidine ring takes two types of conformations and undergoes no chemical exchange within the temperature from −86 to 175 °C. On the other hand, in poly(L-proline) form I the pyrrolidine rings in the crystalline state was found to be undergoing chemical exchange [173e].

Poly(β-benzyl L-aspartate) (PBLA) in the solid state adopts various conformations such as a right-handed $\alpha(\alpha_R)$-helix, a left-handed α (α_L)-helix, a left-handed $\omega(\omega_L)$-helix, and a β-sheet form, depending on the temperature. The VT ^{13}C CP/MAS NMR spectra of PBLA are shown in Figure 4.63 [182]. From this, it is clear that the conformational changes such as α_R-helix $\Rightarrow$ ω_L-helix $\Rightarrow$ β-sheet occur with increasing temperature.

Poly(γ-*n*-octadecyl L-glutamate) (POLG) forms a crystalline phase composed of paraffin-like crystallites together with the α-helical main chain, packing into a characteristic layer structure. The polymer forms thermotropic cholesteric liquid crystals by the melting of the side-chain crystallites. Various temperature of the conformational changes of POLG are studied readily by high-resolution solid-state NMR [183–187]. Figure 4.64 shows the ^{13}C CP/MAS NMR spectrum POLG at room temperature together with the assignment of peaks. At room temperature, ^{13}C chemical shift values of the C=O and Cα carbons are 176.0 and 57.6 ppm, respectively. These values show that the main chain takes an α-helix conformation. On the other hand, the interior CH_2 signal splits into two peaks. From the reference data of *n*-alkanes [188, 189] and polyethylene, it is seen that the peak at 33.4 ppm is the carbon signal in the all-*trans* zigzag conformation in the crystalline state and the peak at 30.6 ppm is the carbon signal in the noncrystalline phase or the liquid phase.

Figure 4.65 shows the VT ^{13}C CP/MAS NMR spectra of POLG at a wide temperature range from −40 to 230 °C. The *n*-alkyl peaks change observably as the temperature is increased. The peak of the carbon signal in the noncrystalline state disappears above 50 °C, and the intensity of the interior CH_2 carbon in the noncrystalline state increases noticeably. This is due to the melting of side-chain crystallities. On the other hand, ^{13}C chemical shift values of the C=O and Cα carbons are independent of temperature. It is shown that the main chain takes on a right-handed α-helix conformation within the temperature range from −40 to 230 °C.

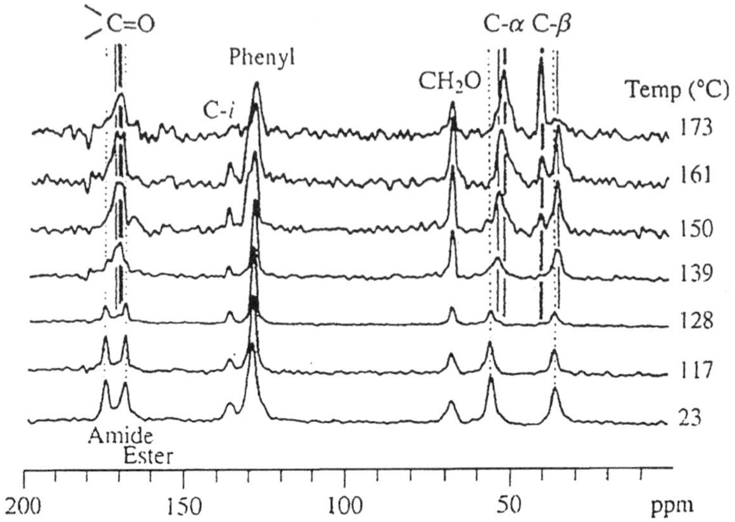

Figure 4.63 Variable-temperature TOSS ^{13}C CP/MAS NMR spectrum of PBLA in the solid state at various temperatures: dashed line-α_R-helix form, light solid line-ω_L-helix form, and bold solid line-β-sheet form.

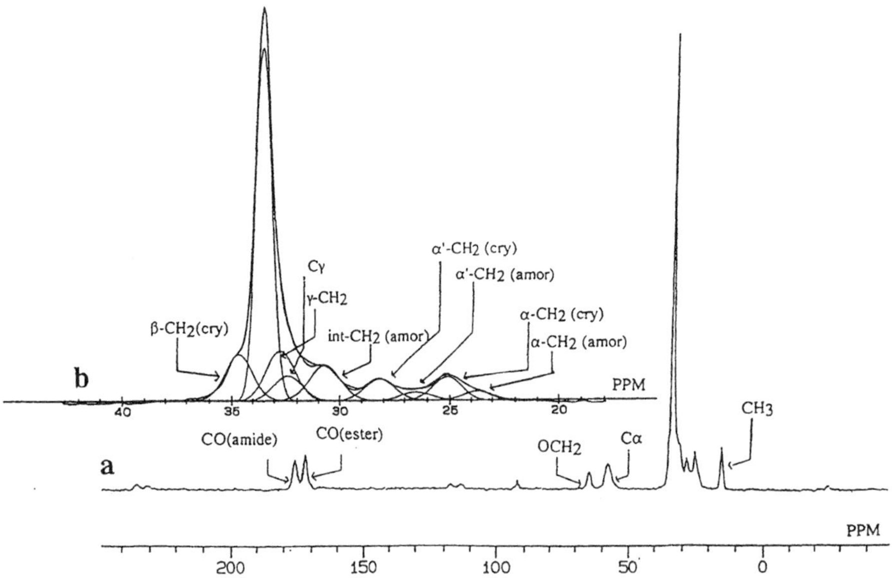

Figure 4.64 (a) ^{13}C CP/MAS NMR spectrum of POLG in the solid state at room temperature. (b) The methylene region is expanded and is deconvoluted by computer-fitting with the Gaussian function (cry and amor indicate crystalline and noncrystalline states, respectively).

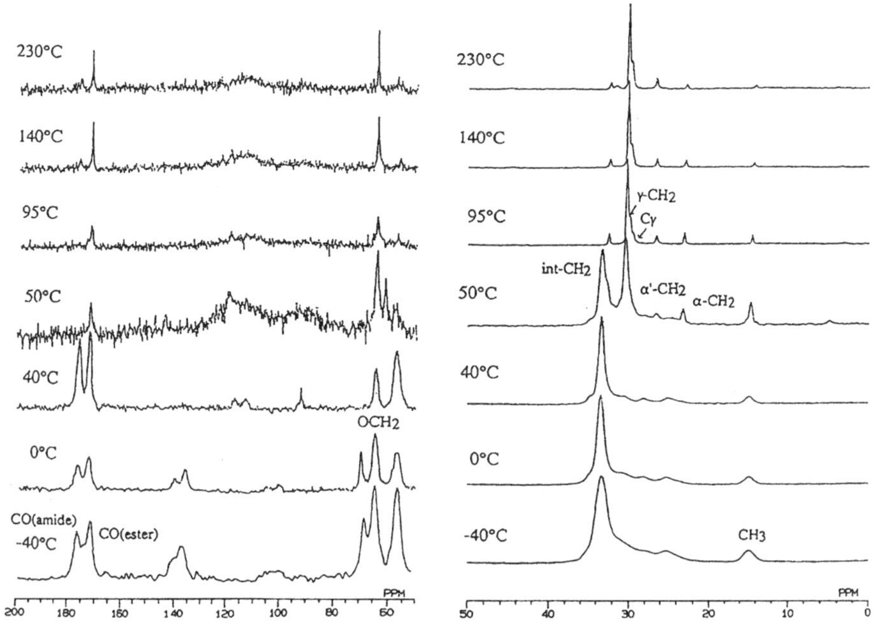

Figure 4.65 ^{13}C CP/MAS NMR spectra of POLG as a function of temperature.

For the samples' varied molecular motion with temperature, it is useful to measure the T_1 value using the inversion-recovery method with PST pulse sequence and Torchia's pulse sequence according to its molecular motions. Katoh *et al.* have investigated the molecular motion of POLG within the temperature range from −40 to 230 °C using both methods [186]. Figure 4.66 shows the temperature dependence of T_1 values for some typical carbons. According to the BPP theory, T_1 decreases, passes through a minimum (correlation time: τ_c), and increases again when molecular motion increases; that is, temperature is decreased. The ^{13}C T_1 values of POLG are very short even though, as seen from the chemical shifts, the *n*-alkyl side chains of POLG are in the crystalline state in the temperature range from −40 to 40 °C. If the *n*-alkyl side chains are in the crystalline state, ^{13}C T_1 values must be very long by analogy with the *n*-alkane data [188, 189]. However, the ^{13}C T_1 values for individual carbons of the *n*-alkyl side chains are very short and very close to the T_1 values of *n*-alkane in the rotator phase; further, the N ^{13}C T_1 values for individual carbons of *n*-alkyl side chains are almost the same as in the case of *n*-alkane in the rotator phase. From these results, it can be said that *n*-alkyl side chains of POLG are in a phase like the rotator phase, in which they undergo fast rotation such as libration, along the *trans* zigzag chain axis. They also have investigated the transition changes at 45, 60, and between 90 and 110 °C for the individual carbons of the side chains. The transitional change at 45 °C comes

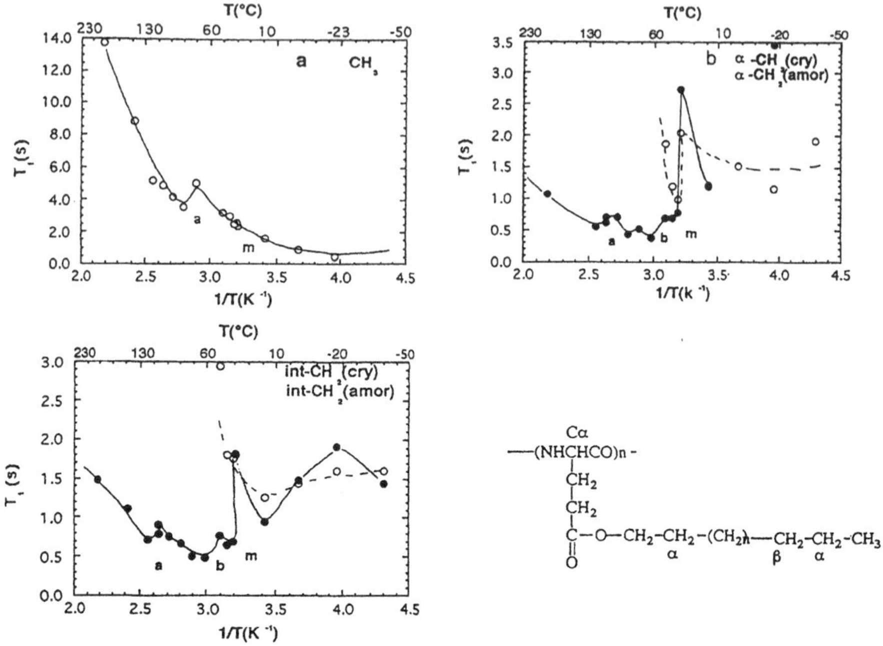

Figure 4.66 Temperature dependence of ^{13}C T_1 for individual carbons of POLG: (a) CH_3, (b) *int*-CH_2 and (c) α-CH_2 (*a* and *b* indicate transition temperature and *m* indicates melting point). For T_1 measurement in the temperature range −40 to 50 °C Torchia's pulse sequence was used, and in the temperature range 50 to 230 °C, the inversion recovery pulse sequence was used.

from the melting of the side chains, and, on the other hand, the other changes come from transitional change of the liquid crystalline phase, which has not been proved but only observed by the ^{13}C chemical shifts.

It has been demonstrated that the introduction of fluorine atoms into polymers induces new physical properties by changes of intramolecular and intermolecular interactions. Katoh and Ando have studied poly(γ-glutamate)s with *n*-fluoroalkyl side chains using high-resolution solid-state NMR [190]. From these experimental results, it was found that the conformation of the main chain depends on the carbon number of the *n*-fluoroalkyl side chains.

The main chain of poly(γ-alkyl L-glutamate) with short *n*-fluoroalkyl side chains takes the helix form, and the poly(γ-alkyl L-glutamate) with long *n*-fluoroalkyl side chain takes the β-sheet form. These conformational behaviors of poly(γ-alkyl L-glutamate) with *n*-fluoroalkyl side-chains in the solid state are different from those of poly(γ-alkyl L-glutamate).

4.10.2.1.2.2 Fibrous Proteins Because fibrous proteins generally have periodical amino acid sequence and higher-order structure, the clarification of their fine structure in the solid state becomes very important not only when discussing the physical and chemical properties but when obtaining information about the molecular design of synthetic polypeptides. The conformation-dependent ^{13}C CP/MAS NMR chemical shifts are particularly useful for the determination of the conformational features of fibrous proteins such as silk fibroin [191], collagen [192], and wool keratin [193–195]. For example, Yoshimizu *et al.* have endeavored to estimate the conformational transition of *S*-carboxymethyl keratin, which has low-sulphur fractions (SCMK-low sulphur) [193–195]. For SCMK-low sulphur heated at 200 °C for 3 h *in vacuo*, the ^{13}C MAS NMR spectrum shows that each signal becomes broader than those of other treated specimen. This indicates the existence of various conformations, different microenvironments, or both, in the heated SCMK-low sulphur. Thus, it can be said that the random coil form appears by heating. On the other hand, from the X-ray diffraction, the α-helix form completely vanishes in the SCMK-low sulphur under same condition. The difference between the results from X-ray diffraction and NMR spectroscopy suggests that only the packing of the ordered structure (α-helix form) in the SCMK-low sulphur is disrupted by heating but that the secondary structure is retained.

4.10.2.1.2.3 Membrane Proteins Membranes provide a vital interface between a biological organism and its environment. Furthermore, they provide a means for forming compartments within a cell. In every sense, they are vital to living systems. Indeed, because membranes and membrane proteins cannot easily be crystallized, NMR spectroscopy is a particularly powerful method for probing their structure and function. Tuzi *et al.* have demonstrated that [1-^{13}C]Ala, [3-^{13}C]Ala, and [1-^{13}C]Val selected labeling of bacteriorhodopsin (bR) is a very convenient probe for conformation and dynamics [196, 197]. Figure 4.67 shows the ^{13}C CP MAS NMR spectrum of [3-^{13}C]Ala and [1-^{13}C]Val bR. The peaks of [3-^{13}C]Ala bR have been assigned to the α_{II}-helix form at 16.3 ppm (60%), α_{I}-helix form at 14.9 ppm (20%), and the loops at 17.2 ppm (20%) by taking a variety of turn structures by deconvolution of the spectrum. Further, from the VT CP/MAS ^{13}C–NMR experiment of [3-^{13}C]Ala-labeled bR, well-resolved ^{13}C–NMR signals were observed either at ambient temperature or at −20 °C but were broadened considerably at temperatures below −40 °C [198]. This situation was interpreted in several conformations slightly different from each other. Researchers found that an exchange process was strongly influenced by the manner of organization of the lipid bilayers, depending upon the presence or absence of cations responsible for electric shielding of negative charge at the polar head groups. The secondary

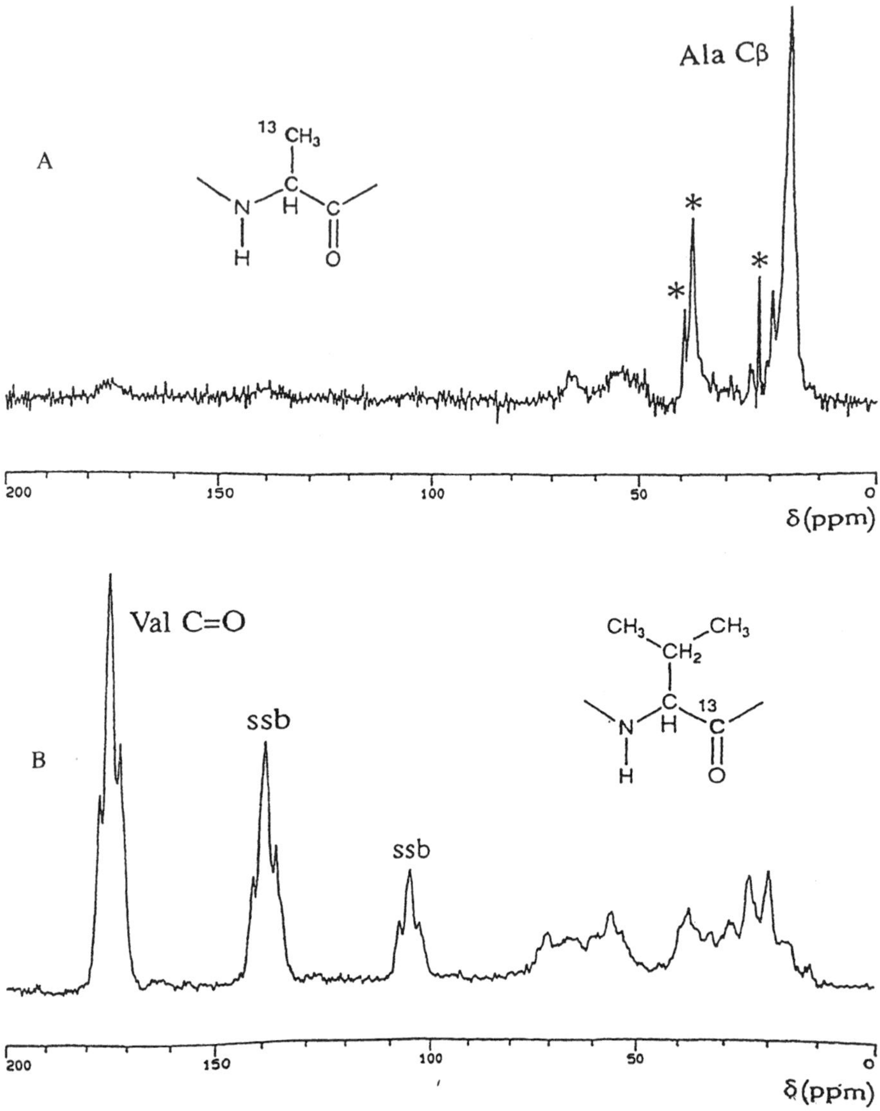

Figure 4.67 ^{13}C CP/MAS NMR spectra of (a) [3-^{13}C]Ala bR and (b) [1-^{13}C]Val bR in the purple membranes (* denotes that peaks are signals of ^{13}C-labeled lipids).

structures of [3-^{13}C]Ala-labeled bR was not always identical at temperatures between ambient and lower temperatures because the ^{13}C chemical shifts and relative peak intensities form purple membrane preparations containing these salts changed with temperature in the range $-110\,°C$ to $23\,°C$. In particular, they found that some residues involving Ala residues at the α_{II}-helix and loop region were converted at temperatures below $-60\,°C$ to a conformation of bR involving α_I-helix.

Most recently, Opella showed that it is the ability of solid-state NMR experiments to give completely resolved spectra of immobile proteins that enables the structures of larger membrane proteins to be determined in the definitive environment of lipid bilayers [199]. Further, Naito *et al.* recorded high-resolution solid-state ^{13}C–NMR spectra of [3-^{13}C]Ala-labeled fragments of bR incorporated into dimyristoylphosphatidylcholine bilayers [200]. This approach has proved to be useful for the assignment of peaks as well as dynamic features of the transmembrane helixes in lipid bilayers.

In addition to synthetic polymers and biopolymers mentioned earlier, solid-state ^{13}C–NMR has been used for studying the structure and dynamics of many materials such as coal [201], wood [202], ceramic materials [203] and foods [204].

4.10.3 APPROACHES BY MULTINUCLEI NMR

^{13}C high-resolution solid-state NMR spectroscopy has provided a deep insight into the structure and dynamics of polymers in the solid state, and many studies have used ^{13}C–NMR. This is because that spin number ($I = 1/2$) and natural abundance (1.1%) are suitable for measuring high-resolution solid-state ^{13}C–NMR spectra. However, it is very important to obtain information about structure and dynamics by using other nuclei. For example, although the study of hydrogen-bonding structure in the solid state has been carried out by ^{13}C–NMR, hydrogen and oxygen atoms play an important role in the determination of the hydrogen-bonding structure, and therefore hydrogen or oxygen atom NMR spectroscopy will provide much information on hydrogen-bonding structure.

Here, some applications of multinuclei (^{15}N, ^{29}Si and ^{17}O) NMR to the studies of the structure and dynamics of polymers are introduced.

4.10.3.1 Structure and Dynamics of Polypyrrole in the Solid State Studied by ^{15}N–NMR Spectroscopy

High-resolution solid-state NMR spectroscopy can provide much information about the structure and dynamics of amorphous polymers. It has generally been

Table 4.7 Ploymerization conditions and electrical conductivity of polypyrrole sample.

Sample	*[Pyrrole] (mol l^{-1})*	*[Sodium p-toluenesulfonate] (mol l^{-1})*	*[^{15}N-labeled Pyrrole] (%)*	*Electrical Conductivity (S cm^{-1})*
a	1.1×10^{-1}	1.8×10^{-3}	20.75	2×10^{-3}
b	1.1×10^{-1}	1.8×10^{-3}	20	4×10^{-6}
c	1.1×10^{-1}	1.8×10^{-3}	20	1×10^{-3}
d	5.3×10^{-2}	5.1×10^{-3}	20	2×10^{-2}
e	1.1×10^{-1}	5.1×10^{-2}	50	1×10^{1}

proposed that polypyrrole can exist in an aromatic form and a quinoid form and that the former corresponds to the undoped state and the latter to the doped state [205–207], as shown in Figure 4.68, However, the actual structure appears to be more complicated. Thus, it is very important to obtain exact structural data for polypyrrole and to elucidate the relationship between structure and physical properties such as electrical conductivity for developing and designing high-performance polypyrroles. The relationship between the structure and electrical conductivity has not been clarified for polypyrrole yet because the polymer is amorphous and insoluble.

^{15}N–NMR spectroscopy in the solid state can provide detailed information about the structure of polypyrrole because an ^{15}N resonance line should correspond to a given structure [230]. Doped and dedoped ^{15}N-labeled polypyrrole films (samples a–e) were prepared by electrochemical polymerization [208] using 20–75% ^{15}N-labeled pyrrole. The details of the experimental conditions for the concentration of pyrrole and sodium *p*-toluene sulfonate in water are listed together with electrical conductivity in Table 4.7. Table 4.8 shows the results of the elemental analysis.

Table 4.8 Elemental analysis data for polypyrrole samples.

	Elemental Composition			
Sample	*Carbon*	*Hydrogen*	*Nitrogen*	*Sulfur*
a	34.8	30.0	7.2	1
b	52.1	39.4	11.9	1
c	104.9	81.2	24.4	1
d	—	—	—	—
e	20.6	19.0	3.5	1

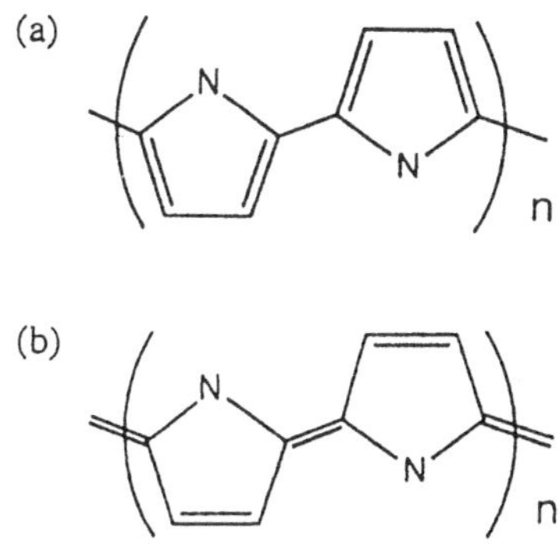

Figure 4.68 (a) Aromatic and (b) quinoid structures for polypyrrole.

The nitrogen-to-sulfur ratio, which is a measure of the number of monomer units per one dopant, varied from 3.5 to 24.4.

Figure 4.69(a) shows the ^{15}N CP/MAS spectrum of sample b, which is measured immediately after polymerization, and Figure 4.69(b) shows the ^{15}N CP/MAS spectrum of sample b′, which has been left in the air for about 1 year. The intensity of the highest-field peak (peak α) at about 90 ppm in the latter spectrum is considerably larger than that in the former spectrum. This means that peak α at about 90 ppm may be assigned to the nitrogen atom in the oxidized structure. In order to make this assignment clear high-resolution solid-state ^{13}C–NMR spectra were measured. Figure 4.70 shows line shape analysis for ^{13}C CP/MAS NMR spectra of sample a and the air-oxidized sample. The main peaks consist of five decomposed peaks and appeared from about 100 to 160 ppm, which can be assigned to a CH carbon of aromatic form, CH carbon of quinoid form, quaternary carbon of aromatic form, quaternary carbon of quinoid form, and quaternary carbon of dopant from high-field to low-field using reference data by Devreux *et al.* [209]. Further, it can be thought that some peaks from the structure such as dopant and oxidized structure are overlapped on these region of the spectra. Previous ^{15}N–NMR study shows [210] that polypyrrole has at least three phases. However, ^{13}C–NMR spectra do not distinguish these structural differences. This shows that ^{15}N–NMR spectroscopy is very useful for structural analysis of these samples. From these ^{13}C–NMR spectra, the chemical shifts of the air-oxidized phases are about 40, 80, and 170 ppm. These spectra were measured by using the TOSS sequence so that these peaks were not spinning sidebands. A recent report by Forsyth *et al.* [211] shows there are spinning sidebands. The oxidized peaks should be overlapped if there is oxidized material. We considered the possible structures of the

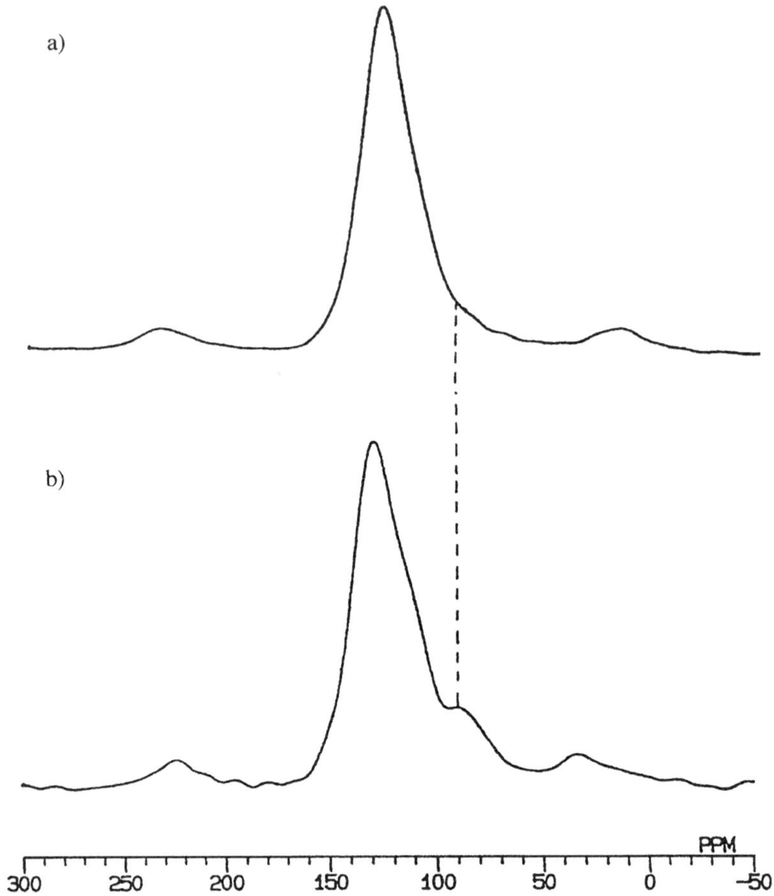

Figure 4.69 50.55 MHz ^{15}N–NMR CP/MAS spectra of polypyrrole (sample B) in the solid state: (a) immediately after polymerization; (b) sample A left for about 1 year in the air.

air-oxidized phase shown in Figure 4.71 and 4.72 for which ^{15}N–NMR chemical shifts are 95 and 94 ppm for structures A and B, respectively. The ^{13}C–NMR chemical shifts of structure A are 176, 128, 148, and 50 ppm for carbon C1, C2, C3, and C4, respectively. The ^{13}C–NMR chemical shifts of structure B are 179, 31, 21, and 43 ppm for carbon C1, C2, C3, and C4, respectively. These assignments can explain both the ^{13}C and ^{15}N–NMR spectra. Further, the peak α can be regarded as an amide prepared by oxidization for which the ^{15}N–NMR chemical shift is about 90 ppm and the ^{13}C–NMR chemical shift is about 168 ppm. Therefore, it

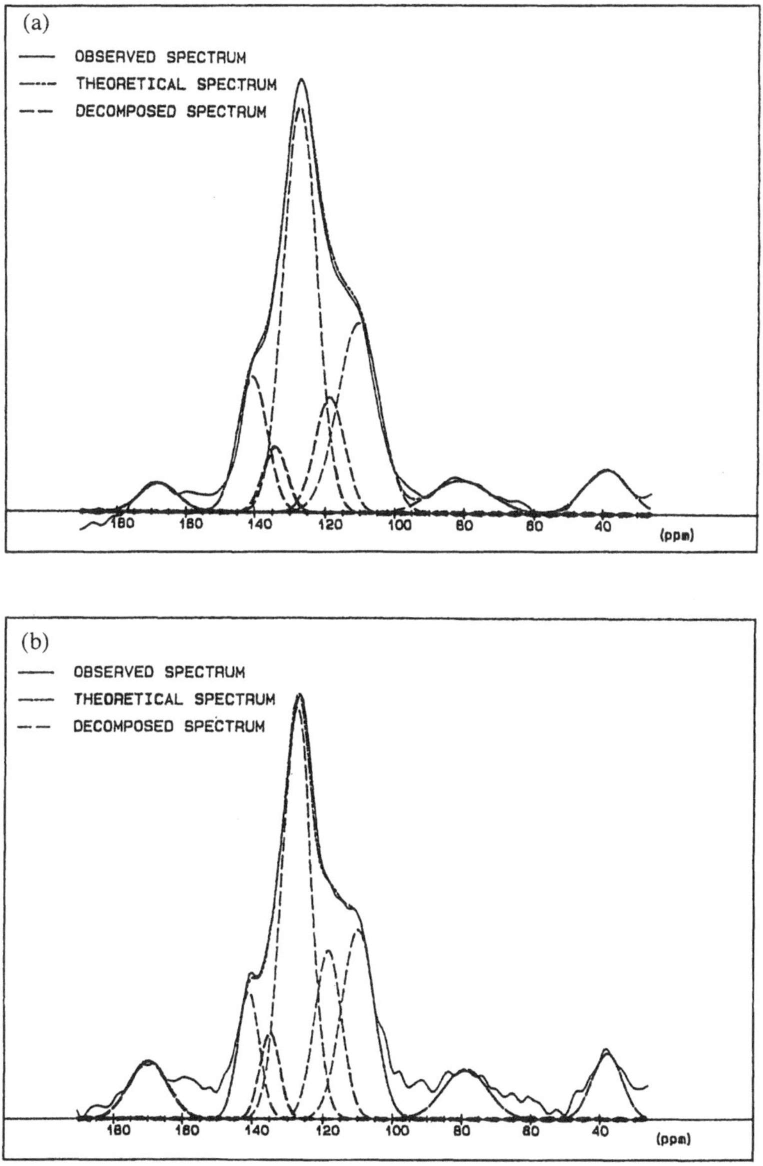

Figure 4.70 The 67.8-MHz ^{13}C CP/MAS spectra and simulated spectra of polypyrrole samples in the solid state: (a) sample A, (b) sample A left for about 1 year in the air.

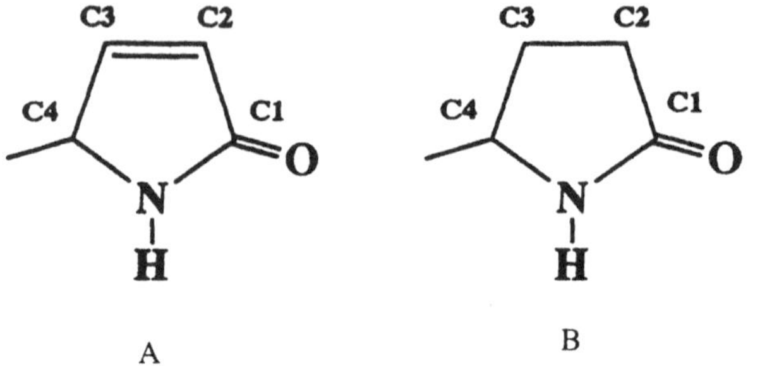

Figure 4.71 Possible models of the air-oxidized structure.

can be concluded that the ^{15}N–NMR chemical shift of 89 ppm is assigned to the air-oxidized structure.

Figure 4.73(a) shows the ^{15}N–NMR spectrum of polypyrrole sample B measured with a contact time of 500 μs to enhance the intensity of peak α. It is clear that this spectrum consists of an intense main peak and three shoulder peaks. These observed ^{15}N–NMR spectra have been deconvoluted into four components designated α, β, γ, and δ in order of decreasing shielding. This indicates that the polypyrrole samples have at least four kinds of structures corresponding to the four

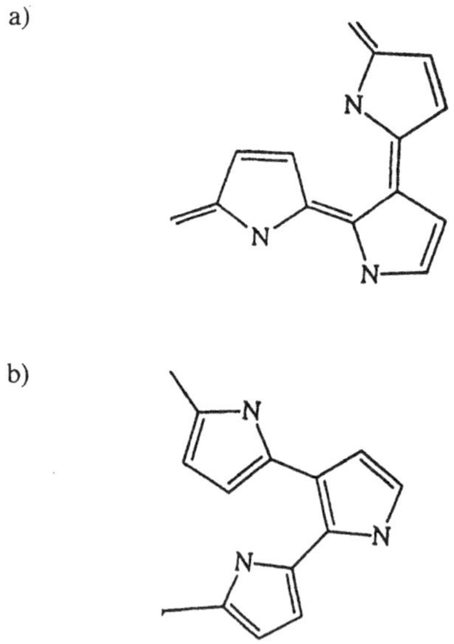

Figure 4.72 The 2,3-bond structure of (a) quinoid and (b) aromatic form.

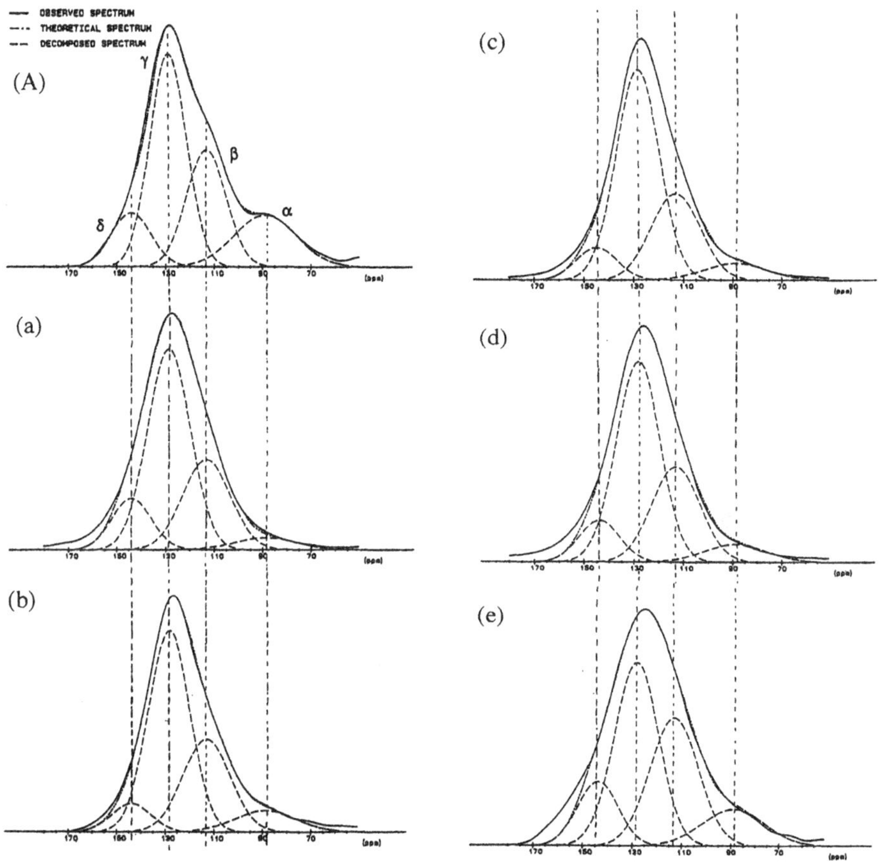

Figure 4.73 50.55-MHz ^{15}N CP/MAS spectra and simulated spectra of polypyrrole samples in the solid state: (a) sample A measured with contact time of 500 μs, (b) sample A, (c) sample B, (d) sample C, (e) sample D and (f) sample E.

peaks. The determined ^{15}N chemical shifts, half-widths, and relative peak intensities of polypyrrole samples are listed in Table 4.9. The spectra of dedoped sample B and iodine doped sample C are shown in Figure 4.5(b) and (c), respectively. The intensity ratios of peak γ to β (I_γ/I_β) for samples B and C are shown in Table 4.10 with the corresponding electrical conductivity data. The relative peak intensity ratio I_γ/I_β is increased by iodine doping. This shows that the peak intensity ratio I_γ/I_β is a measure of the change in electrical conductivity. As mentioned earlier, the aromatic form is the major component in the dedoped sample, and the quinoid form is the major component in the doped sample. For this reason,

Table 4.9 Observed ^{15}N chemical shifts, half-widths, and relative peak intensities of polypyrrole samples.

Sample	*Peak*	*^{15}N Chemical Shift (ppm)*	*Halft-Width (ppm)*	*Relative Peak Intensity (%)*
a	α	89.0	30.0	4.7
	β	133.3	23.0	27.6
	γ	128.65	20.8	54.7
	δ	144.0	19.0	12.9
b	α	89.0	30.0	8.8
	β	113.3	23.0	29.0
	γ	128.60	20.1	54.8
	δ	144.0	19.0	7.4
c	α	89.0	30.0	6.6
	β	113.3	23.0	26.8
	γ	128.55	20.4	58.5
	δ	144.0	19.0	8.2
d	α	89.0	30.0	7.0
	β	113.3	23.0	28.6
	γ	128.60	20.9	54.1
	δ	144.0	19.0	10.3
e	α	89.0	30.0	12.1
	β	113.3	23.0	32.4
	γ	128.6	21.0	42.2
	δ	144.0	19.0	13.4

peak β at about 113 ppm may be assigned to the nitrogen atom in the aromatic form, and peak γ at about 129 ppm to the nitrogen atom in the quinoid form. This assignment was supported by FPT–INDO shielding [210, 213, 214]. Further, the FPT–INDO shielding calculation shows that the minor peak δ at about 144 ppm probably comes from a 2, 3-bond structure, as shown in Figure 4.73 [210]. In

Table 4.10 Electrical conductivity and peak intensity ratios for peaks γ and β.

Sample	*Intensity Ratio of Peak γ to β*	*Electrical Conductivity ($S\ cm^{-1}$)*
b(dedoped)	1.89	4×10^{-6}
c(iodine doped)	2.18	1×10^{-3}

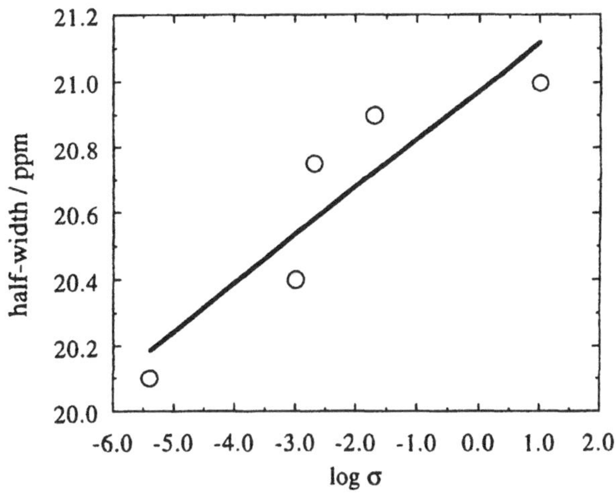

Figure 4.74 Plots of the half-widths of peak γ of polypyrrole samples against the electrical conductivity.

this structure the nitrogen atoms bonded to hydrogen atoms approach other hydrogen atoms bonded to different carbon atoms. From the preceding results, the four peaks α, β, γ, and δ are assigned to the nitrogen atoms in the oxidized structure, the aromatic form, the quinoid form, and the 2, 3-bond structure, respectively. Figure 4.73(a–e) shows 50.55 MHz ^{15}N CP/MAS spectra and simulated spectra of polypyrrole samples (a–e) in the solid state. The four peaks were decomposed by computer fitting, and their ^{15}N chemical shift values are 89.0, 113.3, 128.6, and 144.0 ppm, respectively. The ^{15}N chemical shifts of each samples were almost constant. This shows that the chemical shifts of polypyrrole samples are little affected by the difference in electrical conductivity. However, there is a relationship between the half-widths of peak γ and electrical conductivity. Figure 4.74 shows the plots of the half-widths of peak γ in polypyrrole samples against the electrical conductivity. It is found that an increase of the electrical conductivity leads to an increase of the half-width of peak γ, and an approximately linear relationship between the half-width and the electrical conductivity exists, as shown in Figure 4.74. The half-widths of the other peaks are independent of the degree of doping. As mentioned earlier, the relative peak intensity of the aromatic form is decreased by doping, whereas the relative peak intensity of the quinoid form is increased. It seems that some portion of the aromatic form is converted to the quinoid form in the presence of dopant, and thus the structures of the quinoid form may have some distribution leading to a chemical shift distribution. Therefore, the half-width of peak γ increases as electrical conductivity increases. The half-width of peak γ essentially becomes a measure of electrical conductivity.

To obtain information about the molecular motion in polypyrrole, the ^{15}N spin-lattice relaxation times T_1 of sample A (75% ^{15}N-enriched sample) at room temperature and -100 °C have been measured using Torchia's method. At room temperature, the intensities of peak β as a function of delay time τ show a biexponential form, and thus the peak β has two components with different T_1 values. The T_1 values increase as the temperature is increased. This shows that the molecular motion of these nitrogen atoms is in the slow-motion region. The two T_1 values for peak β are 0.35 and 13.8 s and are very different from each other. This indicates that the mobilities of the two components are quite different. It was reported from an electron diffraction study that polypyrrole has crystalline domains in an amorphous matrix [212]. Therefore, the long T_1 (13.8 s) may be assigned to motion in such domains, and the short T_1 (0.35 s) to motion in the amorphous region. From NMR relaxation time measurements, the ratio of crystalline to amorphous is 42:58. The value of crystalline seems to be large, but it is difficult to compare the cross-polarization intensity of phases with different mobilities. Thus, the real intensity of the crystalline phase is smaller than this value. In an electron diffraction study, the ratio has not been mentioned.

From these results, the following conclusions can be drawn: The peaks α, β, γ, and δ are assigned to the oxidized structure, the aromatic form, the quinoid form, and the 2,3-bond structure, respectively, of polypyrrole. The half-width of peak γ is a measure of electrical conductivity.

4.10.3.2 Structure and Dynamics of Polysiloxane in the Solid State Studied by ^{29}Si NMR Spectroscopy [219]

High-resolution ^{29}Si NMR spectra and ^{29}Si NMR relaxation times of a polysiloxane containing rigid groups, namely, copoly(tetramethyl-*p*-silphenylenesiloxane/dimethyl-siloxane) (CPTMPS/DMS), in the solid state have been measured over a wide range of temperatures to obtain information about the conformation and dynamics [215]. Finite perturbation theory (FPT) calculations of the ^{29}Si shielding constants within the complet neglect of differential overlap (CNDO) MO framework [216, 217] were carried out to understand and discuss the observed ^{29}Si NMR chemical shifts [218].

The ^{29}Si CP/MAS NMR spectra of CPTMPS/DMS at various temperatures are shown in Figure 4.75. At -100 °C, a sharp, intense signal and a broad, weak signal appear at -0.8 ppm (for the TMPS moiety) and at -21.1 ppm (for the DMS moiety), respectively.

The peak intensity of the TMPS moiety is almost independent of temperature in the temperature range -100 to 80 °C, but that of the DMS moiety gradually decreases with an increase of temperature, and at temperatures above -80 °C the

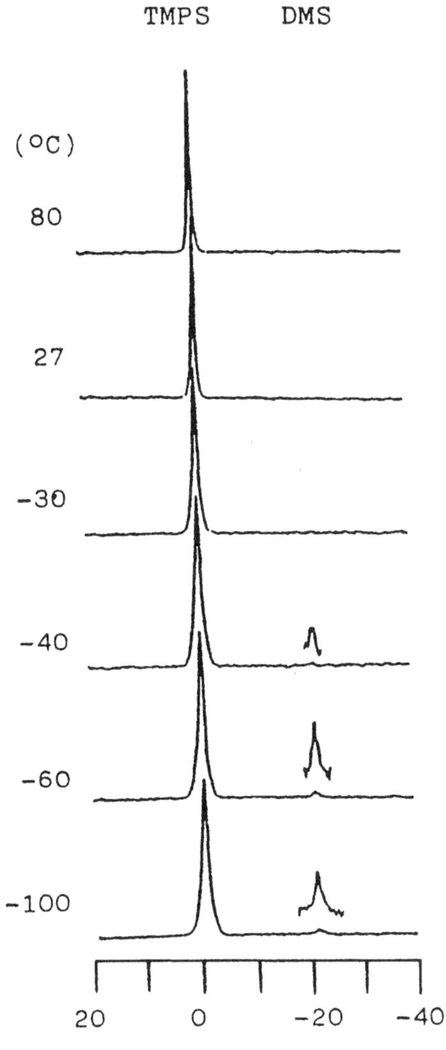

Figure 4.75 VT ^{29}Si CP/MAS NMR spectra of CPTMPS/DMS in the solid state at temperatures from −100 to 80 °C.

signal disappears. This means that the mobility of the DMS moiety is more sensitive compared with that of the TMPS moiety within this temperature range. (The CP efficiency largely depends on the mobility.) The ^{29}Si GHD/MAS NMR spectra of CPTMPS/DMS at various temperatures are shown in Figure 4.76. The ^{29}Si signal of the TMPS moiety shifts upfield ($\Delta\delta = 0.4$ ppm) during the temperature

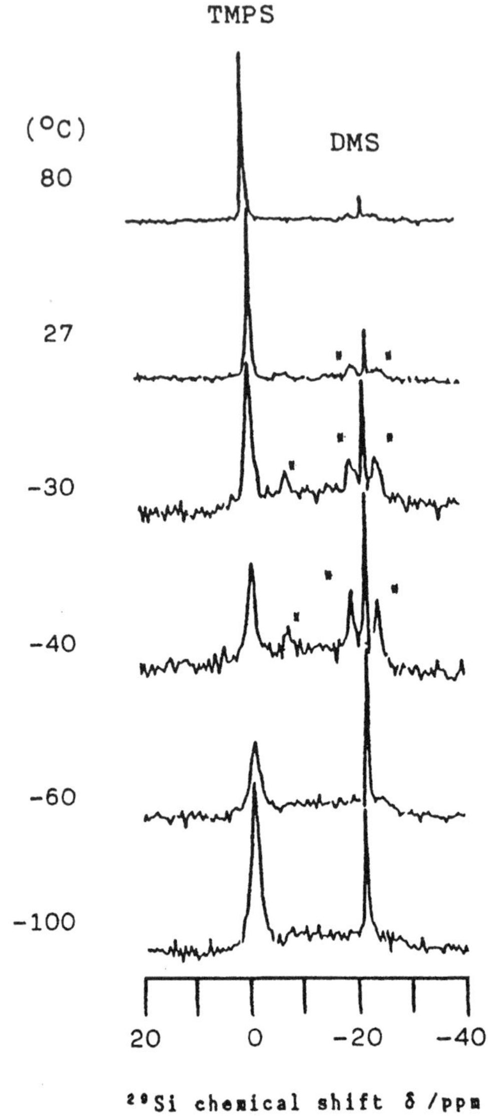

Figure 4.76 VT ^{29}Si GHD/MAS NMR spectra of CPTMPS/DMS in the solid state at temperatures from −100 to 80 °C. Background noise from the NMR probe is labeled by asterisks.

change from −100 to 80 °C. However, the ^{29}Si signal of the DMS moiety shifts further upfield ($\Delta\delta = 1.1$ ppm) with the same temperature change. The observed upfield shifts seem to be due to changes in the main-chain conformation. Besides the TMPS and DMS peaks, some broadening peaks (marked by an asterisk) appear. These peaks come from materials used in the NMR probe.

To understand why the ^{29}Si peaks shift upfield as the temperature is increased, FPT CNDO/2 MO calculations of the ^{29}Si chemical shifts were carried out. The calculated ^{29}Si nuclear shieldings of the model compound for the DMS moiety using the FPT CNDO/2 MO theory are plotted against the dihedral angle θ in Figure 4.77 together with the structure of the model compound (with a dihedral angle θ of 180° relative to the *cis* conformation). The computation for dihedral angles θ less than 60° diverges as a result of greater hindrance. Note that the calculated chemical shift (σ) is a nuclear shielding constant and so the negative sign indicates deshielding. However, the negative sign of the observed chemical

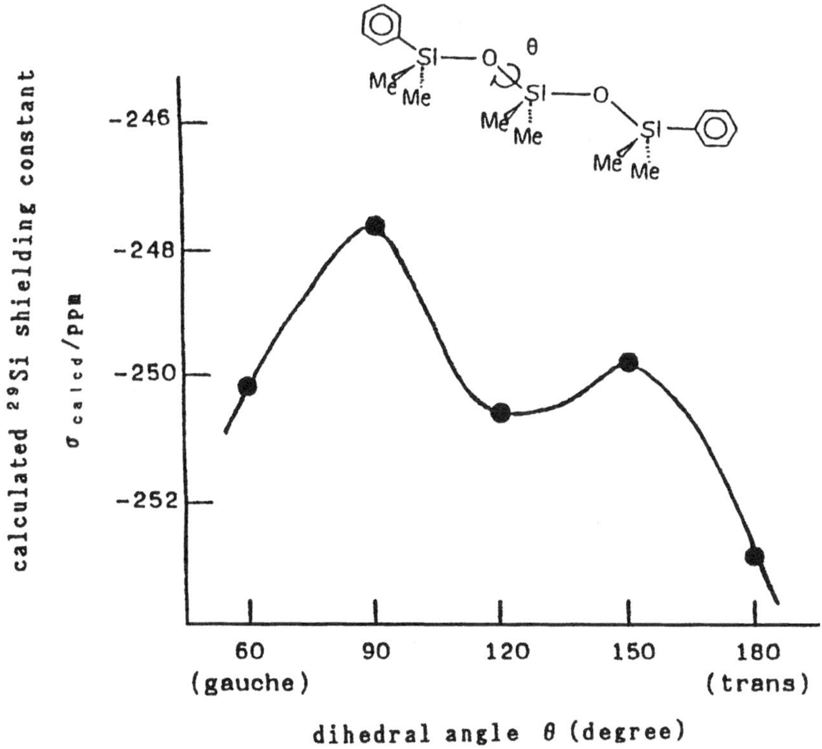

Figure 4.77 The calculated ^{29}Si nuclear shieldings for the central Si atom in the compound as a function of the dihedral angle θ by the FPT CNDO/2 method.

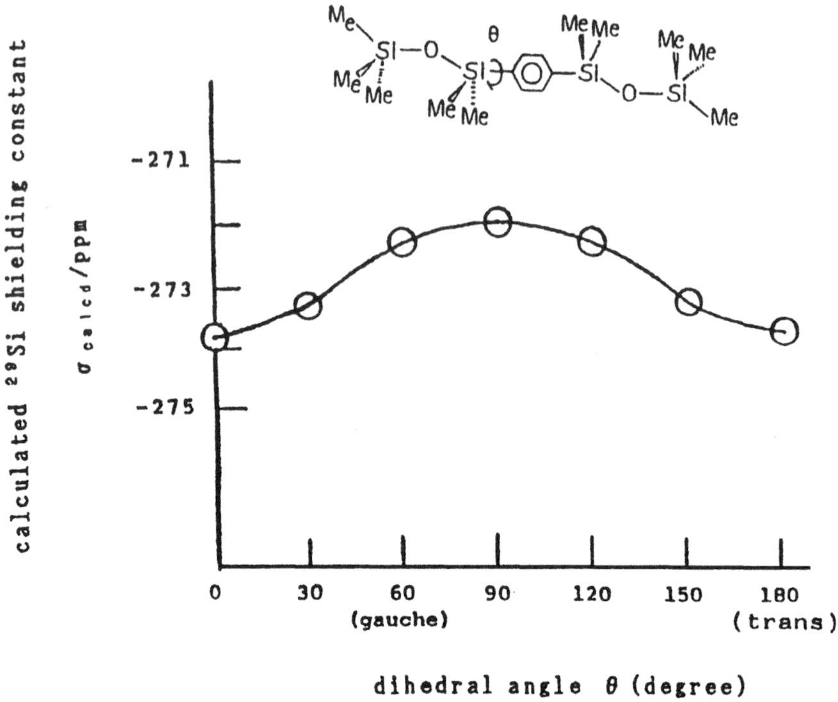

Figure 4.78 The calculated ^{29}Si nuclear shieldings for the silphenylene Si atoms in the model compound as a function of the dihedral angle θ by the FPT CNDO/2 method.

shift (δ) indicates shielding. Therefore, the relative difference in the calculated chemical shifts should be compared with the observed chemical shifts. As can be seen from Figure 4.77, the calculated shielding of the central Si atom in the model compound changes through the maximum upfield shift at $\theta = 90°$ and $150°$ on going from $60°$ to $180°$. It can be said that when θ deviates from $180°$ (trans form), the chemical shift for the central Si atom moves upfield. For the TMPS moiety, the ^{29}Si nuclear shieldings are calculated as a function of θ using the FPT CNDO/2 MO method as shown in Figure 4.78. The calculated ^{29}Si chemical shift of the arylenedisilane group moves through the maximum upfield shift as θ is changed from $0°$ to $180°$. It can be said that as θ deviates from $180°$ (trans form), the ^{29}Si chemical shift moves upfield. This is similar to the case of the DMS moiety, but the magnitudes of moving upfield are smaller compared with those of the DMS moiety. From the NMR experimental finding by Moeller [221] that the ^{29}Si signal for the Si atom of poly(dimethylsiloxane) moves upfield due to a conformational change from the trans to the gauche form with an increase of temperature, it may be expected that at low temperature the DMS moiety predominantly takes the trans form and at high temperature it predominantly takes the gauche form. However, the ^{29}Si signal for the TMPS moiety is almost independent of temperature. This experimental result means that the conformational change of the TMPS moiety is constrained. For the above reason, the DMS moiety is called the "soft" moiety and the TMPS moiety is called the "rigid" moiety. The CPTMPS/DMS copolymer retains rigid and soft properties over a wide range of temperatures.

We are also concerned with the dynamics of the CPTMPS/DMS copolymer through the ^{29}Si T_1 experiment [215]. It is assumed that ^{29}Si T_1 occurs solely by intramolecular dipole–dipole interactions with the attached protons CH_3–Si–CH_3. This is undoubtedly true for motions encountered in the copolymer. As determined from Figure 4.76, the ratio of the peak intensity of the TMPS to the DMS moiety is about 2:1 at −100 °C. However, the intensity of the DMS peak is increased as the temperature is increased from −100 to −60 °C and is gradually decreased as the temperature is further increased from −60 to 80 °C. This means that the ^{29}Si T_1 values of CPTMPS/DMS vary greatly with the temperature change. We measured the ^{29}Si T_1 values of CPTMPS/DMS in the solid state using the inversion recovery method [$(180° - \tau - 90° - 150$ s) pulse sequence] at −90, −60, −30, 27, and 80 °C. Figure 4.79 shows the ^{29}Si GHD/MAS spectra of CPTMPS/DMS at 80 °C. The magnitude of the two signals recovers from the negative value to the equilibrium positive value through the zero value. As seen from this figure, the TMPS signal recovers faster, exponentially, than that of DMS. The determined values of ^{29}Si T_1 for the TMPS and DMS moieties from −90 to 80 °C plotted against $1/T$ (K^{-1}) and are shown in Figure 4.80. The T_1 value for the TMPS moiety decreases as the temperature is increased from −90 to 80 °C. This indicates that the molecular motion is in the slow motion region [11, 215] in this temperature range. However, the T_1 value for the DMS moiety increases as the temperature is increased from −67 to 80 °C, becomes a minimum at −67 °C, and increases again as the temperature is decreased from −67 to −90 °C. This indicates that the molecular motion is in the extreme narrowing region above −67 °C. From the experimental T_1 values and the single correlation-time model based on BPP theory [11], the correlation time τ_c for molecular motion was calculated [215]. From the T_1 minimum, the correlation time τ_c can be obtained for molecular motion at megahertz frequencies. The plot of $\log(\tau_c)$ versus $1/T$ for the DMS moiety in Figure 4.81 shows a good linear relationship. The temperature dependence of the correlation time τ_c usually obeys the Arrhenius form because $\tau_c = \tau_0 \exp(-\Delta E/RT)$, where τ_0 is the prefactor, R is the gas constant, and T is the absolute temperature. The activation energy ΔE, which is considered to correspond to the barrier height for

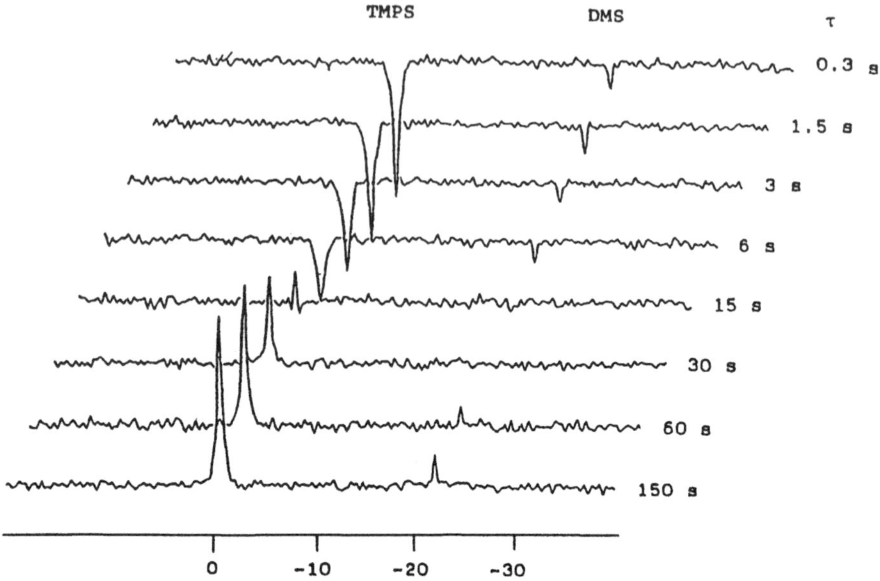

Figure 4.79 ^{29}Si GHD/MAS NMR spectra of CPTMPS/DMS in the solid state at 80 °C using the inversion recovery method (180° − τ − 90° − 150 s) as a function of τ.

the potential hindering rotation, can be estimated from the slope of log(τ_c) versus $1/T$. By this procedure, a ΔE value of 1.8 kcal mol^{-1} is obtained for the DMS moiety. The potential barrier for the hindering rotation in the DMS moiety is only slightly larger than that in poly(dimethylsiloxane) [220], which is itself very small. Thus, it can be said that the soft properties of the DMS moiety come from these low potential barriers for the hindering rotation.

To obtain further information about the molecular motion of the siloxane chain, the ^{29}Si powder pattern spectra were recorded. Figure 4.82 shows the ^{29}Si CP NMR powder pattern spectra for the CPTMPS/DMS copolymer obtained without MAS at −90, 25, and 110 °C. These powder pattern spectra predominantly come from the TMPS moiety because the CP efficiency for the DMS moiety is negligibly smaller compared with that for the TMPS moiety, as seen from the ^{29}Si CP/MAS NMR experiments. The anisotropy widths $|\sigma_{11} - \sigma_{33}|$ obtained from the tentlike powder pattern spectrum were 57.8, 55.0, and 52.0 ppm at −90, 25, and −110 °C, respectively. The anisotropy width decreases as the temperature is increased. This indicates that the mobility for the TMPS moiety increases slightly as the temperature is increased. From the preceding results, it can be concluded

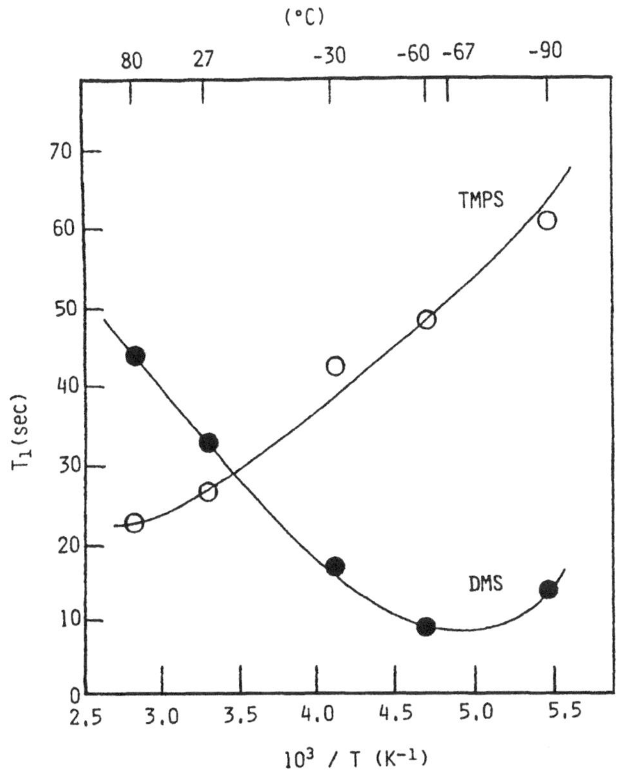

Figure 4.80 Temperature dependence of the observed ^{29}Si T_1 in CPTMPS/DMS in the solid state.

that the DMS moiety predominantly takes the *trans* form at low temperature and at high temperature predominantly takes the *gauche* form, whereas the conformation of the TMPS moiety is almost independent of temperature. The molecular motion of the DMS moiety is fast, whereas that of the TMPS moiety is restricted. This indicates that CPTMPS/DMS retains rigid and soft parts in the main chain.

4.10.3.3 Structure and Dynamics of Polyglycine and Glycylglycine Peptides in the Solid State by ^{17}O NMR Spectroscopy [231]

The oxygen atom is one of the important atoms constituting hydrogen-bonding structure in peptides and polypeptides. Nevertheless, a solid-state ^{17}O NMR study for peptides and polypeptides is very difficult to perform. This is due to the very

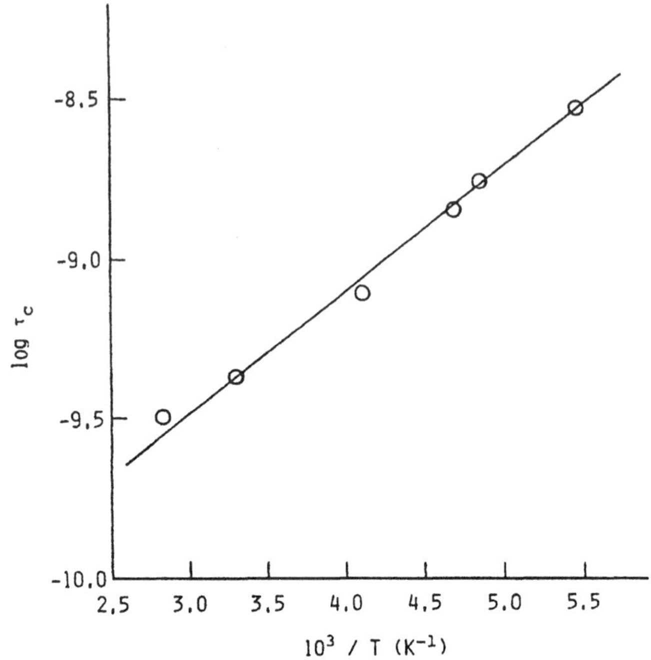

Figure 4.81 Plot of ^{29}Si correlation time for the DMS moiety versus reciprocal absolute temperature.

weak sensitivity of solid-state ^{17}O NMR measurement, which has two causes. One is that the ^{17}O nucleus has a very low natural abundance (0.037%). Another is that the ^{17}O nuclear spin quantum number (I) is 5/2, which means the nucleus is quadrupolar, and thus the ^{17}O signal is broadened by nuclear quadrupolar effects in the solid state. However, solution-state ^{17}O NMR spectroscopy has been successfully employed to elucidate several structural problems in inorganic chemistry [222–225] because the ^{17}O signal becomes very sharp owing to the removal of quadrupolar interactions by isotropic fast reorientation in solution. For example, because the oxygen atom is directly associated with the formation of hydrogen bonds, hydrogen bonding for the carbonyl group in various compounds often results in large upfield shifts of the carbonyl ^{17}O NMR signal [226, 227]. From these results, solution-state ^{17}O NMR has been established as a means of investigating structural characterizations.

From the preceding it can be expected that solid-state ^{17}O NMR would provide a deep insight into the hydrogen-bonding structure in solid peptides and polypeptides. The high-resolution solid-state ^{17}O NMR spectra of polyglycine(PG)I,

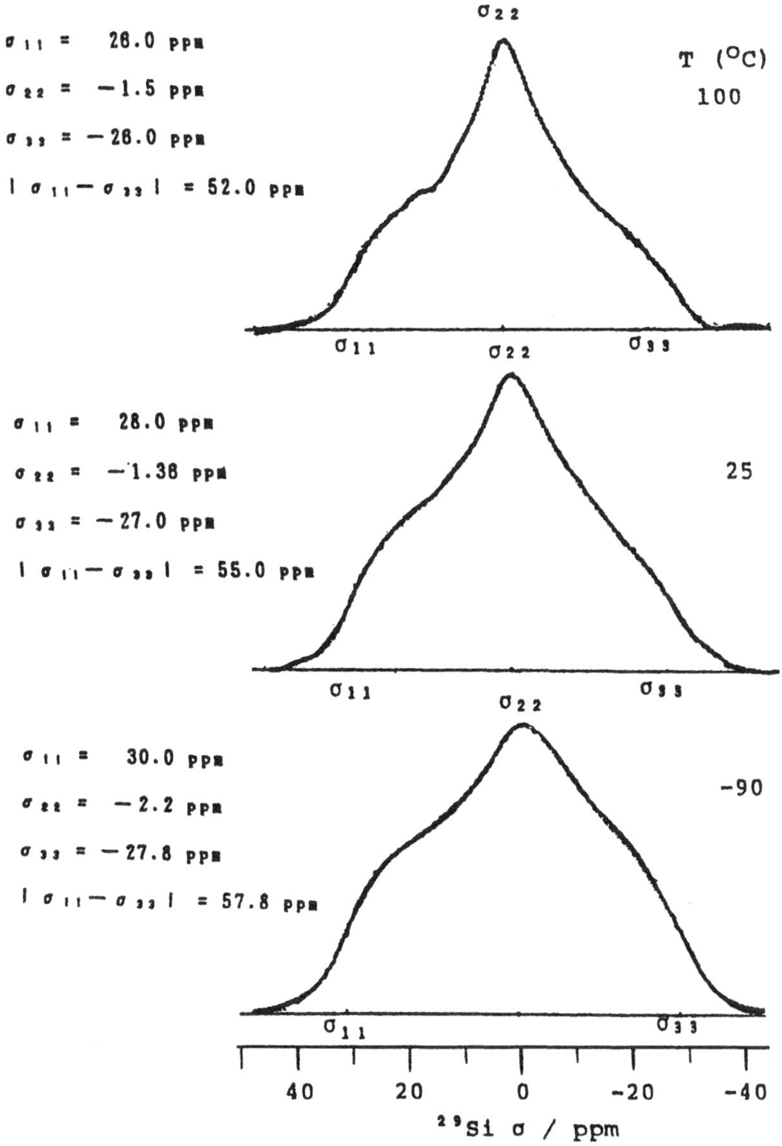

Figure 4.82 The observed ^{29}Si CP NMR powder pattern spectra of CPTMPS/DMS at −90, 25, and 100 °C. The principal values of the shift tensor σ_{11}, σ_{22}, σ_{33} and the anisotropy width $|\sigma_{11} - \sigma_{22}|$ are shown.

PG II, glycylglycine (GlyGly), and glycylglycine nitrate (GlyGly·HNO_3) are observed, and this process covers a wide range of hydrogen bonds [232, 233]. From these results, the relationships between the hydrogen-bonding structures will be understood.

Six-percent ^{17}O-labeled glycine (Gly) was prepared from the glycine methyl ester in Na^{17}OH/methanol solution for which Na^{17}OH was prepared by reacting 11% ^{17}O-labeled water with Na metal. Further, ^{17}O-labeled Gly N-carboxyanhydride (NCA) was prepared using 6% ^{17}O-labeled Gly. Six-percent ^{17}O-labeled polyglycine($(Gly)_n$) was prepared by heterogeneous polymerization of 6%-labeled ^{17}O-glycine-NCA in acetonitrile by using n-butylamine as the initiator.

4.10.3.3.1 Static ^{17}O CP Spectra of the ^{17}O-Labeled Peptides and Polypeptides

Static ^{17}O CP NMR spectra of GlyGly have been observed at 36.6, 54.2, and 67.8 MHz, as shown in Figure 4.83. The spectra at 36.6 and 54.2 MHz consist of two major splitting signals, but that at 67.8 MHz has just one major signal. Such splitting comes from the quadrupolar interaction. By computer simulation, the quadrupolar coupling constant e^2qQ/h is determined to be 8.55 MHz. This value is much larger than that for the amide ^{14}N nitrogen nucleus (1.11 MHz). It is seen that the NMR spectra strongly depend on the measurement frequency because the quadrupolar effect depends on the measurement frequency and the broadening of the ^{17}O NMR signal is further increased by the quadrupolar effect as the measurement frequency is decreased. Therefore, it can be said that high-frequency measurement is needed for a quadrupole nucleus such as ^{17}O.

Figure 4.84 shows 67.8 MHz static ^{17}O CP NMR spectra ofPG I, PG II, GlyGly, and GlyGly·HNO_3. Computer simulations have been performed superimposing theoretically calculated spectra onto the experimental spectra on a computer display. To determine the ^{17}O NMR parameters unequivocally, computer simulations for different frequencies (36.6, 54.2, and 67.8 MHz) were carried out. For example, Figure 4.85 shows 67.8, 54.2, and 36.6 MHz static ^{17}O CP NMR spectra of GlyGly together with theoretically calculated spectra, and Figure 4.86 shows 67.8 MHz static ^{17}O CP NMR spectra of PG I, PG II, GlyGly, and GlyGly·HNO_3 together with theoretically calculated spectra.

The ^{17}O NMR parameters of PG I, PG II, GlyGly, and GlyGly·HNO_3 obtained by computer simulations are shown in Table 4.11. From these results it can be seen that the δ_{11}, δ_{22}, and δ_{33} values of the samples change from 546 to 574 ppm, 382 to 425 ppm, and -132 to -101 ppm, respectively. The magnitude in the chemical shift change is much larger than that of the carbonyl ^{13}C and amide ^{15}N

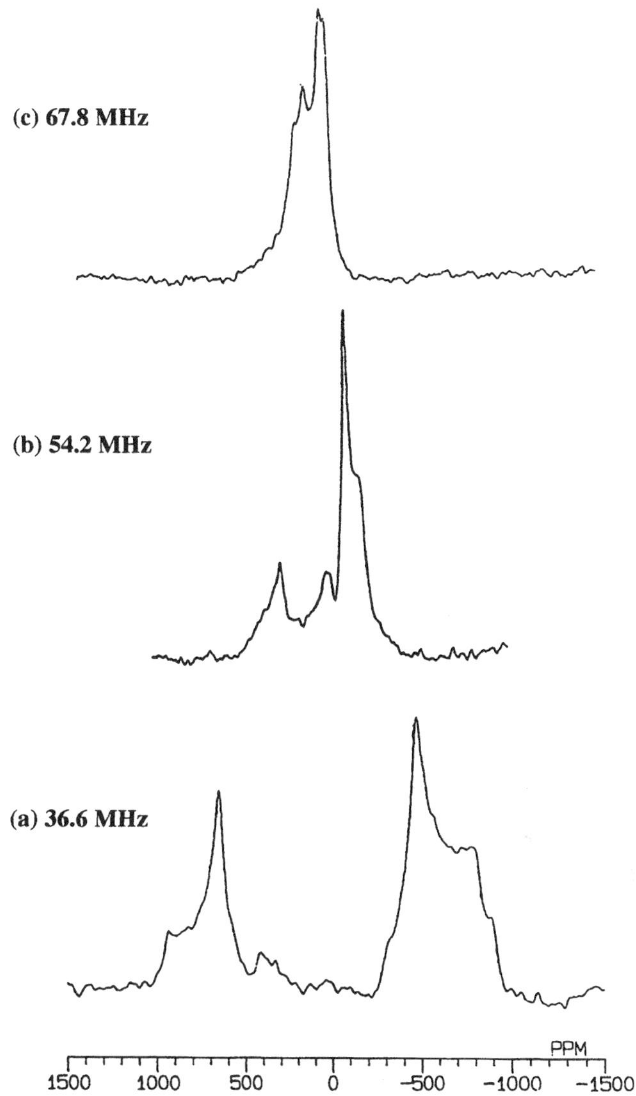

Figure 4.83 (a) 36.6-MHz, (b) 54.2-MHz, and (c) 67.8-MHz ^{17}O CP static spectra of GlyGly in the solid state.

chemical shift. Every $\Delta\delta$ value, defined by $\Delta\delta = \delta_{33} - \delta_{iso}$, is about 400 ppm, which is much larger than that of the carbonyl ^{13}C and amide ^{15}N chemical shifts (about ^{17}O and 100 ppm, respectively). Every η value defined by $\eta = (\delta_{22} - \delta_{11})/\Delta\delta$ is a constant of about 0.4. These values are common to the carbonyl oxygen in peptides. Although the chemical shift tensor is axially symmetric ($\delta_{11} = \delta_{22}$) for

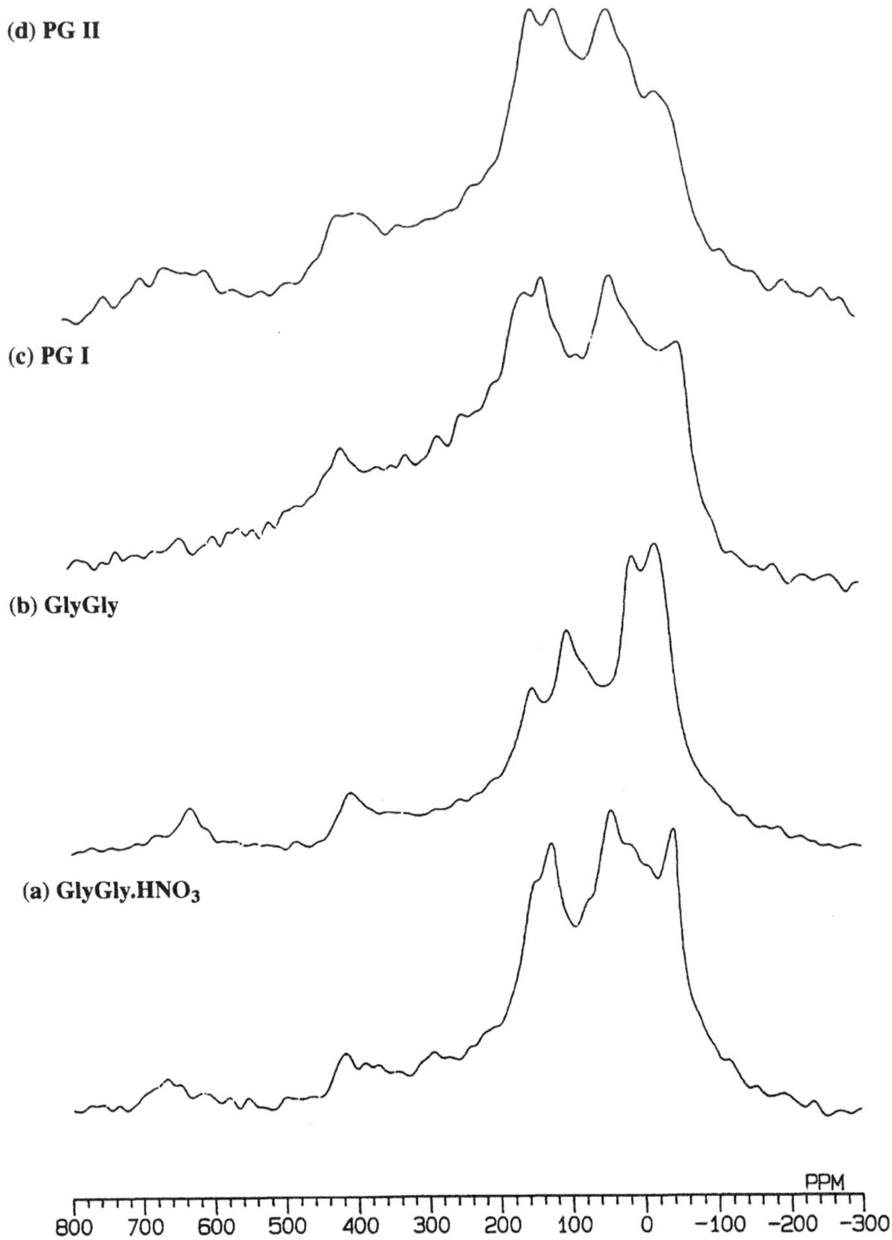

Figure 4.84 The 67.8-MHz ^{17}O CP static NMR spectra of (a) GlyGly·HNO_3, (b) GlyGly, (c) PG I, and (d) PG II.

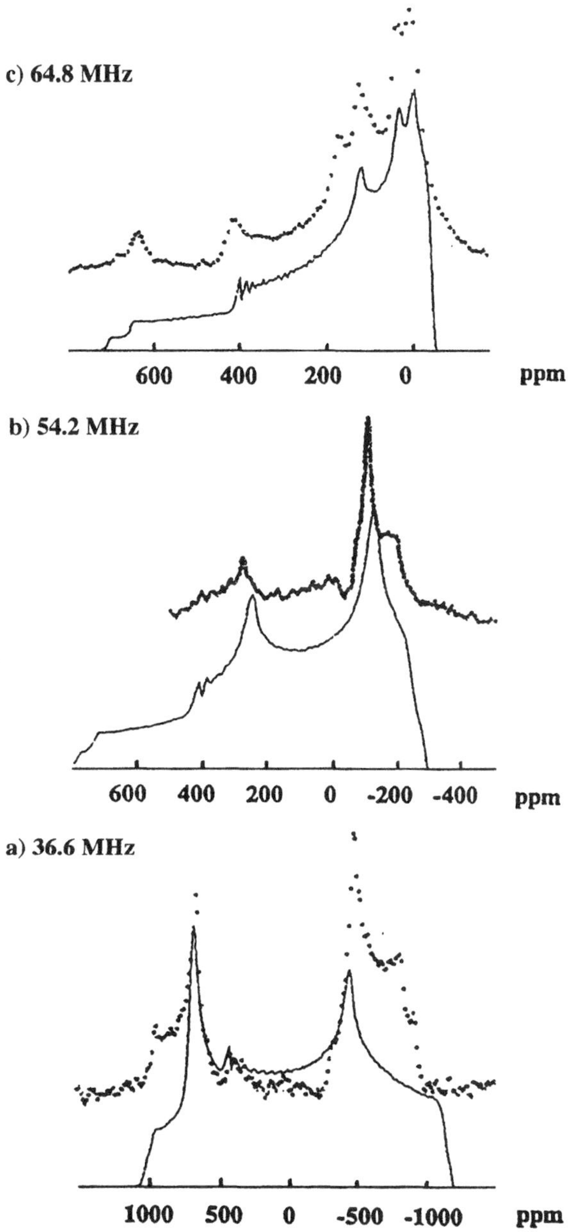

Figure 4.85 (a) 36.6-MHz, (b) 54.2-MHz, and (c) 67.8-MHz ^{17}O CP static spectra of GlyGly together with theoretically simulated spectra.

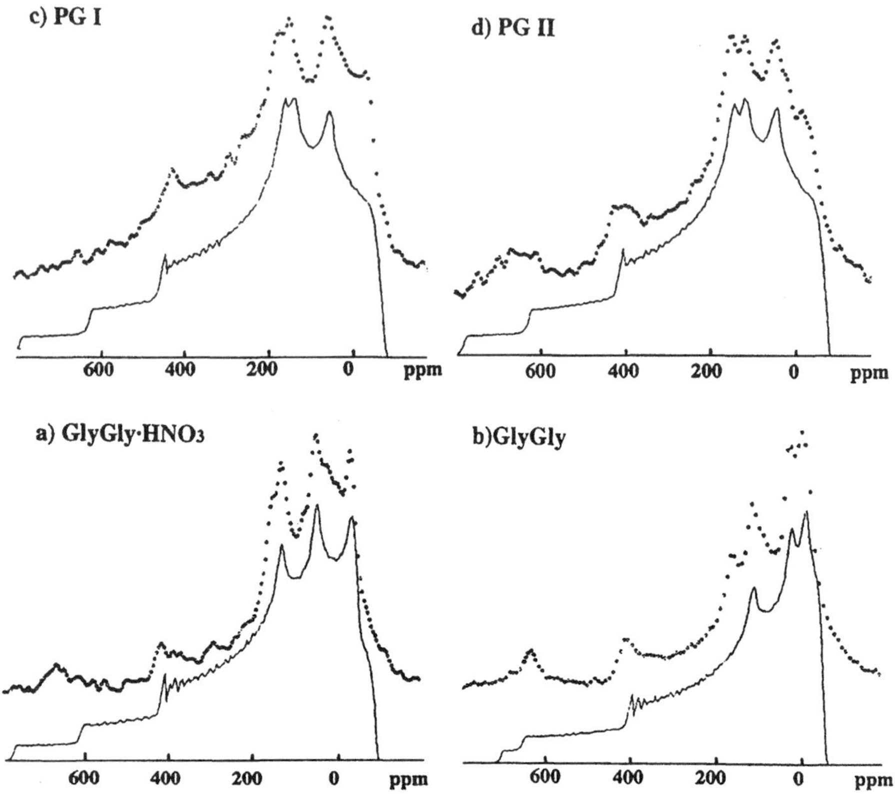

Figure 4.86 67.8-MHz ^{17}O CP static NMR spectra of (a) GlyGly·HNO_3, (b) GlyGly, (c) PG I, and (d) PG II together with theoretically simulated spectra.

the case of the carbonyl oxygen of L-alanine, as reported by Goc *et al.* [228, 229], the chemical shift tensor of carbonyl oxygen in the peptides considered here is not axially symmetric. The e^2qQ/h values change from 8.30 to 8.75 MHz. These values are larger than the e^2qQ/h value of L-leucine, which is 8.0 MHz at 190 K [229]. The η_Q values of polypeptides such as PG I and II are 0.26 and 0.29, and the those of oligopeptides such as GlyGly and GlyGly·HNO_3 are 0.45 and 0.47, respectively. The difference may come from a difference in molecular packing between them. Furthermore, the principal axis of the quadrupolar tensor and the principal axis of the ^{17}O chemical shielding tensor for the carbonyl oxygen of the peptides and polypeptides are not coincident with each other. The relationship between these two principal axes is shown in Figure 4.87 for PG II. From these experimental results it can be said that each of the preceding NMR parameters is strongly influenced by the electronic structure of the molecules.

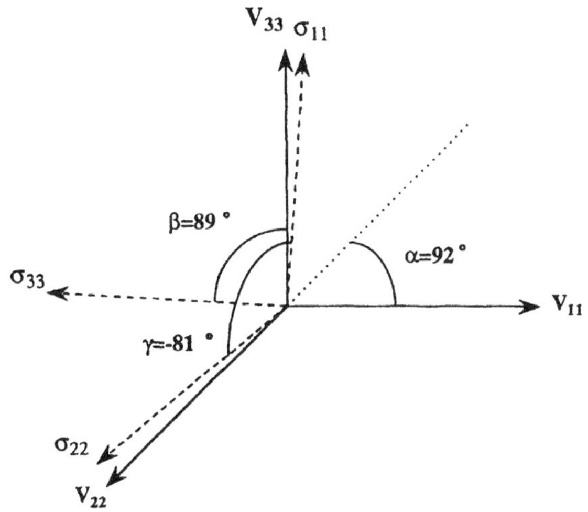

Figure 4.87 Relationship between the principal axes of the quadrupolar tensor and the chemical shift tensor on the carbonyl oxygen for PG II.

4.10.3.3.2 ^{17}O NMR Chemical Shifts of the Carbonyl Oxygen of the Peptides and Polypeptides

From Table 4.11, it can be seen that there is a large difference in the chemical shift value between thepeptides and polypeptides. Figure 4.88 shows the plot of the observed isotropic ^{17}O chemical shift values (σ_{iso}) against the hydrogen bond

Table 4.11 Experimentally determined ^{17}NMR parameters of solid peptides and polypeptides.

			Chemical Shift Tensor (ppm)						Angle Δ^a (deg)		
Sample	e^2qQ/h	η_Q	δ_{11}	δ_{22}	δ_{33}	δ_{iso}	$\Delta\delta^b$	η^c	α	β	γ
Polypeptides											
PG II	8.30	0.29	562	410	−108	288	396	0.38	92	89	−81
PG I	8.55	0.2	574	425	−101	299	400	0.37	100	91	−79
Oligopeptides											
GlyGly	8.55	0.45	546	382	−132	265	397	0.41	94	90	−87
GlyGly·HNO_3	8.75	0.47	559	408	−127	280	407	0.37	94	89	−81

^{a}Angle D indicated the Eular angels between the principal axes of the quadrapolar tensor and chemical shift tensor.

$^b\Delta\delta = \delta_{iso} - \delta_{11}$.

$^c\eta = (\delta_{11} - \delta_{22})/\Delta\delta$.

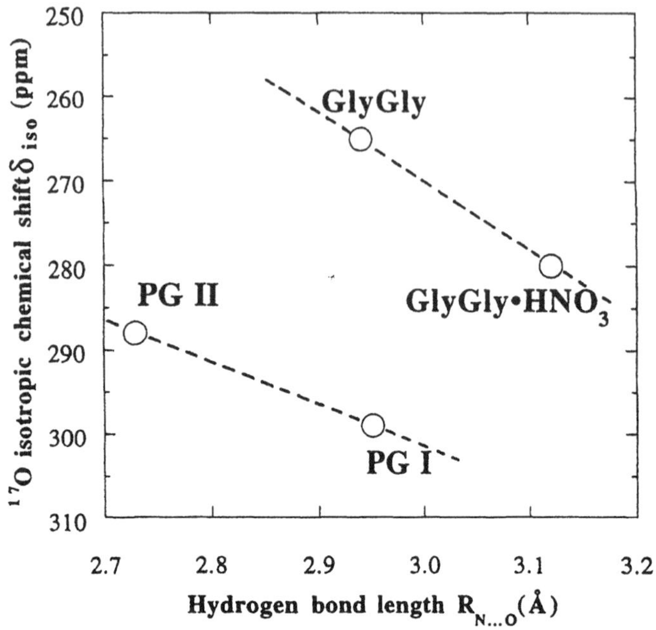

Figure 4.88 Plots of the ^{17}O isotropic chemical shift versus hydrogen bond length ($R_{N\cdots O}$).

length ($R_{N\cdots O}$(Å)). The σ_{iso} values in the peptides and polypeptides move upfield with a decrease in the hydrogen bond length ($R_{N\cdots O}$(Å)). Figures 4.89(a)–(c) show the plot of the observed principal values of the ^{17}O chemical shifts against the hydrogen bond length (($R_{N\cdots O}$(Å)). Every principal value in the peptides and polypeptides moves upfield with a decrease in the hydrogen bond length ($R_{N\cdots O}$(Å)).

What is the origin of the difference in chemical shift between the peptides and polypeptides? Two explanations are considered. The first is that the difference between the peptides and polypeptides comes from the hydrogen bond angle and hydrogen bond dihedral angle, which are closely related to the ^{17}O chemical shift. The second theory is the difference in molecular packing between them.

The dihedral angle N–C=O$\cdots$H (defined by ξ here) shows a measure of the location of the amide group on the peptide plane. In the case of GlyGly ($\xi = -160°$) and GlyGly·HNO_3 ($\xi = 3°$), the amide-group-related hydrogen bonding is nearly in the peptide plane and carbon, oxygen, hydrogen, and nitrogen atoms related to hydrogen bonding are located on the peptide plane. However, in the case of PGI ($\xi = 68°$) and II ($\xi = -47°$), the amide-group-related hydrogen bonding is away

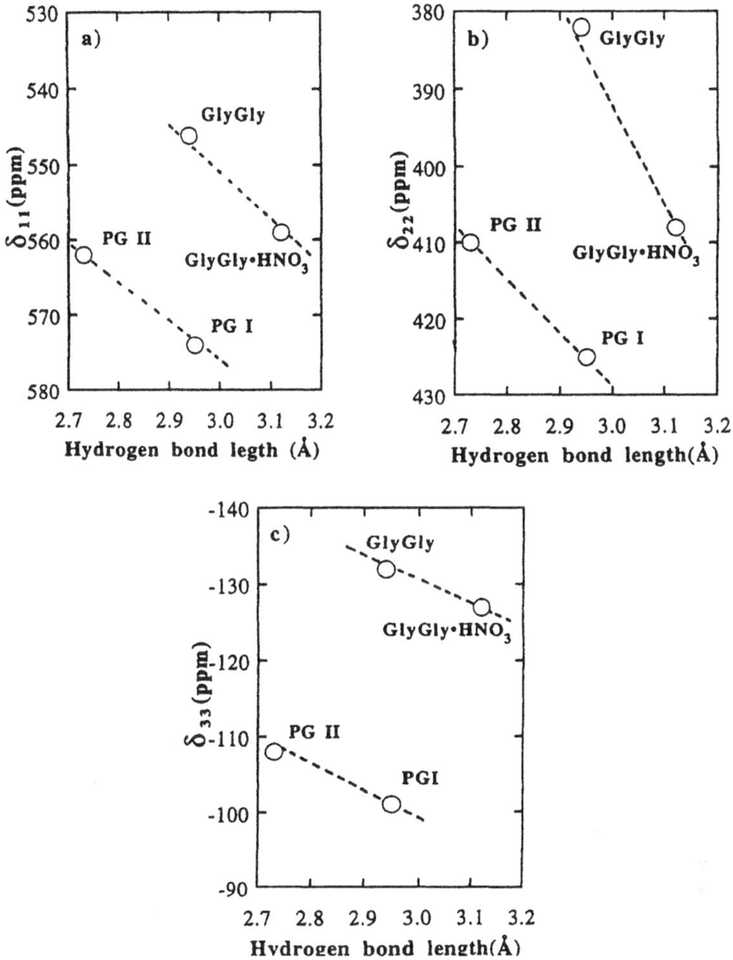

Figure 4.89 Plots of the principal ^{17}O chemical shift values, (a) δ_{11}, (b) δ_{22}, and (c) δ_{33}, versus hydrogen bond length ($R_{N\cdots O}$).

from the peptides plane. The location of the amide group seems to affect the δ value.

The following conclusions can be stated: From the observed carbonyl oxygen ^{17}O NMR spectra of PG I, PG II, GlyGly, and GlyGly·HNO_3 in the solid state, it was found that the e^2qQ/h values decrease linearly with a decrease of the hydrogen bond length. This shows that the hydrogen bond length can be determined through the observation of e^2qQ/h values. Furthermore, it was found that there is

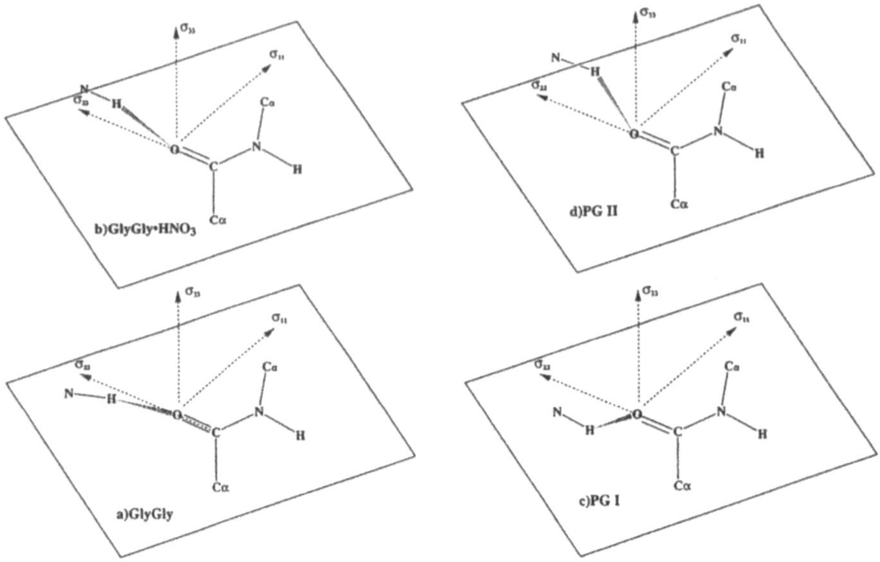

Figure 4.90 Hydrogen bonding spatial location of peptides: (a) GlyGly·HNO_3, (b) GlyGly, (c) PG I, and (d) PG II.

a difference in ηQ between the polypeptides and peptides, and this may stem from the large difference in molecular packing between them. The chemical shift values in the polypeptides and peptides move upfield with a decrease in the hydrogen bond length. However, there is a difference in the chemical shift value between the peptides and polypeptides. These experimental findings demonstrate that ^{17}O NMR spectroscopy is a useful means of elucidating the hydrogen bonding structure in solid peptides and polypeptides.

4.10.4 POLYMER GELS

4.10.4.1 Approaches by ^1H Pulse NMR and Pulsed-Field-Gradient Spin-Echo Techniques

The chemical shift value δ, spin–lattice relaxation time T_1, spin–spin relaxation time T_2, and self-diffusion constant D give information on microscopic structure and dynamics of the components in a gel or a polymer solution such as the network polymer, solvent, ions, and molecules. From δ, information on the chemical structure; the distance between nuclei; the formation of a hydrogen bond; interaction, association, and dissociation of a functional group, and so on, can be obtained.

From T_1 and T_2, the correlation time (υ_c) of the molecular motion can be estimated by using the BPP theory [11] when relaxation mechanisms are predominantly affected by dipole–dipole interaction. The T_1 and T_2 for components in a gel and a polymer solution are closely associated with their microscopic molecular motion and structure. The pulsed-field-gradient spin-echo (PGSE) ^{1}H–NMR method is a powerful technique for determining the self-diffusion coefficients in polymer systems, which give information on the translational mobility of components contained in a gel and a polymer solution.

4.10.4.1.1 Poly(*N*-isopropylacrylamide) Gel

Poly(*N*-isopropylacrylamide) (PNIPAM) in water undergoes a phase transition from solution to gel at around 31 °C [235]. ^{1}H pulse NMR experiments provide useful information about molecular motion of the polymer and solvent in the solution and gel state. The abrupt change in molecular motion for solvent in the gel is caused by the formation of gel structure at the phase transition temperature and, therefore, the change of relaxation times sensitively reflects the change of states.

Figure 4.91 shows the plots of the ^{1}H T_2 values for water in 1, 2.5, 5, and 10 wt/v % PNIPAM/water solutions and for neat water against temperature under atmospheric pressure [236]. The ^{1}H T_2 value increases as the temperature is increased up to 30 °C. The ^{1}H T_2 value of neat water increases linearly with temperature up to 50 °C. At the gelation temperature (about 31 °C), the ^{1}H T_2 value in all of the polymer solutions transitionally decreases. As the temperature is further increased, the T_2 value increases again. The mobility of water increases with increasing temperature up to 30 °C, decreases transitionally at about 31 °C owing to the gelation of the system, and increases again with a further increase of temperature. This trend is very different from the case of neat water. Furthermore, Figure 4.91 shows that the increase of the polymer concentration leads to a decrease in ^{1}H T_2 for water. The motion of water is restricted by an increase in the intermolecular interactions between the polymer chains and water. Because the polymer concentration increases up to 5% at the gelation temperature, the magnitude of the decrease for the ^{1}H T_2 value is larger. As the concentration increases further up to 10% wt/v, the magnitude of the decrease for the ^{1}H T_2 decreases.

Figure 4.92 shows the plot of the ^{1}H T_1 value of water molecules in 5 wt/v % PNIPAM/water solution at atmospheric pressure as a function of temperature. The ^{1}H T_1 value slowly increases as the temperature is increased. At the gelation

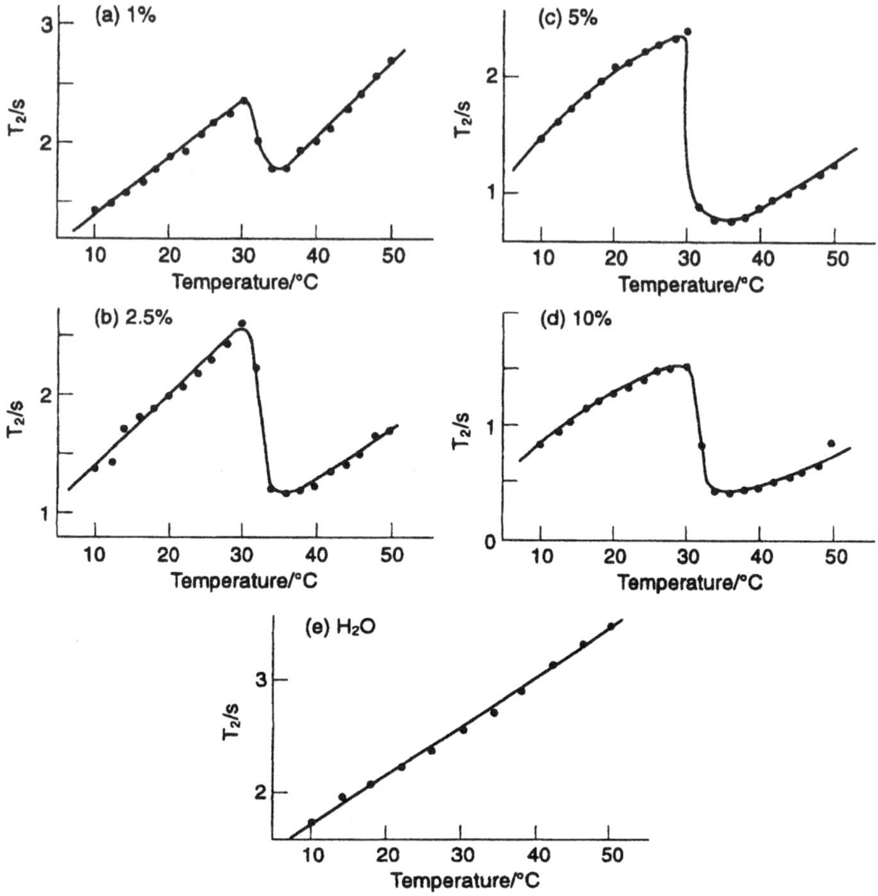

Figure 4.91 Temperature dependence of ^1H T_2 for water in PNIPAM/water solution with the concentrations of (a) 1, (b) 2.5, (c) 5, and (d) 10 wt/v % and for (e) neat water at atmospheric pressure.

temperature, the T_1 value transitionally decreases. As the temperature is further increased, the ^1H T_1 value slowly increases. This behavior is very similar to that for T_2. As seen from the observed ^1H T_1 and T_2 behaviors, the water molecules are in the fast motion region ($\omega\tau_c \ll 1$, where ω is the resonance frequency). At 1, 2.5, and 10 wt/v % solutions, similar results were obtained. Therefore, the ^1H T_1 and T_2 values for water in the PNIPAM/water system are very sensitive to the phase transition and provide dynamic information.

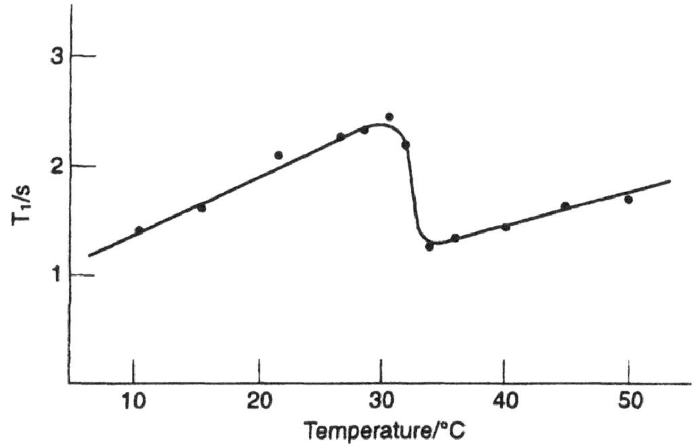

Figure 4.92 Temperature dependence of ^{1}H T_1 for water in 5 wt/v % PNIPAM/water solution at atmospheric pressure.

The application of pressure on the PNIPAM/water system changes the situation around network polymer and solvent. The intermolecular interactions between water and PNIPAM change, and thus the dynamics of phase transition for the system are changed at high pressure. Figure 4.93 shows the temperature dependence of ^{1}H T_2 for water in a 10 wt/v % PNIPAM/water solution at pressures of 100, 200, 300, and 400 kg/cm^2. The temperature for the phase transition is almost independent of pressure. At pressures of 100 and 200 kg/cm^2, the T_2 value decreases transitionally and then sharply at the gelation temperature. At pressures of 300 and 400 kg/cm^2, the T_2 value decreases transitionally but slowly at the gelation temperature. Therefore, the sharpness of the phase transition decreases as the pressure is increased. This means that the gelation of the system occurs slowly under high pressure and that the application of high pressures prevents gelation.

The pressure dependence of the T_2 for water in 5% wt/v PNIPAM/water solution at temperatures of 28, 31, 33, and 35 °C is shown in Figure 4.94. The dependencies change as a function of temperature. At 28 °C the T_2 value is almost independent of pressure. However, at 31 °C the T_2 value slightly increases as the pressure is increased. At 33 °C the T_2 value increases significantly as the pressure is increased, and at 35 °C the pressure dependence of the T_2 value returns to the smaller scale. Therefore, the relatively large T_2 change occurs at the gelation temperature. However, at temperatures higher or lower than the gelation temperature, the T_2 change is very small. Because, at the gelation temperature, the large increase of the T_2 corresponds to a large increase of molecular motion means that the application of

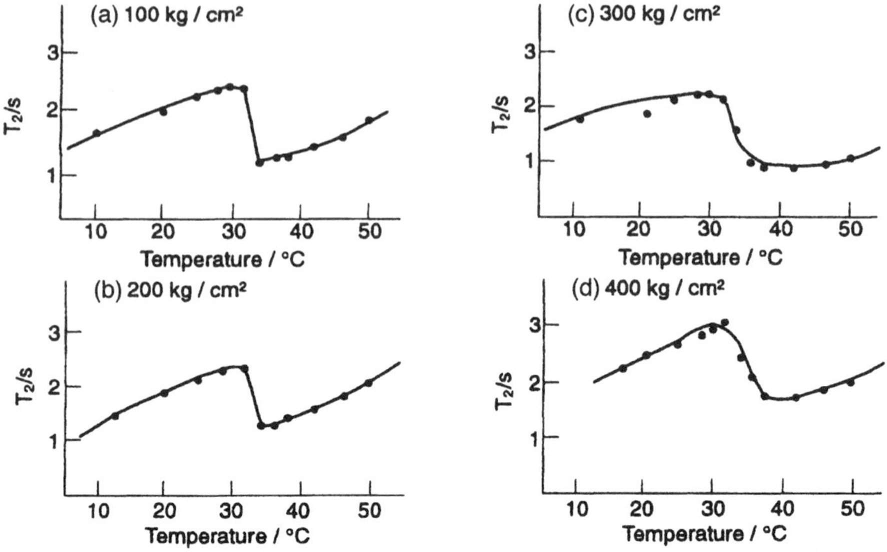

Figure 4.93 Temperature dependence of ^{1}H T_2 for water in 5 wt/v % PNIPAM/water solution under pressures of (a) 100, (b) 200, (c) 300, and (d) 400 kg cm^{-2}.

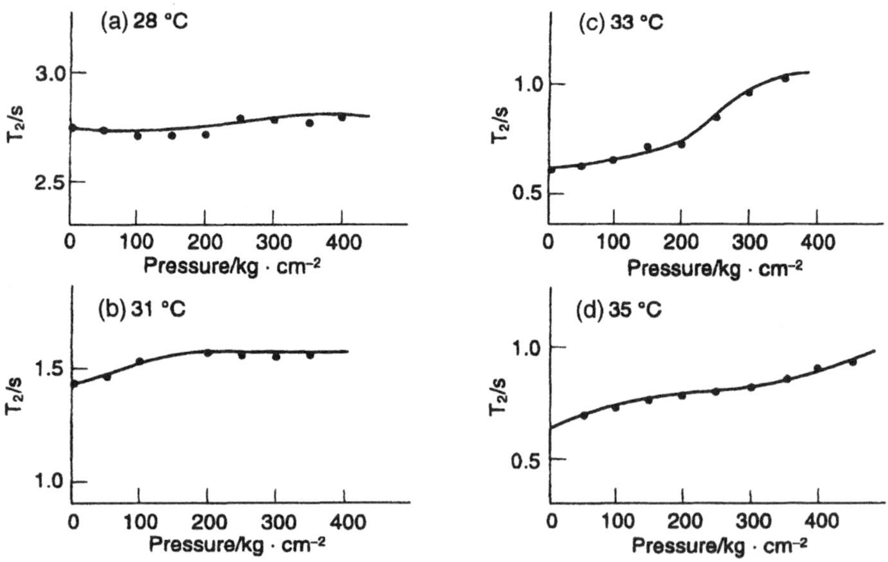

Figure 4.94 Pressure dependence of ^{1}H T_2 for water in 5 wt/v % PNIPAM/water solution at temperatures of (a) 28, (b) 31, (c) 33, and (d) 35 °C.

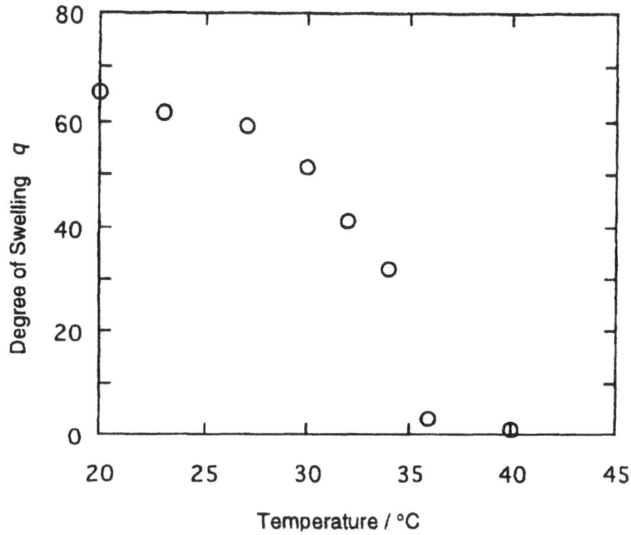

Figure 4.95 Temperature dependence for the volume of PNIPAM gel at the state of equilibrium swelling.

pressure inhibits gelation. At 35 °C, after a stable gel is formed, pressure does not greatly influence gel structure or the molecular motion of water in the gel network.

The degree of swelling q of a PNIPAM gel in water as a function of temperature is shown in Figure 4.95. The q value and volume changes transitionally around 33–34 °C, as has been described already [239, 259]. Such transitions are viewed from dynamics through the observation of ^{1}H T_2. The ^{1}H T_2 values as a function of temperature for a PNIPAM gel ($q = 30$ at 20 °C) are shown in Figure 4.96. The ^{1}H T_2 value increased as the temperature was increased up to 33 °C. At the volume-phase transition temperature (33 °C), the ^{1}H T_2 value transitionally decreased. As the temperature was further increased, the ^{1}H T_2 value increased again. This shows that the mobility of water increases with increasing temperature up to 33 °C like neat water and decreases transitionally at 33 °C due to the abrupt decrease in q.

In the shrinkage process of gels induced by elevating temperature over the transition temperature, a PNIPAM gel often forms the skin structure in which the unshrunk inner layer is surrounded by the shrunk outer layer, and thus it takes a long time to complete the shrinkage process. Figure 4.97 shows the time dependencies of the ^{1}H T_2 values for sample Gel-A with $q = 13$ by rapid change of temperature from 20 to 40 °C and for sample Gel-B with $q = 30$ at 20 °C. The shape of its gel changes as diagrammatically illustrated in Figure 4.98. As the temperature was

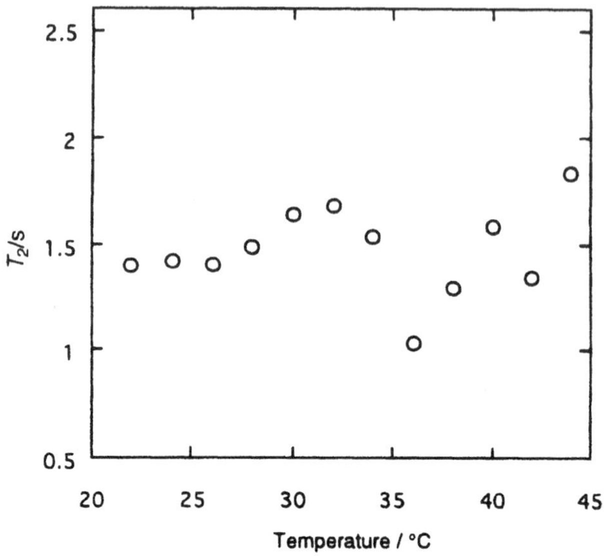

Figure 4.96 Temperature dependence of ^{1}H T_2 for PNIPAM gel.

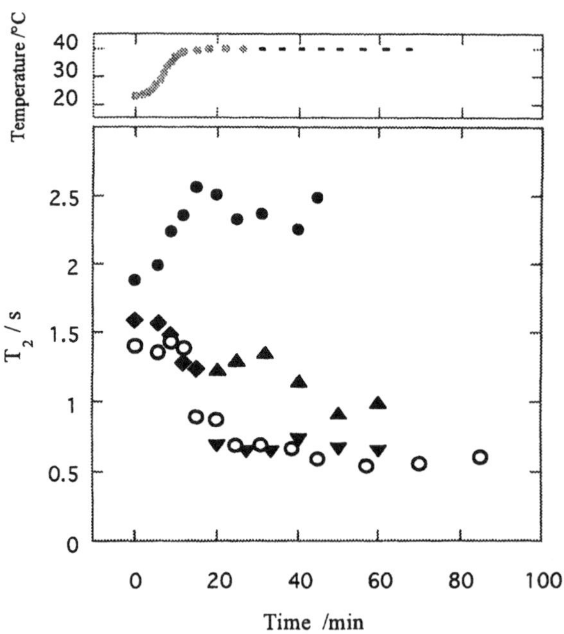

Figure 4.97 Time courses of ^{1}H T_2 for water in neat water (○), in sample Gel-A (●), and in sample Gel-B (□). After elevation of temperature, Gel-A had two components of T2: (▲) and (▼).

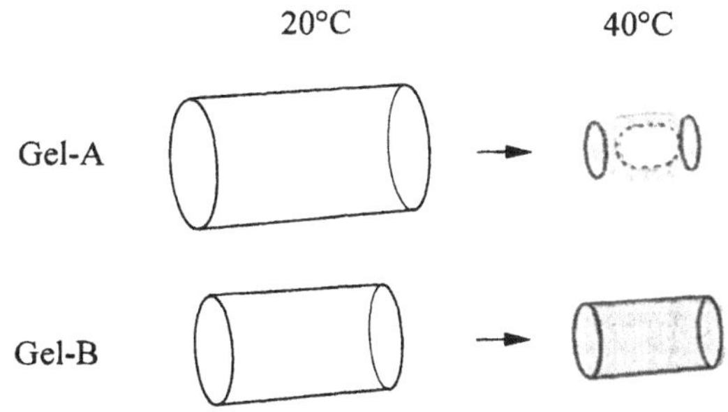

Figure 4.98 Illustration of shape change for samples Gel-A and Gel-B with varying temperature.

increased, the ^{1}H T_2 value for the gel decreased with a decrease in q, whereas the ^{1}H T_2 value of neat water increased linearly. Furthermore, it was found from the ^{1}H T_2 curve that there were two components with different mobility in sample Gel-B after the temperature was raised. A short T_2 component corresponds to the shrunk outer layer, and a long T_2 component to the unshrunk inner layer. This result is supported by the ^{1}H–NMR imaging experiment for a PNIPAM.

The volume phase transition of PNIPAM gel is understood from a dynamical aspect [237]. The D values of HDO (D_{HDO}) in a PNIPAM gel swollen with deuterated water (D_2O), in which a small amount of HDO is contained (<1%), have been determined under the state of equilibrium swelling at 23 °C as a function of q (Figure 4.99). The D_{HDO} value for a PNIPAM gel at high q value (>40) is very close to the D_{HDO} value in neat D_2O. For the gel with q values smaller than 40, the D_{HDO} value of HDO in the gel is decreased as the q value decreases.

The molecular motion of solvent during the volume-phase transition has been clarified through the observation of D_{HDO}. The D_{HDO} values for water in a PNIPAM gel ($q = 66$ at 20 °C) and for neat water are plotted against temperature in Figure 4.100. During the NMR experiments, water is exuded and wiped away. The D_{HDO} value increases as the temperature is increased up to 34 °C. At the volume-phase transition temperature (about 34 °C), the D_{HDO} value transitionally decreases.

The mobility of water in the gel increases with an increase in temperature up to 34 °C in spite of the decrease in q from 66 to about 30 (as shown in Figure 4.100) and decreases transitionally at 34 °C due to the steep decrease of the q value. After the transition, the high-resolution PGSE ^{1}H–NMR spectrum cannot be obtained owing to a large decrease in molecular motion as a result of a large shrinkage of the gel.

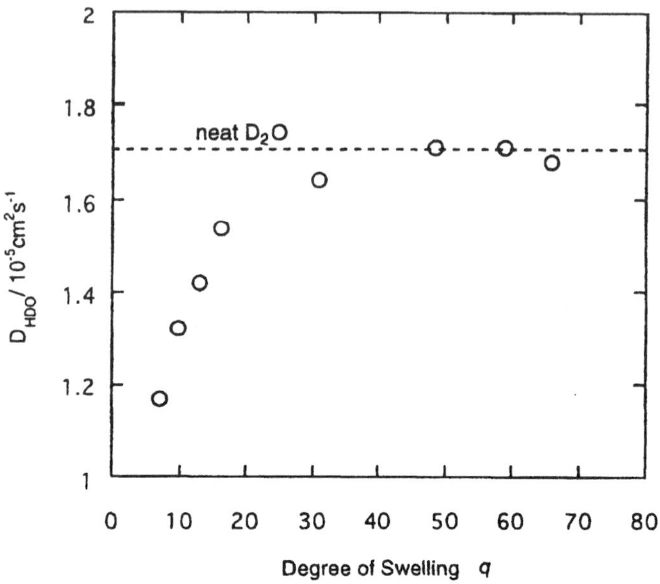

Figure 4.99 Dependence of the diffusion coefficient of water molecule (D_{HDO}) on the degree of swelling (q) in a PNIPAM gel. The dashed line indicates the diffusion coefficient of a small amount of HDO contained in neat D_2O.

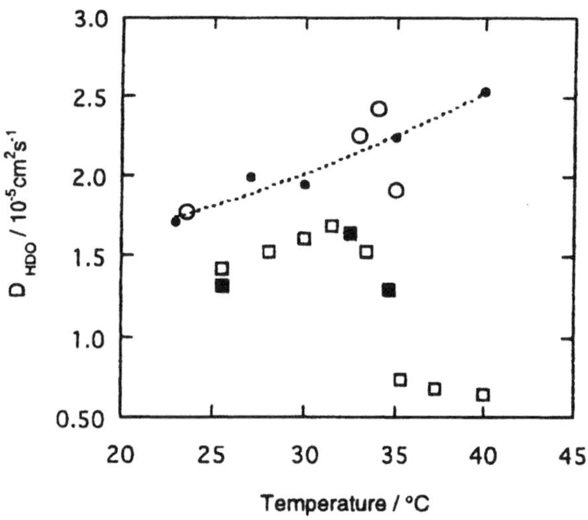

Figure 4.100 Temperature dependence of the diffusion coefficients of water molecule (D_{HDO}) in neat D_2O (●) in a PNIPAM gel at the state of equilibrium swelling ($q = 66$ at 20 °C) (○) and in a PNIPAM gel obtained after heating at 40 °C ($q = 4$) (□) in going from 20 to 40 °C. In the latter sample, the NMR measurements were carried out cooling process from 40 to 20 °C (■).

The change in molecular motion of solvent in the gel over the phase transition temperature without a change of q has been studied. A PNIPAM gel in the shrunk part is put in an NMR tube to determine D_{HDO} at various temperatures by heating or cooling, as shown in Figure 4.100. During the NMR experiment, the volume of the gel is kept constant. As shown in Figure 4.100, D_{HDO} is largely increased at the transition temperature. Such a decrease of D_{HDO} is caused by the large decrease of the mobility of polymer networks that are formed by physical cross-linking at the transition temperature. This process is reversible for heating and cooling, as seen from Figure 4.100.

4.10.4.1.2 Poly(methacrylic acid) Gel

Poly(methacrylic acid) (PMAA) gel is a typical hydropolyelectrolyte gel. In general, hydropolyelectrolyte gels have unique physical and chemical properties. For example, they show a drastic volume change in response to change in solvent composition [238], pH [239], ionic strength [240], and temperature [241], and to the application of an electric field [243, 244]. In these reaction processes, the hydropolyelectrolyte gels change their volume by releasing or absorbing water. In such process, water molecules play a significant role. To reveal the role of water in the gels, it is important to understand the dynamic behavior of water contained in hydropolyelectrolyte gels associated with polymer network structure.

To understand the translational motion of water contained in a PMAA gel by a change in the degree of cross-linking, the diffusion coefficient of H_2O molecules (D_{H_2O}), has been determined by the PGSE ^{1}H–NMR method at 27 °C, as listed in Table 4.12 [245]. As seen from Table 4.12, D_{H_2O} increases as q is increased. Therefore, an increase in q leads to an increase in the mobility of water molecules.

Table 4.12 Determined ^{1}H T_1, ^{1}H T_2, and D_{H_2O} values of water in PMAA gel as a function of the degree of swelling (q).

q	$^1H\ T_1$ (s)	$^1H\ T_2$ (s)	D_{H_2O} (10^{-5} cm^2 s^{-1})
3.4	0.09	0.006	1.37
3.6	0.93	0.010	1.37
4.7	1.03	0.020	1.53
21.7	2.06	0.191	1.90
59.4	3.29	0.577	2.16
105.8	3.78	1.095	2.19

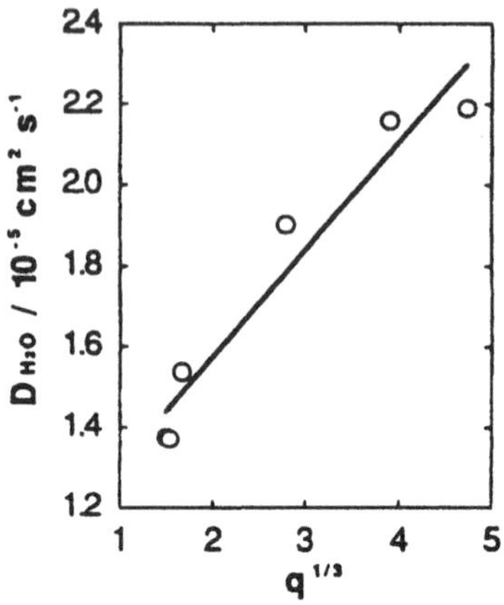

Figure 4.101 Plots of the diffusion coefficients of water (D_{H_2O}) in PMAA gel against $q^{1/3}$ at 27 °C.

As seen from the plot of D_{H_2O} against $q^{1/3}$ in Figure 4.101, D_{H_2O} increases linearly from 1.37×10^{-5} to 2.19×10^{-5} cm^2 s^{-1} with increasing $q^{1/3}$ [244]. This shows that the translational motion of water molecules in the PMAA gel increases with an increase in the degree of swelling. The determined D_{H_2O} value of neat water at 27 °C is 2.5×10^{-5} cm^2 s^{-1}. The translational motion of water molecules in the gel is more strained than that of neat water.

The ^{1}H T_1 and T_2 of water in the PMAA gel may be sensitive to a change in the degree of cross-linking in the gel. The ^{1}H T_1 and T_2 of water in the PMAA gel have been determined at 27 °C as a function of q, as shown in Table 4.12. From this table, it is seen that the ^{1}H T_1 and T_2 increase as q is increased. This means that the molecular motion of water molecules increases with an increase in q. If the ^{1}H T_1 and T_2 values for water in the PMAA gel are plotted against $q^{1/3}$, the linear relationship is obtained (Figure 4.102). The increase in $q^{1/3}$ leads to a linear increase in molecular motion of water molecules in the gel, and the molecular motion is in the fast motion region ($\omega\tau_c \ll 1$) in the BPP theory [11].

To understand the hydrodynamic behavior of water in a PMAA gel, the relationship between q and the radius of a water cluster has been studied [244]. In the case of water molecules, the ^{1}H T_1 is predominantly governed by dipole–dipole interaction [11, 245]. The equation for ^{1}H T_1 is expressed as [246]

$$1/T_1 = (a/b)^2(\gamma^4 h^2/12\pi 2b^4 D_{H_2O})(1 + 3\pi N b^6/5a^3) \qquad (128)$$

where b is the distance between protons of a water molecule ($b = 1.58$), γ is the gyromagnetic ratio of water, h is the Planck constant, N is the number of

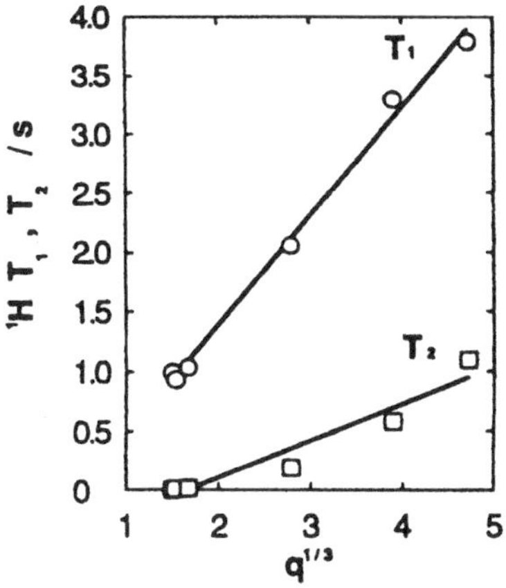

Figure 4.102 Plots of ^{1}H T_1 and T_2 of water in PMAA gel against $q^{1/3}$ at 27 °C.

proton spins of water molecules per 1 cm^3, and a is the radius of a water cluster corresponding to a hard sphere in the Stokes–Einstein model. Figure 4.103 shows the plot of the a values calculated from Eq. (128) against $q^{1/3}$. It is found that a decreases linearly from 3.4×10^{-8} to 1.9×10^{-8} cm with increasing $q^{1/3}$. This shows that the radius of a water cluster in the PMAA gel decreases with an increase of the degree of swelling; i.e. the translational motion of water molecules in the gel is decreased by an increase of the density of the network polymer.

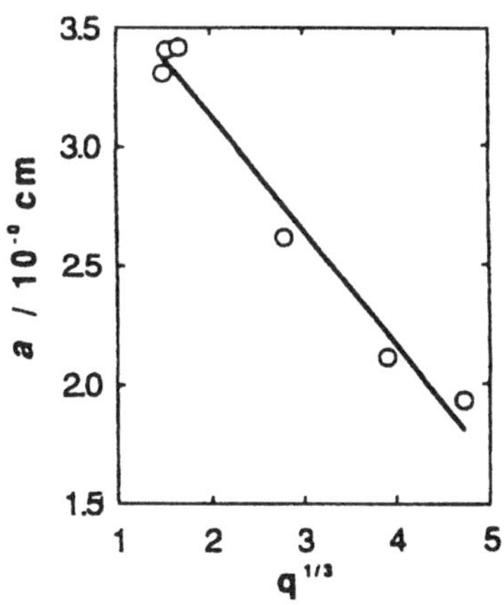

Figure 4.103 Plot of the radius a of a water cluster corresponding to a hard sphere in the Stokes–Einstein model against $q^{1/3}$.

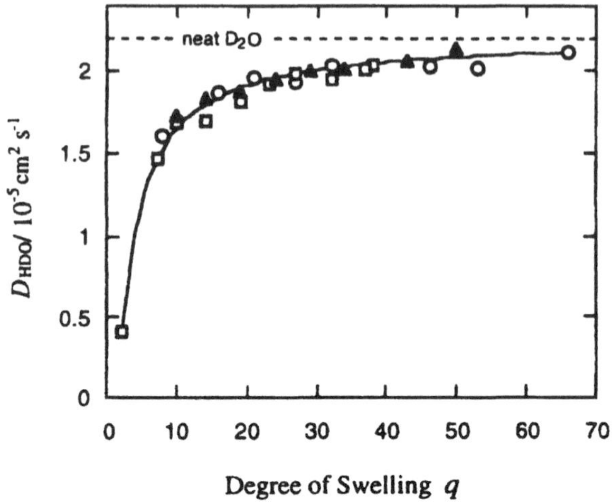

Figure 4.104 Dependence of the diffusion coefficient of the water molecule (D_{HDO}) on the degree of swelling q in a PDMAA gel containing PEG with molecular weights of 4250 (□), 10890 (▲), and 20000 (○) at 30 °C.

4.10.4.1.3 Poly(*N,N*-dimethylacrylamide)

The determined D_{HDO} values in poly(*N,N*-dimethylacrylamide) (PDMAA) gels with various q values were plotted against q, as shown in Figure 4.104 [247]. The D_{HDO} increases with q and is almost independent of the molecular weight of PEG contained in the PDMAA gel. The change in D_{HDO} in the small q region was much larger than that in the large q region. The D_{HDO} value in the large q region asymptotically approached that for HDO contained in neat D_2O (2.2×10^{-5} cm^2 s^{-1}). The intermolecular interactions between water and the polymer network, of which the strength depends on the size of the network, restrain translational motion of water molecules, and there are almost no significant interactions between probe polymer and solvent. This leads to the aforementioned experimental finding that the molecular motion of HDO is independent of the molecular weight of PEG contained in the gel. We can understand more clearly the diffusional behavior of water in the gel system using the following equation obtained from the modified free-volume theory proposed [248]:

$$D_{HDO} = H_{HDO}^{neat} \exp\left[1/\left(f_{solv} - q(f_{solv}^2/\beta)\right)\right] \tag{129}$$

where D_{HDO}^{neat} is the diffusion coefficient for HDO in neat liquid D_2O, f_{solv} is the fractional free volume for neat liquid D_2O, and β is the characteristic constant associated with the volume fraction of solvent. The solid curve obtained from

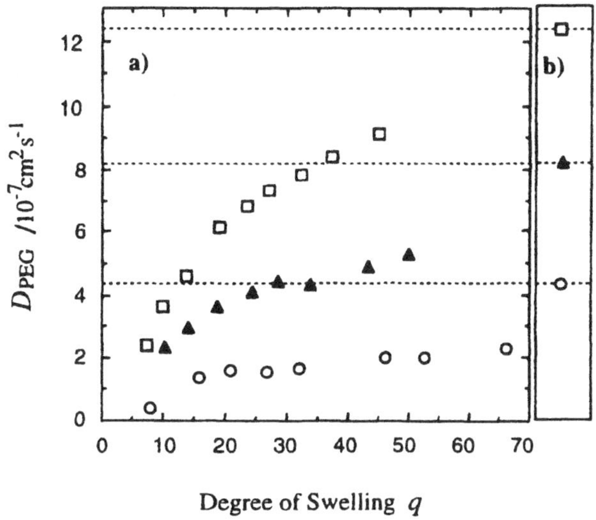

Figure 4.105 (a) Dependence of the diffusion coefficient of PEG in PDMAA gel on the degree of swelling q at 30 °C. (b) The diffusion coefficients of PEG in 1 wt % aqueous solution at 30 °C. The molecular weights of PEGs are 4250 (□), 10890 (▲) and 20000 (○).

a least-squares fitting to the experimental data using Eq. (129) with $D_{\mathrm{HDO}}^{\mathrm{neat}} = 2.2 \times 10^{-5}\ \mathrm{cm^2\ s^{-1}}$ is shown in Figure 4.118. The experimental data lie on the theoretical curve. From this result, the diffusion coefficient for an HDO molecule in bulk PDMAA ($q \rightarrow$ O) is determined to be $3.5 \times 10^{-8}\ \mathrm{cm^2\ s^{-1}}$. The rapid decrease in D_{HDO} in the small q region can be explained by the decrease in the free volume.

To investigate the translational motion of PEG as a probe polymer contained in a PDMAA gel, the diffusion coefficient of PEG (D_{PEG}) was determined at 30 °C by varying q and the molecular weight of PEG contained in the gel. The D_{PEG} values obtained were plotted against q, as shown in Figure 4.105(a) [247]. It was found that D_{PEG} increased as q was increased and depends on the molecular weight of PEG. The D_{PEG} value in gels is smaller than that at 1 wt % solution ($D_{\mathrm{PEG}}^{\mathrm{soln}}$) at 30 °C, as shown in Figure 4.105(b). Therefore, the translational motion of PEG in the gel is more restrained by large intermolecular interactions with the polymer network than that of PEG in solution, and the PEG with higher molecular weight is more strongly restrained.

The relationship between the degree of restraint and the ratio of the screening length of the polymer network to the size of the probe polymer has been studied [249, 250]. If the interchain hydrodynamic interaction in the gel is thought to be more important than the topological constraint, D_{PEG} is expressed by

$$D_{PEG}/D^0_{PEG} = \exp(-\kappa R) \tag{130}$$

where D^0_{PEG} is the diffusion coefficient of an isolated PEG, κ^{-1} is the dynamical screening length of polymer chains, and R is the hydrodynamic radius of the probe polymer [242, 251]. When the D^0_{PEG} value is determined from the diffusion coefficient of PEG in aqueous dilute solution (D^{soln}_{PEG}), the change for local friction of polymer must be taken into account [252]. Then, D^0_{PEG} is given by

$$D^0_{PEG} = D^{soln}_{PEG}/\left(D^{neat}_{HDO}/D_{HDO}\right) \tag{131}$$

The dynamical screening length κ^{-1} is related to the concentration of network polymer c and is expressed by

$$\kappa^{-1} = c^u = q^{-u} \tag{132}$$

where u is expected to be a constant in the range 0.5–1.0 and depends largely on the polymer species [253]. From substitution of Eqs. (131) and (132) into Eq. (130), the following equation is obtained:

$$\ln\{-\ln[(D_{PEG}D^{neat}_{HDO})/(D^{soln}_{PEG}D_{HDO})]\} = u \ln q + \ln R \tag{133}$$

The experimental data shown in Figure 4.105 are plotted on the ln–ln plots of the $-\ln[(D_{PEG}\ D^{neat}_{HDO})/(D^{solv}_{PEG}D_{HDO})]$ against q, as shown in Figure 4.106. The three straight lines for PEG with different weight-average molecular weights (Mw) have

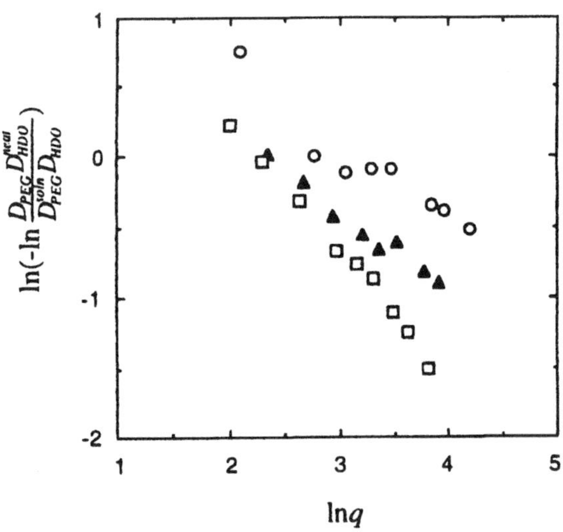

Figure 4.106 Plots of $\ln\{-\ln[(D_{PEG}D^{neat}_{HDO})/(D^{soln}_{PEG}D_{HDO})]\}$ against $\ln q$ at 30 °C. The molecular weights of PEGs are 4250 (□), 10890 (▲), and 20000 (○).

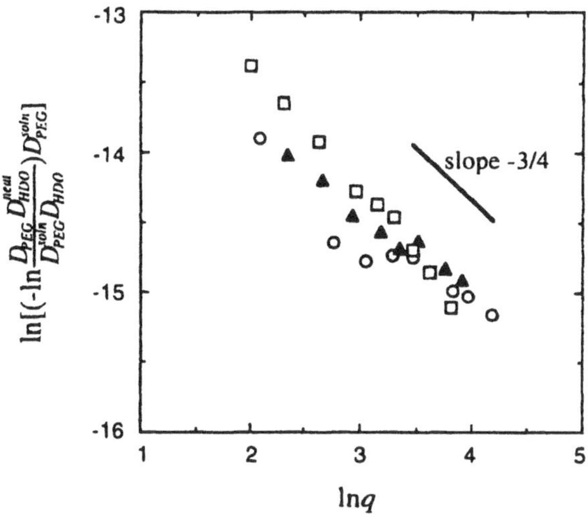

Figure 4.107 Plots of $\ln[\{-\ln[(D_{PEG} D_{HDO}^{neat})/(D_{PEG}^{soln} D_{HDO})]\} D_{PEG}^{soln}]$ against $\ln q$ at 30 °C. The molecular weights of PEGs are 4250 (□), 10890 (▲), and 20000 (○).

almost the same slope and different intercept. Because R in solution is assumed to be the same as that in gel, we have

$$R = kT/6\pi\eta_s D_{PEG}^{soln} \tag{134}$$

where η_s is the viscosity of solvent in solution, k is the Boltzmann constant, and T is the absolute temperature. The substitution of Eq. (134) into Eq. (133) gives

$$\ln\left[\left\{-\ln\left[\left(D_{PEG} D_{HDO}^{neat}\right)/\left(D_{PEG}^{soln} D_{HDO}\right)\right]\right\}\right] D_{PEG}^{soln} = u \ln q + \ln(kT/6\pi\eta_s) \tag{135}$$

In Figure 4.107, the $\ln[\{-\ln[(D_{PEG} D_{HDO}^{neat})/(D_{PEG}^{soln} D_{HDO})]\} D_{PEG}^{soln}]$ values calculated using the experimental data is plotted against $\ln q$. A straight line with a slope of $-3/4$ is obtained, and thus $u = -3/4$. Therefore, we have $\kappa^{-1} = q^{3/4} = c^{-3/4}$. de Gennes [254] proposed that for flexible polymer chains in a gel with good solvent the relation $\kappa^{-1} \approx c^{-3/4}$ is obtained. This prediction is in good agreement with the experimental results.

4.10.4.1.4 Poly(acrylic acid) Gel

Poly(acrylic acid) (PAA) interacts strongly with PEG (255), and a PAA gel shrinks in PEG solution. The diffusion of PEG in a PAA gel may be more restrained than that in the PDMAA gel. In Figure 4.108, the D_{PEG} values obtained for a PAA gel with aqueous 0.05 wt % PEG (Mw = 4250) solution are plotted against q together with those for PDMAA gels [256]. The D_{PEG} value in a PAA gel is much smaller

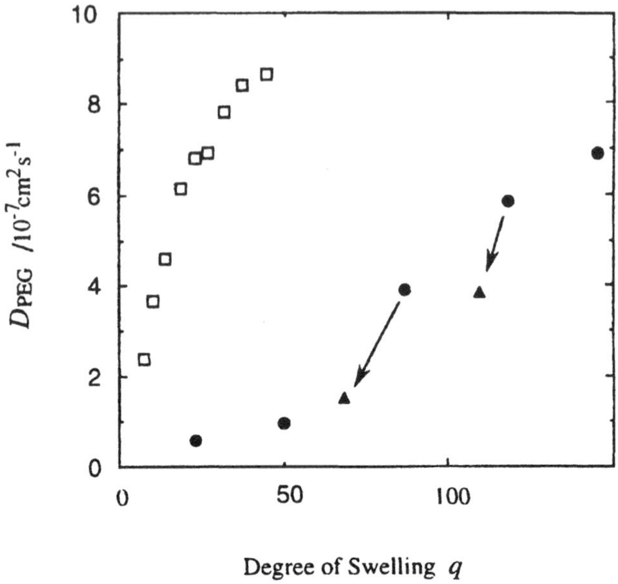

Figure 4.108 Dependence of the diffusion coefficient of PEG in PAA gel (●) and in PDMAA (□) on the degree of swelling q at 30 °C. By an addition of HCl, D_{PEG} is decreased from ● to ▲, as indicated by the arrow.

than that in a PDMAA gel because the translational motion of PEG in PAA gel is more restrained compared with that in the PDMAA gel. From these results, it can be said that there are strong interactions between PAA network and PEG. It was also found that the addition of a small amount of HCl (0.1 mM) leads to the decrease of DPEG together with a slight decrease in q because the dissociation of carboxyl groups in PAA is depressed, and thus the formation of the complex between PEG and PDMAA is facilitated. The ^{1}H T_2 values of PEG in a PDMAA gel and in a PAA gel at the same concentration have been determined to be 0.4 and 0.2 s, respectively. This indicates that the segmental motion of PEG in a PAA gel is also more restrained that in a PDMAA gel. These results agree with the above-mentioned diffusional behavior.

4.10.4.1.5 Poly(*N*,*N*-dimethylacrylamide-*co*-acrylic acid) Gel

The effect of the intermolecular hydrogen bond between a PEG and a network polymer of poly(*N*,*N*-dimethylacrylamide-*co*-acrylic acid) (DMAA-*co*-AA) gel on the dynamic behavior was studied [247, 256]. To investigate the translational

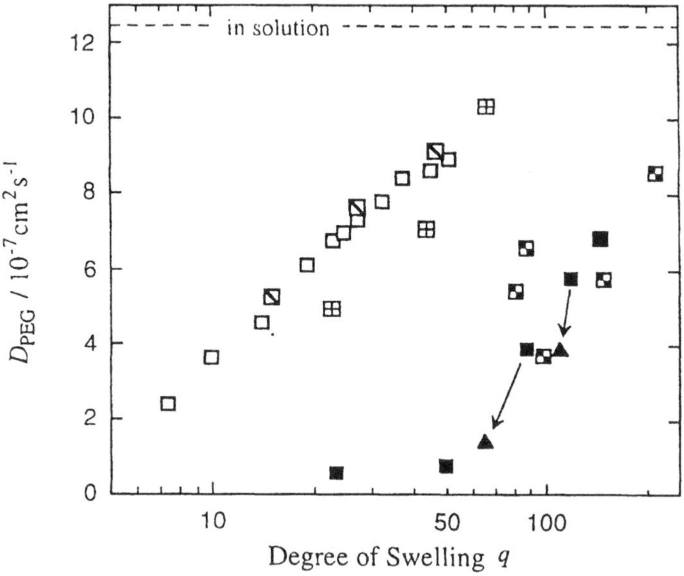

Figure 4.109 Dependence of the diffusion coefficient of the PEG with molecular weight of 4250 (D_{PEG}) in the poly(DMAA-*co*-AA) gels on the degree of swelling q at 30 °C. The diffusion coefficient of the PEG with molecular weight of 4250 in 1 wt % aqueous solution at 30 °C is indicated by the dashed line. The mole fractions of the AA in gels f_{AA} are (□) 0, (□) 0.2, (⊞) 0.5, (◘) 0.9, (■) 1.0, and (▲) 1.0 (soaked in 0.15 mM of HCl).

motion of the PEG in the poly(DMAA-*co*-AA) gels, the diffusion coefficient of the PEG (D_{PEG}) in the gels was determined by the PGSE method. The measurements of D_{PEG} were made for a probe polymer with a molecular weight of 4250 at 30 °C by varying q and the mole fraction of AA in the copolymer gel (f_{AA}). The D_{PEG} values plotted against q are shown in Figure 4.109. The D_{PEG} values in the PDMAA gel ($f_{AA} = 0$) are smaller than those in aqueous 1 wt % PEG solution (dashed line) and increase with increasing q. In the poly(DMAA-*co*-AA) gels, the DPEG is almost constant in the range of $f_{AA} <$ 0.5 and decreases drastically in the range of $f_{AA} \geq 0.9$. These results show that when the AA unit is isolated in the DMAA sequences, the molecular motion of the PEG is not strongly restrained by the intermolecular interaction with AA units. On the other hand, when consecutive AA units are distributed in the network, intermolecular interactions between the PEG and consecutive AA segments are enhanced, which results in the strongly restrained molecular motion of PEG [257].

The addition of a small amount of HCl into the gel causes it to shrink, and the D_{PEG} is decreased, as shown by arrows in Figure 4.109. The complex formation arises from the intermolecular hydrogen bond between oxygen atoms of the PEG and the carboxylic groups of the AA units in the undissociated state shown in Scheme 1. It can be said that the addition of HCl leads to an increase in the amount of undissociated carboxylic groups when going from the left side to the right side in the Scheme 1. This may enhance the formation of the complex. The complex formation and the undissociation induce the shrinkage of the gel and lead to slow diffusion of PEG.

The T_2 value reflects the molecular motion [11, 256]. In the molecular motion of PEG in D_2O solution, the segmental motion is a dominant factor compared with the translational motion and rotational motion of a whole molecule. The ^1H T_2 of PEG (Mw = 4250) was measured by the CPMG method [256] at 30 °C by varying q and f_{AA} of the poly(DMAA-*co*-AA) gel. Figure 4.110 shows the plot of the obtained ^1H T_2 values against q. The ^1H T_2 values for PEG in the gels with $f_{AA} < 0.5$ decrease as q is decreased. This means that the segmental motion of PEG molecule decreases with decreasing size of the network. The segmental

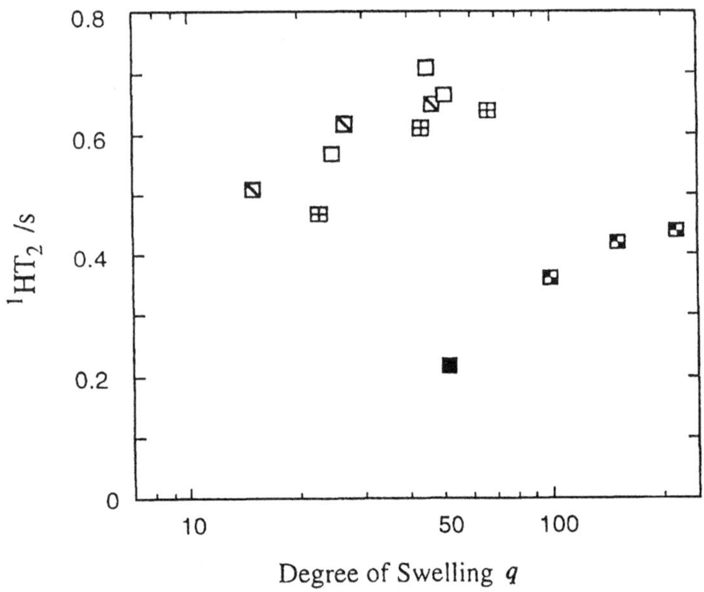

Figure 4.110 Dependence of ^1H T_2 of the PEG with molecular weight of 4250 in the poly(DMAA-*co*-AA) gels on the degree of swelling q at 30 °C. The f_{AA} values are (□) 0, (⧄) 0.2, (⊞) 0.5, (◘) 0.9, (■) 1.0.

motion of PEG may be not strongly restrained by intermolecular hydrogen bonds with the AA units in the gels for $f_{AA} < 0.5$, as indicated by the experimental results on DPEG.

The ^{1}H T_2 values of PEG for $f_{AA} > 0.9$ are much smaller than those for $f_{AA} < 0.5$. This result is similar to that of DPEG. The complex between PEG and AA units in the network polymer restrains the segmental motion of PEG as well as the translational motion. From the preceding results, the complex should be stabilized through hydrogen bondings between the PEG chain and the consecutive AA units in the network polymer.

4.10.4.1.6 Poly(γ-methyl L-glutamate) Gel

Spin-echo ^{1}H–NMR spectra of a poly(γ-methyl L-glutamate) (PMLG) gel containing a mixture of $CHCl_3$ and trifluoroacetic acid (TFA) with the volume fraction of TFA $f_{TFA} = 0.10$ were measured as a function of magnetic field gradient pulse length (δ) [258]. The ^{1}H chemical shift of $CHCl_3$ almost does not change with an increase in TFA content, as shown in Figure 4.111. The ^{1}H chemical shift of TFA is observed at 8.7 to 10.5 ppm with an increase in TFA content.

Figure 4.112 shows the q of PMLG gel in a mixture of TFA–$CHCl_3$ against the f_{TFA}. A PMLG gel in $CHCl_3$ is transparent. The q decreases with increasing

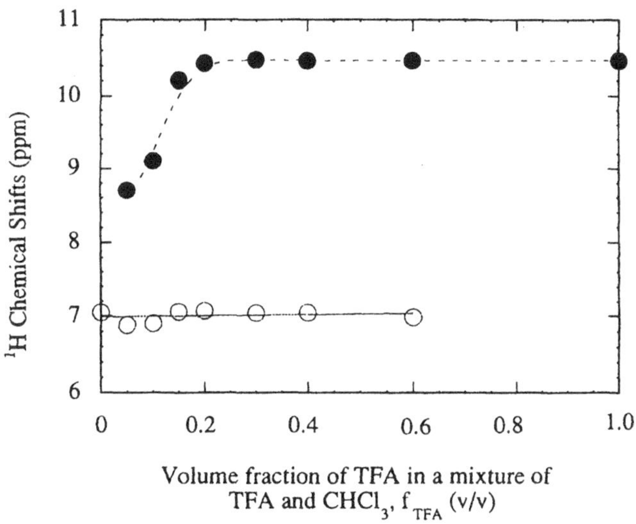

Figure 4.111 Plots of ^{1}H chemical shifts for TFA and $CHCl_3$ in PMLG gel system against the TFA content in a mixture of TFA and $CHCl_3$: TFA (●) and $CHCl_3$ (○).

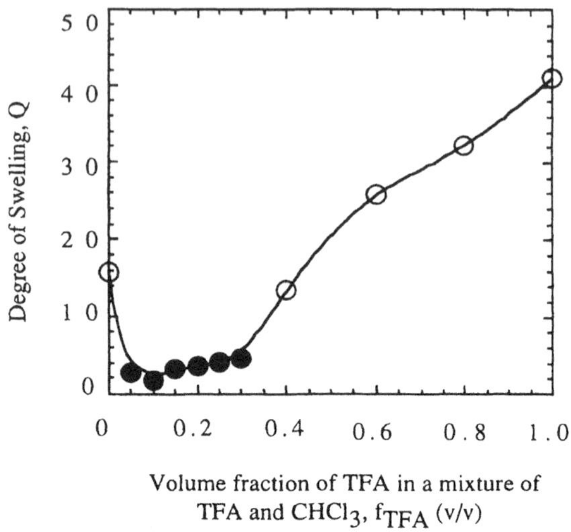

Figure 4.112 Dependence of the degree of swelling q of PMLG gel on the TFA content in a mixture of $CHCl_3$ and TFA: (○) the gel is transparent, and (●) the gel is turbid.

TFA content from $f_{TFA} = 0.05$ to 0.30 while the gel becomes turbid. In the TFA content from $f_{TFA} = 0.30$ to 1.0, the degree of swelling of PMLG gel increases with increasing TFA content; meanwhile the gel becomes transparent again. This shows that the main chain of PMLG gel underwent a conformation transition from the α-helix form to the random coil form. The mixture solvent with $f_{TFA} = 0.05$ to 0.30 acts as a poor solvent for the random coil form, and inhomogeneous aggregation occurs in the gel system. The mixture solvent with higher TFA content such as $f_{TFA} = 0.30$ to 1.0 acts as a good solvent for the random coil form.

The diffusion coefficients of TFA and $CHCl_3$ contained in a PMLG gel were determined by the PGSE ^{1}H–NMR method at 30 °C by varying the TFA content. The decays of TFA and $CHCl_3$ lie on a straight line, as shown in Figure 4.113. This means that the TFA and $CHCl_3$ in PMLG gel have a single diffusion component during the observation time. From the slope of the intensity decay, the diffusion coefficients for TFA (D_{TFA}) and $CHCl_3$ (D_{CHCl_3}) can be determined as 0.37×10^{-5} and 1.00×10^{-5} cm^2 s^{-1}, respectively. The D_{TFA} is smaller than that of $CHCl_3$. This means that the translational motion of TFA is more restrained by the PMLG network compared with $CHCl_3$. The diffusion coefficients of TFA and $CHCl_3$ in PMLG gel were plotted against the TFA content in a mixture of TFA–$CHCl_3$, as shown in Figure 4.114. As seen from this figure, D_{CHCl_3} largely decreases in the f_{TFA} range from 0 to 0.05. This tendency is very similar to that of the degree of

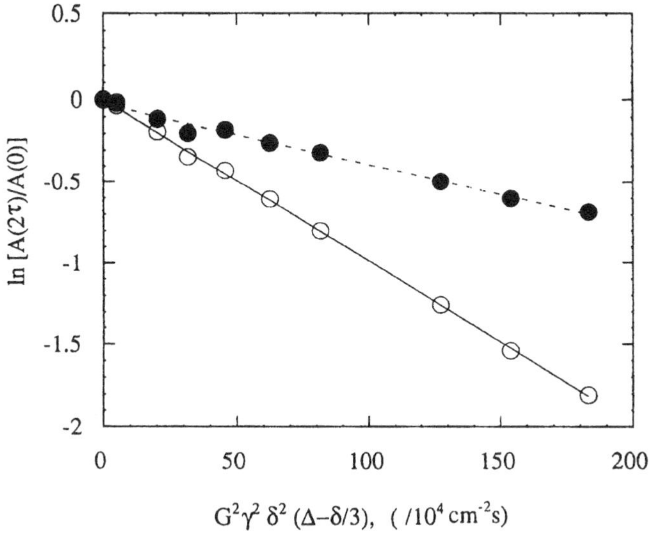

Figure 4.113 Diffusional spin-echo attenuation of TFA (●) and $CHCl_3$ (○) in PMLG gel with $f_{TFA} = 0.10$ by varying field-gradient pulse duration (δ). From the slopes of the solid and dashed straight lines, D_{TFA} and D_{CHCl_3} were determined to be 0.37×10^{-5} and 1.00×10^{-5} cm^2 s^{-1}, respectively.

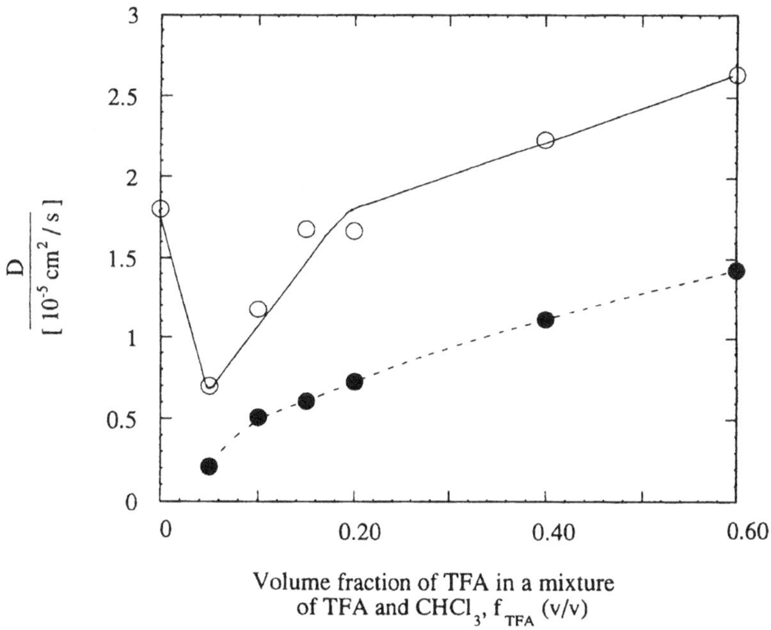

Figure 4.114 Plots of the diffusion coefficients for TFA (●) and $CHCl_3$ (○) in PMLG gel against the TFA content in a mixture of TFA and $CHCl_3$.

swelling in Figure 4.112. This shows that the conformation of PMLG gel changed from the α-helix to the random coil with increasing TFA content. The transition from the α-helix form to the random coil form leads to the shrinkage of the PMLG gel owing to the decrease of the network size and leads to the decrease of diffusion coefficients of $CHCl_3$. In the TFA range from 0.05 to 0.20, D_{TFA} and D_{CHCl_3} increase, although the degree of swelling is almost constant. It can be said that in the f_{TFA} range the helix-coil transition leads to such diffusional behaviors. After the helix-coil transition, the molecular motion of TFA and $CHCl_3$ increases with an increase in f_{TFA}.

The plots of $\ln D$ for TFA and $CHCl_3$ in PMLG gel are shown against the inverse of absolute temperature $(1/T)$ in Figure 4.115. The diffusion coefficients can be expressed against temperature as an Arrhenius-like equation

$$D = \left(a_0^2/\tau_0\right)\exp(-E/kT) \tag{136}$$

where a_0 is the interval of the diffusant, τ_0 is the time of diffusing, k is the Boltzmann constant, and E is an activation energy for diffusion. From the slope for such plots, the activation energies for TFA and $CHCl_3$ were found to be 38.7 and 30.5 kJ/mol, respectively. The activation energies for pure TFA and $CHCl_3$

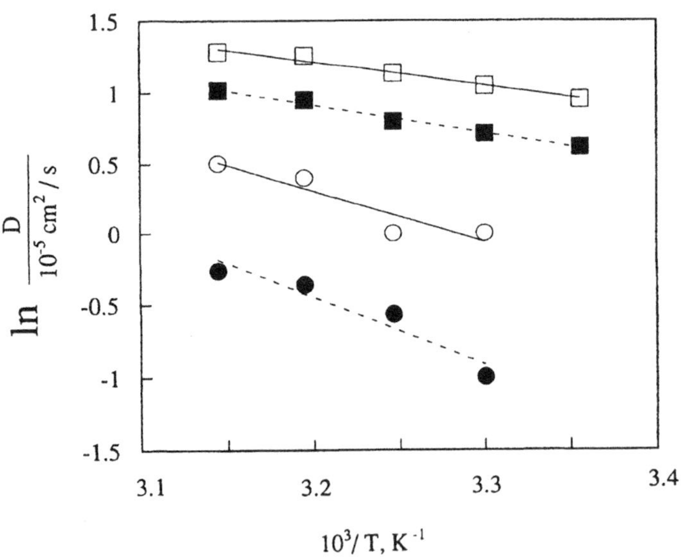

Figure 4.115 Temperature dependence of self-diffusion coefficients of TFA and $CHCl_3$ at $f_{TFA} = 0.10$. In a PMLG gel system, (●) denotes TFA and (○) $CHCl_3$, and in a mixture solvent of $CHCl_3$ and TFA, (■) denotes TFA and (□) $CHCl_3$.

were determined to be 16.2 and 13.5 kJ/mol, respectively. The intermolecular interactions between solvent and polypeptide network are strong enough to restrain the translational motion.

4.10.4.1.7 Poly(vinyl alcohol) Gel

According to the BPP theory [11], T_2 decreases continuously as the correlation time τ_c in molecular motion is increased. Thus, T_2 can provide dynamical information about the gels through the BPP theory. A typical ^{1}H T_2 signal of PVA gel is shown in Figure 4.116 [260]. ^{1}H T_2 values are determined from the slope of the plot of ln M against time t, where M is the amplitude of spin echo signal. This ^{1}H T_2 curve is composed of three components such as a long T_2 component corresponding to the mobile component (component C), a short T_2 component corresponding to the immobile component (component A), and an intermediate T_2 component corresponding to the interfacial component (component B), which exists between the mobile and immobile components. By aid of computer fitting,

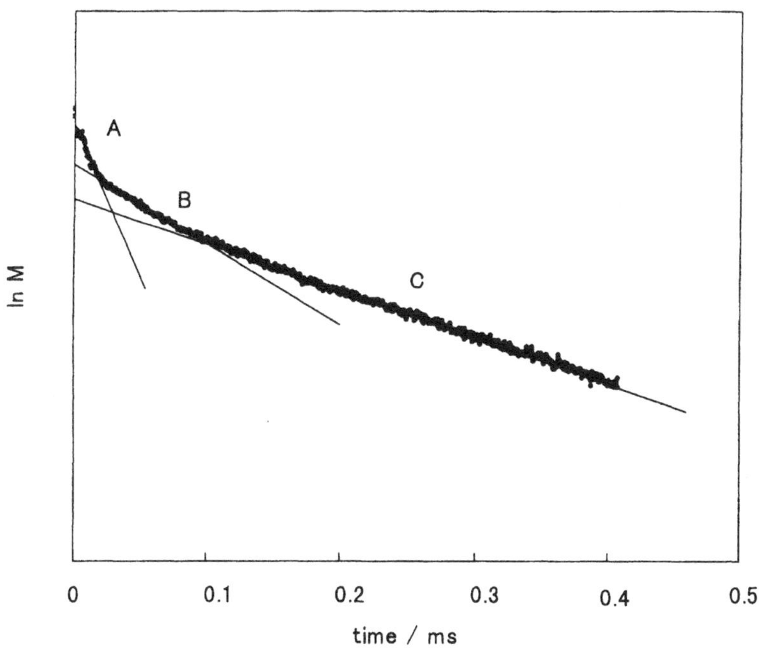

Figure 4.116 ^{1}H T_2 signal of PVA gel at the PVA concentration of 27.4% w/w by the solid echo method.

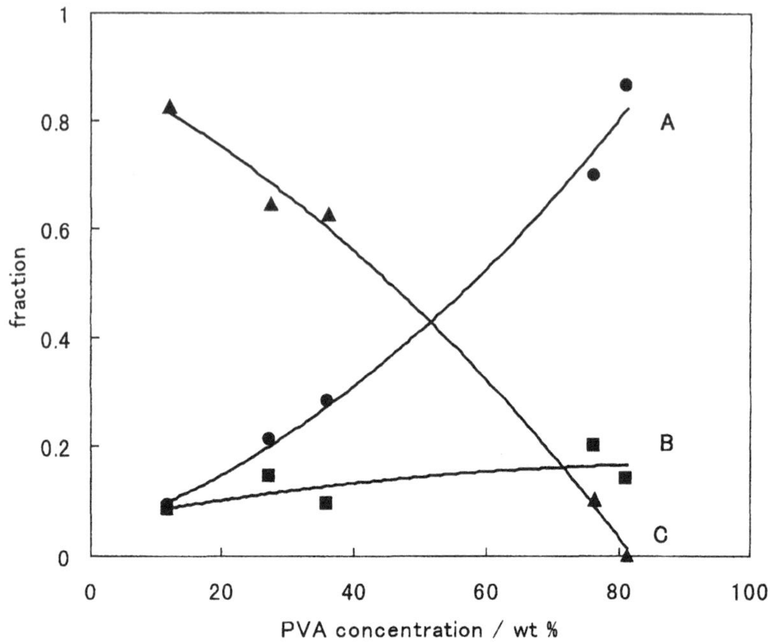

Figure 4.117 The relationship between the fraction of three components with different molecular motion in PVA chains in the gel and the polymer concentration. ● component (A) (immobile), ■ component (B) (interfacial), ▲ component (C) (mobile).

the ^{1}H T_2 value was obtained from the T_2 signal with small experimental error. The obtained ^{1}H T_2 values for the three components were plotted against polymer concentration. With an increase in the polymer concentration, their T_2 values slowly decrease. This means that the molecular motions for the three components are restrained as the polymer concentration is increased. Figure 4.117 shows the fractions of the corresponding three components as a function of polymer concentration. The fractions of the mobile and immobile components largely decrease and increase, respectively, with an increase in the polymer concentration. The fraction of the intermediate component over a wide range of polymer concentrations is very small.

In Figure 4.118 the plots of ^{1}H T_2 values for each of the components A, B, and C against the number of freeze-thaw cycles are shown. Their ^{1}H T_2 value decrease with an increase in the number of freeze–thaw cycles. This shows that the increase in the number of freeze–thaw cycles leads to a slight reduction of their mobilities. As the number of freeze–thaw cycles increases, the fractions of the

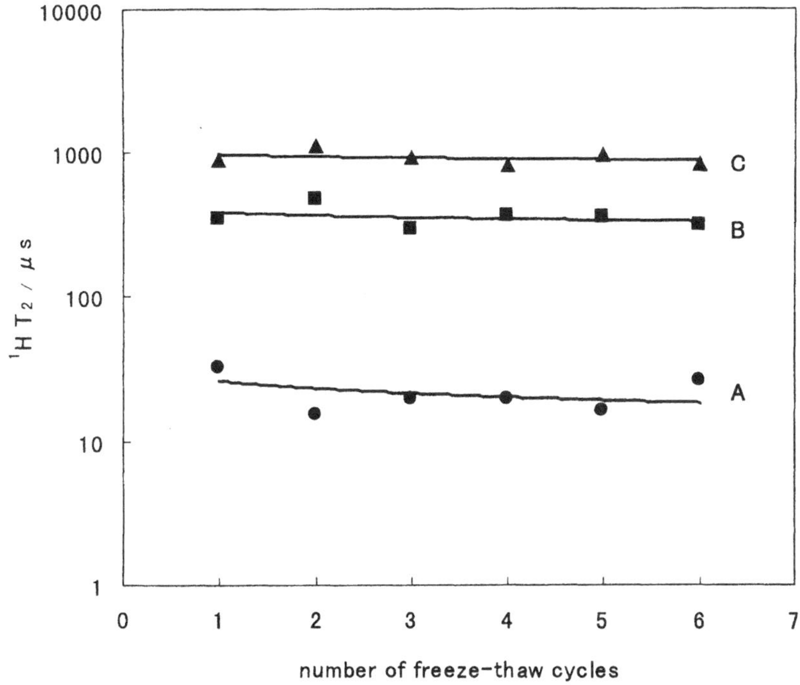

Figure 4.118 The plots of ^{1}H T_2 of PVA gel against the number of freeze–thaw cycles as determined by ^{1}H pulse NMR signal decay: ● denotes component A (immobile); ■ denotes component B (interfacial); ▲ denotes component C (mobile).

immobile component A and the intermediate component B increase and decrease, respectively (Figure 4.119). The fraction of the immobile component C is slightly decreased. This shows that the intermediate component B is largely affected.

4.10.4.2 Approaches by High-Resolution Solid-State ^{13}C–NMR

In Contrast to linear polymers, unique properties of polymer gels arise from the three-dimensional network structure formed by cross-linkages. However, the network of gels has also made their characterization and measurement difficult. Therefore, NMR techniques have been applied to polymer gel systems to characterize them. The studies have demonstrated that NMR spectroscopy is one of the most powerful means for investigating the structure and dynamics of polymer gel systems. The advantages of NMR techniques for characterizing polymer gels

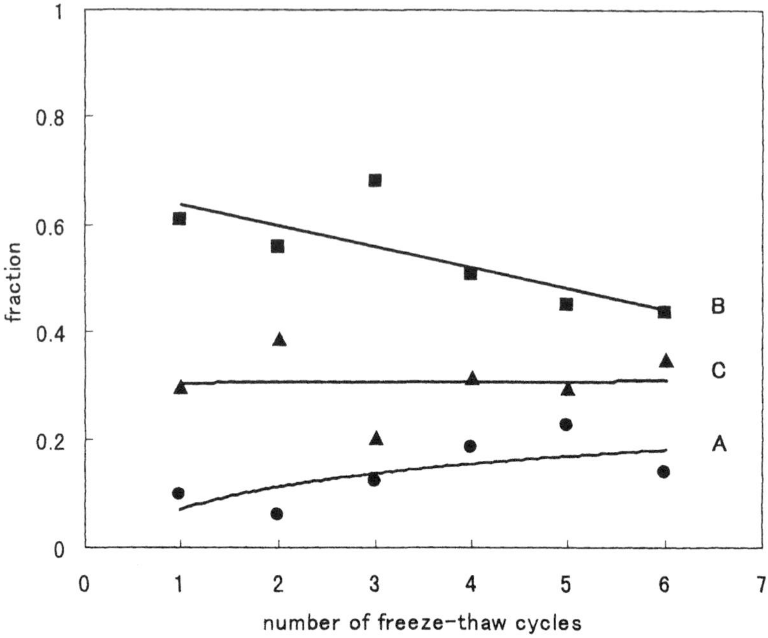

Figure 4.119 The plots of fraction for the three components in PVA gel against the number of freeze–thaw cycles as determined by ^1H pulse NMR signal decay: ● denotes component A (immobile); ■ denotes component B (interfacial); ▲ denotes component C (mobile).

are as follows:

1. They are noninvasive.
2. No probes are needed.
3. Individual information on the main chain, the side chain, and the functional group of a sample is obtained.
4. The change in the structure and the motion of a sample with time is measurable.

The solution NMR techniques are available to characterize the solvent contained in the gels but not the network polymer. We cannot obtain reasonable NMR signals with high intensities, SNR, and resolution for gel samples by normal solution techniques. This is caused by the restraint of motion, the strong dipole–dipole interactions, the chemical shift anisotropy, and the long T_1 and short T_2 of network polymers. The high-resolution solid-state NMR techniques are powerful means for characterizing the network polymer of polymer gels. In this section, research on synthesized polymer gels by high-resolution solid-state NMR techniques will be reviewed.

4.10.4.2.1 Poly(vinyl alcohol) Gel

Cross-linking points in physical gels are formed by strong intermolecular interactions such as coulombic, dipole–dipole, van der Waals, charge transfer, hydrophobic and hydrogen bonding interactions. Some physical gels are thermoreversible and show a reversible transition from the solution to gel state with change in temperature [261].

The poly(vinyl alcohol) (PVA) gel prepared by repeating freeze–thaw cycles from its aqua solution is one of the physical gels. At the beginning of the gelation of PVA aqueous solution, the phase separation into polymer-rich and polymer-poor regions arises with the decrease in temperature [262]. It was suggested that the gel formation is due to the formation of intermolecular hydrogen bonds in the PVA [263]. The structure and the amount of hydrogen bonds may play an important role for the nature of PVA gel.

Figure 4.120 shows ^{13}C–NMR spectra for the PVA in the solution (A), gel (B–D), and solid (E) states as measured by solution and solid-state NMR methods [264]. The NMR spectrum for the PVA aqueous solution obtained by the solution ^{13}C–NMR method (Figure 4.120(a)) shows each of the signals for the CH and CH_2 carbons splits into multiple peaks owing to stereochemical configuration. The three-splitting peaks for the CH carbon are assigned to *mm*, *mr*, and *rr* triads from the down magnetic field [265–267] and further, each of the triad peaks splits into pentad peaks. The splitting peaks for the CH_2 carbon come from tetrad tacticity. In the solution ^{13}C–NMR spectrum for the PVA gel shown in Figure 4.120(b), the ^{13}C signal for the CH carbon splits into three peaks owing to triad tacticity. The signals for the CH and CH_2 carbon become broader as compared with those for the PVA solution. This is caused by the increase in the dipole–dipole interactions between the polymer chains in the network polymer formed in the gel. The ^{13}C PST/MAS NMR spectrum of the PVA gel shown in Figure 4.120(c) is very similar to the solution NMR spectrum of Figure 4.120(b). The NOE enhancement is used to obtain the higher ^{13}C signal in the PST method. The PST method effectively enhances peak intensity for mobile regions in samples as compared with the CP method. This means that the mobile components in the PVA gel are observed in spectra (B) and (C).

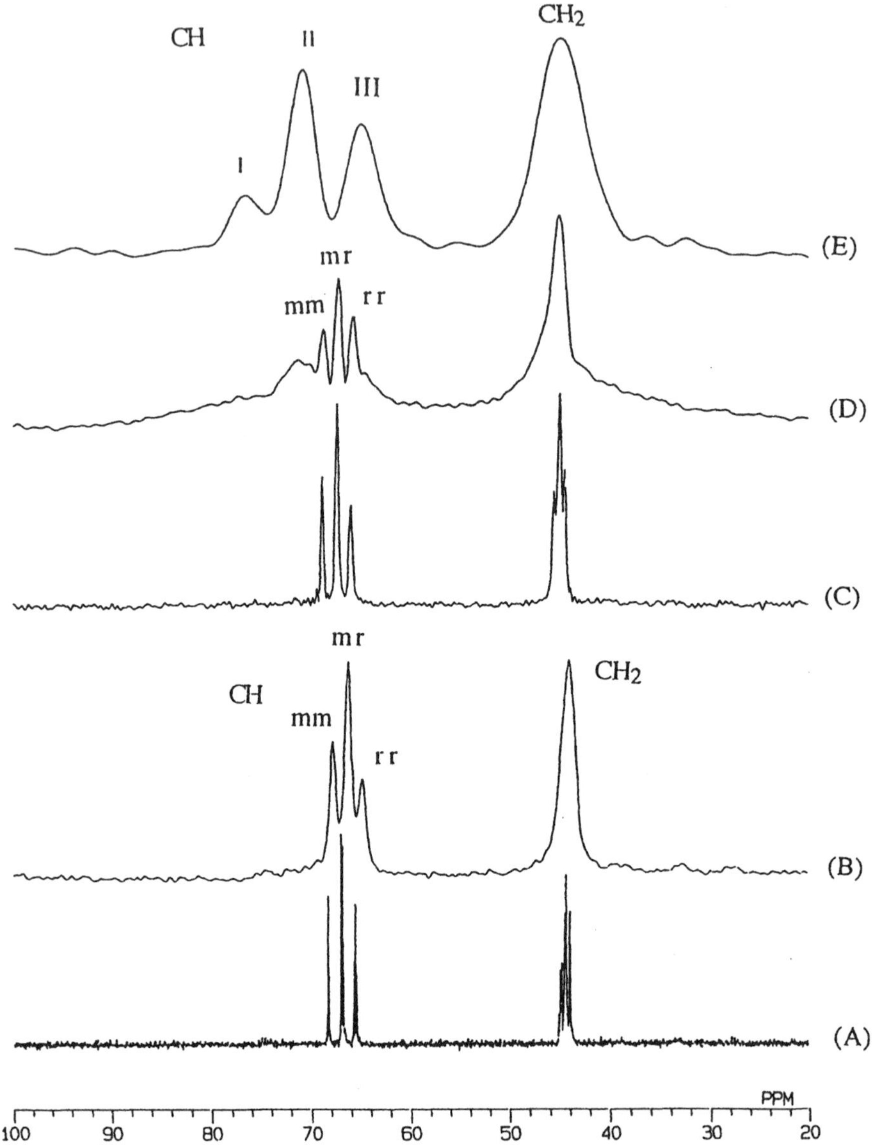

Figure 4.120 ^{13}C–NMR spectra for PVA in the solution, gel, and solid states as measured by several NMR methods. Solution NMR method: the PVA/D_2O solution (A) and the PVA gel (B). Solid-state NMR method: the PVA gel by PST/MAS (C) and by CP/MAS (D), and the solid PVA by CP/MAS (E).

The CP method effectively enhances peak intensity for more rigid regions in samples as compared with the PST method. The ^{13}C signal for the CH carbon of the solid PVA has three split, broad peaks in the ^{13}C CP/MAS NMR spectrum, as shown in Figure 4.120(e). The chemical shift difference between these three peaks is considerably larger than the splitting because of stereochemical configuration. Such a splitting is explained by the number of intramolecular hydrogen bonds formed among neighboring hydroxyl groups [268]. On the assumption that the ^{13}C chemical shift value for the CH carbon shifts down to the magnetic field by 6 ppm per inter- or intramolecular hydrogen bond, the most deshielded peak (peak I) can be assigned to the *mm* triad with two hydrogen bonds, the central peak (peak II) to the *mm* and *mr* triads with one hydrogen bond, and the most shielded peak (peak III) to the *mm*, *mr*, and *rr* triads with no hydrogen bonds, respectively. The proposed schematic crosslinked structure of PVA gel formed by inter- and intramolecular hydrogen bonds is shown in Figure 4.121. In the ^{13}C CP/MAS spectrum for the PVA gel as shown in Figure 4.120(d), the CH signal

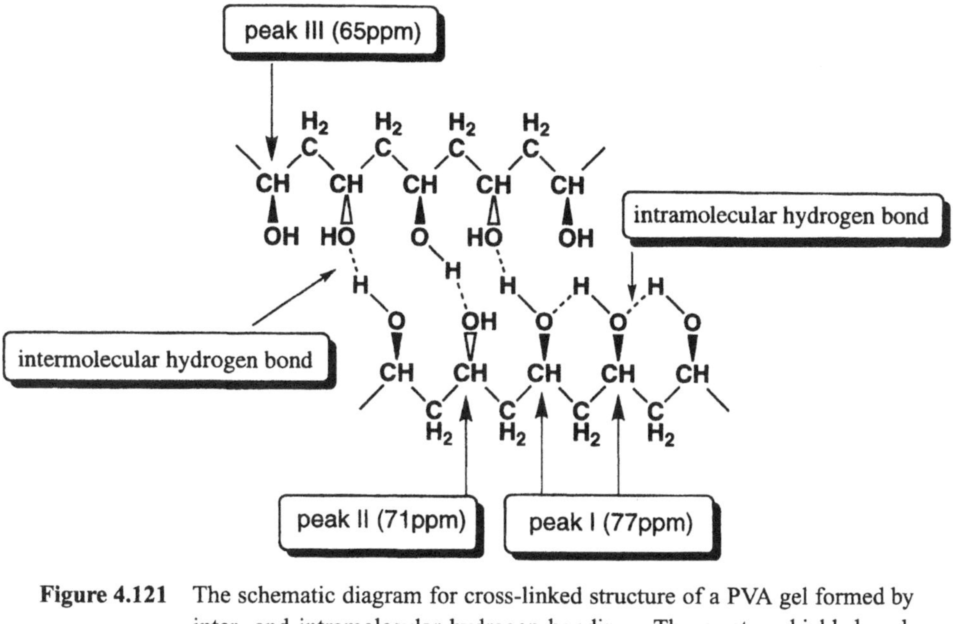

Figure 4.121 The schematic diagram for cross-linked structure of a PVA gel formed by inter- and intramolecular hydrogen bondings. The most unshielded peak (peak I) is assigned to the *mm* triad with two hydrogen bonds, the central peak (peak II) to the *mm* and *mr* triads with one hydrogen bond, and the most shielded peaks (peak III) to the *mm*, *mr*, and *rr* triads with no hydrogen bonds.

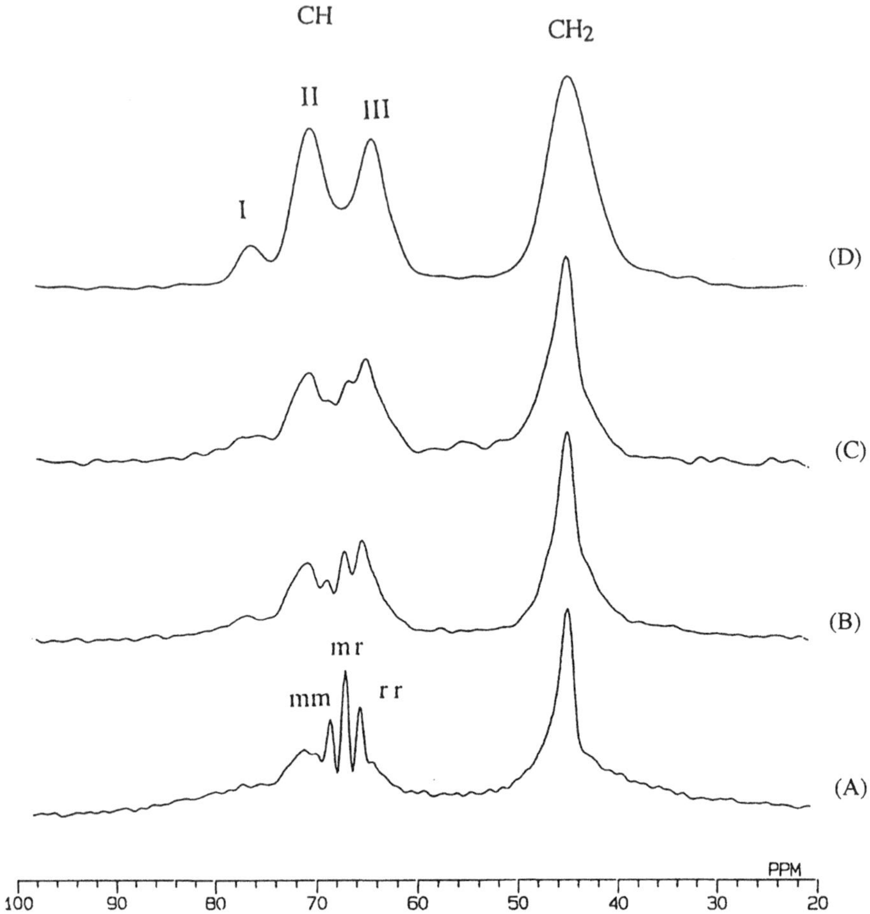

Figure 4.122 ^{13}C CP/MAS NMR spectra for the PVA gels with different polymer concentrations. The polymer concentrations are (a) 9.1, (b) 11.8, (c) 13.8, and (d) 35.0% (wt/wt), respectively.

is composed of the three-split peaks corresponding to the triad tacticity and the broad peaks as found in the solid PVA. This means that both of the immobile and mobile components of the PVA gel are observed by CP/MAS method at the lower polymer concentration.

Figure 4.122 shows ^{13}C CP/MAS NMR spectra of the PVA gels with different polymer concentrations (264). The polymer concentrations of samples A, B, C and D are 9.1, 11.8, 13.8 and 35.0% (wt/wt), respectively. The three peaks due to stereochemical configurations completely disappear in sample D. The peaks I,

II and III can be assigned to the CH carbons with two intra- or intermolecular hydrogen bonds, one hydrogen bond and no hydrogen bonds, respectively. The spectrum for the sample D is very similar to that of solid PVA except the intensity of peak III in the former is somewhat larger than that in the latter. The increase in the fraction of the immobile components and the formation of hydrogen bonds are promoted by the increase in the polymer concentration of the gel.

Figure 4.123 shows ^{13}C PST/MAS NMR spectra of the PVA gels with different polymer concentrations. The signals from mobile components in the gel are observed clearly by the PST/MAS method differing from those by the CP/MAS

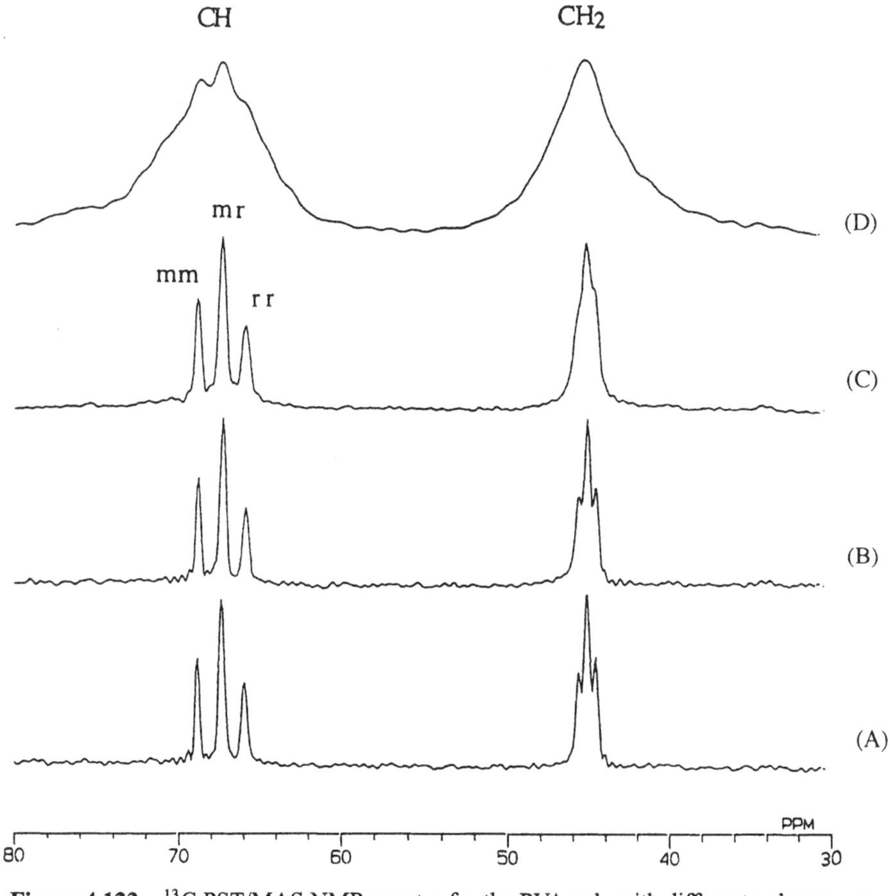

Figure 4.123 ^{13}C PST/MAS NMR spectra for the PVA gels with different polymer concentrations. The polymer concentrations are (a) 9.1, (b) 11.8, (c) 13.8, and (d) 35.0% (wt/wt), respectively.

method. While the spectra for the samples A, B and C are similar to each other and their CH signal splits into three peaks corresponding to the triad configurations, in the spectrum for D, the CH signal shows a broad peak. The broadening is due to the dipole-dipole interactions resulting from an increase in the restraint of molecular motion. This result means that the mobility of the mobile components in the PVA gel is restrained by an increase in the polymer concentration (more than 35.0%) in the gel.

The NMR measurement was made for the PVA solutions at low temperature to clarify the mechanism of gel formation by freeze–thaw cycles [264]. The ^{13}C CP/MAS NMR spectrum for the PVA solution was measured at −50 °C in going from room temperature to −50 °C (Figure 4.124). The signals for the CH group show that, instead of the three peaks that arise from the splitting by tacticity, peaks I, II, and III appear. The result demonstrate that, in the PVA solution at −50 °C, some hydroxyl groups form hydrogen bonds as in the PVA gel, and thus the cross-linked structure is formed in the PVA solution. This means that the gel formation is induced in going from room temperature to −50 °C.

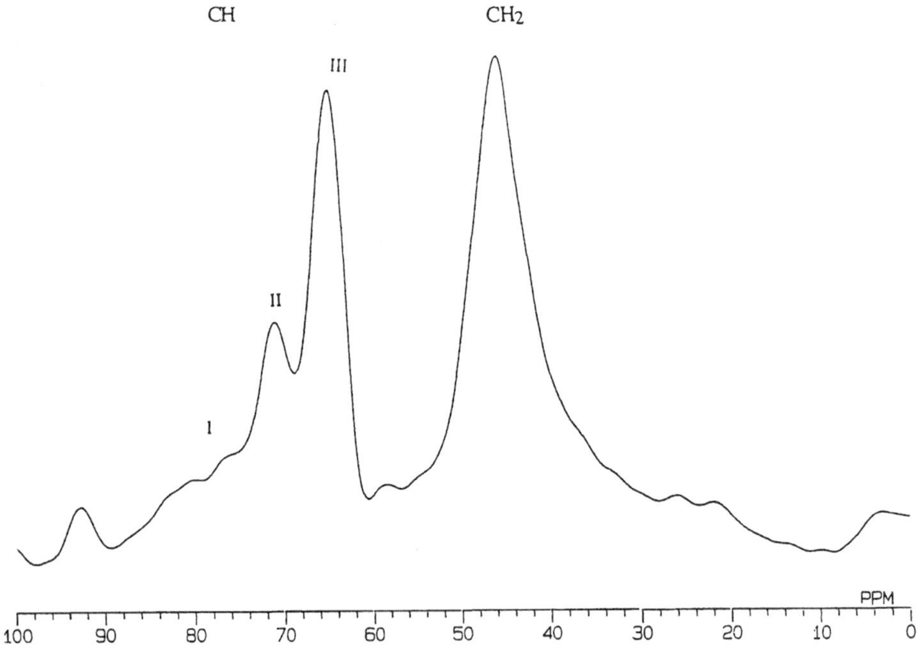

Figure 4.124 ^{13}C CP/MAS NMR spectrum for the PVA solution at −50 °C in the process of cooling.

The ^{13}C CP/MAS NMR spectra for three kinds of the PVA gels with different tacticity (9% of polymer concentration) are shown in Figure 4.125. The CH peaks are composed of both the three sharp peaks corresponding to the triad configurations (*mm*, *mr*, and *rr*) and the three broad peaks (peaks I, II, and III at about 77, 71, and 65 ppm, respectively).

In the ^{13}C CP/MAS NMR spectrum for the *iso*-PVA gel (Figure 4.125(a)), peak I together with three sharp peaks for the triad configurations are obviously observed, whereas peaks II and III are not so obvious. Peaks II and III are observed together with peak I in the spectra for the *at*-PVA and *syn*-PVA gels, as shown in Figure 4.125(b) and (c), respectively.

Peaks I, II, and III were deconvoluted by computer fitting with the Gaussian function, and the *mm*, *mr*, and *rr* peaks with the Lorentzian function. In the *iso*-PVA gel, the *mm* fraction is high, and the *mm* triad with two hydrogen bonds is preferentially formed. The *iso*-PVA gel is much softer than the *at*- and *syn*-PVA gels, although the peak I intensity of the *iso*-PVA gel is higher than those of the *at*- and *syn*-PVA gels. The amount of the intermolecular hydrogen bonds in the *iso*-PVA gel is lower than those in the *at*- and *syn*-PVA gels. The intramolecular hydrogen bonds are mainly formed in the *iso*-PVA. On the other hand, the *rr* fraction is high in the *syn*-PVA gel. In the *syn*-PVA gel, the peak II corresponding to the *mm* and *mr* triads with one hydrogen bond should come from the intermolecular hydrogen bonds that form the cross-linking regions.

The ^{13}C T_1 values for the mobile component in a PVA gel were determined by the IR method combined with the PST/MAS technique [269]. The ^{13}C T_1 values for the immobile component in the gel were also determined by the CP/MAS technique combined with Torchia's pulse sequence [270]. The ^{13}C T_1 values are summarized in Table 4.13. The T_1 values for the triads and the CH_2 are smaller than those for peaks II, III, and the CH_2. The immobile components correspond to the CH carbons in the cross-linked regions formed by inter- and intramolecular hydrogen bonds in the gel, and the mobile ones to the regions away from the cross-linked regions that undergo fast molecular motion in the gel [269, 271]. The mobility of the network polymer in the PVA gel is restrained by hydrogen bonds and that of the cross-linked region, especially, is very low.

4.10.4.2.2 Poly(methacrylic acid) Gel

The poly(methacrylic acid) (PMAA) gel is a typical chemical gel in which cross-linkages are made up of chemical bonds. The hydro-swollen PMAA gel was studied as a function of the degree of cross-linking by ^{13}C high-resolution solidstate

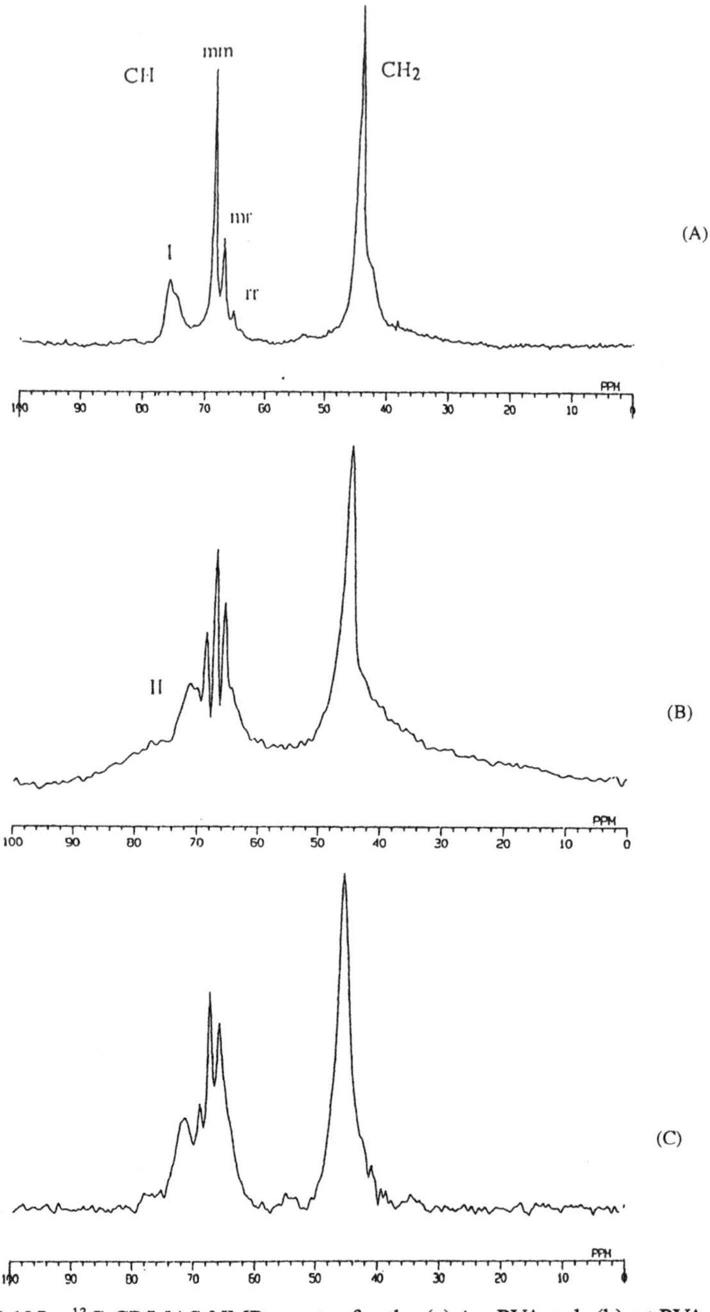

Figure 4.125 ^{13}C CP/MAS NMR spectra for the (a) *iso*-PVA gel, (b) *at*-PVA gel, and (c) *syn*-PVA gel.

Table 4.13 ^{13}C spin-lattice relaxation time T_1 of PVA gel.

	^{13}C T_1 (s)					
	CH					
Method	***mm***	***mr***	***rr***	***II***	***III***	CH_2
Torchia's pulse sequence (CP method)				7.8	7.6	5.2
Inversion-recovery pulse sequence (PST method)	0.37	0.28	0.32			0.16

NMR techniques [271]. The mole ratio of N,N'-methylenebisacrylamide (MBAA), (n_{MBAA}) to methacrylic acid (MAA), (n_{MAA}) in the preparation of the PMAA gel F is defined as

$$F = (n_{MBAA}/n_{MAA}) \tag{137}$$

^{13}C PST/MAS spectra for PMAA gels with three kinds of F and q are shown in Figure 4.126(a), (b), and (c). The chemical shift values for the PMAA gels are listed as a function of q in Table 4.14. Each of the C=O and α-CH_3 signals is split into three peaks by triad configurations of PMAA [272]. The peaks are assigned to isotactic [C=O (mm) and CH_3 (mm)], heterotactic [C=O (mr) and CH_3 (mr)] and syndiotactic [C=O (rr) and CH_3 (rr)] triads, respectively [273, 274]. From the α-CH_3 signals, the triad fraction is evaluated as $rr : mr : mm = 67.4 : 27.1 : 5.5\%$ [275]. The tacticity of the main chain in the PMAA gel prepared by radical polymerization shows mostly rr triads.

Table 4.14 ^{13}C–NMR chemical shift[a] of PMAA gel as a function of the degree of swelling q at 300 K.

	^{13}C Chemical Shift (ppm)							
q	*C=O (rr)*	*C=O (mr)*	*C=O (mm)*	CH_2	-C-	CH_3 (mm)	CH_3 (mr)	CH_3 (rr)
105.8	182.4			54.7	45.3			18.3
59.4	182.9				45.9			17.4
21.7	182.8			54.6	45.9	22.2	19.3	17.9
4.7	182.9			54.1	45.7	21.7	19.2	17.8
3.7	182.7	181.8		54.2	45.8	22.2	19.5	17.9
3.4	182.5	181.5	180.8	54.6	46.0	22.3	19.8	18.2

[a]ppm from tetramethylsilane.

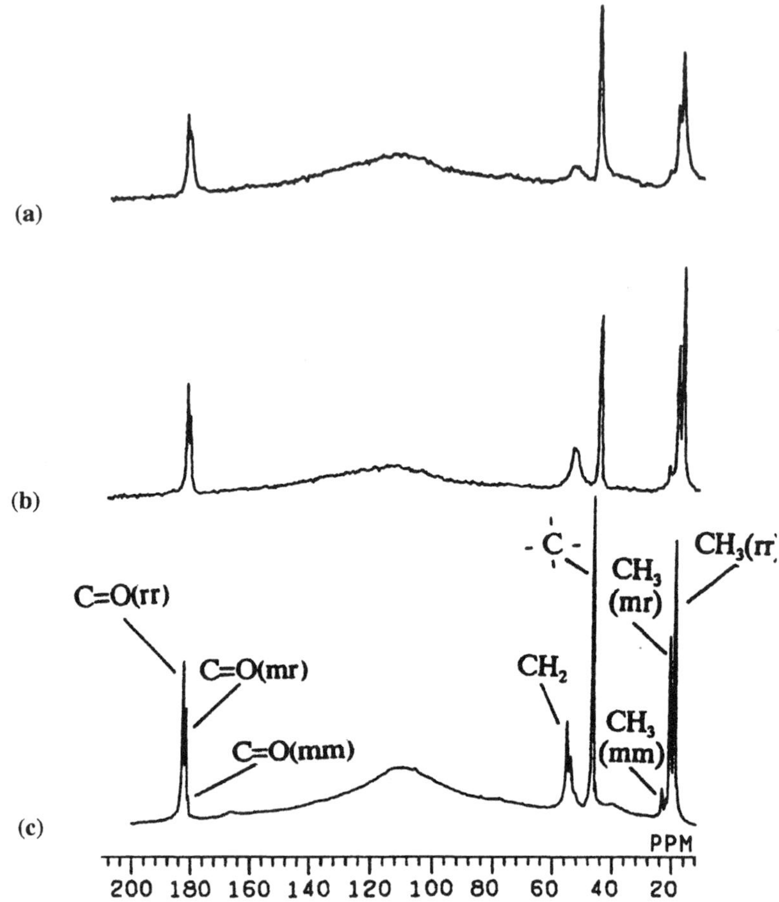

Figure 4.126 ^{13}C PST/MAS spectra for the PMAA gels: (a) $F = 0.001$, $q = 105.8$; (b) $F = 0.0031$, $q = 21.7$; (c) $F = 0.020$, $q = 3.4$.

The ^{13}C chemical shift values for the PMAA gel are almost independent of q. If the formation of hydrogen bonds or dissociation of carboxyl groups is caused by the change in q and is large enough, the chemical shifts for the C=O would change. In fact, the chemical shift changes are found for water in the PMAA gel caused by the change in q [276]. No change in the chemical shifts for the C=O indicate that the COOH groups are surrounded by water molecules and no hydrogen bond formation between the COOH groups is caused by the decrease in q of 3.4–105.8.

The PMAA gel with a low degree of cross-linking and high q is harder than that with a high degree of cross-linking and low q. The increase in q should lead to an

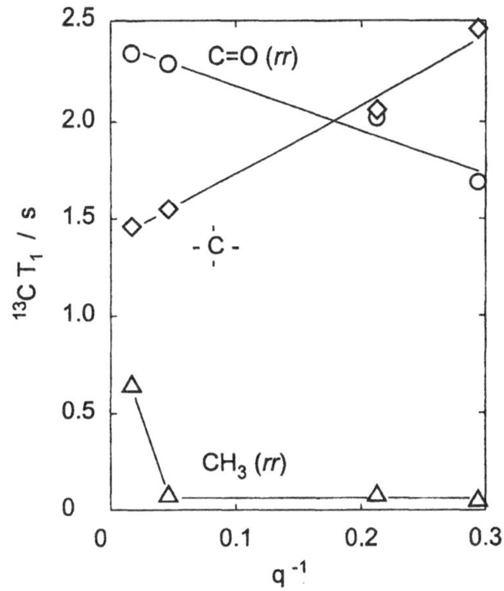

Figure 4.127 Plots of ^{13}C T_1 for the C=O (*rr*), CH_3 (*rr*) and –C– carbons of the PMAA gel against q^{-1} at 27 °C.

increase in the molecular motion of the network polymer in the PMAA gel. The ^{13}C T_1 values for the C=O (*rr*) and CH_3 (*rr*) carbons increase with an increase in q, whereas the T_1 values for the –C– carbons decrease with an increase in q [271]. The ^{13}C T_1 values for the C=O (*rr*), CH_3 (*rr*) and –C– carbons are plotted against q^{-1}, as shown in Figure 4.127. The T_1 for the C=O (*rr*) and –C– carbons decreases and increases linearly with q^{-1}, respectively. The T_1 for the CH_3 (*rr*) carbon decreases abruptly from 0.017 to 0.046 of q^{-1} and is constant from 0.046 to 0.30. The increase in q leads to a decrease in the density of the network polymer in the gel and an increase in the molecular motions of the network. In fact, ^1H T_1, ^1H T_2 and D_{H_2O} of water molecules contained in the PMAA gel increase with an increase in q controlled by the degree of cross-linking, and thus the molecular and translational motion of water molecules in the gel increases with an increase in q [276]. In the PMAA gel system, the spin-lattice relaxation is predominantly governed by dipole–dipole interactions, and the q corresponds to the temperature in the BPP theory on NMR relaxation. Therefore, an increase in q^{-1} means an increase in the inverse of temperature. According to the BPP theory, T_1 decreases with increasing temperature and increases after passing through a minimum as the temperature is increased further. As shown in Figure 4.127, the decrease in ^{13}C T_1 for the C=O (*rr*) and CH_3 (*rr*) carbons of PMAA in the gel with the increase in

q^{-1} means that the motion of side-chain carbons is in the extreme narrowing region at the left side in the BPP curve. However, the increase in T_1 for the –C– carbon means that the motion is in the slow motion region at the right side in the BPP curve. The segmental motion for the side chains of the network polymer in the PMAA gel is faster than that for the main chains. The main chains seem to behave like solid molecules and the side chains liquid ones. This feature of the gel comes from its complex structure composed of three-dimensional network polymer and solvent.

The water content in the PMAA gel contributes to the segmental motion of the network polymer. On the other hand, the ^{13}C T_1 for the PMAA in the gel was also measured as a function of the degree of crosslinking with constant water content [276]. The ^{13}C T_1 values for the main and side chain carbons of the network polymer in the PMAA gel ($q = 3.4$) as a function of the mole ratio of crosslinking reagent to monomer (F) and uncrosslinked PMAA solution (29.4 w%) are plotted in Figure 4.128(a) and (b).

The T_1 values for the C=O (rr) and C=O (mr) carbons decrease in the range of F from 0 to 0.0031 and increase from 0.0031 to 0.020. The efficiency of the NMR relaxation of the carboxyl carbon becomes largest at 0.0031 of F. The motion of the C=O changes from the extreme narrowing region to the slow motion region in the BPP curve with increasing the degree of crosslinking. The T_1 values for the CH_2, CH_3 (mm), CH_3 (mr) and CH_3 (rr) carbons decrease with increasing F. The molecular motion for the CH_3 carbons is in the extreme narrowing region. The order of the T_1 values for the triad methyl carbons is CH_3 (mm) > CH_3 (mr) > CH_3 (rr). This shows that the molecular motion of the CH_3 carbon in the mm triad structure is the fastest and that in the rr triad structure is the slowest.

The T_1 for –C– carbon increases with an increase in F. The motion of –C– that is in the slow motion region is restrained with increasing the degree of crosslinking. The T_1 for the main-chain CH_2 carbon decreases with increasing F in contrast to the –C–. This can come from the difference in their relaxation processes. The two protons and two carbons are attached to the CH_2 carbon, whereas four carbons are attached to the –C–.

This shows that the segmental motion of the network polymer in the PMAA gel is strongly affected by the degree of cross-linking when the water content in the gel is constant.

4.10.4.2.3 Poly(γ-methyl L-glutamate) Gel

The properties of the polymer gels swollen in organic solvents are different from those of hydro-swollen gels. Cross-linked poly(γ-methyl L-glutamate) (PMLG) is prepared by the ester–amide exchange reaction using diaminododecane as a cross-

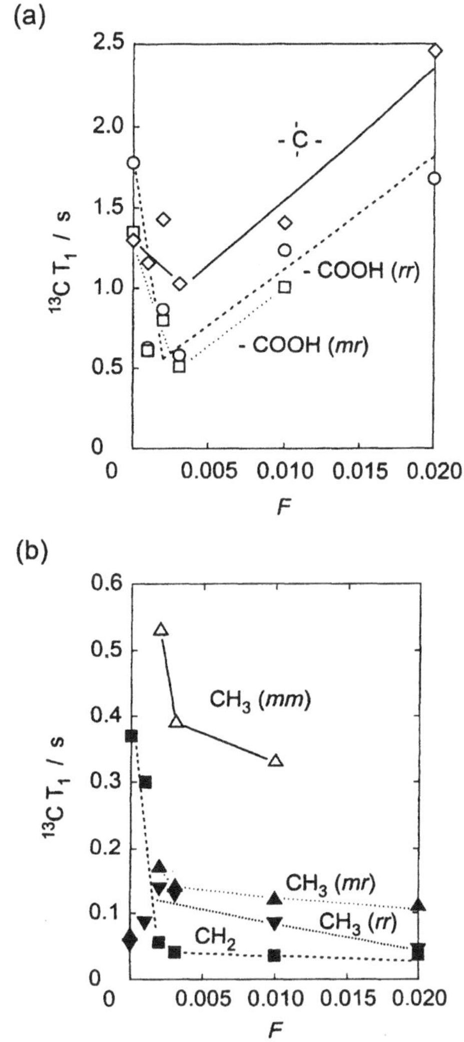

Figure 4.128 Plots of ^{13}C T_1 for C=O (rr), C=O (mr), and –C– carbons (a) and CH_3 (mm), CH_3 (mr), CH_3 (rr), and CH_2 carbons (b) of the PMAA gel ($q = 3.4$) against F at 27 °C.

linker and is swollen in trifluoroacetic acid (TFA) and $CHCl_3$. The PMLG gel was studied by means of high-resolution solid-state ^{13}C NMR spectroscopy [277].

A typical ^{13}C PST/MAS NMR spectrum of PMLG gel swollen in $CHCl_3$ is shown in Figure 4.129. The amide and ester carbonyl carbons of PMLG appear at 176.2 and 173.5 ppm, respectively. The C_α, OCH_3, C_γ, and C_β carbons appear at

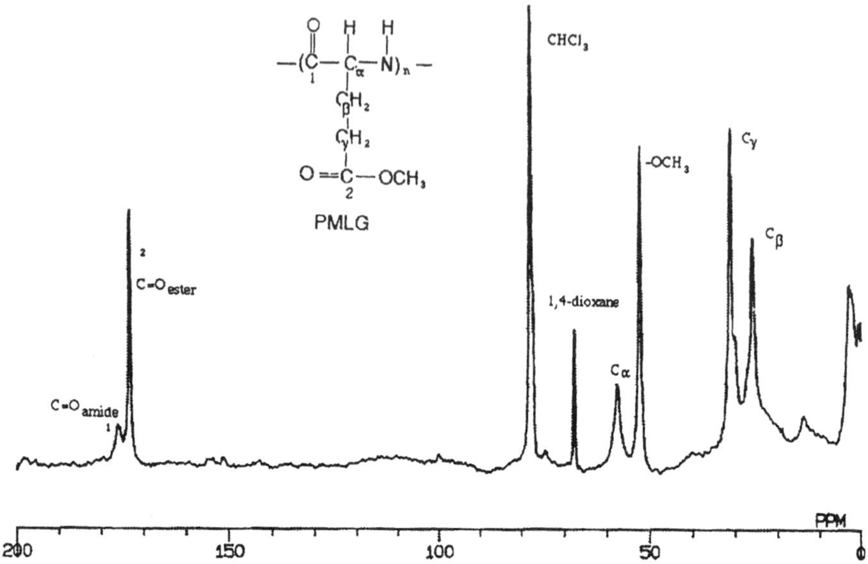

Figure 4.129 ^{13}C PST/MAS NMR spectrum of PMLG gel swollen in $CHCl_3$.

57.6, 52.3, 31.3, and 26.1 ppm, respectively. The peak intensity of the side-chain carbons is more intense than that of the main-chain carbons. This means that the side chains are undergoing faster molecular motion compared with the main chain. The ^{13}C signal of $CHCl_3$ appears at 77.5 ppm. The ^{13}C signal of 1,4-dioxane used in the preparation of the gel appears at 67.8 ppm. The ^{13}C chemical shift values of the amide C=O and C_α carbons are very close to those of solid α-helical polypeptides (176 and 54 ppm, respectively). From these experimental data, it can be concluded that the main chain of the PMLG gel takes the α-helix form.

^{13}C PST/MAS NMR spectra of PMLG gel are shown in Figure 4.130 as a function of solvent composition in a mixture of $CHCl_3$ and TFA at room temperature. The ^{13}C chemical shifts of the individual carbons of PMLG gel are as determined from the TFA content (f_{TFA}) in a mixture of $CHCl_3$ and TFA in Figure 4.131. In the PMLG gel with $CHCl_3$, the ^{13}C chemical shift values of the main-chain amide C=O and C_α carbons are 176.2 and 57.6 ppm, respectively. These are very close to those of α-helical polypeptides in the solid state, which are about 176 and 57 ppm, respectively [173]. The PMLG chain part in the cross-linked network takes the α-helix form. In Figure 4.37, the ^{13}C chemical shifts of all of the main-chain and side-chain carbons are plotted against the TFA content in a mixture of $CHCl_3$ and TFA. The amide C=O and C_α carbon signals appear at 176.5 and 57.6 ppm, respectively, in $CHCl_3$ (Figure 4.130). The PMLG network takes the α-helix conformation as determined by reference data for the solid polypeptide [173]. The amide

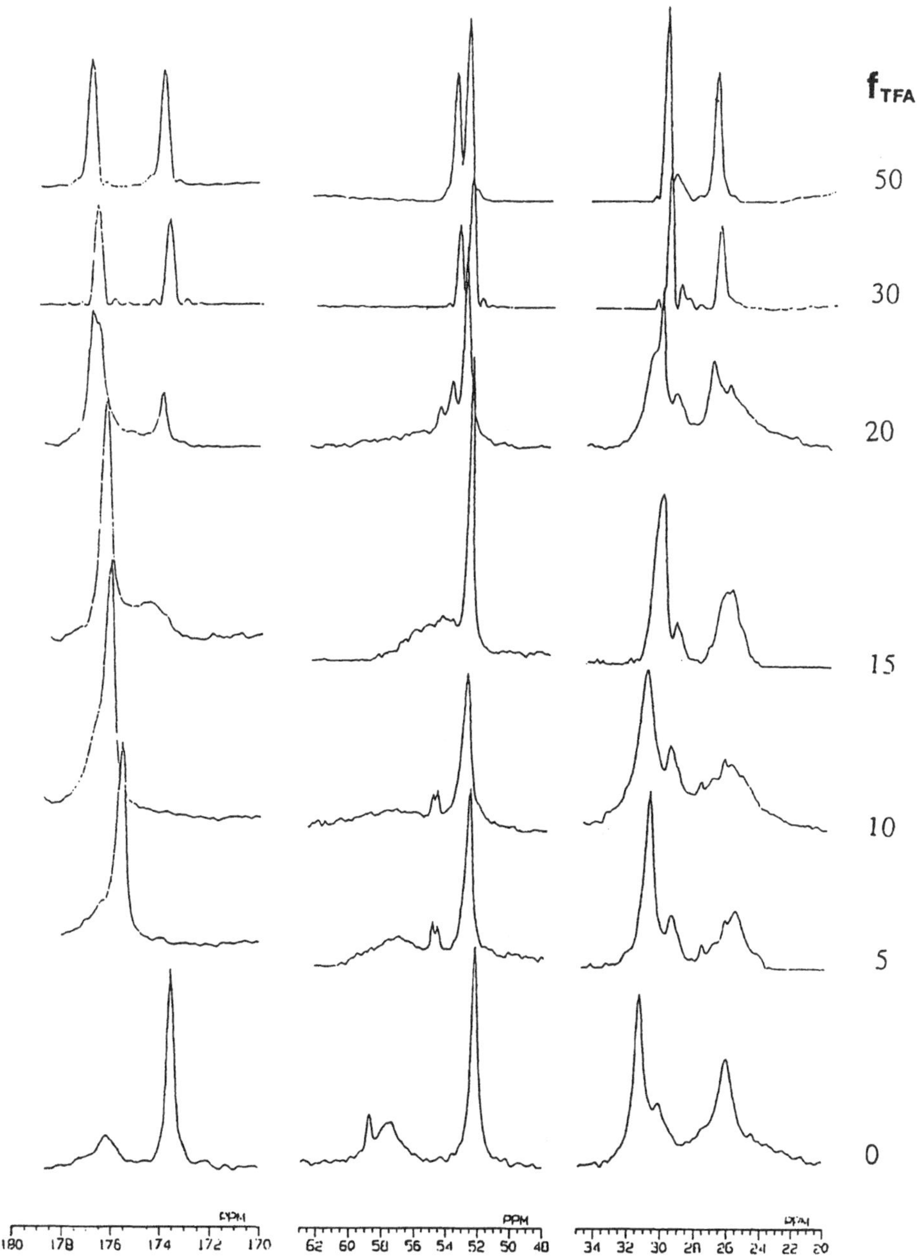

Figure 4.130 ^{13}C PST/MAS NMR spectra of PMLG gel in various TFA contents in a mixture of $CHCl_3$ and TFA.

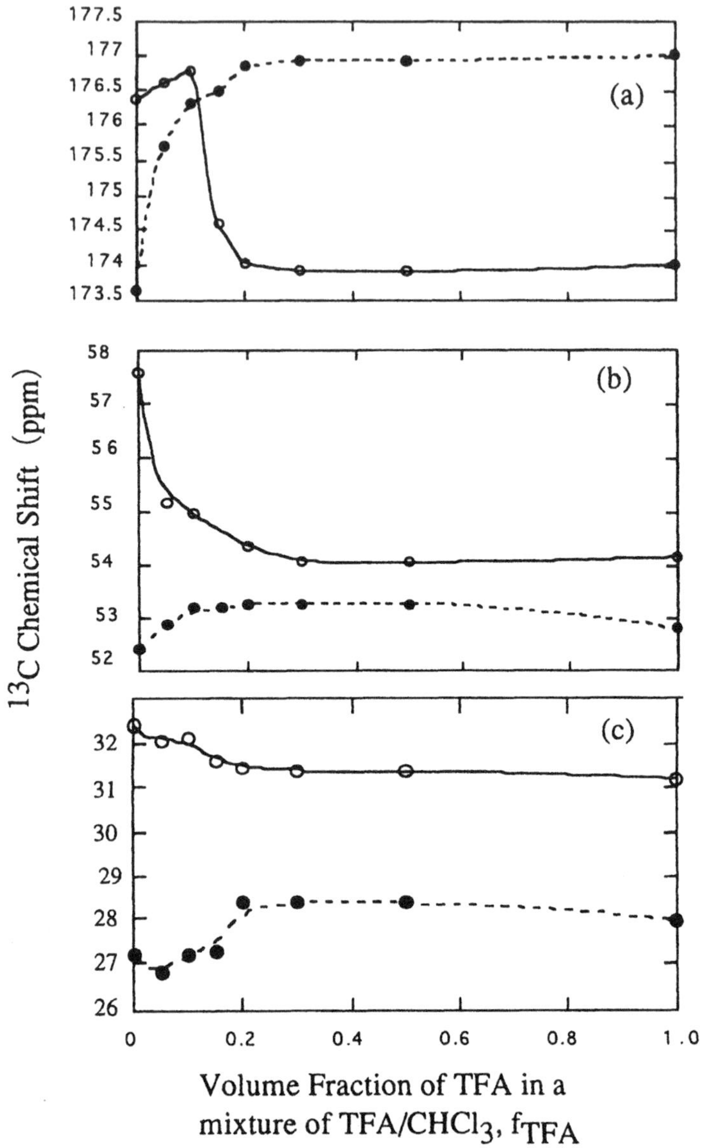

Figure 4.131 ^{13}C chemical shifts for PMLG as a function of the TFA content in a mixture of $CHCl_3$ and TFA: (a) amide carbonyl carbon (○) and ester carbonyl (●), (b) C_α (○) and $-OCH_3$ (●), and (c) C_γ (○) and $C_\square$ (●).

C=O and C_α ^{13}C chemical shifts change transitionally from 176.5 and 57.6 ppm to 174.0 and 54.4 ppm, respectively, and thus move upfield with an increase in TFA content. Such transitional upfield shift shows that the main-chain conformation of the PMLG network changes from the α-helix to the random coil form. The midpoint of the transition is at about $f_{TFA} = 0.15$. On the other hand, the side-chain carbons of the ester C=O carbon shifts downfield at the transition point, but the chemical shifts of the other carbons do not change clearly. The chemical shift of the main chain arises from the conformational change due to the formation of an intramolecular hydrogen bond between the C=O and N–H in the main chain. The main chain chemical shift is considered to be very sensitive to the helix–coil transition. The large shift difference of the ester carbonyl carbons is due to the interaction with TFA. The ^{13}C chemical shifts of the other side-chain carbons also show small changes at $f_{TFA} = 0.15$.

Figure 4.132 shows the TFA content dependence of the ^{13}C T_1 of the amide C=O and ester C=O carbons at 20 and 40 °C. As the temperature decreases from 40 to 20 °C, the ^{13}C T_1 for the amide C=O carbon increases and that of the ester C=O carbon decreases. The molecular motion of the amide C=O carbon is in the slow-motion region ($\omega\tau_c \gg 1$, where ω is the resonance frequency) and that of the ester C=O carbon is in the fast-motion region ($\omega\tau_c \ll 1$) according to the BPP theory. The increase in ^{13}C T_1 of the amide C=O carbon and the decrease in ^{13}C T_1 of the ester C=O carbon with the change in f_{TFA} from 0 to 0.05 show a decrease of the molecular motion of PMLG network. The molecular motion of the PMLG gel is restrained with a rapid decrease of the degree of swelling. In the range of $f_{TFA} = 0.05$ to 0.20, ^{13}C T_1 of the amide C=O carbon decreases with an increase in TFA content, but that of the ester C=O carbon almost does not change. In this range, the fraction of the random-coil form is increased gradually with an increase in f_{TFA} (Figure 4.131), and the degree of swelling is constant [277]. It is reasonable to assume that in the helix-coil transition the conformational exchange occurs. The molecular motion of the main chain increases in going from the α-helix to the random coil in the constant degree of swelling. On the other hand, the molecular motion of the side chains is not affected by the helix-coil transition.

4.10.4.3 Approaches by NMR Imaging

NMR imaging has been widely used for obtaining information on the spatial distribution of the spin density, the relaxation times and the diffusion coefficient in living systems containing mobile components such as water [278, 279, 280]. The NMR imaging technique has been demonstrated to be a powerful means for

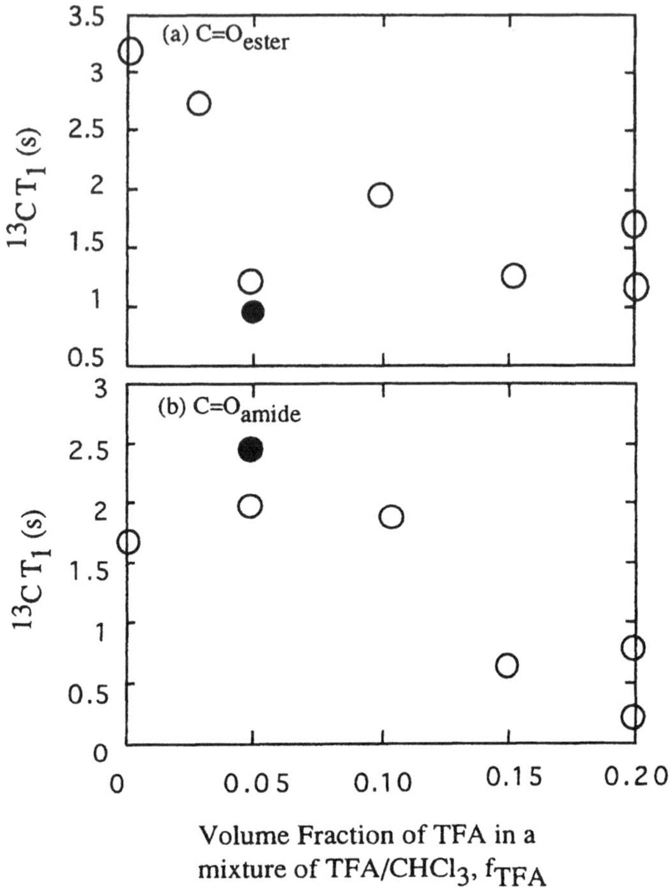

Figure 4.132 ^{13}C T_1 of the ester carbonyl (a) and amide carbonyl carbon (b) of PMLG gel as a function of the TFA content in a mixture of $CHCl_3$ and TFA measured at 20 °C (●) and 40 °C (○). In the gel with $f_{TFA} = 20\%$, two T_1 components are observed.

characterizing the spatial and macroscopic structure and dynamics of polymer samples [281, 282]. This technique may break new ground in characterizing and understanding polymer gels.

4.10.4.3.1 Diffusion in Gels

Polymer gels are microscopically liquid, and mobile components of them such as solvents, counter ions, dissolved molecules, and network polymers migrate in them. It is important to study the diffusion process in gels to understand their dynamics such as structural changes, mobility, and the interaction between molecules and binding phenomena. The diffusion coefficients of mobile components in the gel samples are studied by PGSE techniques. The gels, unlike solution samples, are heterogeneous systems microscopically, macroscopically, or both, and their chemical and physical properties are spatially distributed. The diffusion coefficient of mobile components may be distributed in the gel. It is useful to study the spatial distribution of diffusion coefficients in the gel by the NMR imaging. In this section, the diffusion process in the gel samples studied by NMR imaging is described.

The distribution of the self-diffusion coefficient of methanol and acetone imbibed in cross-linked PMMA was measured [283, 284]. Dry PMMA rods were immersed in methanol or acetone at 30 °C, and images of the solvents absorbed in the PMMA were acquired with the STEAM method. The self-diffusion coefficient of methanol in the swollen PMMA immersed for 20 days was calculated from the images obtained with changing the pulsed-field gradient. The distribution of the self-diffusion coefficient of methanol in the sample was visualized as a 2-D image, and the coefficient values were obtained from the image. The region within 100 mm of the glassy core of the PMMA exhibits a self-diffusion coefficient of $3.2 \pm 0.9 \times 10^{-7}$ cm^2 s^{-1}, and the outer region of the swollen gel exhibits one of $9.2 \pm 0.9 \times 10^{-7}$ cm^2 s^{-1}. The self-diffusion coefficient of methanol measured by the tracer technique at 30 °C is 2.5×10^{-5} cm^2 s^{-1} [285]. The coefficient values for the methanol imbibed in the PMMA network polymer is smaller than that for the free state, and the motion is restrained strongly. The STEAM method is suitable for measuring a self-diffusion coefficient of such small values, which reveal that the self-diffusion of methanol in the swollen PMMA shows a Case II-type character [286–287]. The Case II diffusion is characterized by a sharp diffusion front that is the interface between the swollen and the dry region and the constant solvent concentration in the swollen region. However, the results from the NMR imaging showed that self-diffusion coefficient of methanol is smaller near the glassy core of the PMMA and is larger at the outer region of the swollen gel. The findings disclosed that the decreases in solvent self-diffusion coefficients result from the influence of the polymer motions on the solvent mobility.

4.10.4.3.2 Stress-Strain in Gels

Chemical reactions are caused in polymer gels by the mechanical energy applied. It was reported that ionization of the carboxylic groups of a water-swollen PMAA gel [288] and a PAA gel [289] occurs upon application of stress. These processes involve energy conversion from a mechanical to a chemical form [291]. Such interesting phenomena may come from the viscoelastic characteristics of the gels that constitute an elastic network polymer and liquid solvent. It is important to study the effect of the stress applied on a gel to understand the characteristics of it.

The spatial distribution of the ^{1}H spin density and the ^{1}H T_2 of water molecules in cross-linked water-swollen PMAA gel with the application of stress were studied by ^{1}H–NMR imaging [290]. A cylindrical PMAA gel sample with the diameter of 15.0 mm and a height of 5.3 mm at the degree of swelling, $q = 22$, was placed in a cylindrical sealed glass cell. The stress on the gel sample was applied with a glass piston, the top of which was circular with a diameter of 7.7 mm.

Throughout the experiments, no ^{1}H signals of the network polymer of the PMAA gel swollen with D_2O ($q = 22$) were detectable by spin-warp imaging owing to its short T_2 and great dipolar broadening. On the other hand, ^{1}H signals of water contained in the gel were successfully observed because of its fast molecular motion.

Figure 4.133 shows the observed transverse ^{1}H spin density image of water in the PMAA gel at $q = 22$ without stress (a) and with a stress of 4.8 kPa (b) and (c) where the direction of the cross-sectional profile is perpendicular to the direction of the stress applied. Figure 4.133(c) is the image sliced away from the top surface of the gel compared with (b). The magnitude of ^{1}H spin density is differentiated by 256 steps, and then the observed ^{1}H spin density image is represented by colors from white, representing the highest density, to dark blue, representing the lowest density. The intensity scale indicated by colors is also shown in Figure 4.133.

The ^{1}H spin density distribution of the obtained image of Figure 4.133(a) is approximately homogeneous. This shows that the water molecules that undergo fast molecular motion are homogeneously distributed in the gel. The blue circular portion in Figure 4.133(b) shows the compressed region to which stress is applied. As seen from Figure 4.133(c), there are white regions at the top and the bottom of the image, and colored half-circular stripes appear toward the compressed region as white → light pink → pink → red → orange → bright orange → yellow. The yellow region appears in the compressed region, and the blue one appears at the periphery of the gel. Further, the ^{1}H spin density pattern, as shown in Figure 4.133(c) is asymmetric, and the ^{1}H spin density of water in the right part of the image (the white region) is higher than that in the left one (the yellow and green regions). The application of the stress away from the center of the gel induces such an asymmetric pattern.

The ^{1}H spin density gradient in the gel is induced by stress, and the density of water molecules with higher molecular motion in the compressed region is lower than that in the uncompressed region. On the other hand, the population of water molecules is higher in the white region, and the uncompressed part of the

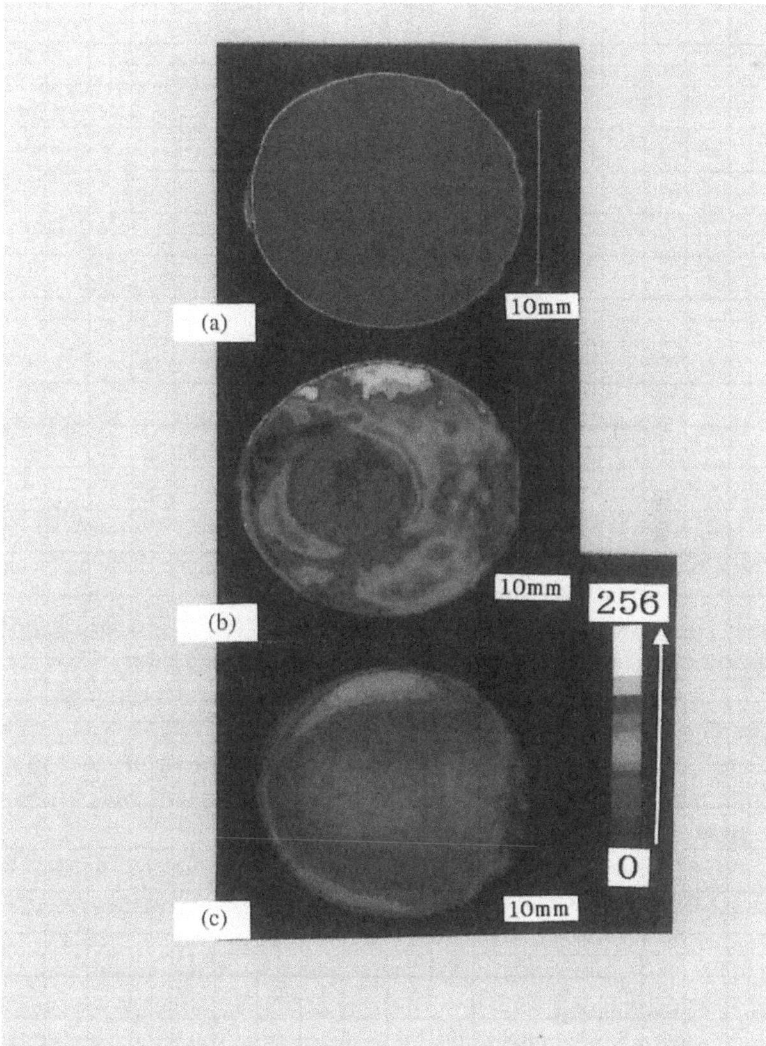

Figure 4.133 Observed transverse ^1H spin-density image for water in the PMAA gel ($q = 22$): (a) without stress at a depth of 1.9 mm from the top surface, (b) with stress of 4.8 kPa at depth of 1.0 mm, and (c) with stress of 4.8 kPa at depth of 1.9 mm. The magnitude of the ^1H spin-density is differentiated by the color signal variation.

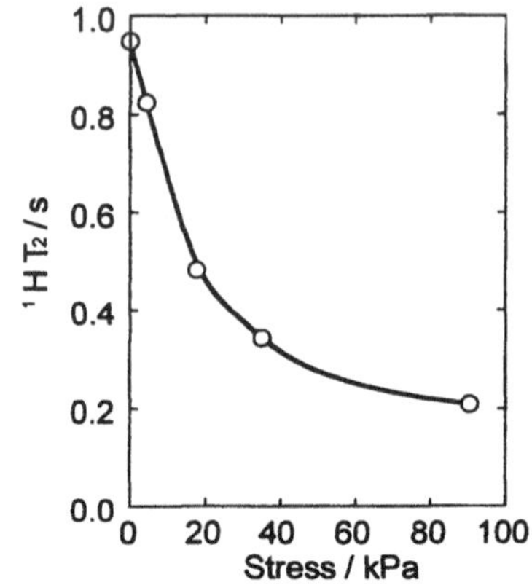

Figure 4.134 Dependence of the ^{1}H T_2 for water in the PMAA gel ($q = 89$) on the applied stress. The ^{1}H T_2 was measured by the CPMG method.

top surface of the PMAA gel rose up slightly owing to deformation of the gel by the stress. This shows that water molecules excluded from the compressed region migrate to the surrounding region.

The ^{1}H T_2 for water in the PMAA gel ($q = 89$) compressed homogeneously was measured by means of 500-MHz ^{1}H-pulse NMR using the CPMG method [39, 40]. The obtained ^{1}H T_2 values are plotted against the strength of the applied stress, as shown in Figure 4.134. This shows that the T_2 decreases with increasing stress, and the T_2 value under a stress of 17.6 kPa is half the value under no stress. The molecular motion of water in the PMAA gel is restricted by the strain induced by the stress.

Figures 4.135(a) to (h) show coronal ^{1}H T_2-enhanced images of water in the PMAA gel at $q = 22$. Here the word *coronal* means that the cross-sectional profile is parallel to the direction of the applied stress. In these images, the relative magnitude of the ^{1}H T_2 value of the water is represented by colors. White indicates the longest T_2 value, and dark blue shows the shortest one according to the color scale shown in Figure 4.135. Although the ^{1}H T_2 distribution was almost homogeneous in the gel under no stress, a gradient pattern in the T_2 distribution was obtained in the gel under a stress of 4.8 kPa when the stress was applied away from the center of the gel. The T_2 value of the top surface is short, as indicated by dark blue or blue. The color in the compressed region changes from the top of the gel toward the bottom in the sequence dark blue → blue → green → light green → yellowish green → yellow → bright orange → orange → red → orange → bright orange → yellow → blue → dark blue. This indicates that the ^{1}H T_2 value of water in the compressed region increases from the top surface of the gel toward the inner part and decreases from the inner part toward the bottom.

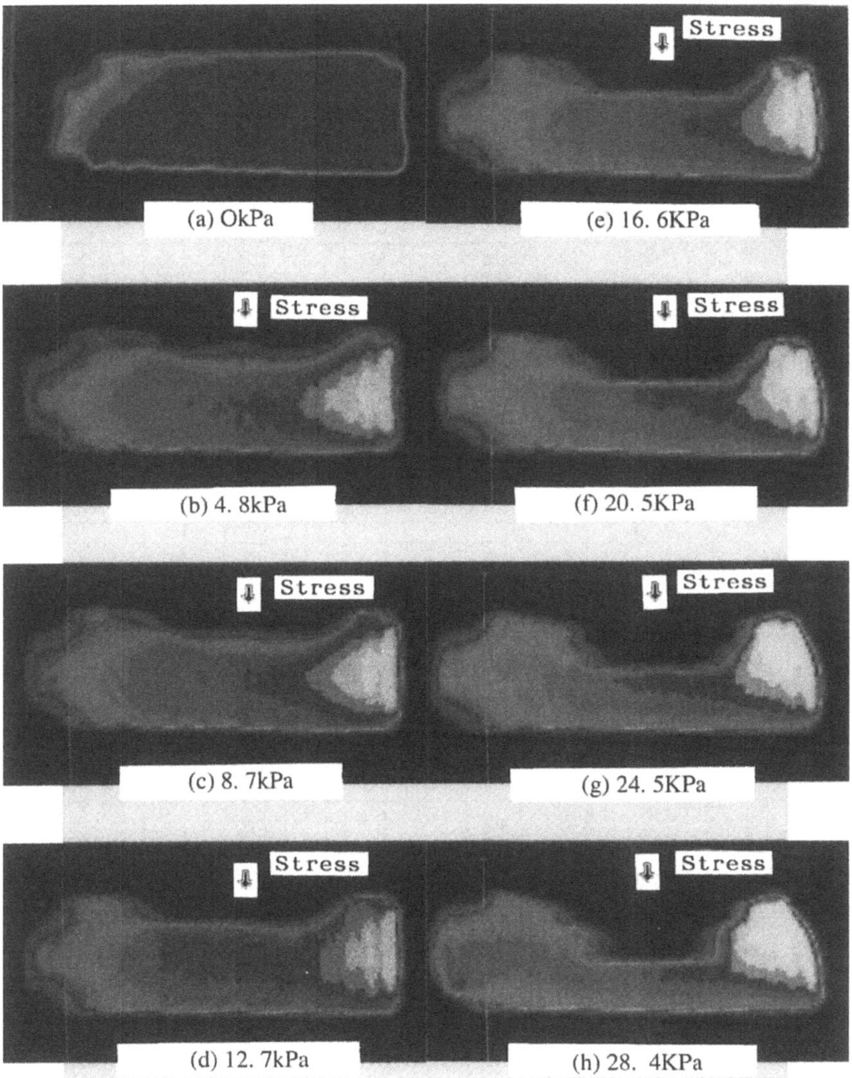

Figure 4.135 Observed coronal ^{1}H T_2-enhanced image for the PMAA gel ($q = 22$) as a function of stress.

The ^{1}H T_2 value increases from the left end of the gel to the right end as indicated by dark blue → blue → green → light green → yellowish green → yellow → bright orange → orange → red → pink → light pink → white sequence. As seen from the results, the propagation of the strain in the gel can be observed as the T_2 distribution.

The gradient pattern of spatial distribution on T_2 changes with increasing stress, as shown in Figure 4.135(a) through (h). The red and orange regions in the compressed region become narrow with increasing stress, and the bright orange and yellow regions become wide. The white region becomes wide with an increase in the strength of the stress from 12.7 to 28.4 kPa. The molecular motion of water molecules in the compressed region is more strongly restrained with an increase in stress. Further, in the uncompressed region (the right- and left-end parts of the gel), the population of water molecules that undergo relatively fast molecular motion increases with increasing stress. This shows that the migration of water in the gel is increased with an increase of stress, and the difference in molecular motion of water in between the compressed and the uncompressed regions becomes large.

The imaging study revealed that the spatial distribution of density and molecular motion of water in the PMAA gel is changed by the mechanical stress applied. This result demonstrates that the NMR imaging technique is useful for studying the response of the gels to stimulation spatially.

4.10.4.3.3 Shrinkage of PMAA Gels Under DC Electric Field

Water-swollen, cross-linked polyelectrolyte gels deform by the application of an electric field [292]. From the first observation of electric–field-induced deformation for a polyelectrolyte gel [293], the phenomena termed chemomechanical reactions were studied from the fundamental or applicational point of view by many researchers [292, 293]. Although a great deal of effort has been made in studying the mechanism, it seems that a fundamental approach at the molecular level has rarely been taken for the gel. Therefore, it is important to clarify the structure and dynamics of small molecules, such as water in the gels, through microscopic observation at the molecular level.

The spatial distribution of ^{1}H spin density and the ^{1}H T_2 of water molecules in a PMAA gel contracted under an electric field was studied to clarify the behavior of water molecules in the gel during the process of shrinkage [294, 295]. The application of an electric field to the PMAA gel leads to its shrinkage accompanied by exhaustion of water. Shrinkage of the gel close to the positive electrode is larger than that at the negative one. The edge of the gel in contact with the negative electrode is swollen with time. Figure 4.136 shows the elapsed-time (T_e) dependence for the shape and the spatial distribution of ^{1}H spin density of the PMAA gel (diameter of 8.0 mm and length of 25.0 mm) with and without the application of an electric field.

The ^{1}H spin-density images showing the gradient at $T_e = 0$ represent the distribution gradient of the water in both gels at first. The density of water molecules

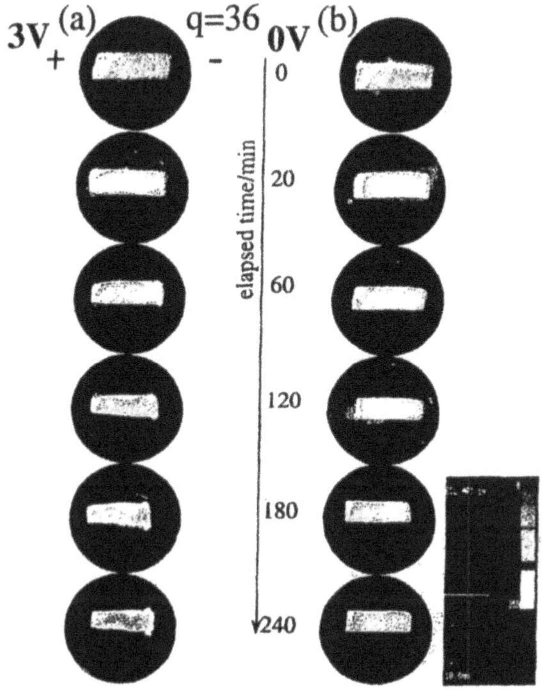

Figure 4.136 Elapsed-time dependence of images for the ^{1}H spin-density distribution of the PMAA gel ($q = 36$) with (a) and without (b) the application of an electric field (3-VDC).

increases in going from the right end of the gels to the left, as indicated by the sequence blue → green → light green → yellowish green → yellow → bright orange → orange → red → white. The region occupied by yellow and red in the gel becomes large from left to right with time under an electric field of 3 V.

At T_e = 20 min, the yellow region in the images of both gels increases. The change in the magnitude of ^{1}H spin density in going from the positive electrode to the negative electrode is indicated by white → red → orange → yellow → white. At T_e = 240 min, the red region in the center area of the gel with the 3-VDC electric field applied becomes large. The ^{1}H spin density changes in going from the positive electrode to the negative electrode as indicated by the sequence white → red → orange → white. To analyze clearly the ^{1}H spin density in detail, the images for the ^{1}H spin density distributions of the PMAA gel with and without the application of 3-VDC of an electric field are represented by a three-dimensional profile, as shown in Figure 4.137. Although water contained in the gel without the application of an electric field decreases homogeneously in the gel with time, water

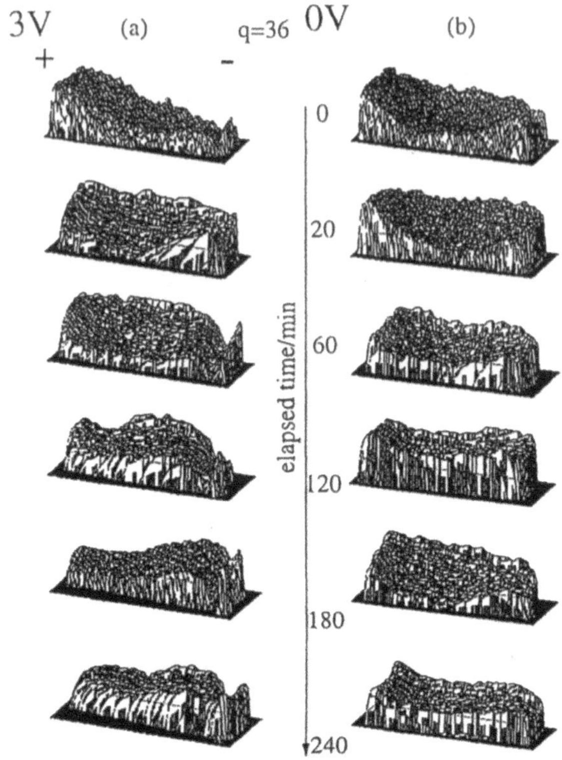

Figure 4.137 Three-dimensional histograms for the ^1H spin-density distribution of the PMAA gel with (a) and without (b) the application of an electric field.

in the gel under 3-VDC of electric field decreases and migrates from the positive side to the negative. These results prove the hypothesis that water molecules in a polyelectrolyte gel migrate from the positive side to a negative one with protonated water molecules under a DC electric field on the basis of electro-osmosis [296]. The water migration is accompanied by the shrinkage of the gel from the positive side and swelling at the edge of the gel near the negative side.

It is important to have information on the ^1H T_2 value of water molecules in the PMAA gel to analyze the dynamical behavior of water molecules under an electric field. The T_2 distributions for the gel with and without the application of an electric field were measured. The T_2 image experiment on the gel without the application of an electric field becomes reference data to clarify the process of gel shrinkage with the application of an electric field. Figure 4.138 shows the ^1H T_2 distribution images for the transverse slice of the PMAA gel under 3-VDC and

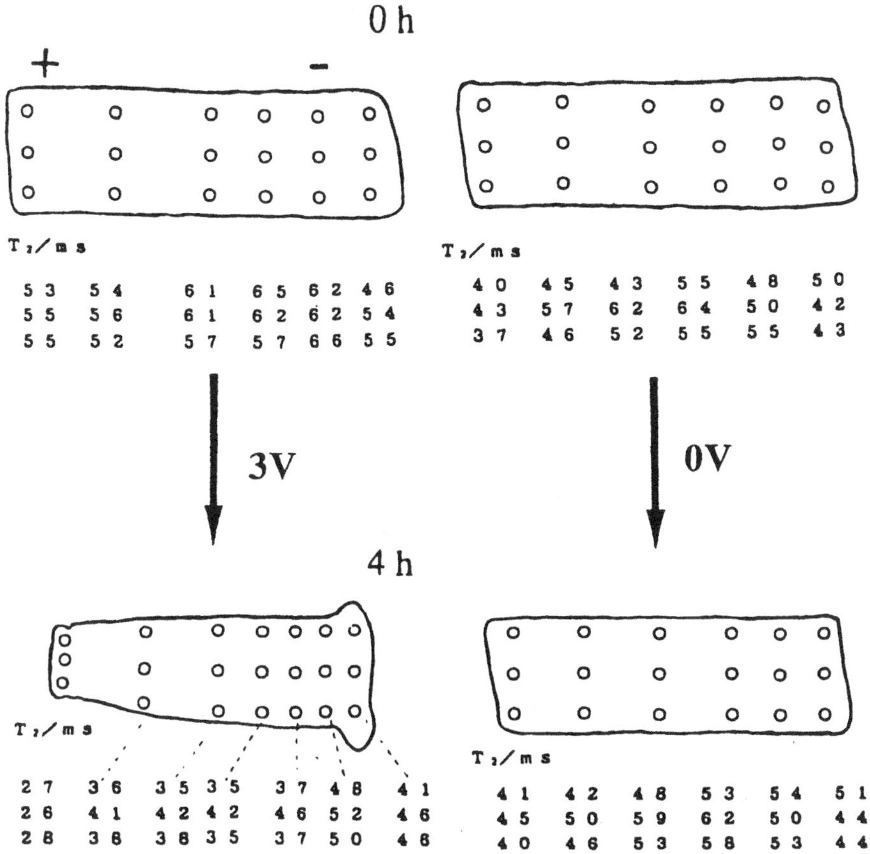

Figure 4.138 ^{1}H T_2 distribution for transverse slice of the PMAA gel ($q = 36$) at $T_e = 0$ h and at 4 h with and without the application of an electric field (3-VDC).

under 0-VDC at elapsed time $T_e = 0$ h (initial state) and 4 h after the application of an electric field; shown are T_2 values for the part indicated by an open circle in the gel. Little change of the T_2 value of water molecules in the gel (0 V) is observed during the image experiment, whereas significant change in the ^{1}H T_2 value of water molecules in the gel (3 V) was observed during the experiment. The ^{1}H T_2 values of water molecules around the positive side are decreased to a larger extent than those around the negative one by the application of an electric field. This means that the molecular motion of water molecules in the gel near the positive electrode was strongly restrained. The elapsed-time dependence of the ^{1}H T_2 distribution change in the gel under the application of an electric field was measured, as shown in Figure 4.139. This figure shows that the molecular motion

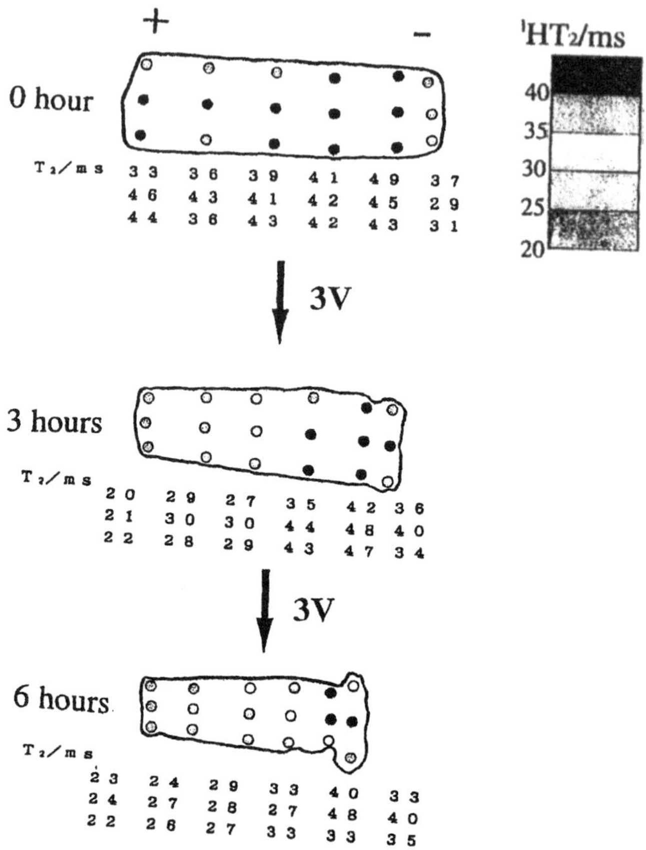

Figure 4.139 ^{1}H T_2 distribution for transverse slice of the PMAA gel at T_e = 0, 3, and 6 h with the application of an electric field (3-VDC).

of water molecules in the gel is restrained with elapsed time, as seen from the decrease of the ^{1}H T_2 value. From these results, the profile of the T_2 distribution image can be divided into four regions in a manner similar to the ^{1}H spin-density distribution image. For convenience, a rough schematic representation for the ^{1}H T_2 distribution image is represented in Figure 4.140. Water molecules in the region with large shrinkage in the gel show low mobility, but at region 3 in Figure 4.140, the mobility of water molecules is high although the density of the network is high. These results are interesting from the viewpoint of the structure and dynamics of the gel and demonstrate that the network polymer in the gel under an electric field may form a complicated structure for which interaction with water molecules is different from that of a normal linear polymer.

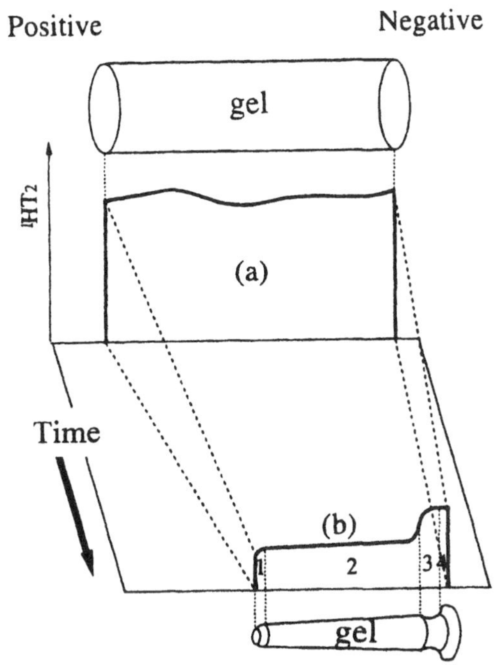

Figure 4.140 Schematic diagram for change in the ^{1}H T_2 distribution image in the PMAA gel at (a) $T_e = 0$ h and (b) $T_e = 6$ h.

It should be also interesting to study the shrinking process of a polyelectrolyte gel that has macroscopically inhomogeneous structure under an electric field. Such a hydro-polymer gel is a good model for investigating the response of a human body to an electric stimulus, because the body is an inhomogeneous hydro-swollen gel.

The shrinking process of a composite hydro-swollen PMAA gel that is a double cylindrical rod under an electric field is shown in Figure 4.141 [295]. The degree of swelling of the outside and inside parts of the gel are different. The amount of cross-linking reagent used during polymerization of the inside network of the gel was different from that of outside one.

The observed ^{1}H spin-density images of the PMAA gel (q for the inside part is 26, and for the outside 28) as a function of elapsed time T_e after the application of an electric field (3-VDC) are shown in Figure 4.142. The q for the outside part of the gel is larger than that for the inside one. At $T_e = 0$ min, most of the inside gel region is represented by yellow, and the outside gel region is represented by red. This indicates that the ^{1}H spin-density of the outside gel is higher than that

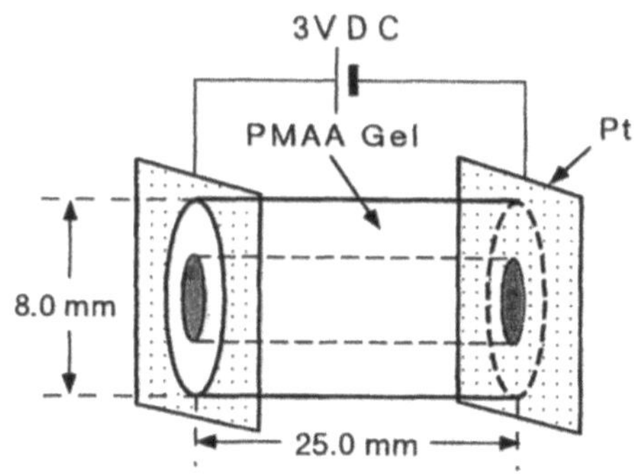

Figure 4.141 Structural scheme of the composite PMAA gel consisting of the inside part (shadowed part) and the outside part (light part). The degrees of swelling of the inside and outside gels are different from each other.

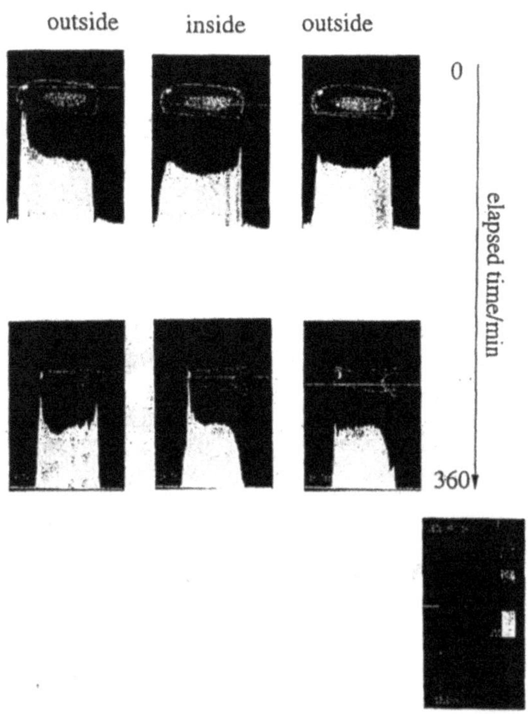

Figure 4.142 Elapsed-time dependence of images for the ^1H spin-density distribution and two-dimensional intensity histogram of the composite PMAA gel with the application of an electric field (3-VDC). The q values of the inside and outside gels are 26 and 28, respectively.

of the inside gel. The experimental result is consistent with degrees of swelling of the inside and outside gels of 26 and 28, respectively. Most of the inside and outside regions are represented by red except for the gel parts in contact with the negative and positive electrodes at $T_e = 360$. The ^{1}H spin-density distributions for the inside and outside of the gel, positions of which are indicated by a white line on the image, are also indicated by a two-dimensional histogram in Figure 4.142. The shrinking rate of the inside gel, of which q is smaller than that of the outside gel, is faster compared with that of the outside gel. This result is very interesting because it is discrepant from the results that the shrinking rate of the gel under an electric field increases with an increase in q [295]. The inhomogeneous structure of the composite PMAA gel may give various effects to the migration of water and ions, the distribution of electric current, and the motion of the network polymer in the gel. It can be predicted easily that the interface between the two kinds of networks in the composite gel also plays an important role in the phenomenon.

The schematic diagrams for the ^{1}H spin-density distribution of the gel undergoing the shrinking process are shown in Figure 4.143(a). The trend of the shrinking process and the ^{1}H spin-density distribution for a homogeneous PMAA gel by the application of an electric field, as described above, are similar to that of the outside gel rather than the inside gel. However, the whole shape of the shrunken composite gel is very different from that of the homogeneous gel that shrinks to form a wine-stopper-like shape.

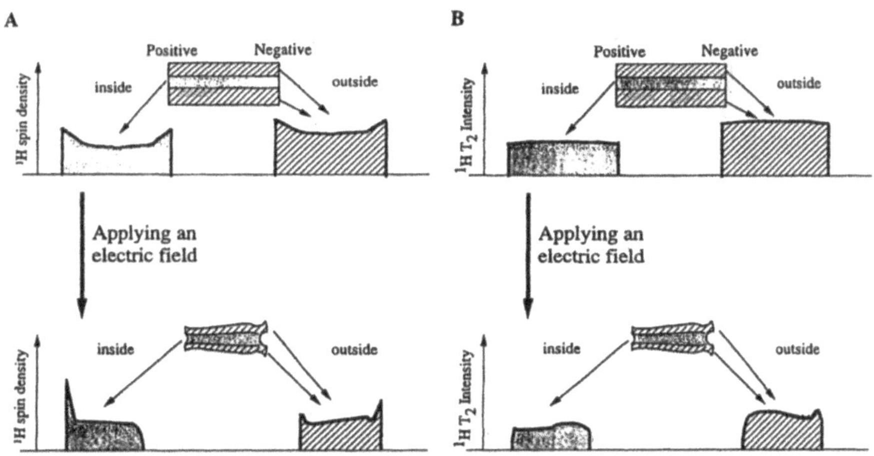

Figure 4.143 Schematic diagrams of the shrinking process of the composite PMAA gel by the application of an electric field as viewed by the ^{1}H spin-density images (a) and the ^{1}H T_2-weighted images (b).

The three-dimensional histograms of elapsed-time dependence of the ^{1}H T_2-weighted image of water in the gel after the application of an electric field are shown in Figure 4.144 with the three-dimensional histograms of ^{1}H spin-density in the gel. The corresponding schematic diagrams are presented in Figure 4.143(b). Before the application of an electric field, the ^{1}H T_2-weighted intensity is almost constant in the whole region of the gel. This means that the mobility of water in the whole region of the inside and outside of the gel is almost the same. However, at a long-elapsed time after the application of an electric field, the ^{1}H T_2-weighted intensity in the region around the negative electrode in the inside and outside gels is somewhat higher than that around the positive electrode. The molecular motion of water in the gel around the negative electrode is faster than that in the other region. The trend is very similar to that for the case of a homogeneous PMAA gel discussed previously. The shrinkage process of a composite gel for which the q values of the inside and outside gels are 28 and 26, respectively, was also studied as the contrasting case with the former one [295].

4.10.4.3.4 Shrinkage of PMAA Gels Containing Mn^{2+}

In order to elucidate the spatial distribution of paramagnetic Mn^{2+} ions in a PMAA gel swollen in an aqueous manganese(II) sulfate solution with the application of a 3-VDC electric field, the images of the spatial distribution of the ^{1}H T_2 value were measured by the ^{1}H–NMR imaging method [297].

The logarithm of the ^{1}H T_2 value of water decreases linearly with an increase in the concentration of paramagnetic Mn^{2+} ion. From the ^{1}H pulse NMR experiments on the gel containing Mn^{2+} ions, the relationship between the ^{1}H T_2 value of water and the concentration of Mn^{2+} ions (C moll^{-1}) in the gel was obtained, as shown in Figure 4.145 and is expressed by

$$T_2(^1\mathrm{H}) = 7.24\ C^{-0.163} \tag{138}$$

This means that the concentration of Mn^{2+} ions in the gel can be determined through the observation of the ^{1}H T_2 value of water in the gel.

The application of an electric field to a PMAA gel leads to shrinkage of the gel with exhaustion of water. Shrinkage around the positive electrode is larger than that around the negative electrode. The gel part in contact with the negative electrode is swollen.

The ^{1}H spin-density images for a transverse slice of a PMAA gel swollen in manganese(II) sulfate solution measured as a function of the elapsed time T_e under a 3-VDC electric field are shown in Figure 4.146 [298]. The ^{1}H spin-density image experiments for PMAA gel (3 V) are represented in Figure 4.146(a). At elapsed

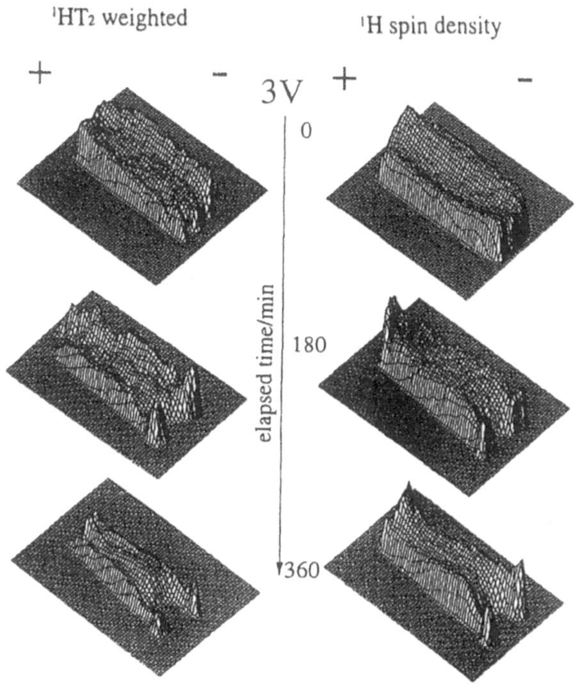

Figure 4.144 Three-dimensional histograms of elapsed-time dependence for the ^{1}H T_2-weighted images and the ^{1}H spin-density images of the composite PMAA gel with the application of an electric field.

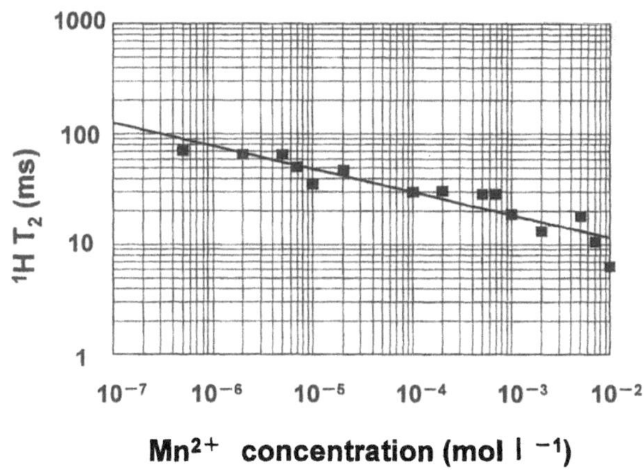

Figure 4.145 The dependence of ^{1}H T_2 of water in the PMAA gel on Mn^{2+} concentration.

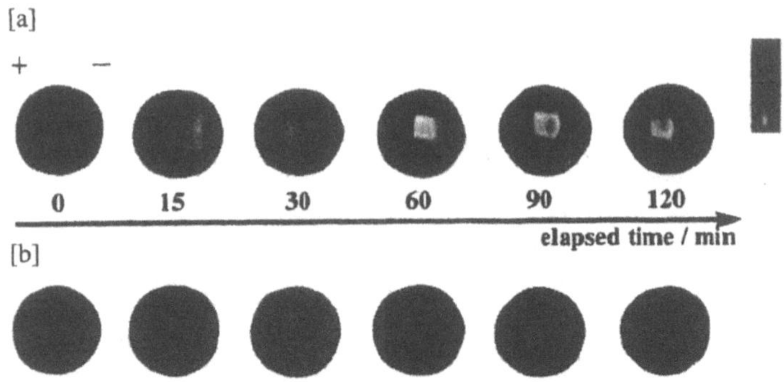

Figure 4.146 Elapsed-time dependence of image for the ^1H spin-density of PMAA gel swollen by an aqueous manganese (II) sulfate solution with (a) and without (b) the application of an electric field.

time $T_e = 0$ min, most of the gel is occupied by the blue region. At $T_e = 15$ min, the magnitude of ^1H spin-density increases in going from the positive electrode to the negative electrode, as indicated by the color sequence blue → green → blue → violet. At $T_e = 30$ min, the ^1H spin density changes as indicated by the progression blue → violet → white → violet → blue → green → blue → red. After $T_e = 60$ min, the white region increases. The ^1H spin-density image experiments for PMAA gel (0 V) are represented in Figure 4.146(b). At elapsed time $T_e = 0$ min, most of the gel is occupied by the blue region. For each image of ^1H spin-density from $T_e = 15$ to 120 min, the magnitude of ^1H spin-density is occupied by green and red regions. There is little change in the ^1H spin-density image for PMAA gel (0 V).

It is important to know information on the ^1H T_2 value of water molecules in a PMAA gel to analyze the distribution of paramagnetic Mn^{2+} ions in the gel under an electric field. The ^1H T_2 value of water molecules in a PMAA gel (3 V) is represented in Figure 4.147(a). At $T_e = 0$ min, the ^1H T_2 value of water molecules in a PMAA gel (3 V) distributes homogeneously in the gel. At $T_e = 15$ min, the ^1H T_2 value of water molecules in a PMAA gel (3 V) decreases from the positive electrode to the negative electrode, as indicated by the sequence 32 → 33 → 32 → 27 → 15 → 10 ms. At $T_e = 30$ min, the ^1H T_2 value of water molecules in a PMAA gel (3 V) changes, as indicated by the progression 41 → 44 → 45 → 39 → 27 → 14 ms; at $T_e = 45$ min, the sequence is 45 → 48 → 45 → 39 → 29 → 14 ms; at $T_e = 60$ min, it is 56 → 73 → 67 → 62 → 42 → 21 ms; at $T_e = 90$ min, 71 → 83 → 93 → 65 → 37 → 20 ms, and at $T_e = 120$ min, 55 → 54 → 48 → 36 → 26 → 17 ms.

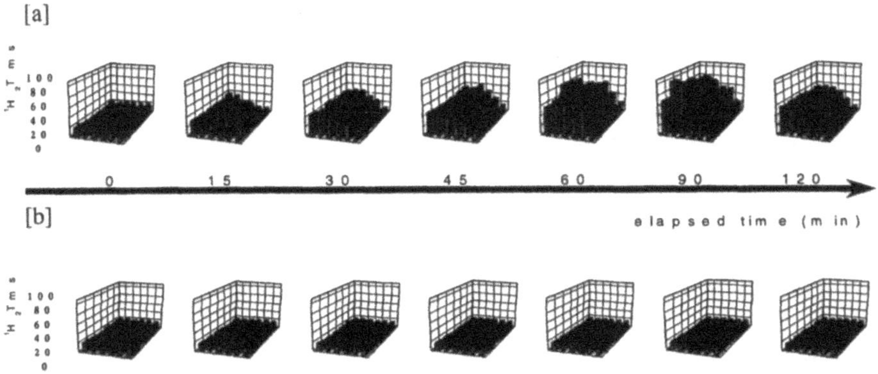

Figure 4.147 Elapsed-time dependence of the ^1H T_2 distribution for a transverse slice of a PMAA gel swollen by an aqueous manganese (II) sulfate solution with (a) and without (b) the application of an electric field.

The ^1H T_2 value of water molecules in the PMAA gel (0 V) is represented in Figure 4.147(b). As seen from this diagram, the ^1H T_2 values of water molecules in a PMAA gel (0 V) are almost about 18 ms without changing from $T_e = 0$ to 120 min. The ^1H T_2 value is influenced by not only the Mn^{2+} ion oncentration but also by the mobility of water molecules in a PMAA gel.

The Mn^{2+} ions in a PMAA gel (3 V) move from the positive electrode side to the negative electrode side. The ^1H T_2 distribution images can be converted to the Mn^{2+} ion distribution images. Their three-dimentional profiles of the spatial distribution of the paramagnetic Mn^{2+} ion concentration in a PMAA gel swollen by an aqueous manganese (II) sulfate solution with (a) and without (b) the application of an electric field are shown in Figure 4.148. At $T_e = 0$ min, the Mn^{2+} ion is distributed homogeneously in the gel. The concentration of the Mn^{2+} ions in the gel is about 3.8×10^{-3} moll^{-1}. As the elapsed time is increased, the Mn^{2+} ions migrate from the positive electrode side to the negative electrode side. The concentration of the Mn^{2+} ions in the gel at the positive electrode side becomes lower compared with that at the negative electrode side. The Mn^{2+} ions exhaust from the negative electrode side of the gel.

The ^1H T_2 of water molecules close to Mn^{2+} ions in the PMAA gel is decreased by the presence of Mn^{2+} ions in a gel. As PMAA gel swollen by an aqueous manganese (II) sulfate solution is applied with an electric field, Mn^{2+} ions in the gel move toward the negative electrode side, and the concentration of Mn^{2+} ions have characteristic spatial distribution. The spatial distribution of the paramagnetic ions in a PMAA gel with application of an electric field is converted by the images

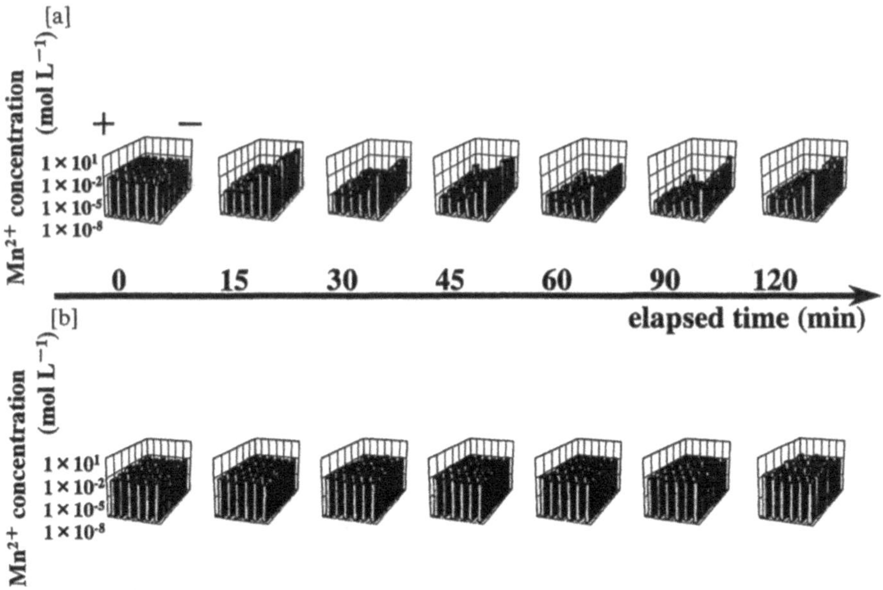

Figure 4.148 Elapsed-time dependence of the spatial distribution of the paramagnetic Mn^{2+} ion concentration in a PMAA gel swollen by an aqueous manganese (II) sulfate solution with (a) and without (b) the application of an electric field.

of the spatial distribution of the ^{1}H T_2 values obtained by the ^{1}H–NMR imaging method.

4.10.4.3.5 Shrinkage of PNIPAM Gel in Response to the Change in Temperature

The PNIPAM gel is a well-known thermo-shrinking gel that collapses at the transition temperature around 33 °C by heating [298, 299]. During the shrinking process, in response to the increase in temperature, PNIPAM gels often form the so-called skin structure in which the inner unshrunken layer is surrounded by the outer shrunken layer [300].

It was shown that two components of ^{1}H T_2 for a PNIPAM gel are observed by heating: a short T_2 component corresponding to the outer shrunken layer and a long T_2 component corresponding to the inner unshrunken layer [301]. It would be interesting to know spatial information about the molecular mobility during the shrinkage process of the gel with skin structure [237]. Figure 4.149 shows a ^{1}H

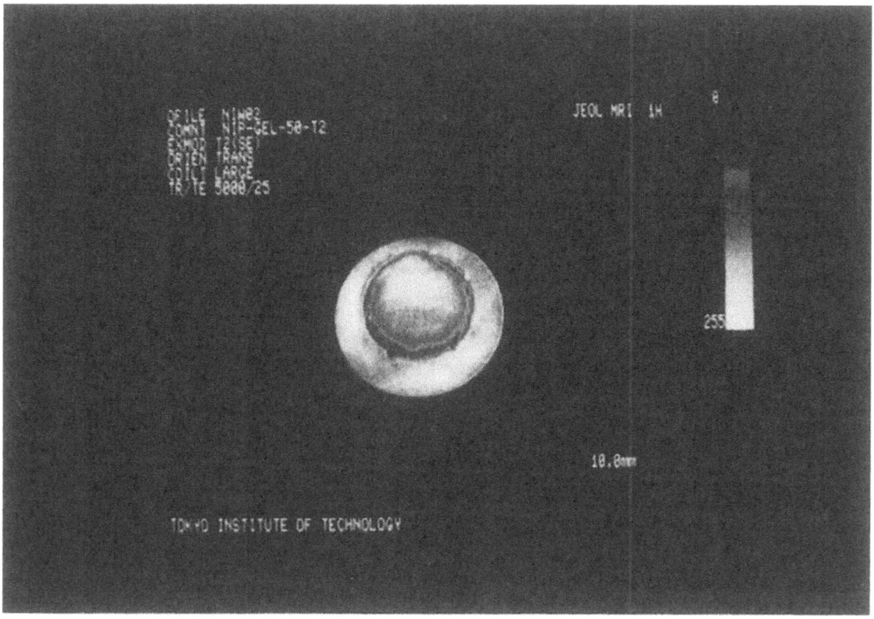

Figure 4.149 Observed ^{1}H T_2-weighted image of the PNIPAM gel ($q = 40$) after rapid temperature change from 20 to 40 °C.

T_2-enhanced image for the gel ($q = 43.6$ at 20 °C) in water after a rapid increase in temperature [301]. The region with short T_2 was observed in the outer layer of the gel. It is clear that the molecular motion in the outer layer is much more restrained than that in the inner layer in which the polymer networks swell.

NMR imaging is a very powerful tool for investigating if such kinds of inhomogeneous reactions occurred in polymer gels, thus giving information on the spatial distribution of molecular motion.

References

1. E. M. Purcell, H. C. Torrey, and R. V. Pound, *Phys. Rev.* **69:** 37 (1946).
2. F. Bloch, W. W. Hansen, and M. Packard, *Phys. Rev.* **69:** 127 (1946).
3. (a) E. R. Andrew, *Nuclear Magnetic Resonance*, Oxford University Press, Oxford, 1954. (b) A. Abragham, *Principles of Nuclear Magnetism*, Oxford, 1961. (c) R. K. Harris, *Nuclear Magnetic Resonance*, Pitman, London, 1983. (d) I. Ando and G. A. Webb, *Theory of NMR Parameters*, Academic Press, London, 1993. (e) M. Mering,

High Resolution NMR in Solids, Springer-Verlag, Berlin, 1985. (f) B. C. Gerstein and C. R. Dybowski, *Transient Techniques in NMR of Solids*, Academic Press, New York, 1985. (g) C. P. Slichter, *Principles of Magnetic Resonance*, Springer-Verlag, Berlin, 1990. (h) E. O. Stejskal and J. D. Memory, *High Resolution NMR in the Solid-State*, Oxford University Press, Oxford, 1994. (i) K. Schmidt-Rohr and H. W. Spiess, *Multidimensional Solid-State NMR and Polymers*, Academic Press, London, 1994.

4. D. M. Grant and R. K. Harris, Eds., *Encyclopedia of NMR*, John Wiley, New York, 1996.
5. (a) F. A. Bovey, *High Resolution NMR of Macromolecules*, Academic Press, New York, 1972. (b) W. M. Pasika, Ed., *Carbon-13 NMR in Polymer Science*, ACS103, American Chemical Society, Washington, D.C., 1979. (c) R. A. Komoroski, Ed., *High Resolution NMR of Synthetic Polymers in Bulk*, VCH Publishers, Berlin, 1986. (d) G. A. Webb and I. Ando, Eds., *Ann. Rept. NMR Spectrosc. (Special Issue: NMR in Polymer Science)*, Vol. 34, Academic Press, London, 1997.
6. I. Ando and T. Asakura, Eds., *Solid-State NMR of Polymers*, Elsevier Science, Amsterdam, 1998.
7. T. C. Farrar and E. D. Becker, *Pulse and Fourier Transform NMR*, Academic Press, New York, 1971.
8. R. R. Ernst and W. A. Anderson, *Rev. Sci. Instrum.* **37:** 93 (1966).
9. (a) P. C. Lauterbur, *Nature* **242:** 190 (1973). (b) P. Mansfield and P. K. Granell, *J. Phys.* **C6:** L422 (1973).
10. (a) M. Karplus, *J. Chem. Phys.* **30:** 11 (1959). (b) M. Karplus, *J. Am. Chem. Soc.* **86:** 5561 (1964).
11. N. Bleombergen, E. M. Purcell, and R. V. Pound, *Phys. Rev.* **73:** 679 (1948).
12. C. Deverell, R. E. Morgen, and J. H. Strange, *Mol. Phys.* **18:** 553 (1970).
13. A. W. Overhauser, *Phys. Rev.* **92:** 411 (1953).
14. R. R. Ernst, G. Bodenhausen, and A. Wokaun, *Principle of Nuclear Magnetic Resonance in One and Two Dimensions*, Clarendon Press, Oxford, 1989.
15. See, for example, (a) I. Ando, in *Encyclopedia of NMR*, John Wiley, New York, p. 176, 1996. (b) T. Yamanobe and H. Kurosu, *Encylopedia of NMR*, John Wiley, New York, p. 4468, 1996.
16. I. Ando, A. Nishioka, and M. Kondo, *J. Magn. Reson.* **21:** 429 (1976).
17. I. Ando, Y. Kato, M. Kondo, and A. Nishioka, *Makromol. Chem.* **178:** 803 (1977).
18. M. Kondo, S. Watanabe, and I. Ando, *Mol. Phys.* **37:** 1521 (1979).
19. I. Ando and G. A. Webb, *Org. Magn. Reson.* **15:** 111 (1981).
20. I. Ando, in *Encyclopedia of NMR*, John Wiley, New York, p. 2512, 1996.
21. M. Witanowski, L. Stefaniak, and G. A. Webb, *Ann. Rept. NMR Spectrosc.* **25:** 42 (1993).
22. M. Arshadi, D. Johnels, U. Edlund, C. H. Ottoss, and D. Cremer, *J. Am. Chem. Soc.* **118:** 5120 (1996).
23. R. Bonaccorsi, P. Pola, and J. Tomasi, *J. Am. Chem. Soc.* **106:** 1945 (1984).
24. P. J. Flory, *Statistical Mechanics of Chain Molecules*, Interscience, New York, 1969.

25. I. Ando, A. Nishioka, and T. Asakura, *Makromol. Chem.* **176:** 411 (1975).
26. I. Ando and A. Nishioka, *Makromol.Chem.* **176:** 3089 (1975).
27. I. Ando, Y. Kato, and A. Nishioka, *Makromol. Chem.* **177:** 2759 (1976).
28. T. Asakura, I. Ando, and A. Nishioka, *Makromol. Chem.* **176:** 1151 (1975).
29. T. Asakura, I. Ando, and A. Nishioka, *Makromol. Chem.* **177:** 1493 (1977).
30. T. Yamanobe, R. Chujo, and I. Ando, *Mol. Phys.* **50:** 123 (1983).
31. R. Yamanobe and I. Ando, *J. Chem. Phys.* **83:** 3154 (1985).
32. T. Ishii, H. Kurosu, T. Yamanobe, and I. Ando, *J. Chem. Phys.* **89:** 7315 (1988).
33. H. Kurosu, T. Yamanobe, and I. Ando, *J. Chem. Phys.* **89:** 6261 (1988).
34. I. Ando, H. Kurosu, T. Yamanobe, and G. A. Webb, *Ann. Rept. NMR Spectrosc.* **22:** 205 (1990).
35. J. J. Ladik, *Quantum Theory of Polymers as Solids*, Plenum Press, New York, 1988.
36. H. Kurosu, M. Kikuchi, and I. Ando, *J. Polym. Sci., Polym. Phys. Ed.* **33:** 769 (1995).
37. (a) T. C. Farrar and E. D. Becker, *Pulse and Fourier Transform NMR. Introduction to Theory and Methods*. Academic Press, New York, pp. 20–2, 97–9, 1971. (b) F. Bloch, *Phys. Rev.* **70:** 460 (1946). (c) F. Bloch, H. H. Hansen, and M. Packard, *Phys. Rev.* **70:** 474 (1946).
38. E. L. Harn, *Phys. Rev.* **80:** 580 (1950).
39. H. Y. Carr and E. M. Purcell, *Phys. Rev.* **94:** 630 (1954).
40. S. Meiboom and D. Gill, *Rev. Sci. Instrum.* **29:** 688 (1958).
41. J. E. Tanner and E. O. Stejskal, *J. Chem. Phys.* **49:** 1768 (1968).
42. (a) J. Karger, H. Pfeier, and W. Heink, *Adv. Magn. Reson.* **12:** 1 (1988). (b) W. S. Price, *Annual Report on NMR Spectroscopy*; G. A. Webb, Ed., Academic Press, London, p. 140, Vol. 35, 1997. (c) T. Nose, ibid., p. 217, Vol. 27, 1993.
43. A. E. Derome, *Modern Techniques for Chemistry Research*, Pergamon Press, New York, 1987.
44. J. W. Emsley, J. Feeney, and L. H. Sutcliffe, *High Resolution Nuclear Magnetic Resonance Spectroscopy*, Pergamon Press, New York, 1965.
45. P. L. Corio, *Structure of High Resolution NMR Spectra*, Academic Press, London, 1966.
46. R. J. Abraham, *Analysis of High Resolution NMR Spectra*, Elsevier, Amsterdam, 1971.
47. M. Karplus, *J. Am. Chem. Soc.* **85:** 2870 (1963).
48. L. M. Jackman and S. Shernhell, *Applications of Nuclear Magnetic Resonance Spectroscopy in Organic Chemistry*, 2nd Ed., Pergamon, New York, 1969.
49. J. H. Noggle and R. E. Shirmer, The Nuclear Overhauser Effect, Academic Press, New York, 1971.
50. G. C. Levy and G. L. Nelson, *Carbon-13 Nuclear Magnetic Resonance for Organic Chemistry*, John Wiley & Sons, Inc., New York, 1972.
51. J. B. Stothers, *Carbon-13 NMR Spectroscopy*, Academic Press, New York, 1972.
52. R. R. Ernst, *J. Chem. Phys.* **45:** 3845 (1966).
53. J. K. Sanders and J. D. Mersh, *Prog. NMR Spectrosc.* **15:** 353 (1982).
54. G. A. Morris and R. Freeman, *J. Am. Chem. Soc.* **102:** 428 (1980).

55. D. M. Doddrell, D. T. Pegg, and M. R. Bendall, *J. Magn. Reson.* **48:** 323 (1982).
56. A. Bax, *Two-Dimensional Nuclear Magnetic Resonance in Liquids*, Delft University Press, Delft, The Netherlands, 1982.
57. J. K. M. Sanders and B. K. Hunter, *Modern NMR Spectroscopy*, Oxford University Press, Oxford, 1988.
58. W. P. Aue, E. Bartholdi, and R. R. Ernst, *J. Chem. Phys.* **64:** 2229 (1976).
59. A. Bax, R. Freeman, and G. A. Morris, *J. Magn. Reson.* **42:** 164 (1981).
60. S. Macura and R. R. Ernst, *Mol. Phys.* **41:** 95 (1980).
61. A. A. Maudsley, L. Muller, and R. R. Ernst, *J. Magn. Reson.* **28:** 463 (1977).
62. A. Bax and S. Subramanian, *J. Magn. Reson.* **67:** 565 (1986).
63. A. Bax, R. Freeman, and S. P. Kempsell, *J. Am. Chem. Soc.* **102:** 4849 (1980).
64. W. P. Aue, J. Karhan, and R. R. Ernst, *J. Chem. Phys.* **64:** 4226 (1976).
65. (a) E. R. Andrew, A. Bradbury, and R. G. Eades, *Arch. Sci. (Geneva)* **11:** 233 (1958). (b) E. R. Andrew, *Prog. NMR Spectroscopy* **8:** 1 (1971).
66. S. R. Hartmann and E. L. Hahn, *Phys. Rev.* **128:** 2042 (1962).
67. J. S. Waugh, L. M. Huber, and U. Haeberlen, *Phys. Rev.* **20:** 180 (1968).
68. J. Jeener, Ampere International Summer School, Basko Polje, Yugoslavia, 1971.
69. A. Hagemeyer, K. Schmidt-Rohr, and H. W. Spiess, *Adv. Magn. Reson.* **13:** 85 (1989).
70. M. Linder, A. Hohener, and R. R. Ernst, *J. Chem. Phys.* **73:** 4959 (1980).
71. T. Terao, T. Fujii, T. Onodera, and A. Saika, *Chem. Phys. Lett.* **107:** 145 (1984).
72. J. E. Roberts, G. S. Harbison, M. G. Munowitz, J. Herzfeld, and R. G. Griffin, *J. Am. Chem. Soc.* **109:** 4163 (1987).
73. T. Gullion and J. Schaefer, *J. Magn. Reson.* **81:** 196 (1989).
74. T. Gullion and J. Schaefer, *Adv. Magn. Reson.* **13:** 58 (1989).
75. Y. Pan, T. Gullion, and J. Schaefer, *J. Magn. Reson.* **90:** 330 (1990).
76. A. M. Christensen and J. Schaefer, *Biochem.* **32:** 2868 (1993).
77. A. Schmidt, R. A. MaKay, and J. Schaefer, *J. Magn. Reson.* **96:** 644 (1992).
78. M. H. Levitt, D. P. Raleigh, F. Creuzet, and R. G. Griffin, *J. Chem. Phys.* **92:** 6347 (1990).
79. R. G. S. Spencer, K. J. Halverson, M. Auger, A. E. McDermott, R. G. Griffin, and P. T. Lansbury, *Biochem.* **30:** 10382 (1991).
80. R. Tycko and G. Dabbagh, *Chem. Phys. Lett.* **173:** 461 (1990).
81. R. Tycko and S. O. Smith, *J. Chem. Phys.* **98:** 932 (1992).
82. R. R. Ernst, *Quarterly Review of Biophysics* **19** (3/4): 183 (1987).
83. P. Mansfield and E. L. Hahn, Eds., *Phil. Trans. R. Soc. Lond. A* **333:** 401 (1990).
84. W. A. Edelstein, J. M. S. Hutchison, G. Johnson, and T. Redpath, *Phys. Med. Biol.* **25:** 751 (1980).
85. G. Johnson, J. M. S. Hutchison, T. Redpath, and L. M. Eastwood, *J. Magn. Reson.* **54:** 374 (1983).
86. A. Kumer, D. Welti, and R. R. Ernst, *J. Magn. Reson.* **18:** 69 (1975).
87. F. A. Bovey, L. Jelinsky, and P. A. Mirau, *Nuclear Magnetic Resonance Spectroscopy*, Academic Press, New York, 1988.

88. The Society of Polymer Science, Japan, Ed., *Sinkobunsi Zikkengaku 5: Kobunsi no Kozo*, 338–341, Kyoritu, Tokyo (1995).
89. E. O. Stejskal and J. E. Tanner, *J. Chem. Phys.* **42:** 288 (1965).
90. H. Yasunaga, M. Kobayashi, S. Matsukawa, H. Kurosu, and I. Ando, *Annual Reports on NMR Spectroscopy* **34:** 39 (1997).
91. K.-D. Merboldt, W. H. Nicke, and J. Frahm, *J. Magn. Reson.* **64:** 479 (1985).
92. W. Sattin, T. H. Mareci, and K. N. Scott, *J. Magn. Reson.* **65:** 298 (1985).
93. D. G. Cory, J. C. de Boer, and W. S. Veeman, *Macromolecules* **22:** 1618 (1989).
94. S. Matsui, *J. Magn. Reson.* **98:** 618 (1992).
95. D. G. Cory, J. B. Miller, and A. N. Garroway, *J. Magn. Reson.* **90:** 205 (1990).
96. P. Mansfield and B. Chapman, *J. Phys. E.: Sci. Instrum.* **19:** 540 (1986).
97. P. Mansfield and B. Chapman, *J. Magn. Reson.* **72:** 211 (1987).
98. K. Hikichi and M. Yasuda, *J. Polymer* **19:** 1003 (1987).
99. M. W. Crowther, N. M. Szeverenyi, and G. C. Levy, *Macromolecules* **19:** 1333 (1986).
100. M. D. Bruck, F. A. Bovey, and R. E. Cais, *Macromolecules* **17:** 2547 (1984).
101. T. Asakura, M. Demura, and Y. Nishiyama, *Macromolecules* **24:** 2334 (1991).
102. K. Hikichi, M. Ikuda, and M. Yasuda, *J. Polymer* **20:** 851 (1988).
103. K. Hikichi, T. Hirai, M. Ikura, K. Higuchi, and M. Ohuchi, *J. Polymer* **19:** 1317 (1987).
104. K. Beshah, *Macromolecules* **25:** 5597 (1992).
105. H. Ketels, J. Beulen, and G. V. Velden, *Macromolecules* **21:** 2032 (1988).
106. G. H. J. Doremaela, A. L. German, N. K. Veries, and G. P. M. Valden, *Macromolecules* **23:** 4206 (1990).
107. G. M. Benedisk, *Pure Appl. Chem.* A29(2), 85 (1992).
108. K. Hatada, K. Ute, K. Tanaka, M. Imanari, and N. Fujii, *J. Polymer* **21:** 447 (1989).
109. T. Asakura, N. Nakayama, M. Demura, and A. Asano, *Macromolecules* **25:** 4876 (1992).
110. S. Amiya and M. Uetsuki, *Anal. Sci.* **1:** 91 (1985).
111. K. Nagayama, A. Kumar, K. Wurlblich, and P. R. Ernst, *J. Magn. Reson.* **40:** 321 (1980).
112. A. Bax and D. G. Davis, *J. Magn. Reson.* **65:** 355 (1985).
113. G. P. Gippert and L. R. Brown, *Polym. Bull.* **11:** 585 (1984).
114. R. K. Tubbs, *J. Polym. Sci.*, PartA-1 **4:** 623 (1966).
115. T. Moritani and Y. Fujiwara, *Macromolecules* **10:** 532 (1977).
116. G. Velden and J. Beulen, *Macromolecules* **15:** 1071 (1982).
117. S. Toppet, P. Lemstra, and G. Velden, *Polymer* **24:** 507 (1983).
118. D. C. Bugada and A. Rudin, *Polymer* **25:** 1759 (1984).
119. J. D. Roberts, F. J. Weigert, J. I. Kroschwitz, and H. J. Reich, *J. Am. Chem. Soc.* **92:** 1388 (1970).
120. L. P. Lindeman and J. Q. Adams, *Anal. Chem.* **43:** 1245 (1971).
121. I. Sakurada, *Koubunshi* **17:** 21 (1968).
122. I. Sakurada, Y. Sakaguchi, and S. Shima, *Koubunshi Kakagu* **13:** 348 (1956).
123. R. S. Spencer, *J. Polym. Sci.* **4:** 606 (1948).

124. I. Sakurada and T. Osugi, *Gosei Seni Kenkyu* **2:** 192 (1944).
125. S. Imoto, J. Ukida, and T. Kominami, *Koubunshi Kagaku* **14:** 214 (1957).
126. R. L. Adelman and R. C. Ferguson, *J. Polym. Sci., Polym. Chem. Ed.* **13:** 891 (1975).
127. S. Amiya and M. Uetsuki, *Macromolecules* **15:** 166 (1982).
128. P. J. Flory and F. S. Leutner, *J. Polym. Sci.* **3:** 880 (1948).
129. S. Imoto, *Koubunshi Kagaku* **16:** 125 (1959).
130. S. Amiya, S. Tsuchiya, R. Qian, and A. Nakajima, *Pure Appl. Chem.* **62:** 2139 (1990).
131. Y. Morishima, H. Iimuro, and S. Nozakura, *J. Polym. Sci., Polym. Chem. Ed.* **13:** 89 (1975).
132. S. Nozakura, Y. Morishima, H. Iimuro, and Y. Irie, *J. Polym. Sci., Polym. Chem. Ed.* **14:** 759 (1976).
133. M. T. Roedel, *J. Am. Chem. Soc.* **75:** 6110 (1953).
134. K. Maruyama, Y. Kuramoto, M. Yagi, and Y. Tanizaki, *Polymer* **29:** 24 (1988).
135. I. Ando, T. Sorita, T. Yamanobe, T. Komoto, H. Sato, K. Deguchi, and M. Imanari, *Polymer* **26:** 1864 (1985).
136. D. L. VanderHart and F. Khoury, *Polymer* **25:** 1589 (1984).
137. W. L. Earl and W. L. VanderHart, *Macromolecules* **12:** 762 (1979).
138. T. Yamanobe, T. Sorita, T. Komoto, and I. Ando, *J. Mol. Struct.* **31:** 267 (1985).
139. H. Kurosu, T. Yamanobe, and I. Ando, *J. Mol. Struct.*, (Theochem.) **201:** 239 (1989).
140. S. Akiyama, T. Komoto, I. Ando, and H. Sato, *J. Polym. Sci.: Part B: Polym. Phys.* **28:** 587 (1990).
141. I. Ando, T. Yamanobe, S. Akiyama, T. Komoto, H. Sato, T. Fujito, K. Deguchi, and M. Imanari, *Solid State Commun.* **62:** 785 (1987).
142. I. Ando and G. A. Webb, *Theory of NMR Parameters*, Academic Press, London, p. 47, 1983.
143. Q. Chen, T. Yamada, H. Kurosu, I. Ando, T. Shiono, and Y. Doi, *J. Polym. Sci.: Part B: Polym. Phys.* **30:** 591 (1992).
144. J. J. Dechter, R. A. Komoroski, D. E. Axelson, and L. Mandelkern, *J. Polym. Sci.: Part B: Polym. Phys.* **19:** 631 (1981).
145. B. Schroter and A. Posern, *Makromol. Chem.* **182:** 675 (1981).
146. R. Kitamaru, F. Horii, and K. Murayama, *Polym. Bull.* **7:** 583 (1982).
147. D. E. Axelson, *High Resolution NMR Spectroscopy of Synthetic Polymers in Bulk*, R. A. Komoroski, Ed., Ch. 5, p. 197, VCH Publishers, Deerfield Beach, Florida, 1986.
148. A. Hagemeyer, K. Schmidt-Rohr, and H. W. Spiess, *Adv. Magn. Reson.* **13:** 85 (1989).
149. D. R. Holmes, C. W. Bunn, and D. T. Smith, *J. Polym. Sci.* **19:** 401 (1956).
150. I. Ando and T. Asakura, in *Solid State NMR of Polymers*, Elsevier Science, Amsterdam, p. 445, 1998.
151. I.-S. Chuang and G. E. Maciel, *Ann. Rept. NMR Spectrosc.* **29:** 169 (1994).
152. H. Miura, J. Hirschinger, and A. D. English, *Macromolecules* **23:** 2169 (1990).
153. K. Kubo, T. Yamanobe, T. Komoto, I. Ando, and T. Shiibashi, *J. Polym. Sci.: Part B: Polym. Phys.* **27:** 929 (1989).
154. K. Kubo, I. Ando, T. Shiibashi, T. Yamanobe, and T. Komoto, *J. Polym. Sci.: Part B: Polym. Phys.* **29:** 57 (1991).

155. T. L. Weeding, W. S. Veeman, H. A. Guaurand, and H. G. B. Huysmans, *Macromolecules* **21:** 2028 (1988).
156. A. Okada, M. Kawasumi, I. Kurauchi, and O. Kamigaito, *J. Appl. Polym. Sci.* **37:** 1363 (1989).
157. C. G. Johnson and L. Mathias, *Polymer* **35:** 66 (1994).
158. P. Holstein, J. Spevacek, D. Geshke, and V. Thiele, *Prog. Colloid & Polym. Sci.* **85:** 60 (1991).
159. J. Shaefer and E. O. Stejskal, *J. Am. Chem. Soc.* **98:** 1031 (1976).
160. J. Shaefer and E. O. Stejskal, *Macromolecules* **10:** 384 (1997).
161. M. D. Sefcik, E. O. Stejskal, R. A. McKay, and J. Shaefer, *Macromolecules* **12:** 423 (1979).
162. S. A. Swanson, W. W. Fleming, and D. C. Hofer, *Macromolecules* **25:** 582 (1992).
163. A. F. Fry and A. C. Lind, *New Polym. Mater.* **2:** 235 (1990).
164. (a) R. K. Harris and P. Jackson, *Chem. Rev.* **91:** 1427 (1991). (b) S. F. Dec, R. A. Wind, and G. E. Maciel, *Macromolecules* **20:** 349 (1987). (c) E. Katoh, H. Sugimoto, Y. Kita, and I. Ando, *J. Mol. Struct.* **355:** 21(1995). (d) E. Katoh, H. Sugisawa, A. Oshima, Y. Tabata, T. Seguchi, and T. Yamazaki, *Rad. Phys. Chem.* 1998, **54:** 165 (1998).
165. E. W. Hagaman, *J. Magn. Reson.* **83:** 1397 (1983).
166. P. Holstein, U. Scheler, and R. K. Harris, *Magn. Reson. Chem.* **35:** 647 (1997).
167. B. H. Klein Douwel, W. E. J. R. Maas, W. S. Veeman, G. H. Werumeus Buning, and J. M. J. Vankan, *Macromolecules* **23:** 406 (1990).
168. W. E. J. R. Maas, W. A. C. Heijden, W. S. Veeman, J. M. J. Vankan, and G. H. Werumeus Buning, *J. Chem. Phys.* **95:** 4698 (1991).
169. A. P. A. M. Eijkelenboom, W. E. J. R. Maas, W. S. Veeman, G. H. Werumeus Buning, and J. M. J. Vankan, *Macromolecules* **25:** 4511 (1992).
170. J. L. White and P. A. Mirau, *Macromolecules* **26:** 3049 (1993).
171. P. Tang, J. A. Reimer, and M. M. Denn, *Macromolecules* **26:** 4269 (1993).
172. I. Ando, T. Yamanobe, H. Kurosu, and G. A. Webb, *Ann. Rept. NMR Spectrosc.* **22:** 205 (1989).
173. See, for example, (a) H. Saito, T. Tabeta, A. Shoji, I. Ando, and T. Asakura, in *Magnetic Resonance in Biology and Medicine*, G. Govil, C. L. Kheterapal, and A. Saran, Eds., Tata McGraw-Hill, New Delhi, 1985, p. 195. (b) H. Saito and I. Ando, *Ann. Rept. NMR Spectrosc.* **21:** 210 (1989) (1990). (c) A. Shoji, S. Ando, S. Kuroki, I. Ando, and G. A. Webb, *Ann. Rept. NMR Spectrosc.* **26:** 55 (1993). (d) H. Kurosu, S. Ando, H. Yoshimizu, and I. Ando, *Ann. Rept. NMR Spectrosc.* **28:** 189 (1994). (e) I. Ando and S. Kuroki, in *Encyclopedia of NMR*, D. M. Grant and R. K. Harris, Eds., John Wiley, New York, 1996, p. 4458. (f) I. Ando, *Encyclopedia of NMR*, p. 176. (h) I. Ando, *Macromol. Symp.* **101:** 371 (1996). (g) N. Asakawa, T. Kameda, H. Kurosu, S. Ando, and I. Ando, *Ann. Rept. NMR Spectrosc.* **35:** 55 (1997).
174. H. Saito, R. Tabeta, A. Shoji, T. Ozaki, and I. Ando, *Macromolecules* **16:** 1050 (1983).
175. H. Saito, R. Tabeta, I. Ando, T. Ozaki, and A. Shoji, *Chem. Lett.*: 1437 (1983).
176. I. Ando and G. A. Webb, *Theory of NMR Parameters*, Academic Press, London, p. 115, 1983.

177. I. Ando, T. Yamanobe, H. Kurosu, and G. A. Webb, *Ann. Rept. NMR Spectrosc.* **22:** 205 (1989).
178. I. Ando, H. Saito, R. Tabeta, A. Shoji, and T. Ozaki, *Macromolecules* **17:** 457 (1984).
179. N. Asakawa, H. Kurosu, and I. Ando, *J. Mol. Struct.* **323:** 279 (1994).
180. H. Saito, R. Tabeta, T. Asakura, Y. Iwanaga, A. Shoji, T. Ozaki, and I. Ando, *Macromolecules* **17:** 1405 (1984).
181. N. Asakawa, H. Kurosu, I. Ando, A. Shoji, and T. Ozaki, *J. Mol. Struct.* **317:** 119 (1994).
182. T. Akieda, H. Mimura, S.Kuroki, H. Kurosu, and I. Ando, *Macromolecules* **25:** 5794 (1992).
183. M. Tsukahara, T. Yamanobe, T. Komoto, J. Watanabe, and I. Ando, *J. Mol. Struct.* **159:** 345 (1987).
184. T. Yamanobe, M. Tsukahara, T. Komoto, J. Watanabe, I. Ando, I. Uematsu, K. Deguchi, T. Fujito, and M. Imanari, *Macromolecules* **21:** 48 (1988).
185. T. Yamanobe, H. Tsukamoto, Y. Uematsu, I. Ando, and I. Uematsu, *J. Mol. Struct.* **295:** 25 (1993).
186. E. Katoh, H. Kurosu, and I. Ando, *J. Mol. Struct.* **318:** 123 (1994).
187. E. Katoh, H. Kurosu, S. Kuroki, and I. Ando, *J. Mol. Struct.* **326:** 145 (1994).
188. S. Ishikawa, H. Kurosu, and I. Ando, *J. Mol. Struct.* **248:** 361 (1991).
189. S. Ishikawa and I. Ando, *J. Mol. Struct.* **273:** 227 (1992).
190. E. Katoh and I. Ando, *J. Mol. Struct.* **378:** 225 (1996).
191. See, for example, T. Asakura, H. Kashiba, and H. Yoshimizu, *Macromolecules* **21:** 644 (1988).
192. See, for example, (a) Y. Hiyama and D. A. Torcha, *Basic Life Sci.* **56:** 273 (1990). (b) R. Fujiwara and Y. Kuboi, *Biochem. Biophys. Res. Commun.* **167:** 761 (1990).
193. H. Yoshimizu and I. Ando, *Macromolecules* **23:** 2908 (1990).
194. H. Yoshimizu, H. Mimura, and I. Ando, *Macromolecules* **24:** 862 (1991).
195. H. Yoshimizu, H. Mimura, and I. Ando, *J. Mol. Struct.* **246:** 367 (1991).
196. S. Tuzi, A. Naito, and H. Saito, *Biochemistry* **33:** 15046 (1994).
197. S. Tuzi, A. Naito, and H. Saito, *Eur. J. Biochem.* **239:** 294 (1996).
198. S. Tuzi, S. Yamaguchi, A. Naito, R. Needleman, J. K. Lanyi, and H. Saito, *Biochemistry* **35:** 7520 (1996).
199. S. J. Opella, *Nat. Struct. Biol.* **4:** 845 (1997).
200. A. Naito, S. Kimura, S. Tuzu, and H. Saito, *Pept. Chem.* **34:** 369 (1996).
201. See, for example, F. P. Miknis, *Ann. Rept. NMR Spectrosc.* **33:** 207 (1997).
202. See, for example, M. Bardet, L. Emsley, and M. Vincendon, *Solid State Nucl. Magn. Reson.* **8:** 25 (1997).
203. See, for example, D. Derouet, S. Forgeard, J. C. Brosse, J. Emery, and J. Y. Buzare, *J. Polym. Sci., Part A: Polym. Chem.* **36:** 437 (1998).
204. See, for exmaple, M. Ha, W. Jardine, J. Gordon, and C. Micheal, *J. Agric. Food. Chem.* **45:** 117 (1997).
205. J. L. Bredas and G. B. Street, *Acc. Chem. Res.* **18:** 309 (1985).

206. M. G. Kanatzidis, *Chem. Eng. News* December **3:** 36 (1990).
207. J. L. Bredas, B. Themans, J. G. Fripiat, and J. M. Andre, *Phys. Rev. B* **29:** 6761 (1984).
208. S. Yoshimura and H. Yasujima, *Koubinshi* **37:** 886 (1988).
209. F. Devreux, G. Bidan, A. A. Syed, and C. Tsintavis, *J. Phys.* **46:** 1595 (1986).
210. M. Kikuchi, H. Kurosu, and I. Ando, *J. Mol. Struct.* **269:** 183 (1992).
211. M. Forsyth, V.-T. Tmong, and M. E. Smith, *Polymer* **86:** 1593 (1994).
212. R. H. Geiss, G. B. Street, W. Volksen, and J. Economy, *IBM J. Res. Devel.* **27:** 321 (1983).
213. I. Ando, A. Nishioka, and M. Kondo, *J. Magn. Reson.* **21:** 429 (1976).
214. I. Ando and G. A. Webb, *Org. Magn. Reson.* **15:** 111 (1981).
215. T. Takayama and I. Ando, *J. Mol. Struct.* **271:** 75 (1992).
216. T. Takayama and I. Ando, *Bull. Chem. Soc. Japan* **60:** 3125 (1987).
217. T. Takayama and I. Ando, *Bull. Chem. Soc. Japan* **62:** 1233 (1989).
218. T. Takayama and I. Ando, *J. Mol. Struct.* **222:** 275 (1990).
219. T. Takayama, in *Solid State NMR of Polymers*, I. Ando and T. Asakura, Eds., Elsevier Science, Amsterdam, p. 613, 1998.
220. T. Takayama and I. Ando, *J. Mol. Struct.* **241:** 101 (1991).
221. (a) G. Kogler, A. Hasenhindl, and M. Moeller, *Macromolecules* **22:** 4190 (1989). (b) V. M. Litvinov, A. K. Wittaker, A. Hagemeyer, and H. W. Spiess, *Colloid Polym. Sci.* **267:** 681 (1989).
222. D. W. Boykin, Ed., *^{17}O NMR Spectroscopy in Organic Chemistry*, CRC Press, Boca Raton, FL, 1991.
223. A. L. Baumstark and D. W. Boykin, *^{17}O NMR Spectroscopy: Applications to Structural Problems in Organic Chemistry*, in A. L. Baumstark, Ed., *Advances in Oxygenated Processes*, Vol. III, JAI Press, London, pp. 141–176, 1991.
224. W. G. Klemperer, in J. B. Lambert and F. G. Riddell, Eds., *The Multinuclear Approach to NMR Spectroscopy*, D. Reidel, Dordrecht, The Netherlands, pp. 245–260, 1983.
225. J. P. Kintzinger, in P. Laszlo, Ed., *Newly Accessible Nuclei*, Vol. 2, Academic Press, New York, pp. 79–104, 1983.
226. F. H. C. Crich and A. Rich, *Nature (London)* **176:** 780 (1955).
227. T. H. Walter, G. L. Turner, and E. Oldfield, *J. Magn. Reson.* **76:** 106 (1988).
228. R. Goc, E. Ponnusomy, J. Tritt-Goc, and D. Fiat, *Int. J. Pept. Protein Res.* **31:** 130 (1988).
229. R. Goc, J. Tritt-Goc, and D. Fiat, *Bull. Magn. Reson.* **11:** 238 (1989).
230. H. Kurosu, M. Kikuchi, and I. Ando, *J. Polym. Sci.: Part B: Polym. Phys.* **33:** 769 (1995).
231. S. Kuroki, in *Solid-State of NMR of Polymers*, I. Ando and T. Asakura, Eds., Elsevier Science, Amsterdam, p. 236, 1998.
232. (a) S. Kuroki, A. Takahashi, I. Ando, A. Shoji, and T. Ozaki, *J. Mol. Struct.* **323:** 197 (1994). (b) S. Kuroki, I. Ando, A. Shoji, and T. Ozaki, *J. Chem. Soc. Chem. Commun.* 433 (1992).

233. A. Takahashi, S. Kuroki, I. Ando, T. Ozaki, and A. Shoji, *J. Mol. Struct.* **422:** 196 (1998).
234. B. Bluemich, P. Bluemer, and K. Saito, in *Solid-State NMR of Polymers*, Elsevier Science, Amsterdam, 1998.
235. M. Heskins and J. E. Guillet, *J. Macromol. Sci. Chem.* **A2:** 1441 (1986).
236. H. Ohta, I. Ando, S. Fujishige, and K. Kubota, *J. Polym. Sci.* **B29:** 963 (1991).
237. S. Hirotu, Y. Hirokawa, and T. Tanaka, *J. Chem. Phys.* **87**(2)**:** 1392 (1987).
238. N. Tanaka, S. Matsukawa, H. Kurosu, and I. Ando, *Polymer* **39:** 4703 (1998).
239. T. Tanaka, *Phys. Rev. Lett.* **40:** 820 (1978).
240. T. Tanaka, *Sci. Am.* **244:** 110 (1981).
241. I. Omine and T. Tanaka, *J. Chem. Phys.* **77:** 5725 (1982).
242. Y. Hirokawa and T. Tanaka, *J. Chem. Phys.* **81:** 6379 (1984).
243. T. Tanaka, I. Nishio, and S.-T. Sun, S. Ueno-Nishio, *Science* **218:** 467 (1982).
244. Y. Osada and M. Hasebe, *Chem. Lett.* **1285** (1985).
245. H. Yasunaga and I. Ando, *Polym. Gels Networks* **1:** 83 (1993).
246. P. S. Hubbard, *Phys. Rev.* **131:** 275 (1963).
247. S. Matsukawa and I. Ando, *Macromolecules* **22:** 7136 (1996).
248. H. Fujita, *Adv. Polym. Sci.* **3:** 1 (1961).
249. H. Yasuda, C. E. Lamaze, and A. Peterlin, *J. Polym. Sci.* **A9:** 1117 (1971).
250. J. A. Barrie, *Diffusion in Polymer*, J. Crank and G. S. Parks, Eds., p. 259, Academic Press, New York, 1968.
251. H. Yoon, H. Kim, and H. Yu, *Macromolecules* **22:** 848 (1989).
252. N. Nemoto, M. R. Landry, I. Noh, T. Kitano, J. Wesson, and H. Yu, *Macromolecules* **18:** 308 (1985).
253. R. Cukier, *Macromolecules* **17:** 252 (1984).
254. P.-G. de Gennes, *Macromolecules* **9:** 594 (1976).
255. Y. Osada, *J. Polym. Sci.* **A17:** 3485 (1979).
256. S. Matsukawa and I. Ando, *Macromolecules* **30:** 8310 (1997).
257. I. Iliopoulos and R. Audebert, *Macromolecules* **24:** 2566 (1991).
258. C. Zhao, S. Matsukawa, H. Kurosu, and I. Ando, *Macromolecules* **31:** 3139 (1998).
259. J. M. Guenet, *Thermoreversible Gelation of Polymers and Biopolymers*, Academic Press, London, 1992.
260. M. Kanekiyo, M. Kobayashi, I. Ando, H. Kurosu, T. Ishii, and S. Amiya, *J. Mol. Struct.* **447:** 49 (1998).
261. D. R. Paul, *J. Appl. Polym. Sci.* **11:** 439 (1976).
262. A. Takahashi and S. Hiramitsu, *Polymer J.* **6:** 103 (1974).
263. M. Kobayashi, I. Ando, T. Ishii, and S. Amiya, *Macromolecules* **28:** 6677 (1995).
264. T. K. Wu and D. W. Ovenall, *Macromolecules* **6:** 582 (1973).
265. Y. Inoue, R. Chujo, and A. Nishioka, *J. Polym. Sci., Polym. Phys. Ed.* **11:** 393 (1973).
266. Y. Inoue, R. Chujo, A. Nishioka, S. Nozakura, and H. Iimuro, *Polymer J.* **4:** 244 (1973).
267. T. Terao, S. Maeda, and A. Saika, *Macromolecules* **16:** 1535 (1983).

268. M. Kobayashi, I. Ando, T. Ishii, and S. Amiya, *J. Mol. Struct.* **440:** 155 (1998).
269. D. A. Torchia, *J. Magn. Reson.* **30:** 613 (1978).
270. H. Yasunaga, M. Kobayashi, S. Matsukawa, H. Kurosu, and I. Ando, *Ann. Rept. NMR Spectrosc.* **34:** 39 (1997).
271. H. Yasunaga and I. Ando, *J. Mol. Struct.* **301:** 129 (1993).
272. C. Pichot, A. Hamoudi, Q. T. Pham, and A. Guyot, *Eur. Polym. J.* **14:** 109 (1978).
273. J. Schaefer, *Macromolecules* **4:** 98 (1971).
274. E. Klesper, A. Johnsen, W. Gronski, and F. W. Wehrli, *Makromol. Chem.* **176:** 1071 (1975).
275. H. Yasunaga and I. Ando, *J. Mol. Struct.* **301:** 125 (1993).
276. H. Yasunaga and I. Ando, *Polym. Gels Networks* **1:** 267 (1993).
277. C. Zhao, S. Matsukawa, M. Kobayashi, and I. Ando, *J. Mol. Struct.* **442:** 235 (1998).
278. P. Mansfield and P. G. Morris, *NMR Imaging in Biomedicine. Advances in Magnetic Resonance Supplement 2*, Academic Press, New York, 1982.
279. V. R. McCready, M. Leach, and P. J. Ell, Eds., *Functional Studies Using NMR*, Springer-Verlag, Berlin, 1987.
280. P. Mansfield and E. L. Hahn, Eds., *Phil. Trans. R. Soc. Lond. A* **333:** 401 (1990).
281. P. T. Callaghan, *Principles of Nuclear Magnetic Resonance Microscopy*, Oxford Science Pub., Oxford, 1991.
282. L. A. Weisenberger and J. L. Koenig, *J. Polym. Sci., Polym. Lett.* **27:** 55 (1989).
283. L. A. Weisenberger and J. L. Koenig, *Macromolecules* **23:** 2445 (1990).
284. R. E. Rathbum and A. L. Babb, *J. Phys. Chem.* **65:** 1072 (1961).
285. N. L. Thomas and A. H. Windle, *Polymer* **23:** 529 (1982).
286. C.-Y. Hui, K.-C. Wu, R. C. Lasky, and E. J. Kramer, *J. Appl. Phys.* **61:** 5129 (1987).
287. C.-Y. Hui, K.-C. Wu, R. C. Lasky, and E. J. Kramer, *J. Appl. Phys.* **61:** 5137 (1987).
288. Y. Osada, R. Kishi, and M. Hasebe, *J. Polym. Sci.: Part C: Polym. Lett.* **25:** 481 (1987).
289. K. Sawahata, J.-P. Gong, and Y. Osada, *Macromol. Rapid Commun.* **16:** 713 (1995).
290. H. Yasunaga, H. Kurosu, and I. Ando, *Macromolecules* **25:** 6505 (1992).
291. Y. Osada, *Adv. Polym. Sci.* **82:** 1 (1987).
292. D. DeRossi, K. Kajiwara, Y. Osada, and A. Yamauchi, Eds., *Polymer Gels—Fundamentals and Biomedical Applications*, Plenum Press, New York, 1991.
293. T. Tanaka, Y. Osada, S. B. Ross-Murphy, and R. A. Siegel, Eds., *Polym. Gels Networks* **4** (1994).
294. T. Shibuya, H. Yasunaga, H. Kurosu, and I. Ando, *Macromolecules* **28:** 4377 (1995).
295. H. Kurosu, T. Shibuya, H. Yasunaga, and I. Ando, *Polymer J.* **28:** 80 (1996).
296. J. P. Gong, T. Nitta, and Y. Osada, *J. Phys. Chem.* **98:** 9583 (1994).
297. A. Yamazaki, Y. Hotta, and H. Kurosu, and I. Ando, *Polymer* **39:** 1511 (1998).
298. S. Fujishige, K. Kubota, and I. Ando, *J. Phys. Chem.* **93:** 3311 (1989).
299. Y. H. Bae, T. Okano, and S. W. Kim, *J. Polym. Sci. Polym. Phys. Ed.* **28:** 923 (1990).
300. R. Yoshida, K. Uchida, Y. Kaneko, K. Sakai, A. Kikuchi, Y. Sakurai, and T. Okano, *Nature* **374:** 240 (1995).
301. N. Tanaka, S. Matsukawa, H. Kurosu, and I. Ando, *Polymer* **39:** 4703 (1998).

Chapter 5 | Mechanical Spectroscopy of Polymers

M. Mours
BASF AG, Ludwigshafen
Germany

H. H. Winter
Department of Chemical Engineering
University of Massachusetts at Amherst
Amherst, Massachusetts

5.1 Introduction

We all have bounced balls and can remember differences between the bouncing of balls for soccer, basketball, tennis, golf, or billiards. Balls bounce differently when thrown hard at a surface or when just dropped. Bouncing is an informative experiment but not very controlled in its course. Mechanical spectroscopy (MS) can be viewed as a sophisticated bouncing experiment in which a material sample, most commonly in the shape of a small disk or a rectangular sheet, is deformed under a periodic load. The material can be elastic (like the bouncing ball), but the test also applies to viscous fluids because deformation and restoration of the original shape are both prescribed by the experiment.

Mechanical spectroscopy is able to probe the dynamics of relaxation processes for liquids and solids in their equilibrium state. The applied stress or strain needs to be small enough so that perturbations of the equilibrium structure can safely be neglected.

Relaxation processes in polymers occur on a wide range of time scales. Fast dynamics is associated with small-scale relaxation processes (small molecules, molecular strands, subunits of molecules) and, vice versa, slow dynamics belongs to macromolecular motions or long-range correlation of supermolecular motions (physical aggregation, for instance). To appreciate the wide interest in MS, it is important to realize that the time dependence of isochoric rheology is completely described by a single material function, the relaxation time spectrum $H(\lambda)$, and that MS is the most advanced method of determining $H(\lambda)$. The shape of $H(\lambda)$ is often correlated with specific molecular architectures. The spectrum's sensitivity to small changes in molecular connectivity makes it a powerful tool to distinguish small differences in otherwise undistinguishable materials. Systematic patterns have been

EXPERIMENTAL METHODS
IN POLYMER SCIENCE

ISBN 0-12-683265-X/$30.00.

found in $H(\lambda)$ that are useful for developing polymers of novel molecular architecture or for tailoring polymers to specific applications. Recent successes include discoveries of the relaxation time spectrum of the critical gel (polymer at its gel point) and of the spectrum of long, linear, flexible molecules of uniform molecular weight.

Mechanical spectroscopy is a mature experimental technique. Excellent commercial instruments are available for measuring the necessary data to determine $H(\lambda)$, and computer methods are readily available to analyze the data. However, the high sensitivity of the experiment makes it prone to artifacts. It is necessary to understand MS to gain full access to its potential.

5.1.1 RHEOLOGY OF THE EQUILIBRIUM STATE

Before proceeding, several basic viscoelastic parameters of polymeric liquids and solids need to be defined. This can be done most easily with the stress relaxation experiment (Figure 5.1) in which a small shear strain γ_0 is imposed on an

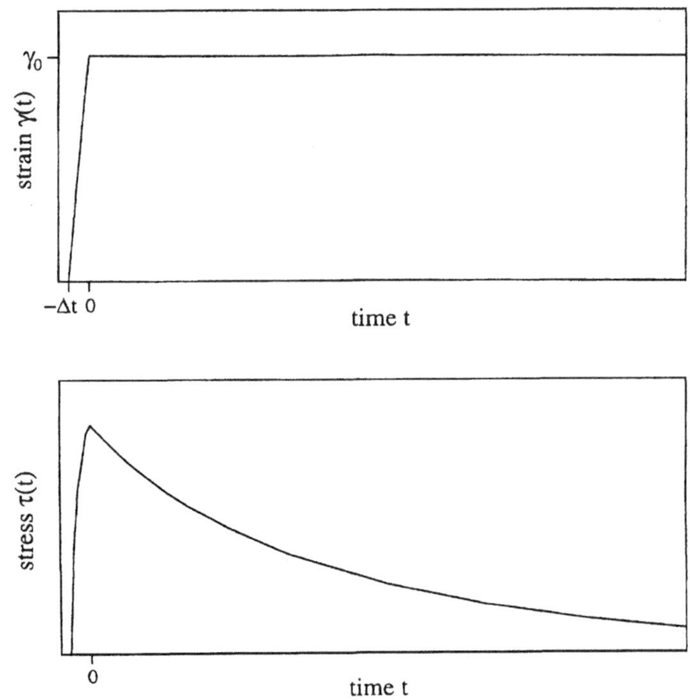

Figure 5.1 Dynamics of the stress relaxation experiment. A finite strain γ_0 is applied rapidly by shearing during a short time interval Δt. The resulting stress $\tau(t)$ is measured over a long time.

equilibrated material sample while the shear stress response as a function of time is measured. The stress in the sample rises when γ_0 is applied at the beginning of the experiment, but then it relaxes as γ_0 is kept constant. Instead of plotting the decaying shear stress $\tau(t)$ (Figure 5.1), one most commonly plots the ratio of stress and strain, called the relaxation modulus $G(t, \gamma_0)$. Data are shown for a polyethylene melt in Figure 5.2. Stress relaxation data at very small strain (linear viscoelastic limit) collapse into a single time function

$$\overset{0}{G}(t) = \tau(t)/\gamma_0 \tag{1}$$

represented by the top curve in Figure 5.2. The linear relaxation modulus $\overset{0}{G}(t)$ is independent of the applied strain.

The dynamics of the equilibrium state has significance for the nonequilibrium state (large strain). For a polymeric liquid, as shown here, Einaga *et al.* [2] discovered that large shear strains cause the time function $\overset{0}{G}(t)$ to shift vertically

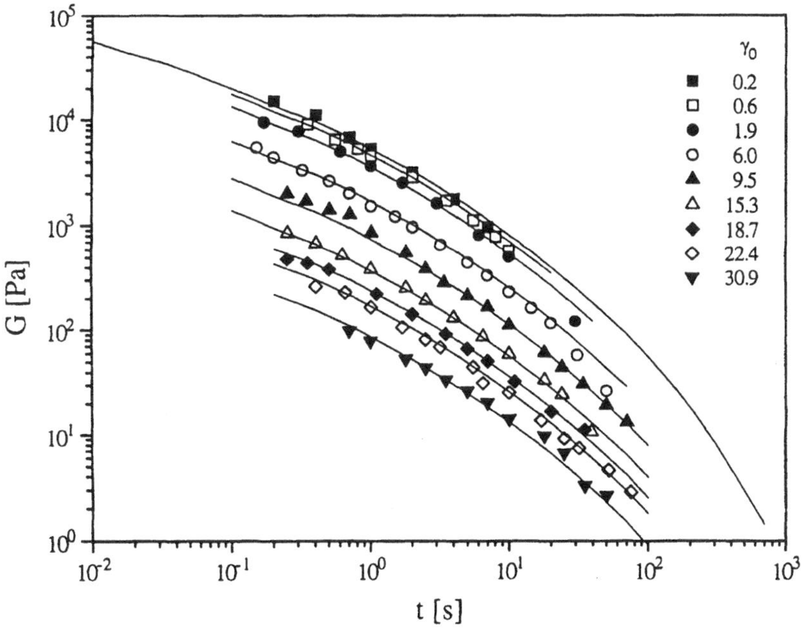

Figure 5.2 Strain dependence of the relaxation modulus $\overset{0}{G}(t)$ of a molten polyethylene. Data replotted from [1]. The top line represents the relaxation modulus calculated from the spectrum. The other lines were shifted vertically to obtain the best possible fit for each experimental data set.

to lower values

$$G(t, \gamma_0) = \overset{0}{G}(t)\, h(\gamma_0), \tag{2}$$

however, without changing its shape: time effects and strain effects can be expressed as a product of two separate functions. Many materials have been found to exhibit this phenomenon (when probed at long time scales [3]). This observation has important consequences for the modeling of molecular dynamics. The magnitude of the shift $h(\gamma_0)$ describes the effect of large strains [4, 5]. This chapter will only discuss the small strain behavior, i.e., the viscoelasticity of the equilibrium state.

Rheological experiments are quite limited in the range of time scales they can access at once. The polyethylene melt (Figure 5.2) was probed at relatively large time scales at which all its short modes had already relaxed. A more complete relaxation modulus would include glassy modes, as shown schematically in Figure 5.3, in which curve (a) represents a typical polymer melt. Distinct regions can be identified that are characterized by their own molecular dynamics. At very short times, the polymer can only relax on a very small scale. It is in its glassy state. Given more time, a gradual transition (glass transition) occurs towards the

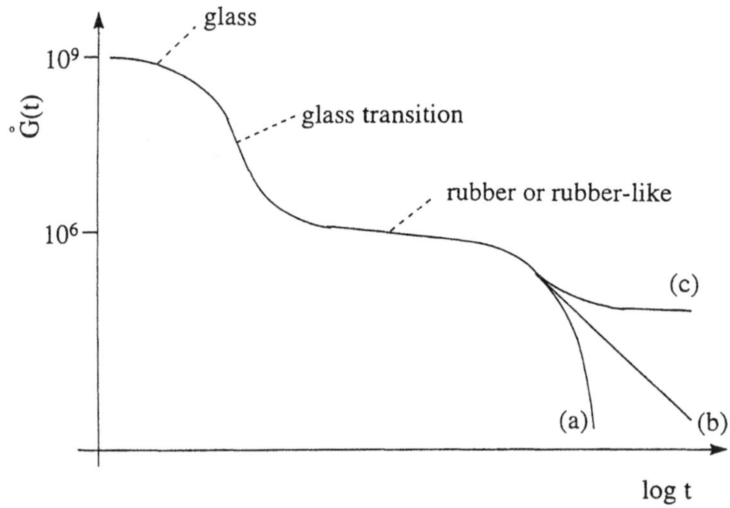

Figure 5.3 Schematic of the relaxation modulus of polymer melts (a), polymer critical gels (b), and solid polymers (c). The critical gel is the material at the transition from liquid to solid.

rubbery state. For polymer melts, the time scale of rubbery behavior is called the entanglement region. At even longer times, the polymer melt, curve (a), has time to flow. The flow region is also called the terminal region because no longer time scales beyond λ_{max} are present in the polymer.

5.1.2 *SOLIDS VERSUS LIQUIDS*

The relaxation modulus is defined for liquids and for solids. The stress of the polymeric liquid of Figure 5.2 relaxes to zero, as is typical for a liquid. In the case of a solid (curve (c) in Figure 5.3), the stress can only decay to a finite value that corresponds to the equilibrium modulus G_e. The nonrelaxing stress contributions of a solid in the preceding stress relaxation experiment would be

$$\lim_{t\to\infty} \tau = G_e \gamma_0 \tag{3}$$

Even solids have transient relaxation modes. The longest relaxation time of these modes is finite. In this chapter on MS, we are mostly concerned with transient stress contributions $\overset{0}{G}(t) - G_e$ and treat G_e as an independent material parameter, which is zero for liquids and finite for solids.

An ideal solid would show no stress relaxation at all, that is, $\tau(t) = G_e\gamma_0 =$ constant. The longest relaxation time is vanishingly small.

At the transition from liquid to solid, stress relaxes in a power law (see curve (b) in Figure 5.3). The transition state (critical gel) has its own relaxation pattern, which is universal [6, 7].

5.1.3 *BOLTZMANN EQUATION OF LINEAR VISCOELASTICITY*

The relaxation modulus $\overset{0}{G}(t)$ is the only material function required to calculate the stress $\tau(t)$ as a function of any strain history $\dot{\gamma}(t')$ for $-\infty < t' < t$ as long as the strain is small enough to avoid perturbation of the equilibrium state of the material. The extra stress in the equilibrium material is fully defined by the classical theory of linear viscoelasticity of Boltzmann (see, for example, [8]):

$$\tau(t) = \int_{-\infty}^{t} dt' \, \overset{0}{G}(t - t') \, \dot{\gamma}(t') \tag{4}$$

The variable $t - t'$ denotes the distance into the past, where t is the current time and t' is an instant in the past.

5.1.4 RELAXATION TIME SPECTRUM

The relaxation modulus can be expressed as a Laplace transform of a continuous relaxation time spectrum $H(\lambda)$:

$$\overset{0}{G}(t) = G_e + \int_0^\infty \frac{d\lambda}{\lambda} H(\lambda)\, e^{-t/\lambda} \tag{5}$$

$H(\lambda)$, called the spectrum from here on, is an alternative to the directly observable relaxation modulus $\overset{0}{G}(t)$. Each of these material functions alone is sufficient for describing the viscoelastic behavior of a material. The equivalent Boltzmann equation for $H(\lambda)$ is given in Appendix 5.A.

The $H(\lambda)$ expresses the distribution of relaxation strengths as a function of the relaxation time λ. It is a positive-valued function that is finite in the range $0 < \lambda < \lambda_{max}$ and zero above the longest relaxation time λ_{max}. The longest relaxation time λ_{max} belongs to the largest correlation length in the material (= entire molecule in a polymer melt or largest defect in a polymeric solid).

The spectrum $H(\lambda)$ cannot be measured directly. It is preferably obtained through mechanical spectroscopy and it requires extensive data analysis, resulting in a discrete spectrum (see Section 5.3). Most physical models of materials are expressed in discrete spectra, and it is not clear yet whether $H(\lambda)$ is really continuous or discrete.

Molecular models for relaxation time spectra based on molecular theory were introduced by Rouse [9], de Gennes [10], and Doi and Edwards [11]. The models start out with the most simple molecular geometry, that of long, linear, flexible chains of uniform length. Predictions from these initial theories compare well with experimental data of polymers within narrow time domains, but theories need to become quite sophisticated for covering a wider range of relaxation times [11–14].

Many empirical models have also been proposed [71]. A recent proposal for the relaxation of long, linear, flexible molecules of nearly uniform length gives very good agreement with measured data [15]. This so-called BSW spectrum (of Baumgärtel, Schausberger and Winter)

$$H(\lambda) = \begin{Bmatrix} n_e G_N^0 \left[\left(\dfrac{\lambda}{\lambda_c} \right)^{-n_g} + \left(\dfrac{\lambda}{\lambda_{max}} \right)^{n_e} \right] & \text{for } \lambda < \lambda_{max} \\ 0 & \text{for } \lambda_{max} < \lambda \end{Bmatrix} \tag{6}$$

comprises a superposition of two power laws, the first one describing the self-similarity of the glass transition and the second one expressing the self-similarity of the entanglement behavior. The material parameter G_N^0 is the plateau modulus,

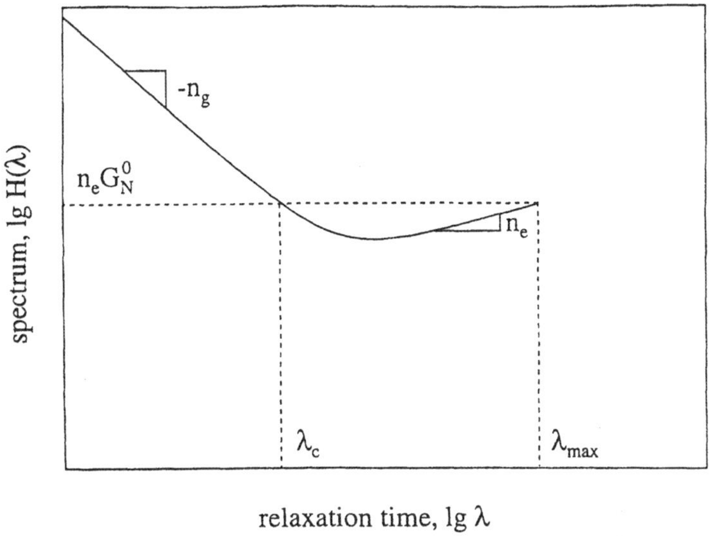

Figure 5.4 Schematic of the BSW spectrum. The two power laws with slopes $-n_g$ and n_e correspond to the glass transition and entanglement relaxation regime, respectively. The λ_c is the crossover relaxation time from entanglement to glass transition. The longest relaxation time λ_{max} belongs to the relaxation of the entire chain. The power laws are an expression of the self-similarity of the polymer chain's, n_e, and of the glassy superstructure, n_g.

n_g and n_e are the relaxation exponents in the glass transition and entanglement regime, respectively, λ_c is the characteristic time of crossover to the glass transition, and λ_{max} is the longest relaxation time. A schematic of the BSW spectrum is presented in Figure 5.4. Generic parameter sets can be evaluated for materials with similar molecular architecture but different molecular weights M [16]. The only molecular-weight-dependent parameter is the longest relaxation time

$$\lambda_{max} = \lambda_c \left(\frac{M}{M_c} \right)^z \tag{7}$$

which compares M to the polymer-specific crossover molecular weight M_c for the onset of entanglements. Exponent z is the scaling exponent that also describes the dependence of zero-shear viscosity on molecular weight [8]; experimentally z was found to have values of about 3.4 for large molecules $M \gg M_c$ [17]. The BSW spectrum is the simplest spectrum known to be able to express relaxation data of polymers with linear flexible molecules of uniform molecular weight. It has been serving as reference when studying more complicated materials.

5.1.5 DISCRETIZED FORM OF MODULUS AND SPECTRUM

Discrete spectra result in a relaxation modulus

$$\overset{0}{G}(t) = G_e + \sum_{i=1}^{N} g_i e^{-t/\lambda_i} \tag{8}$$

where coefficients g_i and relaxation times λ_i are material parameters. Here, we chose the sum of N exponential decays (Maxwell modes) for the time dependence of the material. This does not pose any lack of generality because any continuous function $\overset{0}{G}(t)$ may be approximated by such a discrete set of Maxwell relaxation modes. It should be noted that the discrete relaxation time spectrum

$$H(\lambda) = \sum_{i=1}^{N} g_i\, \delta(\lambda/\lambda_i - 1) \tag{9}$$

is governed by the same parameter set $g_i, \lambda_i, i = 1, 2, 3, \ldots N$ as the relaxation modulus. The expression $\delta(y)$ is the Dirac delta function.

Figure 5.5 shows the discrete relaxation and retardation time spectrum for a linear polybutadiene with a weight-average molecular weight Mw = 70,000. At the same time, the calculated material functions $\overset{0}{G}(t)$ and $\overset{0}{J}(t)$ are displayed as solid lines. The compliance $\overset{0}{J}(t)$ will be defined next.

5.1.6 RETARDATION TIME SPECTRUM

The stress relaxation experiment is only one of many ways of defining and measuring viscoelasticity in materials. Another useful approach requires application of a constant stress τ_0 on the material while the resulting strain $\gamma(t)$ is measured. This is the so-called creep experiment. The resulting strain is commonly expressed as compliance

$$\overset{0}{J}(t) = \gamma(t)/\tau_0 \tag{10}$$

or as retardation time spectrum $L(\lambda)$ [8]. All such related viscoelastic functions can be converted into each other [18]. If, for instance, the relaxation time spectrum is known, it is possible to convert it to the retardation time spectrum; there is a one-to-one relation between these two types of spectra [8]:

$$L(\lambda) = \frac{H(\lambda)}{\left(G_e - \displaystyle\int_0^{\infty} \frac{du}{u} \frac{H(u)}{\lambda/u - 1}\right)^2 + \pi^2 H(\lambda)^2} \tag{11}$$

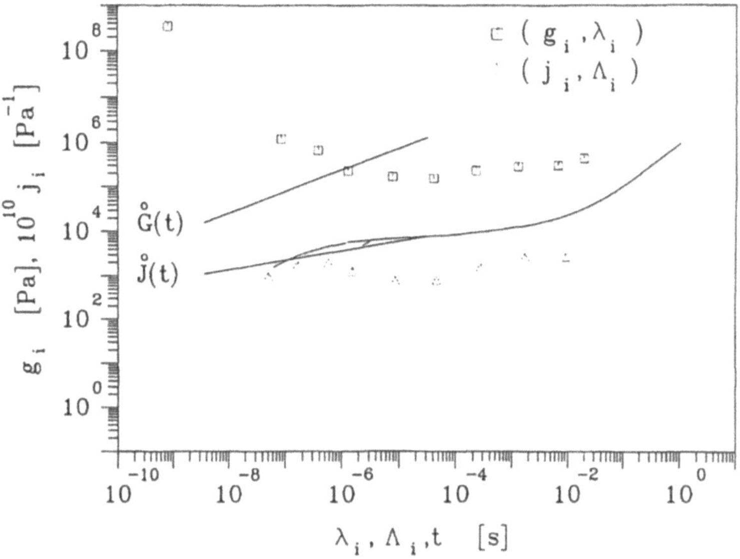

Figure 5.5 Discrete relaxation time spectra g_i, λ_i and retardation time spectra j_i, Λ_i for a polybutadiene (symbols). These spectra were used to calculate the relaxation modulus $\overset{0}{G}(t)$ and the compliance $\overset{0}{J}(t)$ (lines). The values of g_i and j_i depend on the spacing of the discrete modes.

It would exceed the scope of this chapter to establish the wide range of such relations. The focus will be on the relaxation modulus and the relaxation time spectrum because either one, by itself, already defines the viscoelasticity polymeric liquids and solids.

5.1.7 SCOPE OF MECHANICAL SPECTROSCOPY

Mechanical spectroscopy (MS) determines the above material properties by probing specific material samples in small amplitute oscillation at prescribed frequency ω. MS quantifies the strength, time scale, and temperature dependence of the characteristic relaxation modes of a material. Most advantageous is its ability to detect and resolve small differences between materials and to express changes in materials during cross-linking or melting (to name some typical examples). Mechanical spectroscopy is used for developing new materials and for tailoring their processing behavior.

Mechanical spectroscopy applies a periodic strain to the material. This has several important advantages compared with other techniques:

1. Each data point is dominated by the specific relaxation modes with relaxation times near $\lambda = 1/\omega$. Other relaxation modes have less influence.
2. The experimental time for taking a single data point (sampling time) is about $t_S = 2\pi/\omega$ independent of any longer relaxation modes that might be present in the sample.
3. Each of the dynamic moduli independently carries all the information about the spectrum.

As an introduction we used step strain behavior to demonstrate the relaxation properties of materials in a most obvious way. For determining relaxation properties, however, the time-periodic experiment of MS is preferable. It will be discussed next.

5.2 Mechanical Spectroscopy Experiment

5.2.1 SHEAR FLOW NOTATION

For the purpose of MS, which conventionally is performed in shear rheometers, it is sufficient to consider $\tau(t)$ to be the shear stress in response to a shearing deformation at a shear rate

$$\dot{\gamma}(t') = \frac{\partial}{\partial t'}\gamma(t;\, t') \tag{12}$$

over the history of the material, $-\infty < t' < t$. The restriction to shear flow does not cause any loss of generality (in isotropic materials at small strains, equilibrium state) because the same spectrum governs all flows and deformations in the linear viscoelastic range, i.e. the spectrum $H(\lambda)$ is equally applicable to other types of deformations (extension, for instance) within the linear viscoelastic range, as will be shown in Section 5.6.

The stress components for the shear experiment are shown in Figure 5.6. These are the shear stress

$$\tau(t) \equiv \tau_{21} \tag{13}$$

which we will simply call τ in the context of this discussion of MS, and the normal stress differences

$$N_1 = \tau_{11} - \tau_{22} \tag{14}$$

$$N_2 = \tau_{22} - \tau_{33} \tag{15}$$

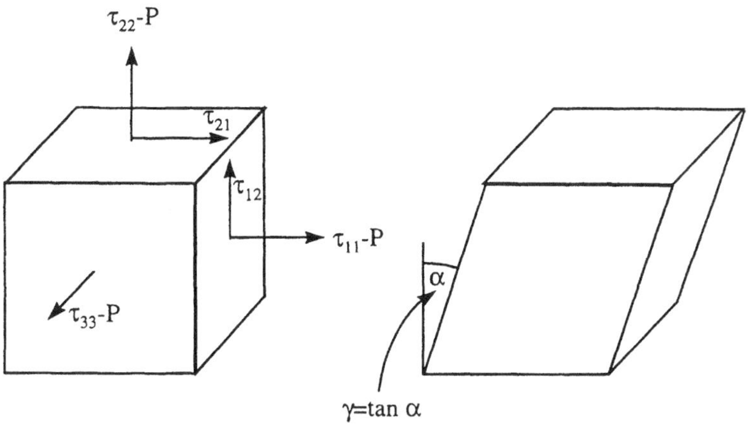

Figure 5.6 Infinitesimal material element in shear, where 1 is the velocity direction, 2 is the velocity gradient direction, and 3 is the vorticity direction. The material element is shown at two times, t' and t; $\gamma(t;\ t')$ is the shear strain.

Normal stresses and the shear stress are governed by the same relaxation modulus and spectrum, i.e., MS can concentrate on measurement of the shear stress only. Normal stresses and pressure P do not contribute directly.

5.2.2 OSCILLATORY SHEAR

The MS experiment is performed by placing the viscoelastic material (test sample) in a mechanical device (rheometer) that allows small-amplitude sinusoidal shearing at a strain

$$\gamma(t) = \gamma_0 \sin(\omega t); \quad \text{or} \quad \text{a strain rate } \dot{\gamma}(t) = \omega\gamma_0 \cos(\omega t) \tag{16}$$

After some start-up time (in the order of $1/\omega$), the stress in the sample responds sinusoidally

$$\begin{aligned} \tau(t) &= \tau_0 \sin(\delta + \omega t) \\ &= G'(\omega)\,\gamma_0 \sin(\omega t) + G''(\omega)\,\gamma_0 \cos(\omega t) \end{aligned} \tag{17}$$

(see Figure 5.7). The frequency of the sinusoidal stress $\tau(t)$ is the same as that of the strain $\gamma(t)$; however, $\tau(t)$ is shifted by a phase angle $\delta(\omega)$. The factor $\tau_0(\omega, \gamma_0)$ is the amplitude of the sinusoidal stress. The angular frequency $\omega = 2\pi f$ [rad/s] is defined by the number of cycles per time given by the frequency f [Hz]. For the data analysis, it does not matter whether the strain is prescribed and the stress is responding, or vice versa. Both types of mechanical spectrometers are equally common.

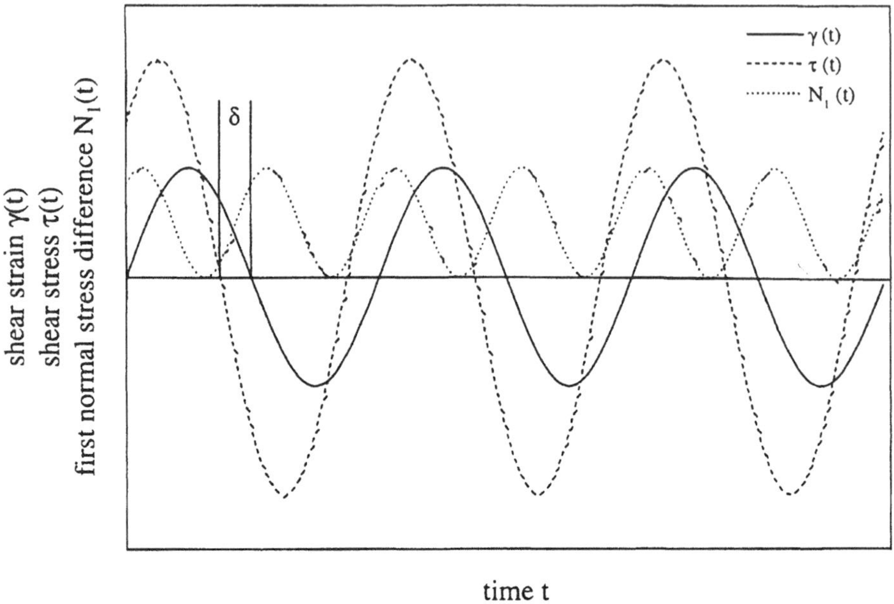

Figure 5.7 Schematic of the imposed strain function $\gamma(t)$ and the material response (shear stress $\tau(t)$ and first normal stress difference $N_1(t)$) in a small-amplitude oscillatory shear experiment.

Mechanical spectroscopy derives its name from the fact that the sample is probed at prescribed time scales $t \approx 1/\omega$ that depend on the choice of the experimental frequency. The experiment is repeated for a range of frequencies, $\omega_{\min} < \omega < \omega_{\max}$. Because of experimental constraints (e.g., weak torque values at low frequencies or large slip and inertial effects at high frequencies), it is usually impossible to measure $G'(\omega)$ and $G''(\omega)$ with a commercial rheometer directly over more than three or four decades of frequency.

It is customary to decompose the stress (Eq. (17)) into a part that is in phase with the sinusoidal strain (Eq. (16)) governed by the storage modulus

$$G'(\omega) = \frac{\tau_0}{\gamma_0} \cos\delta \tag{18}$$

and a 90° out-of-phase part governed by the loss modulus

$$G''(\omega) = \frac{\tau_0}{\gamma_0} \sin\delta \tag{19}$$

The objective of the oscillatory shear experiment is to determine these two material-specific moduli over a wide range of frequency, temperature, pressure, or other

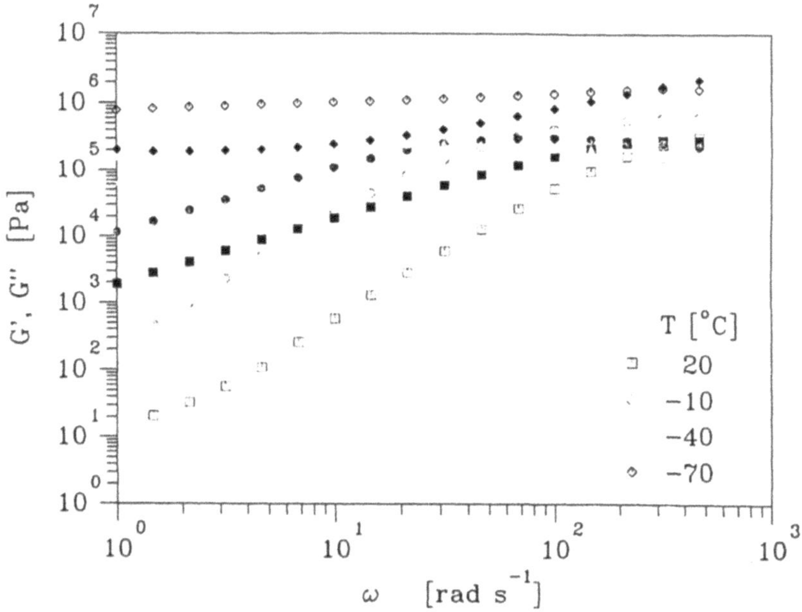

Figure 5.8 Dynamic moduli measured at different temperatures as used for the construction of the master curve shown in Figure 5.21 by time–temperature superposition. Open symbols represent G'; filled symbols represent G''. Data from Mours and Winter 1994 [19].

material-affecting parameters. The frequency-dependent moduli are generally valid for liquids and solids, i.e., for any isotropic material as long as the strain amplitude γ_0 is sufficiently small. G' is a measure of the stored energy and G'' of the dissipated energy per cycle.

A typical data set (see Figure 5.8), spans about three to four decades in frequency. The frequency range can be extended by time–temperature superposition [8], which will be discussed in Section 5.5.1.

5.2.3 MULTIWAVE TECHNIQUES [20]

Instead of probing samples with a single frequency at a time (or a sequence of single frequencies), it is possible to apply m frequencies simultaneously

$$\gamma(t) = \sum_{j=1}^{m} \gamma_j \sin(\omega_j t) \tag{20}$$

and decompose the stress response

$$\tau(t) = \sum_{j=1}^{m} \tau_j \sin(\delta_j + \omega_j t) \tag{21}$$

into m individual components. The frequency is the same for corresponding components of strain and stress. The result of the experiment is a set of m dynamic moduli $G'(\omega_j)$, $G''(\omega_j)$ that are then evaluated as if each of them were being measured in a conventional (single-frequency) experiment. The number of probing modes m cannot be very high, for the sum of the strain amplitudes

$$\gamma_{\max} = \sum_{j=1}^{m} \gamma_j \tag{22}$$

should not exceed the linear viscoelastic limit. Typically one may choose $m = 3$.

The multiwave technique saves experimental time because, while measurements are being made at the lowest frequency appropriate for a specific test sample, several higher frequencies may be run simultaneously. Preferably, the lowest experimental frequencies are probed together to save the most time. Such time savings are especially important for the study of transient materials that change because of cross-linking, degradation, or crystallization, for instance.

5.2.4 *UNIVERSAL TERMINAL BEHAVIOR*

The ratio of longest relaxation time $\lambda_{\max}$ and characteristic experimental time defines the Deborah number [21] for the MS experiment as follows:

$$N_{\mathrm{De}} = \frac{\lambda_{\max}}{t_{\mathrm{S}}} = \frac{\omega \lambda_{\max}}{2\pi} \tag{23}$$

Here, the characteristic experimental time is the sampling time t_{S} (Eq. (36)). For very low frequencies, the Deborah number becomes very small, $N_{\mathrm{De}} \ll 1$, and the polymer shows its distinct behavior as liquid or solid.

5.2.4.1 Polymeric Liquids

The low-frequency asymptotes of the moduli define the zero shear viscosity of the liquid

$$\eta_0 = \lim_{\omega \to 0} \frac{G''}{\omega} \tag{24}$$

and the recoverable compliance

$$J_e^0 = \frac{1}{\eta_0^2} \lim_{\omega \to 0} \frac{G'}{\omega^2} \tag{25}$$

For such low-frequency experiments, the polymer behavior is dominated by a representative longest mode $\langle \lambda \rangle$

$$\overset{0}{G}(t) \approx g_0 e^{-t/\langle \lambda \rangle} \tag{26}$$

with

$$\langle \lambda \rangle = \frac{\sum_{i=1}^{\infty} g_i \lambda_i}{\sum_{i=1}^{\infty} g_i} \tag{27}$$

in discrete representation. The dynamic moduli approach the classic terminal behavior for $N_{\text{De}} \ll 1$:

$$G'(\omega) \sim \omega^2, \qquad G''(\omega) \sim \omega \tag{28}$$

(see [8]). No new material parameters need to be introduced.

5.2.4.2 Polymeric Solids

The low-frequency asymptotes of the moduli define the equilibrium modulus

$$G_e = \lim_{\omega \to 0} G' \tag{29}$$

and a transient shear viscosity

$$\eta_s = \lim_{\omega \to 0} \frac{G''}{\omega} \tag{30}$$

which determines viscous losses in solids during transient deformations.

5.2.5 TYPICAL EXPERIMENTAL RESULTS

A wide range of dynamic mechanical data on liquids and solids can be found in the classical textbooks of Ferry [8] and Tschoegl [22]. Here we can only show a few typical examples.

The typical features of dynamic mechanical data of polymeric fluids express themselves most distinctly in the special case of narrowly distributed, linear, flexible polymers of high molecular weight. Polystyrene is a good example (see Figure 5.9 in which the data shown are the starting point for finding the BSW

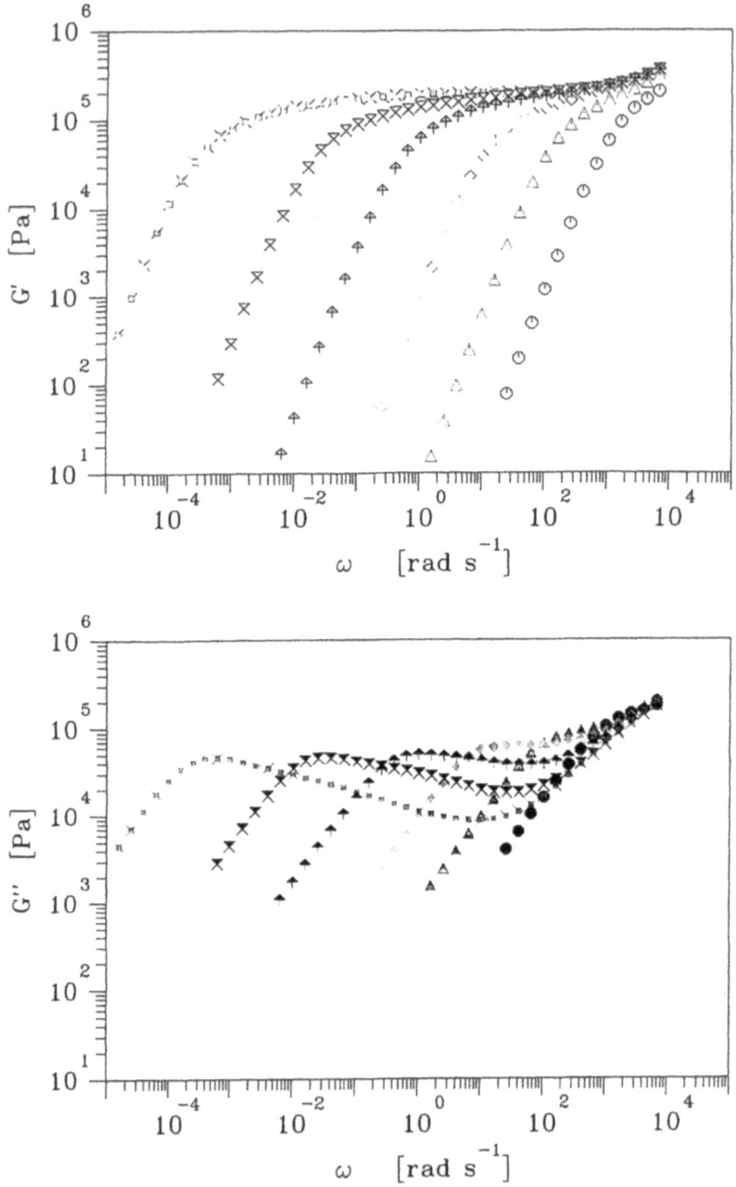

Figure 5.9 Storage modulus $G'(\omega)$, loss modulus $G''(\omega)$, and phase shift $\delta(\omega)$ of narrowly distributed polystyrene at 180 °C. The molecular weights are 34,000; 65,000; 125,000; 292,000; 757,000; 2,540,000. Data from Schausberger *et al.* [23].

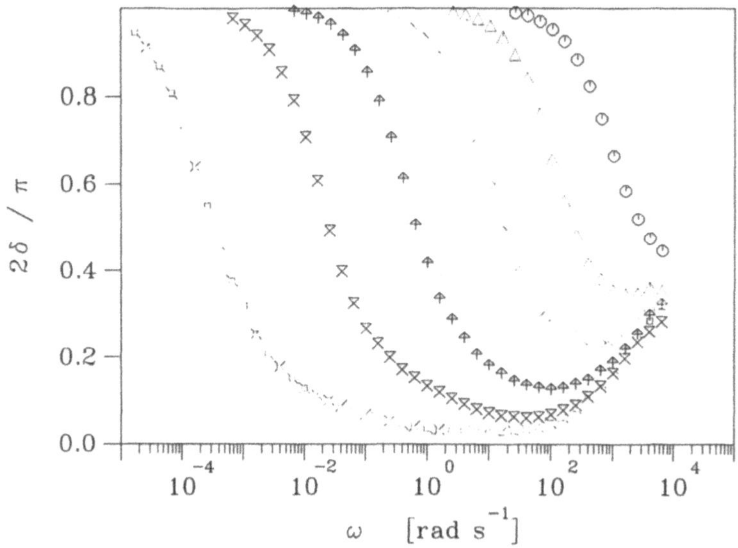

Figure 5.9 Continued.

spectrum, Eq. (6)). At low frequencies, the typical liquid behavior of $G'(\omega_j) \sim \omega^2$ and $G''(\omega_j) \sim \omega$ can be seen. The crossover from terminal behavior to entanglement occurs at a characteristic frequency that decreases with molecular weight

$$\omega_c \sim M_w^{-z} \quad \text{with} \quad z \approx 3.4 \tag{31}$$

For model polymeric liquids of Figure 5.9, the crossover from flow to entanglement behavior results in an actual intersect of G' and G''; the intersect frequency ω_x often serves as the inverse of a representative longest relaxation time $\lambda^*_{max} = 1/\omega_x$. This representative time constant is somewhat smaller than the longest relaxation time $\lambda^*_{max} < \lambda_{max}$ but has the advantage that it can be read directly from G', $G''(\omega)$ graphs.

The entanglement plateau of $G''(\omega)$, called the plateau modulus G_N^0, is known to be independent of molecular weight M_w. The upturn of the moduli in Figure 5.9. at high frequencies marks the crossover to the glass transition, which does not significantly depend on M_w.

A mixture of two polymers of different molecular weights exhibits two distinct entanglement regions [25, 24]. An example is shown in Figure 5.10. Each of the components is narrowly distributed; broadening of the molecular weight

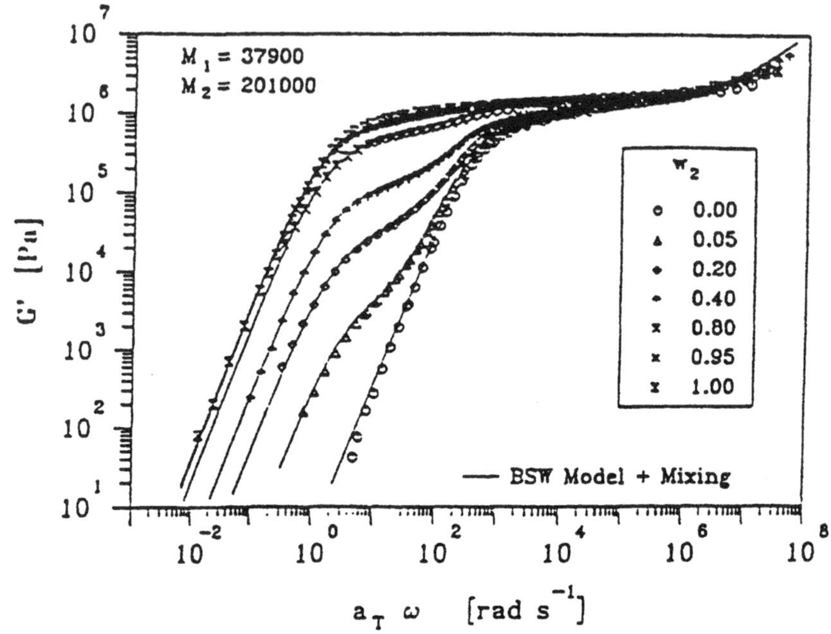

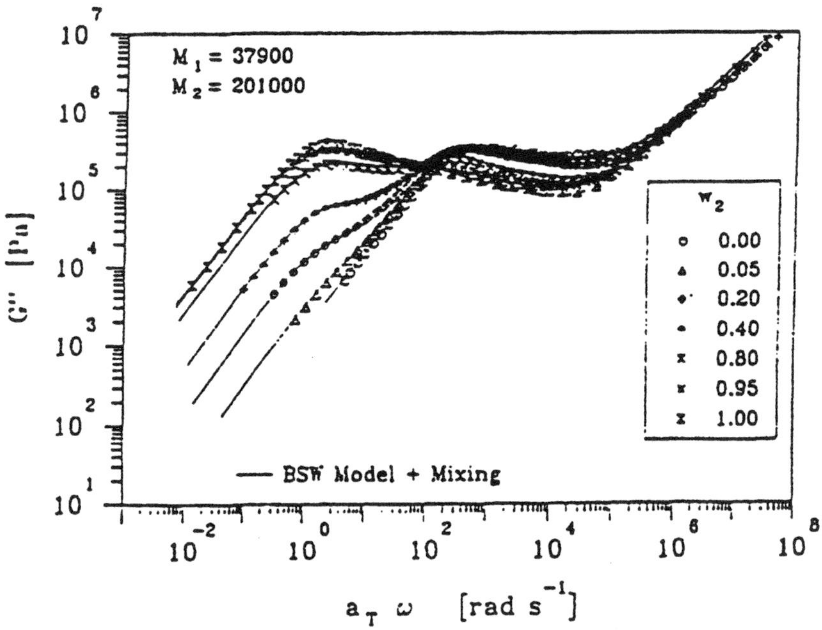

Figure 5.10 The (a) storage and (b) loss moduli of bidisperse polybutadiene (PBD) blends measured at 28 °C. The line through the data represents the model spectrum described by Jackson and Winter [24]. Both M_1 and M_2 are highly entangled.

distribution would widen the crossover regions, and the distinct features of the dynamic moduli would get lost [26].

5.2.6 INTERRELATION WITH RELAXATION MODULUS AND SPECTRUM

Experimental data of dynamic moduli $G'(\omega)$ and $G''(\omega)$ carry all the information needed to determine the relaxation modulus $\overset{0}{G}(t)$. However, extraction of the information requires the inversion of integrals

$$\text{storage modulus} \quad G'(\omega) = \omega \int_0^\infty dt\ \overset{0}{G}(t) \sin(\omega t) \tag{32}$$

$$\text{loss modulus} \quad G''(\omega) = \omega \int_0^\infty dt\ \overset{0}{G}(t) \cos(\omega t), \tag{33}$$

which is a nontrivial task. It will be discussed in Section 5.3. The above equations can be derived by inserting Eq. (16) into Boltzmann's equation of linear viscoelasticity, Eq. (4) (see, for example, Bird *et al.* [27]). The dynamic moduli may also be expressed with the continuous spectrum $H(\lambda)$ by

$$G'(\omega) - G_e = \int_0^\infty \frac{d\lambda}{\lambda} H(\lambda) \frac{(\omega\lambda)^2}{1 + (\omega\lambda)^2} \tag{34}$$

$$G''(\omega) = \int_0^\infty \frac{d\lambda}{\lambda} H(\lambda) \frac{(\omega\lambda)}{1 + (\omega\lambda)^2} \tag{35}$$

The spectrum covers the transient part of the relaxation process. This is why the equilibrium modulus needs to be subtracted when performing spectroscopy on solids.

5.2.7 SPECTROSCOPIC NATURE OF EXPERIMENT

The basic advantages of small-amplitude oscillatory strain (shear or extension) come through its spectroscopic nature: The experimental time for taking a single data point, called sampling time t_s, is roughly equal to the period of the strain wave

$$t_s = 2\pi/\omega \tag{36}$$

This allows measurement of specific relaxation modes with time constants on the order of $1/\omega$ independent of any longer or shorter modes that might be present in the polymer, i.e., only a small fraction of the spectrum is actually sampled in an individual experiment. Each data point ω_s, $G'(\omega_s)$, $G''(\omega_s)$ is dominated by the specific relaxation modes with relaxation times near $\lambda = 1/\omega_s$. Other relaxation modes have less influence, which can be demonstrated most efficiently with a polymer at its gel point for which the longest relaxation time diverges to infinity, $\lambda_{max} \to \infty$ [7]. Any measurement that depends on λ_{max} will necessarily fail at the gel point. Mechanical spectroscopy, however, is able to probe intermediate modes of the spectrum without interference from the diverging λ_{max}. The spectrum at the gel point is known to be self-similar at times larger than some material characteristic crossover time λ_0 [6, 7].

$$H(\lambda) = a\lambda^{-n} \quad \text{for} \quad \lambda_0 < \lambda < \infty \tag{37}$$

For the purpose of demonstrating the spectroscopic nature of MS, we select a sampling frequency ω_s within the self-similar region of the spectrum ($\omega_s \ll 1/\lambda_0$) and evaluate the corresponding dynamic moduli by inserting Eq. (37) into Eqs. (34) and (35):

$$G'(\omega_s) = a\omega_s^n \int_0^\infty dx\, F(x) \quad \text{with} \quad F(x) = x^{-(n+1)} \frac{x^2}{1+x^2} \tag{38}$$

$$G''(\omega_s) = a\omega_s^n \int_0^\infty dx\, F''(x) \quad \text{with} \quad F''(x) = x^{-(n+1)} \frac{x}{1+x^2} \tag{39}$$

The areas under the F and F'' curves determine the values of the complex moduli G', G'' (see Figure 5.11). The graph shows that the relaxation modes near $\lambda\omega_s = 1$ determine the outcome of the experiment. Note that G_e is equal to zero at the gel point and needs not to be considered in this example. The self-similar spectrum of the critical gel allows calculation of dynamic moduli

$$G' = a\omega^n \pi/(2\,\sin(n\pi/2)) \tag{40}$$

$$G'' = a\omega^n \pi/(2\,\cos(n\pi/2)) \tag{41}$$

for $\omega < 1/\lambda_0$.

This spectroscopic property needs to be contrasted with nonspectroscopic methods that measure steady-state properties (the steady shear viscosity, for instance). They simultaneously probe an integral over all relaxation modes.

5.2.8 *RHEOMETER GEOMETRIES*

For the most widely used configuration, the plate-and-plate rheometer (Figure 5.12), the polymer is molded into a disk-shaped sample of, for instance, radius R = 12.5 mm and h = 1-mm height. The sample is sandwiched between disk-shaped rheometer fixtures. The configuration has a common symmetry axis. When one of

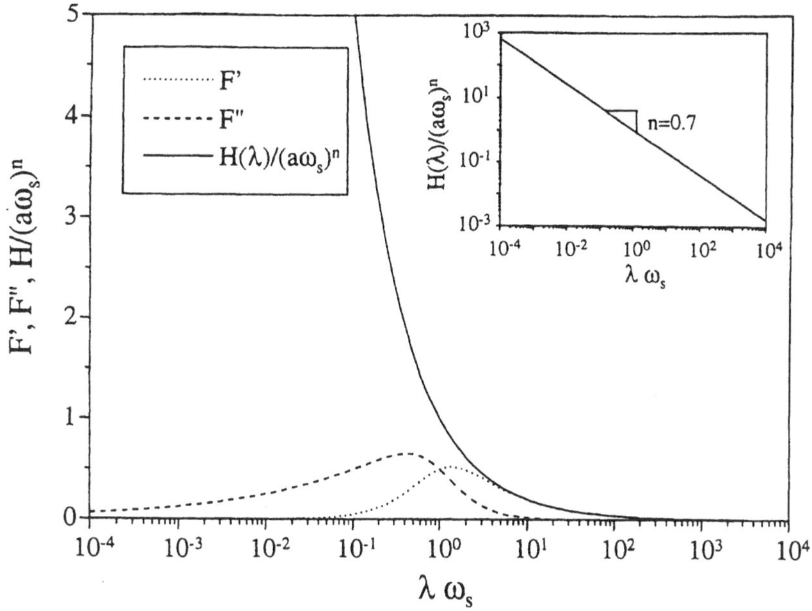

Figure 5.11 $F(\lambda\omega_s)$ and $F''(\lambda\omega_s)$ for the evaluation of G' and G''. Inset: reduced power law spectrum $H(\lambda) = a\lambda^{-n}$.

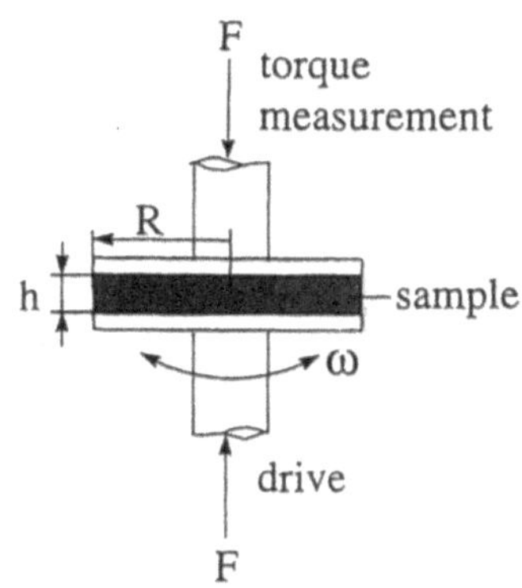

Figure 5.12 Plate and plate rheometer with schematic of a shearing experiment with disk-shaped sample. Drive and torque measurement are on opposite sides (in this example). An axial force F is required to maintain constant sample thickness h.

the fixtures is rotated (at constant h) while the other fixture is held stationary, the sample deforms and transmits a torque that can be measured. The angle of rotation and the torque can be recalculated into shear strain $\gamma(t)$ and shear stress $\tau(t)$ for further analysis. Stress and strain vary across the sample, but this does not pose a problem for experiments in the linear viscoelastic range because a linear relation exists between $\gamma(t)$ and $\tau(t)$. Large strain experiments, which we do not address here, would require Rabinowitch-type corrections or switching to other flow geometries.

Other common geometries are cone-and-plate and annulus between concentric cylinders (Couette rheometer). Detailed descriptions of the many possible geometries and related data analyses are well documented in textbooks [27–31]. Data analysis requires corrections for instrument compliance and inertia.

5.2.9 EXPERIMENTAL PROBLEMS

It seems to be the general property of polymers that they slip on surfaces [32]. However, the slip is negligibly small in most experiments. Data analysis proceeds under the assumption that there is no relative velocity between the polymer and the wall of the rheometer ("stick"). Problems arise when the slip velocity is large enough to alter the strain in the sample. Discussion of this subject would exceed the task of this chapter.

Another common problem is the lack of chemical stability of the sample. Rheological experiments require time during which the polymer may change and standard data analysis will no longer be possible. This subject is of great importance and will be discussed in Section 5.4.

The experimental frequencies range over $\omega_{\min} < \omega < \omega_{\max}$. Sample inertia and elastic instabilities limit the experiment at high frequencies ($\rightarrow\omega_{\max}$). Transducer sensitivity (or lack thereof) and increasing sampling time $t_s \sim 1/\omega$, limit the experiment at low frequencies ($\rightarrow\omega_{\min}$).

The large strain limits of the linear viscoelastic range of the experiment can be detected by Fourier analysis of the stress response to the sinusoidal strain [33–35]. While at small strain (linear viscoelastic regime), the stress response is of the same frequency as the input, and higher frequencies are not represented except for the noise in the experiment. However, at larger strain amplitudes additional frequencies start to appear in the stress response. These frequencies are multiples of the sampling frequency ω_s. Most pronounced are the odd harmonies with $3\,\omega_s$, $5\,\omega_s$, $7\,\omega_s$, and higher. Discussion of these nonlinear effects is beyond the scope of this chapter.

5.2.10 PROBING NONEQUILIBRIUM STATES

Mechanical Spectroscopy also allows the study of material samples that were forced away from their equilibrium state. Such nonequilibrium conditions may arise, for instance, during large strains. For the purpose of MS, one modulates the large strain by superimposing small perturbations of sinusoidal [36] or step [37] dynamics. The only requirement for MS is a linear response of the stress increments to the small strain perturbations (or vice versa), whereas the main part of the experiment is still nonlinear. The experiments are difficult to perform with sufficient accuracy, and not much data are available. The main problem seems to be that the perturbations have to be extremely small in order to maintain linear modulation conditions [38]. The probing of nonequilibrium states with MS is very appealing; more such work will be seen in the near future. The basic details of data analysis are the same as those entailed in probing equilibrium states except that one has to account for the out-of-equilibrium condition in which the Boltzmann equation, Eq. (4), does not apply any more. No separate discussion will be needed here. In the remainder of this chapter, we will assume that the material has fully equilibrated because this is the most common experimental condition.

5.3 Data Analysis for Mechanical Spectroscopy

In this section, we will assume that oscillatory experiments have been completed successfully, that the experiments have been analyzed, and that a $G'(\omega)$, $G''(\omega)$ data set is ready for further analysis. Many different methods have been proposed for converting the measured dynamic moduli G', G'' into the relaxation time spectrum of the test sample [8, 1, 39–45]. Orbey and Dealy [46] have presented an overview of several of these methods. The spectrum may also be extracted from any other material function besides G' and G''. Advanced methods have been proposed for inverting creep data (see, for example, [47, 48]). However, the focus here will continue to be on the G', G'' data.

The solution of the inversion problem will always be a compromise. Important criteria to consider are as follows [74]:

1. Good fit of the experimental data.
2. Avoidance of overfitting. The algorithm should be able to find the optimum amount of detail that can be extracted from experimental data without producing artifacts.

3. The format of g_i and λ_i should be freely adjustable during the inversion of the data.
4. The resulting material parameters should have physical meaning.
5. Minimization of the truncation error. The data are always cut off on both sides of the frequency scale. This truncation may be the source of substantial error [49].
6. Verification of experimental data quality. The data are always inconsistent to some degree because the signal-to-noise ratios differ for G' and G''. This may cause a substantial discrepancy between fit and data, which can be detected during data analysis.
7. For practical considerations, the relaxation modulus or the spectrum should be expressed by a function or a sum of functions that can be integrated easily in the various linear viscoelastic model calculations. This again leads us to an exponential form (Eq. (8)).
8. For practical considerations, the number of parameters should be small (for ease of use in modeling calculations and for storing material data in a database).
9. For practical considerations, the computation time should be short. All conversion methods in the literature seem to satisfy this criterion owing to the high computation speed of desktop computers.

Most problematic of these criteria is truncation. *Low* frequency data should extend into the terminal region to avoid truncation, but no characteristic limiting behavior exists at *high* frequencies; high frequency truncation is unavoidable.

5.3.1 METHOD OF EVALUATING $H(\lambda)$ FROM G', G'' DATA

Conversion methods are fairly established by now, and practitioners can choose from several options. The robust numerical algorithm of Baumgärtel *et al.* [43, 50] satisfies most of the preceding criteria. It avoids ill-posedness, which seems to be inherent in the curve fitting when too many parameters are chosen. The method is especially useful when working with complex polymeric materials because the data do not need to be extrapolated beyond the experimental frequency window (as with a Fourier transform); no material-specific empiricisms are needed. A nonlinear regression simultaneously adjusts $g_i, \lambda_i, i = 1, 2, 3 \ldots, N$ to obtain a best fit of G', G''. A type of regularization is achieved by systematically searching for the spectrum with the smallest number of Maxwell modes that still represents the data within the experimental error margin (parsimonious modeling). A closer fit could

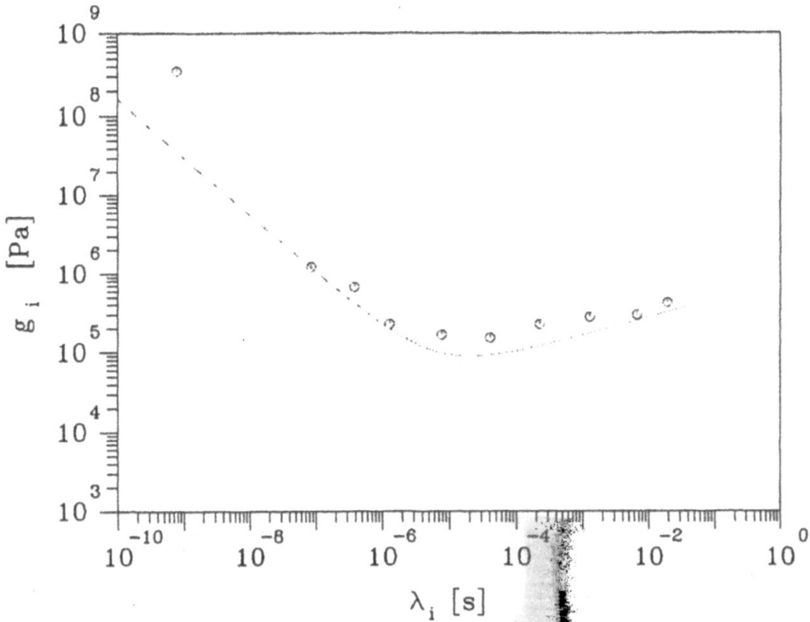

Figure 5.14 Discrete relaxation time spectrum as calculated from dynamic mechanical data of a nearly monodisperse linear polybutadiene with M_w = 70,000. The line represents the continuous BSW spectrum for the same material.

results that long for explanation just as in a real experiment. We start out by searching for general trends in the behavior of the algorithm and then try to abstract these trends into general statements ("rules") and check the limits at which they start to fail. Some of these properties will be discussed in the next paragraphs, and often it will become clear that our observations will need more rigorous proof before they will be acceptable in a more general context.

5.3.2.1 Ill-Posedness

The inversion of a *single* integral equation such as Eq. (22) for G' (or Eq. (23) for G'') is known to be an ill-posed problem [54, 40]. A sign of ill-posedness would be large variations in the solution when adding (or omitting) some data or even a single data point to the experimental set. Little is known about the simultaneous inversion of *two* interrelated integrals. A surprising observation, contrary to statements in the literature, is that the typical characteristics of ill-posedness completely vanish when the preceding inversion is simultaneously performed on both integrals for G' and G''. This lack of ill-posedness is most valuable for our purposes of finding the spectrum.

Ill-posedness has been avoided with discrete relaxation spectra by keeping N small [54]. The algorithm of Baumgärtel *et al.* also becomes more robust when reducing N. However, robustness does not seem to be the problem, for the algorithm behaved well-posed up to high N values as long as data were artificially smoothed (smoothing for testing purposes only), see Figure 16 in Ref. [50]. An upper limit occurred naturally at very high N at which the modes are so dense that rounding errors become important in the minimization procedure. This suggests that it is not the ill-posedness that creates an upper limit on N, but it is the over-fitting of noise in the data.

5.3.2.2 Uniqueness

It will be necessary to show uniqueness before further physical conclusions can be drawn from any experimental spectrum. For checking uniqueness, one needs to transform the discrete spectra g_i, λ_i into continuous form to see that, indeed, various very different looking discrete spectra reduce to the *same continuous function* $H(\lambda)$. If this is satisfied, we call the discrete solutions unique because they are all equivalent representations of one and the same $H(\lambda)$.

Several additional observations let us believe that the parsimonious spectrum poses a "unique" solution to the problem (within experimental error). The various observations entail the following:

1. The starting set $g_{i,0}$, $\lambda_{i,0}$ does not affect the final parsimonious spectrum.
2. Random omission of several data points from the experimental input does not change the calculated continuous $H(\lambda)$.
3. Addition of some artificial, random noise to the data does not affect the calculated continuous $H(\lambda)$ provided that we maintain the approximate overall noise level.

In addition to these observations on real data, Baumgärtel *et al.* [43, 50] produced artificial G', G'' data from a known spectrum $H(\lambda)$ and, to close the circle of thought, used the algorithm as described above and always returned to the original spectrum. This is an important test that needs to be satisfied by a conversion algorithm before considering it further.

5.3.3 *COMPUTER-AIDED CONSISTENCY TESTS*

5.3.3.1 Prediction of Other Material Functions and Comparison with Experiment

The newly determined $H(\lambda)$ may be expressed in any of the other linear viscoelastic material functions [8] and the result should be consistent with experimental observations and with theory. The most obvious tests are

plotting of G', G'' data against calculated G', G'' (from $H(\lambda)$),
plotting of deviations between the preceding functions versus frequency,
plotting of relaxation modulus (from $H(\lambda)$),
plotting of creep compliance (from $H(\lambda)$), and
predicting of stress during start-ups and comparison with experiments (see equations in Section 5.6).

All of these tasks are commonly performed by computer, and the result is presented in the interactive graphics mode. Specific tests are discussed in the following paragraphs.

5.3.3.1.1 Kramers–Kronig Check [74]

The storage modulus G' and the loss modulus G'' are not independent of each other. They are interrelated by the Kramers–Kronig relation

$$\frac{G'(\omega)}{\omega^2} = \frac{2}{\pi}\int_0^\infty \frac{dx}{x}\frac{G''(x)}{\omega^2 - x^2}, \tag{49}$$

which has been described by Bird *et al.* [27] and by Boij *et al.* [55]. The relation is difficult to apply to experimental data because the data are truncated at low and high frequency. This is where discretization helps. The discrete form of Eq. (49)

$$\sum_{i=1}^{N} \frac{g_i\lambda_i^2}{1+(\omega\lambda_i)^2} = \frac{2}{\pi}\int_0^\infty \frac{dx}{x}\frac{1}{\omega^2 - x^2}\sum_{i=1}^{N}\frac{g_i\lambda_i}{1+(\omega\lambda_i)^2} \tag{50}$$

may be rearranged into

$$\sum_{i=1}^{N} g_i\lambda_i \left\{\frac{\lambda_i}{1+(\omega\lambda_i)^2} - \frac{2}{\pi}\int_0^\infty \frac{dx}{x}\frac{1}{\omega^2 + x^2}\frac{1}{1+(x\lambda_i)^2}\right\} = 0 \tag{51}$$

It can be shown that the function in the bracket is equal to zero for any λ_i because each individual Maxwell mode g_i, λ_i satisfies the Kramers–Kronig relation. The equilibrium modulus G_e is included in these calculations as a Maxwell mode with infinitely long relaxation time. The conclusion is that the discrete relaxation modulus, Eq. (8), will always satisfy the Kramers–Kronig relation.

This result has far-reaching implications for the evaluation of dynamic mechanical experiments because the spectrum calculation tells us when a data set went bad: If the computer algorithm is not able to fit a given set of G', G'' data, then one can assume that the data set violates the Kramers–Kronig relation. Such violation can always happen for various experimental reasons. We have used that criterion effectively for detecting faulty data sets. The Kramers–Kronig check is one of the very strengths of the parsimonious spectrum calculation.

5.3.4 COMPUTER CODE

The preceding data analysis has been written as computer code. The code is used by many laboratories worldwide. Detailed descriptions can be found through the Internet at http://members.tripod.com/~Rheology/.

5.4 Time-Resolved Mechanical Spectroscopy (TRMS)

Mechanical spectroscopy is the preferred tool to study materials with changing structure such as polymers during gelation (an example will be given later), phase transition, decomposition, polymerization, and so forth. Time-resolved techniques were developed to study the dynamic properties of such mutating materials [56, 57, 58, 19]. We use the term mutation as a general expression for changes that affect the molecular mobility and thus the relaxation behavior of the investigated material. The change in chemically cross-linking polymers depends on the extent of reaction p, which changes from $p = 0$ to $p \to 1$ as the reaction proceeds. For want of a generally applicable variable, we will use symbol p to identify the structural state of a changing material without specifying the type of structural change.

The linear viscoelastic behavior of a transient material may be expressed as

$$\tau(t) = \int_{-\infty}^{t} dt' \, \underset{t'}{\overset{t}{G}}(t, t', p(t'))\, \dot{\gamma}(t') \tag{52}$$

where $\tau(t)$ is the shear stress and $\dot{\gamma}(t)$ is the shear rate. The relaxation function $\underset{t'}{\overset{t}{G}}(t, t', p(t'))$ accounts for mutation during the relaxation processes. It is positively valued and decays monotonically with $s = t - t'$. The equation is based on the linear superposition principle of stress and strain, i.e., its range of validity is restricted to sufficiently small strains and strain rates.

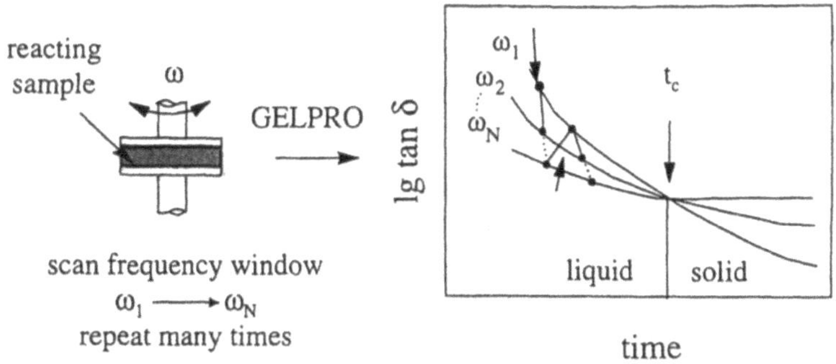

Figure 5.15 Schematic of time-resolved mechanical spectroscopy and expected data for a cross-linking material. The data analysis is performed with a software called GELPRO. Reference [19].

Mechanical spectroscopy is applicable to transient samples, especially because of its spectroscopic nature. However, a wide range of frequencies needs to be applied either simultaneously [20] or sequentially. Simultaneous probing at several frequencies (multiwave method, Fourier transform mechanical spectroscopy) has the advantage of shortening the experimental time and accommodating higher rates of mutation. However, it has the disadvantage that the overall strain amplitude may quickly exceed the limits of linear viscoelasticity. For this reason, we recently used sequentially increasing frequencies, one after the other, just as in conventional MS. At each frequency a single measurement was performed. This frequency cycle was then repeated again and again until the rate of mutation ceased. A schematic of this technique is shown in Figure 5.15 for a cross-linking material. In this example, the point of liquid–solid transition can be determined from the crossover of $\tan\delta(\omega)$, as described by Chambon and Winter [6]. A real data set is shown in Figure 5.16.

The sampling time t_s in time-resolved MS (TRMS) is again the duration of a single measurement, i.e., the time required by the rheometer to take a single data point. For an oscillatory shear experiment, t_s is approximately given by the period of the imposed shear wave (Eq. (36)). During this period the shear stress response is measured continuously and then transformed into a single data point for the properties of interest in such an experiment; for example, the dynamic moduli G' and G''.

5.4.1 *QUASI-STABILITY*

The sample changes during each sampling. If the total change is large during t_s, the function $\overset{t}{\underset{t'}{G}}$ of Eq. (52) needs to be known for the data analysis (which rarely is the case). To avoid such problems of sample mutation, the mutation during

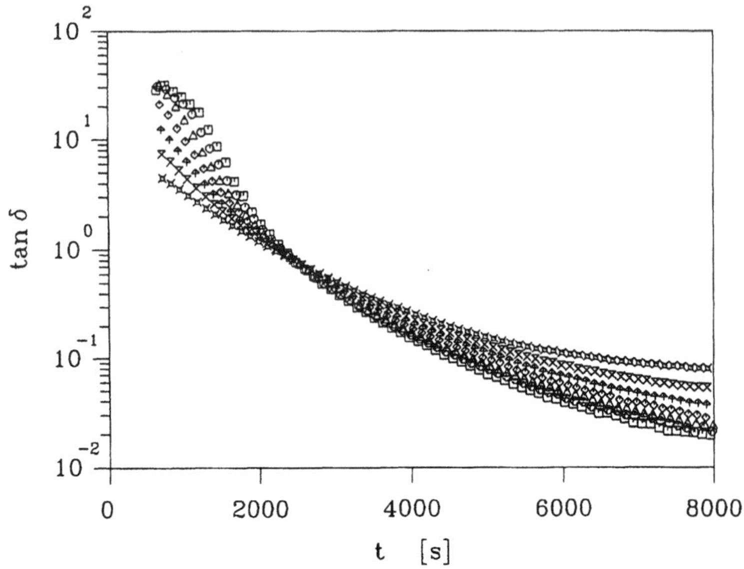

Figure 5.16 Evolution of loss tangent ($\tan\delta$) during cross-linking of PBD38 at 28 °C for several frequencies (1–100 rad/s, 3 frequencies/decade). The gel point is marked by the $\tan\delta$ crossover, i.e., the frequency independence of $\tan\delta$. Reference [73].

t_s needs to be small enough so that it can be neglected. The rate of mutation is shown as slope in Figure 5.17. The inverse of the slope has been used to define a mutation time λ_{mu}, which may refer to the properties G' or G''

$$\lambda'_{mu} = \left|\frac{1}{G'}\frac{\partial G'}{\partial t}\right|^{-1}, \qquad \text{or} \quad \lambda''_{mu} = \left|\frac{1}{G''}\frac{\partial G''}{\partial t}\right|^{-1} \tag{53}$$

The total change during the sampling time $t_s = 2\pi/\omega$ has been expressed in the mutation number [59] by

$$N_{mu} = \frac{\text{experimental time}}{\text{mutation time}} = \text{experimental time} * \text{rate of change},$$

which needs to be evaluated for both G' and G'' as follows:

$$N'_{mu} = \frac{2\pi}{\omega G'}\frac{\partial G'}{\partial t}, \qquad \text{or} \quad N''_{mu} = \frac{2\pi}{\omega G''}\frac{\partial G''}{\partial t}, \tag{54}$$

the larger value determines the admissibility of the experiment. Usually, G' responds more sensitively for transient materials. The mutation number may be

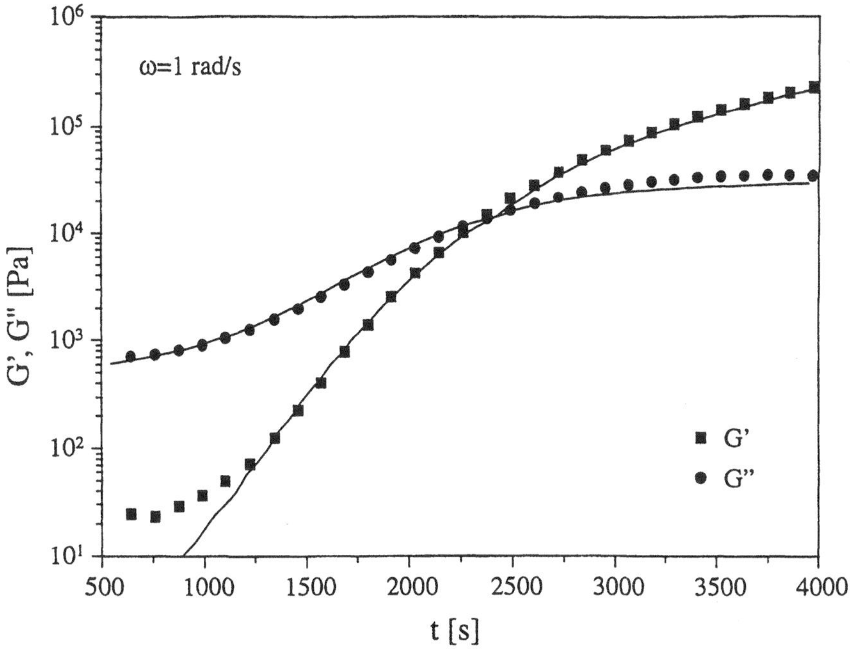

Figure 5.17 Evolution of dynamic moduli during cross-linking of PBD38 at $\omega = 1$ rad/s ($T = 28$ °C). Lines represent the model calculations as outlined in this chapter, and symbols depict the experimental data.

combined with the Deborah number (Eq. (23)) by

$$N_{\mathrm{De}}N'_{\mathrm{mu}} = \frac{\lambda}{G'}\frac{\partial G'}{\partial t}, \quad \text{or} \quad N_{\mathrm{De}}N''_{\mathrm{mu}} = \frac{\lambda}{G''}\frac{\partial G''}{\partial t} \tag{55}$$

This dimensionless group is important because dynamic mechanical experiments become nonlinear when N_{mu} or N_{mu} N_{De} exceed certain values. Winter *et al.* [59] found, for instance, that nonlinear effects in a gelation experiment could be observed for N_{mu} greater than 0.15. On the other hand, Mours and Winter [19] determined a critical mutation number of 0.9 for a constant heating rate experiment. For N_{mu} or N_{mu} N_{De} below the critical value, the sample can be assumed to behave quasi-stably during the sampling of single experimental points.

The characterization of a quasi-stable sample can be understood by looking at the schematic plot (property G', G'' versus time on different time scales) in Figure 5.18. The material properties change during the entire mutation. If the probing time or the rate of overall property change is small, the sample behaves quasi-stably during the experimental time t_s, resulting in a sinusoidal stress response

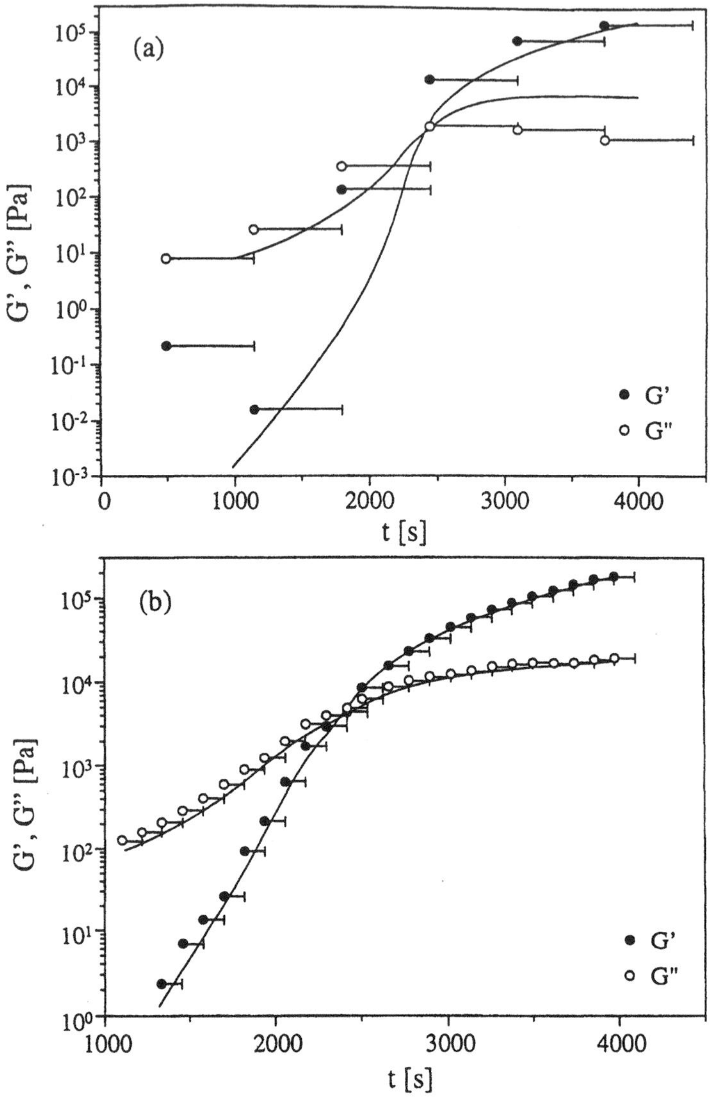

Figure 5.18 Evolution of dynamic moduli during cross-linking of PBD38 at (a) $\omega = 0.01$ rad/s and (b) $\omega = 0.1$ rad/s ($T = 28\,°\mathrm{C}$). Moduli calculated from the modeled shear stress are depicted by symbols. The displayed range is equivalent to the experimental time t_s. Lines correspond to the phenomenological model of Mours and Winter [73].

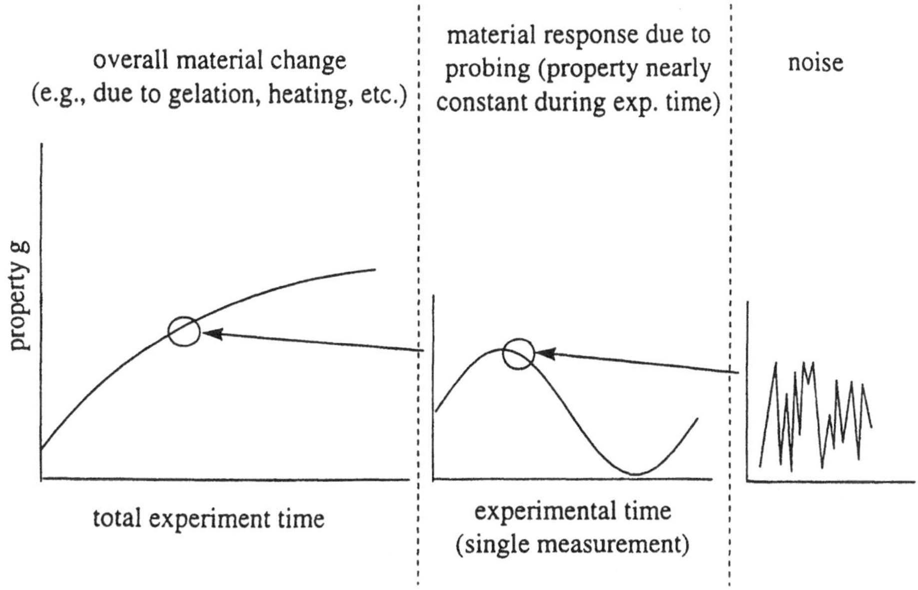

Figure 5.19 Change of interesting property g on different time scales (left: change of property during overall experiment; middle: response to the probing at prescribed frequency (during experimental time t_s); right: rapid variations of experimental signal (noise)) [19].

to an imposed sinusoidal strain. In this case, each data point represents a distinct (quasi-stable) state of the material.

5.4.2 TRMS DATA ANALYSIS [19]

Each TRMS data point represents a different state of the material, and interpolation is necessary to obtain the desired data as a function of frequency at distinct material states. The interpolation requires the following sequence of procedures, which have been assigned to a personal computer:

5.4.2.1 Sorting

The data points $G'(\omega_j, p(t))$ and $G''(\omega_j, p(t))$ ($j = 1, 2, \ldots M$, where M is the number of discrete frequencies), initially sorted in terms of increasing time, must be rearranged in terms of increasing frequency.

5.4.2.2 Smoothing

Continuous curves $f_j'(p)$ and $f_j''(p)$ are calculated for each frequency ω_j representing the evolving moduli $G'(\omega_j, p(t))$ and $G''(\omega_j, p(t))$ as a function of the variable p, which describes the ongoing mutation process. Data smoothing in conjunction with the curve fitting gives the most consistent results. This must be repeated for each discrete frequency ω_j of the G' and G'' data set.

5.4.2.3 Interpolation

Using these fitted curves, dynamic mechanical material functions ($G'(\omega, t_i)$, $G''(\omega, t_i)$, $\tan\delta(\omega, t_i)$, $2\delta(\omega, t_i)/\pi$, etc.) can be interpolated at discrete material states (represented by the time t_i) for each of the frequencies. This results in a comprehensive rheological description of each material state during the structural or other development that influences the flow properties.

5.4.2.4 Further Analysis

The interpolated data sets can now be used to study the material behavior at different structural states. One example is the use of superposition principles, if applicable, such as time–temperature superposition or time–cure superposition, to combine the data into master curves. As an example for a typical TRMS experiment, data on a cross-linking polybutadiene [73] are shown in Figure 5.20.

5.4.3 CROSS-LINKING OF POLYBUTADIENE (PBD 38)

As an example, we consider cross-linking of a polybutadiene precursor with $M_w = 37{,}900$ (PBD38). The evolution of the relaxation time spectrum during cross-linking was described using a phenomenological model and corresponding parameters, as given by Mours and Winter [73]. Mutation numbers for all frequencies were calculated using

$$N_{mu} = \frac{t_s}{G'}\frac{\partial G'}{\partial t},$$

as given in Table 5.1. The very low mutation numbers ($N_{mu} < 0.07$) at the experimental frequencies ($\omega > 1$ rad/s) ensured that effects of the material change on the measured dynamic moduli were negligible. In these cases, agreement between the model calculations and the experimental dynamic moduli was satisfactory, as displayed in Figure 5.17, for the case of $\omega = 1$ rad/s. The deviations at short times in the storage modulus are due to the transducer insensitivity at low torque values.

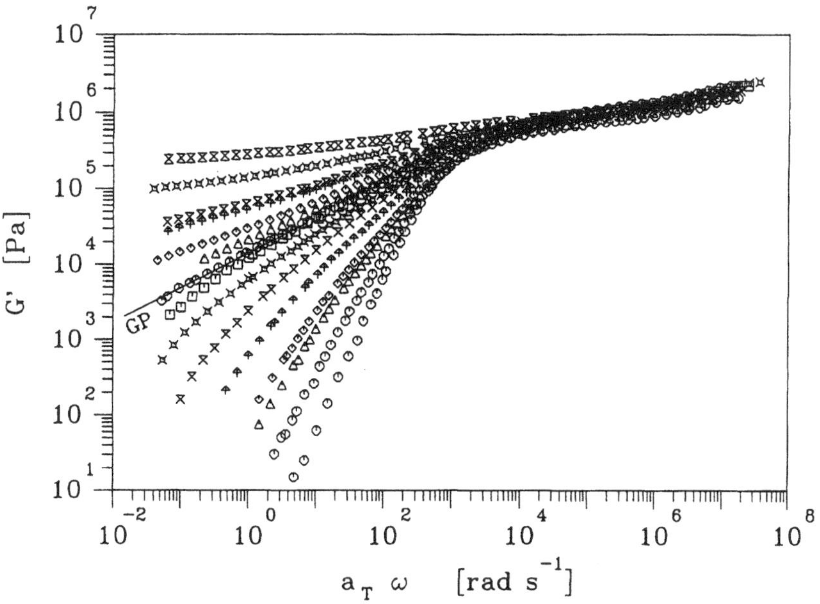

Figure 5.20 Storage modulus data of PBD38 samples at different extents of reaction ($T_{ref} = 28$ °C). Time-resolved mechanical spectroscopy data in combination with data on chemically stopped samples extend the frequency window, which, for TRMS alone, would be quite small. The gel point is marked by a line [73].

The resulting dynamic moduli for the lower frequencies ($\omega = 0.01$ rad/s and $\omega = 0.1$ rad/s) are shown in Figure 5.18. Two sets of symbols are displayed for each frequency connected by horizontal lines. The reason for this is the ambiguity in determining the time for each data point. Because the experimental time t_s can be rather long, depending on the frequency (see Table 5.1), the calculated

Table 5.1 Average and maximum mutation numbers and experimental times at several frequencies during cross-linking of PBD38.

ω (rad/s)	$t_s(s)$	$N_{mu,max}$	$N_{mu,average}$
0.01	650	3.7	2.1
0.03	300	3.0	1.2
0.05	200	2.1	0.9
0.1	120	1.1	0.5
1	10	0.07	0.03
10	3	0.02	0.01

values represent an average over this time period t_s. In an actual experiment, the rheometer usually uses the initial time of $t_{initial}(t_s = t_{end} - t_{initial})$ as the actual time for the corresponding data point. This results in an overestimation of the moduli because they keep increasing during the cross-linking reaction. Similarly, if one takes t_{end} as time for the data point, the moduli are generally too low. The lines in the preceding figures represent model calculations of the dynamic moduli using the phenomenological model developed by Mours and Winter [73].

Calculated moduli agree reasonably well with experimental data, except for the lowest frequency cases ($\omega = 0.01$ rad/s). The average and maximum mutation numbers are very high (above 2) in this case. For the three higher frequencies ($\omega = 0.03$ rad/s to $\omega = 0.1$ rad/s), the moduli increase monotonically with time and follow along the expected values. However, because of the ambiguity in the determination of the time t for each data point, it is impossible to find actual agreement between simulation and model. Lower mutation numbers (at higher frequencies) tend to give closer agreement. This is especially true beyond the gel point, where mutation numbers are usually lower than before the gel point. The criterion N_{mu} tends to take on the highest value shortly before the gel point for all studied frequencies.

In conclusion, the preceding modeling basically agrees with experimental findings that the mutation number during time-resolved rheometry experiments of cross-linking samples should stay below a value of 0.2. The average value of 0.5 at $\omega = 0.1$ rad/s is still too high to achieve good agreement with the expected data without arbitrarily changing the time stamp of each data point. However, the lower mutation number at frequencies above $\omega = 1$ rad/s results in good agreement between experimental data and simulation.

5.5 Temperature Effects and Time–Temperature Superposition

5.5.1 *TIME–TEMPERATURE SUPERPOSITION (tTS)*

Thermorheologically simple materials obey the so-called time–temperature superposition principle for which time and temperature changes are equivalent [8]. Frequency-dependent data at different temperatures can be superimposed (by simultaneous horizontal and vertical shifting) to yield a so-called master curve at a reference temperature T_{ref} within the experimental temperature range. This property was found with homopolymers (and some miscible blends) over wide frequency or time windows. Commercial software packages are available that can easily perform this type of shifting on a personal computer (see, for instance, http://members.tripod.com/ ~Rheology/). The superposition typically yields plots of

reduced storage modulus $b_T G'$	versus reduced frequency $a_T \omega$
reduced loss modulus $b_T G''$	versus reduced frequency $a_T \omega$
loss tangent G''/G'	versus reduced frequency $a_T \omega$
reduced relaxation modulus $b_T \overset{0}{G}$	versus reduced time t/a_T
reduced dynamic viscosity $b_T \eta^*/a_T$	versus reduced frequency $a_T \omega$

The temperature shift factors a_T and b_T represent the horizontal and vertical shift, respectively, of data at the experimental temperature and onto T_{ref}.

Time–temperature superposition allows shifting of discrete spectra to new temperatures. This requires that groups of discrete relaxation modes

$$\lambda_i(T) = \lambda_i(T_{\mathrm{ref}})a_T, \qquad \text{and} \quad g_i(T) = g_i(T_{\mathrm{ref}})/b_T \tag{56}$$

shift with common shift factors a_T and b_T. Time–temperature superposition, if applicable, is able to extend the available frequency window (see Figure 5.21). This extension of the frequency window is one of the main reasons for performing time–temperature superposition.

5.5.2 *LIMITS OF tTS*

An understanding of tTS behavior in MS experiments helps to define experimental criteria for measuring the limits when tTS starts to break down. For example, temperature changes may induce transitions in morphology that cause sudden changes in the molecular mobility. Appropriate methods for MS have been discussed in Section 5.4. Time–temperature superposition is not possible for these materials.

It should be noted that even the long-chain dynamics (entanglement and terminal region) of homopolymers do not necessarily shift with the same shift factor as the short segmental dynamics (glass transition, glass), as shown by Inoue *et al.* [60]. This is not very obvious because most data sets do not cover a wide enough frequency or time window. In general, miscible polymer blends do not obey the tTS. Exceptions are polymer blends for which (a) the glass transition temperatures of the components are very close to each other, or (b) the molecules of the blend components interact very strongly, or both [61, 62].

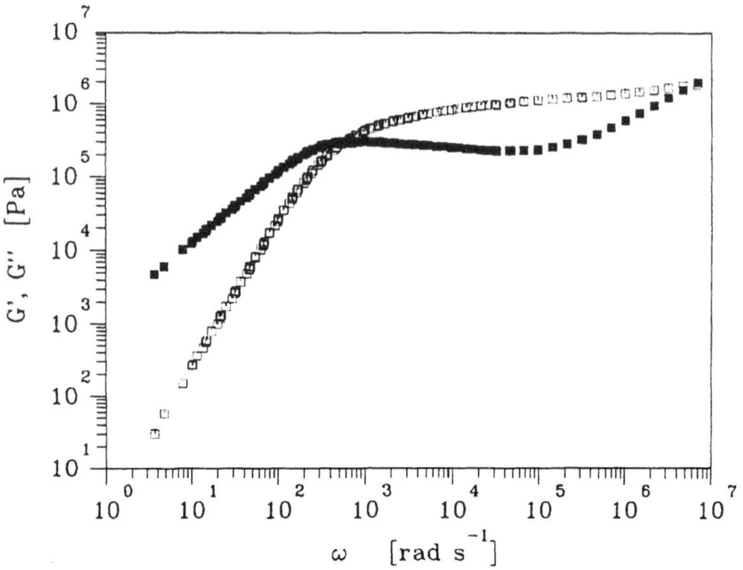

Figure 5.21 Storage and loss modulus master curves of a nearly monodisperse linear polybutadiene with Mw = 37,700. The data cover the terminal region at low frequencies and show the entanglement plateau in G' at intermediate frequencies and the crossover to the glass transition at high frequencies [72].

5.5.3 TEMPERATURE SHIFT FACTORS

Relaxation rates are highest at high temperatures. There have been different attempts in the literature to describe the temperature dependence of the horizontal shift factors. Today, the three-parameter Williams–Landel–Ferry equation (WLF) is widely accepted [8]:

$$\log a_T = \frac{-C_1(T - T_{\text{ref}})}{C_2 + T - T_{\text{ref}}}, \qquad T_g < T < (T_g + 200\,\text{K}), \tag{57}$$

where T_g is the glass transition temperature of the specific polymer. Williams *et al.* [63] proposed universal values of $C_1 = 7.60$ and $C_2 = 227.3$ K if T_{ref} is chosen somewhat close to T_g.

Some researchers prefer the format of the three-parameter Vogel equation [64, 65]:

$$\ln a_T = \frac{E_a}{R}\left(\frac{1}{T - T_V} - \frac{1}{T_{\text{ref}} - T_V}\right), \tag{58}$$

which is equivalent to the WLF equation. The Vogel temperature T_V is much below the glass temperature. Parameters in Eqs. (57) and (58) are convertible into each other:

$$C_1 = \frac{E_a}{R(T_{ref} - T_V)} \quad \text{and} \quad C_2 = T_{ref} - T_V \tag{59}$$

Far above the Vogel temperature, $T \gg T_V$, a simpler two-parameter Arrhenius-type expression is sufficient to trace experimental data:

$$\ln a_T = \frac{E_a}{R}\left(\frac{1}{T} - \frac{1}{T_{ref}}\right) \quad \text{for } (T_g + 200\,\text{K}) < T. \tag{60}$$

The Arrhenius parameter E_a/R is usually extracted from a plot of $\log a_T$ versus reciprocal temperature $1/T$. If the Arrhenius-type dependence is a valid approximation, this plot shows a straight line with a slope of $E_a/(2.303R)$. In the case of WLF-behavior, the temperature dependence of a_T deviates from this straight line (see Figure 5.22 for polybutadiene with $M_w = 37{,}900$).

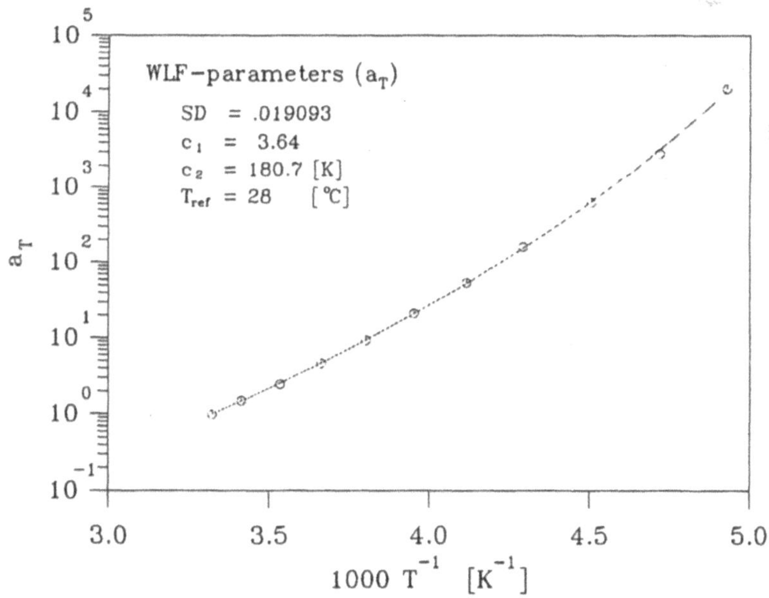

Figure 5.22 Horizontal temperature shift factors a_T and WLF-fit for PBD38. Reference [73].

Temperature-induced density changes result in a vertical shift factor [8]

$$b_T(T; T_{\text{ref}}) = \frac{T_{\text{ref}}\rho(T_{\text{ref}})}{T\rho(T)} \tag{61}$$

Its value is usually close to unity for moderate temperatures differences.

Time–temperature superposition of experimental data yields values of shift factors a_T and b_T at different temperatures so that the necessary parameters of Eqs. (57), (58), and (60) can be evaluated by regression techniques. If the parameters E_a/R, C_1, C_2, T_V, and $\rho(T_{\text{ref}})/\rho(T)$ are known, the constructed master curve can be shifted to any temperature within the experimental range. Extrapolation of shift factors to temperatures outside the experimental window is possible but should be viewed with caution.

5.5.4 *NONISOTHERMAL FLOW*

Processing flows are rarely isothermal, and the temperature variations severely influence the stress. Hopkins [66] and Morland and Lee [67] suggested a simple modification of the memory integral equations to account for nonisothermal histories $T(t') \neq T_{\text{ref}}$. The Boltzmann's equation, Eq. (4), may be rewritten as

$$\tau(t) = \int_{-\infty}^{t} dt' \dot{\gamma}(t') \sum_{i=1}^{N} g_i \, e^{-f(t;t')/\lambda_i} \tag{62}$$

with an exponent

$$f(t;\, t') = \int_{t'}^{t} dt'' \frac{1}{a_T(t'')} \tag{63}$$

The temperature shift factor changes along the path line of material elements according to their temperature history $a_T(t'') = a_T(T_{\text{ref}}; T(t''))$. The nonisothermal effects result in a stretching or compressing of the time scale. Conveniently, one uses $f(t;\, t')$ as the new time. For isothermal flow, f reduces to $f = t - t'$. It should be noted that Eqs. (62) and (63) only apply to polymers for which the tTS principle holds.

5.5.5 *TEMPERATURE SCAN*

The temperature dependence of molecular mobility in polymers can be measured by gradually increasing (or decreasing) temperature during small-amplitude oscillatory shear. This so-called temperature sweep or temperature scan (TS) is a commonly used experiment in polymer rheometry. The frequency ω is held constant throughout and, hence, the experiment has been called isochronal TS (although it is only isochronal with respect to the experimental time scale; the probed material time scale changes with temperature). The choice of frequency is quite arbitrary and limits the observation to molecular motions in the vicinity of a single mode instead of a wider spectrum of relaxation times. The TS experiment is not very informative for this reason.

In extension of this common procedure, the rheometer should be programmed to probe, while heating a polymer sample, a sequence of frequencies ω_i within a frequency window $\omega_{\min} < \omega_i < \omega_{\max}$. The sequence of frequencies may be repeated cyclically during the temperature scan. The method requires that the sample be in thermal equilibrium at all times, which seems to be the case for many rheometers even at moderate-to-high heating rates [19]. In this fashion, a single experiment provides the temperature and frequency dependence of the relaxation spectrum. The problem is that every data point in this experiment, which can be called frequency–temperature scan (FTS), is taken at a new temperature, and data analysis requires interpolation. However, such data interpolation methods and their limits of application are well established (time-resolved rheometry, see Section 5.4). Modern rheometers allow a similar procedure, the FTS. In this experiment, the polymer sample is probed at a cyclically repeated sequence of frequencies. After each frequency sequence, the temperature is increased or decreased in preprogrammed steps. Furthermore, one can perform several independent frequency scans at different temperatures to achieve the desired frequency and temperature-dependent data.

5.6 Applications of the Relaxation Time Spectrum

The relaxation spectrum describes the underlying features of all linear viscoelastic material functions (except for the equilibrium modulus G_e in solids). It contains all necessary information about the relaxation behavior and therefore represents a complete rheological picture of the material within the linear viscoelastic regime. Many different material functions can be directly calculated from the spectrum. For example, a dynamic mechanical experiment with data sets $G'(\omega)$ and $G''(\omega)$ may be converted into $\overset{0}{G}(t)$ and compared with the results from a stress relaxation experiment, $\overset{0}{G}(t)$. Similarly, MS and creep experiments can be compared and scrutinized for self-consistency. The MS data may be converted into a spectrum that defines the zero-shear viscosity η_0 and first normal stress coefficient ψ_1:

$$\eta_0 = \int_0^{\infty} d\lambda H(\lambda) = \sum_{i=1}^{N} g_i \lambda_i \tag{64}$$

$$\psi_1 = 2\int_0^{\infty} d\lambda\, \lambda H(\lambda) = 2\sum_{i=1}^{N} g_i \lambda_i^2 \tag{65}$$

5.6.1 *LINEAR VISCOELASTIC SIMULATIONS*

The spectrum allows for the simulation of start-up of shear or extensional flow in a sample that had been equilibrated at rest. Simulation involves the tensorial form of the Boltzmann equation, see Appendix 5.A, which can be found in rheology text books. For example, in a start-up and stress relaxation experiment, a material is sheared initially with a constant strain rate $\dot{\gamma}_0$ for a time t_0. The strain is then kept at a constant value. During this experiment, the stress is recorded. In the linear viscoelastic region, the shear stress $\tau(t)$ and the first normal stress difference $N_1(t)$ response during the initial shear, $0 < t < t_0$, is given by [1]:

$$\tau(t) = \dot{\gamma}_0 \left[G_e t + \sum_{i=1}^{N} g_i (1 - e^{-t/\lambda_i}) \right] \tag{66}$$

$$N_1(t) = \dot{\gamma}_0^2 \left[G_e t_0^2 + 2\sum_{i=1}^{N} g_i \lambda_i^2 (1 - (1 + 1/\lambda_i) e^{-t/\lambda_i}) \right]$$

and during stress relaxation ($t_0 < t$)

$$\tau(t) = \dot{\gamma}_0 \left[G_e t_0 + \sum_{i=1}^{N} g_i \lambda_i \left(e^{-(t-t_0)/\lambda_i} - e^{-t/\lambda_i} \right) \right] \tag{67}$$

$$N_1(t) = \dot{\gamma}_0^2 \left[G_e t_0^2 + 2\sum_{i=1}^{N} g_i \lambda_i^2 (1 - (1 + t_0/\lambda_i) e^{-t_0/\lambda_i}) e^{-(t-t_0)/\lambda_i} \right]$$

Another example is the start-up of uniaxial extension at constant rate $\dot{\varepsilon}_0$. The first normal stress difference is predicted as [68]:

$$\tau_{11} - \tau_{22} = \sum_{i=1}^{N} g_i \left[\frac{1 - 2\dot{\varepsilon}_0 \lambda_i e^{-(1-2\dot{\varepsilon}_0\lambda_i)t/\lambda_i}}{1 - 2\dot{\varepsilon}_0 \lambda_i} - \frac{1 + \dot{\varepsilon}_0 \lambda_i e^{-(1+\dot{\varepsilon}_0\lambda_i)t/\lambda_i}}{1 + \dot{\varepsilon}_0 \lambda_i} \right] \tag{68}$$

5.6.2 *FLOW MODELLING*

The spectrum has been used in commercially available programs to simulate the flow behavior of viscoelastic materials at large strains (for example, simulation software using finite element methods to study the flow in complex geometries). The spectrum is an important part of large-strain constitutive equations [27, 69].

Flow modeling of polymer processing often relies on the steady shear viscosity $\eta(\dot{\gamma})$ when only the relaxation time spectrum is available. Fortunately, for many polymers heuristic relations exist that predict the viscosity from the linear relaxation time spectrum. An estimate of the flow curve $\eta(\dot{\gamma})$ may be based on the Cox–Merz relation [70] between the steady and dynamic viscosity:

$$\eta(\dot{\gamma}) = \eta^*(\omega)\Big|_{\omega=\dot{\gamma}} = \sqrt{G'^2 + G''^2}/\omega\Big|_{\omega=\dot{\gamma}}$$

$$= \sqrt{\left[\sum_{i=1}^{N} \frac{g_i \lambda_i^2 \dot{\gamma}}{1 + (\lambda_i \dot{\gamma})^2}\right]^2 + \left[\sum_{i=1}^{N} \frac{g_i \lambda_i}{1 + (\lambda_i \dot{\gamma})^2}\right]^2} \tag{69}$$

5.6.3 *RELAXATION PATTERNS*

A long-standing quest in polymer physics is the search for a relation of molecular architecture and linear viscoelastic properties [8, 71, 72]. Ideal molecular architectures are believed to generate some simple relaxation patterns. One example for this simplicity in pattern detection is the BSW spectrum, which describes the relaxation of a linear flexible polymer of uniform length (Eq. (6)). From all the commonly used rheological material functions, this spectrum was preferred because it allows the most simple expression for the observed relaxation phenomena. A look at the dynamic moduli G' and G'' (as, for example, shown in Figure 5.9) does not easily reveal this simple pattern of the spectrum (Figure 5.4) that completely describes the linear viscoelastic relaxation behavior. Another example for such an ideal spectrum is found with cross-linking polymers at their gel point, Eq. (37).

Appendix 5.A Boltzmann Equation

The linear constitutive equation of a viscoelastic material following the Boltzmann superposition principle is given by

$$\tau(t) = \int_{-\infty}^{t} dt' \, \overset{0}{G}(t - t')\dot{\gamma}(t') \tag{A1}$$

Various forms of the Boltzmann equation have been in use. It often is written in terms of strain γ between states at t and t'

$$\gamma(t;t') = \int_{t'}^{t} dt''\dot{\gamma}(t'') \qquad \text{(A2)}$$

The Boltzmann equation is then transformed to

$$\tau(t) = \int_{-\infty}^{t} dt'\overset{0}{m}(t-t')\,\gamma(t;t') \qquad \text{(A3)}$$

with the memory function $\overset{0}{m}(s)$ given by

$$\overset{0}{m}(s) = -\frac{d\overset{0}{G}(s)}{ds} \qquad \text{(A4)}$$

with $s = t - t'$. The continuous spectrum may be introduced into the Boltzmann equation, Eq. (4), to obtain

$$\tau(t) = G_e \int_0^{\infty} ds\,\dot{\gamma}(t-s) + \int_0^{\lambda_{max}} \frac{d\lambda}{\lambda} H(\lambda) \int_0^{\infty} ds\, e^{-s/\lambda}\dot{\gamma}(t-s) \qquad \text{(A5)}$$

Calculations with the spectrum are as convenient as with the relaxation modulus. Both need to be considered on an equal footing.

Tensorial forms of the Boltzmann equation are required for flow simulations (as shown in Section 5.6.1). These use the velocity gradient tensor $\nabla\boldsymbol{\nu}$ and the extra stress tensor $\boldsymbol{\tau}$

$$\boldsymbol{\tau} = \int_{-\infty}^{t} dt'\overset{0}{G}(t-t')(\nabla\boldsymbol{\nu}(\boldsymbol{x},t') + \nabla\boldsymbol{\nu}(\boldsymbol{x},t')^T) \qquad \text{(A6)}$$

and the Finger strain tensor $\boldsymbol{B}(t;t')$ [75, 69]

$$\boldsymbol{\tau} = G_e\boldsymbol{B}(t;t_0) + \int_{-\infty}^{t} dt'\overset{0}{m}(t-t')\boldsymbol{B}(t;t') \qquad \text{(A7)}$$

No new material properties need to be introduced. $\boldsymbol{B}(t;t_0)$ denotes the strain with respect to the stress-free state at equilibrium.

Appendix 5.B Related Material Functions

The small amplitude oscillatory shear experiments can be expressed in many different ways that are all related to G' and G''. The most common of these functions are

complex modulus $$G^*(\omega) = \sqrt{G'^2 + G''^2} \tag{B1}$$

complex viscosity $$\eta^*(\omega) = \sqrt{G'^2 + G''^2}/\omega \tag{B2}$$

loss tangent $$\tan\delta = G''/G' \tag{B3}$$

dynamic viscosities $$\eta'(\omega) = G''/\omega \quad \text{and} \quad \eta''(\omega) = G'/\omega \tag{B4}$$

stiffness $$S'(\omega) = \frac{G'(\omega)}{\Gamma(2\delta/\pi)\cos\delta\,\omega^{2\delta/\pi}} \tag{B5}$$

relaxation exponent $$n(\omega) = 2\delta/\pi \tag{B6}$$

dynamic compliances $$J'(\omega) = \frac{G'}{G'^2 + G''^2} \text{ and } J''(\omega) = \frac{G''}{G'^2 + G''^2} \tag{B7}$$

where $\Gamma(x)$ is the gamma function.

References

1. H. M. Laun, *Rheol. Acta* **17:** 1–15 (1978).
2. Y. Einaga, K. Osaki, M. Kurata, S. Kimura, and M. Tamura, *Polym. J.* **2:** 550 (1971).
3. K. Osaki, *Rheol. Acta* **32:** 429–437 (1993).
4. K. Osaki, *Proc. 7th Int. Congr. Rheology*, Gothenburg: 104–109 (1976).
5. M. H. Wagner, *Rheol. Acta* **15:** 136–142 (1976).
6. F. Chambon and H. H. Winter, *Polym. Bull.* **13:** 499–503 (1985).
7. H. H. Winter and M. Mours, *Adv. Polym. Sci.* **134:** 165–234 (1997).
8. J. D. Ferry, *Viscoelastic Properties of Polymers*, 3rd ed., John Wiley & Sons, New York, 1980.
9. P. E. Rouse, *J. Chem. Phys.* **21:** 1272–1280 (1953).
10. P. G. de Gennes, *J. Chem. Phys.* **55:** 572 (1971).
11. M. Doi and S. Edwards, *The Theory of Polymer Dynamics*, Clarendon Press, Oxford, 1986; M. Doi, *Chem. Phys. Lett.* **26:** 269–272 (1974); M. Doi and S. F. Edwards, *J. Chem. Soc. Faraday Trans.* II **74:** 1789–1832 (1978).
12. J. des Cloizeaux, *Euro. Phys. Lett.* **5:** 437 (1988); *Macromolecules* **23:** 4678 (1990).
13. K. S. Schweizer, *J. Chem. Phys.* **91:** 5802 (1989); *J. Chem. Phys.* **91:** 5822-5839 (1989); *J. Chem. Phys.* **103:** 1934 (1995); *J. Chem. Phys.* **103:** 6296 (1995); *J. Chem. Phys.* **106:** 347 (1997).
14. S. T. Milner and T. C. B. McLeish, *Phys. Rev. Lett.* (1998).

15. M. Baumgärtel, A. Schausberger, and H. H. Winter, *Rheol. Acta* **29:** 400–408 (1990).
16. J. K. Jackson, M. E. De Rosa, and H. H. Winter, *Macromolecules* **27:** 2426–2431 (1994).
17. G. C. Berry and T. G. Fox, *Adv. Polym. Sci.* **5:** 261 (1968).
18. B. Gross, *Mathematical Structure of the Theories of Viscoelasticity*, Hermann & Cie., Paris (1953).
19. M. Mours and H. H. Winter, *Rheol. Acta* **33:** 385–397 (1994).
20. E. E. Holly, S. K. Venkataraman, F. Chambon, and H. H. Winter, *J. Non-Newtonian Fluid Mech.* **27:** 17–26 (1988).
21. M. Reiner, *Physics Today*, Jan: 62 (1964).
22. N. W. Tschoegl, *The Phenomenological Theory of Linear Viscoelastic Behavior*, Springer-Verlag, Heidelberg, 1989.
23. A. Schausberger, G. Schindlauer, and H. Janeschitz-Kriegl, *Rheol. Acta* **24:** 220 (1985).
24. J. K. Jackson and H. H. Winter, *Macromolecules* **28:** 3146–3155 (1995).
25. H. Watanabe and T. Kotaka, *Macromolecules* **17:** 2316 (1984).
26. S. H. Wassermann and W. W. Graessley, *J. Rheol.* **36:** 543 (1992); S. H. Wassermann, *J. Rheol.* **39:** 601 (1995).
27. R. B. Bird, R. C. Armstrong, and O. Hassager, *Dynamics of Polymeric Liquids*, Vol. 1, John Wiley & Sons, New York, 1987.
28. K. Walters, *Rheometry*, Chapman and Hall, London, 1975.
29. J. M. Dealy and K. F. Wissbrun, *Melt Rheology and its Role in Plastics Processing*, Van Nostrand Reinhold, New York, 1990.
30. C. W. Macosko, *Rheology: Principles, Measurements, and Applications*, VCH Publ., Inc., New York, 1994.
31. G. Marin, *Rheological Measurement*, 2nd ed., A. A. Collyer and D. W. Clegg, Eds., Chapman & Hall, London, 1998.
32. H. M. Laun, Characterization of Wall Slip of Polymer Melts and its Relevance to Polymer Processing, Proceedings of IUPAC Conference, p. 11, World Polymer Congress, Australia, 1998.
33. M. Wilhelm, D. Maring, and H. W. Spiess, *Rheol. Acta* **37:** 399–405 (1998).
34. J. A. Yosick, A. J. Giacomin, W. E. Stewart, and F. Ding, *Rheol. Acta* **37:** 365 (1998).
35. A. J. Giacomin and J. M. Dealy, *Rheological Measurement*, A. A. Collyer and D. W. Clegg, Eds., Chapman & Hall, London, 1998.
36. J. M. Simmons, *Rheol. Acta* **7:** 184–188 (1968) and R. I. Tanner and G. Williams, *Rheol. Acta* **10:** 528–538 (1971).
37. G. B. McKenna and L. J. Zapas, *Polym. Engr. Sci.* **26:** 725–729 (1986); *J. Polym. Sci., Phys. Ed.* **23:** 1647–1656 (1985).
38. L. Walker, J. Vermant, P. Moldernaers, and J. Mewis, *Rheol. Acta* (1998).
39. C. Friedrich and B. Hofmann, *Rheol. Acta* **22:** 425–434 (1983).

40. J. Honerkamp and J. Weese, *Macromolecules* **22:** 4372–4377 (1989); *Rheol. Acta* **32:** 65–73 (1993).
41. V. M. Kamath, M. R. Mackley, *J. Non-Newtonian Fluid Mech.* **32:** 119–144 (1989).
42. F. Schwartzl and A. J. Staverman, *Appl. Sci. Res. A* **4:** 127–141 (1953).
43. M. Baumgärtel and H. H. Winter, *Rheol. Acta* **28:** 511–519 (1989).
44. D. W. Mead, *J. Rheol.* **38:** 1769–1795 (1994).
45. I. Emri and N. W. Tschoegl, *Rheol. Acta* **31:** 161–174 (1993); *Rheol. Acta* **32:** 322–327 (1993); *Rheol. Acta* **33:** 60–70 (1994).
46. N. Orbey and J. M. Dealy, *J. Rheol.* **35:** 1035–1049 (1991).
47. J. Kaschta and F. R. Schwarzl, *Rheol. Acta* **33:** 530–541 (1994).
48. A. J. Staverman, F. Schwarzl, In: *Die Physik der Hochpolymeren*, Vol. 4, Springer-Verlag, Berlin, 1956.
49. J. K. Jackson, C. Garcia-Franco, and H. H. Winter, ANTEC **38:** 2438–2442 (1992).
50. M. Baumgärtel and H. H. Winter, *J. Non-Newtonian Fluid Mech.* **44:** 15–36 (1992).
51. H. Nyquist, *Trans. Amer. Inst. Electr. Eng.* **47:** 617–644 (1928).
52. H. H. Winter, M. Mours, M. Baumgärtel, and P. Soskey, *Rheological Measurement*, 2nd ed., A. A. Collyer and D. W. Clegg, Eds., Chapman & Hall, London, 1998.
53. V. A. Mozorov, *Methods for Solving Incorrectly Posed Problems*, Springer, Berlin, 1984.
54. R. I. Tanner, *J. Appl. Polym. Sci.* **12:** 1649–1652 (1968).
55. H. C. Booij and G. P. J. M. Thoone, *Rheol. Acta* **21:** 15–24 (1982).
56. K. te Nijenhuis and H. Dijkstra, *Rheol. Acta* **14:** 71–84 (1975).
57. G. Marin, J. J. Labaig, and P. Monge, *Rheol. Acta* **16:** 527 (1975).
58. J. C. Scanlan and H. H. Winter, *Makrom. Chem., Makrom. Symp.* **45:** 11–21, (1991).
59. H. H. Winter, P. Morganelli, and F. Chambon, *Macromolecules* **21:** 532–535 (1988).
60. T. Inoue, H. Hayashihara, H. Okamoto, and K. Osaki, *J. Polym. Sci.* **30:** 409–414 (1992).
61. R. H. Colby, *Polymer* **30:** 1275–1278 (1989).
62. J. A. Pathak, R. H. Colby, S. Y. Kamath, S. K. Kumar, and R. Stadler, *Macromolecules* **31:** 8988–8997 (1998).
63. M. L. Williams, R. F. Landel, and J. D. Ferry, *J. Amer. Chem. Soc.*\ **77:** 3701–3707 (1955).
64. H. Vogel, *Physik Z.* **22:** 645–646 (1921).
65. G. S. Fulcher, *J. Amer. Cer. Soc.* **8:** 339–355, 789–794 (1925).
66. I. L. Hopkins, *J. Polym. Sci.* **28:** 631 (1958).
67. L. W. Morland and E. H. Lee, *Trans. Soc. Rheol.* **4:** 233–263 (1960).
68. H. Chang and A. S. Lodge, *Rheol. Acta* **11:** 127–129 (1972).
69. R. G. Larson, *The Structure and Rheology of Complex Fluids.* Oxford University Press, New York, 1999.
70. W. P. Cox and E. H. Merz, *J. Poly. Sci.* **28:** 619–622 (1958); D. Doraiswamy, A. M. Mujumdar, I. Tsao, A. N. Beris, S. C. Danforth, and A. B. Metzner, *J. Rheol.* **35:** 647–685 (1991); Y. G. Lin, P. W. Jin, J. C. W. Chien, and H. H. Winter, *Polymer* **30:** 831–834 (1989).

71. W. W. Graessley, *Adv. Polym. Sci.* **16:** 1–179 (1974); G. Marin, W. W. Graessley, *Rheol. Acta* **16:** 527 (1977); V. R. Raju, E. V. Menezes, G. Marin, W. W. Graessley, L. J. Fetters, *Macromolecules* **14:** 1668–1676 (1981); W. W. Graessley, *Adv. Polym. Sci.* **47:** 68 (1982); D. Pearson, *Rubber Chem. Tech.* **60:** 437 (1987).
72. M. Rubinstein, E. Helfand, and D. Pearson, *Macromolecules* **20:** 822 (1989).
73. M. Mours and H. H. Winter, *Macromolecules* **29:** 7221–7229 (1996).
74. H. H. Winter, *J. Non-Newtonian Fluid Mech.* **68:** 225–239 (1997).
75. A. S. Lodge, *Elastic Liquids*, Academic Press, New York, 1964.

Chapter 6 | Polymer Hydrogel Phase Transitions

Anthony E. English

Department of Biomedical Engineering
University of Iowa
Iowa City, Iowa

Elazer R. Edelman

Harvard–MIT Division of Health Sciences and Technology
Massachusetts Institute of Technology
Cambridge, Massachusetts

Toyoichi Tanaka

Department of Physics and Center for Materials Science and Engineering
Massachusetts Institute of Technology
Cambridge, Massachusetts

Abstract

This chapter is an introduction to the experimental study of neutral, polyelectrolyte, and polyampholyte hydrogel volume phase transitions. In each case, the necessary thermodynamic background required to understand the effects of changing various experimental parameters is presented. In particular, the similarity between hydrogel phase transitions and other materials is emphasized by illustrating the analogous behavior in the van der Waals fluid. Throughout this chapter, detailed experimental protocols for fabricating hydrogels are provided. Solvent-mediated phase transitions are first treated for neutral hydrogels. Polyelectrolyte hydrogel phase transitions resulting from changes in bath pH and ionic strength are then examined. The formation of first-order transitions in polyelectrolyte hydrogels that result from competing solvent and ionic interactions are also presented. The penultimate section of this chapter discusses the competing effects of polyelectrolyte repulsion and polyampholyte attraction.

6.1 Introduction

Synthetic hydrogels consist of monomers that are polymerized into linear chains and cross-linked into a nonlinear three-dimensional network. In their swollen state, hydrogels consist of a large liquid solvent fraction and a very small polymer component. As Figure 6.1 shows, changes in some ambient control variable, such

ISBN 0-12-683265-X/$30.00.

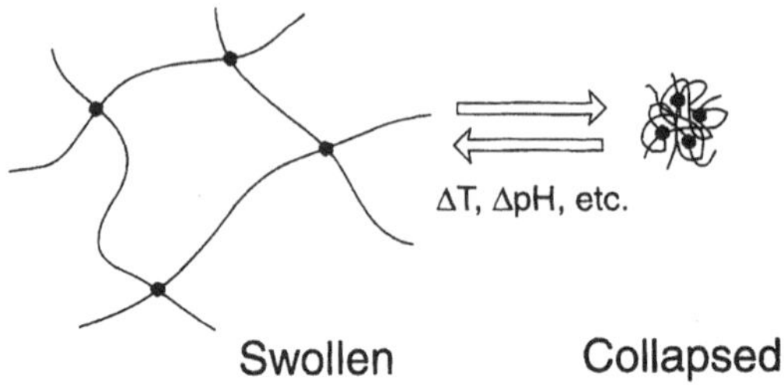

Figure 6.1 Hydrogels can undergo large discontinuous volume phase transitions in response to incrementally small changes in ambient conditions. Changes in temperature, pH, ionic strength, applied electric fields, and solvent composition have been shown to produce large first-order volume transitions.

as temperature, pH, ionic strength, or solvent composition can induce a large reversible collapse transition. In some cases, incrementally small changes in an external control variable can produce a volume transition of over three orders of magnitude. The ability to alter the physical response of these materials dramatically by varying their composition makes them extremely important in many technological and scientific applications.

The physical characteristics of hydrogel phase transitions are universally shared by a large number of other materials. The swelling transition illustrated in Figure 6.1 is analogous to the gas-to-liquid phase transition and has been observed in synthetic vinyl hydrogels and hydrogels made of natural polymers such as DNA, polysaccharides, and polypeptides. Synthetic and natural hydrogels have been observed to undergo critical point phenomena and first-order phase transitions similar to magnetic systems and other fluids [1, 2]. Some of the original concepts first proposed by van der Waals for understanding the liquid-to-gas phase transformation provide a conceptual foundation for understanding hydrogel phase transitions. Furthermore, the thermodynamics of ionic solutions have also provided valuable insight into understanding ionic hydrogel phase transitions [3].

In recent years, the unique physical properties of synthetic hydrogels have been exploited in a number of interesting and novel applications [4]. Because hydrogels have the capacity to contain a large volume of solvent in their swollen state, they have been used in a variety of superabsorbent materials. Separation and filtration devices for small ions and large-molecular-weight macromolecules have also been successfully fabricated from these materials. Hydrogels made of hydrophobic monomers that undergo transitions near 37 °C have been considered as drug delivery devices, materials for tissue reconstruction, and materials for preventing surgical adhesions [5]. In addition, hydrogel phase transitions driven by enzymatic reactions have been used in some creative controlled drug delivery strategies [6–8]. Porous hydrogels have been prepared for three-dimensional cellular scaffoldings, soft contact lenses, and artificial lenses. The application of these materials in medicine, agriculture, food preparation, and several other industries continues to grow [9–19].

This chapter is an introduction to the experimental study of neutral, polyelectrolyte, and polyampholyte hydrogel volume phase transitions. The relevant thermodynamics for the experimental study of hydrogel phase transitions are briefly reviewed. In particular, the similarity between hydrogel phase transitions and other materials is emphasized by illustrating the analogous behavior in the van der Waals fluid. Throughout this chapter, detailed experimental protocols for fabricating hydrogels are provided. General experimental considerations are given in Section 6.2. Section 6.3 describes the role of solvent-mediated phase transitions in neutral hydrogels, and Sections 6.4 and 6.5 discuss the effects of introducing ionization into a hydrogel network.

6.2 General Experimental Considerations

The equilibrium and nonequilibrium swelling transitions of synthetic and natural hydrogels are sensitive functions of the monomer composition and cross-linking density. Changing the ratio of hydrophobic to hydrophilic monomers can be used to adjust the network response to changes in temperature or solvent composition. Hydrogels can also be fabricated to respond to ionic solutions by incorporating acidic and basic monomeric groups into the network. The monomer ion dissociation constants and the relative proportion of acidic and basic monomers will determine how the network swells in response to changing the bath ionic strength and pH.

6.2.1 MONOMERS, CROSS-LINKERS, INITIATORS, AND ACCELERATORS

When purchased from a manufacturer, monomers and cross-linkers generally contain inhibitors, partially polymerized molecules, and other impurities. Although synthetic hydrogels can be formed from chemicals as received from a manufacturer in many cases, purification is often necessary to improve the quality of the

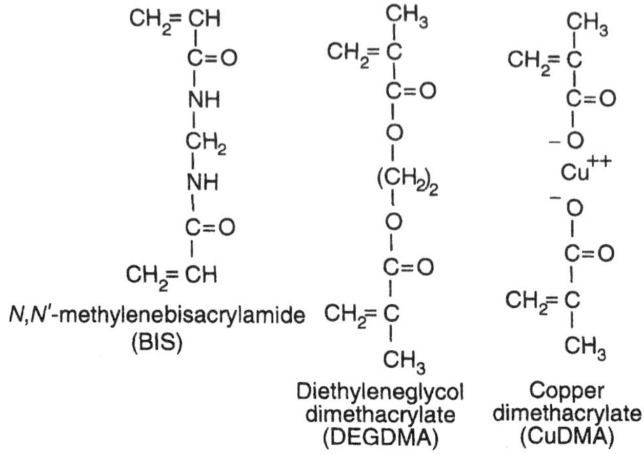

Figure 6.2 Molecular structures of the cross-linkers *N,N'*-methylenebisacrylamide, diethyleneglycol dimethacrylate, and copper dimethacrylate. The two vinyl groups cross-link different polymer strands into a three-dimensional mesh.

resulting hydrogel. Recrystallization and distillation are the two most common methods for purifying monomers and cross-linkers. Perrin and Armarego have prepared an excellent reference for the purification of a wide variety of chemicals used for fabricating hydrogels [20].

Figure 6.2 shows three representative cross-linkers used in the free-radical polymerization of hydrogel networks. *N,N'*-methylenebisacrylamide is commonly used in aqueous solutions, whereas diethyleneglycol dimethacrylate is more soluble for cross-linking polymer networks in organic solvents such as dimethyl sulfoxide. Ionic cross-linking can also be achieved using multivalent ions. Alginate hydrogels, for example, can be cross-linked using calcium ions [21]. Other bivalent ions complexed with anionic vinyl monomers can also be used as cross-linkers. Copper dimethacrylate or magnesium dimethacrylate, for example, can be purchased from Polysciences and used as cross-linkers.

The choice of an initiator for free-radical polymerization depends on the solvent and monomers being used. Ammonium persulfate (APS) is commonly used to initiate a polymerization in aqueous solutions. In organic solvents, 2,2'-azobisisobutyronitrile (AIBN) is often the initiator of choice. Most free-radical polymerizations can also be accelerated by immersing the pregel solution in a hot water bath typically at 60 °C. An elevated temperature or an ultraviolet light source can be used in conjunction with 2,2'-azobisisobutyronitrile. At lower temperatures, the polymerization can be accelerated using *N,N,N',N'*-tetramethylethylenediamine (TEMED) or sodium metabisulfite (SMBS). When using *N,N,N',N'*-tetramethylethylenediamine to accelerate the polymerizing of acidic monomers, however, it is necessary to neutralize the solution before adding the accelerator.

6.2.2 DEGASSING

Oxygen can interfere with a free-radical polymerization, and it is therefore advantageous to remove as much of it as possible by degassing the pregel solution before starting the polymerization. A small magnetic stirrer or glass pipette should be in the solution to avoid uncontrolled boiling while degassing under a vacuum. In addition, a filter trap should be used to avoid contaminating a vacuum system with toxic monomers. Dry ice in an isopropyl alcohol bath can be used to cool a filter trap and enhance the capture of monomers. Once the pregel solution is degassed, purging the system with nitrogen or argon gas helps reduce the rate of oxygen uptake in the pregel solution. Degassing is particularly important for aqueous solutions but may not be necessary for organic solvents. Caution should also be exercised when degassing volatile solvents.

Dissolved carbon dioxide can influence the swelling measurements of ionic hydrogels—particularly acidic hydrogels at low ionic strengths. Carbon dioxide dissolves readily in aqueous solutions and forms carbonic acid, which can lower the bath pH to about 5.5. The Donnan inclusion of hydrogen ions from a bath containing carbonic acid can thus significantly lower the internal hydrogel pH. The swelling equilibrium of weakly acidic hydrogels, such as acrylic acid or methacrylic acid for example, can therefore be a sensitive function of the amount of dissolved carbon dioxide. During experiments, the effects of dissolved carbon dioxide can be minimized by continuously flushing the hydrogel bath solution with nitrogen or argon gas.

6.2.3 CASTING

Volume phase transitions can be monitored experimentally by either measuring the size of a macroscopic hydrogel [22] or by measuring the wet mass [23]. Planar hydrogel geometries are generally desirable for membrane transport studies. Thin teflon spacers with punched-out holes sealed between two glass slides can serve as an effective mold. The pregelation solution can be injected into the model using a hypodermic needle, and binder clips can be used to hold the glass slides together. A tight seal allows the mold to be immersed in a temperature-controlled bath and ensures that oxygen is excluded during polymerization.

Kinetic considerations favor smaller-sized hydrogels. Many phase transitions are difficult to observe in large hydrogels because of the lengthy swelling times.

Small cylindrical hydrogels can be cast in micropipettes that are added after degassing the pregel solution. After polymerization, one side of the pipette can be nicked with a file or glass cutter and carefully broken to remove a small cylindrical length of hydrogel. Often it is necessary to experiment with several different cross-linker and monomer concentrations to obtain hydrogels that are easy to remove from small pipettes. High cross-linking densities produce hydrogels that are very brittle and hard to remove from pipettes. Too low a cross-linking concentration results in hydrogels that are too weak to remove.

Hydrogels with positive charges tend to adhere to glass surfaces and can be difficult to manage unless the surface is coated with a repellent. A glass repellent consisting of a 20-g/liter solution of dimethyldichlorosilane in 1,1,1-trichloroethane can be purchased from Pharmacia (Repel–Silane) and other chemical companies. Repellents, however, may influence the physical properties of hydrogels and thus should be used with caution.

6.3 Neutral Hydrogels

Fabricating neutral hydrogels to respond to changes in solvent composition or temperature in a prescribed manner requires some knowledge of the forces that determine the equilibrium swelling behavior. A detailed molecular understanding of hydrogel swelling forces, however, represents a very formidable task. Nevertheless, a great deal of physical insight can be obtained by considering neutral hydrogel swelling equilibrium as a simple balance between elastic and solvent forces. The following sections present an analogy between nonideal gas transitions and the thermodynamics of hydrogel phase transitions. Detailed experimental protocols for the fabrication of neutral hydrogel networks are then presented. Later sections consider the additional effects resulting from including ionizable monomers in the network.

6.3.1 THERMODYNAMICS OF FLUID AND NEUTRAL HYDROGEL TRANSITIONS

6.3.1.1 The van der Waals Fluid

The van der Waals fluid represents the classic model for qualitatively describing the phase transitions of materials with molecular interactions [24, 2]. Unlike the ideal gas, the van der Waals fluid includes both attractive and repulsive interactions capable of producing critical-point phenomena and first-order or discontinuous transitions. For a fixed molar number N of fluid particles confined to a volume V

in contact with a thermal reservoir at temperature T, the properties of the van der Waals fluid follow from the Helmholtz free energy

$$\Delta F = -NRT \ln(V - Nb) - \frac{aN^2}{V}. \tag{1}$$

In Eq. (1), ΔF represents the Helmholtz free energy, R is the universal gas constant, and the two terms a and b represent the van der Waals constants. Evaluating the pressure P from the thermodynamic relation

$$P = -\left(\frac{\partial \Delta F}{\partial V}\right)_{T,N} \tag{2}$$

yields the van der Waals equation of state

$$P = \frac{NRT}{(V - Nb)} - \frac{aN^2}{V^2}. \tag{3}$$

Figure 6.3 shows a series of isotherms corresponding to the van der Waals equation of state. At high temperatures, the pressure follows the ideal gas law. An inflection point, however, occurs at the critical temperature T_c. Below the critical temperature a kink in the isotherm implies an unphysical negative compressibility. The straight lines in Figure 6.3 represent a correction to the unphysical region of the phase

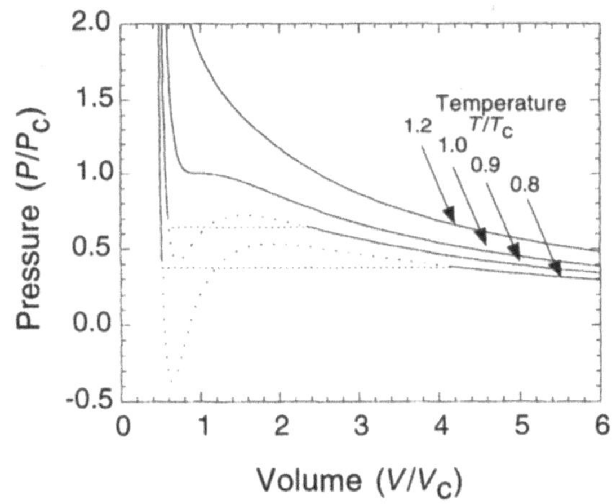

Figure 6.3 Pressure versus volume isotherms for the van der Waals fluid. The pressure, volume, and temperature values are normalized to the critical values. Below the critical temperature T_c, a kink in the isotherms results. On the basis of Maxwell construction and minimization of the free energy, two phases are predicted to coexist in equilibrium.

diagram based on the Maxwell construction. In this region, two phases are predicted to coexist in equilibrium.

6.3.1.2 Neutral Hydrogels

The state of a neutral hydrogel network immersed in a solvent can be defined by the volume V, solvent quality χ, and the monomer density n. The solvent quality plays the same role as the temperature does in the van der Waals fluid. Classically, the free energy of a neutral hydrogel consists of the two terms

$$\Delta F = \Delta F_{\mathrm{el}} + \Delta F_{\mathrm{M}}, \tag{4}$$

where ΔF_{el} and ΔF_{M} represent the elastic and solvent contributions, respectively. The solvent and elastic components of the free energy are given by the relations [25–27]

$$\Delta F_{\mathrm{el}} = \frac{3k_{\mathrm{B}}T}{2}\frac{n_0 V_0}{N_x}\left[\left(\frac{n_0}{n}\right)^{\frac{2}{3}} - 1 - \frac{1}{3}\ln\left(\frac{n_0}{n}\right)\right] \tag{5}$$

and

$$\Delta F_{\mathrm{M}} = k_{\mathrm{B}}T n_0 V_0 (1 - n_0 V_0)[\ln(1 - nv) + \chi nv], \tag{6}$$

where v is the lattice site volume, n the monomer density, n_0 the reference state monomer density, V_0 the reference state polymer volume, and N_x is the number of monomers between cross-links. The first term includes the elasticity of the polymer chains, and the second represents the solvent interaction.

Similar to the van der Waals gas, the hydrogel osmotic pressure follows from the relation

$$\pi = -\left(\frac{\partial \Delta F}{\partial V}\right)_{\chi} = -n\left(\frac{\partial \Delta F}{\partial n}\right)_{\chi}. \tag{7}$$

The resulting osmotic pressure

$$\frac{\pi}{k_{\mathrm{B}}T} = \frac{\pi_{\mathrm{el}} + \pi_{\mathrm{M}}}{k_{\mathrm{B}}T} = \frac{n_0}{N_x}\left[\frac{1}{2}\left(\frac{n}{n_0}\right) - \left(\frac{n}{n_0}\right)^{1/3}\right]$$
$$- \frac{1}{v}[\ln(1 - nv) + nv + \chi n^2 v^2], \tag{8}$$

is plotted in Figure 6.4. In equilibrium, the zero osmotic pressure condition, $\pi = 0$, is satisfied.

Decreasing the solvent quality, or increasing χ, results in a smaller equilibrium volume. A closer analogy between the van der Waals gas and the Flory theory can

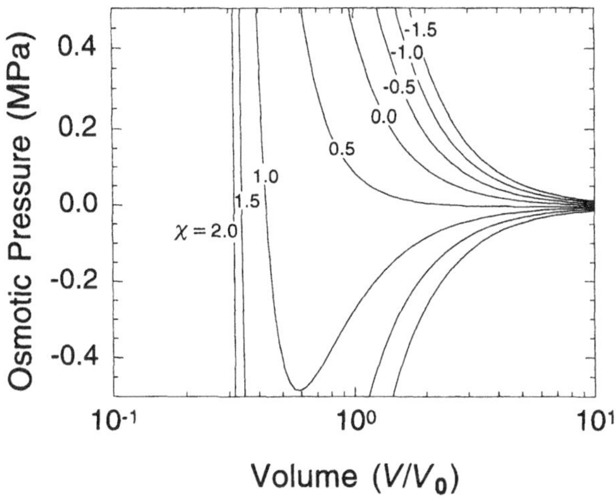

Figure 6.4 Osmotic pressure versus hydrogel volume for different solvent qualities. Decreasing the solvent quality, or increasing χ, results in lower osmotic pressures and hence a smaller equilibrium swelling.

be appreciated by interpreting the osmotic pressure in terms of the reduced temperature $\tau = 1/2 - \chi$. Experimentally, changes in solvent quality can be brought about by changing the solvent composition, hydrogel monomer composition, or temperature. Furthermore, changing the cross-linking density or efficiency alters the number of monomers between cross-links N_x.

In the absence of any externally applied forces, the equilibrium osmotic pressure is zero. On the basis of the family of curves shown in Figure 6.4, changes in solvent quality, or the equivalent reduced temperature, are predicted to produce corresponding continuous volume changes. If a small tensile force or negative pressure is applied to the system, a discontinuous volume transition can result, as seen by the Maxwell loop that occurs at lower pressures. This important fact was first recognized by Dusek and Patterson [28]. Additional intermolecular forces, produced by ionized monomers for example, that bias the isotherms to higher pressures can produce first-order phase transitions with changes in solvent quality under free-swelling conditions [29–31].

6.3.2 PREPARATION

Neutral hydrogels can be fabricated from a large number of different monomers. Figure 6.5 shows three representative monomers that are commonly used for

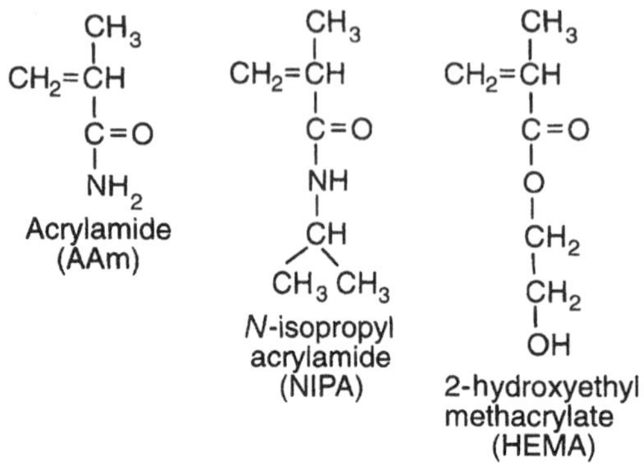

Figure 6.5 Molecular structures of acrylamide, *N*-isopropylacrylamide, and 2-hydroxyethyl methacrylate monomers. Hydrogel networks formed from these neutral monomers are sensitive to the ambient solvent composition or temperature.

forming neutral hydrogel networks. Acrylamide (AAm) is used to make hydrogels for electrophoretic studies and readily undergoes swelling transitions with variations in solvent composition. *N*-isopropylacrylamide (NIPA) has been used extensively to form hydrogels that respond to temperature. The monomer 2-hydroxyethyl methacrylate (HEMA) has found many applications in the fabrication of biocompatible hydrogels.

Acrylamide (AAm) hydrogels can be formed using the chemicals listed in Table 6.1. Acrylamide, *N,N'*-methylenebisacrylamide (BIS), ammonium persulfate (APS), and *N,N,N',N'*-tetramethylethylenediamine can be purchased from BioRad, Aldrich, Sigma, or Mallinckrodt. To 100 ml of distilled deionized water add 4.98 g of acrylamide, 0.133 g of *N,N'*-methylenebisacrylamide, and 1 ml of a 4 wt % ammonium persulfate. After stirring thoroughly and degassing, the polymerization can be accelerated using either *N,N,N',N'*-tetramethylethylenedia-

Table 6.1 Chemical composition of acrylamide hydrogels.

Chemical	*Formula Weight*	*Concentration*	*Mass or Volume*
Water			100 ml
AAm (monomer)	71.08 g/mol	700 mM	4.98 g
BIS (cross-linker)	154.17 g/mol	8.6 mM	0.133 g
APS (initiator)	228.20 g/mol	1.76 mM	1 ml 4 wt % stock[a]
TEMED (accelerator)			240 μL

[a] A 4 wt % APS stock solution is prepared by dissolving 0.4 g APS into 10 ml water.

mine or by immersing the pregelation solution in a 60 °C bath. Immediately after adding *N,N,N′,N′*-tetramethylethylenediamine or before immersing in a 60 °C bath, micropipettes should be added or the pregel solution should be injected into the required mold.

Spherical acrylamide hydrogels can be formed by inverse suspension polymerization using paraffin oil as a continuous phase [32]. Paraffin oil can be purchased from Aldrich, Sigma, or Mallinckrodt. A stock monomer solution is prepared by dissolving 4.98 g acrylamide (700 mM) and 0.133 g *N,N′*-methylenebisacrylamide (8.6 mM) in 100 ml of deionized water. After degassing and purging 10 ml of the stock solution with nitrogen, add 100 μl of 4 wt % ammonium persulfate. Before the pregel solution polymerizes, use a pipette to drop small aliquots of the prepared stock solution into purified, degassed, and nitrogen-saturated paraffin oil. Hydrogels are formed in about 5 minutes when the polymerization temperature is 20 °C. After the polymerization is complete, it is necessary to wash the hydrogels with copious amounts of distilled deionized water to remove unreacted monomers and short polymer chains.

As Table 6.2 indicates, *N*-isopropylacrylamide monomers can be polymerized at 4 °C using *N,N′*-methylenebisacrylamide as a cross-linker, ammonium persulfate as an initiator, and *N,N,N′,N′*-tetramethylethylenediamine as an accelerator. *N*-isopropylacrylamide, however, generally requires purification prior to polymerization. In a fume hood *N*-isopropylacrylamide can be purified by recrystallization using the following procedure: Dissolve approximately 30 g of unpurified *N*-isopropylacrylamide in 180 mM of toluene. Filter the toluene and *N*-isopropylacrylamide solution using a 25–50 μm glass filter under aspiration. Pour the filtered toluene and *N*-isopropylacrylamide solution into a 2000-mM beaker and add approximately 1500 mM of petroleum ether. If the *N*-isopropylacrylamide does not precipitate out of solution, continue adding more petroleum ether until it does. Filter out the precipitated *N*-isopropylacrylamide using a 25–50 μm glass filter under aspiration. Dry the recrystallized *N*-isopropylacrylamide

Table 6.2 Chemical composition of *N*-isopropylacrylamide hydrogels.

Chemical	***Formula Weight***	***Concentration***	***Mass or Volume***
Water			100 ml
NIPA (monomer)	113.16 g/mol	700 mM	7.92 g
BIS (cross-linker)	154.17 g/mol	8.6 mM	0.133 g
APS (initiator)	228.20 g/mol	1.76 mM	1 ml 4 wt % stock
TEMED (accelerator)			240 μL

under a vacuum connected to a filter trap immersed in a dry ice and isopropyl alcohol bath for 24 h.

Hydroxyethyl methacrylate hydrogels can be formed by polymerization in dimethyl sulfoxide solvent using ethyleneglycol dimethacrylate as a cross-linker and 2,2′-azobisisobutyronitrile as a free radical initiator. Hydroxyethyl methacrylate, ethyleneglycol dimethacrylate, and 2,2′-azobisisobutyronitrile can be purchased from Aldrich or Polysciences. Polymerization can be carried out in a 60 °C bath for 12 h. If one attempts to polymerize a 0.7-M solution of 2-hydroxyethyl methacrylate using an aqueous solvent with *N,N′*-methylenebisacrylamide and ammonium persulfate, a weak spongelike polymer hydrogel is formed unless higher 2-hydroxyethyl methacrylate concentrations are used. This can be avoided if a mixture of ethylene glycol and water is used as the solvent. When 2-hydroxyethyl methacrylate is polymerized at 0.7 M in dimethyl sulfoxide and washed in an aqueous solvent, it undergoes a collapse transition.

6.3.3 EQUILIBRIUM SWELLING TRANSITIONS

Figure 6.6 shows the swelling equilibrium values of an acrylamide hydrogel, as described in the previous section. In an aqueous solution, the hydrogel exists in a swollen state. Increasing relative concentrations of acetone in the surrounding bath

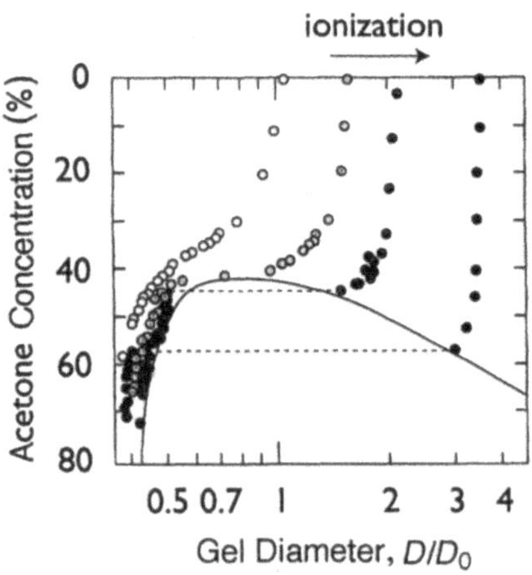

Figure 6.6 Swelling equilibria of acrylamide versus acetone concentration. Increasing the acetone concentration lowers the solvent quality and produces a collapsed transition. Continuous changes in acetone concentration or solvent quality produce continuous volume transitions.

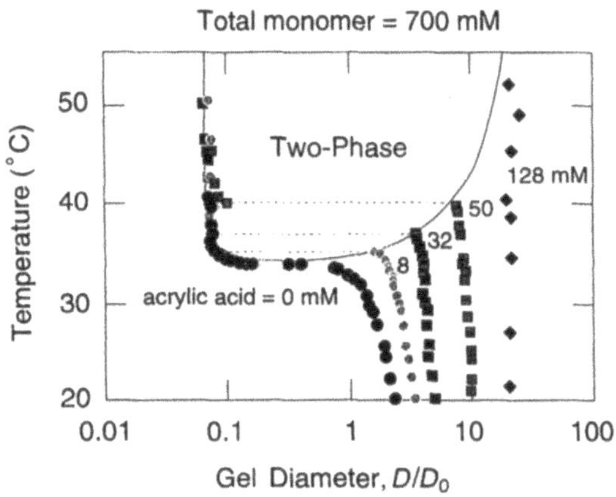

Figure 6.7 Swelling equilibria of *N*-isopropylacrylamide versus temperature. Near the transition temperature, *N*-isopropylacrylamide undergoes a sharp but continuous swelling transition as a function of temperature.

lowers the solvent quality and produces a collapse transition. A similar swelling transition occurs when the bath temperature is varied and the solvent composition is fixed [33]. These swelling transitions exhibit a sharp but continuous swelling pattern. As will be discussed in later sections, discontinuous first-order swelling transitions occur when ionized monomers are added to the network.

Figure 6.7 shows the equilibrium swelling of *N*-isopropylacrylamide as a function of temperature. At low temperatures, *N*-isopropylacrylamide is swollen, and at higher temperatures it is collapsed. At about 34.4 °C, *N*-isopropylacrylamide undergoes a sharp but continuous volume transition. By increasing the monomer concentration or decreasing the cross-linking density, a first-order or discontinuous volume phase transition has also been observed at the transition point [34]. The transition from continuous to discontinuous swelling behavior in hydrogel systems is one of the hallmarks of critical-point phenomena [33].

More complex phase transitions of *N*-isopropylacrylamide and other neutral hydrogels can be observed in mixed solvents. Reentrant volume-phase transitions of *N*-isopropylacrylamide hydrogels can be observed, for example, in water and methanol mixtures [35]. When the solvent composition is varied systematically, *N*-isopropylacrylamide hydrogels undergo a discontinuous collapse transition followed by a discontinuous swelling transition. These two phase transitions indicate the presence of a closed-loop instability phase boundary having both upper and lower critical points. The closed-loop phase boundary was also observed to depend on temperature and diminished to zero as the temperature approached 0 °C. Reentrant phase behavior of N-isopropylacrylamide has also been observed when the relative concentrations of water and dimethyl sulfoxide are varied [36].

6.3.4 NONEQUILIBRIUM SWELLING TRANSITIONS

An understanding of hydrogel swelling kinetics is indispensable to the experimentalist undertaking the study of hydrogel phase transitions. Very slow swelling kinetics can occur in neutral and ionic hydrogels. Not being aware of very slow transitions can lead to unrepeatable measurements of metastable states. The following paragraphs, therefore, review the kinetic response arising from network diffusion and critical-point phenomena. In subsequent sections, the effects of ionization are considered. Ionized hydrogels can have very slow kinetics that are limited by stagnant layers and diffusion reaction mechanisms.

Studies of the critical divergence and critical slowing down of polymer hydrogel thermal fluctuations provided a foundation for understanding the nonequilibrium network swelling response. Tanaka, Hocker, and Benedek [37, 38, 32, 39, 40] examined the density fluctuations of acrylamide hydrogel networks and showed that the longitudinal elastic modulus can be determined from the intensity of light scattered from the density fluctuations. This was followed by a series of experimental and theoretical studies that examined the critical density fluctuations in acrylamide hydrogels [41, 42, 33]. Microscopic light-scattering experiments based on photon correlation spectroscopy provided a direct measurement of the network diffusion coefficient. Near a critical point, the network diffusion coefficient was observed to diverge [43].

The collective diffusion of a hydrogel network was reexamined by measuring the macroscopic nonequilibrium swelling of acrylamide hydrogels [44, 32]. In these studies, the network diffusion is associated with the bulk counterflow of water through the network. The local motion of the polymer network obeys a diffusion equation in which the diffusion coefficient is defined by the ratio of the network elastic modulus K to the frictional coefficient f between the network and the fluid. The characteristic swelling time is proportional to the square of the linear hydrogel size and inversely proportional to the network diffusion coefficient.

When a hydrogel undergoes a small volume transition, linearization of the elastic response is an acceptable approximation. During large first-order phase transitions, however, the elastic moduli will be concentration-dependent, and the swelling response will be nonlinear. During a first-order phase transition, the mechanical instability associated with large osmotic pressure gradients within the hydrogel further complicates the swelling kinetics and the evolution pattern [45].

6.4 Polyelectrolyte Hydrogels

Polyelectrolyte hydrogels are polymer networks consisting of charged monomers of only one sign. At the outset, it is imperative that two very different system definitions for polyelectrolyte hydrogels be made and understood. If the surrounding bath contains no ions, the system is completely defined by a hydrogel phase with a given volume and set of ion particle numbers. When ions are present in the surrounding bath, the ion chemical potentials in the bath and hydrogel phases must be equal in equilibrium. This is an important consideration even in distilled deionized water having no added salt. Although the pH of water is 7 under ideal conditions, the electrostatic inclusion of hydrogen or hydroxide ions can create very low or high internal pH values, depending on the polymer charge density. Hence, polyelectrolyte hydrogels made of very strong acidic or basic monomers can exhibit ion association effects at low bath salt concentrations.

6.4.1 EQUILIBRIUM THERMODYNAMICS IN NONIONIC SOLUTIONS

If the surrounding bath is free of ions, counterions are confined to the hydrogel phase. The electrostatic confinement of the counterions maintains charge electroneutrality and a constant total number of each type of particle. As Figure 6.8 shows, the hydrogel state is defined by the volume V, solvent quality χ, and the set of particle numbers $\{N_i\}$. The Helmholtz free energy therefore becomes the natural state function to consider in describing the system thermodynamics.

A polyelectrolyte hydrogel immersed in a large reservoir with no added ions can be treated in the same way as a neutral hydrogel if the translational freedom of the counterions within the hydrogel is included. It will be assumed that the charged monomers remain completely dissociated. The free energy of a polyelectrolyte

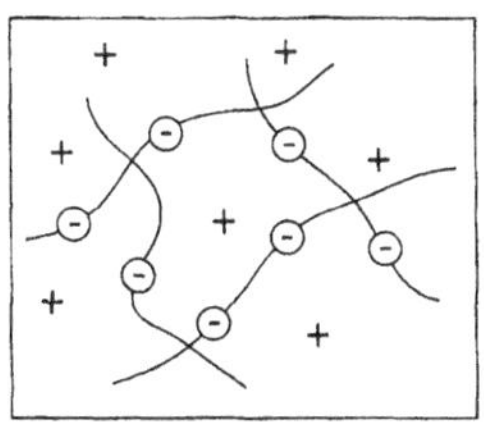

Figure 6.8 In the absence of bath ions, the hydrogel state is defined by the volume V, solvent quality χ, and the set of particle of numbers $\{N_i\}$. The Helmholtz free energy becomes the natural thermodynamic state function to consider.

hydrogel is given by

$$\Delta F^{\mathrm{H}} = \Delta F_{\mathrm{el}} + \Delta F_{\mathrm{M}} + \Delta F_{0}, \tag{9}$$

where ΔF_{el} and ΔF_{M} are as previously defined for a neutral hydrogel. The third term on the right-hand side of Eq. (9) represents the entropic contribution of the mobile counterions,

$$\Delta F_0 = k_{\mathrm{B}} T \sum_{i=1}^{\sigma} N_i \left[\ln\left(\frac{N_i}{V}\right) - 1 \right] + \sum_{i=1}^{\sigma} z_i e N_i \varphi, \tag{10}$$

where z_i is the ith ion valance, N_i the number of ith particles in the hydrogel phase, e the unit electrostatic charge, σ is the number of ion species, and φ represents the uniform contribution to the internal potential arising from the electrostatic double layer at the hydrogel boundary.

The osmotic pressure, which follows from Eq. (9),

$$\frac{\pi}{k_{\mathrm{B}} T} = \frac{n_0}{N_x}\left[\frac{1}{2}\left(\frac{n}{n_0}\right) - \left(\frac{n}{n_0}\right)^{1/3}\right] - \frac{1}{v}[\ln(1 - nv) + nv + \chi n^2 v^2] + \sum_{i=1}^{\sigma} c_i \tag{11}$$

is plotted in Figure 6.9. Compared with Figure 6.4, the pressure versus volume curves for a given solvent quality are shifted to higher osmotic pressures. As a

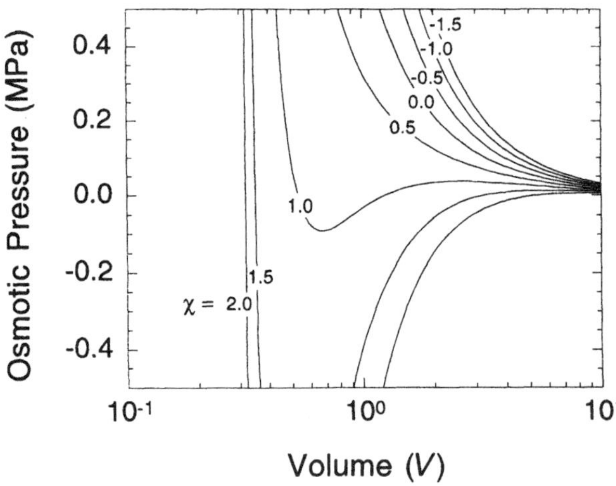

Figure 6.9 Osmotic pressure versus hydrogel volume for different solvent qualities. This model assumes that the charged monomers remain completely ionized and the surrounding bath possesses no ions. In comparison with Figure 6.4 ionization biases the constant solvent quality curves to higher osmotic pressures.

result, the Maxwell loop can be reached if the zero osmotic pressure equilibrium condition is satisfied. Ionization in this system can thus give rise to a first-order volume phase transition under free-swelling conditions.

6.4.2 *EQUILIBRIUM THERMODYNAMICS IN IONIC SOLUTIONS*

6.4.2.1 System Definition

When ions are present in the surrounding bath, it is no longer valid to define the system as a confined volume with no capacity for ion exchange. In this case, a bath phase exists in equilibrium with the hydrogel phase, and the internal ion concentrations are set by the chemical potentials of the surrounding bath. As shown in Figure 6.10, the volumes, solvent quality, and the set of particle numbers in the bath and hydrogel phase are required to define the system state. The effects of monomeric ion dissociation will also be included at this point.

6.4.2.2 Free-Energy Model

The total system free energy ΔF^{T} can be decomposed into a sum of hydrogel and bath components such that

$$\Delta F^{\mathrm{T}} = \Delta F^{\mathrm{H}} + \Delta F^{\mathrm{B}}, \tag{12}$$

where ΔF^{H} and ΔF^{B} represent the hydrogel and bath contributions, respectively. The hydrogel free energy is defined as it is in Eq. (9). The bath component includes the translational freedom of the bath ions and the electrostatic contribution from a

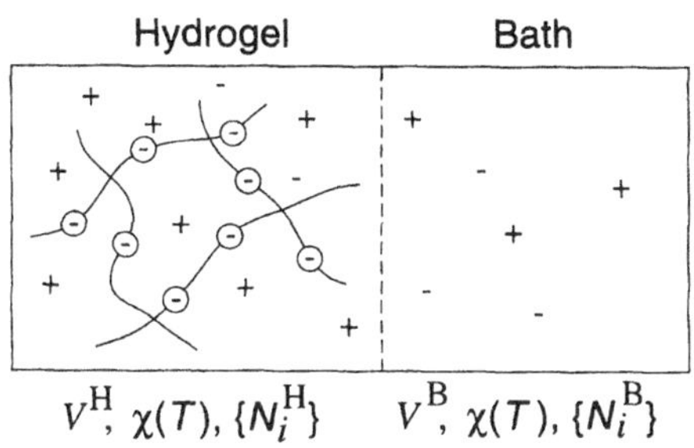

Figure 6.10 In the presence of bath ions, the system definition includes the volume, solvent quality, and the particle numbers in the hydrogel and bath phases. In equilibrium the ion chemical potentials in each phase are equivalent.

uniform potential such that

$$\Delta F^{\mathrm{B}} = k_{\mathrm{B}}T \sum_{i=1}^{\sigma} N_i^{\mathrm{B}} \left[\ln\left(\frac{N_i^{\mathrm{B}}}{V^{\mathrm{B}}}\right) - 1 \right] + \sum_{i=1}^{\sigma} z_i e N_i^{\mathrm{B}} \varphi^{\mathrm{B}}, \tag{13}$$

where the superscript indicates the corresponding bath volume, particle numbers, and uniform potential φ. Normally, the internal uniform contribution to the electrostatic potential of the hydrogel is referenced to that of the surrounding bath.

6.4.2.3 Swelling and Ionic Equilibrium

In equilibrium, the chemical potentials of the solvent and mobile ion species in the hydrogel and bath phase balance. Balancing the solvent chemical potentials, for example, is equivalent to balancing the osmotic pressures according to the relation

$$-\frac{1}{v}[\ln(1 - nv) + nv + \chi n^2 v^2] + \frac{n_0}{N_x}\left[\frac{1}{2}\left(\frac{n}{n_0}\right) - \left(\frac{n}{n_0}\right)^{1/3}\right] + \sum_{i=0}^{\sigma} c_i^{\mathrm{H}} - \sum_{i=0}^{\sigma} c_i^{\mathrm{B}} = 0, \tag{14}$$

where c_i^{H} and c_i^{B} represent the ith ion concentration in the hydrogel and bath phases, respectively. Balancing the ion chemical potentials gives rise to the Donnan relations,

$$\left(\frac{C_i}{C_i^{\mathrm{B}}}\right)^{\frac{1}{z_i}} = \left(\frac{C_j}{C_j^{\mathrm{B}}}\right)^{\frac{1}{z_j}}, \tag{15}$$

where capital letters are used to represent molar quantities.

6.4.2.4 Acid and Base Dissociation Equilibrium

The swelling response of a polyelectrolyte hydrogel network to changes in bath ionic strength or pH can be strongly dependent on the monomer acid and base dissociation contants. Sulfonic acid groups or quaternary ammonium halides remain dissociated over a very wide pH range. In contrast, carboxylic acid and amino groups are much more dependent on the ambient pH. At very low ionic strengths, however, the internal pH of ionizable hydrogels can reach extreme values. As a result, even highly acidic or basic monomers can exhibit swelling patterns consistent with ion dissociation.

Equality of chemical potentials between unbound and bound hydrogen ion states is equivalent to the relations

$$K_{\mathrm{a}} = \frac{C_{\mathrm{A^-}} C_{\mathrm{H^+}}}{C_{\mathrm{AH}}} \tag{16}$$

and

$$K_{\mathrm{b}} = \frac{C_{\mathrm{BOH}} C_{\mathrm{H^+}}}{C_{\mathrm{B^+}}} \tag{17}$$

for acidic and basic polyelectrolyte, respectively. The subscripts $\mathrm{A^-}$ and AH refer to the dissociated and undissociated acidic monomeric states, and the subscripts $\mathrm{B^+}$ and BOH refer to the corresponding dissociated and undissociated basic monomeric states.

6.4.3 NONEQUILIBRIUM THERMODYNAMICS

As is true with neutral hydrogels, network diffusion and critical slowing down can play an important role in determining the nonequilibrium swelling behavior of ionic hydrogels. Ionic hydrogel swelling transitions, however, are often associated with the transport of ions into or out of the network. Ion tranport can be rate limited by a number of different mechanisms [46, 47]. The time it takes to establish charge electroneutrality on a Debye length scale can be estimated using the relation $\tau = \sigma/\varepsilon$, where σ and ε are the medium conductivity and permittivity, respectively. This time is on the order of 1 ns and is therefore negligible. Coupled ion diffusion will scale directly as the hydrogel dimensions and inversely as the coupled ion diffusion coefficient.

When a high concentration of charged monomeric groups capable of binding ions is present, the diffusive transport of ions can be greatly slowed [48]. In acidic polyelectrolyte hydrogels, for example, the effective diffusivity of the hydrogen ion becomes

$$D_{\mathrm{H_{eff}}} = \frac{D_{\mathrm{H^+}}}{1 + \dfrac{C_{\mathrm{A}} K_{\mathrm{a}}}{(K_{\mathrm{a}} + C_{\mathrm{H^+}})^2}} \tag{18}$$

where C_{A} is the total concentration of acidic monomers, $C_{\mathrm{H^+}}$ the concentration of hydrogen ions, K_{a} the acid dissociation constant, and $D_{\mathrm{H^+}}$ the diffusion constant of hydrogen ions in solution. Given the $\mathrm{p}K_{\mathrm{a}}$ of acrylic acid, for example, the transport of hydrogen ions can be retarded by several orders of magnitude.

In the absence of diffusion reaction effects, ion transport in polyelectrolyte hydrogels can be limited by either coupled ion transport or by film boundary effects [47]. A criterion for estimating whether film boundary effects or coupled ion diffusion will be rate limiting is given by Helferich. Consider a hydrogel with a concentration ρ of ionic groups immersed in a bath with ion concentration C_s. If D represents the coupled ion diffusion constant, D_f the film diffusion constant, and R_0 the hydrogel dimension, δ the film layer thickness, and α is a separation factor, then

$$\frac{\rho D\delta}{C_s D_f R_0}(5 + 2\alpha) \gg 1 \tag{19}$$

implies the film diffusion limit and

$$\frac{\rho D\delta}{C_s D_f R_0}(5 + 2\alpha) \ll 1 \tag{20}$$

the particle diffusion limit. Film boundary effects thus become an important consideration when equilibrating polyelectrolyte hydrogels at low ionic strengths.

6.4.4 PREPARATION

6.4.4.1 Acidic or Anionic Polyelectrolytes

Several different acidic or anionic monomers are available commercially for fabricating polyelectrolyte hydrogels. Figure 6.11 shows the two common acidic monomers, methacrylic acid and acrylic acid. These are weakly acidic monomers having pK_a values approximately equal to 5.5. Hydrogels can also be fabricated with strong acidic monomers. The sulfonic acidic group, for example, is a much stronger acid than the carboxyl group. The sodium salts of styrene sulfonic acid and 2-acrylamido-2-methyl-1-propane sulfonic acid, shown in Figure 6.12, are common monomers used to make strongly dissociated polyelectrolytes. When

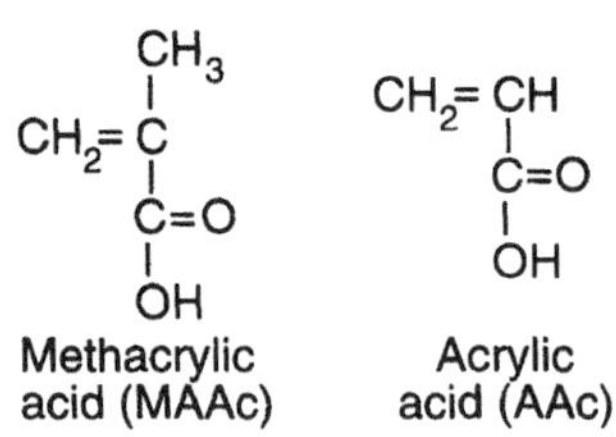

Figure 6.11 Molecular structures of methacrylic acid and acrylic acid monomers. These monomers are weak acids, and thus polyelectrolyte hydrogels formed from either of them tend to be sensitive functions of the ambient bath pH and ionic strength.

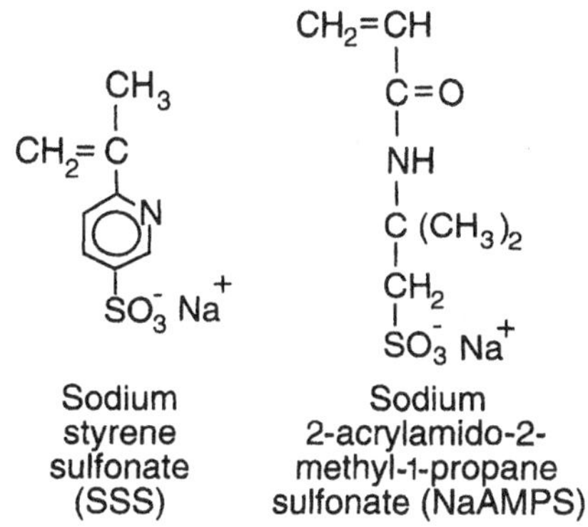

Figure 6.12 Sodium salts of styrene sulfonic acid and 2-acrylamido-2-methyl-1-propanesulfonic acid monomers. Unlike the carboxyl group, the sulfonate group remains dissociated over a much wider pH range.

fabricating hydrogels from acidic monomers, it is often necessary to neutralize the pregel solution with a strong base, such as sodium hydroxide, to improve the quality of the hydrogel.

6.4.4.2 Basic or Cationic Polyelectrolytes

Weakly basic polyelectrolyte hydrogels can be prepared using tertiary amine monomers such as 2-vinylpryridine and 2-diethylaminoethyl (methyl) acrylate that are shown in Figure 6.13. Tertiary amines, such as 2-vinylpryridine and 2-diethylaminoethyl (methyl) acrylate, become charged when the amine functional group is protonated. Strongly basic polyelectrolyte hydrogels can be prepared using

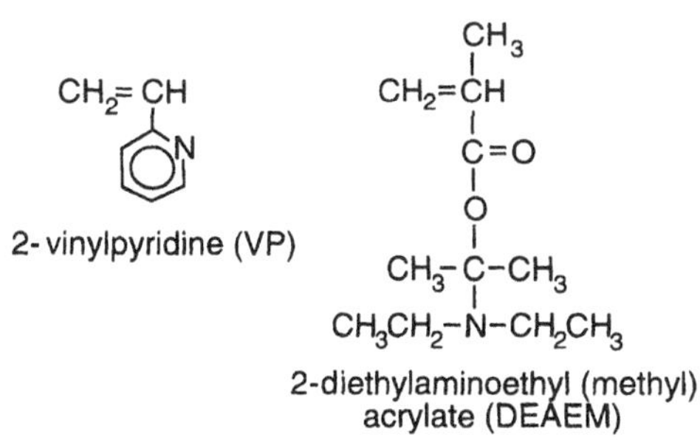

Figure 6.13 Molecular structures of 2-vinylpyridine and 2-diethylaminoethyl (methyl) acrylate. The two monomers 2-vinylpyridine and 2-diethylaminoethyl (methyl) acrylate are weak bases that become charged by protonation of the nitrogen group.

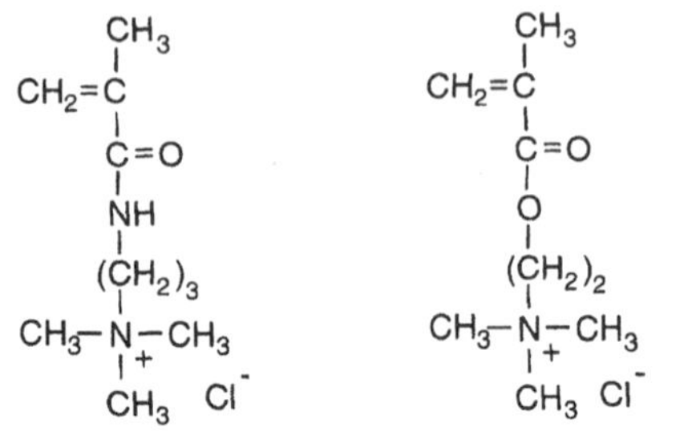

Figure 6.14 Molecular structures of [3-(methacryloylamino)propyl]trimethylammonium chloride and [2-(methacryloyloxy)ethyl]trimethylammonium chloride monomers. These monomers are strong bases that remain dissociated over a wide pH range.

the quaternary ammonium salts [3-(methacryloylamino)propyl]trimethylammonium chloride or [2-(methacryloyloxy) ethyl] trimethylammonium chloride, shown in Figure 6.14.

6.4.4.3 Sample Experimental Protocols

Methacrylic acid (MAAc) hydrogels can be prepared using the chemical concentrations listed in Table 6.3. Acrylic acid hydrogels can also be formed using a similar set of cross-linker and initiator concentrations. Purification of these monomers

Table 6.3 Chemical composition of methacrylic acid hydrogels.

Chemical	*Formula Weight*	*Concentration*	*Mass of Volume*
Water			94.063 ml
MAAc	86.09 g/mol	700 mM	[a]5.937 ml
BIS	154.17 g/mol	8.6 mM	0.1326 g
APS	228 g/mol	1.76 mM	1 ml 4 wt % stock
TEMED			[b]240 μm

[a]The density of MAAc is 1.015 g/ml.
[b]The pregelation solution should be neutralized with NaOH before adding TEMED. As an alternative, TEMED can be omitted if a 60 °C bath is used to accelerate the polymerization.

generally produces better-quality hydrogels. Acrylic acid can be distilled at 60–70 °C under a 20–25 mmHg vacuum, and methacrylic acid can be distilled at 79 °C under a 1.3 mmHg vacuum. Anhydrous copper sulfate should be added to the predistillation solution to absorb water. The polymerization can be accelerated using a 60 °C bath or by using *N,N,N′,N′*-tetramethylethylenediamine (TEMED) when the polymerization temperature is 25 °C or less. If *N,N,N′,N′*-tetramethylethylenediamine is used, it is necessary to neutralize the pregelation solution before adding *N,N,N′,N′*-tetramethylethylenediamine.

Spherical hydrogels of *N*-isopropylacrylamide and sodium acrylate can be formed by inverse suspension polymerization using paraffin oil as a continuous phase. A 700-mM neutral monomer stock solution can be made as previously described by dissolving 7.8 g *N*-isopropylacrylamide, 1.33 g *N,N′*-methylenebisacrylamide, and 40 mg ammonium persulfate in 100 ml of distilled deionized water. A similar ionic monomer stock solution can be made by neutralizing 4.8 ml of acrylic acid with NaOH in 100 ml of distilled deionized water containing the same cross-linker and initiator concentrations. Pregel solutions with varying proportions of ionic monomers can be made by mixing the stock solutions in varying proportions. Before polymerization, the solutions must be degassed and purged with nitrogen. Immediately after adding 24 μl *N,N,N′,N′*-tetramethylethylenediamine to 10 ml of the pregel solution, a 1-ml aliquot of the pregel solution can be injected into 50 ml of purified, degassed, and nitrogen-saturated paraffin oil solution. Spherical droplets are formed by the mechanical agitation of a stir bar. When the preparation temperature is 20 °C, hydrogels are formed in 30 min. After polymerization is complete, it is necessary to wash the hydrogels with copious amounts of distilled water to remove unreacted monomers and short polymer chains.

Submicron hydrogel beads have also been prepared by inverse emulsion polymerization at room temperature. In one series of studies, hexane was used as the solvent with the nonionic surfactant sorbitan monolaurate (SPAN 20) [49]. Hydrogels with diameters ranging from 0.2 to 1 μm were successfully produced and observed using photon correlation spectroscopy. Sharp but continuous hydrogel volume transitions were observed in these studies.

6.4.5 ION DISSOCIATION AND POLYELECTROLYTE SCREENING

The swelling response of acidic polyelectrolytes to changes in bath salt concentration and pH is illustrated in Figure 6.15. The solid line in Figure 6.15 represents a

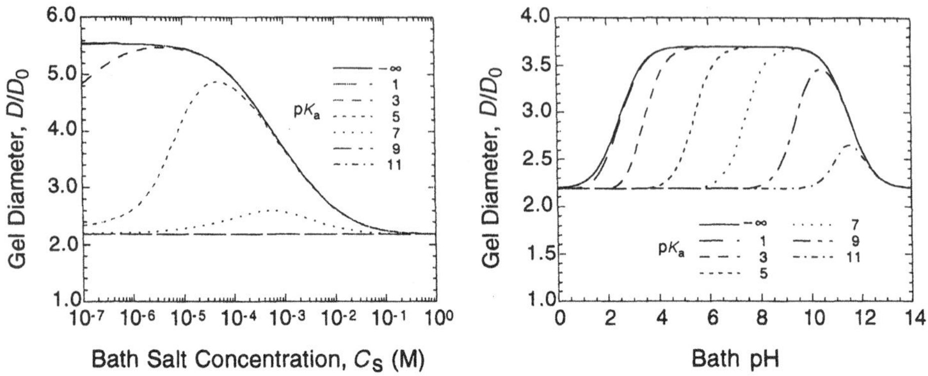

Figure 6.15 Swelling equilibrium profiles of polyelectrolyte hydrogels with different acid dissociation constants. With increasing pK_a values, hydrogen ion association produces a deswelling transition.

polyelectrolyte hydrogel that remains in its dissociated state even at extremely low pH values. Strongly acidic polyelectrolytes remain swollen at very low bath salt concentrations. As the bath salt concentration increases, the osmotic pressure difference between the hydrogel and the surrounding bath decreases, and a deswelling transition occurs.

As the acid dissociation constants decrease, a deswelling transition occurs at low ionic strengths. This deswelling transition is the result of the electrostatic inclusion of hydrogen counterions. At low ionic strengths, the hydrogen ion is the only counterion available to maintain charge electroneutrality within the hydrogel. As a result, at low ionic strengths the internal pH drops to very low values and acid association occurs. The subsequent loss in charge density gives rise to the deswelling pattern. Adding small amounts of low-molecular-weight salts to the bath results in the exchange of hydrogen ions for the bath cation. This increases the internal pH and causes the acidic groups to dissociate and thereby increase the internal charge density. The deswelling transition that follows with further increases in bath salt concentration is the result of losing the osmotic pressure difference between the hydrogel and the surrounding bath.

The swelling response of polyelectrolyte hydrogels with bath pH and acid dissociation constant can be similarly understood. When the acidic monomeric groups remain completely dissociated at all pH values, the swelling response with pH is shown by the solid line in Figure 6.15. At very low and very high pH values, the deswelling pattern is the result of losing the osmotic pressure difference between the hydrogel and the bath because of the high bath ionic strength. This is similar

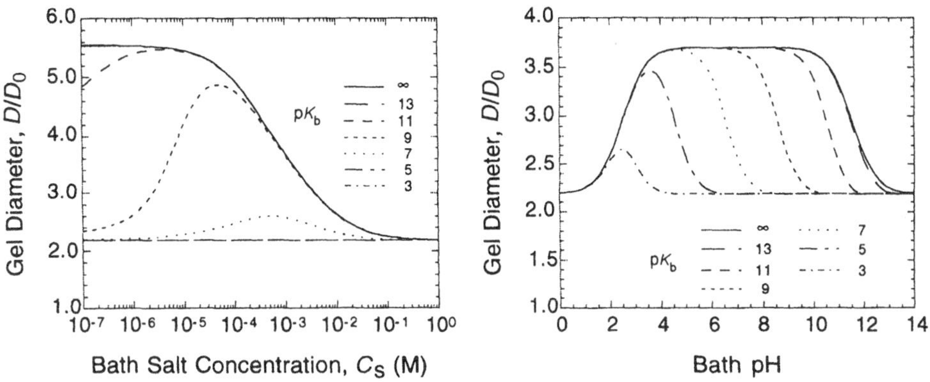

Figure 6.16 Swelling equilibrium profiles of polyelectrolyte hydrogels with different base dissociation constants. In parallel with the behavior of acidic polyelectrolytes, as the base strength of the monomer becomes smaller, charge neutralization reduces the osmotic pressure.

to the deswelling pattern that occurs when the bath salt concentration increases. As the acid dissociation constant decreases, the swelling pattern shifts to the right as a result of the decreasing network ionization.

Figure 6.16 shows a similar set of swelling patterns for basic or cationic polyelectrolyte hydrogels. Just as in Figure 6.15, completely dissociated basic polyelectrolytes are swollen at low bath ionic strengths. Increasing the bath salt concentration leads to a deswelling transition for the same reason as it does for a completely dissociated acidic polyelectrolyte. With decreasing monomer base strength, a deswelling pattern occurs at low ionic strengths similar to that of an acidic polyelectrolyte. In this case, however, the internal pH at low ionic strengths becomes very high because the inclusion of hydroxide ions is now required to maintain charge electroneutrality. The response to changes in pH follows a symmetric pattern opposite to that of the acidic polyelectrolyte case.

Copolymer hydrogels with varying proportions of N-isopropylacrylamide, acrylamide and acrylic acid illustrate the competition between solvent and repulsive polyelectrolyte forces. As shown in Figure 6.17, at low ionic strengths polyelectrolytes collapse from the hydrogen ion association. The equilibrium swelling at low ionic strengths is dominated by the different solvent qualities of the acrylamide and N-isopropylacrylamide monomers. At very high ionic strengths, a collapse transition is observed in pure N-isopropylacrylamide hydrogels resulting from a short-range solvent-mediated effect associated with high ionic strengths.

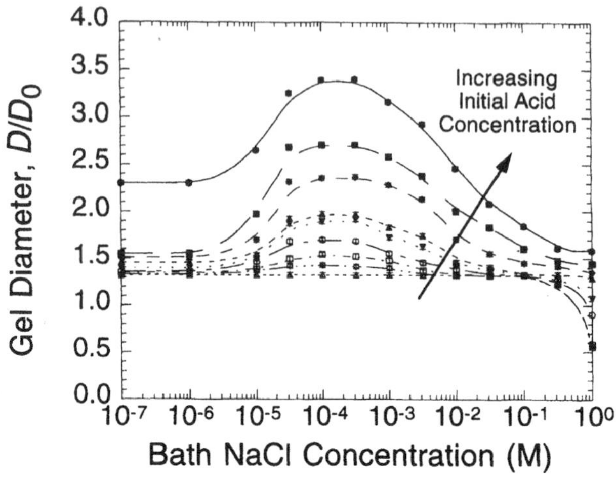

Figure 6.17 Swelling equilibria of N-isopropylacrylamide, acrylamide, and acrylic acid copolymer hydrogels. The lowest swelling curve represents pure N-isopropylacrylamide. Successively higher swelling curves represent hydrogels polymerized with acrylamide and acrylic acid concentrations increasing in powers of 2. Reprinted with permission from ref. [46]. Copyright 1997 American Institute of Physics.

6.4.6 FIRST-ORDER TRANSITIONS

6.4.6.1 Poor Solvent Quality with Strong Dissociation

First-order swelling transitions in polyelectrolyte hydrogels can be brought about in both cationic and anionic hydrogels. Theoretically, it has been shown that a hydrogel will undergo a discontinuous volume phase transition when enough positive internal osmotic pressure is present and the solvent quality is sufficiently poor. A positive internal osmotic pressure is produced by ionizing the network. Strongly dissociable acidic and basic polyelectrolytes have been observed to undergo these transitions. Sodium styrene sulfonate and [3-(methacryloylamino)propyl] trimethylammonium chloride are two strongly ionized polyelectrolytes that have been investigated as a function of charge and solvent quality [50–53]. The formation of a first-order swelling transition therefore does not depend on the sign of the charge.

The formation of first-order or discontinuous phase transitions that result from the addition of charged monomers has also been shown experimentally in studies of N-isopropylacrylamide. As discussed earlier, N-isopropylacrylamide hydrogels undergo sharp but continuous or weakly discontinuous volume phase transitions at a particular temperature. With increasing relative concentrations of acrylic acid monomers, first-order or discontinuous phase transitions result. The formation of a first-order transition with increasing ionization has been illustrated by a series of experiments by Hirotsu *et al.* [49]. With increasing relative concentrations of acrylic acid monomers, the discontinuity in the swelling transition increases.

6.4.6.2 Poor Solvent Quality with Weak Dissociation

When the solvent quality becomes poor, first-order transitions can occur in ionic hydrogels as a result of ion dissociation. Figure 6.18, for example, illustrates the formation of a first-order phase transformation in a copolymer hydrogel of 2-hydroxyethyl methacrylate and methacrylic acid. The lowest set of data points in Figure 6.18 represents the equilibrium swelling measurements of pure 2-hydroxyethyl methacrylate hydrogels. The successively higher swelling points are those with increasing proportions of methacrylic acid. The arrows denote regions with discontinuous volume phase transitions. In this case, when the bath salt concentration reaches a critical value the dissociation of the acidic monomers leads to a first-order swelling transition.

Theoretical swelling patterns of polyelectrolyte hydrogels with variations in bath ionic composition and solvent quality are shown in Figure 6.19. In good

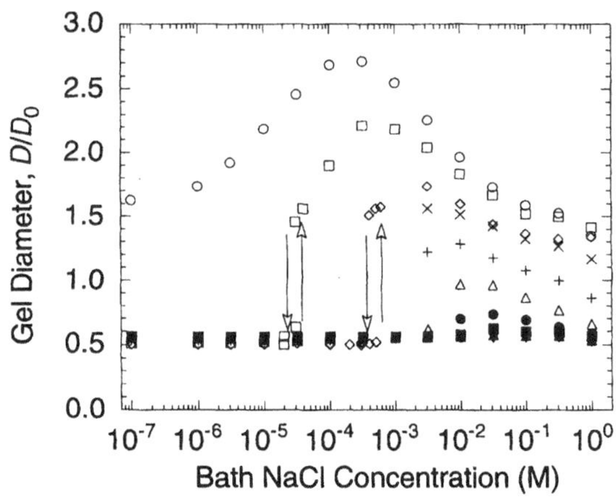

Figure 6.18 Swelling response of copolymer 2-hydroxyethyl methacrylate and methacrylic acid hydrogels with changing bath salt concentration. The first-order swelling transition that occurs at low ionic strengths is consistent with ion dissociation. Reprinted with permission from *J. Chem. Phys.* **105** (23), 10606. Copyright 1996 American Institute of Physics.

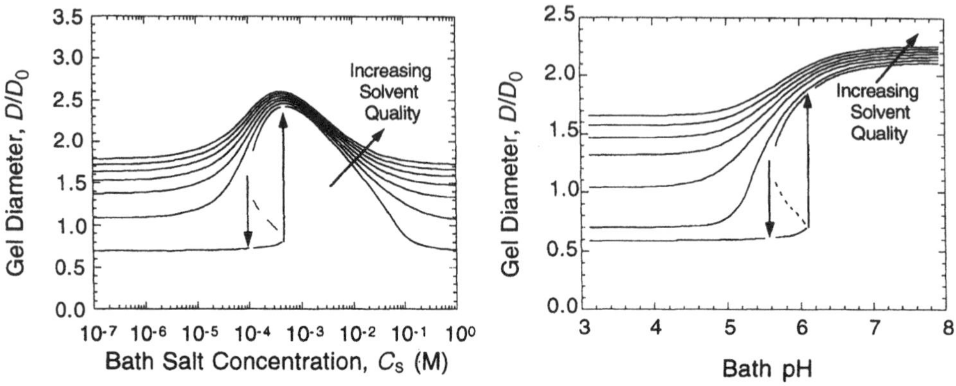

Figure 6.19 Theoretical swelling response of hydrogels with acidic monomers in a bath with varying solvent quality and salt concentration. With a low-quality solvent, ion dissociation can produce a first-order swelling transition when the bath ionic strength or pH is varied. Reprinted with permission from ref. [46]. Copyright 1997 American Institute of Physics.

solvents at neutral pH, deswelling transitions at low ionic strengths, are produced by ion association. At high ionic strengths, deswelling occurs as a result of losing the osmotic pressure difference between the hydrogel and the surrounding bath. Increasing the bath pH from 3 to 8 leads to a dissociation of the acidic monomers and, hence, a swelling transition. Decreasing the solvent quality leads to the formation of first-order swelling transitions in both cases that are the result of ion dissociation equilibrium.

In retrospect, the first-order swelling behavior of ionized *N*-isopropylacrylamide and acrylic acid copolymers can be understood in greater detail. If *N*-isopropylacrylamide and acrylic acid copolymers are washed in distilled deionized water for long periods, most of the acrylic acid monomers will have hydrogen ions bound to them. If, however, the hydrogel is initially prepared with sodium acrylate and measurements are made on a time scale that does not allow complete exchange of the sodium ions for hydrogen ions, the network can be assumed completely dissociated. Diffusion reaction kinetics and the effects of interfacial boundary layers that occur at low ionic strengths make the assumption of complete ionization valid if measurements are made on a short time scale relative to the ion exchange kinetics. In this case, the observed first-order behavior is the result of fixed ion groups. When ions are allowed to equilibrate completely at very low ionic strengths, first-order transitions are highly dependent on the monomer dissociation constants.

6.4.6.3 Enzymatically Driven Transitions

Phase transitions in synthetic hydrogels can be driven in a very novel fashion using enzymatic reactions within the hydrogel. This offers the potential for a number of important applications in drug delivery and biochemical to mechanical transduction. First-order swelling transitions have been produced enzymatically in a series of hydrogel systems examined by Kokufuta and Tanaka [54]. Their systems consisted of *N*-isopropylacrylamide hydrogels in which liver esterase was imbedded. Phase transitions were driven by the hydrolyzation of ethyl butyrate into ethanol and butyric acid. Changes in the substrate-to-product composition triggered the phase transition.

Seigel *et al.* [8] examined the swelling transitions in hydrogels with dissociable acidic groups that were triggered by the enzymatic conversion of glucose to gluconic acid. As previously discussed, hydrogels with acidic monomers swell with increasing pH and deswell with decreasing pH. By imbedding glucose oxidase in an acidic polyelectrolyte network it was possible to catalyze the conversion of glucose to hydrogen ions and thereby decrease the pH. When these hydrogels were exposed to a glucose solution, they underwent deswelling transitions. In some cases, swelling oscillations were observed.

6.5 Polyampholyte Hydrogels

Polyampholyte hydrogels are cross-linked polymer networks consisting of positive, negative, and even neutral monomers. One of the most important factors that determines the physical properties of a polyampholyte hydrogel is the relative concentration of positive and negative monomers. The charge of a monomeric group also depends on the acid and base dissociation constants. As a result, it is possible to fabricate polyampholytes with a wide range of physical properties, depending on the relative composition of the network.

6.5.1 EQUILIBRIUM THERMODYNAMICS

6.5.1.1 System Definition

A great deal of interest in polyampholyte hydrogel applications concerns their physicochemical properties when immersed in electrolyte baths. Unlike polyelectrolytes, balanced polyampholytes have the unique capacity to swell at high bath salt concentrations. For many practical applications, a thermodynamic system

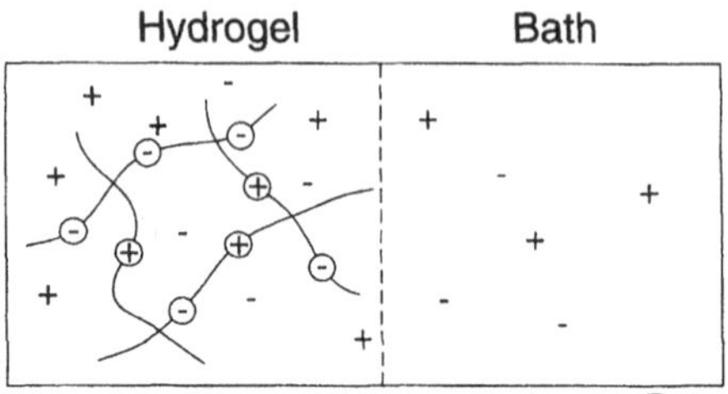

$V^{\mathrm{H}}, \chi(T), \{N_i^{\mathrm{H}}\}$ $V^{\mathrm{B}}, \chi(T), \{N_i^{\mathrm{B}}\}$

Figure 6.20 System definition for a polyampholyte hydrogel immersed in an electrolyte bath. The total system consists of a hydrogel and bath phase in equilibrium with each other. In equilibrium, the exchangeable ion chemical potentials in the hydrogel and bath phase are equal to each other.

definition of polyampholyte behavior must therefore include a bath phase containing low-molecular-weight ions. Similar to the case of a polyelectrolyte in equilibrium with a surrounding bath, we define the polyampholyte system as shown in Figure 6.20. The polyampholyte hydrogel network contains varying concentrations of acidic and basic monomers, and the system is defined by the volume, solvent quality, and particle numbers in each phase.

6.5.1.2 Free-Energy Model

Just as in the case of a polyelectrolyte hydrogel in equilibrium with a surrounding bath, the total system free energy ΔF^{T} can be decomposed into a sum of hydrogel and bath components such that

$$\Delta F^{\mathrm{T}} = \Delta F^{\mathrm{H}} + \Delta F^{\mathrm{B}}, \tag{21}$$

where ΔF^{H} and ΔF^{B} represent the hydrogel and bath contributions, respectively. The free energy of a polyampholyte hydrogel phase is given by

$$\Delta F^{\mathrm{H}} = \Delta F_{\mathrm{el}} + \Delta F_{\mathrm{M}} + \Delta F_0 + \Delta F_{\mathrm{ex}}, \tag{22}$$

where ΔF_{el}, ΔF_{M}, and ΔF_0 are as previously defined. The fourth term on the right-hand side of Eq. (22) represents the excess free-energy term necessary to include the attractive interactions of the polyampholyte effect.

The excess free-energy term has been the subject of recent studies [55, 56]. Most studies have used the Debye–Hückel approximation to compute the excess free-energy correction. At high bath ionic strengths, however, the Debye-Hückel approximation fails catastrophically. Higher-order correction terms have been used to obtain qualitative agreement with theory. The excess free energy using the Debye–Hückel plus second virial coefficient approximation, for example, is defined as

$$(\Delta F_{\text{ex}})_{\text{DHLL+B}_2} = -k_{\text{B}}TV\left[\frac{\kappa^3}{12\pi} + \sum_{i=1}^{\sigma}\sum_{j=1}^{\sigma} c_i c_j B_{ij}(\kappa)\right], \tag{23}$$

where κ is the inverse Debye length and B_{ij} is the second virial coefficient in the Mayer ionic solution theory [57, 58].

The bath free energy for the polyampholyte system can be defined as

$$\Delta F^{\text{B}} = \Delta F_0^{\text{B}} + \Delta F_{\text{ex}}^{\text{B}}, \tag{24}$$

where

$$\Delta F^{\text{B}} = k_{\text{B}}T\sum_{i=1}^{\sigma} N_i^{\text{B}}\left[\ln\left(\frac{N_i^{\text{B}}}{V^{\text{B}}}\right) - 1\right] + \sum_{i=1}^{\sigma} z_i e N_i^{\text{B}} \varphi^{\text{B}}, \tag{25}$$

and the definition of excess bath free energy is similar to that of Eq. (23).

6.5.1.3 Swelling and Ionic Equilibrium

Just as in the equilibrium definition of a polyelectrolyte in contact with a thermodynamic bath, polyampholyte hydrogel equilibrium occurs when

$$-\frac{1}{v}[\ln(1 - nv) + nv + \chi n^2 v^2] + \frac{n_0}{N_x}\left[\frac{1}{2}\left(\frac{n}{n_0}\right) - \left(\frac{n}{n_0}\right)^{1/3}\right] + \left(\sum_{i=0}^{\sigma} c_i^{\text{H}} + \frac{\pi_{\text{ex}}^{\text{H}}}{k_{\text{B}}T}\right) - \left(\sum_{i=0}^{\sigma} c_i^{\text{B}} + \frac{\pi_{\text{ex}}^{\text{B}}}{k_{\text{B}}T}\right) = 0, \tag{26}$$

where $\pi_{\text{ex}}^{\text{H}}$ and $\pi_{\text{ex}}^{\text{B}}$ are the excess osmotic pressure corrections for the hydrogel and bath phases, respectively [59–64].

Balancing the ion chemical potentials gives rise to a relation between ions i and j of the form

$$\left(\frac{C_i \gamma_i}{C_i^{\text{B}} \gamma_i^{\text{B}}}\right)^{\frac{1}{z_i}} = \left(\frac{C_j \gamma_j}{C_j^{\text{B}} \gamma_j^{\text{B}}}\right)^{\frac{1}{z_j}}, \tag{27}$$

where capital letters are used to represent molar quantities and the γ terms represent the activity coefficients. In the limit of small ion concentrations, the activity coefficients approach unity, and the familiar Donnan relations

$$\left(\frac{C_i}{C_i^{\mathrm{B}}}\right)^{\frac{1}{z_i}} = \left(\frac{C_j}{C_j^{\mathrm{B}}}\right)^{\frac{1}{z_j}} \tag{28}$$

are recovered.

6.5.1.4 Acid and Base Dissociation Equilibrium

Including the higher-order effects needed to describe the polyampholyte effect with acid and base dissociation gives rise to the relations

$$K_{\mathrm{a}} = \frac{\gamma_{\mathrm{A}^-} C_{\mathrm{A}^-} \gamma_{\mathrm{H}^+} C_{\mathrm{H}^+}}{\gamma_{\mathrm{AH}} C_{\mathrm{AH}}} \tag{29}$$

and

$$K_{\mathrm{b}} = \frac{\gamma_{\mathrm{BOH}} C_{\mathrm{BOH}} \gamma_{\mathrm{H}^+} C_{\mathrm{H}^+}}{\gamma_{\mathrm{B}^+} C_{\mathrm{B}^+}}. \tag{30}$$

At low ion strengths the activity coefficients approach unity and the relations

$$K_{\mathrm{a}} = \frac{C_{\mathrm{A}^-} C_{\mathrm{H}^+}}{C_{\mathrm{AH}}} \tag{31}$$

and

$$K_{\mathrm{b}} = \frac{C_{\mathrm{BOH}} C_{\mathrm{H}^+}}{C_{\mathrm{B}^+}} \tag{32}$$

are recovered.

6.5.2 *PREPARATION*

6.5.2.1 Basic Principles

The physical properties of polyampholytic hydrogels can be appreciated by first considering the behavior of polyampholyte polymers. Historically, polyampholyte polymers were examined as a model system for proteins and other biomolecules. Polyampholytes formed from the two sets of monomers shown in Figures 6.21 and 6.22, for example, were examined by Alfrey and coworkers [65–68, 1]. These polyampholytes have a net positive charge at low pH, a net negative charge at high pH, and are neutral at some intermediate pH. Far from the isoelectric point, or charge balance state, these polymers behave like polyelectrolytes and exist in a swollen state that can be forced to collapse by increasing the ionic strength.

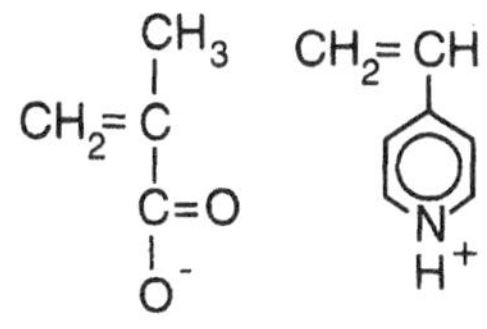

Figure 6.21 Molecular structures of methacrylic acid and 4-vinylpyridine monomers used to form polyampholytes. Polyampholytes formed with these two monomers behave like polyelectrolytes at high and low pH values. At some intermediate pH, the concentration of positive and negative monomers is equal.

The acid and base dissociation constants play an important role in determining the background of positive and negative charges at the isoelectric point. Because the difference in the methacrylic acid and 4-vinylpyridine dissociation constants is small, the background charge concentration at the isoelectric point is insignificant. The difference in the dissociation constants of the methacrylic acid and diethyl (or dimethyl)aminoethyl methacrylate pair, however, is much greater. Hence, a large background of positive and negative charges can exist at the isoelectric point. When a significant background of charges exists at the isoelectric point, attraction between oppositely charged side chains leads to a relatively tight coiling of the polyions at low bath ionic strengths. Polyampholytes in a balanced charge state are also observed to undergo a swelling transition with increasing bath ionic strength. This antipolyelectrolyte swelling with increasing salt ion screening is one of the classic hallmarks of polyampholytes.

The swelling response of a polyampholyte is highly dependent on the relative concentrations and microstructural arrangement of charged and neutral monomers established during polymerization. The microstructure, for example, has been

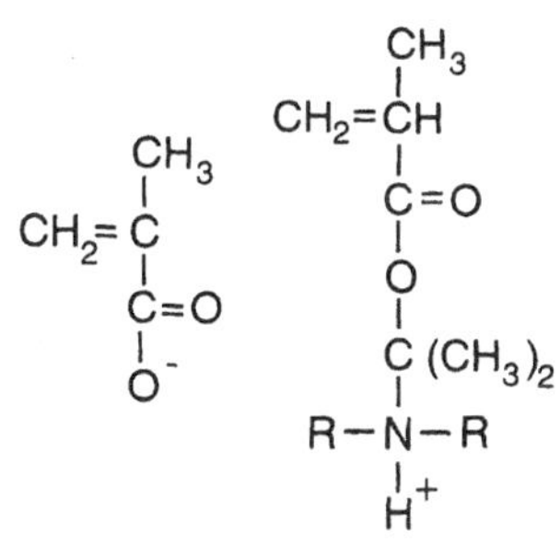

Figure 6.22 Molecular structures of methacrylic acid and diethyl (or dimethyl) aminoethyl methacrylate monomers used to fabricate polyampholytes. The R group can be either a methyl or an ethyl group.

shown to depend on the type of monomers used in the feedstock and the conditions under which the polymerization occurs [69, 70]. Balanced or unbalanced polyampholytes can be prepared by varying the acidic and basic monomer feedstock concentrations during copolymerization [22, 55]. The charge density can also be manipulated by incorporating neutral monomers into the polymer backbone [55].

6.5.2.2 Counterion Removal

Although counterions are normally present in unbalanced polyampholytes to maintain charge electroneutrality, even balanced polyampholytes can bear residual counterions. Counterion screening during polymerization normally leads to a random monomer distribution with no preferential arrangement based on monomer charges. Microemulsion techniques also appear to produce polyampholytes with near balanced states and random structures [71]. The preparation of balanced polyampholytes without counterions and the formation of polymers with smaller-length-scale charge fluctuations has been investigated by a number of groups. Extensive dialysis of balanced polyampholytes, for example, has been used to remove residual counterions and reduce the length scale of the charge fluctuations successfully.

Counterions have also been removed from stoichiometrically balanced solutions of cationic and anionic monomers using precipitation techniques. The use of counterion precipitation in the synthesis of balanced polyampholytes was pioneered by Salamone and coworkers [72–76]. Recently, a method based on the precipitation of counterions from 2-acrylamido-2-methyl-1-propanesulfonic acid and [3-(methacryloylamino)propyl] trimethylammonium chloride solutions has been proposed [22]. As shown in Figure 6.23, the silver salt of 2-acrylamido-2-methyl-1-propanesulfonic acid is prepared by reacting the acidic monomer with silver carbonate to produce silver 2-acrylamido-2-methyl-1-propanesulfonate, water, and carbon dioxide gas. Mixing an equimolar concentration of silver 2-acrylamido-2-methyl-1-propanesulfonate, and [3-(methacrylolamino)propyl] trimethylammonium chloride results in an insoluble silver chloride precipitate and the two charged monomers free of counterions.

6.5.2.3 Polyzwitterions

Zwitterions are electrically neutral molecules that carry an equal number of positive and negative charges covalently linked to each other [77]. Polyampholytes with dissociable acidic and basic monomeric groups at the isoelectric point can be

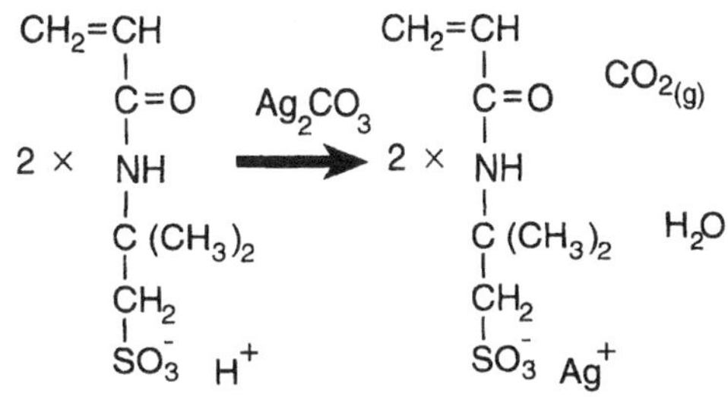

Figure 6.23 Reacting 2-acrylamido-2-methyl-1-propanesulfonic acid with silver carbonate produces silver 2-acrylamido-2-methyl-1-propanesulfonate, water, and carbon dioxide gas. By reacting silver 2-acrylamido-2-methyl-1-propanesulfonate with [3-(methacryloylamino)propyl]trimethylammonium chloride an insoluble silver chloride precipitate and the two charged monomers free of counterions are produced.

considered zwitterions. However, the unique properties of the zwitterionic structure are manifested when the cationic and anionic functional groups are located on the same monomer. Polybetaines, for example, have positive and negative charges separated by a single molecular group. Polyzwitterions, therefore, have associated charge fluctuations on the length scale of a monomer and exist as a dipolar structure.

Synthetic routes to zwitterionic polyampholytes based on pyridine derivatives were first given by various groups towards the end of the 1950s. Ladenheim and Morawetz presented the synthesis of poly(4-vinylpyridine) betaine [78], and the synthesis of polymeric tertiary amine sulfobetaines is described by Hart and Timmermann [79].

6.5.2.4 Interpenetrating Polyelectrolyte Complexes

Polyampholytes can also be formed from ionic complexes of oppositely charged polyelectrolyte polymers. Unlike polyzwitterions in which charge fluctuations occur on the length scale of a single monomer, polyelectrolyte complexes possess charge fluctuations on much greater length scales. The physical properties of polyelectrolyte complexes are dependent on the acid and base dissociation constants of the monomers, and the ionic stoichiometry of the polyelectrolyte chains [80, 81].

The mechanical stability and swelling response of polyelectrolyte complexes depend on the acid and base strength of the ionic monomers. Fuos *et al.* [82] have examined polyelectrolyte complexes formed from poly(sodium styrene sulfonate) and poly(vinylmethyl pyridinium chloride). Deuel and coworkers [83] have examined the polyelectrolyte complexes formed from weak acidic and basic monomers. Polyelectrolyte complexes formed with strong acid and base groups are generally stronger mechanically and more stable than complexes formed with weaker acidic and basic monomers.

Kudaibergenov has recently investigated cross-linked interpenetrating polyelectrolyte networks [84]. In this synthesis, a second polyelectrolyte network is formed in another prepolymerized polyelectrolyte network of the opposite charge. Compared with random polyelectrolytes, the high density of charges and the formation of ionic cross-links were observed to give rise to greater swelling transitions.

6.5.3 ION DISSOCIATION AND POLYAMPHOLYTE SCREENING

Figure 6.24 illustrates the effect of varying the acidic or basic monomer concentration of polyampholytes prepared near the balance point [22]. Each point represents the swelling equilibrium of a 2-acrylamido-2-methyl-1-propanesulfonate and [3-(methacryloylamino)propyl] trimethylammonium polyampholyte copolymer hydrogel prepared with a total monomer concentration of 0.7 M. Increasing the relative proportions of either acidic or basic monomers away from the balance point produces a swelling transition. The differences in the acid and base

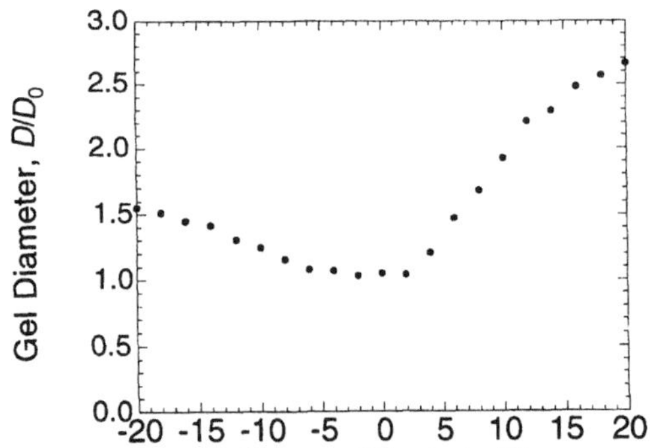

Figure 6.24 Equilibrium swelling of polyampholytes with varying acidic and basic monomer concentration offsets. Near the charge balance point, a minimum in the equilibrium swelling occurs. With increasing acidic or basic monomer offset, a swelling transition is observed. Reprinted with permission from ref. [22]. Copyright 1996 American Institute of Physics.

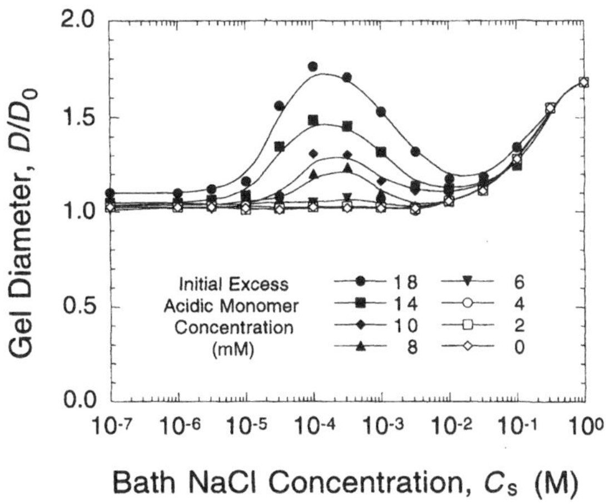

Figure 6.25 Polyampholyte hydrogel equilibrium swelling as a function of bath salt concentration. Balanced polyampholytes monotonically swell with increasing bath salt concentration. Unbalanced polyampholytes show polyelectrolyte swelling patterns at low and intermediate ionic strengths and polyampholyte swelling patterns at high ionic strengths. Reprinted with permission from ref. [55]. Copyright 1998 American Chemical Society.

dissociation equilibrium of the two charged monomers produces the swelling pattern asymmetry.

The swelling response of polyampholyte hydrogels near the balance point to changes in bath salt concentration illustrates the competing effects of ion dissociation, polyelectrolyte screening, and polyampholyte screening. Figure 6.25 shows the swelling response of 1.4 M polyampholyte hydrogels prepared near the balance point with varying excess concentrations of acidic monomers. At low ionic strengths, all polyampholytes undergo a deswelling transition. Increasing the bath salt concentration from 1 to 100 μM results in the swelling transitions of unbalanced polyampholytes. When the salt concentration increases from 0.1 to 10 mM, unbalanced polyampholytes undergo a deswelling transition. Further increases in the bath salt concentration produce swelling transitions in all hydrogels. Despite the complexity of the swelling response, it can be decomposed and understood as a sum of different contributions.

If polyampholyte effects are ignored, varying the excess charge offset produces the series of swelling curves shown in Figure 6.26. The deswelling transitions observed at intermediate bath salt concentrations are simply the result of polyelectrolyte screening. This effect can be recognized in the data shown in Figure 6.25 at intermediate bath ionic strengths. Including the effects of ion dissociation produce the series of swelling curves shown in Figure 6.27. At low ionic strengths,

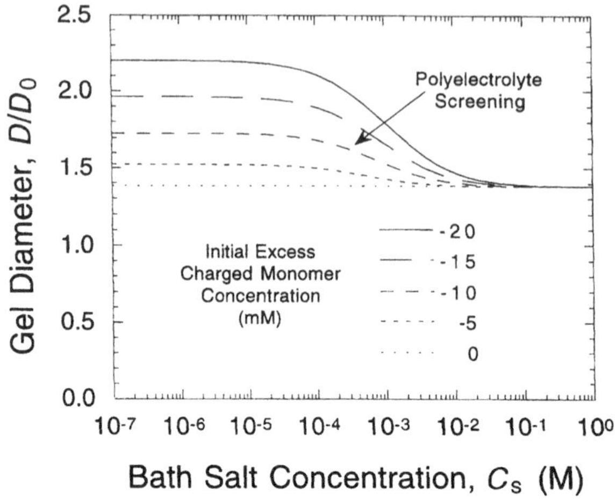

Figure 6.26 Nondissociable polyelectrolyte swelling versus bath salt concentration and excess charge. At low ionic strengths the hydrogel swelling equilibrium reaches a maximum value determined by the solvent and elastic network pressures. The deswelling transition that occurs at intermediate ionic strengths is the result of polyelectrolyte shielding. Reprinted with permission from ref. [55]. Copyright 1998 American Chemical Society.

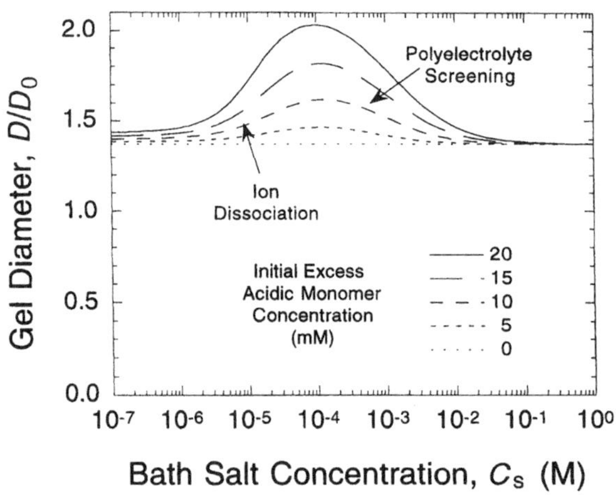

Figure 6.27 Acidic polyelectrolyte swelling versus bath salt concentration and excess acidic monomer offset. At low ionic strengths hydrogen ions bind to the acid monomer groups and produce a deswelling transition. The deswelling transition at intermediate ionic strengths is the result of polyelectrolyte shielding. Reprinted with permission from ref. [55]. Copyright 1998 American Chemical Society.

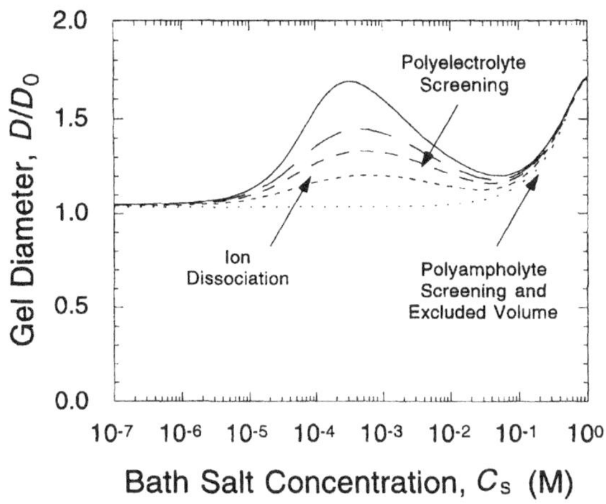

Figure 6.28 Polyelectrolyte swelling versus bath salt concentration and excess acidic monomer offset. At low ionic strengths hydrogen ions bind to the acid monomer groups and produce a deswelling transition. The deswelling transition at intermediate ionic strengths is the result of polyelectrolyte shiedling. The swelling transition at high ionic strengths is the result of polyampholyte shielding and the excluded volume effect. Reprinted with permission from ref. [55]. Copyright 1998 American Chemical Society.

the electrostatic inclusion of hydrogen ions results in a very low internal pH and hence leads to the neutralization of the acidic monomers. The loss of charge and osmotic pressure therefore produce a deswelling transition. Increasing the bath salt concentration leads to the exchange of hydrogen ions for sodium ions, and thus the acidic monomers dissociate.

Figure 6.28 shows the composite effect of including ion dissociation, polyelectrolyte repulsion, and polyampholyte attraction. At low ionic strengths, ion dissociation equilibrium and attractive polyampholyte effects dominate the swelling response. At intermediate bath ionic strengths, polyelectrolyte shielding produces the deswelling transition in unbalanced polyampholytes. At very high bath ionic strengths, polyampholyte shielding and excluded volume effects result in a swelling transition.

6.6 Conclusion

This chapter has discussed the thermodynamic behavior and synthesis of polymer hydrogels. The objective of this chapter has been to provide an introduction to the experimental study of hydrogels. The universality of hydrogel volume phase transitions was illustrated by describing the analogous behavior in the van der Waals fluid. Detailed experimental protocols for fabricating neutral, polyelectrolyte, and polyampholyte hydrogels have also been presented.

Neutral hydrogels can be fabricated from a wide variety of different monomers. Neutral hydrogel networks can undergo large volume phase transition when the solvent composition or temperature is varied. Acrylamide hydrogels, for example, undergo a collapse transition in a water and acetone solvent mixture as the relative concentration of acetone increases. Similarly, *N*-isopropylacrylamide hydrogels collapse with increasing bath temperature in an aqueous solvent. Temperature-dependent reentrant phase transitions can also be observed with *N*-isopropylacrylamide in water and methanol mixtures.

Introducing ionization into a polymer network makes it possible to fabricate hydrogels that respond to changes in ambient ionic strength and pH. When only one type of charge is present in the network, polyelectrolyte behavior is observed. Polyelectrolyte hydrogels typically undergo deswelling transitions at low ionic strengths at neutral pH as a result of the Donnan inclusion of hydrogen or hydroxide ions. At high ionic strengths, polyelectrolyte hydrogels collapse as a result of losing the osmotic pressure difference between the hydrogel and the surrounding bath. The competition between repulsive ionic interactions and solvent-mediated forces can lead to first-order phase transitions in polyelectrolyte hydrogels.

Polyampholyte hydrogels possess positive and negative charges. Polyampholyte hydrogels with a balanced concentration of positive and negative charges remain collapsed at low ionic strengths and swell at high ionic strengths. With increasing charge offset, polyampholyte hydrogels exhibit properties similar to polyelectrolyte hydrogels and collapse at intermediate bath ionic strengths. The acid and base dissociation constants of the monomers play an important role in determining the swelling response of the network to changes in bath ionic strength and pH.

References

1. S.-K. Ma, *Modern Theory of Critical Phenomenon*, Advanced Book Program, Reading, MA (1976).
2. H. E. Stanley, *Introduction to Phase Transitions and Critical Phenomenon*, Oxford University Press, New York, 1971.
3. T. Tanaka 1977.
4. Y. Osada, *Adv. Polym. Sci.* **82:** 1–46 (1987).
5. C. Yu and J. Kohn, *Inverse Temperature Transition Materials*, Fifth World Biomaterials Congress, p. 565, 1996.

6. Y. H. Bae, T. Okano, and S. W. Kim, *Makromol. Chem. Rapid Commun.* **8:** 481 (1987).
7. T. Okano, Y. H. Bae, and S. W. Kim, in *Modulated Control Release System,* J. Kost, Ed., CRC Press, Boca Raton, FL.
8. R. A. Siegel and C. G. Pitt, *J. Controlled Release* **33:** 173–188 (1995).
9. D. De Rossi, *Trans. Am. Soc. Artif. Inter. Organs* **XXXII:** 157 (1986).
10. A. S. Hoffman, *J. Controlled Release* **6:** 297 (1987).
11. A. S. Hoffman, *Artif. Organs* **19:** 458–467 (1995).
12. A. S. Hoffman and B. D. Ratner, in *Synthetic Hydrogels for Biomedical Applications*, ACS Symposium Series 31, p. 1, American Chemical Society, Washington, DC, 1976.
13. A. S. Hoffman, A. Afrasiabi, and L. C. Dong, *J. Controlled Release* **4:** 213 (1986).
14. Y. Osada, K. Umezawa, and A. Yamauchi, *Makromol. Chem.* **189:** 597 (1988).
15. R. A. Siegel and B. A. Firestone, *Macromolecules* **21:** 3254 (1988).
16. M. Suzuki and Y. Sawada, *J. Appl. Phys.* **51:** 5667 (1980).
17. M. Suzuki, T. Tateishi, T. Ushida, and S. Fujishige, *Biorheol.* **23:** 274 (1986).
18. P. Verdugo, *Biophys. J.* **49:** 231a (1986).
19. P. Verdugo, *Rev. Physiol.* **52:** 157 (1990).
20. D. D. Perrin and W. L. F. Armarego, *Purification of Laboratory Chemicals*, Pergamon Press, New York, 1988.
21. E. R. Edelman, E. Mathiowitz, R. Langer, and M. Klagsbrun, *Biomaterials* **12:** 619–626 (1991).
22. A. E. English, S. Mafe, J. A. Manzanares, X. Yu, A. Y. Grosberg, and T. Tanaka, *J. Chem. Phys.* **104:** 8713–8720 (1996).
23. T. M. Quinn and A. J. Grodzinsky, *Macromolecules* **26:** 4332–4338 (1993).
24. K. Huang, *Statistical Mechanics,* John Wiley & Sons, New York, 1963.
25. P. Flory, *Principles of Polymer Chemistry*, Cornell University Press, Ithaca, NY, 1953.
26. P. J. Flory, *Proc. Roy. Soc. Lond. A.* **351:** 351–380 (1976).
27. P.-G. de Gennes, *Scaling Concepts in Polymer Physics,* Cornell University Press, Ithaca, NY, 1979.
28. K. Dusek and D. Patterson, *J. Polym. Sci. A-2* **6:** 1209–1216 (1968).
29. A. R. Khokhlov, *Polymer* **21:** 376 (1980).
30. A. R. Khokhlov and S. K. Nechaev, *Phys. Lett.* **112-A:** 156 (1985).
31. A. Khokhlov and I. Nyrkova, *Macromolecules* **25:** 1493 (1992).
32. T. Tanaka and D. J. Fillmore, *J. Chem. Phys.* **70:** 1214–1218 (1979).
33. T. Tanaka, S. Ishiwata, and C. Ishimoto, *Phys. Rev. Lett.* **38:** 771–774 (1977).
34. Y. Li and T. Tanaka, *J. Chem. Phys.* **90:** 5161–5166 (1989).
35. T. Amiya, Y. Hirokawa, Y. Hirose, and Y. Li, *J. Chem. Phys.* **86:** 2375–2379 (1987).
36. S. Katayama, Y. Hirokawa, and T. Tanaka, *Macromolecules* **17:** 2641–2643 (1984).

37. T. Tanaka, L. Hocker, and G. Benedek, *J. Chem. Phys.* **59:** 5151–5159 (1973).
38. J. P. Munch, S. J. Candau, R. Duplessix, C. Picot, J. Herz, and R. Benoit, *J. Polym. Sci.* **14:** 1097 (1976).
39. A. Peters and S. J. Candau, *Macromolecules* **19:** 1952, **21:** 2278 (1988).
40. Y. Li and T. Tanaka, *J. Chem. Phys.* **92:** 1365–1371 (1990).
41. A. Hochberg, T. Tanaka, and D. Nicoli, *Phys. Rev. Lett.* **43:** 217–219 (1979).
42. T. Tanaka, *Phys. Rev. Lett.* **40:** 820–823 (1978).
43. E. Matsuo and T. Tanaka, *Phase Transitions* **46:** 217–228 (1994).
44. T. Tanaka, *Physica* **140A:** 261–268 (1986).
45. T. Tanaka, *Nature* **325:** 796–798 (1987).
46. A. E. English, T. Tanaka, and E. R. Edelman, *J. Chem. Phys.* **107:** 1645–1654 (1997).
47. F. Helfferich, *Ion Exchange.* McGraw-Hill, New York, 1962.
48. J. Nussbaum and A. J. Grodzinsky, *J. Membrane Sci.* **8:** 193–219 (1981).
49. S. Hirotsu, Y. Hirokawa, and T. Tanaka, *J. Chem. Phys.* **87:** 1392–1395 (1987).
50. Y. Hirokawa, T. Tanaka, and E. Sato, *Macromolecules* **18:** 2782–2784 (1985).
51. M. Ilavsky, J. Hrouz, and K. Ulbrich, *Polym. Bull.* **7:** 107 (1982).
52. H. Inomata, S. Goto, and S. Saito, *Langmuir* **8:** 1030 (1992).
53. K. Otaka, H. Inomata, M. Konno, and S. Saito, *Macromolecules* **23:** 283 (1990); A. E. English *et al.*, *J. Chem. Phys.* **105:** 10606 (1996).
54. E. Kokufuta and T. Tanaka, *Macromolecules* **24:** 1605–1607 (1991).
55. A. E. English, T. Tanaka, and E. Edelman, *Macromolecules* (1998).
56. A. E. English, T. Tanaka, and E. R. Edelman, *Polymer* (1998).
57. J. E. Mayer, *J. Chem. Phys.* **18:** 1426–1436 (1950).
58. W. G. McMillan and J. E. Mayer, *J. Chem. Phys.* **13:** 276–305 (1945).
59. J. P. Baker, H. W. Blanch, and J. M. Prausnitz, *Polymer* **36:** 1061–1069 (1995).
60. A. V. Dobrynin and M. Rubinstein, *Phys.* II, **5:** 677–695 (1995).
61. Y. Kantor and M. K. Kardar, *Phys. Rev. E* **51:** 1299–1312 (1995).
62. Y. Kantor, M. K. Kardar, and H. Li, *Phys. Rev. E* **49:** 1383–1392 (1994).
63. P. G. Higgs and J. F. Joanny, *J. Chem. Phys.* **94:** 1543–1554 (1991).
64. J. Wittmer, J. Johner, and J.-F. Joanny, *Europhysics Lett.* **24:** 263–268 (1993).
65. T. Alfrey, R. M. Fuoss, H. Morawetz, and H. Pinner, *J. Am. Chem. Soc.* **74:** 438–441 (1952).
66. T. Alfrey and H. Morawetz, *J. Am. Chem. Soc.* **72:** 1864 (1950).
67. T. Alfrey, C. G. Overberger, and S. H. Pinner, *J. Am. Chem. Soc.* **75:** 4221–4223 (1953).
68. T. Alfrey and S. H. Pinner, *J. Polym. Sci.* **23:** 533–547 (1957).
69. C. L. McCormick and C. B. Johnson, *Macromolecules* **21:** 694–699 (1988).
70. C. L. McCormick and L. C. Salzar, *Macromolecules* **25:** 1896–1900 (1992).
71. J. Corpart and F. Candau, *Macromolecules* **26:** 1333–1343 (1993).
72. J. C. Salamone, I. Ahmed, E. L. Rodriguez, L. Quach, and A. C. Watterson, *J. Macromol. Sci. Chem.* **A25:** 811–837 (1988).
73. J. C. Salamone, L. Quach, A. C. Watterson, S. Krauser, and M. U. Mahmud, *J. Macromol. Sci. Chem.* **A22:** 653–664 (1985).

74. J. C. Salamone, C.-C. Tsai, and A. C. Watterson, *J. Macromol. Sci. Chem.* **A13:** 665–672 (1979).
75. J. C. Salamone, C. C. Tsai, and A. C. Watterson, in *Advances in Chemistry* A. Eisenberg, Ed., pp. 337–346, American Chemical Society, Washington, DC, 1980.
76. J. C. Salamone, W. Volksen, A. P. Olson, and S. C. Isreal, *Polymer* **19:** 1157–1162 (1978).
77. J.-C. Galin, in *Polymer Encyclopedia*, J. C. Salamone, Ed., pp. 7189–7201, CRC Press, Boca Raton, FL, 1996.
78. H. Ladenheim and H. Morawetz, *J. Polym. Sci.* **26:** 251–254 (1957).
79. R. Hart and D. Timmerman, *J. Polym. Sci.* **28:** 638–640 (1958).
80. H. J. Bixler and A. S. Michaels, in *Encyclopedia of Polymer Science and Technology*, N. M. Bikales, Ed., pp. 765–780, John Wiley & Sons, New York, 1969.
81. E. Tsuchida, Y. Osada, and H. Ohno, *J. Macromol. Sci. Phys.* **B17:** 683–714 (1980).
82. R. M. Fuoss and H. Sadek, *Science* **110:** 552–554 (1949).
83. H. Deuel, J. Solms, and A. Dengler, *Helv. Chim. Acta* **36:** 1671–1680 (1953).
84. S. E. Kudaibergenov, *Ber. Bunsenges. Phys. Chem.* **100:** 1079–1082 (1996).

Index

A

B

G

H

I

J

K

L

FSC
www.fsc.org
MIX
Papier | Fördert
gute Waldnutzung
FSC® C083411

Kolibri 360 GmbH
Ferdinand-Jühlke-Straße 7
99095 Erfurt, Deutschland
produktsicherheit@kolibri360.de